PHYSICAL-CHEMICAL TREATMENT OF WATER AND WASTEWATER

PHYSICAL-CHEMICAL TREATMENT OF WATER AND WASTEWATER

Arcadio P. Sincero Sr., D.Sc., P.E.
Morgan State University
Baltimore, Maryland

Gregoria A. Sincero, M. Eng., P.E.
Department of the Environment
State of Maryland

CRC PRESS

Boca Raton   London   New York   Washington, D.C.

Library of Congress Cataloging-in-Publication Data

Sincero, Arcadio P. (Arcadio Pacquiao)
  Physical–chemical treatment of water and wastewater / Arcadio Pacquiao Sincero, Sr.,
Gregoria Alivio Sincero.
     p. cm.
  Includes bibliographical references and index.
  ISBN 1-58716-124-9 (alk. paper)
  1. Water—Purification. 2. Sewage—Purification. I. Sincero, Gregoria A. (Gregoria Alivio)
  II. Title

  TD430 .S47 2002
  628.1′62—dc21   2002023757

Visit the CRC Press Web site at www.crcpress.com

© 2003 by A. P. Sincero and G. A. Sincero

Co-published by IWA Publishing, Alliance House, 12 Caxton Street, London, SW1H 0QS, UK
Tel. +44 (0) 20 7654 5500, Fax +44 (0) 20 7654 5555
publications@iwap.co.uk
www.iwapublishing.com
ISBN 1-84339-028-0

No claim to original U.S. Government works
International Standard Book Number 1-58716-124-9
Library of Congress Card Number 2002023757
Printed in the United States of America 1 2 3 4 5 6 7 8 9 0
Printed on acid-free paper

Preface

This textbook is intended for undergraduate students in their junior and senior years in environmental, civil, and chemical engineering, and students in other disciplines who are required to take the course in physical–chemical treatment of water and wastewater. This book is also intended for graduate students in the aforementioned disciplines as well as practicing professionals in the field of environmental engineering. These professionals include plant personnel involved in the treatment of water and wastewater, consulting engineers, public works engineers, environmental engineers, civil engineers, chemical engineers, etc. They are normally employed in consulting firms, city and county public works departments, and engineering departments of industries, and in various water and wastewater treatment plants in cities, municipalities, and industries. These professionals are also likely to be employed in government agencies such as the U.S. Environmental Protection Agency, and state agencies such as the Maryland Department of the Environment.

The prerequisites for this textbook are general chemistry, mathematics up to calculus, and fluid mechanics. In very few instances, an elementary knowledge of calculus is used, but mostly the mathematical treatment makes intensive use of algebra. The entire contents of this book could be conveniently covered in two semesters at three credits per semester. For schools offering only one course in physical–chemical treatment of water and wastewater, this book gives the instructor the liberty of picking the particular topics required in a given curriculum design.

After the student has been introduced to the preliminary topics of water and wastewater characterization, quantitation, and population projection, this book covers the unit operations and unit processes in the physical–chemical treatment of water and wastewater. The unit operations cover flow measurements and flow and quality equalization; pumping; screening, sedimentation, and flotation; mixing and flocculation; conventional filtration; advanced filtration and carbon adsorption; and aeration, absorption, and stripping. The unit processes cover water softening, water stabilization, coagulation, removal of iron and manganese, removal of phosphorus, removal of nitrogen, ion exchange, and disinfection.

The requirements for the treatment of water and wastewater are driven by the Safe Drinking Water Act and Clean Water Act, which add more stringent requirements from one amendment to the next. For example, the act relating to drinking water quality, known as the Interstate Quarantine Act of 1893, started with only the promulgation of a regulation prohibiting the use of the common cup. At present, the Safe Drinking Water Act requires the setting of drinking water regulations for some 83 contaminants. The act relating to water quality started with the prohibition of obstructions in harbors as embodied in the Rivers and Harbors Act of 1899. At present, the Clean Water Act requires that discharges into receiving streams meet water quality standards; in fact, regulations such as those in Maryland have an antidegradation policy. In recent years, problems with *Cryptosporidium parvum* and *Giardia lamblia* have come to the fore. Toxic substances are being produced by industries every day which could end up in the community water supply. These acts are technology forcing, which means that as we continue to discover more of the harmful effects of pollutants on public health and welfare and the environment, advanced technology will continue to be developed to meet the needs of treatment.

The discipline of environmental engineering has mostly been based on empirical knowledge, and environmental engineering textbooks until recently have been written in a descriptive manner. In the past, the rule of thumb was all that was necessary. Meeting the above and similar challenges, however, would require more than just empirical knowledge and would require stepping up into the next level of sophistication in treatment technology. For this reason, this textbook is not only descriptive but is also analytical in nature. It is hoped that sound concepts and principles will be added to the already existing large body of empirical knowledge in the discipline. These authors believe that achieving the next generation of treatment requirements would require the next level of sophistication in technology. To this end, a textbook written to address the issue would have to be analytical in nature, in addition to adequately describing the various unit operations and processes.

This book teaches both principles and design. Principles are enunciated in the simplest way possible. Equations presented are first derived, except those that are obtained empirically. Statements such as "It can be shown..." are not used in this book. These authors believe in imparting the principles and concepts of the subject matter, which may not be done by using "it-can-be-shown" statements. At the end of each chapter, where appropriate, are numerous problems that can be worked out by the students and assigned as homework by the instructor.

The question of determining the correct design flows needs to be addressed. Any unit can be designed once the flow has been determined, but how was the flow determined in the first place? Methods of determining the various design flows are discussed in this book. These methods include the determination of the average daily flow rate, maximum daily flow rate, peak hourly flow rate, minimum daily flow rate, minimum hourly flow rate, sustained high flow rate, and sustained low flow rate.

What is really meant when a certain unit is said to be designed for the average flow or for the peak flow or for any flow? The answer to this question is not as easy as it may seem. This book uses the concept of the probability distribution to derive these flows. On the other hand, the loss through a filter bed may need to be determined or a deep-well pump may need to be specified. The quantity of sludge for disposal produced from a water softening process may also be calculated. This book uses fluid mechanics and chemistry without restraint to answer these design problems.

Equivalents and equivalent mass are two troublesome and confusing concepts. If the chemistry and environmental engineering literature were reviewed, these subjects would be found to be not well explained. Equivalents and equivalent mass in a unified fashion are explained herein using the concept of the reference species. Throughout the unit processes section of this book, reference species as a method is applied. Related to equivalents and equivalent mass is the dilemma of expressing concentrations in terms of calcium carbonate. Why, for example, is the concentration of acidity expressed in terms of calcium carbonate when calcium carbonate is basic and acidity is acidic? This apparent contradiction is addressed in this book.

As in any other textbook, some omissions and additions may have produced some error in this book. The authors would be very grateful if the reader would bring them to our attention.

Acknowledgments

First, I acknowledge Dr. Joseph L. Eckenrode, former Publisher, Environmental Science & Technology, Technomic Publishing Company, Inc. Dr. Eckenrode was very thorough in determining the quality, timeliness, and necessity of the manuscript. It was only when he was completely satisfied through the strict peer review process that he decided to negotiate for a contract to publish the book. Additionally, I acknowledge Brian Kenet and Sara Seltzer Kreisman at CRC Press.

This book was written during my tenure at Morgan State University. I acknowledge the administrators of this fine institution, in particular, Dr. Earl S. Richardson, President; Dr. Clara I. Adams, Vice President for Academic Affairs; Dr. Eugene M DeLoatch, Dean of the School of Engineering; and Dr. Reginald L. Amory, Chairman of the Department of Civil Engineering. I make special mention of my colleague, Dr. Robert Johnson, who was the acting Chairman of the Department of Civil Engineering when I came on board. I also acknowledge my colleagues in the department: Dr. Donald Helm, Prof. A. Bert Davy, Dr. Indranil Goswami, Dr. Jiang Li, Dr. Iheanyi Eronini, Dr. Gbekeloluwa B. Oguntimein, Prof. Charles Oluokun, and Prof. Neal Willoughby.

This acknowledgment would not be complete if I did not mention my advisor in doctoral studies and three of my former professors at the Asian Institute of Technology (A.I.T.) in Bangkok, Thailand. Dr. Bruce A. Bell was my advisor at the George Washington University where I earned my doctorate in Environmental–Civil Engineering. Dr. Roscoe F. Ward, Dr. Rolf T. Skrinde, and Prof. Mainwaring B. Pescod were my former professors at A.I.T., where I earned my Masters in Environmental–Civil Engineering.

I acknowledge and thank my wife, Gregoria, for contributing Chapter 2 (Constituents of Water and Wastewater) and Chapter 7 (Conventional Filtration). I also acknowledge my son, Roscoe, for contributing Chapter 17 (Disinfection). Gregoria also contributed a chapter on solid waste management when I wrote my first book *Environmental Engineering: A Design Approach*. This book was published by Prentice Hall and is being adopted as a textbook by several universities here and abroad. This book has been recommended as a material for review in obtaining the Diplomate in Environmental Engineering from the Academy of Environmental Engineers.

Lastly, I dedicate this book to members of my family: Gregoria, my wife; Roscoe and Arcadio Jr., my sons; the late Gaudiosa Pacquiao Sincero, my mother; Santiago Encarguiz Sincero, my father; and the late Aguido and the late Teodora Managase Alivio, my father-in-law and my mother-in-law, respectively. I also dedicate this book to my brother Meliton and to his wife Nena; to my sister Anelda and to her husband Isidro; to my other sister Feliza and to her husband Martin; and to my brother-in-law Col. Miguel M. Alivio, MD and to his wife Isabel. My thoughts also go to my other brothers-in-law: the late Tolentino and his late wife Mary, Maximino and his wife Juanita, Restituto and his wife Ignacia, the late Anselmo and his wife Silvina, and to my sisters-in-law: the late Basilides and her late husband Dr. Alfonso Madarang, Clarita and her late husband Elpidio Zamora, Luz and her husband Perpetuo Apale, and Estelita.

Arcadio P. Sincero
Morgan State University

About the Authors

Arcadio P. Sincero is Associate Professor of Civil Engineering at Morgan State University, Baltimore, MD. He was also a former professor at the Cebu Institute of Technology, Philippines. He holds a Bachelor's degree in Chemical Engineering from the Cebu Institute of Technology, a Master's degree in Environmental–Civil Engineering from the Asian Institute of Technology, Bangkok, and a Doctor of Science degree in Environmental–Civil Engineering from the George Washington University. He is a registered Professional Engineer in the Commonwealth of Pennsylvania and in the State of Maryland and was a registered Professional Chemical Engineer in the Philippines. He is a member of the American Society of Civil Engineers, a member of the American Institute of Chemical Engineers, a member of the Water Environment Federation, a member of the American Association of University Professors, and a member of the American Society of Engineering Education.

Dr. Sincero has a wide variety of practical experiences. He was a shift supervisor in a copper ore processing plant and a production foreman in a corn starch processing plant in the Philippines. He was CPM (Critical Path Method) Planner in a construction management firm and Public Works Engineer in the City of Baltimore. In the State of Maryland, he was Public Health Engineer in the Bureau of Air Quality and Noise Control, Department of Health and Mental Hygiene; Water Resources Engineer in the Water Resources Administration, Department of Natural Resources; Water Resources Engineer in the Office of Environmental Programs, Department of Health and Mental Hygiene; Water Resources Engineer in the Water Management Administration, Maryland Department of the Environment. For his positions with the State of Maryland, Dr. Sincero had been Chief of his divisions starting in 1978. His last position in the State was Chief of Permits Division of the Construction Grants and Permits Program, Water Management Administration, Maryland Department of the Environment. These practical experiences have allowed Dr. Sincero to gain a wide range of environmental engineering and regulatory experiences: air, water, solid waste, and environmental quality modeling.

Gregoria A. Sincero is a senior level Water Resources Engineer at the Maryland Department of the Environment. She was also a former professor at the Cebu Institute of Technology, Philippines. She holds a Bachelor's degree in Chemical Engineering from the Cebu Institute of Technology and a Master's degree in Environmental–Civil Engineering from the Asian Institute of Technology, Bangkok, Thailand. She is a registered Professional Environmental Engineer in the Commonwealth of Pennsylvania and was a registered Professional Chemical Engineer in the Philippines. She is a member of the American Institute of Chemical Engineers.

Mrs. Sincero has practical experiences both in engineering and in governmental regulations. She was Senior Chemist/Microbiologist in the Ashburton Filters of

Baltimore City. In the State of Maryland, she was Water Resource Engineer in the Water Resources Administration, Department of Natural Resources and Water Resources Engineer in the Office of Environmental Programs, Department of Health and Mental Hygiene, before joining her present position in 1988 at MDE. At MDE, she is a senior project manager reviewing engineering plans and specifications and inspecting construction of refuse disposal facilities such as landfills, incinerators, transfer stations, and processing facilities. Also, she has experiences in modeling of surface waters, groundwaters, and air and statistical evaluation of groundwater and drinking water data using EPA's Gritstat software.

Contents

PART III
Unit Processes of Water
and Wastewater Treatment 465

Appendices and Index *789*

Background Prerequisites

The background prerequisites for this textbook are general chemistry, mathematics up to calculus, and fluid mechanics. In very few instances, an elementary knowledge of calculus is used, but mostly the mathematical treatment makes intensive use of algebra. In fluid mechanics, the only sophisticated topic used is the Reynolds transport theorem. Although this topic is covered in an undergraduate course in fluid mechanics, it was thought advantageous to review it here. The other background prerequisite is general chemistry. Environmental engineering students and civil engineering students, in particular, seem to be very weak in chemistry. This part will therefore also provide a review of this topic. Depending upon the state of knowledge of the students, however, this part may or may not be discussed. This state of knowledge may be ascertained by the instructor in the very first few days of the course, and he or she can tailor the discussions accordingly.

The contents of this "Background Prerequisites" section are not really physical–chemical treatment but just background knowledge to comfortably understand the method of approach used in the book. This book is analytical and therefore must require extensive use of the pertinent chemistry, mathematics, and fluid mechanics. This section contains two chapters: "Introduction" and "Background Chemistry and Fluid Mechanics."

Introduction

This book is titled *Physical–Chemical Treatment of Water and Wastewater*. This chapter begins by defining wastewater and physical–chemical treatment of water and wastewater and treats briefly the coverage. It also addresses the unit operations and unit processes. In addition, in the environmental engineering field, construction of water and wastewater treatment plants and the requirements of their levels of performance are mostly driven by government laws and regulations. For example, the Clean Water Act mandates construction of wastewater treatment facilities that, at least, must produce the secondary level of treatment. The Safe Water Drinking Water Act also requires performance of treatment plants that produce drinking water of quality free from harmful chemicals and pathogens. For these reasons, the Clean Water Act and the Safe Drinking Water Act are discussed at length, detailing their developments and historical perspectives.

WASTEWATER

Wastewater is the spent water after homes, commercial establishments, industries, public institutions, and similar entities have used their waters for various purposes. It is synonymous with *sewage*, although sewage is a more general term that refers to any polluted water (including wastewater), which may contain organic and inorganic substances, industrial wastes, groundwater that happens to infiltrate and to mix with the contaminated water, storm runoff, and other similar liquids. A certain sewage may not be a spent water or a wastewater.

The keyword in the definition of wastewater is "used" or "spent." That is, the water has been used or spent and now it has become a *waste water*. On the other hand, to become a sewage, it is enough that water becomes polluted whether or not it had been used. When one uses the word wastewater, however, the meaning of the two words is blended such that they now often mean the same thing. *Wastewater* equals *sewage*.

PHYSICAL–CHEMICAL TREATMENT OF WATER AND WASTEWATER

What is physical–chemical treatment of water and wastewater? The dictionary defines *physical* as having material existence and subject to the laws of nature. *Chemical*, on the other hand, is defined as used in, or produced by chemistry. Being used in and produced by chemistry implies a material existence and is subject to the laws of nature. Thus, from these definitions, chemical is physical. The fact that chemical is physical has not, however, answered the question posed.

To explore the question further, we go to the definition of chemistry. *Chemistry* is defined as the branch of science that deals with the composition, structure, and properties of substances and the transformation that they undergo. Now, from this definition can be gleaned the distinguishing feature of chemical—the transformation that the substance undergoes. The transformation changes the original substance into an entirely different substance after the transformation. Chemical transformation can be distinguished from physical transformation. In physical transformation, although also involving a change, the change is only in appearance but not in substance. For example, FeO in the beginning is still FeO in the end; the size of the particles may have changed, however, during the process. We now define physical treatment of water and wastewater as a process applied to water and wastewater in which no chemical changes occur. We also define chemical treatment of water and wastewater as a process applied to water and wastewater in which chemical changes occur. Gleaning from these definitions of physical and chemical treatments, in the overall sense, physical–chemical treatment of water and wastewater is a process applied to water and wastewater in which chemical changes may or may not occur.

UNIT OPERATIONS AND UNIT PROCESSES

Figure 1 shows the schematic of a conventional wastewater treatment plant using primary treatment. Raw wastewater is introduced either to the screen or to the comminutor. The grit channel removes the larger particles from the screened sewage, and the primary clarifier removes the larger particles of organic matter as well as inorganic matter that escapes removal by the grit channel. Primary treated sewage is then introduced to a secondary treatment process train downstream (not shown) where the colloidal and dissolved organic matter are degraded by microorganisms.

The scheme involves mere physical movement of materials, no chemical or biological changes occur. In addition, the function of the various operations in the scheme, such as screening, may be applied not only to the primary treatment of sewage as the figure indicates but to other plant operations as well. For example, bagasse may be screened from sugar cane juice in the expression of sugar in a sugar mill, or the larger particles resulting from the cleaning of pineapples in a pineapple factory may be screened from the rest of the wastewater. To master the function of screening, it is not necessary that this be studied in a wastewater treatment plant, in

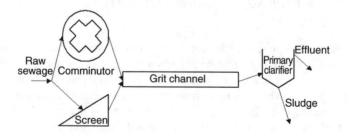

FIGURE 1 A primary treatment system.

a sugar mill, or in a pineapple factory. It can be studied in any setting where screens are used.

Furthermore, the functions of the operation of the primary clarifier may be applied not only to the treatment of sewage as indicated in the schematic but also to the clarification in a water treatment plant, as well as in the clarification of tailings in a mining operation. Similar statements may be made about the operation of the grit channel. In other words, to master the function of clarification and grit removal, it is not necessary that these be studied in a sewage treatment plant or in a water treatment plant. It can be studied in any setting where clarifiers and grit removal are used.

The foregoing operations are physical; they are therefore physical treatments. These physical treatments are called *unit operations* and as gleaned from the previous discussion they may be defined as physical treatments that are identified only according to their functions without particular reference to the location of the units utilizing the functions. For example, screening may be studied without particular reference to any sugar mills or pineapple factories. The unit operations of clarification may be studied without particular reference to any wastewater or water treatment plants or mining operations. The unit operations of grit removal may also be studied without particular reference to any sewage plant but only to any setting where grit removal is involved.

The unit operations discussed previously are all physical operations. In the biological or chemical scene where materials are changed, unit operations have counterparts called *unit processes*. Examples of unit processes are *coagulation* and *biological oxidation*. In coagulation, a chemical called coagulant undergoes a chemical reaction. This chemical reaction may occur in any plant or factory or any location at all where the function of the chemical reaction in coagulation is utilized. For example, coagulation is employed in water treatment to enhance the settling of the turbidity of raw waters. Coagulation may also be used in the clarification of sugar cane juice to remove the fibers that the juice may contain.

The other unit process, biological oxidation, is used in sewage treatment; it may also be used in biofiltration applied in water treatment. The biological reaction that occurs in either sewage treatment or biofiltration are the same. In other words, the function of the biological reaction is the same whether the reaction occurs in sewage treatment or in biofiltration.

Coagulation and biological oxidation are identified on the basis of the function of their characteristic chemical or biological reactions irrespective of the plant, factory, or any other location that uses the reactions. The function of a coagulation reaction is coagulation whether the reaction occurs in a water treatment plant or in a sugar plant; and the function of a biological oxidation reaction is biological oxidation whether the reaction occurs in a sewage treatment plant or in a water treatment plant. The setting is immaterial; what is of concern is the function of the chemical reaction. Unit processes may therefore be defined as chemical (or biological) treatments that are identified only according to the functions of the chemical (or biological) reactions irrespective of where the units utilizing the reactions are occurring.

COVERAGE

This book is divided into four general parts and addresses important topics on the physical–chemical treatment of water and wastewater. The "first" part is titled Background Prerequisites. Its contents are not really physical–chemical treatment but just background knowledge to comfortably understand the method of approach used in the book. This book is analytical and therefore requires the use of pertinent chemistry, mathematics, and fluid mechanics. Relevant prerequisite topics are discussed in this part.

Part I, Characteristics of Water and Wastewater, covers the chapters on quantity and constituent physical and chemical characteristics of water and wastewater. Part II, Unit Operations of Water and Wastewater Treatment, includes the chapters on flow measurements and flow and quality equalization; pumping; screening, sedimentation, and flotation; mixing and flocculation; conventional filtration; advanced filtration and carbon adsorption; and aeration and stripping. Part III, Unit Processes of Water and Wastewater Treatment, includes the chapters on coagulation; water softening; chemical stabilization; removal of iron and manganese; removal of phosphorus; removal of nitrogen by nitrification–denitrification; ion exchange; and disinfection. Removal of nitrogen by nitrification–denitrification, on the surface, is a biological process. To control the process, however, its intrinsic chemical reactions must be unraveled and totally understood. The treatment (as used in this textbook) therefore turns toward being chemical in nature. The organisms only serve as mediators (i.e., the producer of enzymes needed for the reaction). Thus, on the most fundamental level, nitrogen removal by nitrification–denitrification is a chemical process (more, accurately, a biochemical process), which is subsequently included as a chapter in Part III of this book.

CLEAN WATER ACT

To gain a broader perspective of the Clean Water Act, it is important to know about the United States Code (USC).* This code is a consolidation and codification of the *general* and *permanent laws* of the United States. It contains several titles; in one of these titles, the Clean Water Act is codified. Because many of the general and permanent laws required to be incorporated into the code are inconsistent, redundant, and obsolete, the Office of the Law Revision Counsel of the House of Representatives revises for enactment into law each title of the code. This process is called *positive law codification*. Positive law codification is the process of preparing a bill and enacting it into law, one title at a time, a revision and restatement of the general and permanent laws of the United States. This codification does not change the meaning or legal effect of the statute but removes ambiguities, contradictions, and other imperfections from the law. Certain titles of the code have now been enacted into positive law, and pursuant to Section 204 of Title 1 of the code, the text of these titles is legal evidence of the law. The other titles of the code that have not been

---

* The United States Code itself is public domain. Portions of the code can be used and redistributed without permission from the U.S. Government Printing Office.

TABLE 1
The United States Code

Title 1	General Provisions	Title 26	Internal Revenue Code
Title 2	The Congress	Title 27	Intoxicating Liquors
Title 3	The President	Title 28	Judiciary and Judicial Procedure
Title 4	Flag and Seal, Seat of Government, and the States	Title 29	Labor
		Title 30	Mineral Lands and Mining
Title 5	Government Organization and Employees	Title 31	Money and Finance
		Title 32	National Guard
Title 6	Surety Bonds (repealed)	Title 33	Navigation and Navigable Waters
Title 7	Agriculture	Title 34	Navy (repealed)
Title 8	Aliens and Nationality	Title 35	Patents
Title 9	Arbitration	Title 36	Patriotic Societies and Observances
Title 10	Armed Forces	Title 37	Pay and Allowances of the Uniformed Services
Title 11	Bankruptcy		
Title 12	Banks and Banking	Title 38	Veterans' Benefits
Title 13	Census	Title 39	Postal Service
Title 14	Coast Guard	Title 40	Public Buildings, Property, and Works
Title 15	Commerce and Trade	Title 41	Public Contracts
Title 16	Conservation	Title 42	The Public Health and Welfare
Title 17	Copyrights	Title 43	Public Lands
Title 18	Crimes and Criminal Procedure	Title 44	Public Printing and Documents
Title 19	Customs Duties	Title 45	Railroads
Title 20	Education	Title 46	Shipping
Title 21	Food and Drugs	Title 47	Telegraphs, Telephones, and Radiotelegraphs
Title 22	Foreign Relations and Intercourse		
		Title 48	Territories and Insular Possessions
Title 23	Highways	Title 49	Transportation
Title 24	Hospitals and Asylums	Title 50	War and National Defense
Title 25	Indians		

enacted into positive law are *prima facie* evidence of the laws contained in these titles. Table 1 shows the listing of all the titles of the USC.

Refer to Table 1. The codification of the Clean Water Act is within Title 33, Navigation and Navigable Waters. This title contains 41 chapters, the first chapter being Navigable Waters Generally and the 41st chapter being National Coastal Monitoring. Chapter 26 of this title is Water Pollution Prevention and Control. This is the codification of the Clean Water Act.

Each title of the USC is divided into several sections. Title 33, of course, starts with Section 1, which is under Chapter 1, Navigable Waters Generally. Chapter 26 starts at Section 1251 and ends at Section 1387. A portion of the USC is cited by specifying the title number before the USC and the section or the range of sections after the USC. For example, 33 USC 1251-1387 is the codification of the Clean Water Act.

Environmental pollution is global. Both developed and developing countries all experience this problem. These countries therefore enact laws and regulations to

TABLE 2
Clean Water Act (Chapter 26) and Major Amendments

Year	Act	Public Law
1948	Federal Water Pollution Control Act	P. L. 80-845 (June 30, 1948)
1956	Water Pollution Control Act of 1956	P. L. 84-660 (July 9, 1956)
1961	Federal Water Pollution Control Act Amendments	P.L. 87-88
1965	Water Quality Act of 1965	P.L. 89-234
1966	Clean Water Restoration Act	P.L. 89-753
1970	Water Quality Improvement Act of 1970	P.L. 91-224, Part I
1972	**Federal Water Pollution Control Act Amendments**	**P.L. 92-500**
1977	Clean Water Act of 1977	P.L. 95-217
1981	Municipal Wastewater Treatment Construction Grants Amendments	P.L. 97-117
1987	Water Quality Act of 1987	P.L. 100-4

control the continuing pollution of the environment. In the United States, the passage of the Rivers and Harbors Act of 1899 may be considered as the first law controlling water pollution. Subsequently, the Federal Water Pollution Control Act, codified as 33 U.S.C. 1251-1387, was passed in 1948. The Federal Water Pollution Control Act is the full name of the Clean Water Act mentioned in the previous paragraphs. This Act is the principal law governing pollution of surface waters of this country. The following years show various amendments to this act: 1956, 1961, 1965, 1966, and 1970; but it was in 1972 that the act was totally revised by amendments to have taken practically its current shape. The 1972 legislation spelled out ambitious programs for water quality improvement that have since been expanded by various additional amendments. Congress made certain fine-tuning amendments in 1977, revised portions of the law in 1981, and enacted further amendments in 1987, the last year that the act was amended. Table 2 traces the dates of the major amendments to the Clean Water Act and Table 3 summarizes the major sections of the Act (indicated in the second column) as codified in the corresponding sections of the USC (first column).

The following events were common before 1972: Lake Erie was dying; the Chesapeake Bay was deteriorating rapidly; and the Potomac River was clogged with blue-green algae blooms. Many of the rivers were like open sewers, and sewage frequently washed up on shores. Fish kills were a common sight. Wetlands were disappearing. The Cuyahoga River in Cleveland, Ohio, was burned because of gross pollution due to a discharge of oil. To stop this widespread water pollution, in 1972, Congress enacted the first comprehensive national clean water legislation—**the Federal Water Pollution Control Act Amendments of 1972**.

The Clean Water Act focuses on improving the water quality of the nation. It provides for establishment of standards, development of technical tools, and financial assistance to address the causes of pollution and poor water quality, including municipal and industrial wastewater discharges, nonpoint source runoff pollution

TABLE 3
Major U.S. Code Sections of the Clean Water Act

Code Section	Subchapter
	Subchapter I—Research and Related Programs
1251	Congressional declaration of goals and policy, Sec. 101
1252	Comprehensive programs for water pollution control, Sec. 102
1253	Interstate cooperation and uniform laws, Sec. 103
1254	Research, investigations, training and information, Sec. 104
1255	Grants for research and development, Sec. 105
1256	Grants for pollution control programs, Sec. 106
1257	Mine water pollution demonstrations, Sec. 107
1258	Pollution control in the Great Lakes, Sec. 108
1259	Training grants and contracts, Sec. 109
1260	Applications for training grants and contracts, allocations, Sec. 110
1261	Scholarships, Sec. 111
1262	Definitions and authorization, Sec. 112
1263	Alaska village demonstration project, Sec. 113
1265	In-place toxic pollutants, Sec. 115
1266	Hudson River reclamation demonstration project, Sec. 116
1267	Chesapeake Bay, Sec. 117
1268	Great Lakes, Sec. 118
1269	Long Island Sound, Sec. 119
1270	Lake Champlain management conference, Sec. 120
	Subchapter II—Grants for Construction of Treatment Works
1281	Congressional declaration of purpose, Sec. 201
1282	Federal share, Sec. 202
1283	Plans, specifications, estimates, and payments, Sec. 203
1284	Limitations and conditions, Sec. 204
1285	Allotment of grant funds, Sec. 205
1286	Reimbursement and advanced construction, Sec. 206
1287	Authorization of appropriations, Sec. 207
1288	Areawide waste treatment management, Sec. 208
1289	Basin planning, Sec. 209
1290	Annual survey, Sec. 210
1291	Sewage collection system, Sec. 211
1292	Definitions, Sec. 212
1293	Loan guarantees, Sec. 213
1294	Wastewater recycling and reuse information and education, Sec. 214
1295	Requirements for American materials, Sec. 215
1296	Determination of priority, Sec. 216
1297	Guidelines for cost effective analysis, Sec. 217
1298	Cost effectiveness, Sec. 218
1299	State certification of projects, Sec. 219

(*continued*)

TABLE 3 (*Continued*)
Major U.S. Code Sections of the Clean Water Act

Code Section	Subchapter
	Subchapter III—Standards and Enforcement
1311	Effluent Limitations, Sec. 301
1312	Water quality-related effluent limitations, Sec. 302
1313	Water quality standards and implementation plans, Sec. 303
1314	Information and guidelines, Sec. 304
1315	State reports on water quality, Sec. 305
1316	National standards of performance, Sec. 306
1317	Toxic and pretreatment effluent standards, Sec. 307
1318	Records and reports, inspections, Sec. 308
1319	Enforcement, Sec. 309
1320	International pollution abatement, Sec. 310
1321	Oil and hazardous substance liability, Sec. 311
1322	Marine sanitation devices, Sec. 312
1323	Federal facility pollution control, Sec. 313
1324	Clean lakes, Sec. 314
1325	National study commission, Sec. 315
1326	Thermal discharges, Sec. 316
1327	Omitted (alternative financing), Sec. 317
1328	Aquaculture, Sec. 318
1329	Nonpoint source management program, Sec. 319
1330	National estuary study, Sec. 320
	Subchapter IV—Permits and Licenses
1341	Certification, Sec. 401
1342	National pollutant discharge elimination system, Sec. 402
1343	Ocean discharge criteria, Sec. 403
1344	Permits for dredge and fill materials, Sec. 404
1345	Disposal or use of sewage sludge, Sec. 405
	Subchapter V—General Provisions
1361	Administration, Sec. 501
1362	Definitions, Sec. 502
1363	Water pollution control advisory board, Sec. 503
1364	Emergency powers, Sec. 504
1365	Citizen suits, Sec. 505
1366	Appearance, Sec. 506
1367	Employee protection, Sec. 507
1368	Federal procurement, Sec. 508
1369	Administrative procedure and judicial review, Sec. 509
1370	State authority, Sec. 510
1371	Authority under other laws and regulations, Sec. 511
1372	Labor standards, Sec. 513
1373	Public health agency coordination, Sec. 514
1374	Effluent standards and water quality information advisory committee, Sec. 515

TABLE 3 (*Continued*)
Major U.S. Code Sections of the Clean Water Act

Code Section	Subchapter
1375	Reports to Congress, Sec. 516
1376	Authorization of appropriations, Sec. 517
1377	Indian tribes, Sec. 518
	Subchapter VI—State Water Pollution Control Revolving Funds
1381	Grants to states for establishment of revolving funds, Sec. 601
1382	Capitalization grant agreements, Sec. 602
1383	Water pollution control revolving loan funds, Sec. 603
1384	Allotment of funds, Sec. 604
1385	Corrective actions, Sec. 605
1386	Audits, reports, fiscal controls, intended use plan, Sec. 606
1387	Authorization of appropriations, Sec. 607

from urban and rural areas, and habitat destruction. For example, the Clean Water Act requires:

- Establishment of water quality standards by the states and tribes for their waters and development of pollution control programs to achieve them
- Establishment of a minimum level of wastewater treatment for all publicly owned facilities
- Meeting of performance standards by major industries to ensure pollution control
- Funding by the Federal government to states and communities to help meet their wastewater infrastructure needs
- Protection of wetlands and other aquatic habitats through a permitting process that ensures environmentally sound development

The 1972 Clean Water Act declared as its objective the restoration and maintenance of the chemical, physical, and biological integrity of the nation's waters. Two goals were established: (1) zero discharge of pollutants by 1985; and (2) as an interim goal, water quality that is fishable and swimmable by the middle of 1983. These goals were, of course, not being met.

Aside from research and related programs provision, essentially, the Clean Water Act consists of three major parts:

1. Regulatory requirements under the title of Subchapter III
2. Provisions that authorize federal financial assistance for municipal sewage treatment plant construction under the titles of Subchapters II and VI
3. Permits and enforcement under the titles of Subchapters IV and III, respectively.

In terms of historical perspective, these parts are discussed in sequence next.

REGULATORY REQUIREMENTS

As mentioned previously, the act requires each state and tribe to establish water quality standards for all bodies of water in their jurisdictions (Section 303). The Clean Water Act utilizes both water-quality- and technology-based effluent limitations to meet these standards (Sections 301 and 302). Technology-based effluent limitations are normally specified in discharge permits for industries, while water-quality-based effluent limitations are normally specified in discharge permits for publicly owned treatment works (POTWs), although the requirement of secondary treatment in publicly owned treatment works is also technology-based. Water-quality-based effluent limitations are derived by a water quality modeling which, for simple discharges, uses the Streeter–Phelps equation.

Because of strict demands imposed on those who are regulated to achieve higher and higher levels of pollution abatement, the act is a technology-forcing statute. For example, the act started with only requiring the implementation of the secondary and best practicable treatment (BPT) levels of treatment. The requirements, however, gradually increased to requiring the use of the best available technology (BAT) economically achievable, control of toxics, and control of nonpoint pollution sources.

The following scenario depicts the technology-forcing nature of the Clean Water Act. Publicly owned treatment works were once required to meet the secondary treatment level of treatment by July 1, 1977. By this date, industries were also required the equivalent BPT level of treatment. Municipalities that were unable to achieve the secondary treatment by the deadline were allowed to apply for extensions on a case-by-case basis up to July 1, 1988. According to an estimate by the EPA, 86% of all cities met the 1988 deadline with the remainder being put under judicial or administrative schedules requiring compliance as soon as possible.

By no later than March 31, 1989, the act required greater pollutant removal than BPT, generally forcing that industry use BAT technology. Toxic pollutants are generally the target of BAT levels of control. For industrial sources utilizing innovative or alternative technology, compliance extensions of as long as two years were available.

Control of toxic pollutant discharges has now become a key focus in water pollution abatement programs. For waters expected to remain polluted by toxic chemicals even after industrial dischargers have installed the best available cleanup technologies required under the law, the states are required to implement control strategies, in addition to the BAT national standards. In the 1987 Clean Water Act amendments, development of management programs for these post-BAT pollutant problems was a prominent element. This would likely be a key continuing aspect of any Clean Water Act amendments.

It should be realized that all these extensions for compliance are in the nature of forcing the development of technology to achieve compliance. Thus, it is no longer sufficient that the rule of thumb that has been traditionally used in environmental engineering be used in an effort to meet requirements. Instead, the design of unit operations and processes to achieve compliance should be instituted in a more rational and analytical approach.

The original attention of the Clean Water Act prior to the 1987 amendments was primarily directed at point source pollution, which includes wastes discharged from discrete and identifiable sources such as pipes and other outfalls, and did not specifically address control of nonpoint pollution sources. Yet, nonpoint source pollution is equally damaging to water quality. This type of pollution includes stormwater runoff from agricultural lands, forests, construction sites, and urban areas. It is this type of pollution that was the cause of a *Pfiesteria piscicida* outbreak in August 1997 in the Chicamacomico River on the Eastern Shore of Maryland. This is a microorganism that releases toxic substances and is widely believed to be responsible for major fish kills and diseases in several mid-Atlantic states.

As the rain runs off, it picks up pollutant including sediments, toxic materials, nutrients, and conventional wastes that can degrade water quality; these form the nonpoint source pollution. Except for general planning activities, little attention had been paid to these new types of pollution, despite estimates that nonpoint source pollution represents more than 50% of the nation's remaining water pollution problems.

In the 1987 amendments, Section 319 was added. This section authorizes measures to address nonpoint source pollution by directing states to develop and implement nonpoint pollution management programs. In these programs, states were encouraged to pursue groundwater protection activities as part of their overall nonpoint pollution control efforts. Federal grants were authorized to support demonstration projects and actual control activities.

Although the act imposes great technological demands, it also recognizes the need for comprehensive research on water quality problems as stipulated in Title I. Funds were provided for research in the Great Lakes (Section 118) and Chesapeake Bay (Section 117), in-place toxic pollutants in harbors and navigable waterways (Section 115), and water pollution resulting from mine drainage (Section 107). In addition, the act also provides support to train personnel who operate and maintain wastewater treatment facilities (Sections 109 and 110).

FEDERAL FINANCIAL ASSISTANCE

The following treatment traces the history of federal financial assistance to the states. The federal government has been giving grants for planning, design, and construction of municipal sewage treatment facilities since 1956. Beginning with the 1972 amendments, Congress has continued to expand this activity. Since that time, Congress has appropriated $69 billion in funds to aid wastewater treatment plant planning, design, and construction. This appropriation has been made possible through Titles II and VI of the Act. It is estimated that $140 billion more would be required to build and upgrade needed municipal wastewater treatment plants in the United States and for other types of water quality improvement projects that are eligible for funding under the Act.

Under the Title II construction grants program established in 1972, federal grants were made available for several types of projects such as secondary treatment, advanced treatment, and associated sewers. Grants were given based on a priority list established by the states (Section 216). The federal share of the total project cost was generally as much as 55% (Section 202). For projects using innovative or alternative technology such as reuse or recycling of water, as much as 75% federal funds were available. Recipients were not required to repay the federal grants.

Interested parties and policy makers have debated the tension between Title II construction grants program funding needs and the overall federal spending and budget deficits. The 1987 amendments to the Act dealt with this apparent conflict by terminating federal aid for wastewater treatment construction in fiscal year 1994, but providing a transition period toward full state and local government responsibility for financing after this date. Grants under the traditional Title II were continued only through fiscal year 1990.

To allow the states to be self-sustaining in financing their wastewater construction projects, Title VI was created to replace Title II as a federal funding mechanism. This title authorizes grants to capitalize state water pollution control revolving funds or loan programs (Section 603) beginning in fiscal year 1989. States contribute 20% matching funds, and under the revolving loan fund concept, monies used for wastewater treatment construction will be repaid to a state fund, to be available for future construction in other communities. All states now have functioning loan programs, but the shift from grants to loans since fiscal year 1991, after the Title II monies were discontinued in 1990, has not been easy for some. The new financing requirements have been especially a problem for small towns that have difficulty repaying project loans. Because of this problem, however, although statutory authorization for grants to capitalize state loan programs expired in 1994, Congress has continued to provide annual appropriations.

Permits and Enforcement

As mentioned earlier, the 1972 Clean Water Act declared as its objective the restoration and maintenance of the chemical, physical, and biological integrity of the nation's waters. To achieve this objective, the Act assumes that all discharges into the nation's waters are unlawful, except as authorized by a permit. Several sections of the Act require permits and licenses: Sections 402, 404, 405, 403, and 401. Whereas some of these sections will be specifically addressed and explained in the following, the others will not.

Because of discharge permit requirements, some 65,000 industrial and municipal dischargers must obtain permits from the EPA or qualified states. This is required under the National Pollutant Discharge Elimination System (NPDES) program of Section 402 of the Act. An NPDES permit requires the discharger to attain technology-based effluent limits (BPT or BAT for industry and secondary treatment for municipalities), or more stringent limits for water quality-limited waters. Permits specify the control technology applicable to each pollutant, the effluent limitations a discharger must meet, and the deadline for any compliance that must be met. For POTWs with collection systems that receive discharges from industries, they are required to incorporate a pretreatment program for their industrial contributors. POTWs are required to maintain records and to carry out effluent monitoring activities. Permits are issued for 5-year periods and must be renewed thereafter to allow continued discharge.

The NPDES permit incorporates numerical effluent limitations and best management practices. Whereas the BPT and secondary limitations focus on regulating discharges of conventional pollutants, the more stringent BAT limitations emphasize controlling toxic pollutants such as heavy metals, pesticides, and other organic chemicals. BAT limitations apply to categories of industries. In addition, the EPA has issued water quality criteria for more than 115 pollutants, including 65 named priority pollutants. These criteria recommend ambient, or overall, concentration levels for the pollutants and provide guidance to states for establishing water quality standards that will achieve the goals of the Act.

Disposal of dredge or fill materials into receiving bodies of water, including wetlands, is controlled under a separate type of permit program. Authorized by Section 404 of the Act, the U.S. Army Corps of Engineers administers this program subject to and using environmental guidance developed by EPA. Some types of activities such as certain farming, ranching, and forestry practices that do not alter the use or character of the land, are exempt from permit requirements. In addition, some construction and maintenance activities that are deemed not to affect adversely the environment are exempt from permit requirements. As has been done to Michigan and New Jersey, the EPA may delegate certain Section 404 permitting responsibility to qualified states.

Due to the nature of wetlands, the wetlands permit program is a controversial part of the law. Some of the wetlands are privately owned. If the owner wants to develop the area, the law can intrude and impede private decision of what to do with the property. On the other hand, environmentalists seek more protection for remaining wetlands and want limits on activities that can take place there.

The other sections of the Act requiring permits and licenses include using and disposing of sewage sludge (Section 405), ocean discharge (Section 403), and water quality certification (Section 401). The Section 401 certification, obtainable from the state, will show that the project does not violate state water quality standards.

The NPDES permit is the principal enforcement tool of the act. The EPA may issue a compliance order or bring a civil suit in U.S. district courts against persons violating the terms of a permit. The penalty for violation can be as much as $25,000 per day. Stiffer penalties are rendered for criminal violations such as negligent or knowing violations. Penalties of as much as $50,000 per day or a three-year imprisonment, or both may be rendered. For knowing-endangerment violations, such as knowingly placing another person in imminent danger of death or serious bodily injury, a fine of as much as $250,000 and 15 years in prison may be rendered. Finally, the EPA is authorized to assess civil administrative penalties for certain well-documented violations of the law. Section 309 of the Act contains these civil and criminal enforcement provisions. Working with the Army Corps of Engineers, the EPA also has responsibility for enforcing against entities who engage in activities that destroy or alter wetlands.

Similar to other federal environmental laws, enforcement is shared by the EPA and the states. Because of delegation agreements (to be addressed next), however, the majority of actions taken to enforce the law are undertaken by states. This accords the states the primary responsibility, with the EPA having oversight of state enforcement. The EPA also retains the right to bring a direct action when it believes that a state has failed to take timely and appropriate action or where a state or local agency requests EPA involvement.

Finally, individuals may bring a citizen suit in U.S. district courts against persons who violate a prescribed effluent standard or limitation (Section 505). Individuals may also bring citizen suits against the administrator of the EPA or equivalent state official for failure to carry out a nondiscretionary duty under the Act.

FEDERAL AND STATE RELATIONSHIPS

Under the Clean Water Act, federal jurisdiction is broad, especially regarding the establishment of national standards or effluent limitations. For example, the EPA issues regulations containing the BPT and BAT effluent standards applicable to categories of industrial sources such as iron and steel manufacturing, organic chemical manufacturing, and petroleum refining. Certain responsibilities, however, are delegated to the states, and this act stipulates a federal–state partnership in which the federal government sets the agenda and standards for pollution abatement, while states carry out day-to-day activities of implementation and enforcement (Section 510). Delegation agreements are signed between the governor of a state and the EPA.

SAFE DRINKING WATER ACT

Refer to Table 1. The codification of the Safe Drinking Water Act is found in Title 42, "The Public Health and Welfare." The chapters of this title range from Chapter 1 to Chapter 139. Chapter 139, the last chapter, is "Volunteer Protection." Chapter 1, "The Public Health Service," was repealed and renamed as Chapter 6A, but "The Public Health Service" was retained as the chapter title. Subchapter XII of Chapter 6A is "Safety of Public Water Systems;" this is the Safe Drinking Water Act, which, as passed by Congress, is called "Title XIV—Safety of Public Water Systems." It contains Sections 300f through 300j-26. The last section pertains to certification of testing laboratories. Table 4 summarizes the major sections of the act (indicated in the second column) as codified in the corresponding sections of the USC (first column). The USC citation for the Safe Drinking Water Act is 42 USC 300f–300j-26.

In the United States, the passage of the Interstate Quarantine Act of 1893 can be considered as the first law that eventually led to the establishment of drinking water standards. Under this act, the surgeon general of the U.S. Public Health Service was empowered to "... make and enforce such regulations as in his judgment are necessary to prevent the introduction, transmission, or spread of communicable disease from foreign countries into the states or possessions, or from one state or possession into any other state or possession." It was not until 1912, however, that the first water-related regulation was promulgated. This regulation pertains to the simple prohibition of the use of the common cup on carriers of interstate commerce such as trains.

The first act of Congress that had national importance was the Safe Drinking Water Act (SDWA) of 1974. But before this enactment was made, several revisions of the drinking water standards were made in 1914, 1925, 1942, 1946, and 1962.* The following treatment traces the history of the gradually increasing trend of the drinking water standards.

The year 1913 launched the first formal and comprehensive review of drinking water concerns. It was quickly learned that the prohibition of the use of the common

---

* Pontius, F. W. at the American Water Works Association Web site.

TABLE 4
Major U.S. Code Sections of the SDWA (Title XIV—Safety of Public Water Systems)

Code Section	Part
	Part A—Definitions
300f	Definitions, Sec. 1401
	Part B—Public Water Systems
300g	Coverage, Sec. 1411
300g-1	National Drinking Water Regulations, Sec. 1412
300g-2	State Primary Enforcement Responsibility, Sec. 1413
300g-3	Enforcement of Drinking Water Regulations, Sec. 1414
300g-4	Variances, Sec. 1415
300g-5	Exemptions, Sec. 1416
300g-6	Prohibition on Use of Lead Pipes, Solder, and Flux, Sec. 1417
300g-7	Monitoring of Contaminants, Sec. 1418
300g-8	Operator Certification, Sec. 1419
300g-9	Capacity Development, Sec. 1420
	Part C—Protection of Underground Sources of Drinking Water
300h	Regulations for State Programs, Sec. 1421
300h-1	State Primary Enforcement Responsibility, Sec. 1422
300h-2	Enforcement of Program, Sec. 1423
300h-3	Interim Regulation of Underground Injections, Sec. 1424
300h-4	Optional Demonstration by States Relating to Oil or Natural Gas, Sec. 1425
300h-5	Regulation of State Programs, Sec. 1426
300h-6	Sole Source Aquifer Demonstration, Sec. 1427
300h-7	State Programs to Establish Wellhead Protection Areas, Sec. 1428
300h-8	State Ground Water Protection Grants, Sec. 1429
	Part D—Emergency Powers
300i	Emergency Powers, Sec. 1431
300i-1	Tampering with Public Water Systems, Sec. 1432
	Part E—General Provisions
300j	Assurances of Availability of Adequate Supplies of Chemicals Necessary for Treatment, Sec. 1441
300j-1	Research, Technical Assistance, Information, Training of Personnel, Sec. 1442
300j-2	Grants for State Programs, Sec. 1443
300j-3	Special Project Grants and Guaranteed Loans, Sec. 1444
300j-3a	Grants to Public Sector Agencies
300j-3b	Contaminant Standards or Treatment Technique Guidelines
300j-3c	National Assistance Program for Water Infrastructure and Watersheds
300j-4	Records and Inspections, Sec. 1445
300j-5	National Drinking Water Advisory Council, Sec. 1446
300j-6	Federal Agencies, Sec. 1447

(continued)

TABLE 4 (*Continued*)
Major U.S. Code Sections of the SDWA (Title XIV—Safety of Public Water Systems)

Code Section	Part
300j-7	Judicial Review, Sec. 1448
300j-8	Citizen's Civil Action, Sec. 1449
300j-9	General Provisions, Sec. 1450
300j-10	Appointment of Scientific, etc., Personnel by Administrator of Environmental Protection Agency for Implementation of Responsibilities; Compensation
300j-11	Indian Tribes
300j-12	State Revolving Loan Funds, Sec. 1452
300j-13	Source Water Quality Assessment
300j-14	Source Water Petition Program
	Part F—Additional Requirements to Regulate Drinking Water
300j-15	Water Conservation Plan
300j-16	Assistance to Colonials
300j-17	Estrogenic Substances Screening Program
300j-18	Drinking Water Studies
300j-21	Definitions
300j-22	Recall of Drinking Water Coolers with Lead-Lined Tanks
300j-23	Drinking Water Coolers Containing Lead
300j-24	Lead Contamination in School Drinking Water
300j-25	Federal Assistance for State Programs Regarding Lead Contamination in School Drinking Water
300j-26	Certification of testing laboratories

cup was of no value if the water placed in it was not safe in the first place. In 1914, the first drinking water standard was adopted. This required a limit for the total bacterial plate count of 100 organisms/mL and stipulated that not more than one of five 10-cc portions of each sample examined could contain *B. coli* (now called *Escherichia coli*). The standards were legally binding only on water supplies used by interstate carriers.

By 1925, the technology of filtration and chlorination was already established, and large cities encountered little difficulty in producing treated water in the range of 2 coliforms per 100 mL. Conforming to the technology-forcing nature of regulations, the standard was therefore changed to 1 coliform per 100 mL, establishing the principle of attainability. The standard now also contained limits on lead, copper, zinc, and excessive soluble mineral substances.

In 1942, significant new initiatives were stipulated into the standards. Samples for bacteriological examination from various points in the distribution system were to be obtained, requiring a minimum number of bacteriological samples for examination each month. The laboratories and the procedures used in making the examinations became subject to state and federal inspection at any time. Maximum permissible concentrations were also established for lead, fluoride, arsenic, and selenium. Salts of barium, hexavalent chromium, heavy metals, and other substances having deleterious physiological effects were not allowed in the water system. Maximum concentrations were also set for copper, iron, manganese, magnesium, zinc, chloride, sulfate, phenolic compounds, total solids, and alkalinity.

In 1946, a maximum permissible concentration was added for hexavalent chromium. The use of the salts of barium, hexavalent chromium, heavy metal glucosides, and other substances were prohibited in water treatment processes. In addition, the 1946 standards authorized the use of the membrane filter procedure for bacteriological examination of water samples.

In the early 1960s, over 19,000 municipal water systems had been identified. These systems drew surface waters for treatment into drinking water. At this time, however, federal water pollution control efforts revealed that chemical and industrial wastes had polluted many surface waterways. Thus, the 1962 standards provided the addition of more recommended maximum limiting concentrations for various substances such alkyl benzene sulfonate, barium, cadmium, carbon-chloroform extract, cyanide, nitrate, and silver.

All in all, the standards covered 28 constituents. The 1962 standards were the most comprehensive pre-SDWA federal drinking water standards. Mandatory limits were set for health-related chemical and biological impurities and recommended limits for impurities affecting appearance, taste, and odor. The implementing regulations, however, were legally binding only at the federal level, and were applicable on only about 700 water systems that supplied common carriers in interstate commerce representing fewer than 2% of the nation's water supply systems.* For the vast majority of consumers, as an enforcement tool, the 1962 standards were of limited use in ensuring safe drinking water.

The U.S. Public Health Service undertook a comprehensive survey of water supplies in the United States, known as the Community Water Supply Study (CWSS). Released in 1970, the study found that 41% of the systems surveyed did not meet the 1962 standards. Many systems were deficient with respect to various aspects of source protection, disinfection, clarification, and pressure in the distribution system. The study also showed that, although the water served to the majority of the U.S. population was safe, about 360,000 people were being supplied with potentially dangerous drinking water.

The results of the CWSS generated congressional interest in federal safe drinking water legislation. Subsequently, more studies revealed that dangerous substances were contaminating drinking water. Thirty-six organic compounds were found in the raw water supply of the Carrollton Water Treatment Plant in New Orleans, Louisiana. This plant drew raw water from the Mississippi River, a river heavily polluted with industrial wastes. Many of these wastes came from the manufacture of synthetic organic chemicals (SOCs) and had found their way into the New Orleans raw water supply. Three of the organic chemicals found in the raw water supply were chloroform, benzene, and bis-chloroethyl ether. These chemicals are known carcinogens. Additionally, trihalomethanes, a by-product of chlorination in drinking water, were discovered in The Netherlands. Trihalomethanes are suspected carcinogens.

---

* Train, R. S. (1974). Facing the real cost of clean water, *J. AWWA*, 66, 10, 562.

TABLE 5
Safe Drinking Water Act and Major Amendments

Year	Act	Public Law
1974	Safe Drinking Water Act	P.L. 93-523 (Dec. 16, 1974)
1977	SDWA Amendments of 1977	P.L. 95-190 (Nov. 16, 1977)
1979	SDWA Amendments of 1979	P.L. 96-63 (Sept. 6, 1979)
1980	SDWA Amendments of 1980	P.L. 96-502 (Dec. 5, 1980)
1986	SDWA Amendments of 1986	P.L. 99-339 (Jun. 16, 1986)
1988	Lead Contamination Control Act	P.L. 100-572 (Oct. 31, 1988)
1996	SDWA Amendments of 1996	P.L. 104-182 (Aug. 6, 1996)

The occurrence of these chemicals in the drinking water supplies heightened public awareness. As a result, Congress passed the Safe Drinking Water Act of 1974. The drinking water regulations resulting from this Act were the first to apply to all public water systems in the United States, covering both chemical and microbial contaminants. Recall that, except for the coliform standard under the Interstate Quarantine Act of 1893, drinking water standards were not legally binding until the passage of the Safe Drinking Water Act of 1974. As shown in Table 5, the Act was amended several times, the last being 1996.

Highlights of the Safe Drinking Water Act

The most important of the safe drinking water acts are the Safe Drinking Water Act of 1974, the Safe Drinking Water Act Amendments of 1986, and the Safe Drinking Water Act of 1996. The Safe Drinking Water Act was first enacted on December 16, 1974 to protect public drinking water systems in the United States from harmful contaminants. The major provision of this Act requires the development of:

1. National primary drinking water regulations (Section 1412)
2. Underground injection control regulations to protect underground sources of drinking water (Section 1428)
3. Protection programs for sole-source aquifers (Section 1427)

Most notably, the 1986 amendments include:

1. Setting drinking water regulations for 83 specified contaminants by 1989
2. Establishment of requirements for disinfection and filtration of public water supplies and providing related technical assistance to small communities
3. Banning the use of lead pipes and lead solder in new drinking water distribution systems
4. Establishing an elective wellhead protection program around public water supply wells

5. Establishing an elective demonstration grant program for states and local authorities having designated sole-source aquifers to develop ground water protection programs
6. Issuing rules for monitoring wells that inject wastes below a drinking water source

Finally, the most notable highlights of the 1996 amendments include:

1. Requiring community water systems serving more than 10,000 customers to notify customers annually of the levels of federally regulated contaminants in the drinking water. These notifications must include information on the presence of suspicious but still unregulated substances. If a violation of the standard occurs, the notifications must contain information about the health effects of the contaminants in question.
2. Establishment of programs to train and certify competent water treatment plant operators
3. Establishment of key drinking water standards for *Cryptosporidium*, certain carcinogens, and other contaminants that threaten drinking water in the United States.

DEVELOPMENT OF MCLs AND MCLGs

The Safe Drinking Water Act directs the EPA to develop national primary drinking water standards. These standard are designed to protect human health. In addition, secondary drinking water standards are developed to protect public welfare that deal primarily with contaminants affecting drinking water aesthetics such as odor, taste, and appearance. These standards are not federally enforceable and are issued only as guidelines. The primary drinking water standards are enforceable on all public water systems serving at least 25 persons.

With respect to setting standards, two terms have been invented: maximum contaminant level goals (MCLGs) and maximum contaminant levels (MCLs). MCLGs are health goals that are not enforceable. MCLs are the enforceable counterpart of the MCLGs. They are set as close to the MCLGs as feasible and are based upon treatment technologies, costs, and feasibility factors such as availability of analytical methods and treatment technology. For lead and copper, MCLGs and MCLs are not used; instead of specifying standards, water treatment is required.

The process of determining an MCL starts with an evaluation of the adverse effects caused by the chemical in question and the doses needed to cause such effects. The final result of this process is a safe dose that includes a margin of safety thought to provide protection against adverse effects. This dose is called a reference dose (RfD) and is established based on the results of animal experiments. The research results are extrapolated to humans using standard EPA methods. This extrapolation varies depending upon whether the chemical is not a carcinogen, a known or probable carcinogen, or a possible carcinogen.

For chemicals that do not cause cancer, an MCLG is established by first converting the RfD to a water concentration. This number is then divided by five. The number five is based on the assumption that exposure to the chemical through drinking water represents only one-fifth of all the possible exposures to this substance. Other sources of exposure may include air, soil, and food. In most cases, the MCL established is the same value as the MCLG.

For a known or probable carcinogen, (EPA Class A or B), the MCLG is set at zero (i.e., no amount of chemical is acceptable). Because no analytical methods can measure zero, however, the MCL is based on the lowest concentration that can be measured on a routine basis. This is known as the practical quantitation limit (PQL). Thus, it is obvious that for known or probable carcinogens, the MCL is not guaranteed to be a safe level but instead is the lowest measurable level.

For possible cancer-causing chemicals (EPA Class C—some evidence exists that they may cause cancer, but it is not very convincing), a value equivalent to the MCLG is calculated as if they were not carcinogens. This value is then divided by a factor of ten to give the final MCL. Division by ten provides a margin of safety in case the chemical is later determined to be a carcinogen.

DRINKING WATER REGULATIONS UNDER THE ACT

In support of each regulation, Section 1412(b) of the Act requires that the EPA must make available to the public a document that specifies, to the extent practicable, the population addressed by the regulation. The document must state the upper, central, and lower estimates of risk and significant uncertainties and studies that would help resolve uncertainties. Finally, the document must include peer-reviewed studies that support or fail to support estimates.

Section 1412(b) further requires that whenever the EPA proposes a national primary drinking water regulation, it must publish a cost-benefit analysis. In the analysis for alternative MCLs, the effects on sensitive subpopulation must be considered; and in the analysis for treatment technique proposed for regulations, the cost and benefit factors required for an MCL regulation must be taken into account, as appropriate.

Section 1412(b)(6) requires that when the EPA proposes an MCL, it must publish a determination as to whether the costs of the standard are justified by the benefits. If the EPA determines that the costs of an MCL are not justified by the benefits, the law allows the EPA to set an MCL that maximizes health risk reduction benefits at a cost that is justified by the benefits. This section further limits the authority of the EPA to adjust the MCL from the feasible level if the benefits are justified for systems that serve 10,000 or more persons and for systems that are unlikely to receive a variance. This section further provides that the determination by the EPA as to whether or not the benefit of an MCL justifies the cost is judicially reviewable only as part of a court's review of the associated primary drinking water regulation.

Section 1412(b)(5) authorizes the EPA to consider "risk–risk" tradeoffs when setting an MCL. In other words, an MCL may be set at a level other than the feasible level if the technology to meet the MCL would increase health risk by (1) increasing concentration of other contaminants in drinking water, or (2) interfering with treatment used to comply with other primary drinking water regulations. When establishing such an MCL, the EPA shall (1) minimize overall risk by balancing both the risk reductions from treating the individual contaminant and possible side effects of such treatment on concentrations of other contaminants, and (2) assure that the combination of treatments for the individual contaminant and other contaminants shall not be more stringent than the "feasible."

FEDERAL FINANCIAL ASSISTANCE

As discussed earlier under the Clean Water Act, the federal government has been giving grants for planning, design, and construction of municipal sewage treatment facilities in the form of revolving funds. Based on a similar concept, a State Revolving Fund (Section 1452) has been created to provide low interest loans to assist community and nonprofit noncommunity water systems in installing and upgrading treatment facilities. Part of this loan fund can be used to provide loan subsidies and loan forgiveness to poor communities. Also, based on seriousness of health risk, compliance needs, and system economic need, each year the states prepare plans identifying eligible projects and their priorities.

FEDERAL AND STATE RELATIONSHIPS

As in the Clean Water Act, federal jurisdiction is broad, but the states have the primary responsibility of enforcing the law (Section 1413) provided, however, that the EPA has determined that the state to enforce the law has adopted drinking water regulations that are no less stringent than the national primary drinking water regulations promulgated by the Administrator of the EPA. As a condition of primacy, states have the authority to impose administrative penalties. For example, for systems serving more than 10,000 persons, states are able to assess not less than $1,000 per day per violation. For smaller systems, states only have the authority adequate to ensure compliance. The EPA has the authority to take over the primacy, if the state fails to implement the authority given to it in the delegation.

RELATIONSHIP OF THIS BOOK TO THE ACTS

As we learned from the previous discussions of the Clean Water Act and Safe Drinking Water Act, more advanced and sophisticated technologies are needed for the more stringent requirements in meeting the water quality standards and drinking water standards. It is no longer sufficient that empirical knowledge be used in treating water and wastewater. It is true that, armed with an empirical knowledge, an engineer can proceed to design unit operations and unit processes that will treat a certain contaminant. But to advance to the next level of sophistication in treatment, an understanding of the underlying concept of the processes is important and, certainly, would be necessary. For example, how do we economically remove trihalomethanes? How about radionuclides and the 83 contaminants specified in the 1986 amendments of the Safe Drinking Water Act? This book therefore presents the fundamental concepts of the unit operations and unit processes used in the treatment of water and wastewater. The authors hope to enlighten engineers and other professionals, who are engaged in water and wastewater treatment practice, with the ability to answer not only the *how* but also the *why* of the physical and chemical treatment of water and wastewater. It is, therefore, by necessity, analytical in nature.

GLOSSARY

Chemical treatment—A process applied to water and wastewater in which chemical changes occur.

Physical treatment—A process applied to water and wastewater in which no chemical changes occur.

Physical–chemical treatment—A process applied to water and wastewater in which chemical changes may or may not occur.

Unit operations—Physical treatments that are identified only according to their functions without particular reference to the location of the units utilizing the functions.

Unit processes—Chemical (or biological) treatments that are identified only according to the functions of the chemical (or biological) reactions, irrespective of where the units utilizing the reactions are occurring.

Wastewater—The spent water after homes, commercial establishments, industries, public institutions, and similar entities have used their waters for various purposes.

Background Chemistry and Fluid Mechanics

As mentioned, this chapter discusses the background knowledge needed in order to understand the subsequent chapters of this book. The student must have already gained this knowledge, but it is presented here as a refresher. Again, the background knowledge to be reviewed includes chemistry and fluid mechanics. Before they are discussed, however, units used in calculations need to be addressed, first. This is important because confusion may arise if there is no technique used to decipher the units used in a calculation.

UNITS USED IN CALCULATIONS

In calculations, several factors may be involved in a term and it is important to keep tract of the units of each of the factors in order to ascertain the final unit of the term. For example, consider converting 88 kilograms to micrograms. To make this conversion, several factors are present in the term for the calculation. Suppose we make the conversion as follows:

$$88 \, kg = 88(1000)(1000)(1000) = 88,000,000,000 \, \mu g \qquad (1)$$

where $88(1000)(1000)(1000)$ is called the *term* of the calculation and 88 and the 1000s are the *factors* of the term. As can be seen, it is quite confusing what each of the 1000s refers to. To make the calculation more tractable, it may be made by putting the units in each of the factors. Thus,

$$88 \, kg = 88 \, kg\left(1000\frac{g}{kg}\right)\left(1000\frac{mg}{g}\right)\left(1000\frac{\mu g}{mg}\right) = 88,000,000,000 \, \mu g \qquad (2)$$

This second method makes the calculations more tractable, but it makes the writing long, cumbersome, and impractical when several pages of calculations are done. For example, in designs, the length of the calculations can add up to the thickness of a book. Thus, in design calculations, the first method is preferable with its attendant possible confusion of the units. Realizing its simplicity, however, we must create some method to make it tractable.

This is how it is done. Focus on the right-hand side of the equation of the first calculation. It is known that the unit of 88 is kg, and we want it converted to μg. Remember that conversions follow a sequence of units. For example, to convert kg to μg, the sequence might be any one of the following:

$$kg \Rightarrow g \Rightarrow mg \Rightarrow \mu g \qquad (3)$$

$$kg \Rightarrow \mu g \qquad (4)$$

In Sequence (3), the conversion follows the *detailed steps*: first, from kg to g, then from g to mg and, finally, from mg to μg. Looking back to Equation (1), this is the sequence followed in the conversion. The first 1000 then refers to the g; the second 1000 refers to the mg; and the last 1000 refers to the μg. Note that in this scheme, the value of a given unit is exactly the equivalent of the previous unit. For example, the value 1000 for the g unit is exactly the equivalent of the previous unit, which is the kg. Also, the value 1000 for the mg unit is exactly the equivalent of its previous unit, which is the g, and so on with the μg.

In Sequence (4), the method of conversion is a *short cut*. This is done if the number of micrograms in a kilogram is known. Of course, we know that there are 10^6 micrograms in a kilogram. Thus, using this sequence to convert 88 kg to micrograms, we proceed as follows:

$$88 \, kg = 88(10^6) = 88,000,000,000 \, \mu g$$

Example 1 Convert 88,000,000,000 μg to kg using the detailed-step and the short-cut methods.

Solution: Detailed step:

$$88,000,000,000 \, \mu g = 88,000,000,000(10^{-3})(10^{-3})(10^{-3}) = 88 \, kg \quad \textbf{Ans}$$

Note that the first (10^{-3}) refers to the mg; the second (10^{-3}) refers to the g; and the last (10^{-3}) refers to the kg. There is no need to write the units specifically.
Short cut:

$$88,000,000,000 \, \mu g = 88,000,000,000(10^{-6}) = 88 \, kg \quad \textbf{Ans}$$

Note that there are 10^{-6} kg in one μg.

Example 2 Convert 10 cfs to m^3/d.

Solution: cfs is cubic feet per second.

$$1 \, ft = 1/3.281 \, m$$

$$1 \, s = 1/60 \, min; \; 1 \, min = 1/60 \, hr; \; 1 \, hr = 1/24 \, d$$

Therefore,

$$10 \, cfs = 10 \, ft^3/s = 10\left(\frac{1}{3.281}\right)^3 \frac{1}{\left(\frac{1}{60}\right)\left(\frac{1}{60}\right)\left(\frac{1}{24}\right)} = 24,462 \, m^3/d \quad \textbf{Ans}$$

Now, let us turn to a more elaborate problem of substituting into an equation. Of course, to use the equation, all the units of its parameters must be known. Accordingly, when the substitution is done, these units must be satisfied. For example, take the following equation:

$$I = \frac{96,494[C_o]Q_o\eta(m/2)}{M} \tag{5}$$

It is impossible to use the above equation if the units of the factors are not known. Thus, any equation, whether empirically or analytically derived, must always have its units known. In the above equation, the following are the units of the factors:

I	amperes, A
$[C_o]$	gram equivalents per liter, geq/m^3
Q_o	cubic meters per second, $\text{m}^3\text{/s}$
η	dimensionless
m	dimensionless

Now, with all the units known, it is easy to substitute the values into the equation; the proper unit of the answer will simply fall into place.

The values can be substituted into the equation in two ways: *direct substitution* and *indirect substitution*. Direct substitution means substituting the values directly into the equation and making the conversion into proper units while already substituted. Indirect substitution, on the other hand, means converting the values into their proper units outside the equation before inserting them into the equation. These methods will be elaborated in the next example.

Example 3   In the equation $I = \frac{96,494[C_o]Q_o\eta(m/2)}{M}$, the following values for the factors are given: $[C_o] = 4000$ mg/L of NaCl; $Q_o = 378.51$ m^3/d; $\eta = 0.77$, $m = 400$, and $M = 0.90$. Calculate the value of I by indirect substitution and by direct substitution.

Solution: Indirect substitution:
In indirect substitution, all terms must be in their proper units before making the substitution, thus:

NaCl = 23 + 35.45 = 58.45, molecular mass of sodium chloride

Therefore,

$$[C_o] = 4000\,\text{mg/L} = \frac{4000(10^{-3})}{58.45} = 0.068\,\text{gmols/L}$$

$$= 0.068\,\text{geq/L} = 68\,\text{geq/m}^3$$

$$Q_o = 378.51\,\text{m}^3\text{/d} = 378.51\left[\frac{1}{24(60)(60)}\right] = 0.0044\,\text{m}^3\text{/s}$$

And, now, substituting, therefore,

$$I = \frac{96,494[C_o]Q_o\eta(m/2)}{M} = \frac{96,494(68)(0.0044)(0.77)(400/2)}{0.90}$$

$$= 122.84 \ A \quad \textbf{Ans}$$

Direct substitution:

$$I = \frac{96,494[C_o]Q_o\eta(m/2)}{M}$$

$$= \frac{96,494\left\{\frac{4000(10^{-3})}{58.45}(1000)\right\}\left\{378.51\left[\frac{1}{24(60)(60)}\right]\right\}(0.77)/(400/2)}{0.90}$$

$$= 122.84 \ A \quad \textbf{Ans}$$

Note that the conversions into proper units are done inside the equation, and that the conversions for $[C_o]$ and Q_o are inside pairs of braces { }. Once accustomed to viewing these conversions, you may not need these braces anymore.

One last method of ascertaining units in a calculation is the use of consistent units. If a system of units is used consistently, then it is not necessary to keep track of the units in a given calculation. The proper unit of the answer will automatically fall into place.

The system of units is based upon the general dimensions of space, mass, and time. Space may be in terms of displacement or volume and mass may be in terms of absolute mass or relative mass. An example of absolute mass is the gram, and an example of relative mass is the mole. The mole is a relative mass, because it expresses the ratio of the absolute mass to the molecular mass of the substance. When the word mass is used without qualification, absolute mass is intended.

The following are examples of systems of units: meter-kilogram-second (mks), meter-gram-second (mgs), liter-gram-second (lgs), centimeter-gram-second (cgs), liter-grammoles-second (lgmols), meter-kilogrammoles-second (mkmols), centimeter-grammoles-second (cgmols), etc. Any equation that is derived analytically does not need to have its units specified, because the units will automatically conform to the general dimension of space, mass, and time. In other words, the units are automatically specified by the system of units chosen. For example, if the mks system of units is chosen, then the measurement of distance is in units of meters, the measurement of mass is in units of kilograms, and the measurement of time is in seconds. Also, if the lgs system of units is used, then the volume is in liters, the mass is in grams, and the time is in seconds. **To repeat, if consistent units are used, it is not necessary to keep track of the units of the various factors, because these units will automatically fall into place by virtue of the choice of the system of units**. The use of a consistent system of units is illustrated in the next example.

Example 4   The formula used to calculate the amount of acid needed to lower the pH of water is

$$[A_{cadd}]_{geq} = [A_{cur}]_{geq} + \frac{10^{-pH_{to}} - 10^{-pH_{cur}}}{\phi}$$

Calculate the amount of acid needed using the lgmols system of units.

Solution: Of course, to intelligently use the above equation, all the factors should be explained. We do not need to do it here, however, because we only need to make substitutions. Because the lgmols units is to be used, volume is in liters, mass is in gram moles, and time is in seconds. Therefore, the corresponding concentration is in gram moles per liter (gmols/L). Another unit of measurement of concentration is also used in this equation, and this is equivalents per liter. For the lgmols system, this will be gram equivalents per liter (geq/L).

Now, values for the factors need to be given. These are shown below and note that no units are given. Because the lgmols system is used, they are understood to be either geq/L or gmols/L. Again, it is not necessary to keep track of the units; they are understood from the system of units used.

$$[A_{cur}]_{geq} = 2.74(10^{-3}) \qquad pH_{to} = 8.7 \qquad pH_{cur} = 10.0$$

Therefore,

$$[A_{cadd}]_{geq} = 2.74(10^{-3}) + \frac{10^{-8.7} - 10^{-10}}{0.00422} = 2.74(10^{-3}) \text{ geq/L} \quad \textbf{Ans}$$

Note that the values are just freely substituted without worrying about the units. By the system of units used, the unit for $[A_{cadd}]_{geq}$ is automatically geq/L.

GENERAL CHEMISTRY

Chemistry is a very wide field; however, only a very small portion, indeed, of this seemingly complex subject is used in this book. These include equivalents and equivalent mass, methods of expressing concentrations, activity and active concentration, equilibrium and solubility product constants, and acids and bases. This knowledge of chemistry will be used under the unit processes part of this book.

EQUIVALENTS AND EQUIVALENT MASSES

The literature shows confused definitions of equivalents and equivalent masses and no universal definition exists. They are defined based on specific situations and are never unified. For example, in water chemistry, three methods of defining equivalent mass are used: equivalent mass based on ionic charge, equivalent mass based on acid–base reactions, and equivalent mass based on oxidation–reduction reactions (Snoeyink and Jenkins, 1980). This section will unify the definition of these terms by utilizing the concept of *reference species*; but, before the definition is unified, the aforementioned three methods will be discussed first. The result of the discussion, then, will form the basis of the unification.

Equivalent mass based on ionic charge. In this method, the equivalent mass is defined as (Snoeyink and Jenkins, 1980):

$$\text{Equivalent mass} = \frac{\text{molecular weight}}{\text{ionic charge}} \tag{6}$$

For example, consider the reaction,

$$Fe(HCO_3)_2 + 2Ca(OH)_2 \rightarrow Fe(OH)_2 + 2CaCO_3 + 2H_2O$$

Calculate the equivalent mass of $Fe(HCO_3)_2$.

When this species ionizes, the Fe will form a charge of plus 2 and the bicarbonate ion will form a charge of minus 1 but, because it has a subscript of 2, the total ionic charge is minus 2. Thus, from the previous formula, the equivalent mass is $Fe(HCO_3)_2/2$, where $Fe(HCO_3)_2$ must be evaluated from the respective atomic masses. The positive ionic charge for calcium hydroxide is 2. The negative ionic charge of OH^- is 1; but, because the subscript is 2, the total negative ionic charge for calcium hydroxide is also 2. Thus, if the equivalent mass of $Ca(OH)_2$ were to be found, it would be $Ca(OH)_2/2$. It will be mentioned later that $Ca(OH)_2/2$ is not compatible with $Fe(HCO_3)_2/2$ and, therefore, Equation (6) is not of universal application, because it ought to apply to all species participating in a chemical reaction. Instead, it only applies to $Fe(HCO_3)_2$ but not to $Ca(OH)_2$, as will be shown later.

Equivalent mass based on acid–base reactions. In this method, the equivalent mass is defined as (Snoeyink and Jenkins, 1980):

$$\text{Equivalent mass} = \frac{\text{molecular weight}}{n} \tag{7}$$

where n is the number of hydrogen or hydroxyl ions that react in a molecule.

For example, consider the reaction,

$$H_3PO_4 + 2NaOH \rightarrow 2Na^+ + HPO_4^{2-} + 2H_2O$$

Now, calculate the equivalent mass of H_3PO_4. It can be observed that H_3PO_4 converts to HPO_4^{2-}. Thus, two hydrogen ions are in the molecule of H_3PO_4 that react and the equivalent mass of H_3PO_4 is $H_3PO_4/2$, using the previous equation. The number of hydroxyl ions that react in NaOH is one; thus, using the previous equation, the equivalent mass of NaOH is NaOH/1.

Equivalent mass based on oxidation–reduction reactions. For oxidation–reduction reactions, the equivalent mass is defined as the mass of substance per mole of electrons involved (Snoeyink and Jenkins, 1980).

For example, consider the reaction,

$$4Fe(OH)_2 + O_2 + 2H_2O \rightarrow 4Fe(OH)_3$$

The ferrous is oxidized to the ferric form from an oxidation state of +2 to +3. The difference between 2 and 3 is 1, and because there are 4 atoms of Fe, the amount of electrons involved is $1 \times 4 = 4$. The equivalent mass of $Fe(OH)_2$ is then $4Fe(OH)_2/4$. Note that the coefficient 4 has been included in the calculation. This is so, because in order to get the total number of electrons involved, the coefficient must be included. The electrons involved are not only for the electrons in a molecule but for the electrons in all the molecules of the balanced chemical reaction, and, therefore must account for the coefficient of the term. For oxygen, the number of moles electrons involved will also be found to be 4; thus, the equivalent mass of oxygen is $O_2/4$.

Now, we are going to unify this equivalence using the concept of the reference species. The positive or negative charges, the hydrogen or hydroxyl ions, and the moles of electrons used in the above methods of calculation are actually references species. They are used as references in calculating the equivalent mass. Note that the hydrogen ion is actually a positive charge and the hydroxyl ion is actually a negative charge. From the results of the previous three methods of calculating equivalent mass, we can make the following generalizations:

1. The mass of any substance participating in a reaction per unit of the number of reference species is called the equivalent mass of the substance, and, it follows that
2. The mass of the substance divided by this equivalent mass is the number of equivalents of the substance.

The expression *molecular weight/ionic charge* is actually mass of the substance per unit of the reference species, where ionic charge is the reference species. The expression *molecular weight n* is also actually mass of the substance per unit of the reference species. In the case of the method based on the oxidation–reduction reaction, no equation was developed; however, the ratios used in the example are ratios of the masses of the respective substances to the reference species, where 4, the number of electrons, is the number of reference species.

From the discussion above, the reference species can only be one of two possibilities: the electrons involved in an oxidation–reduction reaction and the positive (or, alternatively, the negative) charges in all the other reactions. These species (electrons and the positive or negative charges) express the combining capacity or valence of the substance. The various examples that follow will embody the concept of the reference species.

Again, take the following reactions, which were used for the illustrations above:

$$4Fe(OH)_2 + O_2 + 2H_2O \rightarrow 4Fe(OH)_3$$

$$Fe(HCO_2)_3 + 2Ca(OH)_2 \rightarrow Fe(OH)_2 + 2CaCO_3 + 2H_2O$$

In the first reaction, the ferrous form is oxidized to the ferric form from an oxidation state of +2 to +3. Thus, in this reaction, electrons are involved, making them the reference species. The difference between 2 and 3 is 1, and since there are 4 atoms of Fe, the amount of electrons involved is $1 \times 4 = 4$. For the oxygen molecule, its atom has been reduced from 0 to -2 per atom. Since there are 2 oxygen atoms in the molecule, the total number of electrons involved is also equal to 4 (i.e., $2 \times 2 = 4$). In both these cases, the number of electrons involved is 4. This number is called the number of reference species, combining capacity, or valence. (*Number of reference species* will be used in this book.) Thus, to obtain the equivalent masses of all the participating substances in the reaction, each term must be divided by 4: $4Fe(OH)_2/4$, $O_2/4$, $2H_2O/4$, and $4Fe(OH)_3/4$. We had the same results for $Fe(OH)_2$ and O_2 obtained before.

If the total number of electrons involved in the case of the oxygen atom were different, a problem would have arisen. Thus, if this situation occurs, take the convention of using the smaller of the number of electrons involved as the number of reference species. For example if the number of electrons involved in the case of oxygen were 2, then all the participating substances in the chemical reaction would have been divided by 2 rather than 4. For any given chemical reaction or series of related chemical reactions, however, whatever value of the reference species is chosen, the answer will still be the same, provided this number is used consistently. This situation of two competing values to choose from is illustrated in the second reaction to be addressed below. Also, take note that the reference species is to be taken from the reactants only, not from the products. This is so, because the reactants are the ones responsible for the initiation of the interaction and, thus, the initiation of the equivalence of the species in the chemical reaction.

In the second reaction, no electrons are involved. In this case, take the convention that if no electrons are involved, either consider the positive or, alternatively, the negative oxidation states as the reference species. For this reaction, initially consider the positive oxidation state. Since the ferrous iron has a charge of +2, ferrous bicarbonate has 2 for its number of reference species. Alternatively, consider the negative charge of bicarbonate. The charge of the bicarbonate ion is -1 and because two bicarbonates are in ferrous bicarbonate, the number of reference species is, again, $1 \times 2 = 2$. From these analyses, we adopt 2 as the number of reference species for the reaction, subject to a possible modification as shown in the paragraph below. (Notice that this is the number of reference species for the whole reaction, not only for the individual term in the reaction. In other words, all terms and each term in a chemical reaction must use the same number for the reference species.)

In the case of the calcium hydroxide, since calcium has a charge of +2 and the coefficient of the term is 2, the number of reference species is 4. Thus, we have now two possible values for the same reaction. In this situation, there are two alternatives: the 2 or the 4 as the number of reference species. As mentioned previously, either can be used provided, when one is chosen, all subsequent calculations are based on the one particular choice; however, adopt the convention wherein the number of reference species to be chosen should be the smallest value. Thus, the number of reference species in the second reaction is 2, not 4—and all the equivalent masses of the participating substances are obtained by dividing each balanced term of the reaction by 2: $Fe(HCO_3)_2/2$, $2Ca(OH)_2/2$, $Fe(OH)_2/2$, $2CaCO_3/2$ and $2H_2O/2$.

Note that $Ca(OH)_2$ has now an equivalent mass of $2Ca(OH)_2/2$ which is different from $Ca(OH)_2/2$ obtained before. This means that the definition of equivalent mass in Equation (6) is not accurate, because it does not apply to $Ca(OH)_2$. The equivalent mass of $Ca(OH)_2/2$ is not compatible with $Fe(HCO_3)_2/2$, $Fe(OH)_2/2$, $2CaCO_3/2$, or $2H_2O/2$. Compatibility means that the species in the chemical reaction can all be related to each other in a calculation; but, because the equivalent mass of $Ca(OH)_2$ is now made incompatible, it could no longer be related to the other species in the reaction in any chemical calculation. In contrast, the method of reference species that is developed here applies in a unified fashion to all species in the chemical reaction: $Fe(HCO_3)_2$, $Ca(OH)_2$, $Fe(OH)_2$, $CaCO_3$, and H_2O and the resulting equivalent masses are therefore compatible to each other. This is so, because all the species are using the same number of reference species.

Take the two reactions of phosphoric acid with sodium hydroxide that follow. These reactions will illustrate that the equivalent mass of a given substance depends upon the chemical reaction in which the substance is involved.

$$H_3PO_4 + 2NaOH \rightarrow 2Na^+ + HPO_4^{2-} + 2H_2O$$

$$H_3PO_4 + 3NaOH \rightarrow 3Na^+ + PO_4^{3-} + 3H_2O$$

Consider the positive electric charge and the first reaction. Because Na^+ of NaOH (remember that only the reactants are to be considered in choosing the reference species) has a charge of +1 and the coefficient of the term is 2, two positive charges $(2 \times 1 = 2)$ are involved. In the case of H_3PO_4, the equation shows that the acid breaks up into HPO_4^{2-} and other substances with one H still "clinging" to the PO_4 on the right-hand side of the equation. This indicates that two H^+'s are involved in the breakup. Because the charge of H^+ is +1, two positive charges are accordingly involved. In both the cases of Na^+ and H^+, the reference species are the two positive charges and the number of reference species is 2. Therefore, in the first reaction, the equivalent mass of a participating substance is obtained by dividing the term (including the coefficient) by 2. Thus, for the acid, the equivalent mass is $H_3PO_4/2$; for the base, the equivalent mass is $2NaOH/2$, etc.

In the second reaction, again, basing on the positive electric charges and performing similar analysis, the number of reference species would be found to be +3. Thus, in this reaction, for the acid, the equivalent mass is $H_3PO_4/3$; for the base, the equivalent mass is $3NaOH/3$, etc., indicating differences in equivalent masses with the first reaction for the same substances of H_3PO_4 and NaOH. Thus, the equivalent mass of any substance depends upon the chemical reaction in which it participates.

In the previous discussions, the unit of the number of reference species was not established. A convenient unit would be the mole (i.e., mole of electrons or mole of positive or negative charges). The mole can be a milligram-mole, gram-mole, etc. The mass unit of measurement of the equivalent mass would then correspond to the type of mole used for the reference species. For example, if the mole used is the gram-mole, the mass of the equivalent mass would be expressed in grams of the substance per gram-mole of the reference species; and, if the mole used is the milligram-mole, the equivalent mass would be expressed in milligrams of the substance per milligram-mole of the reference species and so on.

Because the reference species is used as the standard of reference, its unit, the mole, can be said to have a unit of one equivalent. From this, the equivalent mass of a participating substance may be expressed as the mass of the substance per equivalent of the reference species; but, because the substance is equivalent to the reference species, the expression "per equivalent of the reference species" is the same as the expression "per equivalent of the substance." Thus, the equivalent mass of a substance may also be expressed as the mass of the substance per equivalent of the substance.

Each term of a balanced chemical reaction, represents the mass of a participating substance. Thus, the general formula for finding the equivalent mass of a substance is

$$\text{Equiv. mass} = \frac{\text{mass of substance}}{\text{number of equivalents of substance}}$$
$$= \frac{\text{term in balanced reaction}}{\text{number of moles of reference species}}$$

Example 5   Water containing 2.5 moles of calcium bicarbonate and 1.5 moles of calcium sulfate is softened using lime and soda ash. How many grams of calcium carbonate solids are produced **(a)** using the method of equivalent masses and **(b)** using the balanced chemical reaction? Pertinent reactions are as follows:

$$Ca(HCO_3)_2 + Ca(OH)_2 \rightarrow 2CaCO_3\downarrow + 2H_2O$$
$$CaSO_4 + Na_2CO_3 \rightarrow CaCO_3 + Na_2SO_4$$

Solution:

(a) $Ca(HCO_3)_2 + Ca(OH)_2 \rightarrow 2CaCO_3\downarrow + 2H_2O$

number of reference species = 2

Therefore, eq. mass of $CaCO_3 = \dfrac{2CaCO_3}{2} = 100$

eq. mass of $Ca(HCO_3)_2 = \dfrac{Ca(HCO_3)_2}{2} = \dfrac{40 + 2[1 + 12 + 3(16)]}{2}\dfrac{162}{2} = 81$

qeq of $Ca(HCO_3)_2 = \dfrac{2.5(162)}{81} = 5 = $ geq of $CaCO_3$

g of $CaCO_3$ solids $= 5(100) = 500$ g   **Ans**

$$CaSO_4 + Na_2CO_3 \rightarrow CaCO_3 + Na_2SO_4$$

number of reference species = 2

$$\text{Therefore, eq. mass of } CaCO_3 = \frac{CaCO_3}{2} = 50$$

$$\text{eq. mass of } CaSO_4 = \frac{CaSO_4}{2} = \frac{40 + 32 + 4(16)}{2} = \frac{136}{2} = 68$$

$$\text{qeq of } CaSO_4 = \frac{1.5(136)}{68} = 3 = \text{geq of } CaCO_3$$

$$\text{g of } CaCO_3 \text{ solids } = 3(50) = 150 \text{ g} \quad \textbf{Ans}$$

$$\text{Total grams of } CaCO_3 = 500 + 150 = 650 \quad \textbf{Ans}$$

(b) $Ca(HCO_3)_2 + Ca(OH)_2 \rightarrow 2CaCO_3\downarrow + 2H_2O$

$$\text{g of } CaCO_3 \text{ solids } = \frac{2CaCO_3}{Ca(HCO_3)_2}(2.5)(162) = 500$$

$$CaSO_4 + Na_2CO_3 \rightarrow CaCO_3 + Na_2SO_4$$

$$\text{g of } CaCO_3 \text{ solids } = \frac{CaCO_3}{CaSO_4}(1.5)(136) = 150$$

$$\text{Total grams of } CaCO_3 = 500 + 150 = 650 \quad \textbf{Ans}$$

METHODS OF EXPRESSING CONCENTRATIONS

Several methods are used to express concentrations in water and wastewater and it is appropriate to present some of them here. They are molarity, molality, mole fraction, mass concentration, equivalents concentration, and normality.

Molarity. Molarity is the number of gram moles of solute per liter of solution, where from the general knowledge of chemistry, gram moles is the mass in grams divided by the molecular mass of the substance in question. Take note that the statement says "per liter of solution." This means the solute and the solvent are taken together as a mixture in the liter of solution. The symbol used for molarity is M.

For example, consider calcium carbonate. This substance has a mass density ρ_{CaCO_3} equals 2.6 g/cc. Suppose, 35 mg is dissolved in a liter of water, find the corresponding molarity.

The mass of 35 mg is 0.035 g. Calcium carbonate has a molecular weight of 100 g/mol; thus, 0.035 g is 0.035/100 = 0.00035 gmol. The volume corresponding to 0.35 g is 0.035/2.6 = 0.0135 cc = 0.0000135 L and the total volume of the mixture then is 1.0000135 L. Therefore, by the definition of molarity, the corresponding molarity of 35 mg dissolved in one liter of water is 0.00035/1.0000135 = 0.00035 M.

Molality. Molarity is the number of gram moles of solute per 1000 g of solvent. Take note of the drastic difference between this definition of molality and the difinition of molarity. The solvent is now "separate" from the solute. The symbol used for molality is m.

Now, consider the calcium carbonate example above, again, and find the corresponding molality. The only other calculation we need to do is to find the number of grams of the liter of water. To do this, an assumption of the water temperature must be made. Assuming it is 5°C, its mass density is 1.0 g/cc and the mass of one liter is then 1000 g. Thus, the 0.00035 gmol of calcium carbonate is dissolved in 1000 gm of water. This is the very definition of molality and the corresponding molality is therefore 0.00035m. Note that there is really no practical difference between molarity and molality in this instance. Note that the mass density of water does not vary much from the 5°C to 100°C.

Mole fraction. Mole fraction is a method of expressing the molar fractional part of a certain species relative to the total number of moles of species in the mixture. Letting n_i be the number of moles of a particular species i, the mole fraction of this species x_i is

$$x_i = \frac{n_i}{\sum_1^N n_i} \tag{8}$$

N is the total number of species in the mixture.

Example 6   The results of an analysis in a sample of water are shown in the table below. Calculate the mole fractions of the respective species.

Ions	Conc (mg/L)
$Ca(HCO_3)_2$	150
$Mg(HCO_3)_2$	12.0
Na_2SO_4	216.0

Solution: The calculations are shown in the table, which should be self-explanatory.

Ions	Conc (mg/L)	Molecular Mass	Moles/Liter	Mole Fraction
$Ca(HCO_3)_2$	150	$40.1(2) + 2\{1 + 12 + 3(16)\} = 202.2$	0.74[a]	0.32[b]
$Mg(HCO_3)_2$	12.0	$24.3(2) + 2\{1 + 12 + 3(16)\} = 170.6$	0.07	0.03
Na_2SO_4	216.0	$23(2) + 32.1 + 4(16) = 142.1$	1.52	0.65
		Sum = Σ	2.33	1.00

[a] $\dfrac{150}{202.2} = 0.74$

[b] $\dfrac{0.74}{2.33} = 0.32$

Mass concentration. Generally, two methods are used to express mass concentration: mass of solute per unit volume of the mixture (m/v basis) and mass of the solute per unit mass of the mixture (m/m basis). In environmental engineering, the most common expression in the m/v basis is the mg/L. The most common in the m/m basis is the ppm, which means parts per million. In other words, in a ppm, there is one mass of the solute in 10^6 mass of the mixture.

One ppm for a solute dissolved in water can be shown to be equal to one mg/L. This is shown as follows: 1 ppm = $(1 \text{ mg})/(10^6 \text{ mg}) = (1 \text{ mg})/(10^3 \text{ g})$. The mass density of water at 5°C is 1.0 g/cc. Therefore,

$$1 \text{ ppm} = \frac{1 \text{ mg}}{10^6 \text{ mg}} = \frac{1 \text{ mg}}{10^3 \text{ g}} = \frac{1 \text{ mg}}{10^3 \text{ cc}} = \frac{1 \text{ mg}}{L}$$

The mass density of water decreases from 1.0 g/cc at 5°C to 0.96 g/cc at 100°C. Thus, for practical purposes,

$$1 \text{ ppm} = 1 \text{ mg/L}$$

Molar concentration. In concept, molar concentrations can also be expressed on the m/v basis and the m/m basis; however, the most prevalent practice in environmental engineering is the m/v basis. Molar concentration, then, is the number of moles of the solute per unit volume of the mixture. There are several types of moles: milligram-moles, gram-moles, tonne-moles, and so on corresponding to the unit of mass used. In chemistry, the gram moles is almost exclusively used. When the type of moles is not specified, it is understood to be gram moles. So, normally, molar concentration is expressed in gram moles per liter (gmmols/L).

An important application of molar concentration is in a molar mass balance. For example, in the removal of phosphorus using alum, the following series of reactions occurs:

$$Al^{3+} + PO_4^{3-} \rightleftharpoons AlPO_{4(s)}\downarrow$$

$$AlPO_{4(s)}\downarrow \rightleftharpoons Al^{3+} + PO_4^{3-}$$

$$PO_4^{3-} + H^+ \rightleftharpoons HPO_4^{2-}$$

$$HPO_4^{2-} + H^+ \rightleftharpoons H_2PO_4^-$$

$$H_2PO_4^- + H^+ \rightleftharpoons H_3PO_4$$

An indicator of the efficiency of removal is that the concentration of phosphorus in solution must be the minimum. To find this phosphorus in solution, it is necessary to perform a phosphate (PO_4) molar mass balance in solution using the above reactions. $AlPO_{4(s)}$ is a solid; thus, not in solution and will not participate in the balance. The other species, however, are in solution and contain the PO_4 radical. They are $[PO_4^{3-}]$, $[HPO_4^{2-}]$, $[H_2PO_4^-]$, and $[H_3PO_4]$. Note the PO_4 embedded in the respective formulas; thus, the formulas are said to contain the PO_4 radical. Because they contain this PO_4 radical, they will participate in the molar mass balance. Let $[sp_{PO_4Al}]$

represent the total molar concentration of the phospate radical contained in all the species. Then, the molar mass balance on PO_4 becomes

$$[sp_{PO_4Al}] = [PO_4^{3-}] + [HPO_4^{2-}] + [H_2PO_4^-] + [H_3PO_4]$$

The symbol $[sp_{PO_4Al}]$ needs to be explained further, since this mode of subscripting is used in the unit processes part of this book. First, [ ] is read as "the concentration of." The symbol sp stands for species and its first subscript, PO_4, stands for the type of species, which is the PO_4 radical. The second subscript, Al, stands for the "reason" for the existence of the type of species. In other words, the use of alum (the Al) is the reason for the existence of the phosphate radical, the type of species.

Example 7   In removing the phosphorus using a ferric salt, the following series of reactions occurs:

$$Fe^{3+} + PO_4^{3-} \rightleftharpoons FePO_4\downarrow$$
$$FePO_4\downarrow \rightleftharpoons Fe^{3+} + PO_4^{3-}$$
$$PO_4^{3-} + H^+ \rightleftharpoons HPO_4^{2-}$$
$$HPO_4^{2-} + H^+ \rightleftharpoons H_2PO_4^-$$
$$H_2PO_4^- + H^+ \rightleftharpoons H_3PO_4$$

Write the molar mass balance for the total molar concentration of the phosphate radical.

Solution: The solution to this problem is similar to the one discussed in the text, except that $[sp_{PO_4Al}]$ will be replaced by $[sp_{PO_4FeIII}]$. The second subscript, FeIII, stands for the ferric salt. The molar mass balance is

$$[sp_{PO_4FeIII}] = [PO_4^{3-}] + [HPO_4^{2-}] + [H_2PO_4^-] + [H_3PO_4] \quad \textbf{Ans}$$

Example 8   State the symbol $[sp_{PO_4FeIII}]$ in words.

Solution: $[sp_{PO_4FeIII}]$ is the total molar concentration of the PO_4^{3-} radical as a result of using a ferric salt.   **Ans**

Example 9   The complexation reaction of the calcium ion with the carbonate species, OH^-, and SO_4^{2-} are given below:

$$CaCO_3^o \rightleftharpoons Ca^{2+} + CO_3^{2-}$$
$$CaHCO_3^+ \rightleftharpoons Ca^{2+} + HCO_3^-$$
$$CaOH^+ \rightleftharpoons Ca^{2+} + OH^-$$
$$CaSO_4^o \rightleftharpoons Ca^{2+} + SO_4^{2-}$$

Write the molar mass balance for the total molar concentration of Ca.

Solution: Let $[Ca_T]$ represent the total molar concentration of Ca. Therefore,

$$[Ca_T] = [Ca^{2+}] + [CaCO_3^o] + [CaHCO_3^+] + [CaOH^+] + [CaSO_4^o] \quad \textbf{Ans}$$

Equivalent concentration and normality. This method of expressing concentrations is analogous to molar concentrations, with the solute expressed in terms of equivalents. One problem that is often encountered is the conversion of a molar concentration to equivalent concentration. Let $[C]$ be the molar concentration of any substance, where the symbol [ ] is read as "the concentration of." Convert this to equivalent concentration.

To do the conversion, first convert the molar concentration to mass concentration by multiplying it by the molecular mass (MM). Thus, $[C]$ (MM) is the corresponding mass concentration. By definition, the number of equivalents is equal to the mass divided by the equivalent mass (eq. mass). Therefore, the equivalent concentration, $[C]_{eq}$, is

$$[C]_{eq} = \frac{[C](MM)}{\text{eq. mass}}$$

The concentration expressed as geq/L is the normality. The symbol for normality is N.

Example 10   The concentration of $Ca(HCO_3)_2$ is 0.74 gmol/L. Convert this concentration to geq/L.

Solution:

$$[C]_{eq} = \frac{[C](MM)}{\text{eq. mass}}$$

$$[C] = 0.74 \text{ gmol/L}$$

$$MM = 40.1(2) + 2\{1 + 12 + 3(16)\} = 202.2$$

No. of reference species = 2

$$\text{Therefore, eq. mass} = \frac{Ca(HCO_3)_2}{2} = \frac{202.2}{2}$$

$$[C]_{eq} = \frac{0.74(202.2)}{\frac{202.2}{2}} = 1.48 \text{ geq/L} \quad \textbf{Ans}$$

Another problem often encountered in practice is the conversion of equivalent concentration to molar concentration. Let us illustrate this situation using the carbonate system. The carbonate system is composed of the species CO_3^{2-}, HCO_3^-, OH , and H^+. In addition, Ca^{2+} may also be a part of this system. Note that OH and H^+ are always part of the system, because a water solution will always contain these species. To perform the conversion, the respective equivalent masses of the species should first be found. In order to find the number of reference species, the pertinent chemical reactions must all be referred to a common end point. For the carbonate system, this end point is the methyl orange end point when H^+ is added to the system to form H_2CO_3.

The reaction to the end point of HCO_3^- is

$$HCO_3^- + H^+ \rightleftharpoons H_2CO_3$$

Because the number of reference species from this reaction is 1, the equivalent mass of HCO_3^- is $HCO_3/1$. For CO_3^{2-}, the reaction to the end point is

$$CO_3^{2-} = 2H^+ \rightleftharpoons H_2CO_3$$

This reaction gives the number of reference species equal to 2; thus, the equivalent mass of CO_3^{2-} is $CO_3/2$. The reaction of H^+ and OH^- to the end point gives their respective equivalent masses as $H/1$ and $OH/1$. The reaction of Ca^{2+} in the carbonate system is

$$Ca^{2+} + CO_3^{2-} \rightleftharpoons CaCO_3$$

This gives the equivalent of Ca^{2+} as $Ca/2$. The conversion from equivalent to molar concentrations is illustrated in the following example.

Example 11   An ionic charge balance in terms of equivalents for a carbonate system is shown as follows:

$$[CO_3^{2-}]_{eq} + [HCO_3^-]_{eq} + [OH]_{eq} = [H^+]_{eq} + [Ca^{2+}]_{eq}$$

Convert the balance in terms of molar concentrations.

Solution:

$$[C]_{eq} = \frac{[C](MM)}{eq.\ mass}$$

$$(CO_3^{2-})_{eq.\ mass} = \frac{CO_3}{2}; \quad (HCO_3^-)_{eq.\ mass} = \frac{HCO_3}{1}; \quad (OH)_{eq.\ mass} = \frac{OH}{1}$$

$$(H^+)_{eq.\ mass} = \frac{H}{1}; \quad (Ca^{2+})_{eq.\ mass} = \frac{Ca}{2}$$

Therefore, substituting,

$$\frac{[CO_3^{2-}](CO_3)}{\frac{CO_3}{2}} + \frac{[HCO_3^-](HCO_3)}{\frac{HCO_3}{1}} + \frac{[OH^-](OH)}{\frac{OH}{1}} = \frac{[H^+](H)}{\frac{H}{1}} + \frac{[Ca^{2+}](Ca)}{\frac{Ca}{2}}$$

$$2[CO_3^{2-}] + [HCO_3^-] + [OH] = [H^+] + 2[Ca^{2+}] \quad \textbf{Ans}$$

ACTIVITY AND ACTIVE CONCENTRATION

In simple language, *activity* is a measure of the effectiveness of a given species in a chemical reaction. It is an *effective* or *active concentration* and is proportional to concentration. It has the units of concentration. Since activity bears a relationship to concentration, its value may be obtained using the value of the corresponding concentration. This relationship is expressed as follows:

$$\{sp\} = \gamma[sp] \tag{9}$$

where sp represents any species involved in the equilibria such as Ca^{2+}, CO_3^{2-}, HCO_3^- and so on, in the case of the carbonate equilibria. The pair of braces, { }, is read as the "activity of" and the pair of brackets, [ ], is read as "the concentration of;" γ is called activity coefficient.

The activity coefficient expresses the effect of ions on the reactive ability of a species. As the species become crowded, the reactive ability or effectiveness of the species to react per unit individual of the species is diminished. As they become less concentrated, the effectiveness per unit individual is improved. Thus, at very dilute solutions, the activity coefficient approaches unity; at a more concentrated solution, the activity coefficient departs from unity.

Because of the action of the charges upon each other, the activity of the ionized particles is smaller than those of the unionized particles. Ionized particles tend to maintain "relationship" between counterparts, slowing down somewhat their inter-actions with other particles. Thus, Ca^{2+} and CO_3^{2-}, which are ionization products of $CaCO_3$, have activity coefficients less than unity; they tend to maintain relationship with each other rather than with other particles. Thus, their activity with respect to other particles is diminished. On the other hand, a solid that is not ionized such as $CaCO_3$ before ionization, has an activity coefficient of unity; a liquid or solvent such as water (not ionized) has an activity coefficient of unity. Gases that are not disso-ciated have activity coefficients of unity. (Because the activity coefficients are unity, all molar concentrations of these unionized species have a unit of activity.) In other words, unionized particles are free to interact with any other particles, thus having the magnitude of the highest possible value of the activity coefficient. They are not restricted to maintain any counterpart interaction, because they do not have any.

EQUILIBRIUM AND SOLUBILITY PRODUCT CONSTANTS

Let a reactant be represented by the solid molecule A_aB_b. As this reactant is mixed with water, it dissolves into its constituent solute ions. The equilibrium dissolution reaction is

$$A_aB_b \rightleftharpoons aA + bB \tag{10}$$

The ratio $\dfrac{\{A\}^a\{B\}^b}{\{A_aB_b\}}$ is called the *reaction quotient*. Note that in the definition of reaction quotients, the activities of the reactants and products are raised to their respective coefficients. At equilibrium, the reaction quotient becomes the equilibrium constant. Thus, the equilibrium constant is

$$K = \frac{\{A\}^a\{B\}^b}{\{A_aB_b\}} \tag{11}$$

Note the pair of braces denoting activity. Because A_aB_b is a solid, its activity is unity. For this reason, the product $K\{A_aB_b\}$ is a constant and is designated as K_{sp}. K_{sp} is called the *solubility product constant* of the equilibrium dissolution reaction. Equation (11) now transforms to

$$K_{sp} = \{A\}^a\{B\}^b \tag{12}$$

At equilibrium, neither reactants nor products increase or decrease with time. Thus, K_{sp}'s, being constants, can be used as indicators whether or not a given solid will form or dissolve in solution.

For example, $CaCO_3$ has a K_{sp} of $4.8(10^{-9})$ at 25°C. This value decreases to $2.84(10^{-9})$ at 60°C. The equilibrium reaction for this solid is

$$CaCO_{3(s)} \rightleftharpoons Ca^{2+} + CO_3^{2-}. \quad K_{sp} = \{Ca^{2+}\}(CO_3^{2-}\} = 4.8(10^{-9}) \text{ at } 25°C$$
$$= \{Ca^{2+}\}(CO_3^{2-}\} = 2.84(10^{-9}) \text{ at } 60°C$$

Assuming the reaction is at equilibrium at 25°C, the ions at the right side of the reaction will neither cause $CaCO_3$ to form nor dissolve. At 25°C, by the definition of the solubility product constant and, since the reaction is assumed to be in equilibrium, the product of the activities will equal $4.8(10^{-9})$. As the temperature increases from 25°C to 60°C, the product of the activities decreases from $4.8(10^{-9})$ to $2.84(10^{-9})$. This means that there are fewer particles of the ions existing than required to maintain equilibrium at this higher temperature. As a consequence, some of the particles will combine to form a precipitate, the $CaCO_3$ solid.

Table 1 shows values of K_{sp}'s of solids that are of importance in ascertaining whether or not a certain water sample will form or dissolve the respective solids. The subscripts s and aq refers to "solid" and "aqueous," respectively. In writing the chemical reactions for the discussions in this book, these subscripts will not be indicated unless necessary for clarity.

Of all the K_{sp}'s discussed previously, the one involving the carbonate system equilibria is of utmost importance in water and wastewater treatment. This is because carbon dioxide in the atmosphere affects any water body. As carbon dissolves in water, the carbonate system species $H_2CO_3^*$, HCO_3^-, and CO_3^{2-} are formed. Cations will then interact with these species and, along with the H^+ and OH^- that always exist in water solution, complete the equilibrium of the system. $H_2CO_3^*$ is a mixture of CO_2 in water and H_2CO_3 (note the absence of the asterisk in H_2CO_3). The CO_2 in water is written as $CO_{2(aq)}$ and $H_2CO_3^*$ is carbonic acid.

TABLE 1
Solubility Product Constants of Respective Solids at 25°C

K_{sp} reaction of solid	K_{sp}	Significance
$AgCl_{(s)} \rightleftharpoons Ag^+_{(aq)} + Cl^-_{(aq)}$	$1.8(10^{-10})$	Chloride analysis
$Al(OH)_{3(s)} \rightleftharpoons Al^{3+}_{(aq)} + 3OH^-_{(aq)}$	$1.9(10^{-33})$	Coagulation
$BaSO_{4(s)} \rightleftharpoons Ba^{2+}_{(aq)} + SO^{2-}_{4(aq)}$	10^{-10}	Sulfate analysis
$CaCO_{3(s)} \rightleftharpoons Ca^{2+}_{(aq)} + CO^{2-}_{3(aq)}$	$4.8(10^{-9})$	Hardness removal, scales
$Ca(OH)_{2(s)} \rightleftharpoons Ca^{2+}_{(aq)} + 2OH^-_{(aq)}$	$7.9(10^{-6})$	Hardness removal
$CaSO_{4(s)} \rightleftharpoons Ca^{2+}_{(aq)} + SO^{2-}_{4(aq)}$	$2.4(10^{-5})$	Flue gas desulfurization
$Ca_3(PO_4)_{2(s)} \rightleftharpoons 3Ca^{2-}_{(aq)} + 2PO^{2-}_{(aq)}$	10^{-25}	Phosphate removal
$CaHPO_{4(s)} \rightleftharpoons Ca_{(aq)} + HPO^{2-}_{4(aq)}$	$5.0(10^{-6})$	Phosphate removal
$CaF_{2(s)} \rightleftharpoons Ca^{2+}_{(aq)} + 2F^-_{(aq)}$	$3.9(10^{-11})$	Fluoridation
$Cr(OH)_{3(s)} \rightleftharpoons Cr^{3+}_{(aq)} + 3OH^-_{(aq)}$	$6.7(10^{-31})$	Heavy metal removal
$Cu(OH)_{2(s)} \rightleftharpoons Cu^{2+}_{(aq)} + 2OH^-_{(aq)}$	$5.6(10^{-20})$	Heavy metal removal
$Fe(OH)_{3(s)} \rightleftharpoons Fe^{3+}_{(aq)} + 3OH^-_{(aq)}$	$1.1(10^{-36})$	Coagulation, iron removal, corrosion
$Fe(OH)_2 \rightleftharpoons Fe^{2+}_{(aq)} + 2OH^-_{(aq)}$	$7.9(10^{-15})$	Coagulation, iron removal, corrosion
$MgCO_{3(s)} \rightleftharpoons Mg^{2+}_{(aq)} + CO^{2-}_{3(aq)}$	10^{-5}	Hardness removal, scales
$Mg(OH)_2 \rightleftharpoons Mg^{2+}_{(aq)} + 2OH^-_{(aq)}$	$1.5(10^{-11})$	Hardness removal, scales
$Mn(OH)_2 \rightleftharpoons Mn^{2+}_{(aq)} + 2OH^-_{(aq)}$	$4.5(10^{-14})$	Manganese removal
$Ni(OH)_2 \rightleftharpoons Ni^{2+}_{(aq)} + 2OH^-_{(aq)}$	$1.6(10^{-14})$	Heavy metal removal
$Zn(OH)_2 \rightleftharpoons Zn^{2+}_{(aq)} + 2OH^-_{(aq)}$	$4.5(10^{-17})$	Heavy metal removal

From A. P. Sincero and G. A. Sincero (1996). *Enviromental Engineering: A Design Approach.* Prentice Hall, Upper Saddle River, NJ, 42.

Another important application of the equilibrium constant K in general, [Equation (11)], is in the coagulation treatment of water using alum, $Al_2(SO_4)_3 \cdot 14\ H_2O$. (The "14" actually varies from 13 to 18.) In coagulating a raw water using alum, a number of complex reactions are formed by the Al^{3+} ion. These reactions are as follows:

$$Al^{3+} + H_2O \rightleftharpoons Al(OH)^{2+} + H^+$$

$$K_{Al(OH)c} = \frac{\{Al(OH)^{2+}\}\{H^+\}}{\{Al^{3+}\}} = 10^{-5} \tag{13}$$

$$7Al^{3+} + 17H_2O \rightleftharpoons Al_7(OH)_{17}^{4+} + 17H^+$$

$$K_{Al_7(OH)_{17}c} = \frac{\{Al_7(OH)_{17}^{4+}\}\{H^+\}^{17}}{\{Al^{3+}\}^7} = 10^{-48.8} \tag{14}$$

$$13Al^{3+} + 34H_2O \rightleftharpoons Al_{13}(OH)_{34}^{5+} + 34H^+$$

$$K_{Al_{13}(OH)_{34}c} = \frac{\{Al_{13}(OH)_{34}^{5+}\}\{H^+\}^{34}}{\{Al^{3+}\}^{13}} = 10^{-97.4} \tag{15}$$

$$Al(OH)_{3(s)}(\text{fresh precipitate}) \rightleftharpoons Al^{3+} + 3OH^-$$

$$K_{sp,Al(OH)_3} = \{Al^{3+}\}\{OH^-\}^3 = 10^{-33} \tag{16}$$

$$Al(OH)_{3(s)} + OH^- \rightleftharpoons Al(OH)_4^-$$

$$K_{Al(OH)_4c} = \frac{\{Al(OH)_4^-\}}{OH^-} = 10^{+1.3} \tag{17}$$

$$2Al^{3+} + 2H_2O \rightleftharpoons Al_2(OH)_2^{4+} + 2H^+$$

$$K_{Al_2(OH)_2c} = \frac{\{Al_2(OH)_2^{4+}\}\{H^+\}^2}{\{Al^{3+}\}^2} = 10^{-6.3} \tag{18}$$

The equilibrium constants $K_{Al(OH)c}$, $K_{Al_7(OH)_{17}c}$, $K_{Al_{13}(OH)_{34}c}$, $K_{sp,Al(OH)_3}$, $K_{Al(OH)_4c}$, and $K_{Al_2(OH)_2c}$ apply at 25°C. Note that the subscript c is used for the equilibrium constants of the complexes; it stands for "complex." $Al(OH)_{3(s)}$ is not a complex; thus, $K_{sp,Al(OH)_3}$ does not contain the subscript c.

The previous equations can be used to determine the conditions that will allow maximum precipitation of the solid represented by $Al(OH)_3$. The maximum precipitation of $Al(OH)_3$ will produce the utmost clarity of the treated water. To allow for this maximum precipitation, the concentrations of the complex ions $Al(OH)^{2+}$, $Al_7(OH)_{17}^{4+}$, $Al_{13}(OH)_{34}$, $Al(OH)_4^-$, and $Al_2(OH)_2^{4+}$ and Al^{3+} must be held to a minimum. This will involve finding the optimum pH of coagulation. This optimum pH may be determined as follows:

It is obvious that the complex ions contain the Al atom. Thus, the technique is to sum up their molar concentrations in terms of their Al atom content. Once they have been summed up, they are then eliminated using the previous K equations, Eqs. (13) to (18), with the objective of expressing the resulting equation in terms of the constants and the hydrogen ion concentration. Because the K constants are constants, the equations would simply be expressed in terms of one variable, the hydrogen ion concentration, and the equation can then be easily differentiated to obtain the optimum pH of coagulation.

Gleaning from Eqs. (13) to (18), a molar mass balance on the aluminum atom may be performed. Let sp_{Al} represent all the species that contain the aluminum atom standing in solution. Thus, the concentration of all the species containing the aluminum atom, is

$$[sp_{Al}] = [Al^{3+}] + [Al(OH)^{2+}] + 7[Al_7(OH)_{17}^{4+}] + 13[Al_{13}(OH)_{34}^{5+}]$$
$$+ [Al(OH)_4^-] + 2[Al_2(OH)_2^{4+}] \tag{19}$$

Note that, because we are summing the species containing the aluminum atom, the coefficients of the terms of the previous equation contain the number of aluminum atoms in the respective species. For example, $Al_7(OH)_{17}^{4+}$ contains 7 aluminum atoms; thus, its coefficient is 7. Similarly, $Al_{13}(OH)_{34}^{5+}$ contains 13 aluminum atoms; therefore, its coefficient is 13. This explanation holds for the other species as well.

Using Eqs. (13) to (18) and the relations of molar concentration and activity, Equation (9), the following equations are obtained [for the purpose of eliminating the complex ions in Equation (19)]:

$$[Al^{3+}] = \frac{\{Al^{3+}\}}{\gamma_{Al}} = \frac{K_{sp,Al(OH)_3}}{\gamma_{Al}\{OH^-\}^3} = \frac{K_{sp,Al(OH)_3}\{H^+\}^3}{\gamma_{Al}K_w^3} = \frac{K_{sp,Al(OH)_3}\gamma_H^3[H^+]^3}{\gamma_{Al}K_w^3} \tag{20}$$

$$[Al(OH)^{2+}] = \frac{\{Al(OH)^{2+}\}}{\gamma_{Al(OH)c}} = \frac{K_{Al(OH)c}\{Al^{3+}\}}{\gamma_{Al(OH)c}\{H^+\}} = \frac{K_{Al(OH)c}K_{sp,Al(OH)_3}\gamma_H^2[H^+]^2}{\gamma_{Al(OH)c}K_w^3} \tag{21}$$

$$[Al_7(OH)_{17}^{4+}] = \frac{\{Al_7(OH)_{17}\}}{\gamma_{Al_7(OH)_{17}c}} = \frac{K_{Al_7(OH)_{17}c}\{Al^{3+}\}^7}{\gamma_{Al_7(OH)_{17}c}\{H^+\}^{17}} = \frac{K_{Al_7(OH)_{17}c}\gamma_{Al}^7[Al^{3+}]^7}{\gamma_{Al_7(OH)_{17}c}\gamma_H^{17}[H^+]^{17}}$$
$$= \frac{K_{Al_7(OH)_{17}c}K_{sp,Al(OH)_3}^7\gamma_H^4[H^+]^4}{\gamma_{Al_7(OH)_{17}c}K_w^{21}} \tag{22}$$

$$[Al_{13}(OH)_{34}^{5+}] = \frac{\{Al_{13}(OH)_{34}^{5+}\}}{\gamma_{Al_{13}(OH)_{34}c}} = \frac{K_{Al_{13}(OH)_{34}c}\{Al^{3+}\}^{13}}{\gamma_{Al_{13}(OH)_{34}c}(H^+)^{34}} = \frac{K_{Al_{13}(OH)_{34}c}\gamma_{Al}^{13}[Al^{3+}]^{13}}{\gamma_{Al_{13}(OH)_{34}c}\gamma_H^{34}[H^+]^{34}}$$
$$= \frac{K_{Al_{13}(OH)_{34}c}K_{sp,Al(OH)_3}^{13}\gamma_H^5[H^+]^5}{\gamma_{Al_{13}(OH)_{34}c}K_w^{39}} \tag{23}$$

$$[Al(OH)_4^-] = \frac{\{Al(OH)_4^-\}}{\gamma_{Al(OH)_4c}} = \frac{K_{Al(OH)_4c}\{OH^-\}}{\gamma_{Al(OH)_4c}} = \frac{K_{Al(OH)_4c}K_w}{\gamma_{Al(OH)_4c}\gamma_H[H^+]} \tag{24}$$

$$[Al_2(OH)_2^{4+}] = \frac{\{Al_2(OH)_2^{4+}\}}{\gamma_{Al_2(OH)_2c}} = \frac{K_{Al_2(OH)_2c}\{Al^{3+}\}^2}{\gamma_{Al_2(OH)_2c}\{H^+\}^2} = \frac{K_{Al_2(OH)_2c}\gamma_{Al}^2[Al^{3+}]^2}{\gamma_{Al_2(OH)_2c}\gamma_H^2[H^+]^2}$$

$$= \frac{K_{Al_2(OH)_2c}K_{sp,Al(OH)_3}^2\gamma_H^4[H^+]^4}{\gamma_{Al_2(OH)_2c}K_w^6} \tag{25}$$

γ_{Al}, γ_H, $\gamma_{Al(OH)c}$, $\gamma_{Al_7(OH)_{17}c}$, $\gamma_{Al_{13}(OH)_{34}c}$, $\gamma_{Al(OH)_4c}$, $\gamma_{Al_2(OH)_2c}$ are, respectively, the activity coefficients of the aluminum ion and the hydrogen ion and the complexes $Al(OH)^{2+}$, $Al_7(OH)_{17}^{4+}$, $Al_{13}(OH)_{34}^{5+}$, $Al(OH)_4^-$, and $Al_2(OH)_2^{4+}$. $K_{sp,Al(OH)_3}$ is the solubility product constant of the solid $Al(OH)_{3(s)}$ and K_w is the ion product of water. $K_{Al(OH)c}$, $K_{Al_7(OH)_{17}c}$, $K_{Al_{13}(OH)_{34}c}$, $K_{Al(OH)_4c}$, and $K_{Al_2(OH)_2c}$ are, respectively, the equilibrium constants of the complexes $Al(OH)^{2+}$, $Al_7(OH)_{17}^{4+}$, $Al_{13}(OH)_{34}^{5+}$, $Al(OH)_4^-$, and $Al_2(OH)_2^{4+}$.

Let us explain the derivation of one of the above equations. Pick Equation (20). The first part of this equation is

$$[Al^{3+}] = \frac{\{Al^{3+}\}}{\gamma_{Al}} \tag{26}$$

This equation is simply the application of Equation (9), where sp is Al^{3+} and γ is γ_{Al} and where the molar concentration is solved in terms of the activity.

The next part of the equation, $K_{sp,Al(OH)_3}/\gamma_{Al}\{OH^-\}^3$, is derived, in part, from Equation (16), which reads

$$Al(OH)_{3(s)}(\text{fresh precipitate}) \rightleftharpoons Al^{3+} + 3OH^- \quad\quad K_{sp,Al(OH)_3} = 10^{-33} \tag{16}$$

From the definition of K_{sp},

$$K_{sp,Al(OH)_3} = \{Al^{3+}\}\{OH^-\}^3 \tag{27}$$

This equation may be solved for $\{Al^{3+}\}$ to produce

$$\{Al^{3+}\} = \frac{K_{sp,Al(OH)_3}}{\{OH^-\}^3} \tag{28}$$

Substituting $\{Al^{3+}\} = K_{sp,Al(OH)_3}/\{OH^-\}^3$ in Equation (26) produces

$$[Al^{3+}] = \frac{\{Al^{3+}\}}{\gamma_{Al}} = \frac{K_{sp,Al(OH)_3}}{\gamma_{Al}\{OH^-\}^3} \tag{29}$$

The next part of the equation, $[K_{sp,\text{Al(OH)}_3}\{H^+\}^3]/\gamma_{\text{Al}}K_w^3$, should be easy to derive and will no longer be pursued. Note that K_w came from the ion product of water that reads:

$$K_w = \{H^+\}\{OH\} \tag{30}$$

This ion product is an equilbirium constant for the dissociation of water:

$$H_2O \rightleftharpoons H^+ + OH^- \tag{31}$$

Equations (20) to (25) may now be substituted into Equation (19). This produces Equation (32), where, now, the complexes are eliminated and the equation only expressed in terms of the K's and the hydrogen ion concentration. This equation may then be differentiated and equated to zero to obtain the optimum pH. We will not, however, complete this differentiation and equate to zero in this chapter, but will do this in the unit processes part of this book.

$$
\begin{aligned}
[sp_{\text{Al}}] =\ & \frac{K_{sp,\text{Al(OH)}_3}\gamma_H^3[H^+]^3}{\gamma_{\text{Al}}K_w^3} + \frac{K_{\text{Al(OH)}c}K_{sp,\text{Al(OH)}_3}\gamma_H^2[H^+]^2}{\gamma_{\text{Al(OH)}c}K_w^3} \\
& + \frac{7K_{\text{Al}_7(\text{OH})_{17}c}K_{sp,\text{Al(OH)}_3}^7\gamma_H^4[H^+]^4}{\gamma_{\text{Al}_7(\text{OH})_{17}c}K_w^{21}} \\
& + \frac{13K_{\text{Al}_{13}(\text{OH})_{34}c}K_{sp,\text{Al(OH)}_3}^{13}\gamma_H^5[H^+]^5}{\gamma_{\text{Al}_{13}(\text{OH})_{34}c}K_w^{39}} + \frac{K_{\text{Al(OH)}_4c}K_w}{\gamma_{\text{Al(OH)}_4c}\gamma_H[H^+]} \\
& + \frac{2K_{\text{Al}_2(\text{OH})_2c}K_{sp,\text{Al(OH)}_3}^2\gamma_H^4[H^+]^4}{\gamma_{\text{Al}_2(\text{OH})_2c}K_w^6}
\end{aligned}
\tag{32}
$$

Still another important application of the concept of K equilibrium constants is the coprecipitation of $FePO_4$ and $Fe(OH)_3$ in the removal of phosphorus from water. As in the case of coagulation using alum, it is desired to have a final equation that is expressed only in terms of the constants and the hydrogen ion. Once this is done, the equation can then also be manipulated to obtain an optimum pH for the removal of phosphorus.

In phosphorus removal, the phosphorus must be in the phosphate form and, because the reaction occurs in water, the phosphate ion originates a series of reactions with the hydrogen ion. The series is as follows:

$$PO_4^{3-} + H^+ \rightleftharpoons HPO_4^{2-} \Rightarrow HPO_4^{2-} \rightleftharpoons PO_4^{3-} + H^+$$

$$K_{\text{HPO}_4} = \frac{\{PO_4^{3-}\}\{H^+\}}{\{HPO_4^{2-}\}} = 10^{-12.3} \tag{33}$$

$$HPO_4^{2-} + H^+ \rightleftharpoons H_2PO_4^- \Rightarrow H_2PO_4^- \rightleftharpoons HPO_4^{2-} + H^+$$

$$K_{H_2PO_4} = \frac{\{HPO_4^{2-}\}\{H^+\}}{\{H_2PO_4^-\}} = 10^{-7.2} \tag{34}$$

$$H_2PO_4^- + H^+ \rightleftharpoons H_3PO_4 \Rightarrow H_3PO_4 \rightleftharpoons H_2PO_4^- + H^+$$

$$K_{H_3PO_4} = \frac{\{H_2PO_4^-\}\{H^+\}}{\{H_3PO_4\}} = 10^{-2.1} \tag{35}$$

From the previous equations, a molar mass balance on the phosphate radical may be performed. Let $sp_{PO_4,FeIII}$ represent the species in solution containing PO_4 species, using a ferric salt as the precipitant. (Note the FeIII as the second subscript, indicating that a ferric salt is used.) Therefore,

$$[sp_{PO_4,FeIII}] = [PO_4^{3-}] + [HPO_4^{2-}] + [H_2PO_4^-] + [H_3PO_4] \tag{36}$$

To indicate the removal of phosphate using the ferric salt, we use the following precipitation reaction to incorporate the iron into the above equation:

$$FePO_4\downarrow \rightleftharpoons Fe^{3+} + PO_4^{3-} \qquad K_{sp,FePO_4} = 10^{-21.9} \text{ at } 25°C \tag{37}$$

$$K_{sp,FePO_4} = \{Fe^{3+}\}\{PO_4^{3-}\} \tag{38}$$

$[HPO_4^{2-}]$, $[H_2PO_4^-]$, and $[H_3PO_4]$ may be eliminated by expressing them in terms of the K constants using Eqs. (33) to (35). The results are:

$$[HPO_4^{2-}] = \frac{\{HPO_4^{2-}\}}{\gamma_{HPO_4}} = \frac{\{PO_4^{3-}\}\{H^+\}}{\gamma_{HPO_4}K_{HPO_4}} = \frac{\gamma_{PO_4}\gamma_H[PO_4^{3-}][H^+]}{\gamma_{HPO_4}K_{HPO_4}} \tag{39}$$

$$[H_2PO_4^{2-}] = \frac{\{H_2PO_4^-\}}{\gamma_{H_2PO_4}} = \frac{\{HPO_4^{2-}\}\{H^+\}}{\gamma_{H_2PO_4}K_{H_2PO_4}} = \frac{\gamma_{PO_4}\gamma_H^2[PO_4^{3-}][H^+]^2}{\gamma_{H_2PO_4}K_{H_2PO_4}K_{HPO_4}} \tag{40}$$

$$[H_3PO_4] = \{H_3PO_4\} = \frac{\{H_2PO_4^-\}\{H^+\}}{K_{H_3PO_4}} = \frac{\gamma_{PO_4}\gamma_H^3[PO_4^{3-}][H^+]^3}{K_{H_3PO_4}K_{H_2PO_4}K_{HPO_4}} \tag{41}$$

By using Equation (38), $[PO_4^{3-}]$ may also be eliminated in favor of $K_{sp,FePO_4}$ to produce

$$[PO_4^{3-}] = \frac{\{PO_4^{3-}\}}{\gamma_{PO_4}} = \frac{K_{sp,FePO_4}}{\gamma_{PO_4}\{Fe^{3+}\}} = \frac{K_{sp,FePO_4}}{\gamma_{PO_4}\gamma_{FeIII}[Fe^{3+}]} \tag{42}$$

This will incorporate the iron into Equation (36) to indicate the use of the ferric salt. Finally, substituting the whole result into Equation (36), produces

$$[sp_{PO_4,FeIII}] = \frac{K_{sp,FePO_4}}{\gamma_{PO_4}\gamma_{FeIII}[Fe^{3+}]} + \frac{\gamma_H K_{sp,FePO_4}[H^+]}{\gamma_{HPO_4}K_{HPO_4}\gamma_{FeIII}[Fe^{3+}]}$$

$$+ \frac{\gamma_H^2 K_{sp,FePO_4}[H^+]^2}{\gamma_{H_2PO_4}K_{H_2PO_4}K_{HPO_4}\gamma_{FeIII}[Fe^{3+}]} + \frac{\gamma_H^3 K_{sp,FePO_4}[H^+]^3}{k_{H_3PO_4}K_{H_2PO_4}K_{HPO_4}\gamma_{FeIII}[Fe^{3+}]}$$

$$(43)$$

γ_{PO_4}, γ_{Al}, γ_H, γ_{HPO_4}, and $\gamma_{H_2PO_4}$ are, respectively, the activity coefficients of the phosphate, aluminum, hydrogen, hydrogen phosphate, and dihydrogen phosphate ions. K_{HPO_4}, $K_{H_2PO_4}$, and $K_{H_3PO_4}$ are, respectively, the equilibrium constants of the hydrogen phosphate and dihydrogen phosphate ions and phosphoric acid. γ_{FeIII} is the activity coefficient of the ferric ion.

As shown, Equation (43) still expresses $[sp_{PO_4,FeIII}]$ in terms of two variables, $[Fe^{3+}]$ and $[H^+]$. We want it expressed only in terms of the K constants and the hydrogen ion concentration. This is where an expression needs to be found to eliminate $[Fe^{3+}]$. If this expression indicates that some compound of iron is coprecipitating with $FePO_4$, then $[Fe^{3+}]$ could be eliminated using the K constant of this compound.

When two compounds are coprecipitating, they are at equilibrium at this particular instance of coprecipitation. They are at equilibrium, so their equilibrium constants can be used in a calculation. This instance of coprecipitating with $FePO_4$ is, indeed, possible through the use of the compound ferric hydroxide which, of course, must still be investigated whether, in fact, it is possible. This is investigated next.

The dissociation reaction of ferric hydroxide is

$$Fe(OH)_{3(s)} \rightleftharpoons Fe^{3+} + 3(OH^-)$$
$$K_{sp,Fe(OH)_3} = \{Fe^{3+}\}\{OH^-\}^3 = 1.1(10^{-36})$$

$$(44)$$

If the results of this investigation show that the molar concentration of iron needed to precipitate $Fe(OH)_3$ is about the same as the molar concentration of the iron needed to precipitate $FePO_4$, then $Fe(OH)_3$ and $FePO_4$ must be coprecipitating.

Let x be the $\{Fe^{3+}\}$. Because the coefficient of OH^- in the previous reaction is 3, $\{OH^-\}$ is therefore $3x$. Thus, substituting into the K_{sp} equation,

$$K_{sp,Fe(OH)_3} = \{Fe^{3+}\}\{OH^-\}^3 = x(3x)^3 = 1.1(10^{-36})$$
$$x = 4.5(10^{-10}) \text{ gmol/L} = \{Fe^{3+}\}$$

Now, from Equation (37), $K_{sp,FePO_4} = \{Fe^{3+}\}\{PO_4^{3-}\} = 10^{-21.9}$. Letting y equal to $\{Fe^{3+}\}$, $\{PO_4^{3-}\}$ is also equal to y. Substituting into the K_{sp} equation,

$$K_{sp,Fe(PO)_4} = \{Fe^{3+}\}\{PO_4^{3-}\} = y(y) = 10^{-21.9}$$
$$y = 1.1(10^{-11}) \text{ gmol/L} = \{Fe^{3+}\}.$$

These two concentrations, x and y, are practically equal; thus, this justifies the use of ferric hydroxide. We can conclude that $Fe(OH)_{3(s)}$ will definitely coprecipitate with $FePO_4$. Therefore,

$$[Fe^{3+}] = \frac{\{Fe^{3+}\}}{\gamma_{FeIII}} = \frac{K_{sp,Fe(OH)_3}}{\gamma_{FeIII}\{OH^-\}^3} = \frac{K_{sp,Fe(OH)_3}\gamma_H^3[H^+]^3}{\gamma_{FeIII}K_w^3}.$$

Substituting in Equation (43) to eliminate $[Fe^{3+}]$,

$$[sp_{PO_4FeIII}] = \frac{K_{sp,FePO_4}K_w^3}{\gamma_{PO_4}K_{sp,Fe(OH)_3}\gamma_H^3[H^+]^3} + \frac{K_{sp,FePO_4}K_w^3}{\gamma_{HPO_4}K_{HPO_4}K_{sp,Fe(OH)_3}\gamma_H^2[H^+]^2}$$

$$+ \frac{K_{sp,FePO_4}K_w^3}{\gamma_{H_2PO_4}K_{H_2PO_4}K_{HPO_4}K_{sp,Fe(OH)_3}\gamma_H[H^+]} + \frac{K_{sp,FePO_4}K_w^3}{K_{H_3PO_4}K_{H_2PO_4}K_{HPO_4}K_{sp,Fe(OH)_3}}$$

$$\text{(45)}$$

From this equation, the optimum pH for the removal of phosphorus can be determined.

Example 12   In the coagulation treatment of water using copperas, $FeSO_4 \cdot 7H_2O$, the following complex reactions occur.

$$Fe(OH)_{2(s)} \rightleftharpoons Fe^{2+} + 2OH^- \qquad K_{sp,Fe(OH)_2} = 10^{-14.5} \qquad \text{(a)}$$

$$Fe(OH)_{2(s)} \rightleftharpoons FeOH^+ + OH^- \qquad K_{FeOHc} = 10^{-9.4} \qquad \text{(b)}$$

$$Fe(OH)_{2(s)} + OH^- \rightleftharpoons Fe(OH)_3^- \qquad K_{Fe(OH)_3c} = 10^{-5.1} \qquad \text{(c)}$$

A molar mass balance on the ferrous iron produces

$$[sp_{FeII}] = [Fe^{2+}] + [FeOH^+] + [Fe(OH)_3^-] \qquad \text{(d)}$$

where $[sp_{FeII}]$ is the total molar concentration of all species containing the iron atom. Express $[sp_{FeII}]$ in terms of the concentration of hydrogen ion and equilibrium constants.

Solution: In (d), we need to express $[Fe^{2+}]$, $[FeOH^+]$, and $[Fe(OH)_3^-]$ in terms of $[H^+]$ and pertinent equilibrium constants. First, consider $[Fe^{2+}]$. $\{sp\} = \gamma[sp] \Rightarrow [Fe^{2+}] = \{Fe^{2+}\}/\gamma_{FeII}$, where γ_{FeII} is the activity coefficient of ferrous iron, Fe^{2+}. From (a),

$$K_{sp,Fe(OH)_2} = \{Fe^{2+}\}\{OH^-\}^2 \Rightarrow \{Fe^{2+}\} = \frac{K_{sp,Fe(OH)_2}}{\{OH^-\}^2}$$

Therefore,

$$[Fe^{2+}] = \frac{\{Fe^{2+}\}}{\gamma_{FeII}} = \frac{K_{sp,Fe(OH)_2}}{\gamma_{FeII}\{OH^-\}^2}$$

From the iron product of water, $\{OH^-\}\{H^+\} = K_w$, $\{OH^-\} = K_w/\{H^+\} = K_w/\gamma_H[H^+]$, where γ_H is the activity coefficient of the hydrogen ion. Therefore,

$$[Fe^{2+}] = \frac{\{Fe^{2+}\}}{\gamma_{FeII}} = \frac{K_{sp,Fe(OH)_2}}{\gamma_{FeII}\{OH^-\}^2} = \frac{K_{sp,Fe(OH)_2}\{H^+\}^2}{\gamma_{FeII}K_w^2} = \frac{K_{sp,Fe(OH)_2}\gamma_H^2[H^+]^2}{\gamma_{FeII}K_w^2}$$

The steps involved to express $[FeOH^+]$, and $[Fe(OH)_3^-]$ in terms of $[H^+]$ and pertinent equilibrium constants are similar to the ones above. Thus, we will not go through the detailed steps, again, but write the results at once:

$$[FeOH^+] = \frac{\{Fe(OH)^+\}}{\gamma_{FeOHc}} = \frac{K_{FeOHc}}{\gamma_{FeOHc}\{OH^-\}} = \frac{K_{FeOHc}\{H^+\}}{\gamma_{FeOHc}K_w} = \frac{K_{FeOHc}\gamma_H[H^+]}{\gamma_{FeOHc}K_w}$$

$$[Fe(OH)_3^-] = \frac{\{Fe(OH)_3^-\}}{\gamma_{Fe(OH)_3c}} = \frac{K_{Fe(OH)_3c}\{OH^-\}}{\gamma_{Fe(OH)_3c}} = \frac{K_{Fe(OH)_3c}K_w}{\gamma_{Fe(OH)_3c}\{H^+\}}$$

$$= \frac{K_{Fe(OH)_3c}K_w}{\gamma_{Fe(OH)_3c}\gamma_H[H^+]}$$

where γ_{FeOHc} and $\gamma_{Fe(OH)_3c}$ are the activity coefficients of the complexes $FeOH^+$ and $Fe(OH)_3^-$, respectively.

Substituting in **(d)**,

$$[sp_{FeII}] = [Fe^{2+}] + [FeOH^+] + [Fe(OH)_3^-]$$

$$[sp_{FeII}] = \frac{K_{sp,Fe(OH)_2}\gamma_H^2[H^+]^2}{\gamma_{FeII}K_w^2} + \frac{K_{FeOH}\gamma_H[H^+]}{\gamma_{FeOHc}K_w} + \frac{K_{Fe(OH)_3c}K_w}{\gamma_{Fe(OH)_3c}\gamma_H[H^+]} \qquad \textbf{(d)} \quad \textbf{Ans}$$

ACIDS AND BASES

Acids are substances that can donate a proton, H^+, and bases are substances that can accept this proton. The most fundamental reaction of an acid, HA, is its reaction with H_2O,

$$HA + H_2O \rightleftharpoons H_3O^+ + A^- \qquad (46)$$

As seen from this reaction, the acid HA donates its proton to H_2O, producing H_3O^+. H_3O^+ is called a *hydronium ion*. Because H_2O has accepted the proton, it is therefore a base. H_3O^+ that results from the acceptance of this proton is an acid. The water molecule is neutral, but it has the proton, an acid, "clinging" to it. H_3O^+ is called the *conjugate acid* of the base H_2O. Now, when HA donated its proton, it transforms into A^-. Of course, A^- can accept the proton from H_3O^+ to form back HA. For this reason, A^- is a base; it is called the *conjugate base* of the acid HA.

The equilibrium constant for an acid, K_a, is

$$K_a = \frac{\{H_3O^+\}\{A^-\}}{\{HA\}\{H_2O\}} = \frac{\{H_3O^+\}\{A^-\}}{\{HA\}}; \{H_2O\} = 1 \qquad (47)$$

Note that the activity $\{H_2O\}$ is equal to unity, because H_2O is unionized. K_a is also called *ionization constant* for the acid.

As an acid, H_3O^+ reacts in water as follows:

$$H_3O^+ + H_2O \rightleftharpoons H_3O^+ + H_2O \qquad (48)$$

On the left-hand side of the equation, H_3O^+ donates its proton to H_2O, with this H_2O forming the H_3O^+ on the right-hand side of the equation. By virtue of the donation of its proton, the H_3O^+ on the left transforms to the H_2O on the right-hand side of the equation. H_2O on the left is a base and H_3O^+ on the right is its conjugate acid. On the other hand, the H_2O on the right is the conjugate base of the acid H_3O^+ on the left. The net effect of these interactions, however, is that nothing happens, although intrinsically, the mechanism just described is actually happening.

Although the reaction of the hydronium ion H_3O^+ as an acid may seem trivial, it serves an important function. As shown in Table 2, it can serve as an indicator of the lower limit of the strength of strong acids. The strength of an acid can be measured in terms of its production of the hydronium ions. Strong acids ionize completely (or nearly so) producing 100% equivalent of the hydronium ions. Weak acids only ionize partially, which, correspondingly produce a relatively small amount of hydronium ions. They normally ionize to about 10% or less of the original acid. The table shows the relative strengths of acid-conjugate base pairs. Under the column on acids, perchloric acid, sulfuric acid, hydrogen iodide, hydrogen bromide, hydrogen chloride, and nitric acid are the strongest acids. As indicated, they ionize to 100% H_3O^+ and, since they ionize practically completely, the activity of the acid molecule is also practically zero. The denominator in Equation (47) would therefore be zero and K_a would be infinity.

H_2O can also act as an acid. This is shown by the following reaction:

$$H_2O + H_2O \rightleftharpoons H_3O^+ + OH^- \qquad (49)$$

H_3O^+ is the conjugate acid of the second H_2O on the left-hand side of the equation and OH^- is the conjugate base of the first H_2O on the left-hand side of the equation. The second H_2O is a base that reacts with the first H_2O, which is an acid.

Referring to Table 2, the acids from H_3O^+ down to H_2O are weak acids. These are the acids that ionize to only 10% or less to the hydronium ions. The hydronium ion and water then form the boundary limits of the weak acids. Above hydronium, the acids are strong; below water, the compounds do not exhibit any observable acidic

TABLE 2
Relative Strengths of Acid-Conjugate Base Pairs

Acid			Base		
Perchloric acid	$HClO_4$	100% ionized to H_3O^+	Perchlorate ion	ClO_4^-	—
Sulfuric acid	H_2SO_4	100% ionized to H_3O^+	Hydrogen sulfate ion	HSO_4^-	—
Hydrogen iodide	HI	100% ionized to H_3O^+	Iodide ion	I^-	—
Hydrogen bromide	HBr	100% ionized to H_3O^+	Bromide ion	Br^-	—
Hydrogen chloride	HCl	100% ionized to H_3O^+	Chloride ion	Cl^-	—
Nitric acid	HNO_3	100% ionized to H_3O^+	Nitrate ion	NO_3^-	—
Hydronium ion	H_3O^+	↑	Water	H_2O	Weakest base
Hydrogen sulfate ion	HSO_4^-	↑	Sulfate ion	SO_4^{2-}	↓
Phosphoric acid	H_3PO_4	↑	Dihydrogen phosphate ion	$H_2PO_4^-$	↓
Hydrogen fluoride	HF	↑	Fluoride ion	F^-	↓
Nitrous acid	HNO_2	↑	Nitrite ion	NO_2^-	↓
Acetic acid	CH_3CO_2H	↑	Acetate ion	$CH_3CO_2^-$	↓
Carbonic acid	H_2CO_3	↑	Hydrogen carbonate ion	HCO_3^-	↓
Hydrogen sulfide	H_2S	↑	Hydrogen sulfide ion	HS^-	↓
Ammonium ion	NH_4^+	↑	Amonia	NH_3	↓
Hydrogen cyanide	HCN	↑	Cyanide ion	CN^-	↓
Hydrogen carbonate ion	HCO_3^-	↑	Carbonate ion	CO_3^{2-}	↓
Hydrogen sulfide ion	HS^-	↑	Sulfide ion	S^{2-}	↓
Water	H_2O	Weakest acid	Hydroxide ion	OH^-	↓
Ethanol	C_2H_5OH	—	Ethoxide ion	$C_2H_5O^-$	100 ionized to OH^-
Ammonia	NH_3	—	Amide ion	NH_2^-	100 ionized to OH^-
Hydrogen	H_2	—	Hydride ion	H^-	100 ionized to OH^-
Methane	CH_4	—	Methide ion	CH_3^-	100 ionized to OH^-

behavior. Water, then, is the very limit of acidity. Notice the arrows pointing upward from the weakest to the strongest acids.

The most fundamental reaction of a base, B, is its reaction with H_2O,

$$H_2O + B \rightleftharpoons HB + OH^- \tag{50}$$

As seen from this reaction, the base, B, accepts the proton donated by H_2O, producing HB. HB is the conjugate acid of B. The proton donor, H_2O transforms into the hydroxide ion, OH^-. It is this transformation into the hydroxide ion that determines the strength of a base. Strong bases ionize completely (or nearly so) producing 100% equivalent of the hydroxide ions. Weak bases only ionize partially, which, correspondingly produce a relatively small amount of hydroxide ions. As in the case of acids, they normally ionize to about 10% or less of the original base. From the table, under the column on bases, the ethoxide ion, amide ion, hydride ion, and methide ion are the strongest bases. As indicated, they ionize to 100% OH^- and, since they ionize practically completely, the activity of the base molecule is also practically zero.

The equilibrium constant for a base, K_b, is

$$K_b = \frac{\{HB\}\{OH^-\}}{\{B\}\{H_2O\}} = \frac{\{HB\}\{OH^-\}}{\{B\}}; \quad \{H_2O\} = 1 \tag{51}$$

For strong bases, since they are completely ionized, K_b would be equal to infinity. K_b is also called *ionization constant* for the base.

From the table, the bases between water and the hydroxide ion are the weak bases. They are the ones that ionize to 10%. Note that in the case of bases, the other boundary limit is demarcated by the hydroxide ion rather than the hydronium ion. This is so, because the hydronium ion is not a base; thus, it cannot form as a boundary for the bases. One the other hand, H_2O is both an acid and a base. Thus, it consistently forms as a boundary limit in both the acids and the bases. Compounds that act both as an acid and a base are called *amphoteric substances*. H_2O is an amphoteric substance. Above water in the table, the compounds do not exhibit any observable basic behavior. Water, then, is the very limit of basicity. Notice the arrows pointing downward from the weakest to the strongest bases.

Alkalinity and acidity. Related to acids and bases are the concepts of alkalinity and acidity. *Alkalinity* is defined as the capacity of a substance to neutralize an acid, and *acidity* is defined as the capacity of a substance to neutralize a base. Consider one mole per liter of HCl. Its neutralization reaction with OH^- is

$$HCl + OH^- \rightarrow H_2O + Cl^- \tag{52}$$

Note that the symbol "$\rightarrow$" is used instead of "$\rightleftharpoons$." This is so, because HCl is a strong acid and it ionizes completely. From the reaction, one mole per liter of HCl neutralizes one mole per liter of OH^- and its acidity is therefore one mole per liter.

Now, consider one mole per liter of a weak acid HCO_3^-, Its reaction with OH^- is

$$HCO_3^- + OH^- \rightleftharpoons H_2O + CO_3^{2-} \tag{53}$$

Now, note that the symbol "$\rightleftharpoons$" is used instead of "$\rightarrow$." If we look at Table 1, CO_3^{2-} is a stronger base than H_2O. Thus, it can grab H^+ from H_2O forming back the left-hand side of the reaction. Forming back the left-hand side of the reaction means that Reaction (53) is a reversible reaction, so, the symbol "$\rightleftharpoons$" is used. Reversible reactions are a characteristic of weak acids. Because of the reversibility of the reactions of weak acids, a mole of the acid will not neutralize a mole of a base but a quantity less than one mole. This relationship also holds true for weak bases: A mole of a weak base will not neutralize a mole of an acid but a quantity less than one mole. Again, the reason is the reversibility of the reactions of weak bases, as in the case of weak acids. In the case of strong bases, however, one mole of a strong base will neutralize one mole of an acid.

An important application of the concept of alkalinity is the determination of the amount of base needed to raise the pH of a solution. Let us apply this concept to a system composed entirely of the carbonate system. The alkalinity of the system at any instant of time will be given by the concentration of the constituent species. These species are OH^-, HCO_3^-, and CO_3^{2-}.

Call the current pH as pH_{cur} and the pH to be adjusted to or the destination pH as pH_{to}. pH_{cur} is less than pH_{to} and a base is needed. A natural water will always have acidity. Until it is all consumed, this acidity will resist the change in pH when a base is added to the water. Let the current total acidity be $[A_{ccur}]_{geq}$ in gram equivalents per liter. Also, let the total alkalinity to be added as $[A_{add}]_{geq}$ in gram equivalents per liter. Assuming no acidity present, the total base to be added is $10^{-pH_{cur}} - 10^{-pH_{to}}$ gram moles per liter of the equivalent hydroxide ions. But since there is always acidity present, the total alkalinity to be added $[A_{add}]_{geq}$ must include the amount for neutralizing the natural acidity, which would be $[A_{ccur}]_{geq}$. Thus, the total alkalinity to be added is

$$[A_{add}]_{geq} = [A_{ccur}]_{geq} + \frac{10^{-pH_{cur}} - 10^{-pH_{to}}}{\phi_b}$$

where ϕ_b is the fractional dissociation of the hydroxide ion from the base supplied. For strong bases, ϕ_b is unity. Weak bases do not produce 100% hydroxide ions. Thus, the quotient ϕ_b is incorporated to yield the correct amount of base that will produce the amount of base needed to raise the pH.

Example 13   A raw water containing 140 mg/L of dissolved solids is subjected to a coagulation treatment using $Fe_2(SO_4)_3$. For optimum coagulation, the pH should be raised to 8.2. The current acidity is $6.0(10^{-3})$ geq/L and the current pH is 8.0. Calculate the amount of lime needed, assuming $\phi_b = 0.125$.

Solution:

$$[A_{add}]_{geq} = [A_{ccur}]_{geq} + \frac{10^{-pH_{cur}} - 10^{-pH_{to}}}{\phi_b}$$

$$\text{Lime needed} = [A_{add}]_{geq} = 6.0(10^{-3}) + \frac{10^{-6} - 10^{-8.2}}{0.125} = 0.006 \text{ geq/L} \quad \textbf{Ans}$$

FLUID MECHANICS

Of the fluid mechanics applications used in this book, the Reynolds transport theorem is a bit sophisticated. Hence, a review of this topic is warranted. This theorem will be used in the derivation of the amount of solids deposited onto a filter and in the derivation of the activated sludge process. These topics are discussed somewhere in the textbook. Except for these two topics, the theorem is used nowhere else. The general knowledge of fluid mechanics will be used under the unit operations part of this book.

To understand the derivation of the Reynolds transport theorem, there are some prerequisite mathematics needed. The students would have had these mathematics already in their undergraduate training before taking physical-chemical treatment, however, they are presented here as a review. They include the topics on integration symbols, vectors, the Gauss–Green divergence theorem, and partial and total derivatives. These topics are not really used in this book, but they are, however, needed, as mentioned, in the understanding of the derivation of the Reynolds transport theorem. The mathematics used in this book pertains mostly to algebra.

INTEGRATION SYMBOLS

The symbols that we are going to discuss are as follows: $\oint_A$ and $\int_V$. The symbol $\oint_A$ is used to signify integration over a closed surface area and the symbol $\int_V$ is used to signify integration over a volume. But, what does this sentence mean exactly?

Refer to Figure 1. This figure shows a volume of solid, where a differential area dA is shown. This area is on the surface of the solid and, if it is extended around

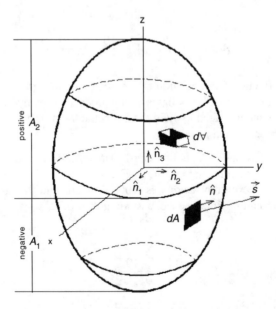

FIGURE 1 Definition of surface and volume integral and Gauss–Green divergence theorem.

the volume, it will form a closed surface around the volume. Now, consider a function such as $f(A) = A + 3A^3$ that applies on the surface of the volume. This function can be integrated over the closed surface and, to do this, the symbol $\oint_A$ is used. The "O" on the integral symbol signifies that the integration is all around the surface area that encloses the volume. Let A_s be the result of this integration. Then,

$$A_s = \oint_A f(A)dA = \oint_A (A + 3A^3)\,dA \tag{54}$$

$\oint_A f(A)\,dA = \oint_A (A + 3A^3)\,dA$ is called a *surface integral*. Because the surface is an area, surface integral is an area integral.

Note that in the surface intergral, we mentioned a function that applies on the surface of the solid. To have a visualization of this concept, consider a tank with water entering and leaving it. The function that would apply on the surface of this tank could be the velocity of the water that enters it and the velocity of the water that leaves it. (The places where the water enters and leaves the tanks are surfaces of the tank, where holes are cut through for the water to enter and leave. The other portions of the tank would not be open to the water; thus, no function would apply on these portions.) The surface integral could then be applied to the velocity functions upon entry and exit of the water.

Now, also, note in the figure the differential volume $d\forall$. This differential volume is not on the surface of the solid but is inside it. Assume a function exists that applies inside the volume. As in the case of the surface integral, this function can be integrated over the entire volume and, to do this, the symbol $\int_\forall$ is used. Let the function be $f(\forall) = \forall + 3\forall^3$ and the result of the integration be $\forall_s$. Thus,

$$\forall_s = \int_\forall f(\forall)d\forall = \int_\forall (\forall + 3\forall^3)\,d\forall \tag{55}$$

$\int_\forall (\forall)d\forall = \int_\forall (\forall + 3\forall)d\forall$ is called a *volume integral*.

An example of a function that applies inside the volume could be anything that transpires inside the volume. Again, consider the tank above, where this time, there is a chemical reaction occurring inside it. This chemical reaction can be expressed as a function; thus, it can be used as the integrand of the volume integral.

VECTORS

A vector is a quantity that has magnitude, direction, and orientation. In this book, we will use the arrow on the roof of a letter to signify a vector. Therefore, let $\vec{S}$ represent any vector. In the Cartesian coordinate system of xyz, this vector may be decomposed into its components as follows:

$$\vec{S} = S_1\hat{n}_1 + S_2\hat{n}_2 + S_3\hat{n}_3 \tag{56}$$

S_1, S_2, and S_3 are the *scalar components* of $\vec{S}$ in the x, y, and z directions, respectively, and $\hat{n}_1, \hat{n}_2$, and $\hat{n}_3$ and are the *unit vectors* in the x, y, and z directions, respectively.

Refer to Figure 1. As shown, dA has the unit vector $\hat{n}$ on its surface. The component of dA on the x-y plane is $dxdy$. As shown in the figure, the unit vector normal to $dxdy$ is $\hat{n}_3$, which is in the direction of the z axis. In vector calculus, the component of dA can be obtained through the scalar product of $\hat{n}$ and $\hat{n}_3$. This product is designated by $\hat{n} \cdot \hat{n}_3$. In other words,

$$dxdy = \hat{n} \cdot \hat{n}_3 \, dA \tag{57}$$

Another vector that we need to discuss is the nabla, $\vec{\nabla}$. This vector may be decomposed in the cartesian coordinates as follows:

$$\vec{\nabla} = \frac{\partial}{\partial x}\hat{n}_1 + \frac{\partial}{\partial y}\hat{n}_2 + \frac{\partial}{\partial z}\hat{n}_3 \tag{58}$$

The dot (or scalar) product is actually the product of the magnitude of two vectors and the cosine between them. The cosine of $90°$ is equal to zero, so the dot product of vectors perpendicular (orthogonal) to each is equal to zero. The unit vectors in the Cartesian coordinates are in an orthogonal axes; therefore, their "mixed dot products" are equal to zero. Thus,

$$\hat{n}_1 \cdot \hat{n}_2 = 0; \quad \hat{n}_1 \cdot \hat{n}_3 = 0; \quad \text{and} \quad \hat{n}_2 \cdot \hat{n}_3 = 0 \tag{59}$$

Now, the cosine of $0°$ is equal to one. Thus, the dot product of a unit vector with itself is equal to one. In other words,

$$\hat{n}_1 \cdot \hat{n}_2 = 1; \quad \hat{n}_2 \cdot \hat{n}_2 = 1; \quad \text{and} \quad \hat{n}_3 \cdot \hat{n}_3 = 1 \tag{60}$$

Using these new-found formulas, we evaluate the dot product of nabla and the vector $\vec{S}$. This is written as follows:

$$\vec{\nabla} \cdot \vec{S} = \left(\frac{\partial}{\partial x}\hat{n}_1 + \frac{\partial}{\partial y}\hat{n}_2 + \frac{\partial}{\partial z}\hat{n}_3\right) \cdot (S_1\hat{n}_1 + S_2\hat{n}_2 + S_3\hat{n}_3) \tag{61}$$

Simplifying the above equation will produce

$$\vec{\nabla} \cdot \vec{S} = \frac{\partial S_1}{\partial x} + \frac{\partial S_2}{\partial y} + \frac{\partial S_3}{\partial z} \tag{62}$$

GAUSS–GREEN DIVERGENCE THEOREM

This theorem converts an area integral to a volume integral and vice versa. Let $\vec{S}$ be any vector and $\hat{n}$ be the unit vector normal to area dA, where A is the area surrounding the domain of volume V. Refer to Figure 1. Now, form the volume integral $\int_V (\partial S_3/\partial z) \, dV$, where S_3 is the scalar component of $\vec{S}$ on the z axis. Thus,

$$\int_V \frac{\partial S_3}{\partial z}dV = \int_V \frac{\partial S_3}{\partial z}dx \, dy \, dz = \int_A \left\{\int_{z_1}^{z_2}\frac{\partial S_3}{\partial z}dz\right\}dx \, dy = \int_A \left\{\int_{z_1}^{z_2}\frac{\partial S_3}{\partial z}dz\right\}[\hat{n} \cdot \hat{n}_3] \, dA$$

where $\hat{n}_3$ is the unit vector in the positive z direction.

$$\int_A \left\{ \int_{z_1}^{z_2} \frac{\partial S_3}{\partial z} dz \right\} [\hat{n} \cdot \hat{n}_3] \, dA = \int_{A_2} \{S_3(x, y, z_2)\}[\hat{n} \cdot \hat{n}_3] \, dA_2$$

$$- \int_{A_1} \{S_3(x, y, z_1)\}[(-\hat{n} \cdot \hat{n}_3)] \, dA_1$$

A_1 and A_2 are the bounding surface areas on the negative z direction and positive z direction, respectively. Thus, $\hat{n} \cdot \hat{n}_3$ is negative on A_1 (because on A_1, $\hat{n}$ is pointing toward the negative z while $\hat{n}_3$ is pointing toward the positive z). $\hat{n} \cdot \hat{n}_3$ is positive on A_2 (by parallel reasoning). Therefore,

$$\int_V \frac{\partial S_3}{\partial z} dV = \int_{A_2} \{S_3(x,y,z_2)\}[\hat{n} \cdot \hat{n}_3] \, dA_2 + \int_{A_1} \{S_3(x,y,z_1)\}[(\hat{n} \cdot \hat{n}_3)] \, dA_1$$

$$= \int_A S_3[\hat{n} \cdot \hat{n}_3] \, dA \tag{63}$$

Considering the x and y directions, respectively, and following similar steps:

$$\int_V \frac{\partial S_1}{\partial x} dV = \int_A S_3[\hat{n} \cdot \hat{n}_1] \, dA \tag{64}$$

$$\int_V \frac{\partial S_2}{\partial y} dV = \int_A S_2[\hat{n} \cdot \hat{n}_2] \, dA \tag{65}$$

S_1 and S_2 are the scalar components of $\vec{S}$ on the x and y axes, respectively, and $\hat{n}_1$ and $\hat{n}_2$ are the unit vectors on the x and y axes, respectively.

Adding Eqs. (63), (64), and (67) produces the *Gauss–Green divergence theorem*. That is,

$$\int_V \frac{\partial S_3}{\partial z} dV + \int_V \frac{\partial S_1}{\partial x} dV + \int_V \frac{\partial S_2}{\partial y} dV = \int_A S_3[\hat{n} \cdot \hat{n}_3] \, dA + \int_A S_1[\hat{n} \cdot \hat{n}_1] \, dA$$

$$+ \int_A S_2[\hat{n} \cdot \hat{n}_2] \, dA \tag{66}$$

$$\int_V (\vec{\nabla} \cdot \vec{S}) \, dV = \oint_A \hat{n} \cdot \{S_3\hat{n}_3 + S_1\hat{n}_1 + S\hat{n}_2\} \, dA \tag{67}$$

$$\int_V (\vec{\nabla} \cdot \vec{S}) \, dV = \oint_A (\hat{n} \cdot \vec{S}) \, dA \tag{68}$$

Note that symbol $\oint_A$ is being used, since we are now integrating around the whole area A surrounding the volume V.

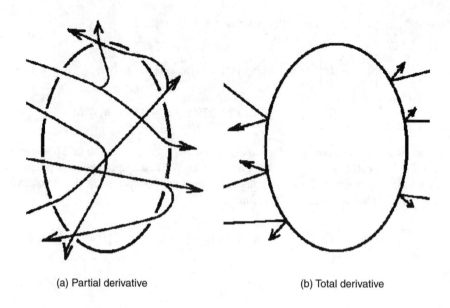

(a) Partial derivative   (b) Total derivative

FIGURE 2 Partial derivative vs. total derivative.

PARTIAL VERSUS TOTAL DERIVATIVE

A world of difference exists between partial derivative and total derivative. Yet, the environmental engineering literature seems not able to distinguish the difference between the two. In a given instance of use of the term, either partial derivative or total derivative is employed when, actually, only one version should be used—not either. To illustrate the difference, first define the word *property*. Property is an observable quality of matter. For example, consider water. Water may contain sodium and chloride ions, and their concentrations are an observable quality. Thus, the concentration of the sodium and chloride ions are properties of water. Water, of course, has temperature, and temperature is an observable quality; therefore, temperature is also a property of water. If the water is flowing, it will have velocity and velocity is an observable quality. Hence, velocity is also a property of water. In other words, to repeat, property is an observable quality of matter.

Now, to demonstrate the difference between partial derivative and total derivative, first, consider the left-hand side of Figure 2. This figure is an oval container which has holes on its sides. As shown, because of these holes, a mass can enter and leave the container. Let Φ represent any value of the property of the mass and let the container move in any direction. As the mass enters and leaves the container, it will be carrying with it its property, so the value of Φ will vary with distance as the container moves. Φ is then said to be a function of distance or space. If the Cartesian coordinate space, xyz, is chosen, then Φ is a function of x, y, and z. In addition, it must also be a function of time, t. In mathematical symbols,

$$\Phi = \Phi(x, y, z, t) \tag{69}$$

The variables x, y, z, and t are called *independent variables*, while Φ is called the *dependent variable* of these independent variables.

Now, if the derivative of Φ with respect to any of the independent variables is to be found, it can only be partial, because it is simultaneously a function of the four independent variables. In other words, it can only be varying partially, say, with respect to x, because it is varying, also partially, with respect to y, z, and t. Partial differentiation uses the symbol "∂." Hence, in mathematical symbols, the partial derivatives are

$$\frac{\partial \Phi}{\partial x} = \frac{\partial \Phi(x, y, z, t)}{\partial x} \qquad y, z, t = \text{constant} \qquad (70)$$

$$\frac{\partial \Phi}{\partial y} = \frac{\partial \Phi(x, y, z, t)}{\partial y} \qquad x, z, t = \text{constant} \qquad (71)$$

$$\frac{\partial \Phi}{\partial z} = \frac{\partial \Phi(x, y, z, t)}{\partial z} \qquad y, x, t = \text{constant} \qquad (72)$$

$$\frac{\partial \Phi}{\partial t} = \frac{\partial \Phi(x, y, z, t)}{\partial t} \qquad y, z, x = \text{constant} \qquad (73)$$

Now, consider the right-hand side of Figure 2. This is also an oval container similar to the one on the left but with no holes on the sides. Because of the absence of holes, no mass can enter the container. Let the container move in any direction. Because no mass is allowed to enter, the value of Φ will not be affected by any outside mass and, therefore, will not be a function of x, y, and z as the container moves in space. It is still, however, a function of time. It is no longer a function of space, so it has no partial derivative with respect to x, y, and z. It is now only a function of one variable t, so its derivative will no longer be partial but total. The symbol d is used for total derivative. In other words,

$$\frac{d\Phi}{dt} = \frac{d\Phi(x, y, z, t)}{dt} = \frac{d\Phi(t)}{dt} \qquad (74)$$

As you can see, there is a world of difference between partial derivative and total derivative. The total derivative applies when the dependent variable is a function of only one independent variable. When it is a function of more than one independent variable, then its derivative with respect to any one of the independent variables can only be partial. From the previous derivation, total derivatives apply to a *closed system*, while partial derivatives apply to an *open system*.

REYNOLDS TRANSPORT THEOREM

The Reynolds transport theorem demonstrates the difference between total derivative (or full derivative) and the partial derivative in the derivation of the material balance equation for the property of a mass. It is important that this distinction be made here because, as previously mentioned, the environmental engineering literature does not appear to distinguish the difference between the two.

For example, the material balance for the microbial kinetics of the activated sludge process is often written as (Metcalf and Eddy, Inc., 1991):

$$\frac{dX}{dt}\mathcal{V} = QX_o - QX + \mathcal{V}\left(\frac{\mu_m XS}{K_s + S} - k_d X\right) \tag{75}$$

where
X = concentration of microorganisms in the reactor
$\mathcal{V}$ = volume of reactor
t = time
Q = flow through the reactor
X_o = influent microbial concentration
μ_m = maximum specific growth rate
S = substrate concentration
K_s = half-velocity constant
k_d = endogenous decay coefficient

Equation 75, however, is incorrect. Note that there is a flow Q into and out of the reactor. This means that the system is open and the full derivative, dX/dt, cannot be used. Nothing in the literature points out this mistake, however, because everything is written the same way. The Reynolds transport theorem distinguishes the difference between the total and partial derivatives, so wrong equations like the previous one will not result in any derivation if using this theorem. After the derivation below of the theorem is complete, we will come back and derive the correct material balance of the microbial kinetics of the activated sludge process.

Two general methods are used in describing fluid flow: *Eulerian* and *Lagrangian* methods. In the Eulerian method, values of the properties of the fluid are observed at several points in space and time. This is analogous to the open container above, where the value of a property varies with time and space and where results are partial derivatives. The Lagrangian method, on the other hand, involves tagging each particle of fluid and observing the behavior of the desired property as the fluid moves, independent of spatial location. For example, if velocity is the desired property, this velocity is measured as the particle moves without regard to location in space. This is analogous to the closed container, where the property is only a function of time, and where the result is a total derivative.

It is possible to convert one method of description to the other using the Reynolds transport theorem attributed to Osborne Reynolds. In other words, the theorem converts the Eulerian method of description of fluid flow to the Lagrangian method of description of fluid flow and vice versa.

To derive the theorem, invent two terms: *control volume* and *control mass*. Refer to Figure 3. This figure is composed of two intersecting solids, where the volume formed due to the intersection of the solids is indicated by the black color, Volume bcde. The volume on the left with the hatching slanting upward to the right plus Volume bcde is the control volume. The volume with the hatching is Volume I.

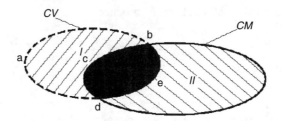

FIGURE 3 Derivation of the Reynolds transport theorem.

The volume on the right with the hatching slanting upward to the left plus Volume bcde is the control mass, and the volume with the hatching is Volume *II*. Control mass is that part of the universe composed of masses identified for analysis; as such, if a boundary is used to enclose the masses, no other masses are allowed to pass through this boundary. In other words, the system is closed. Thus, control mass is also called a *closed system* or simply *system*. The control volume, on the other hand, is a specific volume in space where masses are allowed to pass through, implying that its boundary is permeable to the "traffic" of masses crisscrossing through it. Thus, the control volume is an open system; it is, in fact, also called *open system*.

Imagine the control mass and the control volume coinciding at an initial time *t*, that is, the boundaries of both the control volume and the control mass coincide at this initial time. From this position, let them move at different speeds but with the control mass moving faster. After some time Δt later, the control mass and the control volume take different positions; this is the configuration portrayed in the figure.

Let Φ represent the value of any property common to both the closed and open systems. The change of this property for the closed system when the system goes in time from *t* to $t + \Delta t$ is

$$\Delta \Phi_M = \Phi_{M,t+\Delta t} - \Phi_{M,t} \tag{76}$$

where Δ is a symbol for change and *M* refers to control mass. $\Phi_{M,t+\Delta t}$ is the value of Φ for the control mass at $t + \Delta t$ and $\Phi_{M,t}$ is the value of Φ for the control mass at *t*. At the initial time, because control mass and control volume coincide with each other, the Φ of the control mass $\Phi_{M,t}$ is equal to the Φ of the masses inside the control volume $\Phi_{\mathcal{V},t}$. $\mathcal{V}$ refers to the control volume.

Referring to the figure, at time Δt later, $\Phi_{M,t+\Delta t}$ is given by

$$\Phi_{M,t+\Delta t} = \Phi_{\mathcal{V},t+\Delta t} - \Phi_{I,t+\Delta t} + \Phi_{II,t+\Delta t} \tag{77}$$

where $\Phi_{\mathcal{V},t+\Delta t}$, $\Phi_{I,t+\Delta t}$, and $\Phi_{II,t+\Delta t}$ are, respectively, the values of Φ of the masses in the control volume, Volume *I*, and Volume *II* at time $t + \Delta t$. Substituting Equation (77) in Equation (76), with $\Phi_{\mathcal{V},t}$ equals $\Phi_{M,t}$, produces

$$\Delta \Phi_M = \Phi_{\mathcal{V},t+\Delta t} - \Phi_{I,t+\Delta t} + \Phi_{II,t+\Delta t} - \Phi_{\mathcal{V},t} \tag{78}$$

Rearranging and dividing all terms by Δt produces

$$\frac{\Delta \Phi_M}{\Delta t} = \frac{\Phi_{V,t+\Delta t} - \Phi_{V,t}}{\Delta t} - \frac{\Phi_{I,t+\Delta t}}{\Delta t} + \frac{\Phi_{II,t+\Delta t}}{\Delta t} \tag{79}$$

Because it was derived in the closed system, the function Φ on the left-hand side of Equation (79) is a function of only one independent variable, t. There is only one independent variable, so a partial derivative for Φ is not possible, and there can only be a full or total derivative. Thus, in the limit, the term will produce a total derivative as follows:

$$\lim_{\Delta t \to 0} \frac{\Delta \Phi_M}{\Delta t} = \frac{d\Phi_M}{dt} = \frac{d\int_M \phi\rho \, dV}{dt} \tag{80}$$

where ϕ is the value of the property Φ per unit mass of the system and ρ is the mass density of the system. Note that $\int_M \phi\rho \, dV$ is Φ_M.

Because of the presence of V (space) and t and because it was derived for the control volume, the numerator of the first term on the right-hand side of Equation (79), $\Phi_{V,t+\Delta t} - \Phi_{V,t}$, is a function of both space and time. Therefore, in the limit, the derivative will be a partial derivative. Thus,

$$\lim_{\Delta t \to 0} \frac{\Phi_{V,t+\Delta t} - \Phi_{V,t}}{\Delta t} = \frac{\partial \int_V \phi\rho \, dV}{\partial t} \tag{81}$$

The property, Φ, of the mass inside Volume I results from the masses entering the holes of the control volume. Thus, this property must be a function of space as well as, of course, a function of time. The second term on the right-hand side of Equation (79), $-\Phi_{I,t+\Delta t}/\Delta t$, therefore, in the limit, results in a partial derivative,

$$\lim_{\Delta t \to 0} \left(-\frac{\Phi_{I,t+\Delta t}}{\Delta t} \right) = -\frac{\partial \Phi_I}{\partial t} \tag{82}$$

where $-\partial \Phi_I/\partial t$ is simply the rate of inflow of the property Φ across the boundary, where this boundary is dab. The volume rate of inflow is the product of velocity and cross-sectional area of flow. If the velocity is $\vec{v}$ and the area is A, then the volume rate of inflow is $-\int_{A_{in}} \vec{v} \cdot \hat{n} \, dA$, where $\hat{n}$ is the unit normal vector to dA. Note that the integration is over A_{in}, because the integration pertains to the inflow. Note, also, the negative sign, which is the result of the dot product, $\vec{v}$ and $\hat{n}$. This is explained as follows: The unit normal $\hat{n}$ always points away to the outside of surface (area) A. In this direction, any vector is considered positive, by convention. On the other hand, vector $\vec{v}$, because it is with the mass entering the control volume through the area A, is opposite to the direction of the unit normal. Because $\vec{v}$ and $\hat{n}$ are in opposite directions, their dot product will be negative, thus, $-\int_{A_{in}} \vec{v} \cdot \hat{n} \, dA$.

The mass rate of inflow is, accordingly, the product of the mass density and the volume rate of inflow. If the mass density is ρ, the mass rate of inflow is $-\int_{A_{in}} \rho \vec{v} \cdot \hat{n}\, dA$ Now, if the property per unit mass is ϕ,

$$-\frac{\partial \Phi_I}{\partial t} = -\left\{-\int_{A_{in}} \phi \rho \vec{v} \cdot \hat{n}\, dA\right\} = +\int_{A_{in}} \phi \rho \vec{v} \cdot \hat{n}\, dA \tag{83}$$

The third term of Equation (79), $\Phi_{II,t+\Delta t}/\Delta t$, is a result of mass passing through the boundary of the control volume, out into the control mass. This boundary is bed, the "exit boundary" of the control volume. The velocity at this boundary will be in the same direction as the unit normal vector. Thus, the corresponding dot product will be positive. In the limit, the expression for $\Phi_{II,t+\Delta t}/\Delta t$ will be the same as $\Phi_{I,t+\Delta t}/\Delta t$, except that the sign will be positive. Also, the integration will be over A_{out}, because the integration pertains to the outflow. Therefore,

$$\lim_{\Delta t \to 0}\left(\frac{\Phi_{II,t+\Delta t}}{\Delta t}\right) = \frac{\partial \Phi_{II}}{\partial t}\int_{A_{out}} \phi \rho \vec{v} \cdot \hat{n}\, dA$$

The last two terms of the right-hand side of Equation (79) are then, in the limit,

$$\lim_{\Delta t \to 0}\left(-\frac{\Phi_{I,t+\Delta t}}{\Delta t} + \frac{\Phi_{II,t+\Delta t}}{\Delta t}\right) = \int_{A_{in}} \phi \rho \vec{v} \cdot \hat{n}\, dA + \int_{A_{out}} \phi \rho \vec{v} \cdot \hat{n}\, dA$$

$$= \oint_A \phi \rho \vec{v} \cdot \hat{n}\, dA \tag{84}$$

where $\oint_A$ now represents that the terms inside the parenthesis on the left-hand side are summed over the area A that surrounds the control volume Ψ.

Substituting Eqs. (80), (81), and (84) into Equation (79),

$$\frac{d\int_M \phi \rho\, d\Psi}{dt} = \frac{\partial \int_\Psi \phi \rho\, d\Psi}{\partial t} + \oint_A \phi \rho \vec{v} \cdot \hat{n}\, dA \tag{85}$$

The left-hand side of the above equation is a derivative of the property in the control mass. Hence, in a control mass, the derivative is a total derivative. This is the derivative that would be observed on a given fluid property, irrespective of where the fluid is in space. Remember that the control mass system is closed; no mass can enter. Therefore, the property cannot vary with space, but only with time. This derivative describes what an observer would see if traveling with the mass inside the closed container—the control mass. This is called the Lagrangian method of describing the property. This derivative is also called the *Lagrangian derivative*.

The right-hand side of the equation shows the derivatives of the property inside the control volume, $(\partial \int_\Psi \phi \rho\, d\Psi)/\partial t$, and the property on the boundary of the control volume, $\oint_A \phi \rho \vec{v} \cdot \hat{n}\, dA$. Because $\oint_A \phi \rho \vec{v} \cdot \hat{n}\, dA$ belongs in the same equation as $(\partial \int_\Psi \phi \rho\, d\Psi)/\partial t$, it is also loosely called a derivative and, because this particular derivative is conveyed by a velocity, it is called a convective derivative. In other words, the convective derivative is conveyed from outside of the control volume into the inside of the control volume and vice versa. These derivatives describe what an observer from a distance will see. Standing at a distance, fluid properties as they vary from point to point may be observed (convective derivative), $\oint_A \phi\rho\vec{v} \cdot \hat{n}\, dA$; in addition, standing at the same distance, the same fluid properties as they vary with time may also be observed (the derivative $(\partial\int_v \phi\rho\, d\forall)/\partial t$). This mode of description is called the Eulerian method and the whole right-hand side of the equation may be called the *Eulerian derivative*.

Equation (85) portrays the equivalence of the Lagrangian and the Eulerian views. It is the Reynolds transport theorem. It states the values of the properties in one system, the control mass (the left-hand side), and states the same properties again in another system, the control volume (the right-hand side). These properties are one and the same things, so they must all be equal. They are said, therefore, to have a material balance between the properties in the control mass and the same properties in the control volume. The left-hand side of the equation is also called a *material, substantive*, or *comoving derivative*. The partial derivative on the right-hand side is called a *local derivative*. Because the theorem was derived by shrinking Δt to zero, it must be strongly stressed that the boundaries of both the control mass and the control volume coincide in the application of this theorem.

Kinetics of growth. In order to derive the correct version of Equation (75) using the Reynolds transport theorem, the kinetics of growth needs to be discussed. Let $[X]$ be the concentration of mixed population of microorganisms utilizing an organic waste. The rate of increase of $[X]$ fits the first order rate process as follows:

$$\text{Rate} = \mu[X] \tag{86}$$

where μ is the *specific growth rate* of the mixed population in units of per unit time and t is the time.

Now, the question is, what is that rate? Is it the total derivative $d[X]/dt$ or the partial derivative $\partial[X]/\partial t$? If one wants to determine experimentally the rate of growth of certain microorganisms, an obvious way is to put the culture in a large container reactor, take measurements, and calculate the rate of growth. Because culture will not be introduced and removed continuously to and from the container, the reactor is a closed system—it contains the control mass. Thus, the total derivative is used, instead of the partial derivative, and the rate equation becomes

$$\frac{d[X]}{dt} = \mu[X] \tag{87}$$

When an organism is surrounded by an abundance of food, the growth rate represented by the previous equation is at a maximum. This growth rate is said to be *logarithmic*. When the food is in short supply or depleted, the organisms will cannibalize each other. The growth rate at these conditions is said to be *endogenous*. Monod (1949) discovered that in pure cultures μ is a function of or is limited by the concentration $[S]$ of a limiting substrate or nutrient and formulated the following equation:

$$\mu = \mu_m \frac{[S]}{K_s + [S]} \tag{88}$$

where μ_m is the maximum μ. In Equation (88), if μ is made equal to one-half of μ_m and K_s solved, K_s will be found equal to $[S]$. Therefore, K_s is the concentration of the substrate that will make the specific growth rate equal to one-half the maximum growth rate and, hence, is called the *half-velocity constant*. Although the equation applies only to pure cultures, if average population values of μ_m are used, it can be applied to the kinetics of mixed population such as the activated sludge process, as well.

In the dynamics of any population, some members are born, some members die, or some members simply grow en masse. In addition, in the absence of food, the population may cannibalize each other. The dynamics or kinetics of death and cannibalization may simply be mathematically represented as a mass decay of the population $k_d[X]$, where k_d is the rate of decay. Incorporating Monod's concept and the kinetics of death into Equation (87), the model for the net rate of increase of $[X]$ now becomes

$$\frac{d[X]}{dt} = \mu_m \frac{[S]}{K_s + [S]}[X] - k_d[X] \tag{89}$$

This is the kinetics of growth, which, as can be seen, is expressed in total derivative. This represents the Lagrangian point of view (a closed-system point of view) of the process. Remember that we have used the total derivative, because cultures are not introduced into and withdrawn from the reactor, which means that the reactor is not an open system. The system is closed and, therefore, the total derivative is employed instead of the partial derivative.

Now, we are ready to derive the correct version of Equation (75) using the Reynolds transport theorem. First, let us determine the expression for $(d\int_M \phi\rho \, d\Psi)/dt$ in terms of $[X]$. Using the general dimensions of length L, mass M, and time t, $[X]$ has the dimension of M/L^3. The factors $\phi\rho$ has the dimensions of $(M/M)(M/L^3) = M/L^3$; thus, $\phi\rho$ corresponds to $[X]$. In the theorem $\phi\rho$ is a point value, while $[X]$ is an average value inside the reactor. $[X]$ then correspond to the value of $\phi\rho$ that is averaged. Therefore,

$$\frac{d\int_M \phi\rho \, d\Psi}{dt} = \frac{d\int_M (\phi\rho)_{ave} \, dV}{dt} = \frac{d\int_M [X] \, dV}{dt} = \frac{d[X]\int_M dV}{dt} = \frac{d[X]\Psi}{dt} \tag{90}$$

The volume of the reactor is constant, so Ψ may be taken out of the differential:

$$\frac{d\int_M \phi\rho \, d\Psi}{dt} = \frac{d[X]\Psi}{dt} = \Psi \frac{d[X]}{dt} \tag{91}$$

From Equation (89),

$$\frac{d\int_M \phi\rho \, d\Psi}{dt} = \Psi \frac{d[X]}{dt} = \Psi \left(\mu_m \frac{[S]}{K_s + [S]}[X] - k_d[X] \right) \tag{92}$$

Now, let us determine the expression for $\oint_A \phi \rho \vec{v} \cdot \hat{n} \, dA$. This is the convective derivative and it only applies to the boundary. In the case of the reactor, there are two portions of this boundary: the inlet boundary and the outlet boundary. Let the inflow to the reactor be Q; this will also be the outflow. Note that Q comes from $\vec{v} \cdot \hat{n} \, dA$ and, because the velocity vector and the unit vector are in opposite directions at the inlet, Q will be negative at the inlet. At the outlet, because the two vectors are in the same directions, Q will be positive. The concentration at the outlet will be the same as the concentration inside the tank, which is $[X]$. Thus, letting the concentration at the inflow be $[X_o]$

$$\oint_A \phi \rho \vec{v} \cdot \hat{n} \, dA = (-Q)[X_o] \text{ (at the inlet)} + (+Q)[X] \quad \text{(at the outlet)}$$

(93)

$$\oint_A \phi \rho \vec{v} \cdot \hat{n} \, dA = -Q[X_o] + Q[X]$$

The expression for $(\partial \int_V \phi \rho \, dV)/\partial t$ is straightforward. It is

$$\frac{\partial \int_V \phi \rho \, dV}{\partial t} = \frac{\partial \int_V (\phi \rho)_{ave} \, dV}{\partial t} = \frac{\partial \int_V [X] \, dV}{\partial t} = \frac{\partial [X] \int_V dV}{\partial t} = \frac{\partial [X] V}{\partial t} = V \frac{\partial [X]}{\partial t}$$

(94)

Substituting in Eqs. (92), (93), (94) into the Reynolds transport theorem,

$$V \left(\mu_m \frac{[S]}{K_s + [S]} [X] - k_d [X] \right) = \left\{ V \frac{\partial [X]}{\partial t} \right\} + \{ -Q[X_o] + Q[X] \}$$

(95)

This simplifies to

$$V \frac{\partial [X]}{\partial t} = QX_o - QX + V \left(\frac{\mu_m X S}{K_s + S} - k_d X \right)$$

(96)

This may be compared with Equation (75). As shown, the difference is in the confusion as to whether to use the total derivative or the partial derivative. This difference may be dismissed as trivial, but, of course, the use of the total derivative is very wrong. Compare the two expressions:

$$\frac{d[X]}{dt} = \mu_m \frac{[S]}{K_s + [S]} [X] - k_d [X]$$

(97)

$$\frac{\partial [X]}{\partial t} = \frac{1}{V} \left\{ QX_o - QX + V \left(\frac{\mu_m X S}{K_s + S} - k_d X \right) \right\}$$

(98)

From this comparison, it is very clear that $d[X]/dt$ is not equal to $\partial[X]/\partial t$ and one cannot substitute for the other.

GLOSSARY

Absolute mass—Masses such as grams, kilograms, and tonnes. Masses such as gram-moles, kilogram-moles, and tonne-moles are **relative masses**, because they are normalizations.

Acidity—The capacity of a substance to neutralize a base.

Acids—Substances that donate a proton.

Activity and active concentration—A measure of the effectiveness of a given species in a chemical reaction.

Activity coefficient—Expresses the reactive ability of a species.

Alkalinity—The capacity of a substance to neutralize an acid.

Bases—Substances that accept protons.

Conjugate acid—The acid formed when a base accepts a proton.

Conjugate base—The base formed when an acid loses a proton.

Control mass—That part of the universe composed of masses identified for analysis; as such, if a boundary is used to enclose the masses, no other masses are allowed to pass through this boundary.

Control volume—A specific volume in space where masses are allowed to pass through, implying that its boundary is permeable to the "traffic" of masses crisscrossing through it.

Equilibrium—A state in which the rate of the forward reaction forming the products is equal to the rate of the backward reaction forming the reactants.

Equilibrium constant—The value of the reaction quotient for reactions at equilibrium.

Equivalent mass—The mass of any substance participating in a reaction per unit of the number of reference species or the mass of the substance per equivalent of the substance.

Equivalent concentration—Number of equivalents of solute per unit volume of mixture.

Ionization constant—Equilibrium constant for dissociated ions and their parent compound.

Hydronium ion—H_3O^+ is hydronium ion.

Mass concentration—Mass of solute per unit volume of the mixture (m/v basis) or mass of solute per unit mass of the mixture (m/m basis).

Molar concentration—Number of moles of solute per unit volume of mixture.

Molarity—The number of gram moles of solute per 1,000 g of solvent.

Molarity—The number of gram moles of solute per liter of solution.

Mole fraction—The molar fractional part of a certain species relative to the total number of moles of species in the mixture.

Normality—Number of gram-equivalents of solute per liter of mixture

Number of equivalents—The mass of a substance divided by its equivalent mass.

Number of references species—The number of positive or negative charges or the number of moles of electrons in chemical reactions.

Oxidation—Reaction in which a substance loses electrons.

ppm—One mass of the solute in 10^6 mass of the mixture.

Property—An observable quality of matter.

Reaction quotient—The product of the molar concentration of the product species raised to the respective coefficiecnts divided by the product of the molar concentration of the reactant species raised to the respective coefficients.

Reduction—Reaction in which a substance gains electrons.

References species—The positive or negative charges or the moles of electrons in chemical reactions.

Solubility product contants—Equilibrium constants for dissolving solids.

Unit vector—A vector having a magnitude of unity.

Vector—Quantity that has magnitude, direction, and orientation.

PROBLEMS

1. Convert 10^{-10} kg to μg using the detailed-step and the short-cut methods.
2. Convert 180 million gallons per day to m^3/d.
3. In the equation $[C_o] = I\mathcal{M}/[96,494[C_o]Q_o\eta/(m/2)]$, the following values for the factors are given: $I = 122.84$ A, $Q_o = 378.51$ m^3/d, $\eta = 0.77$, $m = 400$, and $\mathcal{M} = 0.90$. Calculate the value of $[C_o]$ by indirect substitution and by direct substitution.
4. Water containing 3.5 moles of calcium bicarbonate and 2.5 moles of calcium sulfate is softened using lime and soda ash. How many grams of calcium carbonate solids are produced (a) using the method of equivalent masses and (b) using the balanced chemical reaction? Pertinent reactions are as follows:

$$Ca(HCO_3)_2 + Ca(OH)_2 \rightarrow 2CaCO_3\downarrow + 2H_2O$$

$$CaSO_4 + Na_2CO_3 \rightarrow CaCO_3 + Na_2SO_4$$

5. The results of an analysis in a sample of water are shown in the table below. Calculate the equivalent fractions of the respective species.

Ions	Conc (mg/L)
$Ca(HCO_3)_2$	150
$Mg(HCO_3)_2$	12.0
Na_2SO_4	216.0

6.   Calcium hydroxy apatite contains the phosphate and hydroxyl groups. Using calcium hydroxide as the precipitant, the chemical reaction is:

$$5Ca^{2+} + 3PO_4^{3-} + OH^- \rightleftharpoons Ca_5(PO_4)_3OH_{(s)}\downarrow$$

$$Ca_5(PO_4)_3OH_{(s)}\downarrow \rightleftharpoons 5Ca^{2+} + 3PO_4^{3-} + OH^- \qquad K_{sp,apatite} = 10^{-55.9}$$

$$PO_4^{3-} + H^+ \rightleftharpoons HPO_4^{2-}$$

$$HPO_4^{2-} + H^+ \rightleftharpoons H_2PO_4^-$$

$$H_2PO_4^- + H^+ \rightleftharpoons H_3PO_4$$

Write the molar mass balance for the total molar concentration of the phosphate radical.

7.   State the symbol $[sp_{PO_4Ca}]$ in words.

8.   The complexation reactions of the calcium ion with the carbonate species, OH^- and SO_4^{2-} are:

$$CaCO_3^o \rightleftharpoons Ca^{2+} + CO_3^{2-}$$

$$CaHCO_3^+ \rightleftharpoons Ca^{2+} + HCO_3^-$$

$$CaOH^+ \rightleftharpoons Ca^{2+} + OH^-$$

$$CaSO_4^o \rightleftharpoons Ca^{2+} + SO_4^{2-}$$

Write the molar mass balance for the total molar concentration of the CO_3 radical.

9.   The complexation reactions of the calcium ion with the carbonate species, OH^- and SO_4^{2-} are:

$$CaCO_3^o \rightleftharpoons Ca^{2+} + CO_3^{2-}$$

$$CaHCO_3^+ \rightleftharpoons Ca^{2+} + HCO_3^-$$

$$CaOH_3^+ \rightleftharpoons Ca^{2+} + OH^-$$

$$CaSO_4^o \rightleftharpoons Ca^{2+} + SO_4^{2-}$$

Write the molar mass balance for the total molar concentration of the SO_4 radical.

10. The concentration of $Ca(HCO_3)_2$ is 0.74 gmol/L. Convert this concentration to mgeq/L.

11. An ionic charge balance in terms of equivalents for a carbonate system is shown as follows:

$$2[CO_3^{2-}] + [HCO_3^-] + [OH] = [H^+] + 2[Ca^{2+}]$$

Convert the balance in terms of equivalent concentrations.

12.   In the coagulation treatment of water using copperas, $FeSO_4 \cdot 7H_2O$, the following complex reactions occur.

$$Fe(OH)_{2(s)} \rightleftharpoons Fe^{2+} + 2OH^- \qquad K_{sp,Fe(OH)_2} = 10^{-14.5} \qquad (a)$$

$$Fe(OH)_{2(s)} \rightleftharpoons FeOH^+ + OH^- \qquad K_{FeOHc} = 10^{-9.4} \qquad (b)$$

$$Fe(OH)_{2(s)} + OH^- \rightleftharpoons Fe(OH)_3^- \qquad K_{Fe(OH)_3c} = 10^{-5.1} \qquad (c)$$

A molar mass balance on the ferrous iron produces

$$[sp_{FeII}] = [Fe^{2+}] + [FeOH^+] + [Fe(OH)_3^-] \qquad (d)$$

where $[sp_{FeII}]$ is the total molar concentration of all species containing the iron atom. Express $[sp_{FeII}]$ in terms of the concentrations of the ferrous iron, hydrogen ion, and equilibrium constants.

13.   A raw water containing 140 mg/L of dissolved solids is subjected to a coagulation treatment using $Fe_2(SO_4)_3$. For optimum coagulation, the pH should be raised to 8.2. The current acidity is $6.0(10^{-4})$ geq/L. After satisfaction of the existing acidity, the amount of base needed to raise the pH has been calculated to be $1.25(10^{-6})$ geq/L. Calculate the amount of sodium hydroxide needed.

BIBLIOGRAPHY

Holtzclaw, H. F. Jr. and W. R. Robinson (1988). *General Chemistry*. D. C. Heath and Company, Lexington, MA.

Kreyszig, E. (1979). *Advanced Engineering Mathematics*. John Wiley & Sons, New York.

Metcalf & Eddy, Inc. (1991). *Wastewater Engineering: Treatment, Disposal, and Reuse*. McGraw-Hill, New York, 375.

Monod, J. (1949). The growth of bacterial cultures. *Ann Rev. Microbiol*. Vol. 3.

Munson, B. R., D. F. Young, and T. H. Okiishi (1990). *Fundamentals of Fluid Mechanics*. John Wiley & Sons, New York.

Sincero, A. P. and G. A. Sincero (1996). *Environmental Engineering: A Design Approach*. Prentice Hall, Upper Saddle River, NJ, 42.

Snoeyink, V. L. and D. Jenkins (1980). *Water Chemistry*. John Wiley & Sons, New York, 17–20.

Part I

Characteristics of Water and Wastewater

Before any water or wastewater can be treated, it must first be characterized. Thus, characterization needs to be addressed. Waters and wastewaters may be characterized according to their quantities and according to their constituent physical, chemical, and microbiological characteristics. Therefore, Part I is composed of two chapters: "Quantity of Water and Wastewater," and "Constituents of Water and Wastewater."

1 Quantity of Water and Wastewater

Related to and integral with the discussion on quantity are the important knowledge and background on the types of wastewater, sources of water and wastewater, and methods of population projection. The various categories of quantities in the form of design flow rates are also very important. These topics are discussed in this chapter. Because of various factors that have influenced the rate of wastewater generation in recent times, including water conservation and the expanded use of onsite systems, it is critical that designers have more than just typical wastewater generation statistics to project future usage. Thus, a method of determining accurate design flow rates calculated through use of probability concepts are also discussed in this chapter. This method is called *probability distribution analysis*; it is used in the determination of the quantities of water and wastewater, so it will be discussed first.

1.1 PROBABILITY DISTRIBUTION ANALYSIS

Figure 1.1 shows a typical daily variation for municipal sewage, indicating two maxima and two minima during the day. Discharge flows of industrial wastewaters will also show variability; they are, in general, extremely variable and "explosive" in nature, however. They can show variation by the hour, day, or even by the minute. Despite these seemingly uncorrelated variability of flows from municipal and industrial wastewaters, some form of pattern will emerge. For municipal wastewaters, these patterns are well-behaved. For industrial discharges, these patterns are constituted with erratic behavior, but they are patterns nonetheless and are amenable to analysis.

Observe Figure 1.2. This figure definitely shows some form of pattern, but is not of such a character that meaningful values can be obtained directly for design purposes. If enough data of this pattern is available, however, they may be subjected to a statistical analysis to predict design values, or probability distribution analysis, which uses the tools of probability. Only two rules of probability apply to our present problem: the addition rule and the multiplication rule.

1.1.1 ADDITION AND MULTIPLICATION RULES OF PROBABILITY

Before proceeding with the discussion of these rules, we must define the terms events, favorable event, and events not favorable to another event. An *event* is an occurrence, or a happening. For example, consider Figure 1.3, which defines Z as "Going from A to B." As shown, if the traveler goes through path E, he or she arrives at the destination point B. The arrival at B is an event. The travel through path E that causes event Z to occur is also an event.

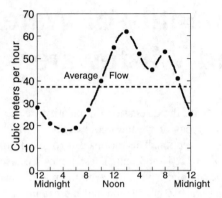

FIGURE 1.1 A typical variation of sewage flow.

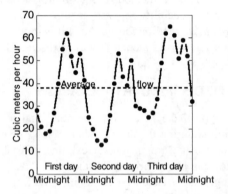

FIGURE 1.2 A three-day variation of sewage flow.

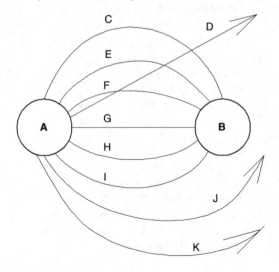

FIGURE 1.3 Definition of event Z as "Going from A to B."

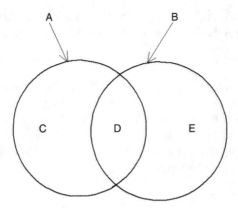

FIGURE 1.4 A Venn diagram for the union and intersection.

The path through E is an event or happening favorable to the occurrence of event Z. The other paths that a traveler could take to reach B are C, F, G, H, and I. Thus, the occurrence of any of these event paths will cause the occurrence of event Z; the occurrence of Z does not, however, mean that all of the event paths E, C, F, G, H, and I have occurred, but that at least one of them has occurred. These events are all said to be *favorable to the occurrence of event Z*.

The paths D, J, and K are events unrelated to Z; if the traveller chooses these paths she or he would never reach the destination point B. The events are *not favorable to the occurrence of Z*. All the events both favorable and not favorable to the occurrence of a given event, such as Z, constitute an event space of a particular domain. This particular domain space is called a *probability space*.

Addition rule of probability. Now, what is the probability that *one event or the other* will occur? The answer is best illustrated with the help of the *Venn diagram*, an example of which is shown in Figure 1.4, for the events A and B. There is D, which contains events from A and B; it is called the *intersection* of A and B, designated as $A \cap B$. This intersection means that D has events or results coming from both A and B. C has all its events coming from A, while E has all its events coming from B.

The sum of the events in A and B constitutes the union of A and B. This is written as $A \cup B$. From the figure,

$$A \cup B = A + B - D = A + B - A \cap B \qquad (1.1)$$

where the subtraction comes from the fact that when A and B "unite," they each contribute to the events at the intersection part of the union (D). This part counted the intersection events twice; thus, the other "half" must be subtracted. The union of A and B is the occurrence of the event: event A *or* event B has occurred—not event A *and* event B have occurred. The event, event A and event B, have occurred is the intersection mentioned previously.

From Equation (1.1), the probability that one event or the other will occur can now be answered. Specifically, what is the probability that the one event A or the other event B will occur? Because the right side of the previous equation is equal to the left side, the probability of the right side must be equal to the probability of the left side. Or,

$$Prob(A \cup B) = Prob(A) + Prob(B) - Prob(A \cap B) \tag{1.2}$$

where *Prob* stands for probability. Equation (1.2) is the probability that one event or the other will occur—the addition rule of probability.

Multiplication rule of probability. The intersection of A and B means that it contains events favorable to A as well as events favorable to B. Let the number of these events be designated as $N(A \cap B)$. Also, let the number of events of B be designated as $N(B)$. Then the expression

$$\frac{N(A \cap B)}{N(B)} \tag{1.3}$$

is the probability of the intersection with respect to the event B.

In the previous formula, event is synonymous with unit event. Unit events are also called outcomes. Probability values are referred to the total number of unit events or outcomes in the probability space, which would be the denominator of the above equation. As shown, however, the denominator of the above probability is referred to $N(B)$. $N(B)$ is smaller than the total number of unit events in the domain space; thus, it is called a *reduced probability space*. Because the reference probability space is that of B and because $N(A \cap B)$ is equal to the number of unit events of A in the intersection, the previous equation is called the *conditional probability* of A with respect to B designated as $Prob(A|B)$, or

$$Prob(A|B) = \frac{N(A \cap B)}{N(B)} \tag{1.4}$$

Let ζ designate the total number of unit events in the domain probability space in which event A is a part as well as event B is a part. Divide the numerator and the denominator of the above equation by ζ. Thus,

$$Prob(A|B) = \frac{N(A \cap B)/\zeta}{N(B)/\zeta} \tag{1.5}$$

The numerator of the previous equation is the intersection probability $Prob(A \cap B)$ and the denominator is $Prob(B)$. Substituting and performing the algebra, the following equation is produced:

$$Prob(A \cap B) = Prob(A|B)Prob(B) \tag{1.6}$$

Equation (1.6) is the *multiplication rule of probability*. If the reduced space is referred to A, then the intersection probability would be

$$Prob(A \cap B) = Prob(B|A)Prob(A) \tag{1.7}$$

In the multiplication rule, if one event precludes the occurrence of the other, the intersection does not exist and the probability is zero. These events are *mutually exclusive*.

If the occurrence of one event does not affect the occurrence of the other, the events are *independent* of each other. They are independent, so their probabilities are also independent of each other and $Prob(A|B)$ becomes $Prob(A)$ and $Prob(A|B)$ becomes $Prob(B)$.

Using the intersection probabilities, the addition rule of probability becomes

$$Prob(A \cup B) = Prob(A) + Prob(B) - Prob(A|B)Prob(B) \qquad (1.8)$$

1.1.2 VALUES EQUALED OR EXCEEDED

One of the values that is often determined is the value equaled or exceeded. The probability of a value equaled or exceeded may be calculated by the application of the addition rule of probability. The phrase "equaled or exceeded" denotes an element equaling a value and elements exceeding the value. Therefore, the probability that a value is equaled **or** exceeded is by the addition rule,

$$\begin{aligned} Prob&(value\ equaled\ or\ exceeded) \\ &= Prob(value\ equaled) + Prob(value\ exceeded) \\ &\quad - Prob(value\ equaled \cap value\ exceeded) \end{aligned} \qquad (1.9)$$

But there are 1, 2, 3,… ψ of the elements *exceeding the value*. Also, for mutually exclusive events, the intersection probability is equal to zero. Thus,

$$\begin{aligned} Prob&(value\ exceeded) \\ &= Prob(value1\ exceeding) + Prob(value2\ exceeding) \\ &\quad + \cdots + Prob(value\,\psi\ exceeding) \end{aligned} \qquad (1.10)$$

Substituting Equation (1.10) into Equation (1.9), assuming mutually exclusive events,

$$\begin{aligned} Prob&(value\ equaled\ or\ exceeded) \\ &= Prob(value\ equaled) + Prob(value1\ exceeding) \\ &\quad + Prob(value2\ exceeding) + \cdots + Prob(value\,\psi\ exceeding) \end{aligned} \qquad (1.11)$$

1.1.3 DERIVATION OF PROBABILITY FROM RECORDED OBSERVATION

In principle, to determine the probability of occurrence of a certain event, the experiment to determine the total number of unit events or outcomes for the probability space should be performed. Then the probability of occurrence of the event is equal to the number of unit events favorable to the event divided by the total possible number of unit events. If the number of unit events favorable to the given event is η and the total possible number of unit events in the probability space is ζ, the probability of the event, $Prob(E)$, is

$$Prob(E) = \frac{\eta}{\zeta} \qquad (1.12)$$

In practical situations, either the determination of the total number of ζs is very costly or the total number is just not available. Assume that the available ζ is ζ_{avail}, then the approximate probability, $Prob(E)_{approx}$, is

$$Prob(E)_{approx} = \frac{\eta}{\zeta_{avail}} \qquad (1.13)$$

The use of the previous equation, however, may result in fallacy, especially if ζ_{avail} is very small. η can become equal to ζ_{total} making the probability equal to 1 and claiming that the event is certain to occur. Of course, the event is not certain to occur, that is the reason why we are using probability. What can be claimed with correctness is that there is a *high degree of probability* that the event will occur (or a high degree of probability that the event will not occur). Because there is no absolute certainty, in practice, a correction of 1 is applied to the denominator of Equation (1.13) resulting in

$$Prob(E)_{approx} = \frac{\eta}{\zeta_{avail} + 1} \qquad (1.14)$$

Take note that for large values of ζ_{avail} the correction 1 in the denominator becomes negligible.

To apply Equation (1.14), recorded data are arranged into arrays either from the highest to the lowest or from the lowest to the highest. The number of values above a given element and including the element is counted and the probability equation applied to each individual element of the array. Because the number of values above and at a particular element is a sum, this application of the equation is, in effect, an application of the probability of the union of events. The probability is called *cumulative*, or *union* probability. After all union probabilities are calculated, an array of probability distribution results. This method is therefore called *probability distribution analysis*. This method will be illustrated in the next example.

Example 1.1   In a facility plan survey, data for Sewer A were obtained as follows:

Week No.	Flow (m^3/wk)	Week No.	Flow (m^3/wk)
1	2900	8	4020
2	3028	9	3675
3	3540	10	3785
4	3300	11	3459
5	3700	12	3200
6	4000	13	3180
7	3135	14	3644

(a) What is the probability that the flow is 3700 m^3/wk? (b) What is the probability that the flow is equal to or greater than 3700 m^3/wk? (c) What is the flow that will never be exceeded?

Solution:

(a) $Prob(3700) = Prob(E)_{approx} = \dfrac{\eta}{\zeta_{avail}+1} = \dfrac{1}{14+1} = 0.07$ **Ans**

(b) From the problem, the values greater than 3700 are 3785, 4000, and 4020. Thus,

$Prob(value\ equaled\ or\ exceeded)$

$\quad = Prob(value\ equaled) + Prob(value\ 1\ exceeding) + \cdots$

$Prob(flow \geq 3700)$

$\quad = Prob(3700) + Prob(3785) + Prob(4000) + Prob(4020)$

$= \dfrac{1}{15} + \dfrac{1}{15} + \dfrac{1}{15} + \dfrac{1}{15} = 0.27$ **Ans**

The problem may also be solved by arranging the data into an array. Because the problem is asking for the probability that is "equal or greater," arrange the data in descending order. This is the probability distribution analysis. The analysis is indicated in the following table.

The values under the column cumulative $\eta = \Sigma\eta$ represent the total number of values above and including the element at a given serial number. For example, consider the serial no. 4. The flow rate at this serial number is indicated as 3700 m^3/wk and, under the column of $\Sigma\eta$, the value is 4. This 4 represents the sum of the number of values equal to or greater than 3700; these values being 3700, 3785, 4000, and 4020. The values 3785, 4000, and 4020 are the values above the element 3700 which numbers 3. Adding 3 to the count of element 3700, itself, which is 1, gives 4, the number under the column $\Sigma\eta$.

The column $\Sigma\eta/(\zeta_{avail}+1) = \Sigma\eta/(14+1)$ is the cumulative probability of the item at a given serial number. For example, the element 3700 has a cumulative probability of 0.27. This cumulative probability is the same $Prob(value\ equaled\ or\ exceeded) = Prob(value\ equaled) + Prob(value\ 1\ exceeding) + \cdots$ used previously.

Serial No.	Flow (m^3/wk)	$\Sigma\eta$	$\dfrac{\Sigma\eta}{\zeta_{total}+1}$
1	4020	1	$\dfrac{1}{14+1} = 0.07$
2	4000	2	$\dfrac{2}{14+1} = 0.13$
3	3785	3	$\dfrac{3}{14+1} = 0.20$
4	3700	4	$\dfrac{4}{14+1} = 0.27$
5	3675	5	$\dfrac{5}{14+1} = 0.33$
6	3644	6	$\dfrac{6}{14+1} = 0.40$
7	3540	7	$\dfrac{7}{14+1} = 0.47$
8	3459	8	$\dfrac{8}{14+1} = 0.53$

(continued)

Serial No.	Flow (m³/wk)	$\Sigma\eta$	$\dfrac{\Sigma\eta}{\zeta_{total}+1}$
9	3300	9	$\dfrac{9}{14+1}=0.60$
10	3200	10	$\dfrac{10}{14+1}=0.67$
11	3180	11	$\dfrac{11}{14+1}=0.73$
12	3135	12	$\dfrac{12}{14+1}=0.80$
13	3028	13	$\dfrac{13}{14+1}=0.87$
14	2900	14	$\dfrac{14}{14+1}=0.93$

(c) The value that will never be exceeded is the largest value. Thus, there will be no value above it and the cumulative count for this element is 1. The data for flows in the previous table are simply for values obtained from a field survey. To answer the question of what is the value that will never be exceeded, we have to obtain this value from an exhaustive length of record and read the value that is never exceeded on that particular record. Of course, the count for this largest value would be 1, as mentioned.

Now, what has the field data to do with the determination of the largest value? The use of the field data is to develop a probability distribution. The resulting distribution is then assumed to model the probability distribution of all the possible data obtainable from the problem domain. The larger the number of data and the more representative they are, the more accurate this model will be.

Obtaining the largest value means that the amount of data used to obtain the probability distribution model must be infinitely large; and, in this infinitely large amount of data, there is only one value that is equaled or exceeded. This means that the probability of this one value is 1/infinity = 0. From the probability distribution, the peak weekly flow rate can be extrapolated at probability 0. This is done as follows (with x representing the weekly flow rate):

x	0
4020	0.07
4000	0.13

Therefore,

$$\frac{x-4020}{4020-4000}=\frac{0-0.07}{0.07-0.13} \qquad x = 4043 \text{ m}^3/\text{wk} \quad \textbf{Ans}$$

1.1.4 VALUES EQUALED OR NOT EXCEEDED

The probability of values equaled or not exceeded is just the reverse of values equaled or exceeded. In the previous example, the values were arranged in descending order. For the case of value equaled or not exceeded, the values are arranged in ascending order.

Deducing from Equation (1.11), the probability of a value equaled or not exceeded is

$$Prob(value\ equaled\ or\ not\ exceeded)$$
$$= Prob(value\ equaled) + Prob(value1\ not\ exceeding)$$
$$+ Prob(value2\ not\ exceeding) + \cdots \qquad (1.15)$$

Example 1.2 In Example 1.1, what is the probability that the flow is equal to or less than 2800? 3000?

Solution: The probability distribution arranged in descending order is as follows:

Serial No.	Flow (m³/wk)	$\Sigma\eta$	$\dfrac{\Sigma\eta}{\zeta_{total}+1} = \dfrac{\Sigma\eta}{14+1}$
1	2900	1	$\dfrac{1}{14+1} = 0.07$
2	3028	2	0.13
3	3135	3	0.2
4	3180	4	$\dfrac{4}{14+1} = 0.27$
5	3200	5	
6	3300	6	
7	3459	7	
8	3540	8	
9	3644	9	
10	3675	10	
11	3700	11	$\dfrac{11}{14+1} = \mathbf{0.73}$
12	3785	12	
13	4000	13	
14	4020	14	

2800	x
2900	0.07
3000	y
3028	0.13

$$\frac{x-0.07}{0.07-0.13} = \frac{2800-2900}{2900-3028}$$

$$\frac{y-0.07}{0.13-0.07} = \frac{3000-2900}{3028-2900}$$

$$x = 0.023 = Prob(flow \le 2800) \quad \textbf{Ans}$$

$$y = 0.12 = Prob(flow \le 3000) \quad \textbf{Ans}$$

Example 1.3 In Example 1.1, calculate the probability that the flow is equal to less than 3700 m³/wk.

Solution: The values less than 3700 are 3675, 3644, 3540, 3459, 3300, 3200, 3180, 3135, 3028, and 2900. Thus,

$$Prob(value\ equaled\ or\ not\ exceeded)$$
$$= Prob(value\ equaled) + Prob(value1\ not\ exceeding)$$
$$+ Prob(value2\ not\ exceeding) + \cdots$$
$$Prob(flow \leq 3700)$$
$$= Prob(3700) + Prob(3675) + Prob(3644) + Prob(3540)$$
$$+ Prob(3459) + Prob(3300) + Prob(3200)Prob(3180)$$
$$+ Prob(3135) + Prob(3028) + Prob(2900)$$
$$= 11\left(\frac{1}{15}\right) = 0.73 \quad \textbf{Ans}$$

Again, probability distribution analysis may also be applied. The procedure is similar to that of the previous one, except that the data values are arranged in ascending order instead of descending order. Thus,

1.2 QUANTITY OF WATER

Any discussion on physical–chemical treatment of water and wastewater is incomplete without knowledge of the quantities involved. How large a volume is being treated? The answer to this question will enable the designer to size the units involved in the treatment.

Two quantities are addressed in this chapter: the quantity of water and the quantity of wastewater. The quantity of water is discussed first. To design water treatment units, the engineer, among other things, may need to know the average flow, the maximum daily flow, and maximum hourly flow. The following information are examples of the use of the design flows:

1. Community water supplies, water intakes, wells, treatment plants, pumping, and transmission lines are normally designed using the maximum daily flow with hourly variations handled by storage.
2. Water distribution systems are designed on the basis of the maximum day plus flow for fighting fires or on the basis of the maximum hourly, whichever is greater. For emergency purposes, standby units are installed.
3. For industrial plants, resort sites, and so on, special studies may be made to determine the various design values of water usage.

In the case of communities, the actual flows that are used in design are affected by the design period. Designs are not normally made on the basis of flow at the end of this period but are spread over the duration. The design period, also called the planning period, is discussed later.

Of the various design flows, we will discuss the average flow first. The average flow in a community is normally taken as the average daily flow computed over a year as follows:

$$x = \frac{y}{365(P) \ \text{or} \ 366(P)} \tag{1.16}$$

where
   x = consumption in units of volume per capita per day
   y = total cubic units of water delivered to the distribution system
   P = the midyear population served by the distribution system

Note that in the previous equation, the consumption has been normalized against population. In industrial plants, commercial, institutional, and other facilities, the normalization is done in some other ways such as per tonnes of product per customer, per student, and so on.

From the previous definition of average flow, it is evident that there would be a number of values depending upon the number of years that the averages are computed. For purposes of design, the engineer must decide which particular value to use. The highest value may not be arbitrarily used because this may result in over design; on the other hand, the lowest value may also not be used for a similar but reverse reasoning.

In reality, only one average value exists, and this value is the long-term value. The reason why we have so many average values is that averages have been taken each year when there should only be one. What should be done is to take all the daily values in the record, sum them up, and divide this sum by the number of daily values. The problem with this approach, however, is that once this one single average is obtained, it is still not certain whether or not this particular one value is **the** average value. In concept, the averaging may be extended for one more additional value and the average recomputed. If the current recomputed average value is equal to or close to the previous value, then it may be concluded that the correct average value has been obtained. If not, then the recomputing of the average may be further extended until the correct average is obtained.

The other way to obtain the average value is to use the probability distribution analysis. From the theory of probability, the long-term average value (which is **the** average value) is the value that corresponds to 50% probability in the probability distribution. Thus, if sufficient data have been gathered, the average can be obtained from the distribution without going through the trial-and-error method of recomputing the average addressed in the previous paragraph.

Depending upon the source of information, average values are vastly different. The following are average usages in cubic meters per capita per day from two communities obtained from two different sources of information:

User	Cubic meters per cap per day	Cubic meters per cap per day
Domestic	0.13	0.24
Commercial and industrial	0.11	0.25
Public use and other losses	0.13	0.08

The previous table shows that you are bound to obtain widely differing answers to the same question from different sources. You, therefore, have to gather your own.

These values may be used for preliminary calculations only. For more accurate values on specific situations, a field determination must have to be made.

As mentioned previously, the maximum day and the maximum hour may also be needed in design. To get the maximum day and the maximum hour, the *proper* average daily flow is multiplied by the ratios of the maximum day and maximum hour to the average day, respectively. The question of what maximum ratios to use is not easy to answer. For any community, industrial plant, commercial establishment, and the like, literally hundreds of maxima are in the record; however, there should only be one maximum.

The following treatment will discuss a method of obtaining the maximum such as the maximum hourly. To obtain these quantities, the record may be scanned for the occurrence of the daily maximum hourly and divided by the corresponding daily average. The results obtained are then arranged serially in decreasing order and probability distribution calculated. The probability obtained from this calculation is a "daily probability" as distinguished from the probabilities of occurrence of storms which are based on years and, hence, are "yearly probabilities." A weekly, monthly, or yearly probability of the maximum hourly may also be calculated. In theory, all these computed probabilities will be equivalent if the records are long enough. Now, what ratio should be obtained from the probability distribution? If the distribution is derived from a very long record, the ratio may be obtained by extrapolating to the probability that the ratio will never be exceeded. This probability would be zero. The definition of the *maximum hourly flow of water use* is the highest of the hourly flows that will ever occur.

The maximum daily flow may be obtained in a similar manner as used for the maximum hourly flow, only the daily values are used rather than the hourly values. Then, in an analogous manner as used to define the maximum hourly flow, we make the following definition of the *maximum daily flow of water use*: the highest of the daily flows that will ever occur.

For an expansion to an existing system, the records that already exist may be analyzed in order to obtain the design values. For an entirely new system, the records of a nearby and similar community, industrial plant, commercial establishment, and the like may be utilized in order to obtain the design values. Of course, in selecting the final design value, there are other factors to be considered. For example, in the case of a community water supply, the practice of metering the consumption inhibits the consumer from wasting water. Therefore, if the record analyzed contains time when meter was used and time when not used, the conclusion drawn from statistical analysis must take this fact into consideration. The record might also contain years when use of water was heavily curtailed because of drought and years when use was unrestricted. This condition tends to make the data "inhomogeneous." A considerable engineering judgment must therefore be exercised to arrive at the final value to use.

1.2.1 DESIGN PERIOD

Generally, it takes years to plan, design, and construct a community water and wastewater facility. Even at the planning stage, the population continues to grow and, along with it, comes the increase of flows during the period. This condition requires that, in addition to determining the ultimate design population, the population must also be predicted during the *initial years* that the project is put into operation.

The flows that the facility is to be designed for are *design flows*. The time from the initial design years to the time that the facility is to receive the final design flows is called the *design* or *planning period*. The facility would not be sized for the initial years nor for the final year; the design must be staged. At the initial years, the facility is smaller, and it gets bigger as it is being expanded during the *staging period* corresponding to the increase in population until finally reaching the end of the planning period. Table 1.1 shows staging periods for expansion of water and wastewater plants, and Table 1.2 shows design periods for various water supply and sewerage components. Tables 1.3 through 1.6 show average rates of water use for various types of facilities.

TABLE 1.1
Staging Periods for Expansion of Water and Wastewater Plants

Ratio, Final Design Fows/ Initial Design Flows	Staging Period, Years
Less than 2.0	15–20
Greater than 2.0	10

TABLE 1.2
Design Periods for Various Water Supply and Sewerage Components

Component	Design Period (years)
Water supply	
Large dams and conduits	25–50
Wells, distribution systems, and filter plants	10–25
Pipes more than 300 mm in diameter	20–25
Secondary mains less than 300 mm in diameter	Full development
Sewerage	
Laterals and submains less than 380 mm in diameter	Full development
Main sewers, outfalls, and interceptors	40–50
Treatment works	10–25

TABLE 1.3
Average Rates of Water Use for Commercial Facilities

User	Unit	Flow (m³/unit.d)
Airport	Passenger	0.02
Apartment house	Person	0.40
Automobile service station	Employee	0.07
Boarding house	Person	0.17
Department store	Employee	0.05
Hotel	Guest	0.20
Lodging house and tourist home	Guest	0.16
Motel	Guest	0.12
Motel with kitchen	Guest	0.16
Laundry	Wash	0.19
Office	Employee	0.06
Public lavatory	User	0.02
Restaurant	Customer	0.03
Shopping center	Parking space	0.01
Theater, indoor	Seat	0.01
Theater, outdoor	Car	0.02

TABLE 1.4
Average Rates of Water Use for Institutional Facilities

User	Unit	Flow (m³/unit.d)
Assembly hall	Seat	0.01
Boarding school	Student	0.28
Day school with cafeteria, gym, and showers	Student	0.09
Day school with cafeteria	Student	0.06
Day school with cafeteria and gym	Student	0.04
Medical hospital	Bed	0.60
Mental hospital	Bed	0.45
Prison	Inmate	0.45
Rest home	Resident	0.34

1.3 TYPES OF WASTEWATER

As mentioned earlier, basically two quantities are addressed in this chapter: those of water and those of wastewater. Quantity of water has already been addressed in the previous treatments, so it is now time to address quantity of wastewaters. Before discussing quantity of wastewaters, however, it is important that the various types of wastewaters be discussed first. The two general types of wastewaters are sanitary and non-sanitary. The *non-sanitary* wastewaters are normally industrial wastewaters. *Sanitary wastewaters* (or *sanitary sewage*) are wastewaters that have been contaminated

TABLE 1.5
Average Rates of Water Use for Recreational Facilities

User	Unit	Flow (m^3/unit.d)
Bowling alley	Alley	0.80
Camp, with central toilet and bath	Person	0.17
Camp, luxury with private bath	Person	0.34
Camp, trailer	Trailer	0.47
Campground, developed	Person	0.11
Country club	Member	0.38
Dormitory (bunk bed)	Person	0.13
Fairground	Visitor	0.01
Picnic park with flush toilets	Visitor	0.03
Swimming pool and beach	Customer	0.04
Resort apartments	Person	0.23
Visitor center	Visitor	0.02

TABLE 1.6
Average Rates of Water Use for Various Industries

Industry	Flow (m^3/tonne product)
Cannery	
Green beans	53
Peaches and pears	15
Vegetables	15
Chemical	
Ammonia	150
CO_2	55
Lactose	640
Sulfur	8
Food and beverage	
Beer	10
Bread	3
Meat packing	15
Milk products	15
Whisky	65
Pulp and paper	
Pulp	415
Paper	128
Textile	
Bleaching	230
Dyeing	40

with human wastes. Sanitary wastewaters generated in residences are called *domestic wastewaters* (*domestic sewages*). *Industrial wastewaters* are wastewaters produced in the process of manufacturing. Thus, because a myriad of manufacturing processes are used, a myriad of industrial wastewaters are also produced. Sanitary wastewaters produced in industries may be called *industrial sanitary wastewaters*. To these wastewaters may also be added infiltration and inflow.

Wastewaters are conveyed through sewers. Various incidental flows can be mixed with them as they flow. For example, *infiltration* refers to the water that enters sewers through cracks and imperfect connections and through manholes. This water mostly comes from groundwater and is not intended to be entering into the sewer. *Inflow* is another incidental flow that enters through openings that have been *purposely* or *inadvertently* provided for its entrance. Inflow may be classified as steady, direct and delayed. *Steady inflows* enter the sewer system continuously. Examples of these are the discharges from cellar and foundation drains that are constantly subjected to high groundwater levels, cooling water and drains from swampy areas, and springs. *Direct inflows* are those inflows that result in an increase of flow in the sewer almost immediately after the beginning of rainfall. The possible sources of these are roof leaders, manhole covers, and yard drains. *Delayed inflows* are those portions of the rainfall that do not enter the sewer immediately but take some days to drain completely. This drainage would include that coming from pumpage from basement cellars after heavy rains and slow entries of water from ponded areas into openings of manholes. Infiltration and inflows are collectively called *infiltration-inflow*.

1.4 SOURCES AND QUANTITIES OF WASTEWATER

The types of wastewaters mentioned above come from various sources. Sanitary wastewaters may come from *residential, commercial, institutional,* and *recreational* areas. Infiltration-inflow, of course, comes from rainfall and groundwater, and industrial wastewaters come from manufacturing industries.

The quantities of these wastewaters as they come from various sources are varied and, sometimes, one portion of the literature would report a value for a quantity of a parameter that conflicts on information of the quantity of the same parameter reported in another portion of the literature. For example, many designers often assume that the amount of wastewater produced is equal to the amount of water consumed, including the allowance for infiltration-inflow, although one report indicates that 60 to 130% of the water consumed ends up as wastewater, and still another report indicates that 60 to 85% ends up as wastewater. For this reason, quantities provided below should not be used as absolute truths, but only as guides. For more accurate values, actual data should be used.

1.4.1 Residential

Flow rates are commonly normalized against some contributing number of units. Thus, flow rates from residences may be reported as liters per person per day (normalized against number of persons), or flow rates from a barber shop may be reported as liters per chair per day (normalized against number of chairs) and so on.

TABLE 1.7
Average Sanitary Wastewater Production in Residential Sources

Type	Production (L per capita per day)
Typical single-family homes	290
Large single-family homes	400
High-rise apartments	220
Low-rise apartments	290
Trailer or mobile home parks	180

TABLE 1.8
Average Sanitary Wastewater Production in Commercial Establishments

Type	Production (L per indicated unit per day)
Hotels and motels, person	200
Restaurants, per employee	115
Restaurants, per customer	35
Restaurants, per meal served	15
Transportation terminals, per employee	60
Transportation terminals, per passenger	20
Office buildings	65
Movie theaters, per seat	15
Barber shop, per chair	210
Dance halls	8
Stores, per employee	40
Shopping centers, per square meter of floor area	8
Commercial laundries, per machine	3000
Car washes, per car	200
Service stations, per employee	190

Table 1.7 shows average sanitary wastewater production in residential areas. Note that the normalization is per person (per capita means per person).

1.4.2 COMMERCIAL

Wastewaters are also produced in the course of doing business. These businesses may include such commercial activities as hotels and motels, restaurants, transportation terminals, office buildings, movie theaters, barber shops, dance halls, stores, shopping centers, commercial laundries, car washes, and service stations. These activities are classified as commercial—year-round activities as opposed to some commercial activities that are seasonal and related to recreational activities, thus, classified as recreational. Table 1.8 shows some average flow rates from these commercial establishments.

TABLE 1.9
Average Sanitary Wastewater Production in Institutional Facilities

Type	Production (L per indicated unit per day)
Boarding schools, per student	300
Day schools, with cafeteria, gym, and showers (per student)	90
Day schools, with cafeteria only (per student)	58
Day schools, without cafeteria, gym, and showers (per student)	40
Mental hospitals, per bed	380
Medical hospitals, per bed	630
Prison, per resident	440
Rest homes, per resident	380

1.4.3 INSTITUTIONAL

Institutional facilities include those from institutions such as hospitals, prisons, schools and rest homes. Table 1.9 shows wastewater production from these and other institutional facilities. Again, note the scheme of normalization: per employee, per customer, per seat, per meal served, etc.

1.4.4 RECREATIONAL

Most wastewaters from recreational facilities use are seasonal. Although strictly commercial, because of their seasonal nature, wastewaters from these facilities are given special classification; they are wastewaters resulting from *recreational* use. Examples are those coming from resort apartments, resort cabins, resort cafeteria, resort hotels, resort stores, visitor centers, campgrounds, swimming pools in resort areas, etc. An example of the seasonal nature of recreational wastewaters is the case of Ocean City, Maryland—a resort town. During summer periods, the wastewater treatment plant is "bursting at the seams;" however, during winter periods, the flow to the treatment plant is very small.

Wastewaters may also be produced from recreational use but on a year-round basis. Hotels and motels in Florida and the Philippines, for example, are not seasonal. They provide year-round services. Wastewaters in these places are best classified as commercial. Hotels in Ocean City are definitely seasonal; thus, their wastewaters are recreational. Table 1.10 shows wastewater production resulting from recreational use.

1.4.5 INDUSTRIAL

The production of industrial wastewaters depends upon the type of processes involved. Table 1.11 shows industrial wastewater productions in some industrial processes.

TABLE 1.10
Average Sanitary Wastewater Production
in Recreational Facilities

Type	Production (L per indicated unit per day)
Luxury resorts hotels, per person	500
Tourist camps, per person	40
Resort motels, per person	210
Resort apartments, per person	230
Resort cabins, per person	145
Resort cafeteria, per customer	8
Resort stores, per employee	40
Visitor centers, per visitor	20

TABLE 1.11
Industrial Wastewater Production
in Some Industries

Industry	Production
Cattle	50 L/head-day
Canning	40 m^3/metric ton
Dairy	80 L/head-day
Chicken	0.5 L/head-day
Pulp and paper	700 m^3/metric ton
Meat packaging	20 m^3/metric ton
Tanning	80 m^3/metric ton raw hides processed
Steel	290 m^3/metric ton

1.5 POPULATION PROJECTION

To determine the flows that water and wastewater treatment facilities need to be designed for, some form of projection must be made. For industrial facilities, production may need to be projected into the future, since the use of water and the production of wastewater are directly related to industrial production. Design of recreational facilities, resort communities, commercial establishments, and the like all need some form of projection of the quantities of water use and wastewater produced. In determining the design flows for a community, the population in the future needs to be predicted. Knowledge of the population, then enables the determination of the flows. This section, therefore, deals with the various methods of predicting population.

Several methods are used for predicting population: arithmetic method, geometric method, declining rate of increase method, logistic method, and graphical comparison method. Each of these methods is discussed.

1.5.1 ARITHMETIC METHOD

This method assumes that the population at the *present time* increases at a constant rate. Whether or not this assumption is true in the past is, of course, subject to question. Thus, this method is applicable only for population projections a short term into the future such as up to thirty years from the present.

Let P be the population at any given year Y and k_a be the constant rate. Therefore,

$$\frac{dP}{dY} = k_a \tag{1.17}$$

Integrating from limits $P = P_1$ to $P = P_2$ and from $Y = Y_1$ to $Y = Y_2$,

$$P_2 = P_1 + k_a(Y_2 - Y_1) \tag{1.18}$$

Solving for k_a,

$$k_a = \frac{P_2 - P_1}{Y_2 - Y_1} \tag{1.19}$$

With k_a known, the population to be predicted in any future year can be calculated using Equation (1.18). Call this population as *Population* and the corresponding year as *Year*. To project the population into the future, the most current values P_2 and Y_2 must be used as the basis for the projection. Thus,

$$Population = P_2 + k_a(Year - Y_2) \tag{1.20}$$

Example 1.4   The population data for Anytown is as follows: 1980 = 15,000, and 1990 = 18,000. What will be the population in the year 2000? What is the value of k_a?

Solution:

$$k_a = \frac{P_2 - P_1}{Y_2 - Y_1} = \frac{18000 - 15000}{1990 - 1980} = 300 \, per \, year \quad \textbf{Ans}$$

$$Population = P_2 + k_a(Year - Y_2)$$
$$= 18,000 + 300(2000 - 1990) = 21,000 \, people \quad \textbf{Ans}$$

1.5.2 GEOMETRIC METHOD

In this method, the population at the present time is assumed to increase in proportion to the number at present. As in the case of the arithmetic method, whether or not this assumption held in the past is uncertain. Thus, the geometric method is also simply used for population projection purposes a short term into the future. Using the same symbols as before, the differential equation is

$$\frac{dP}{dY} = k_g P \tag{1.21}$$

where k_g is the geometric rate constant. Integrating the above equation from $P = P_1$ to $P = P_2$ and from $Y = Y_1$ to $Y = Y_2$, the following equation is obtained:

$$\ln P_2 = \ln P_1 + k_g(Y_2 - Y_1) \tag{1.22}$$

As was the case of the arithmetic method, this equation also needs two data points. Solving for k_g,

$$k_g = \frac{\ln P_2 - \ln P_1}{Y_2 - Y_1} \tag{1.23}$$

By Equation (1.22),

$$\ln(Population) = \ln(P_2) + k_g(Year - Y_2) \tag{1.24}$$

Example 1.5   Repeat Example 1.4 using the geometric method.

Solution:

$$k_g = \frac{\ln P_2 - \ln P_1}{Y_2 - Y_1} = k_g = \frac{\ln 18{,}000 - \ln 15{,}000}{1990 - 1980} = 0.0182 \text{ per year} \quad \textbf{Ans}$$

$$\ln(Population) = \ln P_2 + k_g(Year - Y_2) = \ln(Population)$$
$$= \ln 18{,}000 + 0.0182(2000 - 1990)$$

$$Population = 21{,}593 \text{ people} \quad \textbf{Ans}$$

1.5.3 DECLINING-RATE-OF-INCREASE METHOD

In this method, the community population is assumed to approach a saturation value. Thus, reckoned from the present time, the rate of increase will decline until it becomes zero at saturation. Letting P_s be the saturation population and k_d be the rate constant (analogous to k_a and k_g), the differential equation is

$$\frac{dP}{dY} = k_d(P_s - P) \tag{1.25}$$

where P and Y are the same variable as before. This equation may be integrated twice: the first one, from $P = P_1$ to $P = P_2$ and $Y = Y_1$ to $Y = Y_2$ and the second one, from $P = P_2$ to $P = P_3$ and $Y = Y_2$ to $Y = Y_3$ Thus,

$$\ln(P_s - P_2) = \ln(P_s - P_1) - k_d(Y_2 - Y_1)$$
$$\ln(P_s - P_3) = \ln(P_s - P_2) - k_d(Y_3 - Y_2) \tag{1.26}$$

Solving for k_d,

$$k_d = -\frac{1}{Y_2 - Y_1} \ln \frac{P_s - P_2}{P_s - P_1}$$

$$k_d = -\frac{1}{Y_3 - Y_2} \ln \frac{P_s - P_3}{P_s - P_2}$$

(1.27)

In the previous equations, $Y_2 - Y_1$ may be made equal to $Y_3 - Y_2$, whereupon the value of P_s may be solved for. The final equations including *Population*, are as follows:

$$P_s = \frac{P_1 P_3 - P_2^2}{P_1 + P_3 - 2P_2}$$

(1.28)

$$Population = P_s - (P_s - P_3)e^{-k_d(Year - Y_3)}$$

Example 1.6   The population data for Anytown is as follows: 1980 = 15,000, 1990 = 18,000, and 2000 =20,000. What will be the population in the year 2020? What is the value of k_d?

Solution:

$$P_s = \frac{P_1 P_3 - P_2^2}{P_1 + P_3 - 2P_2} = \frac{15000(20000) - 18000^2}{15000 + 20000 - 2(18000)} = 24{,}000 \text{ people}$$

$$k_d = -\frac{1}{Y_3 - Y_2} \ln \frac{P_s - P_3}{P_s - P_2} = -\frac{1}{10} \ln \frac{24000 - 20000}{24000 - 18000} = 0.04 \text{ per year}   \textbf{Ans}$$

$$Population = P_s - (P_s - P_3)e^{-k_d(Year-Y_3)} = 24000 - (24000 - 20000)e^{-0.04(2020-2000)}$$

$$= 22{,}203 \text{ people}   \textbf{Ans}$$

1.5.4 LOGISTIC METHOD

If food and environmental conditions are at the optimum, organisms, including humans, will reproduce at the geometric rate. In reality, however, the geometric rate is slowed down by environmental constraints such as decreasing rate of food supply, overcrowding, death, and so on. In concept, the factor for the environmental constraints can take several forms, provided, it, in fact, slows down the growth. Let us write the geometric rate of growth again,

$$\frac{dP}{dY} = k_g P$$

(1.29)

This equation states that without environmental constraints, the population grows unchecked, that is, geometric. $k_g P$ is the same as $k_g P(1)$. To enforce the environmental constraint, $k_g P$ should be multiplied by a factor less than 1. This means that the growth rate is no longer geometric but is retarded somewhat. In the logistic method the factor 1 is reduced by P/K, where K is called the *carrying capacity* of the environment and the whole factor, $1 - P/K$, is called *environmental resistance*. The logistic differential equation is therefore

$$\frac{dP}{dY} = k_l P\left(1 - \frac{P}{K}\right) \tag{1.30}$$

where k_g has changed to k_l. Rearranging,

$$\frac{dP}{P\left(1 - \frac{P}{K}\right)} = k_l \, dY \tag{1.31}$$

By partial fractions,

$$\frac{1}{P\left(1 - \frac{P}{K}\right)} = \frac{1}{P} + \frac{1}{K - P} \tag{1.32}$$

Substituting Equation (1.32) in Equation (1.31) and integrating twice: first, between the limits of $P = P_1$ to $P = P_2$ and $Y = Y_1$ to $Y = Y_2$ and second, between the limits of $P = P_2$ to $P = P_3$ and $Y = Y_2$ to $Y = Y_3$ produce the respective equations,

$$\ln \frac{P_2}{P_1}\left(\frac{K - P_1}{K - P_2}\right) = k_l(Y_2 - Y_1)$$

$$\ln \frac{P_3}{P_2}\left(\frac{K - P_2}{K - P_3}\right) = k_l(Y_3 - Y_2) \tag{1.33}$$

Solving for k_l,

$$k_l = \frac{1}{Y_2 - Y_1} \ln \frac{P_2}{P_1}\left(\frac{K - P_1}{K - P_2}\right)$$

$$k_l = \frac{1}{Y_3 - Y_2} \ln \frac{P_3}{P_2}\left(\frac{K - P_2}{K - P_3}\right) \tag{1.34}$$

As in the declining-rate-of-increase method, $Y_2 - Y_1$ may be made equal to $Y_3 - Y_2$, whereupon the value of K may be solved for. The final equations, including *Population*, are as follows:

$$K = \frac{P_2(P_1 P_2 + P_2 P_3 - 2P_1 P_3)}{P_2^2 - P_1 P_3}$$

$$Population = \frac{KP_3 e^{k_l(Year - Y_3)}}{K - P_3[1 - e^{k_l(Year - Y_3)}]} \tag{1.35}$$

Example 1.7 Repeat Example 1.6 using the logistic method.

Solution:

$$K = \frac{P_2(P_1P_2 + P_2P_3 - 2P_1P_3)}{P_2^2 - P_1P_3}$$

$$= \frac{18000[(15000)(18000) + 18000(20000) - 2(15000)(20000)]}{18000^2 - 15000(20000)}$$

$$= 22{,}500 \text{ people}$$

$$k_l = \frac{1}{Y_3 - Y_2}\ln\frac{P_3}{P_2}\left(\frac{K - P_2}{K - P_3}\right) = \frac{1}{10}\ln\frac{20000}{18000}\left(\frac{22500 - 18000}{22500 - 20000}\right) = 0.07 \text{ per year} \quad \textbf{Ans}$$

$$Population = \frac{KP_3 e^{k_l(Year - Y_3)}}{K - P_3[1 - e^{k_l(Year - Y_3)}]} = \frac{22500(20000)e^{0.07(2020 - 2000)}}{22500 - 20000[1 - e^{0.07(2020 - 2000)}]}$$

$$= 21{,}827 \text{ people} \quad \textbf{Ans}$$

1.5.5 GRAPHICAL COMPARISON METHOD

In this method the population of the community of interest is plotted along with those of the populations of larger communities judged to be of similar characteristics as those of the given community. The plotting is such that the populations of the larger communities, which are at the values equal to or greater than those of the given community at the present time, are used to extend the plot of the given community into the future. For example, observe Figure 1.5.

The population data for City A is only known up to 1990, shown by the filled circle and dashed line. Its population needs to be estimated for the year 2020. The population data for two other larger cities considered to be similar to City A are available. At some time in the past, the population of City B had reached 17,000 and in 1990, it was about 18,000. The population of this larger city is therefore

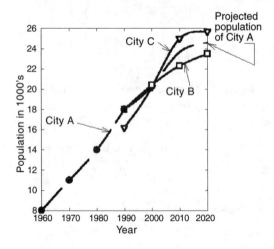

FIGURE 1.5 Population estimate by graphical comparison.

plotted in Figure1.5 starting with 18,000 in 1990. The same treatment is given to the population of the other city, City C, which had a population of about 16,000 in 1990. Using the plotted curves for Cities B and C as a guide (graphical comparison), the curve for City A is extended. Its projected population by the year 2020 is shown by the dashed line and is approximately 24,800.

1.6 DERIVATION OF DESIGN FLOWS
OF WASTEWATERS

The accurate determination of wastewater flows is the first fundamental step in the design of wastewater facilities. To ensure proper design, accurate and reliable data must be available. This entails proper selection of design period, accurate population projection, and the determination of the various flow rates. Again, the discussion that follows addresses the particulars of the community; however, parallel concerns should also be directed to the determination of industrial wastewaters and other types of wastewaters.

1.6.1 DESIGN FLOWS

Different types of flow rates are used in design of wastewater facilities: average daily flow rate, maximum daily flow rate, peak hourly flow rate, minimum daily flow rate, minimum hourly flow rate, sustained high flow rate, and sustained low flow rate. Each of these flow rates has its own use in design. To ensure accurate values for these parameters, a statistically sufficient amount of data should be available for their derivation. Ideally, the amount of data should be infinite or, at least, exhaustive.

The *average daily flow rate* is the mean of all the daily flow rate values obtained from an exhaustive length of flow record. For example, if the total measured flow at a treatment plant for one year (365 days) is 9,000,000 cubic meters, the average daily flow rate is 9,000,000/365 = 24,658 m^3 per calendar day. If the duration of record is more than one year, an average daily flow rate value is established for each year. All these averages are considered together to arrive at a single average daily flow rate. The longer the record, the more accurate the value will be. The ideal value is one in which the length of record is exhaustive.

Note that the average daily flow rate for wastewater is not being given a probability of occurrence. As in water consumption, the average daily flow rate for wastewaters is the average daily flow rate of all daily flow rates available on record. This is the long-term average as was discussed in the case of water. Again, only one value exists for the particular parameter of average daily flow rate. Also, only one value exists for each of the parameters of maximum daily flow rate, maximum hourly flow rate, and so on, with no attached probability of occurrence, although, in concept, one may ask for a particular value corresponding to a certain probability. If a probability of occurrence is attached, this means that some of the wastewaters will not be treated. For example, if a wastewater treatment plant is designed to treat a maximum flow that is equaled or exceeded 10% of the time, this means that 10% of the time, some sewage will not be treated. Sewage, however, must be treated at all times.

As in water consumption, average daily flow rates are usually normalized against the contributing mid-year population. Thus, if the contributing mid-year population for the example given in the previous paragraph is 50,000, the average daily flow rate is also given as 24,658/50,000 = 0.49 m^3 per capita per day.

Analogous to the average design flow rates used for water, average daily flow rate values for wastewater are used in developing ratios in design such as the ratio of maximum daily flow rate to the average daily flow rate, the peak hourly flow rate to the average daily flow rate, and so on. Thus, knowing the average daily flow rate and the ratio, the maximum daily flow rate or the peak hourly flow rate, and the like, can be determined.

The *maximum daily flow rate* is the largest total flow that accumulates over a day as obtained from an exhaustive length of flow record. For example, suppose the flow on March 10 of a certain year was 49,316 m^3 accumulated over the day. If the whole flow record is examined and no one flow summed over the length of one calendar day exceeds this value, then 49,316 m^3/day is the maximum daily flow rate. On the other hand, the *minimum daily flow rate* is the smallest total flow that accumulates over a day as obtained from an exhaustive length of flow record. This is just the reverse of the maximum daily flow rate.

The *peak hourly flow rate* is the largest accumulation of flow in an hour during a particular day as obtained from an exhaustive length of flow record. For practical purposes, the peak hourly flow rate may be considered as the instantaneous peak flow rate. The *minimum hourly flow rate* is the smallest accumulation of flow in an hour during a particular day as obtained from an exhaustive length of flow record. This is just the reverse of the peak hourly flow rate. As with peak hourly flow rates, the minimum hourly flow rate may be considered as the instantaneous minimum flow rate.

The *sustained peak flow rate* is the flow rate that is sustained or exceeded for a specified number of consecutive time periods as obtained from an exhaustive length of flow record. Note the use of the word "or," signifying union of two events. For example, pick any record and take a set of seven-day consecutive flow rates such as 49,300, 48,600, 47,689, 46,000, 46,000, 44,000, and 44,000 all in m^3 per day. The mean of these values (46,513 m^3/day) refers to the "sustained" which, in this case, the flow is being sustained in seven days. The mean of all sets of seven-day consecutive flow rates may be computed similarly. The result of the calculations will then be an array of sustained flow rates, which can be arranged in descending order. As in the previous methods of determining the maximum values using the probability distribution, the sustained peak flow rate is the sustained flow rate that has the probability of being equaled or exceeded by 0%.

The *sustained minimum flow rate* is the flow rate that is sustained or not exceeded for a specified number of consecutive time periods as obtained from an exhaustive length of flow record. Again, note the use of the word "or" for the union of two events. This is just the reverse of the sustained high flow rate; the array, instead of being arranged in descending order, is arranged in ascending order. The sustained minimum flow rate is then the sustained flow rate that is equaled or not exceeded by 0%.

In the previous discussion, the phrases average daily flow rates, maximum daily flow rates, maximum hourly flow rates, minimum daily flow rates, and so on are often simply shortened to average daily flow, maximum daily flow, maximum hourly flow, minimum daily flow and so on, respectively. Again, in the development of these flow rates, sufficient or exhaustive lengths of data should be available to produce statistically acceptable values.

1.7 DERIVING DESIGN FLOWS OF WASTEWATERS FROM FIELD SURVEY

Design flows can be established in several ways: use values in the literature, use the treatment plant record, and conduct a field survey. The use of literature values is the least desirable because flow quantities are dependent on the habits of the people and the characteristics of the land area. Land characteristics affect sewage flow, because the nature of the land affects the depth of groundwater, which, in turn, determines the amount of infiltration into the sewer.

The habits of people also affect the quantity of sewage flow. The habits of the people in Asia are different from the habits of the people in America. Farming communities have different habits from city and urban communities. They produce different quantities of sewage as well as different constituents of sewage. The sewage in Ocean City, MD probably contains a lot of crab shells as compared with the sewage in Salt Lake City, UT. Thus, any literature value may be meaningless, because there may be no way to determine whether or not the value conforms to the land characteristics and habit of the people in the proposed design.

If the use of plant record is futile because of the aforementioned infiltration, a field survey may be conducted. Conducting a field survey is desirable but suffers from two defects: it is very expensive and, by necessity, is conducted over a very short time. Given these drawbacks, however, conducting a field survey is probably the most accurate of the three methods mentioned previously.

For an expansion to an existing system, the flow measurement can be done easily, because the system already exists. For brand-new systems, the metering can be done in a nearby town or village, where the characteristics of the system and the habits of the people are similar to those of the proposed. The requirement of similarity is very important. To have accurate predictions of sewage flows, it is obvious that the reference town should have characteristics similar to those of the proposed. To illustrate the various methods of determining design flows, Figures 1.6 and 1.7 were derived. These figures also show what a flow pattern would look like if a sewer system were monitored in a field survey.

In practice, before any monitoring is done, the as-built sewer system plans must first be obtained to determine where the monitoring apparatus should be set. Once this is established, the contributing population must also be determined. This population is used to normalize the measured flow rates on a per capita basis in order for the data to be transportable; and, most important, the nearby precipitation record should be consulted to determine the months for the dry and wet seasons. If the purpose of the field survey is to determine the dry-weather flows, then the monitoring should be done on the dry-weather months as evidenced from the records of the nearby precipitation station. If the purpose of the field survey is to determine infiltration, then the monitoring should be done on the wet-weather months, again, as evidenced from the records of the nearby precipitation station. In Maryland, the dry-weather months are normally June to August.

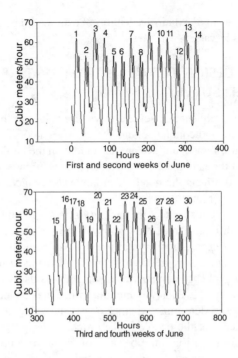

FIGURE 1.6 Hypothetical monitored flow derived for the month of June.

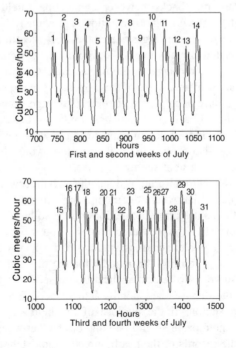

FIGURE 1.7 Hypothetical monitored flow derived for the month of July.

1.7.1 AVERAGE DAILY FLOW RATE

If we had two years worth of data, then the average daily flow rate would be the mean over the two-year period. If we had three, four,... or ζ years worth of data, then the average daily flow rate would be the mean over the three-, four-,..., or ζ-year period. The means obtained from the long years of records would approximate the *long-term means*. If ζ is a very large number such as 30 years, the long-term mean obtained would really be the true mean.

Now, as mentioned, a field survey has a very serious drawback: the length of the survey is very short. Any survey will not last for 30 years, for example. Nonetheless, a field survey can still be accurate compared with other methods of determining design flows. It is to be noted that before any grant for the construction of sewage facilities was given to any community as required by the Clean Water Act of 1972, the Environmental Protection Agency required a field survey.

With this limitation in mind, let us now derive the average daily flow from Figures 1.6 and 1.7 using the probability distribution analysis. Remember that the average corresponds to the 50% probability. The figures contain, all in all, 61 average daily flow rate values. (The total number of days added for the months of June and July.) In order to apply the probability distribution analysis, the average daily flow rates for each of these days must be calculated. This may done easily by graphical integration.

Figure 1.8 shows the general scheme for performing a graphical integration applied to flow rates of the first day of June. As indicated, the integration is performed by

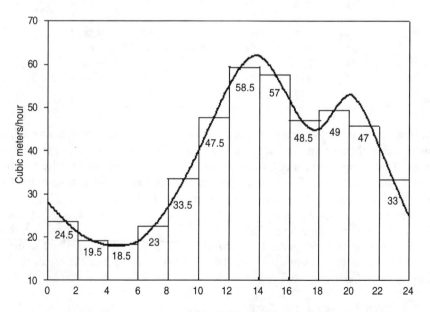

Average daily flowrate= {24.5(2)+19.5(2)+18.5(2)+23(2)+33.5(2)+47.5(2)+
58.5(2)+57(2)+48.5(2)+49(2)+33(2)/24
=459.5(2)/24=38.29 cubic meters/hour

FIGURE 1.8 Graphical integration for the average daily flow rate for Day 1 of June.

TABLE 1.12
Average Daily Flow Rates (m³/hr) for the Months of June and July

Day	Flow Rate	Day	Flow Rate	Day	Flow Rate
		June			
1	38.29	2	30.92	3	45.21
4	38.04	5	31.21	6	31.13
7	38.33	8	30.92	9	45.20
10	38.40	11	38.83	12	30.89
13	45.00	14	37.50	15	30.67
16	29.96	17	42.75	18	38.80
19	30.90	20	45.00	21	37.50
22	31.04	23	44.92	24	45.21
25	38.46	26	30.92	27	37.46
28	38.38	29	30.54	30	38.38
		July			
31	30.92	32	45.20	33	38.42
34	38.83	35	30.92	36	45.21
37	37.50	38	38.29	39	30.92
40	42.21	41	34.13	42	31.79
43	32.00	44	37.38	45	32.58
46	44.92	47	44.56	48	37.96
49	32.00	50	36.79	51	38.58
52	30.33	53	38.83	54	31.04
55	37.92	56	38.00	57	36.83
58	34.50	59	43.67	60	42.54
61	32.71				

drawing small rectangles; each of these rectangles has an area equal to the height multiplied by the time interval (equals 2). The areas of each of these rectangles are then added. This summation divided by 24 represents the average daily flow rate for this first day of June, 38.29 m³/h. This procedure is repeated for each of the days up to the full 61 days duration of the study. The results are given in Table 1.12.

The next step is to arrange the data in either the ascending or descending order. Choosing the descending order, Table 1.13 is produced. Reading at the 0.5 probability, the mean is 37.92 m³/h. Thus, the average daily flow rate is 37.92 m³/h. If the 0.50 probability had not been available from the table, then the mean would have to be interpolated.

1.7.2 PEAK HOURLY FLOW RATE

The peak hourly flow rate will now be derived. Remember that maximum values correspond to a probability of zero. Therefore, to obtain the peak hourly flow rate, read or extrapolate the flow rate corresponding to zero in the probability distribution obtained from an array arranged in descending order.

TABLE 1.13

Table 1.12 Arranged in Descending Order for the Probability Distribution Analysis

Order	Flow Rate	η	$\Sigma\eta$	$\frac{\Sigma\eta}{\zeta_{total}+1}$	Order	Flow Rate	η	$\Sigma\eta$	$\frac{\Sigma\eta}{\zeta_{total}+1}$
1	**45.21**	3	3	**0.048**	23	37.50	3	34	0.55
2	**45.20**	2	5	**0.081**	24	37.46	1	35	0.56
3	45.00	2	7	0.11	25	37.38	1	36	0.58
4	44.92	2	9	0.15	26	36.83	1	37	0.60
5	44.56	1	10	0.16	27	36.79	1	38	0.61
6	43.67	1	11	0.18	28	34.50	1	39	0.63
7	42.75	1	12	0.19	29	34.13	1	40	0.65
8	42.54	1	13	0.21	30	32.71	1	41	0.66
9	42.21	1	14	0.23	31	32.58	1	42	0.68
10	38.83	3	17	0.27	32	32.00	2	44	0.71
11	38.80	1	18	0.29	33	31.79	1	45	0.73
12	38.58	1	19	0.31	34	31.21	1	46	0.74
13	38.46	1	20	0.32	35	31.13	1	47	0.76
14	38.42	1	21	0.34	36	31.04	2	49	0.79
15	38.40	1	22	0.35	37	30.92	6	55	0.89
16	38.38	2	24	0.39	38	30.90	1	56	0.90
17	38.33	1	25	0.40	39	30.89	1	57	0.92
18	38.04	1	26	0.42	40	30.67	1	58	0.94
19	38.29	2	28	0.45	41	30.54	1	59	0.95
20	38.00	1	29	0.47	42	30.33	1	60	0.97
21	37.96	1	30	0.48	43	29.96	1	61	0.98
22	**37.92**	1	31	0.50					

Note: Bold numbers are used in the interpolations and extrapolations discussed in the text.

The daily peak hourly flow rates read from the original data sheets for Figures 1.6 and 1.7 are shown in Table 1.14. These values are then subjected to probability distribution analysis. The results are shown in Table 1.15. From the table, the peak hourly flow rate can be extrapolated at 0 as follows (with x representing the peak hourly flow rate):

x	0
67.0	0.02
66.4	0.03

Therefore,

$$\frac{x-67.0}{67.0-66.4} = \frac{0-0.02}{0.02-0.03} \qquad x = 68.2 \text{ m}^3/\text{h}$$

TABLE 1.14
Daily Peak Hourly Flow Rates (m³/hr) for the
Months of June and July

Day	Peak Flow Rate	Day	Peak Flow Rate	Day	Peak Flow Rate
			June		
1	62.1	2	53.0	3	65.1
4	62.7	5	53.4	6	53.5
7	62.9	8	53.6	9	65.4
10	62.0	11	62.7	12	53.8
13	65.0	14	62.8	15	53.7
16	63.6	17	62.2	18	62.0
19	52.9	20	66.4	21	61.4
22	54.7	23	67.0	24	65.5
25	61.3	26	54.3	27	60.7
28	62.7	29	52.5	30	61.3
			July		
31	50.0	32	65.7	33	62.2
34	62.2	35	54.9	36	66.1
37	62.7	38	62.4	39	53.6
40	65.4	41	62.0	42	53.9
43	53.3	44	62.4	45	53.5
46	65.8	47	65.6	48	62.7
49	53.5	50	62.3	51	62.4
52	53.1	53	62.9	54	53.5
55	63.6	56	62.5	57	62.6
58	53.4	59	65.3	60	62.0
61	53.6				

1.7.3 MAXIMUM DAILY FLOW RATE

Because the maximum daily flow rate is a daily flow rate, the data required for its determination is the same data used in the determination of the average daily flow rate. Instead of using the 50% probability, however, as used for the determination of the average daily flow rate, the 0% probability is used.

From Table 1.13 the maximum daily flow rate can be extrapolated at 0 as follows (with x representing the maximum daily flow rate):

x	0
45.21	0.048
45.20	0.081

TABLE 1.15
Table 1.14 Arranged in Descending Order for the Probability Distribution Analysis

Order	Flow Rate	η	$\Sigma\eta$	$\dfrac{\Sigma\eta}{\zeta_{total}+1}$	Order	Flow Rate	η	$\Sigma\eta$	$\dfrac{\Sigma\eta}{\zeta_{total}+1}$
1	**67.0**	1	1	**0.02**	21	62.1	1	32	0.52
2	**66.4**	1	2	**0.03**	22	62.0	4	36	0.58
3	66.1	1	3	0.05	23	61.4	1	37	0.60
4	65.8	1	4	0.06	24	61.3	2	39	0.63
5	65.7	1	5	0.08	25	60.7	1	40	0.65
6	65.6	1	6	0.10	26	54.9	1	41	0.66
7	65.5	1	7	0.11	27	54.7	1	42	0.68
8	65.4	2	9	0.15	28	54.3	1	43	0.69
9	65.3	1	10	0.16	29	53.9	1	44	0.71
10	65.1	1	11	0.18	30	53.8	1	45	0.73
11	65.0	1	12	0.19	31	53.7	1	46	0.74
12	63.6	2	14	0.23	32	53.6	3	49	0.79
13	62.9	2	16	0.26	33	53.5	4	53	0.85
14	62.8	1	17	0.27	34	53.4	2	55	0.89
15	62.7	5	22	0.35	35	53.3	1	56	0.90
16	62.6	1	23	0.37	36	53.1	1	57	0.92
17	62.5	1	24	0.39	37	53.0	1	58	0.94
18	62.4	3	27	0.44	38	52.9	1	59	0.95
19	62.3	1	28	0.45	39	52.5	1	60	0.97
20	62.2	3	31	0.50	40	50.0	1	61	0.98

Note: Bold numbers are used in the interpolations and extrapolations discussed in the text.

Therefore,

$$\frac{x-45.21}{45.21-45.20} = \frac{0-0.048}{0.048-0.081} \qquad x = 45.22 \text{ m}^3/\text{h}$$

1.7.4 MINIMUM HOURLY FLOW RATE AND MINIMUM DAILY FLOW RATE

The analyses needed for these values are similar to those used for the analyses of the peak hourly flow rate and the maximum daily flow rate, respectively. One difference is that instead of taking the peaks or maxima, the minima are now being used. The other difference is that the cumulative probability used to obtain the values is the union probability of equaled or not exceeded instead of the union probability of equaled or exceeded. This has to be the case, because the equal-or-exceeded is used for the maxima, it must follow that the equal-or-not-exceeded should be used for the minima.

To derive the minimum hourly flow rate from Figures 1.6 and 1.7, scan for the lowest flow each day from the figures. The values obtained are shown in Table 1.16.

TABLE 1.16
Daily Minimum Hourly Flow Rates (m^3/hr)
for the Months of June and July

Day	Flow Rate	Day	Flow Rate	Day	Flow Rate
			June		
1	18.0	2	13.4	3	25.3
4	18.7	5	13.3	6	13.3
7	18.3	8	13.3	9	25.4
10	18.5	11	18.3	12	13.4
13	25.0	14	18.5	15	13.5
16	18.4	17	19.5	18	18.6
19	13.4	20	25.4	21	18.6
22	13.3	23	25.7	24	25.3
25	18.9	26	13.3	27	18.9
28	18.6	29	13.8	30	18.4
			July		
31	13.5	32	26.4	33	18.4
34	18.4	35	13.4	36	25.9
37	18.3	38	18.3	39	13.8
40	25.1	41	18.8	42	13.7
43	13.5	44	18.6	45	13.5
46	25.8	47	18.5	48	15.5
49	13.4	50	18.8	51	18.7
52	13.2	53	18.9	54	13.4
55	18.1	56	18.0	57	13.3
58	15.9	59	18.6	60	23.9
61	13.0				

The probability distribution analysis for the values arranged in ascending order is shown in Table 1.17. From the table, the minimum hourly flow rate can be extrapolated at 0 as follows (with x representing the minimum hourly flow rate):

x	0
13.0	0.02
13.2	0.03

Therefore,

$$\frac{x - 13.0}{13.0 - 13.2} = \frac{0 - 0.02}{0.02 - 0.03} \qquad x = 12.6 \text{ m}^3/\text{h}$$

TABLE 1.17
Table 1.16 Arranged in Ascending Order for the Probability Distribution Analysis

Order	Flow Rate	η	$\Sigma\eta$	$\dfrac{\Sigma\eta}{\zeta_{total}+1}$	Order	Flow Rate	η	$\Sigma\eta$	$\dfrac{\Sigma\eta}{\zeta_{total}+1}$
1	**13.0**	1	1	**0.02**	15	18.6	5	42	0.68
2	**13.2**	1	2	**0.03**	16	18.7	2	44	0.71
3	13.3	6	8	0.13	17	18.8	2	46	0.74
4	13.4	6	14	0.23	18	18.9	3	49	0.79
5	13.5	4	18	0.29	19	19.5	1	50	0.81
6	13.7	1	19	0.31	20	23.9	1	51	0.82
7	13.8	2	21	0.34	21	25.0	1	52	0.84
8	15.5	1	22	0.35	22	25.1	1	53	0.85
9	15.9	1	23	0.37	23	25.3	2	55	0.89
10	18.0	2	25	0.40	24	25.4	2	57	0.92
11	18.1	1	26	0.42	25	25.7	1	58	0.94
12	18.3	4	30	0.48	26	25.8	1	59	0.95
13	18.4	4	34	0.55	27	25.9	1	60	0.97
14	18.5	3	37	0.60	28	26.4	1	61	0.98

Note: Bold numbers are used in the interpolations and extrapolations discussed in the text.

The determination of the minimum daily flow rate refers to daily averages as is the case of the determination of the average daily flow rate and the maximum daily flow rate; thus, the distribution for the daily average is already available in Table 1.12. Instead of the elements in the table being arranged in descending order, however, they will now be rearranged in ascending order, because the union probability to be used is the equal or not exceeded. The reversed probability distribution is shown in Table 1.18. From the table, the minimum daily flow rate can be extrapolated at 0 as follows (with x representing the maximum daily flow rate):

x	0
29.96	0.02
30.33	0.03

Therefore,

$$\frac{x-29.96}{29.96-30.33} = \frac{0-0.02}{0.02-0.03} \qquad x = 29.22 \text{ m}^3/\text{h}$$

TABLE 1.18
Table 1.12 Rearranged in Ascending Order for the Probability Distribution Analysis

Order	Flow Rate	η	$\Sigma\eta$	$\dfrac{\Sigma\eta}{\zeta_{total}+1}$	Order	Flow Rate	η	$\Sigma\eta$	$\dfrac{\Sigma\eta}{\zeta_{total}+1}$
1	**29.96**	1	1	**0.02**	23	37.96	1	32	0.52
2	**30.33**	1	2	**0.03**	24	38.00	1	33	0.53
3	30.54	1	3	0.05	25	38.29	2	35	0.56
4	30.67	1	4	0.06	26	38.04	1	36	0.58
5	30.89	1	5	0.08	27	38.33	1	37	0.60
6	30.90	1	6	0.10	28	38.38	2	39	0.63
7	30.92	6	12	0.19	29	38.40	1	40	0.64
8	31.04	2	14	0.23	30	38.42	1	41	0.66
9	31.13	1	15	0.24	31	38.46	1	42	0.68
10	31.21	1	16	0.26	32	38.58	1	43	0.69
11	31.79	1	17	0.27	33	38.80	1	44	0.71
12	32.00	2	19	0.31	34	38.83	3	47	0.76
13	32.58	1	20	0.32	35	42.21	1	48	0.77
14	32.71	1	21	0.34	36	42.54	1	49	0.79
15	34.13	1	22	0.35	37	42.75	1	50	0.81
16	34.50	1	23	0.37	38	43.67	1	51	0.82
17	36.79	1	24	0.39	39	44.56	1	52	0.84
18	36.83	1	25	0.40	40	44.92	2	54	0.87
19	37.38	1	26	0.42	41	45.00	2	56	0.90
20	37.46	1	27	0.44	42	45.20	2	58	0.94
21	37.50	3	30	0.48	43	45.21	3	61	0.98
22	37.92	1	31	0.50					

Note: Bold numbers are used in the interpolations and extrapolations discussed in the text.

1.7.5 SUSTAINED PEAK FLOW RATE AND SUSTAINED MINIMUM FLOW RATE

Before we proceed to discuss the method of obtaining these flow rates, let us define the word *moving average*. Table 1.19 illustrates the concept. As indicated in the table, a moving average is one in which a constant number of moving elements are averaged in succession such that each succession overlaps each other by one element as the averaging moves forward. The number of moving elements being averaged above is three; therefore, this type of moving average is called a *three-element moving average*. If the number of elements were five, then the moving average would be called a *five-element moving average*.

The number of successive averages in the above table is 6 minus 3 plus 1, where 6 is the total number of elements, 3 is the number of moving elements being averaged in each succession, and 1 is added to get the correct answer. If the total number of elements were 8 and the number of moving elements being averaged were 2, the

TABLE 1.19
Table Illustrating the Concept of Moving Averages

Element number		1	2	3	4	5	6
Element value		23	21	43	32	34	26
First moving average $= \dfrac{23 + 21 + 43}{3} = 29.0$		23	21	43	—	—	—
Second moving average $= \dfrac{21 + 43 + 32}{3} = 32$		—	21	43	32	—	—
Third moving average $= \dfrac{43 + 32 + 34}{3} = 36.3$		—	—	43	32	34	—
Fourth moving average $= \dfrac{32 + 34 + 26}{3} = 30.7$		—	—	—	32	34	26

number of successive moving averages would have been 8 minus 2 plus 1 = 7. In general, if the total number of elements is ζ and the number of moving elements being averaged in each moving average is κ, the number of successive moving averages χ is

$$\chi = \zeta - \kappa + 1 \tag{1.36}$$

One way to derive sustained flow rates is through the use of moving averages. In the field survey data, a number of moving averages or sustained flow rates can be formed. For example, a 7-element, 14-element, and *any*-element sustained flow rate can be formed. The peak or the minimum of these flow rates can be found by the usual method of probability distribution analysis. To be descriptive, substitute the type of element such as hour, day, and so on. Thus, 7-element sustained flow rate becomes 7-day sustained flow rate, 14-element sustained flow rate becomes 14-day sustained flow rate, and so on.

Let us now arbitrarily derive the 14-day sustained flow rates from the field survey data. From Equation (1.36), χ, the number of successive moving averages, is equal to $61 - 14 + 1 = 48$. Each of these averages will have equal likelihood of occurring with a probability of 1/48. Table 1.20 contains the 14-day sustained flow rates derived from Figures 1.6 and 1.7. Tables 1.21 and 1.22 contain the probability distribution analyses for the sustained peak and the sustained minimum flow rates, respectively.

As in previous analyses, the sustained peak flow rate has a probability of zero in an array of descending order; the sustained minimum flow rate also has a probability of zero but in an array of ascending order. From Table 1.21, the sustained peak flow rate may be extrapolated at 0 as follows (with x representing the maximum daily flow rate):

x	
	0
38.24	0.02
38.12	0.04

TABLE 1.20
Table of 14-Day Sustained Flow Rates (m³/hr) Derived from Figures 1.6 and 1.7

Succession Number	Flow Rate	Succession Number	Flow Rate	Succession Number	Flow Rate
1	37.07	2	36.97	3	36.32
4	36.70	5	37.21	6	37.20
7	37.73	8	38.24	9	37.15
10	37.72	11	37.15	12	37.19
13	36.54	14	37.06	15	36.53
16	36.16	17	36.68	18	36.25
19	36.35	20	35.35	21	36.67
22	36.07	23	36.11	24	36.12
25	36.74	26	37.33	27	37.40
28	36.72	29	38.08	30	38.04
31	38.12	32	37.46	33	37.37
34	36.18	35	37.47	36	37.39
37	37.40	38	36.84	39	36.93
40	36.46	41	36.06	42	36.51
43	36.93	44	36.83	45	37.04
46	36.90	47	36.76	48	36.34

Therefore,

$$\frac{x - 38.24}{38.24 - 38.12} = \frac{0 - 0.02}{0.02 - 0.04} \qquad x = 38.36 \text{ m}^3/\text{h}$$

From Table 1.22, the sustained minimum flow rate may also be extrapolated at 0 as follows (with x representing the maximum daily flow rate):

x	0
35.35	0.02
36.06	0.04

Therefore,

$$\frac{x - 35.35}{35.35 - 36.06} = \frac{0 - 0.02}{0.02 - 0.04} \qquad x = 34.64 \text{ m}^3/\text{h}$$

1.7.6 INFILTRATION–INFLOW

Referring back to Figures 1.6 and 1.7, the lowest flows occur during early morning hours. During these hours, the contribution of infiltration to the total flow may be insignificant if the monitoring is done during dry season. During wet-weather seasons, however, the picture would be different. The sanitary flow would now be insignificant

TABLE 1.21
Table 1.20 Arranged in Descending Order for the Probability Distribution Analysis

Order	Flow Rate	η	$\Sigma\eta$	$\dfrac{\Sigma\eta}{\zeta_{total}+1}$	Order	Flow Rate	η	$\Sigma\eta$	$\dfrac{\Sigma\eta}{\zeta_{total}+1}$
1	38.24	1	1	0.02	23	36.72	1	27	0.55
2	38.12	1	2	0.04	24	36.70	1	28	0.57
3	38.08	1	3	0.06	25	36.83	1	29	0.59
4	38.04	1	4	0.08	26	36.76	1	30	0.61
5	37.73	2	6	0.12	27	36.68	1	31	0.63
6	37.72	1	7	0.14	28	36.67	1	32	0.65
7	37.47	1	8	0.16	29	36.54	1	33	0.67
8	37.46	1	9	0.18	30	36.53	1	34	0.69
9	37.40	2	11	0.22	31	36.51	1	35	0.71
10	37.39	1	12	0.24	32	36.46	1	36	0.73
11	37.37	1	13	0.26	33	36.35	1	37	0.76
12	37.21	1	14	0.29	34	36.34	1	38	0.76
13	37.20	1	15	0.32	35	36.32	1	39	0.80
14	37.19	1	16	0.33	36	36.25	1	40	0.82
15	37.15	2	18	0.37	37	36.18	1	41	0.84
16	37.07	1	19	0.39	38	36.16	1	42	0.86
17	37.06	1	20	0.41	39	36.12	1	43	0.88
18	37.04	1	21	0.43	40	36.11	1	44	0.90
19	36.97	1	22	0.45	41	36.90	1	45	0.92
20	36.93	2	24	0.49	42	36.07	1	46	0.94
21	36.84	1	25	0.51	43	36.06	1	47	0.96
22	36.74	1	26	0.53	44	35.35	1	48	0.98

Note: Bold numbers are used in the interpolations and extrapolations discussed in the text.

and infiltration flow will dominate. Infiltration values are therefore monitored at wet-weather conditions during early hours in the morning.

Figure 1.9 shows the result of a monitoring during wet-weather conditions in a sanitary sewer system. As shown, there are several minima and the question is which one should be chosen. The infiltration rate should be a constant value during a particular study period, and the cause of the fluctuation in the figure would be the effect of the sanitary flow. The fluctuation of the sanitary flow has to be above the infiltration rate. The sanitary flow is above the infiltration rate, so the infiltration rate point should be below all else in the figure. Using this criterion, a horizontal straight line is drawn in Figure 1.9 that is below all points in the curve. This horizontal line represents the reading for the infiltration rate which is found to be equal to 24.0 m^3/h.

Direct inflow is important in sizing wastewater engineering structures. This occurs immediately after rainfall and can be determined if the hydrograph of the flow had been recorded. The procedure for determining inflow is as follows: the hydrograph on the times immediately preceding the occurrence of rainfall is extended.

TABLE 1.22
Table 1.20 Arranged in Ascending Order for the Probability Distribution Analysis

Order	Sustained Flow Rate	η	$\Sigma\eta$	$\dfrac{\Sigma\eta}{\zeta_{total}+1}$	Order	Sustained Flow Rate	η	$\Sigma\eta$	$\dfrac{\Sigma\eta}{\zeta_{total}+1}$
1	**35.35**	1	1	**0.02**	23	36.74	1	23	0.47
2	**36.06**	1	2	**0.04**	24	36.84	1	24	0.49
3	36.07	1	3	0.06	25	36.93	2	26	0.53
4	36.90	1	4	0.08	26	36.97	1	27	0.55
5	36.11	1	5	0.10	27	37.04	1	28	0.57
6	36.12	1	6	0.12	28	37.06	1	29	0.59
7	36.16	1	7	0.14	29	37.07	1	30	0.62
8	36.18	1	8	0.16	30	37.15	2	32	0.65
9	36.25	1	9	0.18	31	37.19	1	33	0.67
10	36.32	1	10	0.20	32	37.20	1	34	0.69
11	36.34	1	11	0.22	33	37.21	1	35	0.71
12	36.35	1	12	0.24	34	37.37	1	36	0.73
13	36.46	1	13	0.26	35	37.39	1	37	0.76
14	36.51	1	14	0.29	36	37.40	2	39	0.80
15	36.53	1	15	0.31	37	37.46	1	40	0.82
16	36.54	1	16	0.33	38	37.47	1	41	0.84
17	36.67	1	17	0.35	39	37.72	1	42	0.86
18	36.68	1	18	0.37	40	37.73	2	44	0.90
19	36.76	1	19	0.39	41	38.04	1	45	0.92
20	36.83	1	20	0.41	42	38.08	1	46	0.94
21	36.70	1	21	0.43	43	38.12	1	47	0.96
22	36.72	1	22	0.45	44	38.24	1	48	0.98

Note: Bold numbers are used in the interpolations and extrapolations discussed in the text.

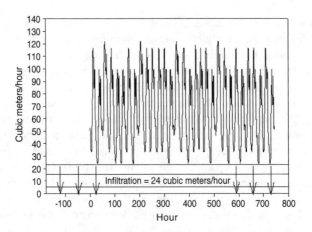

FIGURE 1.9 Wet-weather monitoring during the month of March.

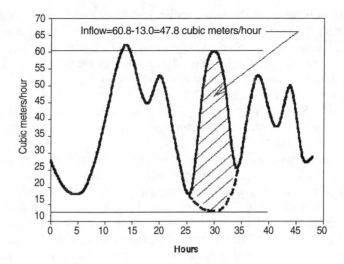

FIGURE 1.10 Hydrograph for deriving inflow.

The time of extension should be at least equal to the time of the hydrograph of the rainfall. Therefore, two hydrographs now exist: the actual hydrograph that contains the rainfall event and the extended hydrograph that assumes there was no rainfall event. The two hydrographs are then superimposed; the hydrograph with rainfall event will be above that of the other. This difference represents the inflow.

Figure 1.10 shows a hydrograph for two days of flow that captures the direct inflow into the sewer. Note the pattern of the hydrograph on the first day. The 0 on the abscissa represents midnight of the first day. At around 25 hours from the previous midnight, a burst of inflow is recorded continuing until approximately 34 hours. The inflow peak is recorded at around 30 hours. The flow pattern of the previous day has been extended as shown by the dotted line. From the indicated construction, the direct inflow is approximately 47.8 m^3/h.

1.7.7 SUMMARY COMMENTS FOR DERIVING FLOW RATES BY THE PROBABILITY DISTRIBUTION ANALYSIS

The technique of obtaining the elements in the probability distribution table is important. Each of these elements should have equal likelihood of occurring. For example, consider element 4 (or order 4) in Table 1.13, which is 44.92. The value of η for this element is 2, which means that there are two members of this element.

Above element 4 are **45.00, 45.20**, and **45.21** numbering a total of 7 members. Thus, 7 element members are above 44.92 making a total of 9 indicated under the column of $\Sigma\eta$. Now, the cumulative probability is calculated as

$$\frac{9}{61+1} = 0.15$$

The use of 9 (which is the total number of element members up to element 4) means that all the members are considered having equal likelihood of occurring. That is, all the element members in a probability distribution analysis have equal probability. Verify this by studying how the calculations are done in the rest of the table and, in fact, all the tables of probability distribution.

The assumption of equal probability is correct. Table 1.13 has 61 members. If you were to determine the probability of element number 16 in this table, for example, how would you do it? Unless it is known that this element is biased, its probability has to be 24 in 62 = 0.39.

The requirement of equal likelihood of occurrence means that the elements must be sampled "uniformly and at equal intervals." For example, for the elements of Table 1.12, if the sampling were done irregularly such as taking measurements in the first two days of a week, skip the next week, resume in the last three days of the fourth week and so on, the resulting element members will not have equal likelihood of occurring and the probability distribution analysis will not apply. The sampling should be done uniformly and at equal intervals. If this is followed, every element will have equal probability. Of course, the more elements there are, the more accurate will be the prediction of the probability distribution analysis.

To repeat, sampling must be done uniformly and at equal intervals for the probability distribution analysis to be applicable. Otherwise, the application of the method will be difficult, since the probability of each element will be undetermined or, if determinable, is calculated with difficulty. Of course, the probability of each element can always be assumed equal, even if the sampling is irregular. If you see this done in practice, the concept upon which the probability distribution analysis is based has been violated and you are guaranteed that whatever is the result, it is wrong.

Example 1.8 Assuming the contributing population in the field survey of Figures 1.6 and 1.7 is 2000, what will be the design flow to size a sewer for a population of 10,000? Sizing of sewers is based on peak flows.

Solution: The peak hourly flow rate was 68.2 m^3/h. Therefore, the design flow to size sewer, $Q_{peak\ hour}$, is

$$Q_{peak\ hour} = \frac{68.2}{2000}(10{,}000) = 341 \ \text{m}^3/\text{h} \quad \textbf{Ans}$$

GLOSSARY

Arithmetic method—A method of predicting population that assumes the current rate of increase is constant.

Average daily flow rate—The mean of all daily flow values obtained from an exhaustive length of flow record.

Carrying capacity—One of the parameters in the factor called environmental resistance in the logistic method of predicting population.

Commercial wastewater—Wastewater produced in the course of conducting business.

Declining-rate-of-increase method—A method of predicting population that assumes the current rate of increase is decreasing and, thus, the population will eventually reach saturation.

Delayed inflow—Portions of rainfall that do not enter the sewer immediately but takes some days for it to be drained completely.

Design flow—The magnitude of flow that a facility is to be designed for.

Design or planning period—The time from the initial design years to the time that the facility is to receive the final design flows.

Direct inflow—An inflow that results in an increase of flow in the sewer almost immediately after the beginning of rainfall.

Domestic wastewater—Sanitary wastewaters generated in residences.

Dry-weather flows—Flows in sewers during dry weather or drought conditions.

Environmental resistance—A factor in the logistic method of predicting population.

Event—An occurrence or a happening.

Geometric method—A method of predicting population that assumes the current rate of increase is proportional to the number of people.

Independent events—Events whose probabilities do not affect the probability of each other.

Industrial wastewater—Wastewaters produced in the process of manufacturing.

Infiltration—Water that enters sewers through cracks and imperfect connections in sewers and manholes.

Infiltration-inflow—Combined infiltration and inflow.

Inflow—Water that enters through openings *purposely* or *inadvertently* provided for its entrance.

Initial years—The first years of a planning period.

Institutional wastewater—Water produced from institutional facilities like hospitals, prisons, schools, and rest homes.

Intersection—The probability space containing events from one or more events happening at the same time.

Intersection probability—The probability of events to occur at the same time.

Logistic method—A method of predicting population that assumes the current rate of increase is affected by environmental factors.

Maximum daily flow rate—The largest total flow that accumulates over a day as obtained from an exhaustive length of flow record.

Minimum daily flow rate—The smallest total flow that accumulates over a day as obtained from an exhaustive length of flow record.

Minimum hourly flow rate—The smallest accumulation of flow in an hour during a particular day as obtained from an exhaustive length of flow record.

Mutually exclusive events—Events that cannot occur at the same time.

Peak hourly flow rate—The largest accumulation of flow in an hour during a particular day as obtained from an exhaustive length of flow record.

Probability—A measure of the likelihood that an event will or will not occur.

Probability distribution analysis—A statistical method applied to a collection of related data for determining likelihood of occurrences using the concept of probability.

Probability space—The domain of all outcomes both favorable and not favorable to the occurrence of an event.

Recreational wastewater—Wastewater produced in the course of recreational activities.

Reduced probability space—The probability space that is a subset of some larger probability space.

Residential wastewater—Wastewater produced in homes and residences.

Sanitary wastewater or sanitary sewage—Wastewaters contaminated with human wastes.

Steady inflow—Water that enters the sewer system continuously.

Storm sewage—Sewage resulting from rainfall.

Sustained minimum flow rate—The flow rate that is sustained or not exceeded for a specified number of consecutive time periods as obtained from an exhaustive length of flow record.

Sustained peak flow rate—The flow rate that is sustained or exceeded for a specified number of consecutive time periods as obtained from an exhaustive length of flow record.

Union—The probability space containing outcomes from events not occurring at the same time.

Union probability—The probability that one event or the other will occur.

Sewage—Contaminated water.

SYMBOLS

k_a	Constant rate in the arithmetic method of predicting population
k_d	Declining rate in the declining-rate method
k_g	Geometric rate in the geometric method of predicting population
k_l	Logistic constant
K	Carrying capacity
P	Population of a community
$Prob$	Probability
P_s	Saturation population
Y	Year
κ	Number of elements in each moving average
η	Number of outcomes favorable to an event space
χ	Number of successive (moving) averages
ζ	Total number of outcomes in a probability space; total number of elements in moving average calculations
ζ_{total}	Approximation of the total number of outcomes in a probability
$\cup$	Union
$\cap$	Intersection

PROBLEMS

1.1   A tourist camp is to be developed for 400 persons. Estimate the volume of the septic tank to be constructed to handle the sewage produced.

1.2   A developer is constructing typical single-family homes. Each home can accommodate a family of four persons. Estimate the size of the drainage pipe to convey the sewage from each home to the wastewater collection system.

1.3   A mobile home park has 40 homes. Assuming that each home has two persons, estimate the design flow for a sewage treatment plant that will be used to service the park.

1.4   A high-rise apartment building complex is 20 stories high. Each floor has 10 apartments occupied by families of 6. Calculate the size of the main drainage pipe to convey the sewage to the collection system.

1.5   A McDonalds restaurant is to be constructed in Easton, Maryland. Estimate the number of meals served in a day and, from this figure, calculate the volume of sewage that is produced.

1.6   A gasoline station is to be constructed along a rural highway. The station will also sell other goods such as candies, soft drinks, and the like. There will be a total of 6 employees. Determine the amount of wastewater produced each day.

1.7   Morgan State University has about 7,000 students. Assuming 60% of these students eat at the school's cafeteria, determine the volume of wastewater produced.

1.8   The School of Engineering at Morgan State University has about 800 students. Design the drainage pipe used to convey the wastewater produced to the collection system of the city of Baltimore.

1.9   Estimate the sewage produced for a typical family of four living in a residential area of a quiet neighborhood.

1.10   Comment on this statement: Two persons estimating sewage production should obtain practically the same values.

1.11   The arithmetic rate constant for a community was previously found to be persons/year. If the population in 2020 is projected to be 21,000, what is the current population in 1997?

1.12   The population in 1990 was 18,000 and the population in 2010 is predicted to be 25,000. What was the population in 2000? Assume the rate of growth at the present time is arithmetic.

1.13   The population in 1990 was 18,000 and the population in 2010 is predicted to be 25,000. What was the population in 1980? Assume the rate of growth at the present time is arithmetic.

1.14   The population in 1970 was 18,000; in 1980, it was 20,000; and in 1990, it was 25,000. Calculate k_a.

1.15   If k_a is 300 persons/year and the population in 1995 was 15,000, what will be the population in 2005?

1.16   The population in 1990 was 18,000, and the population in 2010 is predicted to be 25,000. What was the population in 2000? Assume the rate of growth at the present time is geometric.

1.17   The population in 1990 was 18,000, and the population in 2010 is predicted to be 25,000. What was the population in 1980? Assume the rate of growth at the present time is geometric.

1.18 The population in 1970 was 18,000; in 1980, it was 20,000; and in 1990, it was 25,000. Calculate k_g.

1.19 If k_g is 0.018 ln population/year, and the population in 1995 was 15,000, what will be the population 2005?

1.20 If k_a is 0.041 per year, and the population in 1980 was 15,000, and in 1990, it was 20,000, what will be the saturation population?

1.21 The saturation is 40,000 and k_d is 0.041 per year. If the population in 1990 is 20,000, what will be the population in 2000?

1.22 Determine the saturation population of City C in Figure 1.1. Can you determine the saturation population for City B?

1.23 In Figure 1.5, why are the outcomes in D subtracted when forming the union $A \cup B$? Is D a union or an intersection?

1.24 What is reduced probability space?

1.25 What is intersection probability? What is union probability?

1.26 From the plots in Figures 1.6 and 1.7, determine the lowest rate of flow during the study. What is the corresponding probability of this flow?

1.27 The population data for Anytown is given as follows: 1980 = 15,000, and 1990 = P. If the population in the year 2000 is 21,000, find the value of P. Assume geometric rate of growth.

1.28 The population data for Anytown is as follows: 1980 = 15,000, 1990 = 18,000, and 1995 = 21,000. If the population in some future year is 22,474, what year is it? Assume declining rate of growth.

1.29 The population data for Anytown is as follows: 1980 = 15,000, 1990 = 18,000, and 1995 = P. If the population in the year 2000 is 22,474, find the value of P. Assume declining rate of growth.

1.30 The population data for Anytown is as follows: 1980 = 15,000, 1990 = 18,000, and 1995 = 21,000. If the population in some future year is 22,894, what year is it? Assume logistic rate of growth.

1.31 The population data for Anytown is as follows: 1980 = 15,000, 1990 = 18,000, and 1995 = P. If the population in the year 2000 is 22,894, find the value of P. Assume logistic rate of growth.

1.32 The population data for Anytown is as follows: 1980 = 15,000, 1990 = 18,000, and 1995 = P. If the population in the year 2000 is 22,474, calculate the saturation population.

1.33 The population data for Anytown is as follows: 1980 = 15,000, 1990 = 18,000, and 1995 = P. If the population in the year 2000 is 22,894, calculate the carrying capacity of the environment.

1.34 A portion of a town was monitored during a field survey. During the 30 days that the monitoring was done and using graphical integration, the average daily flow rates were as tabulated in the following table. The population figures obtained from the planning office of the town are as follows: 1980 = 15,000, 1990 = 18,000, and 1995 = 18,500. If the rate of growth is logistic, what is the largest average daily flow expected of the town? Assume that the results of the field survey apply to the whole town.

Average Daily Flow Rates (m³/hr)— Population = 1000

		June		
Day	Avg. Flow Rate		Day	Avg. Flow Rate
1	38.29		2	30.92
3	45.21		4	38.04
5	31.21		6	31.13
7	38.33		8	30.92
9	45.20		10	38.40
11	38.83		12	30.89
13	45.00		14	37.50
15	30.67		16	29.96
17	42.75		18	38.80
19	30.90		20	45.00
21	37.50		22	31.04
23	44.92		24	45.21
25	38.46		26	30.92
27	37.46		28	38.38
29	30.54		30	38.38

Frequency of Flow Rate Values

Avg. Flow Rate	Freq.		Avg. Flow Rate	Freq.
38.29	1		30.92	3
45.21	2		38.04	1
31.21	1		31.13	1
38.33	1		31.04	1
45.20	1		38.40	1
38.83	1		30.89	1
45.00	2		37.50	2
30.67	1		29.96	1
42.75	1		38.80	1
30.90	1		44.92	1
37.46	1		38.38	2
30.54	1		38.46	1

1.35   Using the population data in Problem 1.34 and the result of a field survey below, compute the maximum hourly flow from the town in the year 2050. Use the declining rate of growth.

Daily Peak Hourly Flow Rates (m³/hr)— Population = 1000

		June	
Day	Peak Flow Rate	Day	Peak Flow Rate
1	62.1	2	53.0
3	65.1	4	62.7
5	53.4	6	53.5
7	62.9	8	53.6
9	65.4	10	62.0
11	62.7	12	53.8
13	65.0	14	62.8
15	53.7	16	63.6
17	62.2	18	62.0
19	52.9	20	66.4
21	61.4	22	54.7
23	67.0	24	65.5
25	61.3	26	54.3
27	60.7	28	62.7
29	52.5	30	61.3

Frequency of Flow Rate Values

Peak Flow Rate	Freq.	Peak Flow Rate	Freq.
62.1	1	53.0	1
65.1	1	62.7	3
53.4	1	53.5	1
62.9	1	53.6	1
65.4	1	62.0	2
52.5	1	53.8	1
65.0	1	62.8	1
53.7	1	63.6	1
62.2	1	60.7	1
52.9	1	66.4	1
61.4	1	54.7	1
67.0	1	65.5	1
61.3	2	54.3	1

1.36   Calculate the 7-day moving average in Problem 1.34.

1.37   Calculate the 14-day moving average in Problem 1.34.

BIBLIOGRAPHY

Avnimelech, Y. (1997). Wastewater recycling in Israel: past, present and future. *International Water Irrigation Rev.* 17, 4, 46–48.

Cartwright, P. S. (1997). Municipal wastewater reuse for industrial applications—An Indian experience. *TAPPI Proc.—Environmental Conf. Exhibition Proc. 1997 Environmental Conf. Exhibit, Part 1*, May 5–7, Minneapolis, MN, 1, 65–66, TAPPI Press, Norcross, GA.

Dai, Youyuan, et al. (1998). Comparative study on technical economy for the treatment of wastewater containing phenol between complexation extraction process and biological process. *Zhongguo Huanjing Kexue/China Environmental Science,* August 20, 1998, 18, 4, 293–297.

Harremoes, P. (1997). Integrated water and waste management. *Water Science Technol.* 35, 9, 11–20.

Karneth, A. (1998). Clear Water, Clean DIP. *Pulp Paper Europe.* 3, 4, 28–29.

Masters, G. M. (1991). *Introduction to Environmental Engineering and Science.* Prentice Hall, Englewood Cliffs, NJ.

Metcalf & Eddy, Inc. (1981). *Wastewater Engineering: Collection and Pumping of Wastewater.* McGraw-Hill, New York.

Mitchell, V. G., R. G. Mein, and T. A. McMahon (1998). Stormwater and wastewater use within an urban catchment. *Int. Water Irrigation Rev.* 18, 3, 11–15.

Norris, P. J. (1998). Water reclamation operations in a zero effluent mill. *TAPPI Proc.— Environmental Conf. Exhibit Proc. 1998 TAPPI Int. Environmental Conf. Exhibition, Part 3,* April 5–8 1998, Vancouver, Canada, 3, 1045–1050, TAPPI Press, Norcross, GA.

Qasim, S. R. (1985). *Wasteater Treatment Plants: Planning, Design, and Operation.* Holt, Rinehart & Winston, New York.

Ray, B. T. (1995). *Environmental Engineering.* PWS Publishing Company, Boston.

Roche, T. S. and T. Peterson (1996). Reducing DI water use. *Solid state Technol.* 39, 12, December 1996.

Sincero, A. P. and G. A. Sincero (1996). *Environmental Engineering A Design Approach.* Prentice Hall, Upper Saddle River, NJ.

Singh, Anu and C. E. Adams (1997). Rapid evaluation and installation of a low cost water treatment scheme at a superfund site. *Environmental Progress.* 16, 4, 281–286.

2 Constituents of Water and Wastewater

Given a wastewater, what process should be applied to treat it: biological, chemical, or physical? Should it be treated with a combination of processes? These questions cannot be answered unless the constituents of the wastewater are known. Thus, before any wastewater is to be treated, it is important that its constituents are determined. On the other hand, what are the constituents of a given raw water that make it unfit to drink? Are these constituents simply in the form of turbidity making it unpleasant to the eye, in the form of excessive hardness making it unfit to drink, or in the form bacterial contamination making it dangerous to drink? Water and wastewater may be characterized according to their physical, chemical, and micro-biological characteristics. These topics are discussed in this chapter.

2.1 PHYSICAL AND CHEMICAL CHARACTERISTICS

The constituent physical and chemical characterizations to be discussed include the following: turbidity (physical), color (physical), taste (physical) temperature (physical), chlorides (chemical), fluorides (chemical), iron and manganese (chemical), lead and copper (chemical), nitrate (chemical), sodium (chemical), sulfate (chemical), zinc (chemical), biochemical oxygen demand (chemical), solids (physical), pH (chemical), chemical oxygen demand (chemical), total organic carbon (chemical), nitrogen (chemical), phosphorus (chemical), acidity and alkalinity (chemical), fats and oils and grease (chemical), and odor (physical). The characterization will also include surfactants (physical), priority pollutants (chemical), volatile organic compounds (chemical), and toxic metal and nonmetal ions (chemical). These constituents are discussed in turn in the paragraphs that follow.

2.1.1 TURBIDITY

Done photometrically, *turbidity* is a measure of the extent to which suspended matter in water either absorbs or scatters radiant light energy impinging upon the suspension. The original measuring apparatus that measures turbidity, called the *Jackson turbidimeter*, was based on the absorption principle. A standardized candle was placed under a graduated glass tube housed in a black metal box so that the light from the candle can only be seen from above the tube. The water sample was then poured slowly into the tube until the candle flame was no longer visible. The turbidity was then read on the graduation etched on the tube. At present, turbidity measurements are done conveniently through the use of photometers. A beam of light from a source produced by a standardized electric bulb is passed through a sample vial.

The light that emerges from the sample is then directed to a photometer that measures the light absorbed. The readout is calibrated in terms of turbidity.

The unit of turbidity is the turbidity unit (TU) which is equivalent to the turbidity produced by one mg/L of silica (SiO_2). SiO_2 was used as the reference standard. Turbidities in excess of 5 TU are easily detected in a glass of water and are objectionable not necessarily for health but for aesthetic reasons. A chemical, formazin, that provides a more reproducible result has now replaced silica as the standard. Accordingly, the unit of turbidity is now also expressed as formazin turbidity units (FTU).

The other method of measurement is by light scattering. This method is used when the turbidity is very small. The sample "scatters" the light that impinges upon it. The scattered light is then measured by putting the photometer at right angle from the original direction of the light generated by the light source. This measurement of light scattered at a 90-degree angle is called *nephelometry*. The unit of turbidity in nephelometry is the nephelometric turbidity unit (NTU).

2.1.2 COLOR

Color is the perception registered as radiation of various wavelengths strikes the retina of the eye. Materials decayed from vegetation and inorganic matter create this perception and impart color to water. This color may be objectionable not for health reasons but for aesthetics. Natural colors give a yellow-brownish appearance to water, hence, the natural tendency to associate this color with urine. The unit of measurement of color is the platinum in potassium chloroplatinate (K_2PtCl_6). One milligram per liter of Pt in K_2PtCl_6 is one unit of color.

A major provision of the Safe Drinking Water Act (SDWA) is the promulgation of regulations. This promulgation requires the establishment of primary regulations which address the protection of public health and the establishment of secondary regulations which address aesthetic consideration such as taste, appearance, and color. To fulfill these requirements, the U.S. Environmental Protection Agency (USEPA) establishes maximum contaminant levels (MCL). The secondary MCL for color is 15 color units.

2.1.3 TASTE

Taste is the perception registered by the taste buds. There should be no noticeable taste at the point of use of any drinking water.

The numerical value of taste (or odor to be discussed below) is quantitatively determined by measuring a volume of the sample A (in mL) and diluting it with a volume B (in mL) of distilled water so that the taste (or odor) of the resulting mixture is just barely detectable at a total mixture volume of 200 mL. The unit of taste (or odor) is then expressed in terms of a threshold number as follows:

$$\text{TON or TTN} = \frac{A+B}{A} \qquad (2.1)$$

where
>   TON = threshold odor number
>   TTN = threshold taste number

2.1.4 ODOR

Odor is the perception registered by the olfactory nerves. As in the case of taste, there should be no noticeable odor at the point of use of any drinking water. The secondary standard for odor is 3.

Fresh wastewater odor is less disagreeable than stale wastewater odor but, nonetheless, they all have very objectionable odors. Odors are often the cause of serious complaints from neighborhoods around treatment plants, and it is often difficult for inspectors investigating these complaints to smell any odors in the vicinity of the neighborhood. The reason is that as soon as he or she is exposed to the odor, the olfactory nerves become accustomed to it and the person can no longer sense any odor. If you visit a wastewater treatment plant and ask the people working there if any odor exists, their responses would likely be that there is none. Of course, you, having just arrived from outside the plant, know all the time that, in the vicinity of these workers, plenty of odors exist. The effect of odors on humans produces mainly psychological stress instead of any specific harm to the body. Table 2.1 lists the various odorous compounds that are associated with untreated wastewater.

The determination of odors in water was addressed previously under the discussion on taste. Odors in air are determined differently. They are quantitatively measured by convening a panel of human evaluators. These evaluators are exposed to odors that have been diluted with odor-free air. The number of dilutions required to bring the odorous air to the minimum level of detectable concentration by the panel is the measure of odor. Thus, if three volumes of odor-free air is required, the odor of the air is *three dilutions*. It is obvious that if these evaluators are subjected to the odor several times, the results would be suspicious. For accurate results, the evaluators

TABLE 2.1
Malodorous Compounds Associated with Untreated Wastewater

Compound	Formula	Threshold (ppm)	Odor Quality
Ammonia	NH_3	18	Odor of ammonia
Butyl mercaptan	$(CH_3)_3CSH$	—	Secretion of skunk
Crotyl mercaptan	$CH_3(CH_2)_3SH$	—	Secretion of skunk
Diamines	$NH_2(CH_2)_4NH_2, NH_2(CH_2)_5NH_2$	—	Decayed fish
Ethyl mercaptan	CH_3CH_2SH	0.0003	Decayed cabbage
Hydrogen sulfide	H_2S	<0.0002	Rotten eggs
Indole	C_8H_7N	0.0001	—
Methyl amine	CH_3NH_2	4.6	Fishy
Methyl mercaptan	CH_3SH	0.0006	Decayed cabbage
Methyl sulfide	$(CH_3)_2S$	0.001	Rotten cabbage
Phenyl sulfide	$(C_6H_5)_2S$	0.0001	Rotten cabbage
Skatole	C_9H_9N	0.001	Fecal matter

should be subjected only once, to avoid their olfactory nerves becoming accustomed to the odor thus making wrong judgments.

2.1.5 TEMPERATURE

Most individuals find water at temperatures of 10–15°C most palatable. Groundwaters and waters from mountainous areas are normally within this range. Surface waters are, of course, subject to the effect of ambient temperatures and can be very warm during summer.

The temperature of water affects the efficiency of treatment units. For example, in cold temperatures, the viscosity increases. This, in turn, diminishes the efficiency of settling of the solids that the water may contain because of the resistance that the high viscosity offers to the downward motion of the particles as they settle. Pressure drops also increase in the operation of filtration units, again, because of the resistance that the higher viscosity offers.

2.1.6 CHLORIDES

Chlorides in concentrations of 250 mg/L or greater are objectionable to most people. Thus, the secondary standard for chlorides is 250 mg/L. Whether or not concentrations of 250 mg/L are objectionable, however, would depend upon the degree of acclimation of the user to the water. In Antipolo, a barrio of Cebu in the Philippines, the normal source of water of the residents is a spring that emerges along the shoreline between a cliff and the sea. As such, the fresh water is contaminated by saltwater before being retrieved by the people. The salt imparts to the water a high concentration of chlorides. Chloride contaminants could go as high as 2,000 mg/L; however, even with concentrations this high, the people continue to use the source and are accustomed to the taste.

2.1.7 FLUORIDES

The absence of fluorides in drinking water encourages dental caries or tooth decay; excessive concentrations of the chemical produce mottling of the teeth or dental fluorosis. Thus, managers and operators of water treatment plants must be careful that the exact concentrations of the fluorides are administered to the drinking water. Optimum concentrations of 0.7 to 1.2 mg/L are normally recommended, although the actual amount in specific circumstances depends upon the air temperature, since air temperature influences the amount of water that people drink. Also, the use of fluorides in drinking water is still controversial. Some people are against its use, while some are in favor of it.

2.1.8 IRON AND MANGANESE

Iron (Fe) and manganese (Mn) are objectionable in water supplies because they impart brownish colors to laundered goods. Fe also affects the taste of beverages such as tea and coffee. Mn flavors tea and coffee with a medicinal taste. The SMCLs (secondary MCLs) for Fe and Mn are, respectively, 0.3 and 0.05 mg/L.

2.1.9 LEAD AND COPPER

Clinical, epidemiological, and toxicological studies have demonstrated that lead exposure can adversely affect human health. The three systems in the human body most sensitive to lead are the blood-forming system, the nervous system, and the renal system. In children, blood levels from 0.8 to 1.0 μg/L can inhibit enzymatic actions. Also, in children, lead can alter physical and mental development, interfere with growing, decrease attention span and hearing, and interfere with heme synthesis. In older men and women, lead can increase blood pressure. Lead is emitted into the atmosphere as Pb, PbO, PbO_2, $PbSO_4$, PbS, $Pb(CH_3)_4$, $Pb(C_2H_5)_4$, and lead halides. In drinking water, it can be emitted from pipe solders.

The source of copper in drinking water is the plumbing used to convey water in the house distribution system. In small amounts, it is not detrimental to health, but it will impart an undesirable taste to the water. In appropriate concentrations, copper can cause stomach and intestinal distress. It also causes Wilson's disease. Certain types of PVC (polyvinyl chloride) pipes, called CPVC (chlorinated polyvinyl chloride), can replace copper for household plumbing.

2.1.10 NITRATE

Nitrate is objectionable for causing what is called *methemoglobinemia* (infant cyanosis or blue babies) in infants. The MCL is 10 mg/L expressed as nitrogen.

Before the establishment of stringent regulations, sludges from wastewater treatment plants were most often spread on lands and buried in ditches as methods of disposal. As the sludge decays, nitrates are formed. Thus, in some situations, these methods of disposal have resulted in the nitrates percolating down the soil causing excessive contaminations of the groundwater. Even today, these methods are still practiced. In order for these practices to be acceptable to the regulatory agencies, a material balance of the nitrate formed must be calculated to ascertain that the contamination of the groundwater does not go to unacceptable levels.

2.1.11 SODIUM

The presence of sodium in drinking water can affect persons suffering from heart, kidney, or circulatory ailments. It may elevate blood pressures of susceptible individuals. Sodium is plentiful in the common table salt that people use to flavor food to their taste. It is a large constituent of sea water; hence, in water supplies contaminated by the sea as in the case of Antipolo mentioned earlier, this element would be plentiful.

2.1.12 SULFATE

The sulfate ion is one of the major anions occurring naturally in water. It produces a cathartic or laxative effect on people when present in excessive amounts in drinking water. Its SMCL is 250 mg/L.

2.1.13 ZINC

Zinc is not considered detrimental to health, but it will impart an undesirable taste to drinking water. Its SMCL is 5 mg/L.

2.1.14 BIOCHEMICAL OXYGEN DEMAND

Biochemical oxygen demand (BOD) is the amount of oxygen consumed by the organism in the process of stabilizing waste. As such, it can be used to quantify the amount or concentration of oxygen-consuming substances that a wastewater may contain. Analytically, it is measured by incubating a sample in a refrigerator for five days at a temperature of 20°C and measuring the amount of oxygen consumed during that time.

The substances that consume oxygen in a given waste are composed of carbonaceous and nitrogenous portions. The carbonaceous portion refers to the carbon content of the waste; carbon reacts with the dissolved oxygen producing CO_2. On the other hand, the nitrogenous portion refers to the ammonia content; ammonia also reacts with the dissolved oxygen. Even though the term used is nitrogenous, nitrogen is not referred to in this context. Any nitrogen must first be converted to ammonia before it becomes the "nitrogenous."

Generally, two types of analysis are used to determine BOD in the laboratory: one where dilution is necessary and one where dilution is not necessary. When the BOD of a sample is small, such as found in river waters, dilution is not necessary. Otherwise, the sample would have to be diluted. Table 2.1 sets the criteria for determining the dilution required. This table shows that there are two ways dilution can be made: using percent mixture and direct pipetting into 300-mL BOD bottles. Normally, BOD analysis is done using 300-mL incubation bottles.

Because BOD analysis attempts to measure the oxygen equivalent of a given waste, the environment inside the BOD bottle must be conducive to uninhibited bacterial growth. The parameters of importance for maintaining this type of environment are

TABLE 2.2
Ranges of BOD Measurable with Various Dilutions of Samples

Using Percent Mixtures		By Direct Pipetting into 300-mL Bottles	
% mixture	Range of BOD_5	mL	Range of BOD_5
0.01	35,000–70,000	0.01	40,000–100,000
0.03	10,000–35,000	0.05	20,000–40,000
0.05	7,000–10,000	0.10	10,000–20,000
0.1	3,500–7,000	0.30	4,000–10,000
0.3	1,400–3,500	0.50	2,000–4,000
0.5	700–1,400	1.0	1,000–2,000
1.0	350–700	3.0	400–1,000
3.0	150–350	5.0	200–400
5.0	70–150	10.0	100–200
10.0	35–70	30.0	40–100
30.0	10–35	50.0	20–40
50.0	5–10	100	100
100	0–5	300	0–10

freedom from toxic materials, favorable pH and osmotic pressure conditions, optimal amount of nutrients, and the presence of significant amount of population of mixed organisms of soil origin. Through long years of experience, it has been found that synthetic dilution water prepared from distilled water or demineralized water is best for BOD work, because the presence of such toxic substances as chloramine, chlorine, and copper can be easily controlled. The maintenance of favorable pH can be assured by buffering the dilution water at about pH 7.0 using potassium and sodium phosphates. The potassium and sodium ions, along with the addition of calcium and magnesium ions, can also maintain the proper osmotic pressure, as well as provide the necessary nutrients in terms of these elements. The phosphates, of course, provide the necessary phosphorus nutrient requirement. Ferric chloride, magnesium sulfate, and ammonium chloride supply the requirements for iron, sulfur, and nitrogen, respectively.

A sample submitted for analysis may not contain any organism at all. Such is the case, for example, of an industrial waste, which can be completely sterile. For this situation, the dilution water must be seeded with organisms from an appropriate source. In domestic wastewaters, all the organisms needed are already there; consequently, these wastewaters can serve as good sources of seed organisms. Experience has shown that a seed volume of 2.0 mL per liter of dilution water is all that is needed.

Laboratory calculation of BOD. In the subsequent development, the formulation will be based on the assumption that the dilution method is used. If, in fact, the method used is direct, that is, no dilution, then the dilution factor that appears in the formulation will simply be ignored and equated to 1.

The technique for determining the BOD of a sample is to find the difference in dissolved oxygen (DO) concentration between the final and the initial time after a period of incubation at some controlled temperature. This difference, converted to mass of oxygen per unit volume of sample (such as mg/L) is the BOD.

Let I be the initial DO of the sample, which has been diluted with seeded dilution water, and F be the final DO of the same sample after the incubation period. The difference would then represent a BOD, but since the sample is seeded, a correction must be made for the BOD of the seed. This requires running a blank.

Let I' represent the initial DO of a volume Y of the blank composed of only the seeded dilution water; also, let F' be the final DO after incubating this blank at the same time and temperature as the sample. If X is the volume of the seeded dilution water mixed with the sample, the DO correction would be $(I' - F')(X/Y)$. Letting D be the fractional dilution, the BOD of the sample is simply

$$\text{BOD} = \frac{(I - F) - (I' - F')(X/Y)}{D} \qquad (2.2)$$

In this equation, if the incubation period is five days, the BOD is called the *five-day biochemical oxygen demand*, BOD_5. It is understood that unless it is specified, BOD_5 is a BOD measured at the standard temperature of incubation of 20°C. If incubation is done for a long period of time such as 20 to 30 days, it is assumed that all the BOD has been exerted. The BOD under this situation is the ultimate; therefore, it is called *ultimate BOD*, or BOD_u.

BOD_u, in turn, can have two fractions in it: one due to carbon and the other due to nitrogen. As mentioned before, carbon reacts with oxygen; also, nitrogen in the form of ammonia, reacts with oxygen. If the BOD reaction is allowed to go to completion with the ammonia reaction inhibited, the resulting ultimate BOD is called *ultimate carbonaceous BOD* or CBOD. Because *Nitrosomonas* and *Nitrobacter*, the organisms for the ammonia reaction, cannot compete very well with carbonaceous bacteria (the organisms for the carbon reaction), the reaction during the first few days of incubation up to approximately five or six days is mainly carbonaceous. Thus, BOD_5 is mainly carbonaceous. If the reaction is uninhibited, the BOD after five or six days of incubation also contains the nitrogenous BOD. BOD is normally reported in units of mg/L.

Experience has demonstrated that a dissolved oxygen concentration of 0.5 mg/L practically does not cause depletion of BOD. Also, it has been learned that a depletion of less than 2.0 mg/L produces erroneous results. Thus, it is important that in BOD work, the concentration of DO in the incubation bottle should not fall below 0.5 mg/L and that the depletion after the incubation period should not be less than 2.0 mg/L.

Example 2.1 Ten milliliters of sample is pipetted directly into a 300-mL incubation bottle. The initial DO of the diluted sample is 9.0 mg/L and its final DO is 2.0 mg/L. The initial DO of the dilution water is also 9.0 mg/L, and the final DO is 8.0 mg/L. The temperature of incubation is 20°C. If the sample is incubated for five days, what is the BOD_5 of the sample?

Solution:

$$BOD = \frac{(I-F)-(I'-F')(X/Y)}{D} = \frac{(9-2)-(9-8)([300-10]/300)}{10/300}$$

$$= 183 \text{ mg/L}$$

2.1.15 Nitrification in the BOD Test

The general profile of oxygen consumption in a BOD test for a waste containing oxygen-consuming constituents is shown in Figure 2.1. As mentioned previously, because the nitrifiers cannot easily compete with the carbonaceous bacteria, it takes about 5 days or so for them to develop. Thus, after about 5 days the curve abruptly rises due to the nitrogenous oxygen demand, NBOD. If the nitrifiers are abundant in the beginning of the test, however, the nitrogen portion can be exerted immediately as indicated by the dashed line after a short lag. This figure shows the necessity of inhibiting the nitrifiers if the carbonaceous oxygen demand, CBOD, is the one desired in the BOD test.

The reactions in the nitrification process are mediated by two types of autotrophic bacteria: *Nitrosomonas* and *Nitrobacter*. The ammonia comes from the nitrogen content of any organic substance, such as proteins, that contains about 16% nitrogen. As soon as the ammonia has been hydrolyzed from the organic substance, *Nitrosomonas* consumes it and in the process also consumes oxygen according to

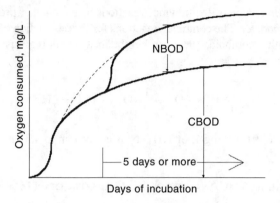

FIGURE 2.1 Exertion of CBOD and NBOD.

the following reactions:

$$\frac{1}{6}NH_4^+ + \frac{1}{3}H_2O \rightarrow \frac{1}{6}NO_2^- + \frac{4}{3}H^+ + e^- \tag{2.3}$$

$$\frac{1}{4}O_2 + H^+ + e^- \rightarrow \frac{1}{2}H_2O \tag{2.4}$$

Adding Eqs. (2.3) and (2.4) produces

$$\frac{1}{6}NH_4^+ + \frac{1}{4}O_2 \rightarrow \frac{1}{6}H_2O + \frac{1}{6}NO_2^- + \frac{1}{3}H^+ \tag{2.5}$$

Equation (2.4) is called an *electron acceptor* reaction. Equation (2.3) is an *elector donor* reaction, that is, it provides the electron for the electron acceptor reaction. Together, these two reactions produce energy for the *Nitrosomonas*.

The NO_2^- produced in Equation (2.5) serves as an electron source for another genus of bacteria, the *Nitrobacter*. The chemical reactions when *Nitrobacter* uses the nitrite are as follows:

$$\frac{1}{6}NO_2^- + \frac{1}{6}H_2O \rightarrow \frac{1}{6}NO_3^- + \frac{1}{3}H^+ + \frac{1}{3}e^- \tag{2.6}$$

$$\frac{1}{12}O_2 + \frac{1}{3}H^+ + \frac{1}{3}e^- \rightarrow \frac{1}{6}H_2O \tag{2.7}$$

Adding Eqs. (2.6) and (2.7) produces

$$\frac{1}{6}NO_2^- + \frac{1}{12}O_2 \rightarrow \frac{1}{6}NO_3^- \tag{2.8}$$

As with *Nitrosomonas*, the previous reactions taken together provide the energy needed by *Nitrobacter*. The combined reactions for the destruction of the ammonium ion, NH_4^+, (or the ammonia, NH_3) can be obtained by adding Eqs. (2.5) and (2.8). This will produce

$$\frac{1}{6}NH_4^+ + \frac{1}{3}O_2 \rightarrow \frac{1}{6}NO_3^- + \frac{1}{6}H_2O + \frac{1}{3}H^+ \tag{2.9}$$

From Equation (2.9), 1.0 mg/L of NH_4–N is equivalent to 4.57 mg/L of dissolved oxygen.

2.1.16 Mathematical Analysis of BOD Laboratory Data

The ultimate carbonaceous oxygen demand may be obtained by continuing the incubation period beyond five days up to 20 to 30 days. To do this, the nitrifiers should be inhibited by adding the appropriate chemical in the incubation bottle. The other way of obtaining CBOD is through a mathematical analysis.

In the incubation process, let y represent the cumulative amount of oxygen consumed (oxygen uptake) at any time t, and let L_c represent the CBOD of the original waste. The rate of accumulation of the cumulative amount of oxygen, dy/dt, is proportional to the amount of CBOD left to be consumed, $L_c - y$. Thus,

$$\frac{dy}{dt} = y' = k_c(L_c - y) \tag{2.10}$$

where k_c is a proportionality constant called *deoxygenation coefficient*.

In the previous equation, if the correct values of k_c and L_c are substituted, the left-hand side should equal the right-hand side of the equation; otherwise, there will be a residual R such that

$$R = k_c(L_c - y) - y' = k_c L_c - k_c y - y' \tag{2.11}$$

At each equal interval of time, the values of y may be determined. For n intervals, there will also be n values of y. The corresponding Rs for each interval may have positive and negative values. If these Rs are added, the result may be zero which may give the impression that the residuals are zero. On the other hand, if the residuals are squared, the result of the sum will always be positive. Thus, if the sum of the squares is equal to zero, there is no ambiguity that the residuals are, in fact, equal to zero.

The n values of y corresponding to n values of time t will have inherent in them one value of k_c and one value of L_c. Referring to Equation (2.11), these values may be obtained by partial differentiation. From the previous paragraph, when the sum of the squares of R is equal to zero, it is certain that the residual is zero. This means that when the sum of the squares is zero, the partial derivative of the sum of the squares must also be zero. Consequently, the partial derivatives of the sum of R^2 with respect to $k_c L_c$ and k_c are zero. Thus, to obtain k_c and L_c, the latter partial derivative of the sum of the squares must be equated to zero to force the solutions. The method just described is called the *method of least squares*, because equating the partial derivatives to zero is equivalent to finding the minimum of the squares. The corresponding equations are derived as follows:

$$\frac{\partial \Sigma R^2}{\partial (k_c L_c)} = 0 = \frac{\partial \Sigma (k_c L_c - k_c y - y')^2}{\partial (k_c L_c)} = \sum 2(k_c L_c - k_c y - y')(1) \quad (2.12)$$

Solving for k_c,

$$k_c = \frac{\Sigma y'}{nL_c - \Sigma y} \quad (2.13)$$

$$\frac{\partial \Sigma R^2}{\partial k_c} = 0 = \frac{\partial \Sigma (k_c L_c - k_c y - y')^2}{\partial k_c} = \sum 2(k_c L_c - k_c y - y')(y) \quad (2.14)$$

In the previous equations, $k_c L_c$ and k_c are the parameters of the partial differentiation. Thus, in Equation (2.14), where the differentiation is with respect to k_c, the partial derivative of $k_c L_c$ with respect to k_c is zero, since the whole expression $k_c L_c$ is taken as a parameter.

Solving Equation (2.14) for k_c,

$$k_c = \frac{\Sigma y y'}{\Sigma y^2 - L_c \Sigma y} \quad (2.15)$$

From Eqs. (2.13) and (2.15), L_c, the ultimate oxygen demand, may finally be solved producing

$$L_c = \frac{\Sigma y' \Sigma y^2 - \Sigma y \Sigma y y'}{\Sigma y' \Sigma y - n \Sigma y y'} \quad (2.16)$$

$$y' = \frac{y_{m+1} - y_{m-1}}{t_{m+1} - t_{m-1}} \quad \text{for } 1 \le m \le n \quad (2.17)$$

The progress of oxygen utilization, y, with respect to time may be monitored by respirometry. Figure 2.2 shows a schematic of an electrolytic respirometer. As the waste is consumed, CO_2 is produced which is then absorbed by a potassium hydroxide solution by a chemical reaction. This absorption causes the pressure inside the bottle to decrease. This decrease is sensed by the electrode triggering the electrolytic decomposition of H_2O to produce O_2 and H_2.

The O_2 is channeled toward the inside of the bottle to recover the pressure and the H_2 is vented to the atmosphere. The amount of oxygen consumed by the waste is correlated with the amount of oxygen electrolytically produced to maintain the pressure inside the bottle.

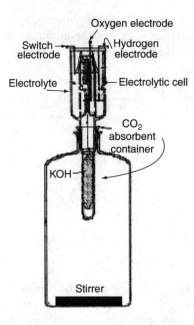

FIGURE 2.2 Electrolytic respirometer for determination of oxygen consumption of a waste.

Example 2.2   The following data represent the cumulative amount of oxygen uptake for a river water receiving waste. Calculate L_c and k_c.

t (day)	2	4	6	8	10
y (mg/L)	10	20	23	25	28

Solution:

$$L_c = \frac{\Sigma y' \Sigma y^2 - \Sigma y \Sigma yy'}{\Sigma y' \Sigma y - n \Sigma yy'}$$

$$k_c = -\frac{\Sigma yy'}{\Sigma y^2 - L_c \Sigma y}$$

$$y' = \frac{y_{m+1} - y_{m-1}}{t_{m+1} - t_{m-1}} \quad \text{for } 1 \le m \le n$$

t (day)	2	4	6	8	10	
y (mg/L)	10	20	23	25	28	$\Sigma y = 68^b$
y'		3.25^a	1.25	1.25		$\Sigma y' = 5.75$
y^2		400	529	625		$\Sigma y^2 = 1554$
yy'		65	28.75	31.25		$\Sigma yy' = 125$

^{a} $3.25 = \frac{23 - 10}{6 - 2}$
^{b} $68 = 20 + 23 + 25$

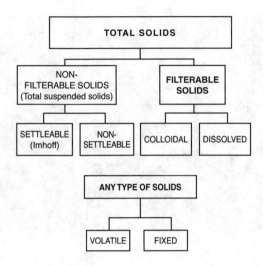

FIGURE 2.3 Components of total solids.

$$L_c = \frac{5.75(1554) - 68(125)}{5.75(68) - 3(125)} = \frac{435.5}{16} = 27.22 \text{ mg/L} \quad \textbf{Ans}$$

$$k_c = -\frac{125}{1554 - 27.22(68)} = 0.42 \text{ per day} \quad \textbf{Ans}$$

2.1.17 SOLIDS

Solids that find their way into wastewaters include the solids on the kitchen table: corn, vegetables, crab, rice, bread, chicken, fish, egg, and so on. In short, these are the solids flushed down the toilet. In addition, there are also solids coming from the bathroom such as toilet paper and human wastes. In the old combined sewer systems, solids may include the soils from ground eroded by runoff. Figure 2.3 shows a pictorial representation of the various components of total solids.

The *total solids* content of a wastewater are the materials left after water has been evaporated from the sample. The evaporation is normally done at 103–105°C. Total solids may be classified as *filtrable* and *nonfiltrable*. The filtrable fraction contains the *colloidal particles* and the *dissolved solids* that pass through the filter in a prescribed laboratory procedure. The nonfiltrable fraction contain the *settleable* and the *nonsettleable* fractions that did not pass through the filter.

The nonsettleable fraction of the nonfiltrable fraction is in true suspension; it is composed of suspended solids. On the other hand, the settleable fraction does not suspend in the liquid and, thus, these component solids are not suspended solids; they are settleable fractions because they settle. Solids retained on the filter (nonfiltrable solids) are, however, collectively (and erroneously) called suspended solids while those that pass the filter are collectively (and, also, erroneously) called dissolved solids.

The solids that pass through the filter are not all dissolved, because they also contain colloidal particles. Also, the solids retained on the filter are not all suspended,

FIGURE 2.4 Imhoff cones.

because they also contain settleable solids; however, the use of these terms have persisted. More accurately, the *nonfiltrable-nonsettleable* fraction should be the one called suspended solids. Since the nonfiltrable solids are composed of the true suspended solids and the "nonsuspended" suspended solids, nonfiltrable solids are also called *total suspended solids.*

The *settleable fraction* is the volume of the solids after settling for 30 minutes in a cone-shaped vessel called an *Imhoff cone.* The volume of solids that settled, in milliliters, divided by the corresponding grams of solids mass is called the *sludge volume index, SVI.* Settleable solids are an approximate measure of the volume of sludge that will settle by sedimentation. Figure 2.4 shows a photograph of Imhoff cones.

All the types of solids described previously can have fixed and volatile portions. The *fixed* portions of the solids are those that remain as a residue when the sample is decomposed at 600°C. Those that disappear are called *volatile solids.* Volatile solids and fixed solids are normally used as measures of the amount of organic matter and inorganic matter in a sample, respectively. Magnesium carbonate, however, decomposes to magnesium oxide and carbon dioxide at 350°C. Thus, the amount of organic matter may be overpredicted and the amount of inorganic may be underpredicted if the carbonate is present in an appreciable amount.

Example 2.3   A suspended solids analysis is run on a sample. The tared mass of the crucible and filter is 55.3520 g. A sample of 260 mL is then filtered and the residue dried to constant mass at 103°C. If the constant mass of the crucible, filter, and the residue is 55.3890 g, what is the suspended solids (SS) content of the sample?

Solution:

$$SS = \frac{55.3890 - 55.3520}{260}\frac{g}{mL} = \frac{55.3890 - 55.3520}{260}(1000)(1000)$$

$$= 142.3 \text{ mg/L}   \textbf{Ans}$$

2.1.18 pH

Even the purest water exhibits ionization. Kohlrausch, a German physical chemist, demonstrated this property by measuring the electrical conductivity of water using a very sensitive instrument. The existence of the electrical conductivity is a result of the chemical reaction between two water molecules as shown below:

$$H_2O + H_2O \rightleftharpoons H_3O^+ + OH^- \tag{2.18}$$

The first term on the right-hand side of Equation (2.18) is called the hydronium ion; the second is called the hydroxide ion. These ions are responsible for the electrical conductivity of water. The concentrations of these ions are very small. At 25°C, for pure water, there is a concentration of 1×10^{-7} mole per liter of the hydronium ion and of the hydroxide ion, respectively. When the water is not pure, these concentrations would be different. In a large number of environmental engineering textbooks, the hydronium is usually written as H^+. Also the hydronium ion is usually referred to as the hydrogen ion. In essence, the hydronium ion can be looked at as a hydrated hydrogen ion.

Let the symbol $\{H^+\}$ be read as "the effective concentration or activity of H^+" and the symbol $[H^+]$ be read as "the concentration of H^+." The effective concentration $\{H^+\}$ refers to the ions of H^+ that actually participate in a reaction. This is different from the concentration $[H^+]$, which refers to the actual concentration of H^+, but not all the actual concentration of this H^+ participate in the chemical reaction. Effective concentration is also called activity. The effective concentration or activity of a solute is obtained from its actual concentration by multiplying the actual concentration by an activity coefficient, f (i.e., $\{H^+\} = f[H^+]$).

When the concentration of a solute such as H^+ is dilute, the solute particles are relatively far apart behaving independently of each other. Because the concentration is dilute (particles far apart), the particles participating in a reaction are essentially the concentration of the solute. Therefore, for dilute solutions, $\{H^+\}$ is equal to $[H^+]$.

The existence of the hydronium ion is the basis for the definition of pH as originated by Sorensen. pH is defined as the negative logarithm to the base 10 of the hydrogen ion activity expressed in gmols per liter as shown below.

$$pH = -\log_{10}\{H^+\} \tag{2.19}$$

The product of the activities of H^+ and OH^- at any given temperature is constant. This is called the ion-product of water, K_w, which is equal to 1×10^{-14} at 25°C.

Sorensen also defined a term pOH as the negative of the logarithm to the base 10 of the hydroxide ion activity expressed in gmols per liter.

$$pOH = -\log_{10}\{OH^-\} \tag{2.20}$$

In Equation (2.19), when $[H^+]$ is equal to one mole per liter, the pH is equal to zero. When the concentration is 1×10^{-14} mole per liter, the pH is 14. Although the pH could go below 0 and be greater than 14, in practice, the practical range is considered to be from 0 to 14. Low pH solutions are acidic while high pH solutions are basic. A pH equal to 7 corresponds to a complete neutrality. The range of pH from 0 to 14 corresponds to a range of pOH from 14 to 0.

The ion product of water, K_w, is

$$\{H^+\}\{OH^-\} = K_w \tag{2.21}$$

Taking the logarithm of both sides to the base 10,

$$pH + pOH = pK_w$$
$$pH = pK_w - pOH \tag{2.22}$$

where

$$pK_w = -\log_{10} K_w \tag{2.23}$$

pH is an important parameter both in natural water systems and in water and wastewater engineering. The tolerable concentration range for biological life in water habitats is quite narrow. This is also the case in wastewater treatment. For example, nitrification plants are found to function at only a narrow pH range of 7.2 to 9.0. In water distribution systems, the pH must be maintained at above neutrality of close to 8 to prevent corrosion. Above pH 8, the water could also cause scaling, which is equally detrimental when compared with corrosion.

Example 2.4 10^{-2} mole of HCl is added to one liter of distilled at 25°C. After completion of the reaction, the pH was found to be equal to 2. **(a)** What is the solution reaction? **(b)** What are the concentrations of the hydrogen and hydroxide ions?

Solution:
(a) The solution reaction is acidic.

(b) $pH = -\log_{10}\{H^+\} \Rightarrow 2 = -\log_{10}\{H^+\}$

$\{H^+\} = 10^{-2}$ gmol/L Ans $\Rightarrow \{OH^-\} = 10^{-2}$ gmol/L   **Ans**

2.1.19 CHEMICAL OXYGEN DEMAND

The chemical oxygen demand (COD) test has been used to measure the oxygen-equivalent content of a given waste by using a chemical to oxidize the organic content of the waste. The higher the equivalent oxygen content of a given waste, the higher is its COD and the higher is its polluting potential. Potassium dichromate has been found to be an excellent oxidant in an acidic medium. The test must be conducted at elevated temperatures. For certain types of waste, a catalyst (silver sulfate) may be used to aid in the oxidation.

The COD test normally yields higher oxygen equivalent values than those derived using the standard BOD_5 test, because more oxygen equivalents can always be oxidized by the chemical than can be oxidized by the microorganisms. In some types of wastes, a high degree of correlation may be established between COD and BOD_5. If such is the case, a correlation curve may be prepared such that instead of analyzing for BOD_5, COD may be analyzed, instead. This is practically advantageous, since it takes five days to complete the BOD test but only three hours for the COD test. The correlation may then be used for water plant control and operation.

The chemical reaction involved in the COD test for the oxidation of organic matter is as follows:

$$\text{Organic matter} + Cr_2O_7^{2-} + H^+ \xrightarrow[\text{heat}]{\text{catalyst}} Cr^{3+} + CO_2 + H_2O \qquad (2.24)$$

From this reaction, chromium as reduced from an oxidation state of +6 to an oxidation state of +3. The oxidation products are carbon dioxide and water. The *oxidation state* is a measure of the degree of affinity of the atom to the electrons it shares with other atoms. A negative oxidation state of an atom indicates that the electrons spend more time with the atom, while a positive oxidation state indicates that the electrons spend more time with the other atom.

2.1.20 TOTAL ORGANIC CARBON

The polluting potential and strength of a given waste may also be assessed by measuring its carbon content. Because carbon reacts with oxygen, the more carbon it contains, the more polluting and stronger it is. The carbon content is measured by converting the carbon to carbon dioxide. The test is performed by injecting a known quantity of sample into an oxidizing furnace. The amount of carbon dioxide formed from the reaction of C with O_2 inside the furnace is quantitatively measured by an infrared analyzer. The concentration of the total organic carbon (TOC) is then calculated using the chemical ratio of C to CO_2.

2.1.21 NITROGEN

Nitrogen is a major component of wastewater. People eat meat and meat contains protein that, in turn, contains nitrogen. Every bite of hamburger is a source of nitrogen and every fried chicken you buy is a source of nitrogen. Nitrogen in protein is needed by humans in order to survive which, in turn, produces wastewater that must be treated.

Protein contains about 16% nitrogen. The nitrogen in protein is an organic nitrogen. *Organic nitrogen*, therefore, is one measure of the protein content of an organic waste. When an organic matter is attacked by microorganisms, its protein hydrolyzes into a type of ammonia called free ammonia. Thus, *free ammonia* is the hydrolysis product of organic nitrogen. The *nitrites* and *nitrates* are the results of the oxidation of ammonia to nitrites by *Nitrosomonas* and the oxidation of nitrites to nitrates by *Nitrobacter*, respectively. The sum of the organic, free ammonia, nitrite, and nitrate nitrogens is called *total nitrogen*. The sum of ammonia and organic nitrogens is called

Kjeldahl nitrogen. Of all the species of nitrogen, ammonia, nitrite, and nitrate are used as nitrogen sources for synthesis. They are to be provided in the correct amount in wastewater treatment. They also cause eutrophication in receiving streams.

The free ammonia may hydrolyze producing the ammonium ion according to the following reaction:

$$NH_3 + H_2O \rightleftharpoons NH_4^+ + OH^- \tag{2.25}$$

At pH levels below 7, the above equilibrium is shifted to the right and the predominant nitrogen species is NH_4^+, the ionized form. On the other hand, when the pH is above 7, the equilibrium is shifted to the left and the predominant nitrogen species is ammonia. The unionized form is most lethal to aquatic life. Ammonia is determined in the laboratory by boiling off with the steam after raising the pH. The steam is then condensed absorbing the ammonia liberated. The concentration is measured by colorimetric methods in the condensed steam.

The nitrite nitrogen is very unstable and is easily oxidized to the nitrate form. Because its presence is transitory, it can be used as an indicator of past pollution that is in the process of recovery. Its concentration seldom exceeds 1 mg/L in wastewater and 0.1 mg/L in receiving streams. Nitrites are determined by colorimetric methods.

The nitrate nitrogen is the most oxidized form of the nitrogen species. Since it can cause *methemoglobinemia*, (infant cyanosis or blue babies), it is a very important parameter in drinking water standards. The maximum contaminant level (MCL) for nitrates is 10 mg/L as *N*. Nitrates may vary in concentrations from 0 to 20 mg/L as *N* in wastewater effluents. A typical range is 15 to 20 mg/L as *N*. The nitrate concentration is usually determined by colorimetric methods.

2.1.22 PHOSPHORUS

Phosphorus can be found in both plants and animals. Thus, bones, teeth, nerves, and muscle tissues contain phosphorus. The nucleic acids DNA and RNA contain phosphorus as well.

The metabolism of food used by the body requires compounds containing phosphorus. The human body gets this phosphorus through foods eaten. These include egg, beans, peas, and milk. Being used by humans, these foods, along with the phosphorus, therefore find their way into wastewaters. Another important source of phosphorus in wastewater is the phosphate used in the manufacture of detergents.

In general, phosphorus occurs in three phosphate forms: orthophosphate, condensed phosphates (or polyphosphates), and organic phosphates. Phosphoric acid, being triprotic, forms three series of salts: dihydrogen phosphates containing the $H_2PO_4^-$ ions, hydrogen phosphate containing the HPO_4^{2-} ions, and the phosphates containing the PO_4^{3-} ions. These three ions collectively are called orthophosphates. As in the case of the nitrogen forms ammonia, nitrite and nitrate, the orthophosphates can also cause eutrophication in receiving streams. Thus, concentrations of orthophosphates should be controlled through removal before discharging the wastewater into receiving bodies of water. The orthophosphates of concern in wastewater engineering are sodium phosphate (Na_3PO_4), sodium hydrogen phosphate (Na_2HPO_4), sodium dihydrogen phosphate (NaH_2PO_4), and ammonium hydrogen phosphate [$(NH_4)_2HPO_4$]. They cause the problems associated with algal blooms.

When phosphoric acid is heated, it decomposes losing molecules of water forming the P–O–P bonds. The process of losing water is called *condensation*, thus the term condensed phosphates and, because they have more than one phosphate group in the molecule, they are also called *polyphosphates*. Among the acids formed from the condensation of phosphoric acid are *dipolyphosphoric acid* or *pyrophosphoric acid* ($H_4P_2O_7$), *tripolyphosphoric acid* ($H_5P_3O_{10}$), and *metaphosphoric acid* ($HPO_3)_n$. Condensed phosphates undergo hydrolysis in aqueous solutions and transform into the orthophosphates. Thus, they must also be controlled. Condensed phosphates of concern in wastewater engineering are sodium hexametaphosphate [$Na(PO_3)_6$], sodium dipolyphosphate ($Na_4P_2O_7$), and sodium tripolyphosphate ($Na_5P_3O_{10}$).

When organic compounds containing phosphorus are attacked by microorganisms, they undergo hydrolysis into the orthophosphate forms. Thus, as with all the other phosphorus species, they have to be controlled before the wastewaters are discharged.

Orthophosphate can be determined in the laboratory by adding a substance that can form a colored complex with the phosphate. An example of such a substance is ammonium molybdate. Upon formation of the color, colorimetric tests may then be applied. The condensed and organic phosphates all hydrolyze to the ortho form, so they can also be analyzed using ammonium molybdate. The hydrolysis are normally done in the laboratory at boiling-water temperatures.

2.1.23 ACIDITY AND ALKALINITY

Acidity and alkalinity are two important parameters that must be controlled in the operation of a wastewater treatment plant. Digesters, for example, will not operate if the environment inside the tank is acidic, since microorganisms will simply die in acid environments. The contents of the tank must be buffered at the proper acidity as well as proper alkalinity.

Acidity is the ability of a substance to neutralize a base. For example, given the base OH^- and a species HCO_3^-, the reaction of the two species in water solution is

$$OH^- + HCO_3^- \rightleftharpoons H_2O + CO_3^{2-}$$

Thus, in the previous reaction, because HCO_3^- has neutralized OH^-, it has acidity and it is an acid.

Alkalinity, on the other hand, is the ability of a substance to neutralize an acid. For example, given the acid HCl and the species HCO_3^-, they react in solution as follows:

$$HCl + HCO_3^- \rightarrow H_2CO_3 + Cl^-$$

In the previous reaction, because HCO_3^- has neutralized the acid HCl, it has alkalinity. Alkaline substances are also called *bases*. From the above two reactions of HCO_3^-, it can be concluded that this species can act both as an acid and as a base. A substance that can act both as an acid and as a base is called an *amphoteric substance.*

Alkalinity in wastewaters results from the presence of the hydroxides, carbonates, and bicarbonates of such elements as calcium, magnesium, sodium, and potassium, and radicals like the ammonium ion. Of the elements, the bicarbonates of calcium and magnesium are the most common. The other alkalinity species that may be found, although not to a major extent as the bicarbonate, are $HSiO_3^-$, $H_2BO_3^-$, HPO_4^{-2}, and $H_2PO_4^-$. Alkalinity helps to resist the change in pH when acids are produced during the course of a biological treatment of a wastewater. Wastewaters are normally alkaline, receiving this alkalinity from the water supply and the materials added during domestic use. Alkalinity is determined in the laboratory by titration using a standard concentration of acid. The reverse is true for the determination of acidity in the laboratory.

2.1.24 FATS, OILS, WAXES, AND GREASE

Organic compounds with the general formula $R\underset{\underset{O}{\|}}{C}{-}OH$ are called organic acids, where R is a hydrocarbon group. When the compound is a long-chain compound, it is called a *fatty acid*. Fatty acids may be saturated or unsaturated. When one or more carbon bonds is a double bond, the fatty acid is said to be unsaturated. An example of a saturated fatty acid is stearic acid which has the formula

$$CH_3CH_2CH_2CH_2CH_2CH_2CH_2CH_2CH_2CH_2CH_2CH_2CH_2CH_2CH_2CH_2CH_2\underset{\underset{O}{\|}}{C}{-}OH$$

containing 18 carbon atoms.

Organic compounds with the general formula R–OH are called *alcohols*. An example of an alcohol is glycerol. It has three OH groups and is called a *triol* and has the formula $OHCH_2CH(OH)CH_2OH$. When glycerol reacts with a saturated fatty acid, *fats* are typically formed. Fats are solid substances. When the product of the reaction is not a solid, it is called *oil.*

When glycerol reacts with unsaturated fatty acids, oils are typically formed. The reaction is similar to that with the saturated fatty acids. Oils are, of course, liquids. The fats and oils formed from the reaction of glycerol with the fatty acids are called *triglycerides* or *triacylglycerides*. Fats and oils are actually esters of glycerol and fatty acids. (Esters have the general formula of $R\underset{\underset{O}{\|}}{C}{-}OR'$, where R' is another hydrocarbon group.) Examples of fats are butter, lard, and margarine; and examples of oils are the vegetable oils cottonseed oil, linseed oil, and palm oil. Fats and oils are abundant in meat and meat products.

Glycerol may also derive, along with phosphoric acid and the fatty acids, a third class of compounds called *phosphoglycerides* or *glycerolphosphatides*. In these glycerides, one of the fatty acids is substituted by organic phosphates attaching to the glycerol backbone at one of the ends. The organic phosphates are the phosphates of choline, ethanolamine, and serine. Phosphoglycerides, fats, and oils are collectively called *complex lipids*. Phosphoglycerides are phospholipids.

Certain alcohols look and feel like lipids or fats; thus, they are called *fatty alcohols*. Fatty alcohols are also called *simple lipids*, which are long-chain alcohols, examples of which are *cetyl alcohol* [$CH_3(CH_2)_{14}CH_2OH$] and *myricyl alcohol* [$CH_3(CH_2)_{29}CH_2OH$]. Therefore, the two types of lipids are: complex lipids and simple lipids. Simple lipids do not have the fatty acid "component" of the complex lipids. The simple lipids can react with fatty acids to form esters called *waxes*. In environmental engineering, waxes and complex lipids (fats, oils, and phospholipids) and mineral oils such kerosene, crude oil, and lubricating oil and similar products are collectively called *grease*. The grease content in wastewater is determined by extraction of the waste sample with trichlorotrifluoroethane. Grease is soluble in trichlorotrifluoroethane.

Grease is among the most stable of the organic compounds that, as such, is not easily consumed by microorganisms. Mineral acids can attack it liberating the fatty acids and glycerol. In the presence of alkali, glycerol is liberated, and the fatty acids, also liberated, react with the metal ion of the alkali forming salts called *soap*. The soaps are equally resistant to degradation by microorganisms.

2.1.25 SURFACTANTS

Surfactants are surface-active agents, which means that they have the property of interacting with surfaces. Grease tends to imbed dirt onto surfaces. In order to clean these surfaces, an agent must be used to loosen the dirt. This is where surfactants come in. Surfactant molecules have nonpolar tails and polar heads. The grease molecules, being largely nonpolar, tend to grasp the nonpolar tail of the surfactant molecules, while the polar water molecules tend to grasp the polar head of the surfactant molecules. Because of the movement during cleansing, a "tug of war" occurs between the water molecules on the one side and the grease on the other with the surfactant acting as the rope. This activity causes the grease to loosen from the surfaces thus effecting the cleansing of the dirt. Detergents are examples of surfactants. They are surface-active agents for cleaning.

Surfactants collect on the air–water interface. During aeration in the treatment of wastewater, they adhere to the surface of air bubbles forming stable foams. If they are discharged with the effluent, they form similar bubbles in the receiving stream.

Before 1965, the type of surfactant used in this country was alkyl benzene sulfonate (ABS). ABS is very resistant to biodegradation and rivers were known to be covered with foam. Because of this and because of legislation passed in 1965, ABS was replaced with linear alkyl benzene sulfonate (LAS). LAS is biodegradable.

The laboratory determination of surfactants involves using methylene blue. This is done by measuring the color change in a standard solution of the dye. The surfactant can be measured using methylene blue, so its other name is methylene blue active substance (MBAS).

2.1.26 PRIORITY POLLUTANTS

Before the 1970s, control of discharges to receiving bodies of water were not very strict. During those times, discharge of partially treated wastewaters were allowed; but although facilities could be built to treat discharges by, at least, partial treatment, several communities and industries were discharging untreated wastewaters. This practice resulted in gross pollution of bodies of water that had to be stopped.

The 1970s show pollution control starting in earnest. Industries were classified into industrial categories. These resulted in the identification of priority pollutants and the establishment of categorical standards for a particular industrial category. These standards apply to commercial and industrial discharges that contain the priority pollutants identified by the EPA. Since industries are allowed to discharge into collection systems, these priority pollutants find their way into publicly owned treatment works (POTWs).

The following are examples of priority pollutants: arsenic, selenium, barium, cadmium, chromium, lead, mercury, silver, benzene, ethylbenzene, chlorobenzene, chloroethene, dichloromethane, and tetrachloroethene. The priority pollutants also include the pesticide and fumigant eldrin, the pesticide lindane, the insecticide methoxychlor, the insecticide and fumigant toxaphene, and the herbicide and plant growth regulator silvex. There are a total of 65 priority pollutants.

2.1.27 VOLATILE ORGANIC COMPOUNDS

Generally, volatile organic compounds (VOCs) are organic compounds that have boiling points of $\leq 100°C$ and/or vapor pressures of >1 mmHg at 25°C. VOCs are of concern in wastewater engineering because they can be released in the wastewater collection systems and in treatment plants causing hazards to the workers. For example, vinyl chloride is a suspected carcinogen and, if found in sewers and released in the treatment plants, could endanger the lives of the workers.

2.1.28 TOXIC METAL AND NONMETAL IONS

Because of their toxicity, certain metals and nonmetal ions should be addressed in the design of biological wastewater treatment facilities. Depending on the concentration, copper, lead, silver, chromium, arsenic, and boron are toxic to organisms in varying degrees. Treatment plants have been upset by the introduction of these metals by killing the microorganism thus stopping the treatment. For example, in the digestion of sludge, copper is toxic in concentrations of 100 mg/L, potassium and the ammonium ion in concentrations of 4,000 mg/L, and chromium and nickel in concentrations of 500 mg/L.

Anions such as cyanides, chromates, and fluorides found in industrial wastes are very toxic to microorganisms. The cyanide and chromate wastes are produced by metal-plating industries. These wastes should not be allowed to mix with sanitary sewage but should be removed by pretreatment. The fluoride wastes are normally produced by the electronics industries.

TABLE 2.3
Typical Composition of Untreated Domestic Wastewater

Constituent	Concentration (mg/L)		
	Strong	Medium	Weak
Biochemical oxygen demand, BOD_5 at 20°C	420	200	100
Total organic carbon (TOC)	280	150	80
Chemical oxygen demand (COD)	1000	500	250
Total solids	1250	700	300
Dissolved	800	500	230
Fixed	500	300	140
Volatile	300	200	90
Suspended	450	200	70
Fixed	75	55	20
Volatile	375	145	50
Settleable solids, mL/L	20	10	5
Total nitrogen	90	50	20
Organic	35	15	10
Free ammonia	55	35	10
Nitrites	0	0	0
Nitrates	0	0	0
Total phosphorus as P	18	10	5
Organic	5	3	1
Inorganic	13	7	4
Chlorides	110	45	30
Alkalinity as $CaCO_3$	220	110	50
Grease	160	100	50

2.2 NORMAL CONSTITUENTS OF DOMESTIC WASTEWATER

The normal constituents of domestic wastewater are shown in Table 2.3. The parameters shown in the table are the ones normally used to characterize organic wastes found in municipal wastewaters. As indicated, untreated domestic wastewater is categorized as weak, medium, and strong.

2.3 MICROBIOLOGICAL CHARACTERISTICS

In addition to the physical and chemical characterization of water and wastewater, it is important that the microbiological constituents be also addressed. The constituent microbiological characterizations to be discussed in this section include the following: bacteria, protozoa, and viruses. In addition, qualitative and quantitative tests for the coliform bacteria will also be addressed. The treatment then proceeds to viruses and protozoa. The treatment on protozoa will include discussion on *Giardia lamblia, Cryptosporidium parvum*, and *Entamoeba histolytica*.

The basic units of classifying living things are as follows: kingdom, phylum, class, order, family, genus, and species. Organisms that reproduce only their own kinds constitute a species. The genera are closely related species. Several genera constitute a family. Several related families form an order and several related orders make a class. A number of classes having common characteristics constitutes a phylum. Lastly, related phyla form the kingdom. Ordinarily, only two kingdoms exist: plant and animal; however, some organisms cannot be unequivocally classified as either a plant or an animal. Haeckel in 1866 proposed a third kingdom that he called *protist* to include protozoa, fungi, algae, and bacteria. Protists do not have cell specialization to perform specific cell functions as in the higher forms of life.

At present, it is known that the protists bacteria and cyanobacteria are different from other protists in terms of the presence or absence of a true nucleus in the cell. This leads to the further division of the protists.

The protists fungi, algae, and protozoa contain a membrane-enclosed organelle inside the cell called a nucleus. This nucleus contains the genetic material of the cell, the DNA (deoxyribonucleic acid) which is arranged into a readily recognizable structure called *chromosomes.* On the other hand, the DNA of bacteria is not arranged in an easily recognizable structure as the chromosomes are. The former organisms are called the eucaryotes; the latter, the procaryotes. Therefore, two types of protists exist: the eucaryotic protists and the procaryotic protists. The eucaryotes are said to have a true nucleus, while the procaryotes do not.

A third type of structure that does not belong to the previous classifications is the virus. Although viruses are not strictly organisms, microorganisms, in general, may be classified as *eucaryotic protists, procaryotic protists*, and *viruses.*

2.3.1 BACTERIA

Bacteria are unicell procaryotic protists that are the only living things incapable of directly using particulate food. They obtain nourishment by transporting soluble food directly from the surrounding environment into the cell. They are below protozoa in the trophic level and can serve as food for the protozoa. Unlike protozoa and other higher forms of life that actually engulf or swallow food particles, bacteria can obtain food only by transporting soluble food from the outside through the cell membrane. The nutrients must be in dissolved form; if not, the organism excretes exoenzymes that solubilize the otherwise particulate food. Because of the solubility requirement, bacteria only dwell where there is moisture. Figure 2.5 shows a sketch of the bacterial cell.

Bacteria are widely distributed in nature. They are found in the water we drink, in the food we eat, in the air we breathe; in fact, they are found inside our bodies themselves (*Escherichia coli*). Bacteria are plentiful in the upper layers of the soil, in our rivers and lakes, in the sea, in your fingernails—they are everywhere.

Bacteria are both harmful and beneficial. They degrade the waste-products produced by society. They are used in wastewater treatment plants—thus, they are beneficial. On the other hand, they can also be pathogenic. The bacteria, *Salmonella typhosa*, causes typhoid fever; *Shigella flexneri* causes bacillary dysentery. *Clostridium tetani* excretes toxins producing tetanus. *Clostridium botulinum* excretes the toxin causing botulism. *Corynebacterium diphtheriae* is the agent for diphtheria.

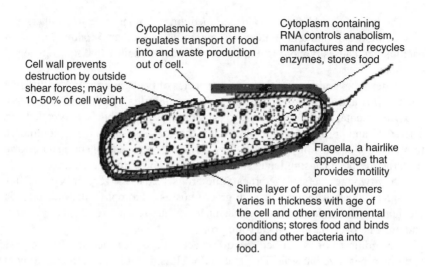

Cytoplasmic membrane regulates transport of food into and waste production out of cell.

Cytoplasm containing RNA controls anabolism, manufactures and recycles enzymes, stores food

Cell wall prevents destruction by outside shear forces; may be 10-50% of cell weight.

Flagella, a hairlike appendage that provides motility

Slime layer of organic polymers varies in thickness with age of the cell and other environmental conditions; stores food and binds food and other bacteria into food.

FIGURE 2.5 Structure of a bacterial cell.

Bacteria come in three shapes: spherical (coccus), rod-shaped (bacillus), and spiral-shaped (vibrio, spirillum, and spirochete). A *vibrio* is a spiral organism shaped like a coma. A *spirillum* is also a spiral organism whose long axis remains rigid when in motion; the *spirochete* is also yet another spiral organism whose long axis bends when in motion.

The cocci range in size from 0.5 μm to 2 μm in diameter. The smallest bacillus is about 0.5 μm in length and 0.2 μm in diameter. In the opposite extreme, bacilli may reach to a diameter of 4 μm and a length of 20 μm. The average diameter and length of pathogenic bacilli are 0.5 μm and 2 μm, respectively. The spirilla are narrow organisms varying in length from 1 μm to 14 μm.

The shape of the bacteria is maintained by a rigid cell wall. The cytoplasm of the cell has a high osmotic pressure that, without a rigid cell wall, can easily rupture by the diffusion of outside water into the cell. The rigidity of the cell wall is due to its chemical makeup of which the chief component is mucocomplex, a polymer of certain amino sugars and short peptide linkages of amino acids. Depending upon the type of bacteria, the bacteria may also contain techoic acid and mucopolysac-charide, or lipoprotein and liposaccharide.

Directly beneath the cell wall is a membrane called the cytoplasmic membrane that surrounds the cytoplasm. In the eucaryotes, an organelle called *mitochondrion*, and, in the photosynthetic eucaryotes, an organelle called *chloroplast* are the sites for the electron-tranport and the respiratory enzyme systems. The bacteria do not have the mitochondrion nor the chloroplast, but the functions of these organelles are embedded within the sites in the cytoplasmic membrane. The *cytoplasm* is the living material which the cell is composed of, minus the nucleus.

Many bacilli and all spirilla are motile when suspended at the proper temperature in a suitable medium. The organ of locomotion is the flagella. The bacterium may have one flagella, few, or many arranged in a tuft. The flagella may protrude at one end or both ends of the organism. True motility is seldom observed in the cocci.

The bacteria may be motile in one medium and nonmotile in another; also, it may be motile at one temperature but nonmotile at another. *Salmonella typhosa*, a bacillus that causes typhoid fever, moves at a rate of about 2,000 times its length in one hour.

Under unfavorable conditions, some species of bacteria, mostly bacillus such as the *Bacillus* and *Clostridium*, assume a form within the cytoplasm that is resistant to environmental influences adverse to bacterial existence. This form is called *spores*. The cocci and spirilla rarely exhibit this behavior—this form being confined only to about 150 species of the bacilli. The most important pathogenic, spore-forming bacteria are those causing tetanus, gas gangrene, botulism, and anthrax.

Only two members of the procaryotic protists exist: the regular bacteria and the blue-green algae. Although the blue-greens possess chlorophyl, their characteristics are those of bacteria possessing no true nucleus. Blue-greens are called *cyanobacteria* and are true bacteria, the procaryotic protists.

According to body surface reaction, bacteria are classified into Gram-positive and Gram-negative bacteria. This method of classification was introduced by Hans Christian Gram, a Danish physician working in Berlin. The reaction of the bacterial surface to staining by crystal violet or gentian violet treated with iodine characterizes this method. For the Gram-positive bacteria, the stain cannot be removed by flooding with alcohol, acetone, or aniline; for the Gram-negative bacteria, on the other hand, the stain is removed by these solvents.

Bacteria reproduce by binary fission. The cell elongates and a constriction is formed; genetic materials are pushed through the constriction and the genetic blueprint is transcribed creating a new daughter cell. The bacterium divides at the constriction. It requires from 15 to 30 minutes for the newborn cells to attain adulthood and restart the cycle of the binary division.

Starting with one cell, let N be the number of cells resulting from the binary fission of one cell at any time. Then

$$N = 2^n \tag{2.26}$$

where n is the number of generations.

Application of the above equation will show that one bacterium will reproduce to $4.7(10^{21})$ bacteria in one day assuming a generation time of 20 minutes. Using the dimension of a bacterium of 2 μm diameter by 12 μm length, the volume of one bacterium is equal to $3.77(10^{-17})$ cubic meter. Thus, assuming that the density of the bacteria is the same as that of water, one bacterial cell will reproduce to $(1000)(3.77)$ $(10^{-17})(5)(10^{21}) = 188,500,000$ kg in 24 hours! This mass is, of course, impossible to be obtained from one bacterium in a day. What this means, though, is, that in a real-world situation, the bacterium does not divide unrestricted, but is influenced by other environmental factors such as crowding and exhaustion of the food supply.

2.3.2 TEST FOR THE COLIFORM GROUP

The coliform group of microorganisms is defined as all aerobic and anaerobic, Gram-negative, nonspore-forming, rod-shaped bacteria that ferment lactose with gas and acid formation within 48 hours at 35°C. The group belongs to the genera *Escherichia*,

Aerobacter, Klebsiella, and *Paracolobacterium* and mostly inhabits the intestinal tract of humans, although they could also be found in the outside environment. Although also found outside the intestinal tract, this group of organisms is used as an indicator for the presence of pathogens in waters. Surface waters can be used for drinking purposes after treatment, so effluents from sewage treatment plant discharges are limited for the acceptable concentrations of these organisms, thus the necessity for testing.

The test for the coliform group may be qualitative or quantitative. As the name implies, the qualitative test simply attempts to identify the presence or absence of the coliform group. The quantitative test, on the other hand, quantifies the presence of the organisms. In all the tests, whether qualitative or quantitative, all media, utensils, and the like must be sterilized. This is to preclude unwanted organisms that can affect the results of the analysis.

Qualitative tests. Three steps are used in the qualitative test: *presumptive, confirmed,* and *completed* tests. These tests rely on the property of the coliform group to produce a gas during fermentation of *lactose*. An inverted small vial is put at the bottom of fermentation tubes. The vial, being inverted, traps the gas inside it forming bubbles indicating a positive gas production.

A serial dilution is prepared such as 10-, 1-, 0.1-, and 0.01-mL portions of the sample. Each of these portions is inoculated into a set of five replicate tubes. For example, for the 10-mL portion, each of five fermentation tubes is inoculated with 10-mL portions of the sample. For the 1-mL portion, each of five fermentation tubes is also inoculated with 1-mL portions of the sample and so on with the rest of the serial dilutions. In this example, because four serial dilutions (10, 1, 0.1, and 0.01) are used, four sets of five tubes each are used, making a total of 20 fermentation tubes.

In the presumptive test, a portion of the sample is inoculated into a number of test tubes containing lactose broths (or lauryl tryptose broth, also containing lactose) and other ingredients necessary for growth. The tubes are then incubated at $35 \pm 0.5°C$. Liberation of gases within 24 to 48 ± 3 hours indicates a positive presumptive test. Organisms other than coliforms can also liberate gases at this fermentation temperature, so a positive presumptive test simply signifies the possible presence of the coliform group.

A further test is done to confirm the presumptive test result; this is called the confirmed test. The theory behind this test uses the property of the coliforms to ferment lactose even in the presence of a green dye. Noncoliform organisms cannot ferment lactose in the presence of this dye. The growth medium is called brilliant green lactose bile broth. A wire loop from the positive presumptive test is inoculated into fermentation tubes containing the broth. As in the presumptive test, the tubes are incubated at $35 \pm 0.5°C$. Liberation of gases within 24 to 48 ± 3 hours indicates a positive confirmed test.

The coliform group of organisms does not come from only the intestinal tract of man but also from the outside surroundings. There are situations, as in pollution surveys of a water supply with a raw water source, that the fecal variety needs to be differentiated from the nonfecal ones. For this and similar situations, the procedure is modified by raising the incubation temperature to $44.5 \pm 0.2°C$. The broth used is the EC medium which still contains lactose. The high incubation temperature precludes the nonfecal forms from metabolizing the EC medium. Positive test results are, again, heralded by the evolution of fermentation gases not within 48 ± 3 hours but within 24 ± 2 hours.

Normally, qualitative tests are conducted only up to the confirmed test. There may be situations, however, where one would want to actually see the organism in order to identify it. Both the presumptive and the confirmed tests are based on circumstantial evidence only—not on actually seeing the organisms. If desired, the completed test is performed.

In the completed test, a wire loop from the positive confirmed test fermentation tube is streaked across either an Endo or MacConkey agar in prepared petri dishes or plates. The dishes are then incubated at $35 \pm 0.5°C$ for 24 ± 2 hours. At the end of this period, the dishes are examined for growth. Typical colonies are growths that exhibit a green sheen; atypical colonies are light-colored and without the green sheen. The typical colonies confirm the presence of the coliform group while the atypical colonies neither confirm nor deny their presence. (From this, Endo agar method can also be used as an alternative to the brilliant green bile broth method.)

The typical or atypical colony from the Endo plate is inoculated into an agar slant in a test tube and to a fermentation tube containing lactose. If no gas evolves within 24 to 48 ± 3 hours after incubating at $35 \pm 0.5°C$, the completed test is negative. If gases had evolved from the tubes, the growth from the agar slant is smeared into a microscopic slide and the Gram-stain technique performed. If spores or Gram-positive rods are present, the completed test is negative. If Gram-negative rods are present, the completed test is positive.

Quantitative tests. In some situations, actual enumeration of the number of organisms present may be necessary. This is the case, for example, for limitations imposed on a discharge permit. Generally, two methods are used to enumerate coliforms: the membrane-filter technique and the multiple-tube technique.

The membrane-filter technique consists of filtering under vacuum a volume of water through a membrane filter having pore openings of 0.45 μm, placing the filter in a Petri dish and incubating, and counting the colonies that developed after incubation. The 0.45-μm opening retains the bacteria on the filter. The volume of sample to be filtered depends on the anticipated bacterial density. Sample volumes that yield plate counts of 20 to 80 colonies on the Petri dish is considered most valid. The standard volume filtered in potable water analysis is 100 mL. If the anticipated concentration is high requiring a smaller volume of sample (such as 20 mL), the sample must be diluted to disperse the bacteria uniformly. The dispersion will ensure a good spread of the colonies on the plate for easier counting.

The M-Endo medium is used for the coliform group and the M-FC medium for the fecal coliforms. The culture medium is prepared by putting an absorbent pad on a petri dish and pipetting enough of the M-Endo medium or M-FC into the pad. After filtration, the membrane is removed by forceps and placed directly on the pad. The culture is then incubated at $35 \pm 0.5°C$ in the case of the coliform group and at $44.5 \pm 0.2°C$ in the case of the fecal coliforms. At the end of the incubation period of 24 ± 2 hours, the number of colonies that developed are counted and reported as number of organisms per 100 mL.

The other method of enumerating coliforms is through the use of the multiple-tube technique. This method is statistical in nature and the result is reported as the most probable number (MPN) of organisms. Hence, the other name of this method is the MPN technique. This technique is an extension of the qualitative techniques of presumptive, confirmed, and completed tests. In other words, MPN results can be a presumptive, confirmed, and completed MPNs. The number of tubes liberating gases is counted from each of the set of five tubes. This information is then used to compute the most probable number of organisms in the sample per 100 mL.

2.3.3 The Poisson Distribution

To use the MPN method, a probability distribution called the *Poisson distribution* is used. From any good book on probability, the probability (also called *probability function*) that a random variable X following the Poisson distribution will have a value y is

$$Prob(X = y) = Prob(y) = e^{-\lambda}\frac{\lambda^y}{y!} \qquad (2.27)$$

where λ is the mathematical expectation. *Mathematical expectation* is the average value obtained if the random variable X is measured exceedingly many times. *Random variable* is a function dependent on chance and whose values are real numbers. A certain probability is attached to the function; this contrasts with an *ordinary variable* in which the probability of its occurrence is not a concern, although it will always have a corresponding probability.

In order to gain an insight into the Poisson distribution, let us derive Equation (2.27). Assume conducting an experiment of sampling microorganisms from a population to obtain a sample consisting of n microorganisms. (A *population* is the total set of all possible measurements in a particular problem. A *sample* is a subset of the population that contains measurements obtained by an experiment.) Let y be the corresponding number of bacteria in the sample. The proportion of bacteria is therefore y/n. This is not, however, the true proportion of bacteria in the entire population. The true proportion is given by the limit $limit_{n \to N} y/n$. In equation form,

$$\underset{n \to N}{limit} \frac{y}{n} = \frac{\lambda}{N} \qquad (2.28)$$

where in the limit y is replaced by the expected number of bacteria, λ, and N is the total number of microorganisms in the population. (As will be learned later, λ eventually determines the most probable number of bacteria.) Because λ/N is the true proportion, it represents the probability that a bacterium can be picked from the sample or from the population. A single picking or observation of a bacterium in the sample is also called a *trial*. In this context, note that the sample can be thought of as comprising of several observations (picking) or trial.

If the probability of picking a bacterium is λ/N, then the probability of *not* picking a bacterium is $1 - \lambda/N$. It was assumed that there are y bacteria in a given sample. Because these y bacteria are occurring at the same time, their simultaneous appearances are an intersection. From the intersection probability of events, the probability of obtaining these y bacteria is $(\lambda/N)(\lambda/N)(\lambda/N)...(\lambda/N) = (\lambda/N)^y$. The occurrence of these bacteria, however, happens at the same time as the occurrence of the "not bacteria." If the number of bacteria in the sample is y, the number of not bacteria is $n - y$ which, from the intersection probability of events, has a probability of occurrence of $(1 - \lambda/N)(1 - \lambda/N)...(1 - \lambda/N) = (1 - \lambda/N)^{n-y}$. Now, the simultaneous appearances of the bacteria and the not bacteria also has intersection probability, and this is given by the product of the probabilities as $(\lambda/N)^y(1 - \lambda/N)^{n-y}$.

There are several ways to obtain the y bacteria in the sample. The number of ways is given by the number of combinations of the n microorganisms taken y at a time, $\binom{n}{y}$. We will not derive the formula for the number of combinations here since this concept is discussed in college algebra. The formula is

$$\binom{n}{y} = \frac{n!}{y!(n-y)!} \tag{2.29}$$

The probability of obtaining y bacteria in any sample is then

$$Prob(y) = \frac{n!}{y!(n-y)!}\left(\frac{\lambda}{N}\right)^y\left(1 - \frac{\lambda}{N}\right)^{n-y} \quad y = 0, 1, 2,..., n \tag{2.30}$$

The previous equation is called the *Bernoulli distribution*, from which we will derive the Poisson distribution.

$n!$ may be expanded as $n(n-1)(n-2)(n-3)...(n-y+1)(n-y)!$. Using this new expression, Equation (2.30) may be rewritten as

$$Prob(y) = \frac{\lambda^y}{y!}\left[\frac{n(n-1)(n-2)(n-3)...(n-y+1)(n-y)!}{(n-y)!N^y}\right]\left(1 - \frac{\lambda}{N}\right)^{n-y} \tag{2.31}$$

where $(n-y)!$ will cancel out from both the numerator and the denominator of the second factor on the right. n in the numerator may be factored out to produce, after canceling out $(n-y)!$.

$$Prob(y) = \frac{\lambda^y}{y!}\left[\frac{n\left[n\left(1-\frac{1}{n}\right)\right]\left[n\left(1-\frac{2}{n}\right)\right]\left[n\left(1-\frac{3}{n}\right)\right]...\left[n\left(1-\frac{y+1}{n}\right)\right]}{N^y}\right]\left(1 - \frac{\lambda}{N}\right)^{n-y} \tag{2.32}$$

When $n!$ is expanded as $n(n-1)(n-2)(n-3)...(n-y+1)(n-y)!$, the number of factors in this expansion is $y + 1$. This is shown as follows: $(n-1)(n-2)(n-3)...(n-y+1)(n-y)!$ has y factors. Including the n, the total number of factors is therefore $y + 1$. Because $(n-y)!$ was canceled out from both the numerator and the denominator to produce Equation (2.32), the numerator in the second factor of this equation contains only y factors. Therefore, factoring out the n's in this numerator will produce n^y. Further manipulations then produces

$$Prob(y) = \frac{\lambda^y}{y!}\left[\frac{n^y\left(1-\frac{1}{n}\right)\left(1-\frac{2}{n}\right)\left(1-\frac{3}{n}\right)\cdots\left(1-\frac{y+1}{n}\right)}{N^y}\right]\left(1-\frac{\lambda}{N}\right)^n\left(1-\frac{\lambda}{N}\right)^{-y} \quad (2.33)$$

In Equation (2.33), as n approaches N, n^y in the numerator will cancel out with N^y in the denominator. Also, after the cancellation, all the factors in the numerator of the second factor will approach 1. Furthermore, as N approaches infinity, $(1 - \lambda/N)^{-y}$ approaches 1. Lastly, as n approaches infinity, we have from calculus $limit_{n\to\infty}$ $(1 - \lambda/N)^n = e^{-\lambda}$. Applying all these to Equation (2.33) produces

$$Prob(y) = \frac{\lambda^y}{y!}e^{-\lambda} \quad y = 0, 1, 2, 3,...\infty \quad (2.34)$$

This distribution is called the *Poisson distribution*. This will be used to derive the method of estimating the most probable number, MPN.

Note the difference between the Binomial distribution and the Poisson distribution. In the former case, the sample is of fixed size, n, and the probability of obtaining y number of bacteria from this sample size is computed. In the Poisson distribution, since n is made to approach infinity, the sample size is the population, itself. In other words, although the sample is of fixed size, the probability is calculated as if the size is that of the population. This is fortunate, since it is impossible to compute the probability using the Binomial distribution; the sample size n is practically not known, unless the sample is painstakingly analyzed to count the number of organisms represented by the value of n, thus the use of the Poisson distribution.

2.3.4 ESTIMATION OF COLIFORM DENSITIES BY THE MPN METHOD

As the phrase "most probable number" implies, whatever value is obtained is just "most probable." It also follows that the techniques of probability will be used to obtain this number.

Consider the 10-mL portions (dilution) inoculated into the five replicate tubes. If α is the expected number of bacteria per mL, the expected number of bacteria in the 10 mL (in one tube) is $\lambda = 10\alpha$. Thus, from the Poisson probability, for any one tube, the probability that there will be no bacteria ($X = y = 0$) is

$$Prob(X = y = 0) = \frac{\lambda^y}{y!}e^{-\lambda} = \frac{(10\alpha)^0}{0!}e^{-10\alpha} = e^{-10\alpha} \quad (2.35)$$

Let q be the number of negative tubes and p be number of positive tubes in the five replicate tubes. (Note that the symbols q and p hold for any number of replicate tubes, not only 5 tubes.) These q negative results occur at the same time, so they are intersection events. Thus, from the probability of intersection events, Equation (2.35) for the q negative results becomes

$$Prob(X = y = 0) = (e^{-10\alpha})^q \tag{2.36}$$

The p positive tubes also are intersection events. If the probability of a negative result is $e^{-10\alpha}$, the probability of a positive result is $1 - e^{-10\alpha}$. By analogy, the intersection probability of the p events is

$$Prob(X = y \neq 0) = [1 - e^{-10\alpha}]^p \tag{2.37}$$

Of course, the q negative results and the p positive results are all occurring at the same time; they are also intersection events to each other. Combining the q and the p results constitutes the dilution experiment. Because the intersection of the q and the p events is the dilution, call the corresponding probability as $Prob(D)$, where D stands for dilution. Thus, from the intersection probabilities of the q and p events, the probability of the dilution is

$$Prob(D) = (1 - e^{-10\alpha})^p (e^{-10\alpha})^q \tag{2.38}$$

Let the number of dilutions be equal to j number, where each dilution has a sample size of m mL (instead of the 10 mL). Each of these j dilutions will also be happening at the same time, and are, therefore, intersection events among themselves. Thus, in this situation, Equation (2.38) becomes

$$Prob(D) = \prod_{i=1}^{i=j} [(1 - e^{-m_i \alpha})^{P_i} (e^{-m_i \alpha})^{q_i}] \tag{2.39}$$

where Π is the symbol for the product factors. The product factors are used, because the j number of dilutions are intersection events.

Now, we finally arrive at what specifically is the quantitative meaning of MPN; this is gleaned from Equation (2.39). In this equation, for a given number of serial dilutions and respective values of q and p, the value of α that will make $Prob(D)$ a maximum is the most probable number of organisms. Thus, to get the MPN, the above equation may be differentiated with respect to α and the result equated to zero to solve for α. This scheme is, however, a formidable task. The easier way would be to program the equation in a computer. For a given number of serial dilutions and corresponding values of q and p, several values of α are inputted into the program. This will generate corresponding values of $Prob(D)$, along with the inputted values of α. The largest of these probability values then gives the value of α that represents the MPN. A table may then be prepared showing serial dilutions and the corresponding MPN.

The American Public Health Association has generated a statistical table for the MPN (APHA, AWWA, WEF, 1992). A modified version is shown in Table 2.4. The first three columns are the combination of positives in only three serial dilutions. Thus,

TABLE 2.4
MPN and 95% Confidence Limits for Coliforms Counts for Various Combinations of Positive and Negative Results when Five 10-mL, Five 1-mL, and Five 0.1-mL Portions are Used

Number of Tubes Giving Positive Reaction Out of			MPN per 100 mL	95% Confidence Level	
5 of 10-mL	5 of 1-mL	5 of 0.1 mL		Lower	Upper
0	0	0	<2	—	—
1	1	1	6	2	18
2	1	1	9	3	24
3	1	1	14	6	35
3	2	1	17	7	40
4	1	1	21	9	55
4	2	1	26	12	65
4	3	1	33	15	77
5	1	1	50	20	150
5	2	1	70	30	210
5	2	2	90	40	250
5	3	1	110	40	300
5	3	2	140	60	360
5	3	3	170	80	410
5	4	1	170	70	480
5	4	2	220	100	580
5	4	3	280	120	690
5	4	4	350	160	820
5	5	1	300	100	1300
5	5	2	500	200	2000
5	5	3	900	300	1900
5	5	4	1600	600	5300
5	5	5	≥1600	—	—

From APHA, AWWA, and WEF (1992). *Standard Methods for the Examination of Water and Wastewater.* 15th ed. American Public Health Association, Washington.

if the analysis had been done on more than three dilutions, only the three highest dilutions can be adopted for use in the table. To have an idea of what is meant by highest dilution, the 0.01 is said to be the highest dilution in a serial dilution of 1, 0.1, and 0.01. Also, the result of the experiment must not show a zero reading.

The reason why a zero reading is unacceptable can be gleaned from the following justification. Serial dilutions are prepared relatively far apart in concentrations. For example, consider the dilution used in Table 2.4, which is 10 mL, 1 mL, and 0.1 mL. Now, suppose the reading is 2, 1, 0 which means 2 positive readings in the 10-mL dilution, 1 positive reading in the 1-mL dilution, and 0 positive reading in the 0.1 mL dilution. Because 0.1 mL is far away from 1 mL, it is possible that the 0 reading would correspond not only to 0.1 mL but also to a dilution between 0.1 mL and 1 mL, such as a dilution of 0.6 mL. Thus, the reading of 0 is uncertain and therefore should not be used. For readings to be valid, they should not contain any zeroes.

Note that if the highest dilution has a reading, then the next higher dilution should have a reading at least equal to the highest dilution. The most accurate would be that it should have a reading greater than the highest dilution. Thus, before the table is used, it should be ascertained that this fact is reflected in the results of the experiment. If this is not the case, the table should not be used and the results of the experiment discarded. Note, however, that if the result of the experiment is erroneous, Table 2.4 cannot be used in the first place, since there will be no entry for this erroneous result in the table.

2.3.5 INTERPOLATION OR EXTRAPOLATION OF THE MPN TABLE

If the reading of a valid experiment cannot be found in the table, the corresponding MPN may be interpolated or extrapolated. For example consider the reading 2, 2, 1 for a certain serial dilution. Referring to Table 2.4, this reading cannot be found; it is, however, between readings 2, 1, 1 and 3, 1, 1. Thus, its corresponding MPN can be interpolated.

The MPN is proportional to the reading. Using this idea to calculate the MPN, assume that there are three dilutions, as are used in Table 2.4. Let MPN_x be the MPN to be interpolated and the corresponding readings be R_{x1}, R_{x2}, R_{x3}. The weighted mean reading R_{xave} corresponding to MPN_x is then

$$R_{xave} = \frac{R_{x1}(D_1) + R_{x2}(D_2) + R_{x3}(D_3)}{D_1 + D_2 + D_3} \qquad (2.40)$$

where D_1, D_2, and D_3 are the three dilutions used in the experiment.

Let MPN_1 and MPN_2 be the MPN of the first and second known MPNs as found in the table, respectively. The corresponding readings of MPN_1 are R_{11}, R_{12}, and R_{13} and those of MPN_2 are R_{21}, R_{22}, and R_{23}. The corresponding mean readings are, respectively, R_{1ave} and R_{2ave} as shown in the equations below:

$$R_{1ave} = \frac{R_{11}(D_1) + R_{12}(D_2) + R_{13}(D_3)}{D_1 + D_2 + D_3} \qquad (2.41)$$

$$R_{2ave} = \frac{R_{21}(D_1) + R_{22}(D_2) + R_{23}(D_3)}{D_1 + D_2 + D_3} \qquad (2.42)$$

From Eqs. (2.40), (2.41), and (2.42), we form the following table:

MPN_1	R_{1ave}
MPN_x	R_{xave}
MPN_2	R_{2ave}

Interpolating from this table, we obtain

$$MPN_x = MPN_1 + (MPN_1 - MPN_2)\left(\frac{R_{xave} - R_{1ave}}{R_{1ave} - R_{2ave}}\right) \qquad (2.43)$$

The previous table was set up for **interpolation**. The table may, however, be setup for extrapolation. In such a case, the following may be obtained:

MPN_x	R_{xave}
MPN_1	R_{1ave}
MPN_2	R_{2ave}

This table may now be **extrapolated**, which will again produce Equation (2.43).

Now, as shown by the previous derivations, whether it is interpolation or extrapolation, the formulas are the same. The difference is only in the placement of the interpolants.

Example 2.5   Samples of river water are analyzed for total and fecal coliforms using the MPN technique. Some of the results are shown below:

Number of Positive Tubes Out of Five			
Serial Dilution	Lactose	BGB	EC
10	5	5	5
1.0	3	3	3
0.1	2	2	2
0.01	1	0	0
0.001	0	0	0

What are the presumptive and confirmed MPNs and the fecal coliform MPN? BGB stands for "brilliant green bile broth" and EC stands for the medium for fecal coliform.

Solution: For the presumptive total coliform MPN, the three highest dilutions are 1.0, 0.1, and 0.01 corresponding to the positive readings of 3, 2, and 1. From Table 2.4, for a serial dilution of 10, 1.0, and 0.1, and positive readings of 3, 2, and 1, the MPN = 17. The serial dilution of 1.0, 0.1, and 0.01 diluted the sample 10 times more than the serial dilution of 10, 1.0, and 0.1 in the table. Because the experiment is diluting the sample 10 times more, the MPN from the reading in the table should be multiplied by 10. Thus,

$$MPN = 17(10) = 170 \text{ organisms/100 mL} \quad \textbf{Ans}$$

$$\text{At 95\% confidence range: } 70 \leq MPN \leq 400 \quad \textbf{Ans}$$

For the confirmed test, the three highest dilutions are 10, 1.0 and 0.1 corresponding to the positive readings of 5, 3, and 2. From the table,

$$MPN = 140 \text{ organisms/100 mL} \quad \textbf{Ans}$$

$$\text{At 95\% confidence range: } 60 \leq MPN \leq 360 \quad \textbf{Ans}$$

For the fecal test, the three highest dilutions are 10, 1.0, and 0.1 corresponding to the positive readings of 5, 2, and 2. From the table,

$$MPN = 90 \text{ fecal coliforms/100 mL} \quad \textbf{Ans}$$

$$\text{At 95\% confidence range: } 40 \le MPN \le 250 \quad \textbf{Ans}$$

Example 2.6   A sample of water is analyzed for the coliform group using three sample portions: 10 mL, 60 mL, and 600 mL. Each of these portions is filtered through five filter membranes using the membrane-filter technique. The results of the colony counts are as follows: 10-mL portions: 6, 7, 5, 8, 6; 60-mL portions: 30, 32, 33, 31, 25; and 500-mL portions: 350, 340, 360, 370, 340. What is the number of coliforms per 100 mL of the sample?

Solution: Using the criterion of 20 to 80 as the most valid count in a membrane-filter technique, the results of the 10- and 500-mL portions can be excluded in the calculation.

$$\text{Average count for the 60-mL portions } = \frac{30 + 32 + 33 + 31 + 25}{5}$$

$$= 30.2 \text{ coliforms per 60 mL}$$

$$\text{Therefore, MPN } = \frac{30.2}{60}(100) = 50.3 \text{ per 100 mL} \quad \textbf{Ans}$$

Example 2.7   Find the MPN for a reading of 2, 2, 1. The experiment is performed on a 10-mL, 1-mL, and 0.1-mL serial dilution.

Solution: The reading cannot be found in Table 2.4. It is, however, between the readings 2, 1, 1 and 3, 1, 1. Therefore,

$$MPN_x = MPN_1 + (MPN_1 - MPN_2)\left(\frac{R_{xave} - R_{1ave}}{R_{1ave} - R_{2ave}}\right)$$

For reading 2, 1, 1, MPN = 9 = MPN_1; for reading 3, 1, 1, MPN = 14 = MPN_2

$$R_{xave} = R_{xave} = \frac{R_{x1}(D_1) + R_{x2}(D_2) + R_{x3}(D_3)}{D_1 + D_2 + D_3} = \frac{2(10) + 2(1) + 1(0.1)}{10 + 1 + 0.1} = 1.99$$

$$R_{1ave} = \frac{R_{11}(D_1) + R_{12}(D_2) + R_{13}(D_3)}{D_1 + D_2 + D_3} = \frac{2(10) + 1(1) + 1(0.1)}{10 + 1 + 0.1} = 1.90$$

$$R_{2ave} = \frac{R_{21}(D_1) + R_{22}(D_2) + R_{23}(D_3)}{D_1 + D_2 + D_3} = \frac{3(10) + 1(1) + 1(0.1)}{10 + 1 + 0.1} = 2.80$$

$$MPN_x = 9 + (9 - 14)\left(\frac{1.99 - 1.90}{1.90 - 2.80}\right) = 9.50, \text{ say } 10 \quad \textbf{Ans}$$

2.3.6 VIRUSES

A virus is a submicroscopic agent of infectious disease that requires a living cell for its multiplication. The two essential components of a virus are protein and nucleic acid. Whereas normal cells contain both RNA (ribonucleic acid) and DNA (deoxyribonucleic acid), a given virus contains only one, not both. A virus cannot multiply on its own as a normal cell does. It has no metabolic enzymes, uses no nutrients, and produces no energy. It is just a particle of protein and nucleic acid. A viral particle is tightly packed inside a protein coat that protects it. This unit is called a *virion*.

A study was conducted on the fate of viruses from the discharge of two sewage treatment plants into a Houston, TX, ship channel (Gaudy and Gaudy, 1980). Significant virus concentrations were detected 8 miles downstream from the point of discharge, and poliovirus was recovered in oysters 21 miles into Galveston Bay. This study shows that viruses can survive in the outside environment for a considerable length of time. In fact, it is known that some viruses may survive for periods of 40 days in water and wastewater at 20°C and for periods of 6 days in streams at normal conditions. For this reason, treated sewage which can contain a multitude of viruses should never be used to irrigate crops for food consumption. Use of treated sewage for spray irrigation can be a hazard from viruses due to droplets carried by the wind.

Viruses excreted by humans may become a major public health hazard. For example, studies have shown that some 10,000 to 100,000 infectious particles of viruses are emitted per gram of feces from people infected with hepatitis. Viruses producing diseases in humans are excreted in feces, so it is the responsibility of the environmental engineer to ensure that viruses in wastewater treatment plants are effectively disinfected. This is usually done by chlorination followed by proper disposal of the effluent.

Requiring a host cell, the virus has a very unique way of propagating itself. Upon entrance into a cell, it disappears. The nucleic acid mingles with the contents of the cytoplasm and loses its identity; nonetheless, its presence is evident from its effects on the host. In one mode of propagation, using the genetic code carried by the nucleic acid, it overpowers the mechanism of the cell to reproduce and diverts the cell materials and energy for viral replication. The virus takes over the machinery, commanding the cell to manufacture hundreds of viral prototypes for further infection of other host cells. The viruses, multiplying in extreme numbers, cause the host cell to swell and burst into pieces. Figures 2.6 and 2.7 show the shapes of the various virus particles and a virus infecting a cell, respectively.

In the second type, the cell tolerates the presence of the virus. The host cell continues to grow and multiply carrying the virus in its interior in a latent or noninfective form. The virus also multiplies at a proportionate rate.

2.3.7 PROTOZOA

Protozoa are single-cell protists one step above the bacteria in the trophic level. They possess the *ectoplasm*, which is a homogeneous portion of the cytoplasm. The ectoplasm helps form the various organs of locomotion, contraction, and prehension, such as cilia, flagella, pseudopods, and suction tubes.

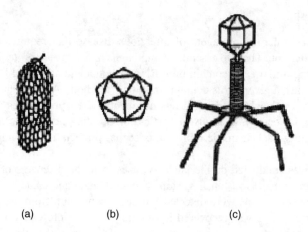

FIGURE 2.6 Structural shapes of viruses: (a) helical, (b) polyhedral, and (c) combination T-even.

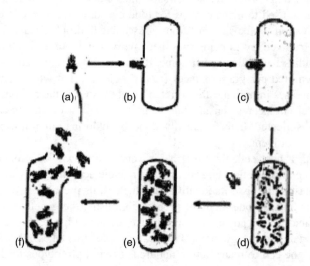

FIGURE 2.7 Life-reproductive cycle of a virus as illustrated by a T-even bacteriophage infecting a bacterial cell: (a) dormancy, (b) adsorption, (c) penetration, (d) replication of new proteins and nucleic acids, (e) maturation, and (f) release and bursting of hold cell.

Protozoa move through the use of the pseudopod or the use of the cilia and flagella. In pseudopodic movement, the ectoplasm flows inside the cell toward the direction of motion. As the mass of ectoplasm moves, the rear of the body is pulled forward. The *flagellum* is a whiplike prolongation of the protoplasm. It propels the organism forward by a lashing action. The *cilium* has a function similar to the flagellum except that it is more numerous, delicate, and shorter. Cilia are attached to the whole outer surface of the body forming a rimlike appearance.

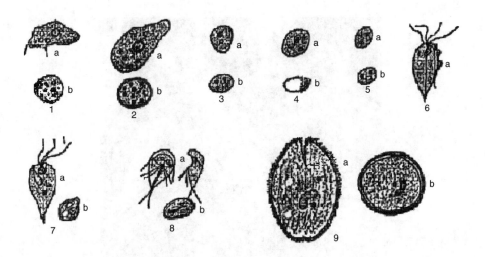

FIGURE 2.8 Vegetative (a) and cyst (b) forms of some representative protozoa: (1) *Entamoeba histolytica*; (2) *Entamoeba coli*; (3) *Endolimax nana*; (4) *Iodomoeba butschlii*; (5) *Dientamoeba fragilis*; (6) *Trichomonas hominis*; (7) *Chilomastix mesnili*; (8) *Giardia lamblia*; (9) *Balantidium coli.*

One class of protozoa called *Telosporidea* does not have any organ of locomotion. The organisms belonging to this class live within cells, tissues, cavities, and fluids of the body. The *Plasmodium*, causing malaria, belongs to this class.

When adverse conditions prevail, the protozoa become inactive, transforming into a rounded form, and eventually developing a protective cell wall. Under the safety inside the cell wall, the organism may live for a long time and resist any destructive insult from the environment. This transformation is called a *cyst*.

Protozoa is capable of sexual and asexual reproduction. Some may have sexual reproduction in one host and asexual in another. Cells capable of sexual reproduction are called *gametes*. The cell formed by the union of two gametes is called *zygote*. Asexual reproduction is achieved by lengthwise or crosswise division of the cell as in the amebas and flagellates. Figure 2.8 shows sketches of some protozoa illustrating the vegetative and cyst forms.

Giardia lamblia. *Giardia lamblia* is a flagellate parasitic protozoan. This parasite is largely confined to the lining of the intestine. It can colonize the lining and feed and grow there. It is shed in feces in the form of cysts. The cyst, however, cannot multiply outside the host.

The vegetative form of *G. lamblia* is called a trophozoite. It is this form that lives and colonizes the lining of the intestine; in particular, the upper small intestine, the parasites attach to the intestinal wall by means of a disc-shaped suction pad on their ventral surface. It is here that the trophozoites actively feed and reproduce. At some time during the trophozoite's life, it releases its hold on the bowel wall and floats in the fecal stream through the intestine. As it makes its journey, it undergoes a morphologic transformation into an egg-like structure, the cyst, mentioned above. Figure 2.9 shows a picture of the trophozoites using a scanning

FIGURE 2.9 Scanning electron microscopy reveals the protozoan parasite *Giardia lamblia* in the small intestine of an animal host.

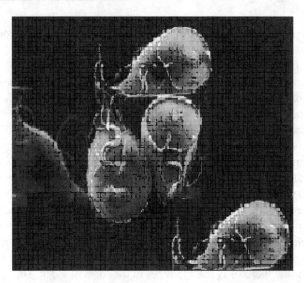

FIGURE 2.10 *Giardia lamblia* trophozoites as they appear with the scanning electron microscope.

electron microscopy that reveals the parasite in the small intestine of an animal host. Figure 2.10 shows another picture of the parasite in the trophozoite stage.

Giardia cysts can be found in water, raw vegetables, and feces of contaminated individuals. Where adequate sanitation cannot be maintained, the cysts can also be found in institutions and day-care centers. The beaver has gained attention as a potential source of *Giardia* contamination of lakes, reservoirs, and streams;

however, human fecal wastes are equally important. Horses have also been implicated as sources of these pathogens. In addition, muskrats have been found to have high infection rates (30 to 40%) (Frost and Liechty, 1980). Studies have shown that they can be infected from cysts obtained from humans and beavers. Occasionally, coyotes, deer, elk, cattle, dogs, and cats have also been found infected. Because all these animals, including humans, can come in contact with surface waters, community water supplies that draw water from impounding reservoirs are a potential problem for diseases caused by the cyst.

The pathogen can cause a disease called giardiasis. Its symptoms include diarrhea, abdominal cramps, fatigue, weight loss, gas, anorexia, and nausea and may persist for 2 to 3 months if untreated. The diarrhea experienced may be mild or severe. Occasionally, some will have chronic diarrhea over several weeks or months, with significant weight loss. Cases may occur sporadically or in clusters or outbreaks. Upon exposure to the cyst, symptoms may appear within 5 to 25 days but usually within 10 days. As a method of treatment, doctors often prescribe antibiotics such as atabrine, metronidazole, or furizolidone to treat giardiasis. Some individuals, however, may recover on their own without medication; fever is rarely present.

The disease is contracted through the following means: The parasite is passed in the feces of an infected person or the feces of animals mentioned previously. The waste may then contaminate water or food. Or, person-to-person transmission may also occur as in day care centers or other settings where handwashing practices are poor.

Although the ingestion of only one *Giardia* cyst is theoretically possible to cause the disease, the minimum number of cysts that can show the symptoms is found to be ten (Rendtorff, 1954). Every 12 hours, trophozoites divide by binary fission. This means that if a person swallowed only a single cyst, reproduction at this rate would result in more than 1 million parasites 10 days later, and 1 billion parasites 15 days later, using the binary-division equation explained earlier for the case of bacteria. From this arithmetic, it would seem that one cyst is enough to cause the disease. The exact mechanism by which *Giardia* causes illness is not yet well understood, however. From this simple calculation, it can be gleaned that contracting the disease is not necessarily related to the number of organisms present. Nearly all of the symptoms, however, are related to dysfunction of the gastrointestinal tract. Also, the parasite rarely invades other parts of the body, such as the gall bladder or pancreatic ducts. In addition, intestinal infection does not result in permanent damage.

The diagnosis of giardiasis is posed by observation of cysts and/or trophozoites in feces or duodenal aspirates. Because of the irregular shedding of parasites, multiple examinations are often required to diagnose the infection.

Measures to prevent transmission include proper disposal of feces, wastewater treatment, and a filtration step before chlorination in water treatment plants that draw from surface water sources. Cooking kills the cysts in contaminated foods and boiling will make water safe for use. For backpackers who walk through the wilderness, iodine has been shown to be a better disinfectant for giardia cyst than chlorine. It should be emphasized that of all the methods of preventing contamination from the cyst, a properly designed and operated water filtration plant is the best line of defense in drinking water supplies.

Analyzing water for *Giardia* involves collecting the cyst through filtration, purifying the collected cyst, detecting, and identifying. The cyst may be identified using a microscope with or without staining. Staining may be accomplished using a method called fluorogenic staining. In this method, live cysts are distinguished by two fluorescent dyes. One dye is fluorescein diacetate (FDA). When absorbed by cysts, FDA produces a fluorescent green only in live cysts. The second dye, either propidium iodide (PI) or ethidium bromide (EB), is excluded efficiently by live cysts but absorbed by dead cysts, resulting in red fluorescence.

Cryptosporidium parvum. *Cryptosporidium parvum* is a single-celled protozoan parasite that infects a wide variety of animals, including humans. It has a life cycle transmission where it can exist as an infective oocyst, usually 5–8 μm in diameter. When ingested, oocysts pass through the stomach into the small intestine where sporozoites are released. Sporozoites invade and penetrate the intestinal epithelial cell membrane and develop into merozoites, impairing the ability of the intestine to absorb water and nutrients. Microgametes and macrogametes are then formed. The former fertilizes the latter and zygotes result. Most of the zygotes pass through the host as very infective, environmentally hardy, thick walled oocysts. These oocysts, if finding a suitable host, release sporozoites, again, completing the cycle. Figures 2.11 and 2.12 show pictures of the oocysts, where the latter shows the parasite using a staining technique.

The oocysts can be found in water, raw vegetables, and feces of contaminated individuals. Where adequate sanitation cannot be maintained, the cysts can also be found in institutions and day care centers. Indeed, the occurrence of the organism is widespread and can be found in both domestic and farm animals, such as cattle, dogs and cats and turkeys; it can also be found in wild animals. The incidence of infection in cattle and sheep is especially high. All these animals when infected excrete vast numbers of oocysts, which may survive in water for up to one year. Because all

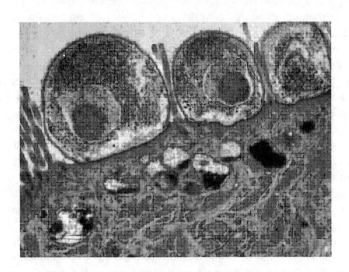

FIGURE 2.11 Electron micrograph of *Cryptosporidium* oocysts.

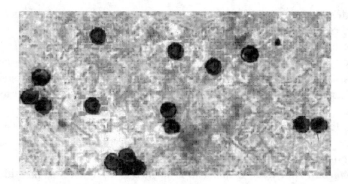

FIGURE 2.12 *Cryptosporidium* oocysts shown by staining.

these animals, including humans, can come in contact with surface waters, as in the case of the *Giardia*, community water supplies that draw water from impounding reservoirs are a potential problem for diseases caused by the oocyst.

The pathogen can cause a disease called cryptosporidiosis. The symptoms may appear within 4 to 6 days, but may appear anytime from 2 to 10 days after infection. While some persons may not have symptoms, others have watery diarrhea, headache, abdominal cramps, nausea, stomach pain, vomiting, and low-grade, flu-like fever. As in giardiasis, these symptoms may lead to weight loss and dehydration. In healthy persons, these symptoms usually last for several days but rarely more than 2 weeks, at which time the immune system is able to stop the infection. In persons with suppressed immune systems, such as persons with cancer and AIDS and persons who have recently had an organ or bone marrow transplant, the infection may continue and become life-threatening. No known cure or effective treatment is available for *Cryptosporidium* infection, although an experimental drug called NTZ shows promise. An infected person will simply just recover as soon as the immune system stops the infection.

Cryptosporidiosis is contracted in practically the same way as giardiasis. That is, the parasite is passed in the feces of an infected person or the feces of animals. The waste may then contaminate water or food. Direct person-to-person transmission may also occur in day-care centers or other settings where handwashing practices are poor. It is estimated that as few as one to 10 oocysts constitute an infective dose. In 1993, waterborne *Cryptosporidium* sickened 403,000 Milwaukee residents and killed more than 100 people, many with HIV/AIDS.

The diagnosis of cryptosporidiosis is posed by observation of the oocysts in the feces of infected individuals. Routine stool examination used for most parasites usually fails to detect *Cryptosporidium*, however. Thus, a stool specimen is examined using staining techniques available for this parasite. See Figure 2.12.

As in giardiasis, measures to prevent transmission include proper disposal of feces, wastewater treatment, and a filtration step before chlorination in water treatment plants that draw from surface water sources. Cooking kills the oocysts in contaminated foods and boiling for at least one minute will make water safe for use. Heating to 160°F is known to kill the oocysts. Chlorination is **ineffective** in killing the parasites. For municipal water systems, the most practical and effective way of getting rid of the parasite is the use of a properly operated and maintained filtration plant. The turbidity of the effluent should be brought down to around 0.1 NTU.

Analysis of water for *Cryptosporidium* uses similar techniques as those used for the analysis of *Giardia*. It involves filtering large quantities of water through a one-micron filter, removing the trapped particles, extracting the oocysts, and refiltering the extract. The sample is then analyzed using an indirect fluorescent antibody procedure that relies heavily on the skill of an experienced microscopist. At this time, there is no accepted standard method for *Cryptosporidium* and the oocyst recovery efficiency and sensitivity are variable.

Entamoeba histolytica. *Entamoeba histolytica* is a single celled parasitic ameba, a protozoan that infects predominantly humans and other primates all around the world. Dogs and cats can become infected but they usually do not shed cysts with their feces, thus do not contribute significantly to transmission. The infectious trophozoite stage exists only in the host and in fresh feces; cysts survive outside the host in water and moist soils and on foods. The ameba are cytotoxic when in contact with human cells. With a rapid influx of calcium into the contacted cell, all membrane movement stops, save for some surface blebbing. Internal organization is disrupted, organelles lyse, and the cell dies. The ameba may eat the dead cell directly or absorb nutrients released from the cell. Following malaria and schistosomiasis, *Entamoeba histolytica* is the third leading cause of morbidity and mortality due to parasitic disease in humans and is estimated to be responsible for infecting one tenth of the world population. Figure 2.13 shows the life cycle of the parasite. Figures 2.14 and 2.15 show pictures of the tropho-zoites, the latter being shown in a section of an intestine.

As depicted in its life cycle in Figure 2.13, the sources of the parasites are the infected humans and other primates. Anyone can harbor this parasite, but it is more frequently found in individuals who live or have visited areas of the world with poor sanitation. It is also more commonly found in individuals living in institutions for the developmentally disabled and in nursing care institutions with infected residents.

Conceptually, the ingestion of only one viable cyst can cause an infection. When swallowed, it causes infections by excysting to the trophozoite stage in the digestive tract and causes the disease called amebiasis. An individual with amebiasis may have no symptoms, or may have a range of symptoms from mild to severe. Mild symptoms, which is the most common and may last for years, include abdominal discomfort and diarrhea that may include blood or mucus alternating with periods of constipation or remission. Other symptoms, such as nausea, weight loss, fever, and chills may also occur. Rarely does the parasite invade other organs of the body such as the liver, lung, and brain. If it does invade these areas, it may cause abscesses there. The symptoms usually appear within 2 to 4 weeks after swallowing the cyst, but they may appear as quickly as a few days, or take as long as several months or even years. Several antibiotics are available through prescription by physicians to treat amebiasis. No over-the-counter medications are available to cure this disease.

As Figure 2.13 shows, the cysts are excreted in feces that may survive in moist environments, which then, again, become available to infect another individual. Thus, the disease is contracted in practically the same way as giardiasis and cryptosporidiosis

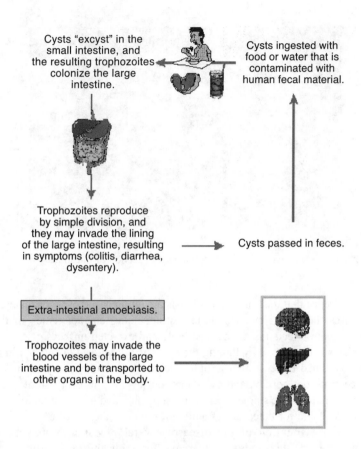

FIGURE 2.13 Life cycle of *Entamoeba histolytica.*

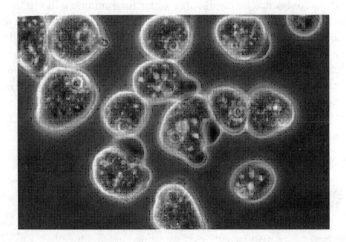

FIGURE 2.14 Phase contrast photomicrograph of cultured *Entamoeba histolytica* trophozoites.

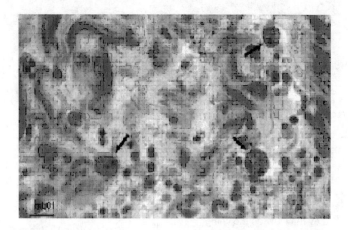

FIGURE 2.15   *Entamoeba histolytica* trophozoites in section of intestine.

with the difference that the hosts in amebiasis are humans and other primates. These three diseases, however, have common distinguishing characteristics in that the causative parasite infects inside the host body and excreted in feces.

Amebiasis is diagnosed by finding the cysts of *Entamoeba histolytica* in a stool specimen examined under a microscope. Because the number of cysts in the stool can change from day to day, and cysts may not even be in every stool specimen, sometimes more than one stool specimen must be obtained for examination. A blood test is also available for detection of antibodies against the parasite.

To avoid transmission of the disease, several precautions may be taken. Always wash hands thoroughly with soap after using the toilet, after changing diapers, before preparing food or drink, and before eating. When traveling in tropical countries where sanitation is poor, drink only bottled water, carbonated water, or canned or bottled sodas. Boiling water for 1 minute will kill the parasite. As in the case of *Cryptosporidium* and *Giardia, Entamoeba histolytica* is not killed by low doses of chlorine or iodine; thus, do not rely on chemical water purification tablets, such as halide tablets, to prevent amebiasis. For municipal water systems, the most practical and effective way of getting rid of the parasite, as in the case of *Cryptosporidium* and *Giardia*, is the use of a properly operated and maintained filtration plant. Again, the turbidity of the effluent should be brought down to around 0.1 NTU.

GLOSSARY

ABS—Alkyl benzene sulfonate.
Acidity—The ability of a substance to neutralize a base.
Alcohols—Organic compounds with the general formula R–OH.
Algae—Eucaryotic protists which may be unicelled, multicelled, or colonial capable of carrying photosynthesis.

Alkalinity—The ability of a substance to neutralize an acid.

Amphoteric substance—A substance that can act both as an acid and as a base.

Bacteria—Unicell procaryotic protists that are the only living things incapable of directly using particulate food.

Biochemical oxygen demand (BOD)—The amount of oxygen consumed by the organism in the process of stabilizing waste.

Chemical oxygen demand (COD)—A measure of the oxygen equivalent content of a given waste obtained by using a chemical such as potassium dichromate to oxidize the organic content of the waste.

Chloroplast—The site for the electron-tranport and the respiratory enzyme systems in photosynthetic eucaryotes.

Chromosomes—The DNA arranged into a readily recognizable structure inside the nucleus.

Cilia—Similar to the flagellum except more numerous, delicate, and shorter.

Clostridium botulinum—A bacterium that excretes the toxin causing botulinism.

Clostridium tetani—A bacterium that excretes toxins producing tetanus.

Coliform group—Aerobic and anaerobic, gram-negative, nonspore-forming, rod-shaped bacteria that ferment lactose with gas and acid formation within 48 hours at 35°C.

Completed test—A further test for the coliform group proceeding from the confirmed test where the test is considered positive when the organisms are stained and considered negative when the organisms are not stained using the Gram staining technique.

Complex lipids—Collective name for phosphoglycerides, fats, and oils.

Condensed phosphates—One of the phosphate forms produced when phosphoric acid is heated and decomposed losing molecules of water and forming the P–O–P bonds. The process of losing water is called *condensation*, thus, the term.

Confirmed test—A further test for the coliform group proceeding from the presumptive test where evolution of gases within 24 to 48 ± 3 hours in the presence of a brilliant green lactose bile broth from samples incubated at 35 ± 0.5°C indicates a positive test.

Corynebacterium diptheriae—A bacterial agent for diptheria.

Cyst—A transformation protozoa undergo when adverse conditions prevail.

Cytoplasm—Living material of which the cell is composed of minus the nucleus.

Deoxygenation coefficient k_c—A proportionality constant in the rate-of-disappearance differential equation for CBOD.

DNA, deoxyribonucleic acid—The genetic material of a cell.

Dipolyphosphoric acid—A polyphosphate of the formula $H_4P_2O_7$.

Dissolved solids—Solids that pass the filter; may also be called filtrable solids.

Electron acceptor—An entity that accepts electors.

Electron donor—An entity that gives (donates) electrons.

Escherichia coli—Bacteria found in the intestinal tract of human beings.

Esters—Organic compounds having the general formula of $R\overset{\text{O}}{\underset{\|}{C}}-OR'$, where R and R' are hydrocarbon groups.

Eucaryotes—Protists that contain chromosomes; they possess true nucleus.

Fat—The product formed from the reaction of glycerol and saturated fatty acid.

Fatty acid—A long-chain organic acid.

Fatty alcohols—Alcohols that look and feel like lipids or fats.

Filtrable solids—The colloidal and dissolved fractions of total solids that pass through a filter in a prescribed laboratory procedure.

Fixed solids—Portions of the various forms of solids that remain as a residue when the sample is decomposed at 600°C.

Five-day biochemical oxygen demand (BOD$_5$)—The BOD for an incubation period of five days.

Flagellum—Whiplike prolongation of the protoplasm.

Free ammonia—Molecular NH_3 dissolved in water formed as the hydrolysis product of organic nitrogen.

Gametes—Protozoal cell capable of sexual reproduction.

Glycerolphosphatides—Another name for phosphoglycerides.

Gram-negative bacteria—Bacteria with a surface stain using crystal violet, or gentian violet treated with iodine that *can* be removed by flooding with alcohol, acetone, or aniline.

Gram-positive bacteria—Bacteria with a surface stain using crystal violet, or gentian violet treated with iodine that *cannot* be removed by flooding with alcohol, acetone, or aniline.

Grease—Waxes, complex lipids (fats, oils, and phospholipids), and mineral oils such kerosene, crude oil, and lubricating oil and similar products.

Imhoff cone—A vessel shaped like a cone used to measure settleable solids.

Kjeldahl nitrogen—The sum of ammonia and organic nitrogens.

K_w—Called the ion-product of water, it is a product of the activities of H^+ and OH^- when any given temperature is constant.

LAS—Linear alkyl benzene sulfonate.

Mathematical expectation—The average value obtained if the random variable is measured exceedingly many times.

Metaphosphoric acid—A condensed phosphate of the formula $(HPO_3)_n$.

Mitochondrion—The site for the electron-transport and the respiratory enzyme systems in eucaryotes.

Multiple-tube or **MPN technique**—A method of estimating the concentration of coliforms using the Poisson distribution.

Nitrobacter—Genus of bacteria responsible for the conversion of nitrite-nitrogen to nitrate-nitrogen.

Nitrosomonas—Genus of bacteria responsible for the conversion of ammonia-nitrogen to nitrite-nitrogen.

Nonfiltrable solids—The settleable and nonsettleable fractions of solids that did not pass through a filter in a prescribed laboratory procedure.

Nucleus—A membrane-enclosed organelle inside the cell containing the genetic material of the cell.

Odor—The perception registered by the olfactory nerves in response to some external stimulus caused by substances.

Oil—The reaction products from the reaction of glycerol and unsaturated fatty acid.

Organic acids—Organic compounds with the general formula $R\overset{\overset{\displaystyle O}{\|}}{C}-OH$, where R is a hydrocarbon group.

Organic nitrogen—The nitrogen content in protein of an organic matter.

Organic phosphorus—The phosphorus content of organic matter.

Orthophosphates—The ions and salts of the three phosphoric acid dissociation products: dihydrogen phosphates containing the $H_2PO_4^-$ ion, hydrogen phosphate containing the HPO_4^{2-} ion, and the phosphates containing the PO_4^{3-} ion.

Oxidation state—A measure of the degree of affinity of the atom to the electrons it shares with other atoms.

pH—The negative of the logarithm to the base 10 of the hydrogen ion activity expressed in gmols per liter.

Phosphoglycerides—Products formed from the reaction of glycerol, phosphoric acid, and fatty acids.

pOH—The negative of the logarithm to the base 10 of the hydroxide ion activity expressed in gmols per liter

Population—The total set of measurements of interest in a particular problem.

POTWs—Publicly owned treatment works.

Polyphosphates—Another term for condensed phosphates.

Presumptive test—A test for the coliform group where evolution of gases within 24 to 48 ± 3 hours when samples are incubated at 35° ± 0.5°C assumes the presence of the microorganisms.

Procaryotes—Protists that do not contain chromosomes; they do not have true nucleus.

Protozoa—Single-cell protist one step above the bacteria in the trophic level.

Protist—A third kingdom of the living things that do not have cell specialization to perform specific cell functions as in the higher forms of life. They include protozoa, fungi, algae, and bacteria.

Pyrophosphoric acid—Another name for dipolyphosphoric acid.

Random variable—A function dependent on chance and whose values are real numbers.

Salmonella typhosa—Pathogenic bacterium that causes typhoid fever.

Sample—A subset of the population that contains measurements obtained by an experiment.

Shigella flexneri—Pathogenic bacterium that causes bacillary dysentery.

Settleable solids—The fraction of solids produced after settling for 30 minutes in a cone-shaped vessel called an *Imhoff cone*.

Simple lipids—The fatty alcohols.

Sludge volume index (SVI)—The volume of solids that settled, in milliliters, divided by the corresponding grams of solids mass.

Spirillum—A spiral organism whose long axis remains rigid when in motion.

Spirochete—An organism whose long axis bends when in motion.

Surfactants—Surface-active agents, which means that they have the property of interacting with surfaces.

Suspended solids—Solids retained on the filter.

Total nitrogen—The sum of the organic, free ammonia, nitrite, and nitrate nitrogens.

Total solids—Materials left after water has been evaporated from a sample.

Total suspended solids—The nonfiltrable solids.

Triglycerides or **triacylglycerides**—Fats and oils formed from the reaction of glycerol with the fatty acids.

Triol—An alcohol with three OH groups.

Tripolyphosphoric acid—A condensed phosphate of the formula $H_5P_3O_{10}$.

Ultimate BOD, BOD_u—BOD obtained after a long period of incubation such as 20 to 30 days.

Ultimate carbonaceous BOD or **CBOD**—A BOD_u with the ammonia reaction inhibited.

Vibrio—A comma-shaped microorganism.

Virion—A unit of viral particle tightly packed inside a protein coat.

Virus—A submicroscopic agent of infectious disease that requires a living cell for its multiplication.

VOCs—Volatile organic compounds.

Volatile solids—Portions of the various forms of solids that disappear when the sample is decomposed at 600°C.

Waxes—Esters of simple lipids and fatty acids.

Zygote—Cell formed by the union of two gametes.

SYMBOLS

BOD	Biochemical oxygen demand
COD	Chemical oxygen demand
D	Fractional dilution of the BOD sample
F	Final DO of the same sample which has been diluted with seeded dilution water after the incubation period
F'	Final DO of the blank after incubating at the same time and temperature as the sample
I	Initial DO of the sample which has been diluted with seeded dilution water
I'	The initial DO of a volume Y of the blank composed of only the seeded dilution water
L_c	CBOD of sample
R	Residual; hydrocarbon group
$RC\text{–}OH$ with O double-bonded to C	General formula are called organic acids
$RC\text{–}OR'$ with O double-bonded to C	General formula for esters where R' is one hydrocarbon group and R is another

R–OH   General formula are called alcohols
TOC   Total organic carbon
VOC   Volatile organic compound
X Volume of the seeded dilution water mixed with the sample
Y Volume of blank

PROBLEMS

2.1   Name as many chemical characteristics as you can for characterizing a wastewater.
2.2   What is biochemical oxygen demand?
2.3   From previous experience, the BOD_5 of a given type of sample is 180 mg/L. What percentage dilution should be used to analyze this sample?
2.4   From previous experience, the BOD_5 of a given type of sample is 180 mg/L. How many milliliters of this sample should be pippetted into a 300-mL BOD bottle?
2.5   In BOD analysis, what ions are used to maintain osmotic pressure?
2.6   From previous experience, the BOD_5 of a given type of sample is 180 mg/L. What percentage dilution should be used to analyze this sample?
2.7   From past experiences on BOD work, what is the lowest concentration of dissolved oxygen allowed in the BOD bottle and what is the lowest oxygen depletion of oxygen allowed to have accurate results?
2.8   A ten-milliliter sample is pipetted directly into a 300-mL incubation bottle. The initial DO of the diluted sample is 9.0 mg/L and its final DO is 2.0 mg/L. The dilution water is incubated in a 200-mL bottle, and the initial and final DOs are, respectively, 9.0 and 8.0 mg/L. If the sample and the dilution water is incubated at 20°C for five days, what is the BOD_5 of the sample at this temperature?
2.9   Solve Problem 2.8 if the volume of the incubation bottle is 200 mL.
2.10   A BOD test is to be conducted on a poultry waste known to contain 650 mg/L of BOD_5. **(a)** What sample portion should be used in the test? **(b)** Estimate the BOD_5 if the initial DO in both the diluted sample and the seed is 8.5 mg/L and at the end of five days, the DOs are 3.0 and 5.0 mg/L for the sample and seed, respectively.
2.11   Calculate the BOD_5 of a domestic sewage with the following laboratory results: sample portion added to 300-mL bottle = 6.5 mL, initial DO = 8.0 mg/L, final DO = 4.5 mg/L. What is the ultimate oxygen demand assuming $k_c = 0.1$ per day?
2.12   The deoxygenation coefficient to the base 10 for a certain waste is 0.1 per day. What fraction is BOD_5 to CBOD of this waste?
2.13   What is the difference between nonfiltrable and suspended solids?
2.14   State all the various fixed and volatile solids.
2.15   Volatile suspended solids analysis is run on a sample. The tare mass of the crucible and filter is 55.3520 g. A sample of 260 mL is then filtered and the residue dried and decomposed at 600°C to drive off the volatile matter. Assume the filter does not decompose at this temperature. After

decomposition, the constant weight of the crucible and the residue was determined and was found to be equal to 30.3415 g. What is the volatile suspended solids if the total suspended solids is 142 mg/L?

2.16 Define pH and pOH.

2.17 What does concentration in terms of activities mean?

2.18 Assuming the temperature is 25°C, calculate the concentration of the other ion indicated by a question mark.

$$[OH^-] = 0.6(10^{-5})M, [H^+] = ?$$

$$[H^+] = 4(10^{-4})M, [OH] = ?$$

2.19 What chemical is normally used in the determination of COD?

2.20 Define oxidation state.

2.21 What is Kjeldahl and total nitrogen?

2.22 What are the three types of phosphorus encountered in wastewater?

2.23 What is an amphoteric substance?

2.24 What are triglycerides?

2.25 What are fatty alcohols?

2.26 What are complex and simple lipids?

2.27 Back River Sewage Treatment Plant of Baltimore, MD, discharges to the Back River estuary. The State of Maryland imposes on its permit a total phosphorus limitation of 2.0 mg/L. Assuming all the total phosphorus converts to orthophosphorus and assuming the dilution effect of the estuary cannot be neglected, calculate the concentration of algae theoretically expected.

BIBLIOGRAPHY

APHA, AWWA, and WEF (1992). *Standard Methods for the Examination of Water and Wastewater.* 15th ed. American Public Health Association, Washington.

Bagdigian, R. M., et al. (1991). Phase III integrated water recovery testing at MSFC: Partially closed hygiene loop and open potable loop results and lessons learned. *SAE (Society of Automotive Engineers) Trans.* 100, 1, 826–841.

Bahri, A. (1998). Fertilizing value and polluting load of reclaimed water in Tunisia. *Water Res.* 32, 11, 3484–3489.

Belkin, S., A. Brenner, and A. Abeliovich (1992). Effect of inorganic constituents on chemical oxygen demand—I. Bromides are unneutralizable by mercuric sulfate complexation. *Water Res.* 26, 12, 1577–1581.

Ball, B. R. and M. D. Edwards (1992). Air stripping VOCs from groundwater. Process design considerations. *Environmental Prog.* 11, 1, 39–48.

Bousher, A., X. Shen, and R. G. J. Edyvean (1997). Removal of coloured organic matter by adsorption onto low-cost waste materials. *Water Res.* 31, 8, 2084–2092.

Frost, F., B. Plan, and B. Liecty (1980). Giardia prevalence in commercially trapped mammals. *J. Environ. Health.* 42, 245–249.

Hammer, M. J. (1986). *Water and Wastewater Technology.* John Wiley & Sons, New York.

Hiraishi, A., Y. Ueda, and J. Ishihara (1998). Quinone profiling of bacterial communities in natural and synthetic sewage activated sludge for enhanced phosphate removal. *Applied Environmental Microbiol.* 64, 3, 992–999.

Isaac, R. A., et al. (1997). Corrosion in drinking water distribution systems: A major contributor of copper and lead to wastewaters and effluents. *Environmental Science Technol.* 31, 11, 3198–3203.

Kim, B. R., et al. (1994). Biological removal of organic nitrogen and fatty acids from metal-cutting-fluid wastewater. *Water Res.* 28, 6, 1453–1461.

Lee, G. and F. A. Jones-Lee (1996). Stormwater runoff quality monitoring: Chemical constituents vs. water quality. *Public Works.* 127, 12, 50–53.

Li, L., N. Crain, and E. F. Gloyna (1996). Kinetic lumping applied to wastewater treatment. *Water Environ. Res.* 68, 5, 841–854.

Noller, B. N., P. H. Woods, and B. J. Ross (1992). Case studies of wetland filtration of mine wastewater in constructed and naturally occurring systems in northern Australia. *Water Science Technol. Proc. IAWQ 3rd Int. Specialist Conf. Wetland Syst. Water Pollut. Control,* Nov. 23–25, 1992, Sydney, Australia, 29, 4, 257–265. Pergamon Press, Tarrytown, NY.

Paxeus, N. and H. F. Schroder (1995). Screening for non-regulated organic compounds in municipal wastewater in Goteborg, Sweden. *Water Science Technol. Proc. 1995 2nd IAWQ Int. Specialized Conf. Hazard Assessment Control Environmental Contaminants Water,* June 29–30, 1995, Lyngby, Denmark, 33, 6, 9–15. Pergamon Press, Tarrytown, NY.

Pawar, N. J., G. M. Pondhe, and S. F. Patil (1998). Groundwater pollution due to sugar-mill effluent, at Sonai, Maharashtra, India. *Environ. Geol.* 34, 2–3, 151–158.

Peavy, H. S., D. R. Rowe, and G. Tchobanoglous (1985). *Environmental Engineering.* McGraw-Hill, New York.

Reemtsma, T. and M. Jekel (1997). Dissolved organics in tannery wastewaters and their alteration by a combined anaerobic and aerobic treatment. *Water Res.* 31, 5, 1035–1046.

Rendtorff, R. C. (1954). The experimental transmission of human intestinal protozoan parasites, II. *Giardia lamblia* cysts given in capsules. *Am. J. Hyg.* 59, 209–220.

Rogers, S. E. and W. C. Lauer (1992). Denver's demonstration of potable water reuse: Water quality and health effects testing. *Water Science Technol. Proc. 16th Biennial Conf. Int. Assoc. Water Pollut. Res. Control—Water Quality Int. '92,* May 24–30, 1992, Washington, 26, 7–8, 1555–1564. Pergamon Press, Tarrytown, NY.

Saari, R. B., J. S. Stansbury, and F. C. Laquer (1998). Effect of sodium xylenesulfonate on zinc removal from wastewater. *J. Environ. Eng.* 124, 10, 939–944.

Scandura, J. E. and M. D. Sobsey (1996). Viral and bacterial contamination of groundwater from on-site sewage treatment systems. *Water Science Technol. Proc. 1996 IAWQ 8th Int. Symp. Health-Related Water Microbiol.,* Oct. 6–10, 1996, Mallorca, Spain, 35, 11–12, 141–146, Elsevier Science, Oxford.

Schnell, A., et al. (1997). Chemical characterization and biotreatability of effluents from an integrated alkaline-peroxide mechanical pulping/machine finish coated (APMP/MFC) Paper Mill. *Water Science Technol. Proc. 1996 5th IAWQ Int. Symp. Forest Industry Wastewaters,* June 10–13, 1996, Vancouver, BC, 35, 2–3, 7–14, Elsevier Science, Oxford.

Sincero, A. P. and G. A. Sincero (1996). *Environmental Engineering. A Design Approach.* Prentice Hall, Upper Saddle River, NJ.

Stober, J. T., J. T. O'Connor, and B. J. Brazos (1997). Winter and spring evaluations of a wetland for tertiary wastewater treatment. *Water Environment Res.* 69, 5, 961–968.

Sztruhar, D., et al. (1997). Case study of combined sewer overflow pollution: Assessment of sources and receiving water effects. *Water Quality Res. J. Canada.* 32, 3, 563–578.

Part II

Unit Operations of Water and Wastewater Treatment

Part II covers the unit operations of flow measurements and flow and quality equalizations; pumping; screening, sedimentation, and flotation; mixing and flocculation; filtration; aeration and stripping; and membrane processes and carbon adsorption. These unit operations are an integral part in the physical treatment of water and wastewater.

3 Flow Measurements and Flow and Quality Equalizations

This chapter discusses the unit operations of flow measurements and flow and quality equalizations. Flow meters discussed include rectangular weirs, triangular weirs, trapezoidal weirs, venturi meters, and one of the critical-flow flumes, the Parshall flume. Miscellaneous flow meters including the magnetic flow meter, turbine flow meter, nutating disk meter, and the rotameter are also discussed. These meters are classified as miscellaneous, because they will not be treated analytically but simply described. In addition, liquid level recorders are also briefly discussed.

3.1 FLOW METERS

Flow meters are devices that are used to measure the rate of flow of fluids. In wastewater treatment, the choice of flow meters is especially critical because of the solids that are transported by the wastewater flow. In all cases, the possibility of solids being lodged onto the metering device should be investigated. If the flow has enough energy to be self-cleaning or if solids have been removed from the wastewater, weirs may be employed. Venturi meters and critical-flow flumes are well suited for measurement of flows that contain floating solids in them. All these flow-measuring devices are suitable for measuring flows of water.

Flow meters fall into the broad category of meters for open-channel flow measurements and meters for closed-channel flow measurements. Venturi meters are closed-channel flow measuring devices, whereas weirs and critical-flow flumes are open-channel flow measuring devices.

3.1.1 RECTANGULAR WEIRS

A *weir* is an obstruction that is used to back up a flowing stream of liquid. It may be of a thick structure or of a thin structure such as a plate. A *rectangular weir* is a thin plate where the plate is being cut such that a rectangular opening is formed in which the flow in the channel that is being measured passes through. The rectangular opening is composed of two vertical sides, one bottom called the *crest*, and no top side. There are two types of rectangular weirs: the suppressed and the fully contracted weir. Figure 3.1 shows a fully contracted weir. As indicated, a *fully contracted rectangular weir* is a rectangular weir where the flow in the channel being measured contracts as it passes through the rectangular opening. On the other hand, a *suppressed rectangular weir* is a rectangular weir where the contraction is absent, that

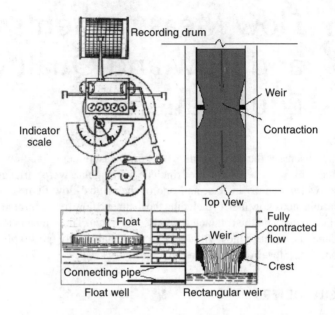

FIGURE 3.1 Rectangular weir measuring assembly.

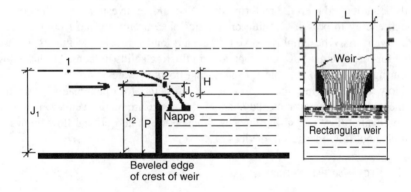

FIGURE 3.2 Schematic for derivation of weir formulas.

is, the contraction is suppressed. This happens when the weir is extended fully across the width of the channel, making the vertical sides of the channel as the two vertical sides of the rectangular weir. To ensure an accurate measurement of flow, the crest and the vertical sides (in the case of the fully contracted weir) should be beveled into a sharp edge (see Figures 3.2 and 3.3).

To derive the equation that is used to calculate the flow in rectangular weirs, refer to Figure 3.2. As shown, the weir height is P. The vertical distance from the tip of the crest to the surface well upstream of the weir at point 1 is designated as H. H is called the *head* over the weir.

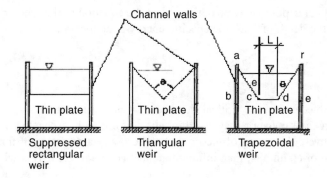

FIGURE 3.3 Various types of weirs.

From fluid mechanics, any open channel flow value possesses one and only one critical depth. Since there is a one-to-one correspondence between this depth and flow, any structure that can produce a critical flow condition can be used to measure the rate of flow passing through the structure. This is the principle in using the rectangular weir as a flow measuring device. Referring to Figure 3.2, for this structure to be useful as a measuring device, a depth must be made critical somewhere. From experiment, this depth occurs just in the vicinity of the weir. This is designated as y_c at point 2. A one-to-one relationship exists between flow and depth, so this section is called a *control section*. In addition, to ensure the formation of the critical depth, the underside of the nappe as shown should be well ventilated; otherwise, the weir becomes submerged and the result will be inaccurate.

Between any points 1 and 2 of any flowing fluid in an open channel, the energy equation reads

$$\frac{V_1^2}{2g} + \frac{P_1}{\gamma} + y_1 + z_1 - h_l = \frac{V_2^2}{2g} + \frac{P_2}{\gamma} + y_2 + z_2 \qquad (3.1)$$

where V, P, y, z, and h_l refer to the average velocity at section containing the point, pressure at point, height of point above bottom of channel, height of bottom of channel from a chosen datum, and head loss between points 1 and 2, respectively. The subscripts 1 and 2 refer to points 1 and 2. g is the gravitational constant and γ is the specific weight of water. Referring to Figure 3.2, the two values of z are zero. V_1 called the *approach velocity* is negligible compared to V_2, the average velocity at section at point 2. The two Ps are all at atmospheric and will cancel out. The friction loss h_l may be neglected for the moment. y_1 is equal to $H + P$ and y_2 is very closely equal to $y_c + P$. Applying all this information to Equation (3.1), and changing V_2 to V_c, produces

$$V_c^2 = 2g(H - y_c) \qquad (3.2)$$

The critical depth y_c may be derived from the equation of the specific energy E. Using y as the depth of flow, the specific energy is defined as

$$E = y + \frac{V^2}{2g} \qquad (3.3)$$

From fluid mechanics, the critical depth occurs at the minimum specific energy. Thus, the previous equation may be differentiated for E with respect to y and equated to zero. Convert V in terms of the flow Q and cross-sectional area of flow A using the equation of continuity, then differentiate and equate to zero. This will produce

$$\frac{Q^2 T}{g A^3} = 1 = \frac{V^2}{gA/T} \Rightarrow \frac{V}{\sqrt{gA/T}} = \frac{V}{\sqrt{gD}} = 1 \qquad (3.4)$$

where T is equal to dA/dy, a derivative of A with respect to y. T is the top width of the flow. A/T is called the *hydraulic depth* D. The expression $V/\sqrt{gD}$ is called the *Froude number*. The flow over the weir is rectangular, so D is simply equal to y_c, thus Equation (3.4) becomes

$$\frac{V_c}{\sqrt{gy_c}} = 1 \qquad (3.5)$$

where V has been changed to V_c, because V is now really the critical velocity V_c. Equation (3.4) shows that the Froude number at critical flow is equal to 1. Equation (3.5) may be combined with Equation (3.2) to eliminate y_c producing

$$V_c = \sqrt{\frac{1}{3}} \sqrt{2gH} \qquad (3.6)$$

The cross-sectional area of flow at the control section is $y_c L$, where L is the length of the weir. This will be multiplied by V_c to obtain the discharge flow Q at the control section, which, by the equation of continuity, is also the discharge flow in the channel. Using Equation (3.5) for the expression of y_c and Equation (3.6) for the expression for V_c, the discharge flow equation for the rectangular weir becomes

$$Q = 0.385 \sqrt{2g} L \sqrt{H^3} \qquad (3.7)$$

Two things must be addressed with respect to Equation (3.7). First, remember that h_l and the approach velocity were neglected and y_2 was made equal to $y_c + P$. Second, the L must be corrected depending upon whether the above equation is to be used for a fully contracted rectangular weir or the suppressed weir.

The coefficient of Equation (3.7) is merely theoretical, so we will make it more general and practical by using a general coefficient K as follows

$$Q = K \sqrt{2g} L \sqrt{H^3} \qquad (3.8)$$

Now, based on experimental evidence Kindsvater and Carter (1959) found that for H/P up to a value of 10, K is

$$K = 0.40 + 0.05\frac{H}{P} \tag{3.9}$$

Due to the contraction of the flow for the fully contracted rectangular weir, the cross-sectional of flow is reduced due to the shortening of the length L. From experimental evidence, for $L/H > 3$, the contraction is $0.1H$ per side being contracted. Two sides are being contracted, so the total correction is $0.2H$, and the length to be used for fully contracted weir is

$$L_{\text{fully contracted weir}} = L - 0.2H \tag{3.10}$$

In operation, the previous flow formulas are automated using control devices. This is illustrated in Figure 3.1. As derived, the flow Q is a function of H. For a given installation, all the other variables influencing Q are constant. Thus, Q can be found through the use of H only. As shown in the figure, this is implemented by communicating the value of H through the connecting pipe between the channel, where the flow is to be measured, and the float chamber. The communicated value of H is sensed by the float which moves up and down to correspond to the value communicated. The system is then calibrated so that the reading will be directly in terms of rate of discharge.

From the previous discussion, it can be gleaned that the meter measures rates of flow proportional to the cross-sectional area of flow. Rectangular weirs are therefore *area meters*. In addition, when measuring flow, the unit obstructs the flow, so the meter is also called an *intrusive flow meter*.

Example 3.1   The system in Figure 3.1 indicates a flow of 0.31 m³/s. To investigate if the system is still in calibration, H, L, and P were measured and found to be 0.2 m, 2 m, 1 m, respectively. Is the system reading correctly?

Solution: To find if the system is reading correctly, the above values will be substituted into the formula to see if the result is close to 0.3 m³/s.

$$Q = K\sqrt{2g}L\sqrt{H^3}$$

$$K = 0.40 + 0.05\frac{H}{P}$$

$$L_{\text{fully contracted weir}} = L - 0.2H = 2 - 0.2(0.2) = 1.96\,\text{m}$$

$$K = 0.40 + 0.05\left(\frac{0.2}{1}\right) = 0.41$$

$$Q = 0.41\sqrt{2(9.81)}(1.96)\sqrt{(0.2)^3}$$

$$= 0.318\,\text{m}^3/\text{s};\ \text{therefore, the system is reading correctly.}\quad \textbf{Ans}$$

Example 3.2 Using the data in the above example, calculate the discharge through a suppressed weir.

Solution:

$$L = 2\text{m}; \ K = 0.41$$

Therefore,

$$Q = 0.41\sqrt{2(9.81)}(2)\sqrt{(0.2)^3} = 0.325 \ \text{m}^3/\text{s} \quad \textbf{Ans}$$

Example 3.3 To measure the rate of flow of raw water into a water treatment plant, management has decided to use a rectangular weir. The flow rate is 0.33 m³/s. Design the rectangular weir. The width of the upstream rectangular channel to be connected to the weir is 2.0 m and the available head H is 0.2 m.

Solution: Use a fully suppressed weir and assume length $L = 0.2$ m. Thus,

$$Q = K\sqrt{2g}L\sqrt{H^3} \Rightarrow 0.33 = K\sqrt{2(9.81)}(2)\sqrt{(0.2)^3} = 0.792K$$

Therefore,

$$K = 0.417 = 0.40 + 0.05\frac{H}{P} = 0.40 + 0.05\left(\frac{0.2}{P}\right)$$

$$P = 0.6 \text{ m}$$

Therefore,

$$\text{dimension of rectangular weir:} \ L = 2.0 \text{ m}, \ P = 0.6 \text{ m} \quad \textbf{Ans}$$

3.1.2 TRIANGULAR WEIRS

Triangle weirs are weirs in which the cross-sectional area where the flow passes through is in the form of a triangle. As shown in Figure 3.3, the vertex of this triangle is designated as the angle θ. When discharge flows are smaller, the H registered by rectangular weirs are shorter, hence, reading inaccurately. In the case of triangular weirs, because of the notching, the H read is longer and hence more accurate for comparable low flows. Triangular weirs are also called *V-notch weirs*. As in the case of rectangular weirs, triangular weirs measure rates of flow proportional to the cross-sectional area of flow. Thus, they are also area meters. In addition, they obstruct flows, so triangular weirs are also intrusive flow meters.

The longitudinal hydraulic profile in channels measured by triangular weirs is exactly similar to that measured by rectangular weirs. Thus, Figure 3.2 can be used for deriving the formula for triangular weirs. The difference this time is that the cross-sectional area at the critical depth is triangular instead of rectangular. From

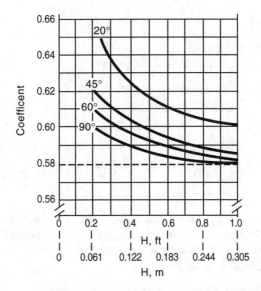

FIGURE 3.4 Coefficient for sharp-crested triangular weirs. (From Lenz, A.T. (1943). *Trans. AICHE*, 108, 759–820. With permission.)

Figure 3.3, the cross-sectional area, A, of the triangle is

$$A = y_c^2 \tan\frac{\theta}{2} \qquad (3.11)$$

Multiplying this area by V_c produces the discharge flow Q.

Now, the Froude number is equal to $V_c/\sqrt{gD}$. For the triangular weir to be a measuring device, the flow must be critical near the weir. Thus, near the weir, the Froude number must be equal to 1. D, in turn, is A/T, where $T = 2y_c \tan\frac{\theta}{2}$. Along with the expression for A in Equation (3.11), this will produce $D = y_c/2$ and, consequently, $V_c = \sqrt{gy_c/2}$ for the Froude number of 1. With Equation (3.2), this expression for V_c yields $y_c = (4/5)H$ and, thus, $V_c = \sqrt{2gH/5}$. $(4/5)H$ may be substituted for y_c in the expression for A and the result multiplied by $\sqrt{2gH/5}$ to produce the flow Q. The result is

$$Q = \frac{16}{25\sqrt{5}} \tan\frac{\theta}{2}\sqrt{2g}H^{5/2} = K\tan\frac{\theta}{2}\sqrt{2g}H^{5/2} \qquad (3.12)$$

where $16/25\sqrt{5}$ has been replaced by K to consider the nonideality of the flow.

The value of the discharge coefficient K may be obtained using Figure 3.4. The coefficient value obtained from the figure needs to be multiplied by $8/15$ before using it as the value of K in Equation (3.12). The reason for this indirect substitution is that the coefficient in the figure was obtained using a different coefficient derivation from the K derivation of Equation (3.12) (Munson et al., 1994).

Example 3.4 A 90-degree V-notch weir has a head H of 0.5 m. What is the flow, Q, through the notch?

Solution:

$$Q = K \tan\frac{\theta}{2}\sqrt{2g}H^{5/2}$$

From Figure 3.4, for an $H = 0.5$ m, and $\theta = 90°$, $K = 0.58$.
Therefore,

$$Q = 0.58\left(\frac{8}{15}\right)(\tan 45°)\sqrt{2(9.81)}(0.5)^{5/2} = 0.24 \text{ m}^3/\text{s} \quad \textbf{Ans}$$

Example 3.5   To measure the rate of flow of raw water into a water treatment plant, an engineer decided to use a triangular weir. The flow rate is 0.33 m³/s. Design the weir. The width of the upstream rectangular channel to be connected to the weir is 2.0 m and the available head H is 0.2 m.

Solution: Because the available head and Q are given, from $Q = K(8/15)\tan \times \theta/2\sqrt{2g} \cdot H^{5/2}$, $K\tan\theta/2$, can be solved. The value of the notch angle θ may then be determined from Figure 3.4.

$$0.33 = K\left(\frac{8}{15}\right)\tan\frac{\theta}{2}\sqrt{2(9.81)}(0.2)^{5/2} \Rightarrow K\left(\frac{8}{15}\right)\tan\frac{\theta}{2} = 4.16$$

From Figure 3.4, for $H = 0.2$ m, we produce the following table:

θ (degrees)	K	$K(\frac{8}{15})\tan\frac{\theta}{2}$
90	0.583	0.31
60	0.588	0.18
45	0.592	0.13
20	0.609	0.06

This table shows that the value of $K(8/15)\tan\frac{\theta}{2}$ is nowhere near 4.16. From Figure 3.4, however, the value of K for θ greater than 90° is 0.58. Therefore,

$$4.16 = 0.58\left(\frac{8}{15}\right)\tan\frac{\theta}{2}, \quad \text{and} \quad \tan\frac{\theta}{2} = 13.45, \quad \theta = 171.49, \text{ say } 171°$$

Given available head of 0.2 m, provide a freeboard of 0.3 m; therefore, dimensions: notch angle = 171°, length = 2 m, and crest at notch angle = 0.2 m + 0.3 m = 0.5 m below top elevation of approach channel.   **Ans**

3.1.3 TRAPEZOIDAL WEIRS

As shown in Figure 3.3, trapezoidal weirs are weirs in which the cross-sectional area where the flow passes through is in the form of a trapezoid. As the flow passes through the trapezoid, it is being contracted; hence, the formula to be used ought to be the contracted weir formula; however, compensation for the contraction may be made by proper inclination of the angle θ. If this is done, then the formula for suppressed rectangular weirs, Equation (3.8), applies to trapezoidal weirs, using the bottom length as the length L. The value of the angle θ for this equivalence to be so is 28°. In this situation, the reduction of flow caused by the contraction is counterbalanced by the increase in flow in the notches provided by the angles θ. This type of weir is now called the *Cipolleti weir* (Roberson et al., 1988). As in the case of the rectangular and triangular weirs, trapezoidal weirs are area and intrusive flow meters.

3.1.4 VENTURI METERS

The rectangular, triangular, and trapezoidal flow meters are used to measure flow in open channels. Venturi meters, on the other hand, are used to measure flows in pipes. Its cross section is uniformly reduced (*converging zone*) until reaching a point called the *throat*, maintained constant throughout the throat, and expanded uniformly (*diverging zone*) after the throat. We learned from fluid mechanics that the rate of flow can be measured if a pressure difference can be induced in the path of flow. The venturi meter is one of the pressure-difference meters. As shown in **b** of Figure 3.5, a venturi meter is inserted in the path of flow and provided with a streamlined constriction at point 2, the throat. This constriction causes the velocity to increase at the throat which, by the energy equation, results in a decrease in pressure there. The difference in pressure between points 1 and 2 is then taken advantage of to measure the rate of flow in the pipe. Additionally, as gleaned from these descriptions, venturi meters are intrusive and area meters.

The pressure sensing holes form a concentric circle around the center of the pipe at the respective points; thus, the arrangement is called a *piezometric ring*. Each of these holes communicates the pressure it senses from inside the flowing liquid to the piezometer tubes. Points 1 and 2 refer to the center of the piezometric rings, respectively. The figure indicates a deflection of Δh. Another method of connecting piezometer tubes are the tappings shown in **d** of Figure 3.5. This method of tapping is used when the indicator fluid used to measure the deflection, Δh, is lighter than water such as the case when air is used as the indicator. The tapping in **b** is used if the indicator fluid used such as mercury is heavier than water.

The energy equation written between points 1 and 2 in a pipe is

$$\frac{P_1}{\gamma} + \frac{V_1^2}{\gamma} + y_1 - h_l = \frac{P_2}{\gamma} + \frac{V_2^2}{\gamma} + y_2 \qquad (3.13)$$

where P is the pressure at a point at the center of cross-section and y is the elevation at point referred to some datum. V is the average velocity at the cross-section and h_l is the head loss between points 1 and 2. γ is the specific weight of water. The subscripts

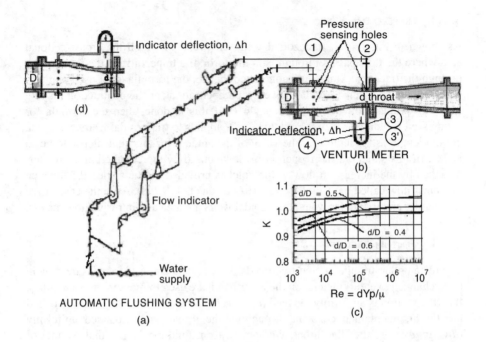

FIGURE 3.5 Venturi meter system: (a) flushing system; (b) Venturi meter; (c) coefficient of discharge. (From ASME (1959). *Fluid Meters—Their Theory and Application,* Fairfield, NJ; Johansen, F. C. (1930). *Proc. R. Soc. London,* Series A, 125. With permission.) (d) Piezometer taps for lighter indicator fluid.

1 and 2 refer to points 1 and 2, respectively. Neglecting the friction loss for the moment and since the orientation is horizontal in the figure, the energy equation applied between points 1 and 2 reduces to

$$\frac{P_1}{\gamma} + \frac{V_1^2}{2g} = \frac{P_2}{\gamma} + \frac{V_2^2}{2g} \tag{3.14}$$

Using the equation of continuity in the form of $(\pi D^2/4)V_1 = (\pi d^2/4)V_2$, where D is the diameter of the pipe and d is diameter of the throat, the above equation may be solved for V_2 to produce

$$V_2 = \sqrt{\frac{2g(P_1 - P_2)}{\gamma(1 - \beta^{41})}} \tag{3.15}$$

where $\beta = d/D$.

Let us now express $P_1 - P_2$ in terms of the indicator deflection, Δh. Apply the manometric equation in **b** in the sequence 1, 4, 3′, 3, 2. Thus,

$$P_1 + \Delta h_{14}\gamma - \Delta h_{3'3}\gamma_{ind} - \Delta h_{32}\gamma = P_2 \tag{3.16}$$

where Δh_{14}, $\Delta h_{3'3}$ $(=\Delta h)$, Δh_{32}, and γ_{ind} refer to the head difference between points 1 and 4, points 3′ and 3, and points 3 and 2, respectively. γ_{ind} is the specific weight of the indicator fluid used to indicate the deflection of manometer levels (i.e., the two levels of the indicator fluid in the manometer tube). Equation (3.16) may be solved for $P_1 - P_2$ producing $P_1 - P_2 = \Delta h(\gamma_{ind} - \gamma)$. However, in terms of an equivalent deflection of water, Δh_{H_2O}, $P_1 - P_2 = \Delta h_{H_2O}\gamma$. Thus,

$$P_1 - P_2 = \Delta h(\gamma_{ind} - \gamma) = \Delta h_{H_2O}\gamma \tag{3.17}$$

and

$$\Delta h_{H_2O} = \frac{\Delta h(\gamma_{ind} - \gamma)}{\gamma} \tag{3.18}$$

If the tapping in **d** is used where the indicator fluid is lighter than water and the above derivation is repeated, $\gamma_{ind} - \gamma$ in Equation (3.18) would be replaced by $\gamma - \gamma_{ind}$. Note that Δh_{H_2O} is not the manometer deflection; it is the water equivalent of the manometer deflection.

$\Delta h_{H_2O}\gamma$ may be substituted for $P_1 - P_2$ in Equation (3.15) and both sides of the equation multiplied by the cross-sectional area at the throat, A_t, to obtain the discharge, Q. The equation obtained by this multiplication is simply theoretical, however; thus, a discharge coefficient, K, is again used to account for the nonideality of actual discharge flows and to acknowledge the fact that the head loss, h_l, was originally neglected in the derivation. The corrected equation follows:

$$Q = KA_t\sqrt{2g\Delta h_{H_2O}} \tag{3.19}$$

where values of K may be obtained from **c** of Figure 3.5 and $A_t = \pi d^2/4$. Because $P_1 - P_2 = \gamma\Delta h_{H_2O}$, Equation (3.19) may also be written in terms of $P_1 - P_2$ as follows

$$Q = KA_t\sqrt{\frac{2g(P_1 - P_2)}{\gamma}} \tag{3.20}$$

Equation (3.20) may be used if the venturi meter is not oriented horizontally. This is done by calculating the pressures at the points directly and substituting them into the equation.

When measuring sewage flows, debris may collect on the pressure sensing holes. Hence, these holes must be cleaned periodically to ensure accurate sensing of pressure at all times. In **a** of Figure 3.5, an automatic cleaning arrangement is designed using an external supply of water. Water from the supply is introduced into the piping system through flow indicator, pipes, valves, and fittings, and into the venturi meter. The design would be such that water jets at high pressure are directed to the pressure sensing holes. These jets can then be released at predetermined intervals of time to wash out any cloggings on the holes. Of course, at the time that the jet is released, erratic readings of the pressure will occur and the corresponding Q should not be used. Line pressure of 70 kN/m^2 in excess over source water supply pressure is satisfactory.

Example 3.6   The flow to a water treatment plant is 0.031 cubic meters per second. The engineer has decided to meter this flow using a venturi meter. Design the meter if the approach pipe to the meter is 150 mm in diameter.

Solution: The designer has the liberty to choose values for the design parameters, provided it can be shown that the design works. Provide a pressure differential of 26 kN/m^2 between the approach to the tube and the throat. Initially assume a throat diameter of 75 mm.

$$Q = KA_t \sqrt{\frac{2g(P_1 - P_2)}{\gamma}}$$

$$Re = \frac{dV\rho}{\mu} \quad V = \frac{0.031}{\pi\left(\frac{0.075^2}{4}\right)} = 7.02 \text{ m/s}$$

From the appendix, $\rho = 997$ kg/m^3 and $\mu = 8.8(10^{-4})$ kg/m·s (25°C); therefore,

$$Re = \frac{0.075(7.02)(997)}{8.8(10^{-4})} = 5.97(10^5)$$

From c of Figure 3.5, at $d/D = 75/150 = 0.5$, $K = 1.02$; therefore,

$$Q = 1.02\left[\pi\left(\frac{0.075^2}{4}\right)\right]\sqrt{\frac{2(9.81)(26000)}{997(9.81)}} = 0.032 \text{ m}^3/\text{s} \approx 0.031 \text{ m}^3/\text{s}$$

Therefore, design values: approach diameter = 150 mm, throat diameter = 75 mm, pressure differential = 26 kN/m^2 **Ans**

3.1.5 PARSHALL FLUMES

Figure 3.6 shows the plan and elevation of a Parshall flume. As indicated, the flow enters the flume through a converging zone, then passes through the throat, and out into the diverging zone. For the flume to be a measuring flume, the depth somewhere at the throat must be critical. The converging and the subsequent diverging as well the downward sloping of the throat make this happen. The invert at the entrance to the flume is sloped upward at 1 vertical to 4 horizontal or 25%. Parshall flumes measure the rate of flow proportional to the cross-sectional area of flow. Thus, they are area meters. They also present obstruction to the flow by making the constriction at the throat; thus, they are intrusive meters.

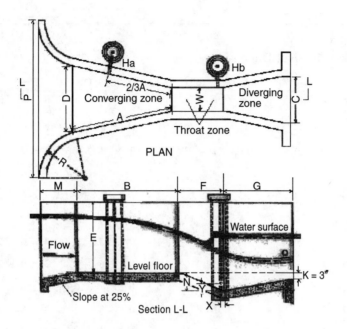

FIGURE 3.6 Plan and sectional view of the Parshall flume.

As defined by Chow (1959), the letter designations for the dimensions are described as follows:

W = size of flume (in terms of throat width)
A = length of side wall of converging section
2/3A = distance back from end of crest to gage point
B = axial length of converging section
C = width of downstream end of flume
D = width of upstream end of flume
E = depth of flume
F = length of throat
G = length of diverging section
K = difference in elevation between lower end of flume and crest of floor
    level = 3 in.
M = length of approach floor
N = depth of depression in throat below crest at level floor
P = width between ends of curve wing walls
R = radius of curved wing walls
X = horizontal distance to H_b gage point from low point in throat
Y = vertical distance to H_b gage point from low point in throat.

The standard dimensions of the Parshall flume are shown in Tables 3.1 and 3.2. If the steps used in deriving the equation for rectangular weirs are applied to the

TABLE 3.1
Standard Parshall Dimensions

W		A		2/3A		B		C		D		E		F	
ft	in.	ft	in.	ft	in.	ft	in.	ft	in.	ft	in.	ft	in.	ft	in.
0	6	2	$\frac{7}{16}$	1	$4\frac{5}{16}$	2	0	1	$3\frac{1}{2}$	1	$3\frac{3}{8}$	2	0	1	0
0	9	2	$10\frac{5}{8}$	1	$11\frac{1}{8}$	2	10	1	3	1	$10\frac{5}{8}$	2	6	1	0
1	0	4	6	3	0	4	$4\frac{7}{8}$	2	0	2	$9\frac{1}{4}$	3	0	2	0
1	6	4	9	3	2	4	$7\frac{7}{8}$	2	6	3	$4\frac{3}{8}$	3	0	2	0
2	0	5	0	3	4	4	$10\frac{7}{8}$	3	0	3	$11\frac{1}{2}$	3	0	2	0
3	0	5	6	3	8	5	$4\frac{3}{4}$	4	0	5	$1\frac{7}{8}$	3	0	2	0
4	0	6	0	4	0	5	$10\frac{5}{8}$	5	0	6	$4\frac{1}{4}$	3	0	2	0
5	0	6	6	4	4	6	$4\frac{1}{2}$	6	0	7	$6\frac{5}{8}$	3	0	2	0
6	0	7	0	4	8	6	$10\frac{3}{8}$	7	0	8	9	3	0	2	0
7	0	7	6	5	0	7	$4\frac{1}{4}$	8	0	9	$11\frac{3}{8}$	3	0	2	0
8	0	8	0	5	4	7	$10\frac{1}{8}$	9	0	11	$1\frac{3}{4}$	3	0	2	0

Parshall flume between any point upstream of the flume and its throat, Equation (3.7) will also be obtained, namely:

$$Q = 0.385\sqrt{2g}L\sqrt{H^3}$$

H may be replaced by H_a, the water surface elevation above flume floor level in the converging zone, and L may also be replaced by W, the throat width. Using a coefficient K, as was done with rectangular weirs, and making the replacements produce

$$Q = K\sqrt{2g}W\sqrt{H_a^3} \tag{3.21}$$

The value of K may be obtained from Figure 3.7 (Roberson et al., 1988). All units should be in MKS (meter-kilogram-second) system.

This equation applies only if the flow is not submerged. Notice in Figure 3.6 that there are two measuring points for water surface elevations: one is labeled H_a, in the converging zone, and the other is labeled H_b, in the throat. These points actually measure the elevations H_a and H_b. The submergence criterion is given by the ratio H_b/H_a. If these ratio is greater than 0.70, then the flume is considered to be submerged and the equation does not apply.

TABLE 3.1
Standard Parshall Dimensions, Continued

W		G		M		N		P		R		Free-Flow Capacity	
												Minimum	Maximum
ft	in.	ft	in.	ft	in.	ft	in.	ft	in.	ft	in.	cfs	cfs
0	6	2	0	1	0	0	$4\frac{1}{2}$	2	$11\frac{1}{2}$	1	4	0.05	3.9
0	9	1	6	1	0	0	$4\frac{1}{2}$	3	$6\frac{1}{2}$	1	4	0.09	8.9
1	0	3	0	1	3	0	9	4	$10\frac{3}{4}$	1	8	0.11	16.1
1	6	3	0	1	3	0	9	5	6	1	8	0.15	24.6
2	0	3	0	1	3	0	9	6	1	1	8	0.42	33.1
3	0	3	0	1	3	0	9	7	$3\frac{1}{2}$	1	8	0.61	50.4
4	0	3	0	1	6	0	9	8	$10\frac{3}{4}$	2	0	1.3	67.9
5	0	3	0	1	6	0	9	10	$1\frac{1}{4}$	2	0	1.6	85.6
6	0	3	0	1	6	0	9	11	$3\frac{1}{2}$	2	0	2.6	103.5
7	0	3	0	1	6	0	9	12	6	2	0	3.0	121.4
8	0	3	0	1	6	0	9	13	$8\frac{1}{4}$	2	0	3.5	139.5

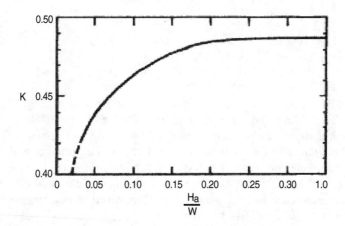

FIGURE 3.7 Discharge coefficient for the Parshall flume.

Example 3.7 **(a)** Design a Parshall flume to measure a rate of flow for a maximum of 30 cfs. **(b)** If the invert of the incoming sewer is set at elevation 100 ft, at what elevation should the invert of the outgoing sewer be set?

Solution: As mentioned previously, in design problems, the designer is at liberty to make any assumption provided she or he can justify it. This means, that two people designing the same unit may not have the same results; however, they must show that their respective designs will work for the purpose intended.

(a) From Table 3.1, for a throat width of 1 ft to 8 ft, the depth E is equal to 3 ft. Thus, allowing a freeboard of 0.5 ft, $H_a = 3.0 - 0.5 = 2.5$ ft. Also, for a throat width of 9 in. $= 0.75$ ft, the depth $E = 2.5$ ft, giving $H_a = 2$ ft for a freeboard of 0.5 ft. From Figure 3.7, the following values of K are obtained for various sizes of throat width, W:

W(ft)	W(m)	E(ft)	H_a(ft)	H_a(m)	H_a/W	K
0.75	0.23	2.5	2.0	0.61	2.67	0.488
2	0.61	3	2.5	0.76	1.25	0.488
3	0.91	3	2.5	0.76	0.83	0.488
4	1.22	3	2.5	0.76	0.63	0.488
5	1.52	3	2.5	0.76	0.50	0.488
6	1.83	3	2.5	0.76	0.42	0.488
7	2.13	3	2.5	0.76	0.34	0.488
8	2.44	3	2.5	0.76	0.31	0.488

Thus, H_a and K are constant for W varying from 0.61 m to 2.44 m. $H_a = 0.61$ and $K = 0.488$ for $W = 0.23$ m. $Q = 30$ cfs $= 30 \, (1/3.283^3) = 0.85$ m³/s. Calculate the values in the following table.

W(m)	$Q = K\sqrt{2g}W\sqrt{H_a^3}$, m³/s
0.23	0.237
0.61	0.874
0.91	1.3
1.22	1.75

The 0.874 m³/s is close to 0.85 m³/s and corresponds to a throat width of 0.61 m $= 2.0$ ft. Since 0.85 m³/s is close to this value, from the table, choose the standard dimensions having a throat width of 2.0 ft $= 0.61$ m.   **Ans**

(b) From the table for a 2-ft flume, M $= 1$ ft 3 in. The entrance to the flume is sloping upward at 25%. Thus, the elevation of the floor level (Refer to Figure 3.6.) is $100 - (1 + 3/12)(0.25) = 99.69$ ft. K, the difference in elevation between lower end of flume and crest of level floor $= 3$ in. Thus, invert of outgoing sewer should be set at $99.69 - 3/12 = 99.44$ ft.   **Ans**

3.2 MISCELLANEOUS FLOW METERS

According to Faraday's law, when a conductor passes through an electromagnetic field, an electromotive force is induced in the conductor that is proportional to the velocity of the conductor. In the actual application of this law in the measurement of the flow of water or wastewater, the salts contained in the stream flow serve as the conductor. The meter is inserted into the pipe containing the flow just as any coupling would be inserted. This meter contains a coil of wire placed around and outside it.

The flowing liquid containing the salts induces the electromotive force in the coil. The induced electromotive force is then sensed by electrodes placed on both sides of the pipe producing a signal that is proportional to the rate of flow. This signal is then sent to a readout that can be calibrated directly in rates of flow. The meter measures the rate of flow by producing a magnetic field, so it is called a *magnetic flow meter*. Magnetic flow meters are nonintrusive, because they do not have any element that obstructs the flow, except for the small head loss as a result of the coupling.

Another flow meter is the *nutating disk meter*. This is widely used to measure the amount of water used in domestic as well as commercial consumption. It has only one moving element and is relatively inexpensive but accurate. This element is a disk. As the water enters the meter, the disk nutates (wobbles). A complete cycle of nutation corresponds to a volume of flow that passes through the disk. Thus, so much of this cycle corresponds to so much volume of flow which can be directly calibrated into a volume readout. A cycle of nutation corresponds to a definite volume of flow, so this flow meter is called a *volume flow meter*. Nutating disk meters are intrusive meters, because they obstruct the flow of the liquid.

Another type of flow meter is the *turbine flow meter*. This meter consists of a wheel with a set of curved blades (turbine blades) mounted inside a duct. The curved blades cause the wheel to rotate as liquid passes through them. The rate at which the wheel rotates is proportional to the rate of flow of the liquid. This rate of rotation is measured magnetically using a blade passing under a magnetic pickup mounted on the outside of the meter. The correlation between the pickup and the liquid rate of flow is calibrated into a readout. Turbine flow meters are also intrusive flow meters; however, because rotation is facilitated by the curved blades, the head loss through the unit is small, despite its being intrusive.

The last flow meter that we will address is the *rotameter*. This meter is relatively inexpensive and its method of measurement is based on the variation of the area through which the liquid flows. The area is varied by means of a float mounted inside the cylinder of the meter. The bore of this cylinder is tapered. With the unit mounted upright, the smaller portion of the bore is at the bottom and the larger is at the top. When there is no flow through the unit, the float is at the bottom. As liquid is admitted to the unit through the bottom, the float is forced upward and, because the bore is tapered in increasing cross section toward the top, the area through which the liquid flows is increased as the flow rate is increased. The calibration in rates of flow is etched directly on the side of the cylinder. Because the method of measurement is based on the variation of the area, this meter is called a *variable-area meter*. In addition, because the float obstructs the flow of the liquid, the meter is an intrusive meter.

3.3 LIQUID LEVEL INDICATORS

Liquid level is a particularly important process variable for the maintenance of a stable plant operation. The operation of the Parshall flume needs the water surface elevation to be determined; for example, how are H_a and H_b determined? As may be deduced from Figure 3.6, they are measured by the float chambers labeled H_a and H_b, respectively. Liquid levels are measured by gages such as floats (as in the case of the Parshall flume), pressure cells or diaphragms, pneumatic tubes and other

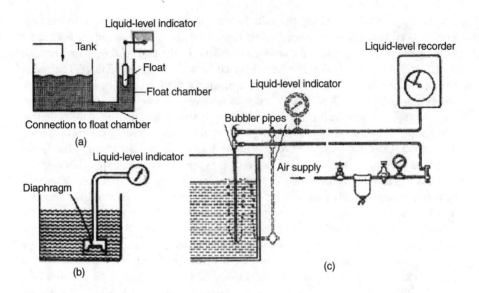

FIGURE 3.8 Liquid-level measuring gages.

devices that use capacitance probes and acoustic techniques. Figure 3.8 shows a float-gaging arrangement (**a**), a pressure cell (**b**), and a pneumatic tube sensing indicator (**c**).

As shown in the figure, float gaging is implemented using a float that rests on the surface of the liquid inside the float chamber. As water or wastewater enters the tank, the liquid level rises increasing the head. The increase in head causes the liquid to flow to the float chamber through the connecting pipe. The liquid level in the chamber then rises. This rise is sensed by the float which communicates with the liquid-level indicator. The indicator can be calibrated to read the liquid level in the tank directly.

In the pressure-cell measuring arrangement, a sensitive diaphragm is installed at the bottom of the tank. As the liquid enters the tank, the increase in head pushes against the diaphragm. The pressure is then communicated to the liquid-level indicator, which can be calibrated to read directly in terms of the level in the tank.

The pneumatic tubes shown in (**c**) relies on a continuous supply of air into the system. The air is purged into the bottom of the tank. As the liquid level in the tank rises, more pressure is needed to push the air into the bottom of the tank. Thus, the pressure required to push the air into the system is a measure of the liquid level in the tank. As shown in the figure, the indicator and recorder may be calibrated to read levels in the tank directly.

3.4 FLOW AND QUALITY EQUALIZATIONS

In order for a wastewater treatment unit to operate efficiently, the loading, both hydraulic and quality should be uniform. An example of this hydraulic loading is the flow rate into a basin, and an example of quality is the BOD in the inflow; however, this kind of condition is impossible to attain under natural conditions. For example, refer to Figure 3.9. The flow varies from a low of 18 m^3/h to a high of 62 m^3/h, and the BOD varies from a low of 27 mg/L to a high of 227 mg/L. To ameliorate the

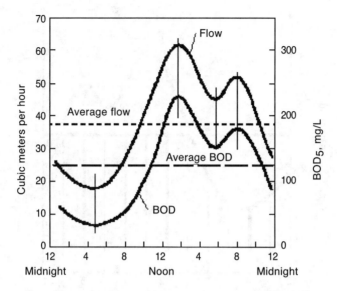

FIGURE 3.9 Long-term extreme sewage flow and BOD pattern in a sewage treatment plant.

difficulty imposed by these extreme variations, an equalization basin should be provided. *Equalization* is a unit operation applied to a flow for the purpose of smoothing out extreme variations in the values of the parameters.

In order to produce an accurate analysis of equalization, a long-term extreme flow pattern for the wastewater flow over the duration of a day or over the duration of a suitable cycle should be established. By *extreme flow pattern* is meant diurnal flow pattern or pattern over the cycle where the values on the curve are peak values—that is, values that are not equaled or exceeded. For example, Figure 3.10 is a flow pattern over a day. If this pattern is an extreme flow pattern then 18 m^3/h is the largest of all the smallest flows on record, and 62 m^3/h is the largest of all the largest flows on record. A similar statement holds for the BOD. To repeat, if the figure is a pattern for extreme values, any value on the curve represents the largest value ever recorded for a particular category. In order to arrive at these extreme values, the probability distribution analysis discussed in Chapter 1 should be used. Remember, that in an array of descending order, extreme values have the probability zero of being equaled or exceeded. In addition to the extreme values, the daily mean of this extreme flow pattern should also be calculated. This mean may be called the *extreme daily mean* and is needed to size the pump that will withdraw the flow from the equalization unit. In the figure, the extreme daily means for the flow and the BOD are identified by the label designated as average.

Now, to derive the equalization required, refer to Figure 3.10. The curve represents inflow to an equalization basin. The unit on the ordinate is m^3/h and that on the abscissa is hours. Thus, any area of the curve is volume. The line identified as average represents the mean rate of pumping of the inflow out of the equalization basin. The area between the inflow curve and this average (or mean) labeled B is the area representing the volume not withdrawn by pumping out at the mean rate;

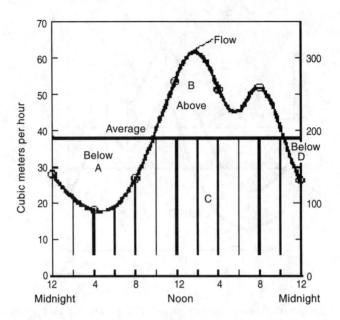

FIGURE 3.10 Determination of equalization basin storage.

it is an excess inflow volume over the volume pumped out at the time span indicated (9:30 a.m. to 10:30 p.m.). The two areas below the mean line labeled A and D represent the excess capacity of the pump over the incoming flow, also, at the times indicated (12:00 a.m. to 9:30 a.m. and 10:30 p.m. to 12:00 a.m.).

The excess inflow volume over pumpage volume, area B, and the excess pumpage volume over inflow volume, areas A plus D, must somehow be balanced. The principle involved in the sizing of equalization basins is that the total amount withdrawn (or pumped out) over a day or a cycle must be equal to the total inflow during the day or the cycle. The total amount withdrawn can be equal to withdrawal pumping at the mean flow, and this is represented by areas A, C, and D. Let these volumes be V_A, V_C, and V_D, respectively. The inflow is represented by the areas B and C. Designate the corresponding volumes as V_B and V_C. Thus, inflow equals outflow,

$$V_A + V_C + V_D = V_B + V_C \qquad (3.22)$$

$$V_A + V_D = V_B \qquad (3.23)$$

From this result, the excess inflow volume over pumpage, V_B, is equal to the excess pumpage over inflow volume, $V_A + V_D$. In order to avoid spillage, the excess inflow volume over pumpage must be provided storage. This is the volume of the equalization basin—volume V_B. From Equation (3.23), this volume is also equal to the excess pumpage over inflow volume, $V_A + V_D$.

Let the total number of measurements of flow rate be ξ and Q_i be the flow rate at time t_i. The mean flow rate, Q_{mean}, is then

$$Q_{mean} = \frac{1}{t_\xi - t_1} \sum_{i=2}^{\xi} \frac{Q_i + Q_{i-1}}{2} (t_i - t_{i-1}) \qquad (3.24)$$

t_ξ = time of sampling of the last measurement. Q_{mean} is the equalized flow rate. Considering the excess over the mean as the basis for calculation, the volume of the equalization basin, V_{basin} is

$$V_{basin} = \sum_{i=2}^{i=\xi} pos \ of \left(\frac{Q_i + Q_{i-1}}{2} - Q_{mean} \right)(t_i - t_{i-1}) \tag{3.25}$$

where $pos \ of \ ((Q_i + Q_{i-1})/2 - Q_{mean})$ means that only positive values are to be summed. By Equation (3.23), using the area below the mean, V_{basin} may also be calculated as

$$V_{basin} = \sum_{i=2}^{i=\xi} \left| neg \ of \left(\frac{Q_i + Q_{i-1}}{2} - Q_{mean} \right)(t_i - t_{i-1}) \right| \tag{3.26}$$

$neg \ of \ ((Q_i + Q_{i-1})/2 - Q_{mean})(t_i - t_{i-1})$ means that only negative values are to be summed. The final volume of the basin to be adopted in design may be considered to be the average of the "posof" and "negof" calculations.

Examples of quality parameters are BOD, suspended solids, total nitrogen, etc. The calculation of the values of quality parameters should be done right before the tank starts filling from when it was originally empty. Let $C_{i-1,i}$ be the quality value of the parameter in the equalization basin during a previous interval between times t_{i-1} and t_i and $C_{i,i+1}$ during the forward interval between times t_i and t_{i+1}. Let the corresponding volumes of water remaining in the tank be $V_{rem_{i-1,i}}$ and $V_{remi,i+1}$, respectively. Also, let C_i be the quality value of the parameter from the inflow at time t_i, C_{i+1} the quality value from the inflow at t_{i+1}, Q_i the inflow at t_i, and Q_{i+1} the inflow at t_{i+1}. Then,

$$
C_{i,i+1} = \frac{C_{i-1,i}(V_{remi-1,i}) + \left(\dfrac{C_i + C_{i+1}}{2}\right)\left(\dfrac{Q_i + Q_{i+1}}{2}\right)(t_{i+1} - t_i) - C_{i,i+1}Q_{mean}(t_{i+1} - t_i)}{(V_{remi-1,i}) + \left(\dfrac{Q_i + Q_{i+1}}{2}\right)(t_{i+1} - t_i) - Q_{mean}(t_{i+1} - t_i)}
$$

$$
= \frac{C_{i-1,i}(V_{remi-1,i}) + \left(\dfrac{C_i + C_{i+1}}{2}\right)\left(\dfrac{Q_i + Q_{i+1}}{2}\right)(t_{i+1} - t_i)}{(V_{remi-1,i}) + \left(\dfrac{Q_i + Q_{i+1}}{2}\right)(t_{i+1} - t_i)} \tag{3.27}
$$

$V_{remi-1,i}$ is the volume of wastewater remaining in the equalization basin at the end of the previous time interval, t_{i-1} to t_i and, thus, the volume at the beginning of the forward time interval, t_i to t_{i+1}. $C_{i-1,i}$ ($V_{remi-1,i}$) is the total value of the quality inside the tank at the end of the previous interval; thus, it is also the total value of the quality at the beginning of the forward interval. $(C_i + C_{i+1})/2$ is the average value of the parameter in the forward interval and $(Q_i + Q_{i+1})/2$ is the average value of the inflow in the interval. Thus, $((C_i + C_{i+1})/2)((Q_i + Q_{i+1})/2)(t_{i+1} - t_i)$ is total value of the quality coming from the inflow during the forward interval. $C_{i,i+1}$ is the equalized quality value during the time interval from t_i to t_{i+1}. $C_{i,i+1} \ Q_{mean} \ (t_{i+1} - t_i)$ is the value of the quality withdrawn from the basin during the interval to t_i to t_{i+1}.

The sizing of the equalization basin should be based on an identified cycle. Strictly speaking, this cycle can be any length of time, but, most likely, would be the length of the day, as shown in Figure 3.10. Having identified the cycle, assume, now, that the pump is withdrawing out the inflow at the average rate of Q_{mean}. For the pump to be able to withdraw at this rate, there must already have been sufficient water in the tank. As the pumping continues, the level of water in the tank goes down, if the inflow rate is less than the average. The limit of the going down of the water level is the bottom of the tank. If the inflow rate exceeds pumping as this limit is reached, the level will start to rise. The volume of the basin during the leveling down process starting from the highest level until the water level hits bottom is the volume V_{basin}.

Let t_{ibot} be this particular moment when the water level hits bottom and the inflow exceeds pumping. Then at the interval between t_{ibot-1} and t_{ibot}, the accumulation of volume in the tank, $V_{remi-1,i} = V_{remibot-1,ibot}$ is 0. At any other interval between t_{i-1} and t_i when the tank is now filling,

$$V_{remi-1,i} = V_{remi-2,i-1} + \left(\frac{Q_{i-1} + Q_i}{2} - Q_{mean} \right)(t_i - t_{i-1}) \qquad (3.28)$$

The value of $V_{remi-1,i}$ will always be positive or zero. It is zero at the time interval between t_{ibot} and t_{ibot-1} and positive at all other times until the water level hits bottom again.

The calculation for the equalized quality should be started at the precise moment when the level hits bottom or when the tank starts filling up again. Referring to Figure 3.10, at around 10:30 p.m., because the inflow has now started to be less than the pumping rate, the tank would start to empty and the level would be going down. This leveling down will continue until the next day during the span of times that the inflow is less than the pumping rate. From the figure, these times last until about 9:30 a.m. Thus, the very moment that the level starts to rise again is 9:30 a.m. and this is the precise moment that calculation of the equalized quality should be started, using Equations (3.27) and (3.28).

Example 3.8 The following table was obtained from Figure 3.10 by reading the flow rates at 2-h intervals. Compute the equalized flow.

Hour Ending	Q (m³/h)	Hour Ending	Q (m³/h)
12:00 a.m.	26	2:00	62
2:00	22	4:00	51
4:00	18	6:00	45
6:00	19	8:00	51
8:00	27	10:00	40
10:00	39	12:00 a.m.	26
12:00 p.m.	52		

Solution:

$$Q_{mean} = \frac{1}{t_\xi - t_1}\sum_{i=2}^{\xi}\frac{Q_i + Q_{i-1}}{2}(t_i - t_{i-1}) = \frac{1}{24 - 0}\left\{\frac{22 + 26}{2}(2) + \frac{18 + 22}{2}(2)\right.$$

$$+ \frac{19 + 18}{2}(2) + \frac{27 + 19}{2}(2) + \frac{39 + 27}{2}(2) + \frac{52 + 39}{2}(2) + \frac{62 + 52}{2}(2)$$

$$+ \frac{51 + 62}{2}(2) + \frac{45 + 51}{2}(2) + \frac{51 + 45}{2}(2) + \frac{40 + 51}{2}(2) + \left.\frac{40 + 51}{2}(2)\right\}$$

$$= 37.7 \text{ m}^3/\text{hr} \quad \textbf{Ans}$$

Example 3.9 Using the data in Example 3.8, design the equalization basin.

Solution:

$$V_{basin} = \sum_{i=2}^{i=\xi}pos\ of\left(\frac{Q_i + Q_{i-1}}{2} - Q_{mean}\right)(t_i - t_{i-2}) = pos\ of\left\{\left(\frac{22 + 26}{2} - 37.7\right)(2)\right.$$

$$+ \left(\frac{18 + 22}{2} - 37.7\right)(2) + \left(\frac{19 + 18}{2} - 37.7\right)(2) + \left(\frac{27 + 19}{2} - 37.7\right)(2)$$

$$+ \left(\frac{39 + 27}{2} - 37.7\right)(2) + \left(\frac{52 + 39}{2} - 37.7\right)(2) + \left(\frac{62 + 52}{2} - 37.7\right)(2)$$

$$+ \left(\frac{51 + 62}{2} - 37.7\right)(2) + \left(\frac{45 + 51}{2} - 37.3\right)(2) + \left(\frac{51 + 45}{2} - 37.7\right)(2)$$

$$+ \left(\frac{40 + 51}{2} - 37.7\right)(2) + \left.\left(\frac{26 + 40}{2} - 37.7\right)(2)\right\}$$

$$= \left(\frac{52 + 39}{2} - 37.7\right)(2) + \left(\frac{62 + 52}{2} - 37.7\right)(2) + \left(\frac{51 + 62}{2} - 37.7\right)(2)$$

$$+ \left(\frac{45 + 51}{2} - 37.3\right)(2) + \left(\frac{51 + 45}{2} - 37.7\right)(2) + \left.\left(\frac{40 + 51}{2} - 37.7\right)(2)\right\}$$

$$= 15.6 + 38.6 + 37.6 + 20.6 + 20.6 + 15.6 = 148.6 \text{ m}^3$$

Use a circular basin at a height of 4 m. Therefore,

$$148.6 = 2\left(\frac{\pi D^2}{4}\right) \Rightarrow D = 9.72 \text{ m, say 10 m}$$

Therefore, dimensions: height = 4 m, diameter = 10 m; use two tanks, one for standby. **Ans**

Example 3.10 The following table shows the BOD values read from Figure 3.9 at intervals of 2 h. Along with the data in Example 3.8, calculate each equalized value of the BOD at every time interval when the tank is filling.

Hour Ending	BOD$_5$ (mg/L)	Hour Ending	BOD$_5$ (mg/L)
12:00 a.m.	75	2:00	235
2:00	50	4:00	175
4:00	42	6:00	151
6:00	42	8:00	181
8:00	52	10:00	135
10:00	100	12:00 a.m.	75
12:00 p.m.	175		

Solution: The tank starts filling at 9:30 a.m.; therefore, calculation will be started at this time.

$$C_{i,i+1} = \frac{C_{i-1,i}\,(V_{remi-1,i}) + \left(\frac{C_i + C_{i+1}}{2}\right)\left(\frac{Q_i + Q_{i+1}}{2}\right)(t_{i+1} - t_i)}{(V_{remi-1,i}) + \left(\frac{Q_i + Q_{i+1}}{2}\right)(t_{i+1} - t_i)}$$

$$V_{remi-1,i} = V_{remi-2,i-1} + \left(\frac{Q_{i-1} + Q_i}{2} - Q_{mean}\right)(t_i - t_{i-1})$$

t_i	BOD = C_i	Q = Q_i	$\frac{C_i + C_{i+1}}{2}$	$\frac{C_{i-1} + C_i}{2}$	$\frac{Q_i + Q_{i+1}}{2}$	$\frac{Q_{i-1} + Q_i}{2}$	$t_{i+1} - t_i$	$V_{remi-1,i}$	$C_{i,i+1}$
8:00	52	27	76	26	33	23	2	3.44	101
10:00	100	39	137.5	76	45.5	33	2	−5.96⇒0	137.5
12:00	175	52	205	137.5	57	45.5	2	15.6[a]	196.88[b]
14:00	235	62	205	205	56.5	57	2	54.2	202.37
16:00	175	51	163	205	48	56.5	2	91.8	182.24
18:00	151	45	166	163	48	48	2	112.24	174.75
20:00	181	51	158	166	45.5	48	2	127.84	167.78
22:00	135	40	105	158	33	45.5	2	143.44	148.0
24:00	75	26	62.5	105	24	33	2	134.04	125.46
2:00	50	22	46	62.5	20	24	2	106.64	103.79
4:00	42	18	42	46	18.5	20	2	71.24	82.67
6:00	42	19	47	42	23	18.5	2	32.84	61.86

[a] $V_{remi-1,i} = V_{remi-2,i-1} + \left(\frac{Q_{i-1} + Q_i}{2} - Q_{mean}\right)(t_i - t_{i-1}) = 0 + (45.5 - 37.7)(2) = 15.6$

[b] $C_{i,i+1} = \dfrac{C_{i-1,i}(V_{remi-1,i}) + \left(\frac{C_i + C_{i+1}}{2}\right)\left(\frac{Q_i + Q_{i+1}}{2}\right)(t_{i+1} - t_i)}{(V_{remi-1,i}) + \left(\frac{Q_i + Q_{i+1}}{2}\right)(t_{i+1} - t_i)} = \dfrac{137.5(15.6) + (205)(57)(2)}{(15.6) + (57)(2)} = 196.88$

GLOSSARY

Cipolleti weir—A trapezoidal weir where the notch angle compensates for the reduction in flow due to contraction.

Control section—A section in an open channel where a one-to-one relationship exists between flow and depth.

Converging zone—The portion in a venturi meter, Parshall flume, or Palmer–Bowlus flume where the cross section is progressively reduced.

Diverging zone—The portion in a venturi meter, Parshall flume, or Palmer–Bowlus flume where the cross section is progressively expanded.

Equalization—A unit operation applied to a flow for the purpose of smoothing out extreme variations in the values of the parameters.

Extreme daily mean—The mean flow rate of the extreme flow pattern.

Extreme flow pattern—Diurnal flow pattern or pattern over the cycle where the values on the curve are peak values—that is, values that are not equaled or exceeded.

Flow meters—Devices used to measure the rate of flow of fluids.

Froude number—Defined as $V/\sqrt{gD}$.

Fully contracted rectangular weir—Rectangular weir where the flow in the channel being measured contracts as it passes through the rectangular opening.

Hydraulic depth—In an open channel, the ratio of the cross-sectional area of flow to the top width.

Parshall flume—Venturi flume invented by Parshall.

Piezometric ring—Pressure sensing holes that form a concentric circle around the center of the pipe.

Rectangular weir—Thin plate where the plate is being cut such that a rectangular opening is formed through which the flow in the channel that is being measured passes through.

Sewer—Pipe that conveys sewage.

Suppressed rectangular weir—Rectangular weir where the contraction is absent, that is, the contraction is suppressed.

Throat—The portion in a venturi meter, Parshall flume, or Palmer–Bowlus flume where the cross section is held constant.

Triangle weirs—Weirs in which the cross-sectional area of flow where the flow passes through is in the form of a triangle, also called V-notch weirs.

Venturi flume—An open-channel measuring device with a longitudinal section that is shaped like a venturi meter.

Venturi meter—Meter used to measure flow in pipes by inducing a pressure differential through reducing the cross section until reaching the throat, maintaining cross section constant throughout the throat, and expanding cross section after the throat.

Weir—Obstruction used to back up a flowing stream of liquid.

SYMBOLS

A	Cross-sectional area of flow
A_c	Cross-sectional area of flow at critical depth
A_t	Cross-sectional area of throat of venturi meter
A_1	Cross-sectional area at section 1
A_2	Cross-sectional area at section 2
b	Bottom width of a trapezoidal section
C_i	Value of inflow of quality into equalization basin at time t_i
C_{i+1}	Value of inflow of quality into equalization basin at time t_{i+1}
$C_{tnki-1,i}$	Value of quality inside equalization basin between time intervals t_{i-1} and t_i
$C_{tnki,i+1}$	Value of quality inside equalization basin between time intervals t_i and t_{i+1}
d	Venturi throat diameter
d_c	Depth at critical section
D	Hydraulic depth; pipe diameter
E	Specific energy at section of open channel
g	Gravitational constant = 9.81 m/s^2
h_l	Head loss, m
Δh	Manometer deflection
$\Delta h_{\mathrm{H_2O}}$	Manometer deflection in equivalent height of water
H	Head over weir
H_a	Upstream head in a Parshall flume
K	Flow coefficient
P_1	Pressure at point 1
P_2	Pressure at point 2
P	Height of weir
Q_i	Inflow rate of flow into an equalization basin at time t_i
Q_{i+1}	Inflow rate of flow into an equalization basin at time t_{i+1}
Q_{mean}	Mean rate of withdrawal from an equalization basin
t_i	Time at index i of calculation
t_{i+1}	Time at index $i + 1$ of calculation
T	Top width of channel
T_c	Top width at critical section
V	Average velocity at cross section of conduit
V_1	Average velocity at cross section at point 1
V_2	Average velocity at cross section at point 2
$(V_{acci-1,i})$	Accumulated volume inside equalization basin between time intervals t_{i-1} and t_i
$V_{acci,i+1}$	Accumulated volume inside equalization basin between time intervals t_i and t_{i+1}
V_{basin}	Volume of equalization basin
V_c	Average velocity at critical section of open channel
W	Throat width of Parshall flume
y	Depth of channel
y_1	Open-channel depth at point 1

y_2	Open-channel depth at point 2
y_c	Depth at control section of open channel (the critical depth)
z	Side slope of a trapezoidal section
z_1	Bottom elevation of open channel at point 1
z_2	Bottom elevation of open channel at point 2
β	Ratio of throat diameter to diameter of pipe, d/D
γ	Specific weight of water
γ_{ind}	Specific weight of manometer indicator fluid
θ	Degrees or the 28-degree angle in the Cipolleti weir

PROBLEMS

3.1 The system in Figure 3.1 is indicating a flow of 0.3 m^3/s. The head over the weir and its length are 0.2 m and 2 m, respectively. Calculate the height of the weir.

3.2 The system in Figure 3.1 is indicating a flow of 0.3 m^3/s. The head over the weir and its height are 0.2 m and 1 m, respectively. Calculate the length of the weir.

3.3 The head over a fully contracted weir of length equal to 2 m is 0.2 m. If the height of the weir is 1 m, what is the discharge?

3.4 A suppressed weir measures a flow in an open channel at a rate of 0.3 m^3/s. The length and head over the weir are 0.2 m and 2.0 m, respectively. Calculate the height of the weir.

3.5 A suppressed weir measures a flow in an open channel at a rate of 0.3 m^3/s. The height and head over the weir are 1 m and 0.2 m, respectively. Calculate the length of the weir.

3.6 The head over a suppressed weir of length equal to 2 m is 0.2 m. If the height of the weir is 1 m, what is the discharge?

3.7 The head over a Cipolleti weir of length equal to 2 m is 0.2 m. If the height of the weir is 1 m, what is the discharge?

3.8 Two triangular weirs are put in series in an open channel. The upstream weir has a notch of 60° while the downstream weir has a notch of 90°. If the head over the 90-degree notched weir is 0.5 m, what is the head over the 60-degree notched weir?

3.9 Two weirs are placed in series in an open channel. The upstream weir is a Cipolleti weir and the other is a 90-degree notched weir. What is the discharge through the Cipolleti weir if the head over the notched weir is 0.5 m?

3.10 A suppressed weir is placed upstream of a V-notch weir in an open channel. The height and head over the weir of the upstream weir are 1 m and 0.2 m, respectively. Calculate the length of the suppressed weir if the head over the V-notch weir is 0.5 m?

3.11 The upstream diameter of a venturi meter is 150 mm and its throat diameter is 75 mm. Determine the pressure at point 1 if the discharge is 0.031 m^3/s. The pressure at point 2 is 170 kN/m^2.

3.12   The diameter of a pipe whose discharge flow of 0.031 m^3/s as measured by a venturi meter is 150 mm. Pressure readings taken at points 1 and 2 are, respectively, 196 kN/m^2 and 170 kN/m^2. Calculate the diameter of the throat.

3.13   In the following figure, the diameter of the pipe D is 35 cm and the throat diameter of the venturi meter d is 15 cm. What is the discharge if H = 5 m?

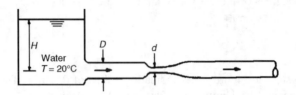

3.14   The manometer deflection of a venturi meter in terms of equivalent water is 205 mm. If the throat diameter is 112.1 mm and the venturi is used to measure flow in a 250-mm pipe, calculate the rate of discharge.

3.15   If the manometer deflection in terms of equivalent water is 112.1 mm, what is the actual deflection if mercury is used as the indicator fluid?

3.16   Water at 20°C flows through a venturi meter that has a throat of 25 cm. The venturi meter is measuring the flow in a 65-cm pipe. Calculate the deflection in a mercury manometer if the rate of discharge is 0.70 m^3/s.

3.17   Design a Parshall flume to measure a range of flows of 20 cfs to 100 cfs.

3.18   The throat width of a Parshall flume is 2.0 ft. If the discharge flow is 30 cfs, calculate the upstream head H_a.

3.19   The invert elevation of the outgoing sewer in Problem 3.17 has been set at 99.44 ft. At what elevation is the invert elevation of the incoming sewer set?

3.20   As an exercise, read off from Figure 3.10 at 1-h intervals and compare your results with the results of Example 3.8.

3.21   Solve Example 3.8 by assuming 12 intervals and compare your results with the values you read from the figure at 2-h intervals.

3.22   In Figure 3.10, how should the values at time 0 and time 24 hours later compare? Why?

3.23   Compute the equalized flow in Example 3.8 if the number of intervals is 48.

3.24   Compute the equalized flow in Example 3.8 if the number of intervals is 12. What is the other name of equalized flow?

3.25   In calculating the equalized values of the quality, the calculation should start when the tank is full. Is this a valid statement? If the calculation is not to start when the tank is full, when then should it be started and why?

3.26   The equalized volume is not the volume of the equalization basin. Is this statement true?

BIBLIOGRAPHY

ASME (1959). *Fluid Meters—Their Theory and Application*. American Society of Mechanical Engineers, Fairfield, NJ.

Babitt, H. E. and E. R. Baumann (1958). *Sewerage and Sewage Treatment*. John Wiley & Sons, New York, 132.

Chow, V. T. (1959). *Open-Channel Hydraulics*. McGraw-Hill, New York, 74–81.

Escritt, L. B. (1965). *Sewerage and Sewage Disposal, Calculations and Design.* p. 95. C. R. Books Ltd., London, 95.

Glasgow, G. D. E. and A. D. Wheatley (1997). Effect of surges on the performance of rapid gravity filtration. *Water Science Technol. Proc. 1997 1st IAWQ-IWSA Joint Specialist Conf. Reservoir Manage. Water Supply—An Integrated System,* May 19–23, Prague, Czech Republic, 37, 2, 8, 75–81. Elsevier Science Ltd., Exeter, England.

Hadjivassilis, I., L. Tebai, and M. Nicolaou (1994). Joint treatment of industrial effluent: Case study of limassol industrial estate. *Water Science Technol. Proc. IAWQ Int. Specialized Conf. Pretreatment of Industrial Wastewaters,* Oct. 13–15, 1993, Athens, Greece, 29, 9, 99–104, Pergamon Press, Tarrytown, NY.

Hanna, K. M., J. L. Kellam, and G. D. Boardman (1995). Onsite aerobic package treatment systems. *Water Res.* 29, 11, 2530–2540.

Hinds, J. (1928). Hydraulic design of flume and siphon transitions. *Trans. ASCE,* 92, 1423.

Johansen, F. C. (1930). *Proc. Roy. Soc. London,* Series A, 125.

Johnson, M. (1997). Remote Turbidity Measurement with a Laser Reflectometer. *Water Science Technol. Proc. 7th Int. Workshop on Instrumentation, Control and Automation of Water and Wastewater Treatment and Transport Syst.,* July 6–9, Brighton, England, 37, 12, 255–261. Elsevier Science Ltd., Exeter. England.

Kindsvater, C. E. and R. W. Carter (1959). Discharge characteristics of rectangular thin-plate weirs. *Trans. ASCE,* 124.

Lenz, A. T. (1943). Viscosity and surface tension effects on V-notch weir coefficients. *Trans. AICHE,* 108, 759–820.

Metcalf & Eddy, Inc. (1981). *Wastewater Engineering: Collection and Pumping of Wastewater.* McGraw-Hill, New York, 89.

Munson, B. R., D. F. Young, and T. H. Okiishi (1994). *Fundamentals of Fluid Mechanics*. John Wiley & Sons, Inc., New York, 676–680.

Roberson, J. A., J. J. Cassidy, and M. H. Chaudry (1988). *Hydraulic Engineering*. Houghton Mifflin, Boston, 68, 211, 215.

Schaarup-Jensen, K., et al. (1998). Danish sewer research and monitoring station. *Water Science Technol. Proc. 1997 2nd IAWQ Int. Conf. Sewer as a Physical, Chemical and Biological Reactor,* May 25–28, 1997, Aalborg, Denmark, 37, 1, 197–204. Elsevier Science Ltd., Exeter, England.

Sincero, A. P. and G. A. Sincero (1996). *Environmental Engineering: A Design Approach.* Prentice Hall, Upper Saddle River, NJ, 71–72.

4 Pumping

Pumping is a unit operation that is used to move fluid from one point to another. This chapter discusses various topics of this important unit operation relevant to the physical treatment of water and wastewater. These topics include pumping stations and various types of pumps; total developed head; pump scaling laws; pump characteristics; best operating efficiency; pump specific speed; pumping station heads; net positive suction head and deep-well pumps; and pumping station head analysis.

4.1 PUMPING STATIONS AND TYPES OF PUMPS

The location where pumps are installed is a *pumping station*. There may be only one pump, or several pumps. Depending on the desired results, the pumps may be connected in parallel or in series. In *parallel connection*, the discharges of all the pumps are combined into one. Thus, pumps connected in parallel increases the discharge from the pumping station. On the other hand, in *series connection*, the discharge of the first pump becomes the input of the second pump, and the discharge of the second pump becomes the input of the third pump and so on. Clearly, in this mode of operation, the head built up by the first pump is added to the head built up by the second pump, and the head built up by the second pump is added to the head built by the third pump and so on to obtain the total head developed in the system. Thus, pumps connected in series increase the total head output from a pumping station by adding the heads of all pumps. Although the total head output is increased, the total output discharge from the whole assembly is just the same input to the first pump.

Figure 4.1 shows section and plan views of a sewage pumping station, indicating the parallel type of connection. The discharges from each of the three pumps are conveyed into a common manifold pipe. In the manifold, the discharges are added. As indicated in the drawing, a *manifold pipe* has one or more pipes connected to it. Figure 4.2 shows a schematic of pumps connected in series. As indicated, the discharge flow introduced into the first pump is the same discharge flow coming out of the last pump.

The word pump is a general term used to designate the unit used to move a fluid from one point to another. The fluid may be contaminated by air conveying fugitive dusts or water conveying sludge solids. Pumps are separated into two general classes: the centrifugal and the positive-displacement pumps. *Centrifugal pumps* are those that move fluids by imparting the tangential force of a rotating blade called an *impeller* to the fluid. The motion of the fluid is a result of the *indirect action* of the impeller. *Displacement pumps*, on the other hand, literally push the fluid in order to move it. Thus, the action is *direct, positively* moving the fluid, thus the name *positive-displacement pumps*. In centrifugal pumps, flows are introduced into the

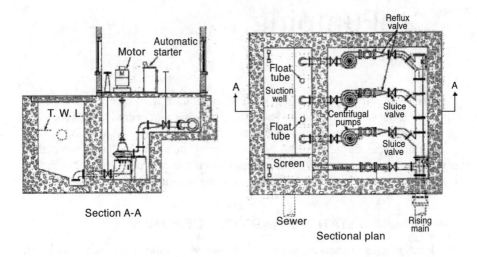

FIGURE 4.1 Plan and section of a pumping station showing parallel connections.

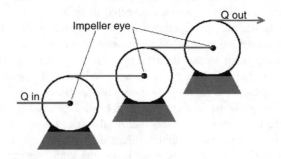

FIGURE 4.2 Pumps connected in series.

unit through the eye of the impeller. This is indicated in Figure 4.2 where the "Q in" line meets the "eye." In positive-displacement pumps, no eye exists.

The left-hand side of Figure 4.3 shows an example of a positive-displacement pump. Note that the screw pump literally pushes the wastewater in order to move it. The right-hand side shows a cutaway view of a deep-well pump. This pump is a centrifugal pump having two impellers connected in series through a single shaft forming a two-stage arrangement. Thus, the head developed by the first stage is added to that of the second stage producing a much larger head developed for the whole assembly. As discussed later in this chapter, this series type of connection is necessary for deep wells, because there is a limit to the depth that a single pump can handle.

Figure 4.4 shows various types of impellers that are used in centrifugal pumps. The one in **a** is used for axial-flow pump. *Axial-flow pumps* are pumps that transmit the fluid pumped in the axial direction. They are also called *propeller pumps*, because the impeller simply propels the fluid forward like the movement of a ship with propellers. The impeller in **d** has a shroud or cover over it. This kind of design can develop more

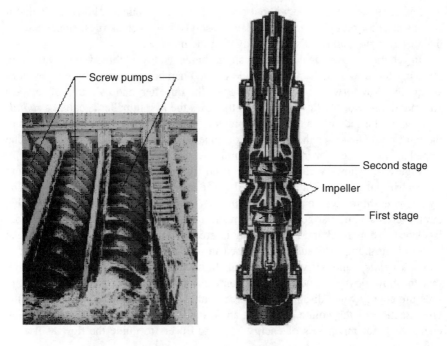

FIGURE 4.3 A screw pump, an example of a positive-displacement pump (left); cutaway view of a deep-well pump (right).

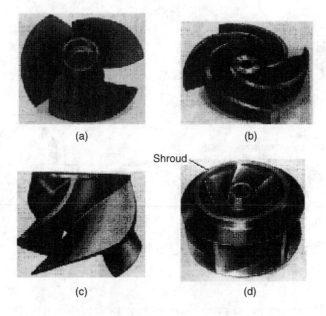

FIGURE 4.4 Various types of pump impellers: (a) axial flow; (b) open type; (c) mix-flow type; and (d) shrouded impeller.

head as compared to the one without a shroud. The disadvantage, however, is that it is not suited for pumping liquids containing solids in it, such as rugs, stone, and the like, because these materials may easily clog the impeller.

In general, a centrifugal impeller can discharge its flow in three ways: by directly throwing the flow radially into the side of the chamber circumscribing it, by conveying the flow forward by proper design of the impeller, and by a mix of forward and radial throw of the flow. The pump that uses the first impeller is called a *radial-flow pump*; the second, as mentioned previously, is called the axial-flow pump; and the third pump that uses the third type of impeller is called a *mixed-flow pump*. The impeller in **c** is used for mixed-flow pumps.

Figure 4.5 shows various impellers used for positive-displacement pumps and for centrifugal pumps. The figures in **d** and **e** are used for centrifugal pumps. The figure in **e** shows the impeller throwing its flow into a discharge chamber that circumscribes a circular geometry as a result of the impeller rotating. This chamber is shaped like a spiral and is expanding in cross section as the flow moves into the outlet of the pump. Because it is shaped into a spiral, this expanding chamber is called a *volute*—another word for spiral. In centrifugal pumps, the kinetic energy that the flow possesses while in the confines of the impeller is transformed into pressure energy when discharged into the volute. This progressive expansion of the cross section of the volute helps in transforming the kinetic energy into pressure energy without much loss of energy. Using diffusers to guide the flow as it exits

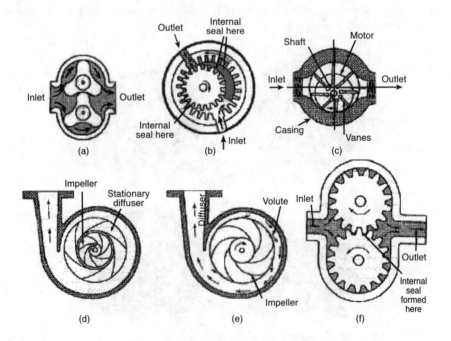

FIGURE 4.5 Various types of pump impellers, continued: (a) lobe type; (b) internal gear type; (c) vane type; (d) impeller with stationary guiding diffuser vanes; (e) impeller with volute discharge; and (f) external gear type impeller.

from the tip of the impeller into the volute is another way of avoiding loss of energy. This type of design is indicated in **d**, showing stationary diffusers as the guide.

The figure in **a** is a *lobe pump*, which uses the lobe impeller. A lobe pump is a positive-displacement pump. As indicated, there is a pair of lobes, each one having three lobes; thus, this is a three-lobe pump. The turning of the pair is synchronized using external gearings. The clearance between lobes is only a few thousandths of a centimeter, thus only a small amount of leakage passes the lobes. As the pair turns, the water is trapped in the "concavity" between two adjacent lobes and along with the side of the casing is positively moved forward into the outlet. The figures in **b** and **f** are gear pumps. They basically operate on the same principle as the lobe pumps, except that the "lobes" are many, which, actually, are now called *gears*. Adjacent gear teeth traps the water which, then, along with the side of the casing, moves the water to the outlet. The gear pump in **b** is an *internal* gear pump, so called because a smaller gear rolls around the inside of a larger gear. (The smaller gear is internal and inside the larger gear.) As the smaller gear rolls, the larger gear also rolls dragging with it the water trapped between its teeth. The smaller gear also traps water between its teeth and carries it over to the crescent. The smaller and the larger gears eventually throw their trapped waters into the discharge outlet. The gear pump in **f** is an *external* gear pump, because the two gears are contacting each other at their peripheries (external). The pump in **c** is called a *vane pump*, so called because a vane pushes the water forward as it is being trapped between the vane and the side of the casing. The vane pushes firmly against the casing side, preventing leakage back into the inlet. A rotor, as indicated in the figure, turns the vane.

Fluid machines that turn or tend to turn about an axis are called *turbomachines*. Thus, centrifugal pumps are turbomachines. Other examples of turbomachines are turbines, lawn sprinklers, ceiling fans, lawn mower blades, and turbine engines. The blower used to exhaust contaminated air in waste-air works is a turbomachine.

4.2 PUMPING STATION HEADS

In the design of pumping stations, the engineer must ensure that the pumping system can deliver the fluid to the desired height. For this reason, energies are conveniently expressed in terms of heights or heads. The various terminologies of heads are defined in Figure 4.6. Note that two pumping systems are portrayed in the figure: pumps connected in series and pumps connected in parallel. Also, two sources of the water are pumped: the first is the source tank above the elevation of the eye of the impeller or centerline of the pump system; the second is the source tank below the eye of the impeller or centerline of the pump system. The flow in flow pipes for the first case is shown by dashed lines. In addition, the pumps used in this pumping station are of the centrifugal type.

The terms *suction* and *discharge* in the context of heads refer to portions of the system before and after the pumping station, respectively. *Static suction lift* h_ℓ is the vertical distance from the elevation of the inflow liquid level below the pump inlet to the elevation of the pump centerline or eye of the impeller. A lift is a negative head. *Static suction head* h_s is the vertical distance from the elevation of the inflow liquid level above the pump inlet to the elevation of the pump centerline. *Static discharge head* h_d is the vertical distance from the centerline elevation of the pump

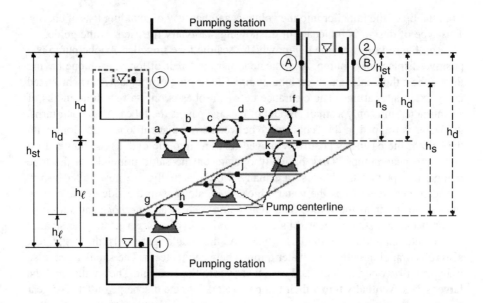

FIGURE 4.6 Pumping station heads.

to the elevation of the discharge liquid level. *Total static head* h_{st} is the vertical distance from the elevation of the inflow liquid level to the elevation of the discharge liquid level. *Suction velocity head* h_{vs} is the entering velocity head at the suction side of the pump hydraulic system. This is not the velocity head at the inlet to a pump such as points a, c, e, etc. in the figure. In the figure, because the velocity in the wet well is practically zero, h_{vs} will also be practically zero. *Discharge velocity head* h_{vd} is the outgoing velocity head at the discharge side of the pump hydraulic system. Again, this is not the velocity head at the discharge end of any particular pump. In the figure, it is the velocity head at the water level in the discharge tank, which is also practically zero.

4.2.1 Total Developed Head

The literature has used two names for this subject: total dynamic head or total developed head (H or TDH). Let us derive TDH first by considering the system connected in parallel between points 1 and 2. Since the connection is parallel, the head losses across each of the pumps are equal and the head given to the fluid in each of the pumps are also equal. Thus, for our analysis, let us choose any pump such as the one with inlet g. From fluid mechanics, the energy equation between the points is

$$\frac{P_1}{\gamma} + \frac{V_1^2}{2g} + z_1 - h_f + h_q + h_p = \frac{P_2}{\gamma} + \frac{V_2^2}{2g} + z_2 \qquad (4.1)$$

where P, V, and z are the pressure, velocity, and elevation head at the indicated points; g is the acceleration due to gravity; h_f is the head equivalent of the resistance loss (friction) between the points; h_q is the head equivalent of the heat added to the flow;

and h_p is the head given to the fluid by the pump impeller. Using the level at point 1 as the reference datum, z_1 equals zero and z_2 equals h_{st}. It is practically certain that there will be no h_q in the physical–chemical treatment of water and wastewater, and will therefore be neglected. Let h_f be composed of the head loss inside the pump h_{lp}, plus the head loss in the suction side of the pumping system h_{fs} and the head loss in the discharge side of the pumping system h_{fd}. Thus, the energy equation becomes

$$\frac{P_1}{\gamma} + \frac{V_1^2}{2g} - h_{lp} - h_{fs} - h_{fd} + h_p = \frac{P_2}{\gamma} + \frac{V_2^2}{2g} + h_{st} \tag{4.2}$$

The equation may now be solved for $-h_{lp} + h_p$. This term is composed of the head added to the fluid by the pump impeller, h_p, and the losses expended by the fluid inside the pump, h_{lp}. As soon as the fluid gets the h_p, part of this will have to be expended to overcome frictional resistance inside the pump casing. The fluid that is actually receiving the energy will drag along those that are not. This dragging along is brought about because of the inherent viscosity that any fluid possesses. The process causes slippage among fluid planes, resulting in friction and turbulent mixing. This friction and turbulent mixing is the h_{lp}. The net result is that between the inlet and the outlet of the pump is a head that has been developed. This head is called the *total developed head* or *total dynamic head* (TDH) and is equal to $-h_{lp} + h_p$.

Solving Equation (4.2) for $-h_{lp} + h_p$, considering that the tanks are open to the atmosphere and that the velocities at points 1 and 2 at the surfaces are practically zero, produces

$$\text{TDH} = \text{TDH0}sd = -h_{lp} + h_p = h_{st} + h_{fs} + h_{fd} \tag{4.3}$$

When the two tanks are open to the atmosphere, P_1 and P_2 are equal; they, therefore, cancel out of the equation. Thus, as shown in the equation, TDH is referred to as TDH0sd. In this designation, 0 stands for the fact that the pressures cancel out. The s and d signify that the suction and discharge losses are used in calculating TDH.

The sum $h_{fs} + h_{fd}$ may be computed as the loss due to friction in straight runs of pipe, h_{fr}, and the minor losses of transitions and fittings, h_{fm}. Thus, calling the corresponding TDH as TDH0rm (rm for run and minor, respectively),

$$\text{TDH} = \text{TDH0}rm = -h_{lp} + h_p = h_{st} + h_{fr} + h_{fm} \tag{4.4}$$

From fluid mechanics,

$$h_{fr} = f\left(\frac{l}{D}\right)\left(\frac{V^2}{2g}\right) \tag{4.5}$$

$$h_{fm} = K\left(\frac{V^2}{2g}\right) \tag{4.6}$$

where f is Fanning's friction factor, l is the length of the pipe, D is the diameter of the pipe, V is the velocity through the pipe, g is the acceleration due to gravity (equals 9.81 m/s^2) and K is the head loss coefficient. $V^2/2g$ is called the velocity head, h_v. That is,

$$h_v = \frac{V^2}{2g} \tag{4.7}$$

If the points of application of the energy equation is between points 1 and B, instead of between points 1 and 2, the pressure terms and the velocity heads will remain intact at point B. In this situation, refering to the TDH as TDH*fullsd* (*full* because velocities and pressure are not zeroed out),

$$\text{TDH} = \text{TDH}fullsd = -h_{lp} + h_p = \frac{P_B - P_{atm}}{\gamma} + \frac{V_B^2}{2g} + z_2 + h_{fs} + h_{fd} \tag{4.8}$$

where z_2 is the elevational head of point B, referred to the chosen datum at point 1. Note that P_{atm} is the pressure at point 1, the atmospheric pressure. When the friction losses are expressed in terms of $h_{fr} + h_{fm}$ and calling the TDH as TDH*fullrm*, the equation is

$$\text{TDH} = \text{TDH}fullrm = -h_{lp} + h_p = \frac{P_B - P_{atm}}{\gamma} + \frac{V_B^2}{2g} + z_2 + h_{fr} + h_{fm} \tag{4.9}$$

If the energy equation is applied using the source tank at the upper elevation as point 1, the same respective previous equations will also be obtained. In addition, if the energy equation is applied to the system of pumps that are connected in series, the same equations will be produced except that TDH will be the sum of the TDHs of the pumps in series. Also the subscripts denoted by B will be changed to A. See Figure 4.6.

4.2.2 INLET AND OUTLET MANOMETRIC HEADS; INLET AND OUTLET DYNAMIC HEADS

Applying the energy equation between an inlet i and outlet o of any pump produces

$$\text{TDH} = \text{TDH}mano = -h_{lp} + h_p = \left(\frac{P_o}{\gamma} + \frac{V_o^2}{2g}\right) - \left(\frac{P_i}{\gamma} + \frac{V_i^2}{2g}\right) \tag{4.10}$$

where TDH$_{mano}$ (*mano* for manometric) is the name given to this TDH. $h_{fs} + h_{fd}$ is equal to zero. $P_i/\gamma = h_i$ is called either the *inlet manometric head* or the *inlet manometric height absolute*; $P_o/\gamma = h_o$ is also called either the outlet *manometric head* or the *outlet manometric height absolute*. The subscripts i and o denote "inlet" and "outlet," respectively.

Manometric head or level is the height to which the liquid will rise when subjected to the value of the gage pressure; on the other hand, manometric height absolute is the height to which the liquid will rise when subjected to the true or absolute pressure in a vacuum environment. The liquid rising that results in the manometric head is under a gage pressure environment, which means that the liquid is exposed to the atmosphere. The liquid rising, on the other hand, that results in the manometric height absolute is not exposed to the atmosphere but under a complete vacuum. Retain h as the symbol for manometric head and, for specificity, use h_{abs} as the symbol for manometric height absolute. Thus, the respective formulas are

$$h = \frac{P_g}{\gamma} \tag{4.11}$$

$$h_{abs} = \frac{P}{\gamma} \tag{4.12}$$

P_g is the gage pressure and P is the absolute pressure. Unless otherwise specified, P is always the absolute pressure.

In terms of the new variables and the velocity heads $V_i^2/2g = h_{vi}$ and $V_i^2/2g = h_{vo}$ for the pump inlet and outlet velocity heads, respectively, TDH, designated as TDH_{abs}, may also be expressed as

$$\text{TDH}_{abs} = h_{abso} + h_{vo} - (h_{absi} + h_{vi}) \tag{4.13}$$

Note that h_{abs} is used rather than h. h is merely a relative value and would be a mistake if substituted into the above equation.

For static suction lift conditions, h_i is always negative since gage pressure is used to express its corresponding pressure, and its theoretical limit is the negative of the difference between the prevailing atmospheric pressure and the vapor pressure of the liquid being pumped. If the pressure is expressed in terms of absolute pressure, then h_{abs} has as its theoretical limit the vapor pressure of the liquid being pumped.

Because of the suction action of the impeller and because the fluid is being lifted, the fluid column becomes "rubber-banded." Just like a rubber band, it becomes stretched as the pressure due to suction is progressively reduced; eventually, the liquid column ruptures. As the rupture occurs, the inlet suction pressure will actually have gone down to equal the vapor pressure, thus, vaporizing the liquid and forming bubbles. This process is called *cavitation*.

Cavitation can destroy hydraulic structures. As the bubbles which have been formed at a partial vacuum at the inlet gradually progress along the impeller toward the outlet, the sudden increase in pressure causes an impact force. Continuous action of this force shortens the life of the impeller.

The sum of the inlet manometric height absolute and the inlet velocity head is called the *inlet dynamic head*, idh (*dynamic* because this value is obtained with fluid in motion). The sum of the outlet manometric height absolute and the outlet velocity head is called the *outlet dynamic head*, odh. Of course, the TDH is also equal to the outlet dynamic head minus the inlet dynamic head.

$$\text{TDH} = \text{TDH}dh = \text{odh} - \text{idh} \qquad (4.14)$$

In general, dynamic head, *dh* is

$$dh = \frac{P}{\gamma} + \frac{V^2}{2g} \qquad (4.15)$$

It should be noted that the correct substitution for the pressure terms in the above equations is always the absolute pressure. Physical laws follow the natural measures of the parameters. Absolute pressures, absolute temperatures, and the like are natural measures of these parameters. Gage pressures and the temperature measurements of Celsius and Fahrenheit are *expedient* or relative measures. This is unfortunate, since oftentimes, it causes too much confusion; however, these relative measures have their own use, and how they are used must be fully understood, and the results of the calculations resulting from their use should be correctly interpreted. If confusion results, it is much better to use the absolute measures.

Example 4.1   It is desired to pump a wastewater to an elevation of 30 m above a sump. Friction losses and velocity at the discharge side of the pump system are estimated to be 20 m and 1.30 m/s, respectively. The operating drive is to be 1200 rpm. Suction friction loss is 1.03 m; the diameter of the suction and discharge lines are 250 and 225 mm, respectively. The vertical distance from the sump pool level to the pump centerline is 2 m. **(a)** If the temperature is 20°C, has cavitation occurred? **(b)** What are the inlet and outlet manometric heads? **(c)** What are the inlet and outlet total dynamic heads? From the values of the idh and odh, calculate TDH.

Solution:

(a) $\dfrac{V_1^2}{2g} + z_1 + \dfrac{P_1}{\gamma} + h_q - h_f + h_p = \dfrac{V_2^2}{2g} + z_2 + \dfrac{P_2}{\gamma}$

Let the sump pool level be point 1 and the inlet to the pump as point 2.

A_d = cross-section of discharge pipe = $\dfrac{\pi D^2}{4} = \dfrac{\pi(0.225^2)}{4} = 0.040$ m^2

Therefore, Q = discharge = $1.3(0.040) = 0.052$ m^3/s

$A_2 = \dfrac{\pi(0.25^2)}{4} = 0.049$ m^2; $V_2 = \dfrac{0.052}{0.049} = 1.059$ m/s

Therefore, $0 + 0 + 0 + 0 - 1.03 + 0 = \dfrac{1.059^2}{2(9.81)} + 2 + \dfrac{P_2}{\gamma}$

$P_2/\gamma = -3.087$ m; because the pressure used in the equation is 0, this value represents the manometric head to the pump

At 20°C, P_v (vapor pressure of water) = 2.34 kN/m^2 = 0.239 m of water

Assume standard atmosphere of 1 atm = 10.34 m of water.

Therefore, theoretical limit of pump cavitation = $-(10.34 - 0.239) = -10.05$ m $\ll$ -3.087

Cavitation has not been reached. **Ans**

(b) Inlet manometric head = -3.087 m of water. **Ans**

Apply the energy to the equation between the sump level and the discharge 30 m above

$$\frac{P_1}{\gamma} + \frac{V_1^2}{2g} - h_{lp} - h_{fs} - h_{fd} + h_p = \frac{P_2}{\gamma} + \frac{V_2^2}{2g} + h_{st}$$

$$-h_{lp} + h_p = TDH = \frac{P_2}{\gamma} - \frac{P_1}{\gamma} + \frac{V_2^2}{2g} - \frac{V_1^2}{2g} + h_{fs} + h_{fd} + h_{st}$$

$$\frac{P_2}{\gamma} - \frac{P_1}{\gamma} = 0 \quad \frac{V_1^2}{2g} = 0 \quad h_{fs} + h_{fd} = 20 \text{ m} \quad h_{st} = 30 \text{ m} \quad \frac{V_2^2}{2g} = \frac{1.3^2}{2(9.81)} = 0.086 \text{ m}$$

Therefore, TDH = $0.086 + 20 + 30 = 50.086$ m of water

Between the inlet and outlet of pump:

$$TDH = TDHvivo = h_{abso} + h_{vo} - (h_{absi} + h_{vi})$$

$$50.086 = h_{abso} + \frac{1.3^2}{2(9.81)} - \left[(-3.087 + 10.34) + \frac{1.059^2}{2(9.81)}\right] = h_{abso} + 0.086 - 7.31$$

$$h_{abso} = 57.31 \text{ m}$$

Therefore, outlet manometric head = $57.31 - 10.34 = 46.97$ m of water. **Ans**

(c) $idh = h_{absi} + h_{vi} = (-3.087 + 10.34) + \frac{1.059^2}{2(9.81)} = 7.31$ m of water **Ans**

$odh = h_{abso} + h_{vo} = 57.31 + \frac{1.3^2}{2(9.81)} = 57.39$ m of water **Ans**

TDH = odh − idh = $57.39 - 7.31 = 50.086$ m of water **Ans**

4.3 PUMP CHARACTERISTICS AND BEST OPERATING EFFICIENCY

It is important that a method be developed to enable the proper selection of pumps to meet specific pumping requirements. Thus, before selecting any particular pump, the designer must consult the characteristics of this pump in order to make an intelligent selection. In fact, manufacturers develop these characteristics for their particular pump. Thus, *pump characteristics* are a set of curves that depict the

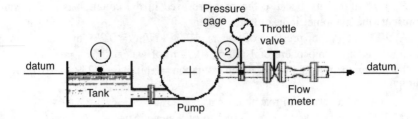

FIGURE 4.7 Setup for developing pump characteristics curves.

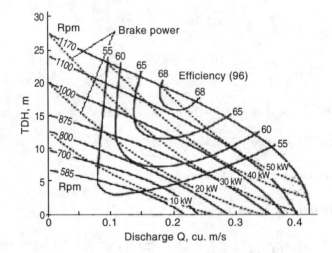

FIGURE 4.8 Pump characteristic curves for a 375-mm impeller. (Courtesy of Smith and Loveless. With permission.)

performance of a given particular pump. Figure 4.7 illustrates the setup used for developing pump characteristics (Hammer, 1986) and Figure 4.8 shows an example of characteristic curves of one particular pump.

Apply the equation for TDH*fullsd* to the figure. For convenience, it is reproduced below.

$$TDH = TDH fullsd = -h_{lp} + h_p = \frac{P_B - P_{atm}}{\gamma} + \frac{V_B^2}{2g} + z_2 + h_{fs} + h_{fd}$$

With point 1 as the datum, z_2 is equal to 0. h_{fs} is the head loss in the suction side from point 1 to the inlet of the pump and h_{fd} is the head loss in the discharge side from the outlet of the pump to point 2. Because the distances are very short, they can be neglected compared to the other terms in the equation. P_B is equal to the gage pressure at point 2, P_{gB}, plus the barometric pressure. In terms of P_{gB}, P_B is then equal to $P_{gB} + P_{atm}$. The most complete treatment will also include the vapor pressure of water. Neglecting vapor pressure since it is negligible, however, the previous equation simply becomes

$$TDH = TDHsetup = -h_{lp} + h_p = \frac{P_{gB}}{\gamma} + \frac{V_B^2}{2g} \qquad (4.16)$$

Note that TDH*fullsd* has been changed to TDH*setup*. This equation demonstrates that the above setup of the unit is a convenient arrangement for determining the TDH. As shown in this equation, TDH can simply be calculated using the pressure gage reading and a measured velocity at point 2.

As depicted in Figure 4.8, the performance of this particular pump has been characterized in terms of total developed head on the ordinate and discharge on the abscissa. The other characteristics are the parameters rpm, power, and efficiency. To illustrate how this chart was developed, consider one curve: when the curve for the 1,170 rpm was developed, the setup in Figure 4.7 was adjusted to 1,170 rpm and the discharge was varied from 0, the shut-off flow, up to the abscissa value depicted in Figure 4.8. The reading of the pressure gage was then taken. This reading converted to head, along with the velocity head obtained from the flow meter reading and the cross-sectional of the discharge pipe, gives the TDH. This was repeated for the other rpm's as well as for the powers consumed (which are indicated in kW).

The relationship of discharge versus total developed head at the shut-off flow cannot be developed for the positive displacement pump operating under a cylinder, without breaking the cylinder head or the cylinder, itself. For pumps operating under a cylinder, the element that pushes the fluid is either the piston or the plunger. This piston or plunger moves inside the cylinder and pushes the fluid inside into the cylinder head located at the end of the forward travel. This pushing creates a tremendous amount of pressure that can rupture the cylinder head or the cylinder body itself, if the piston or plunger has not given way first. In the case of centrifugal pumps, this situation would not be a problem since the fluid will just be churned inside the impeller casing, and testing at shut-off flow is possible.

The activity inside the pump volute incurs several losses: first is the backflow of the flow that had already been acted upon but is slipping back into the suction eye of the impeller or, in general, toward the suction side of the pump. Because energy had already been expended on this flow but failed to exit into the discharge, this backflow represents a loss. The other loss is the turbulence induced as the impeller acts on the flow and swirls it around. Turbulence is a loss of energy. As the impeller rotates, its tips and sides shear off the fluid; this also causes what is called *disk friction* and is a loss of energy. All these losses cause the inefficiency of the pump; h_{lp} is these losses.

During the testing, the power to the motor or prime mover driving the pump is recorded. Multiplying this input power by the prime mover efficiency produces the *brake* or *shaft power*. In the figure, the powers are measured in terms of kilowatts (or kW). Call the head corresponding to the brake power as h_{brake}. The *brake efficiency* of a pump is defined as the ratio of TDH to the brake input power to the pump. Therefore, brake efficiency η is

$$\eta = \frac{TDH}{h_{brake}} \left(= \frac{-h_{lp} + h_p}{h_{brake}} \right) \qquad (4.17)$$

But,

$$\text{TDH} = \frac{P_{gB}}{\gamma} + \frac{V_B^2}{2g} \tag{4.18}$$

as far as the arrangement in Figure 4.7 is concerned. With this equation substituted into Equation (4.17), the efficiency during a trial run can be determined. This new equation for efficiency is

$$\eta = \frac{1}{h_{brake}} \left(\frac{P_{gB}}{\gamma} + \frac{V_B^2}{2g} \right) \tag{4.19}$$

As shown in the figure, along a certain curve there are several values of efficiencies determined. Among these efficiencies is one that is the highest of all. This particular value of the efficiency corresponds to the best operating performance of the pump; hence, this point is called the *best operating efficiency*. For example, for the characteristic curve determined at a brake kilowatt input of 40 kW, the best operating efficiency is approximately 67%. This corresponds to a TDH of approximately 16 m and approximately a discharge of 0.18 m³/s. To operate this pump, its discharge should be set at 0.18 m³/s to take advantage of the best operating efficiency. In practice, the operating performance is normally set anywhere from 60 to 120% of the best operating efficiency.

Note that the brake power has been given in terms of its head equivalent h_{brake}. To obtain h_{brake} from a given brake power expressed in *horsepower*, h_p, use the equivalent that $h_p = 745.7$ N·m/s. If Q is the rate of flow in m³/s, and γ is the specific weight in N/m³, then h_{brake} in meters is

$$h_{brake} = \frac{745.7 h_p}{Q\gamma} \tag{4.20}$$

Example 4.2 Pump characteristics curves are developed in accordance with the setup of Figure 4.7. The pressure at the outlet of the pump is found to be 196 kN/m² gage. The discharge flow is 0.15 m³/s and the outlet diameter of the discharge pipe is 375 mm. The motor driving the pump is 50 hp. Calculate TDH.

Solution:

$$\text{TDH} = \frac{P_{gB}}{\gamma} + \frac{V_B^2}{2g}$$

$$V_B = \frac{0.15}{\text{cross sectional area of pipe}} = \frac{0.15}{\pi(0.375)^2/4} = 1.36 \text{ m/s}$$

Assume temperature of water = 25°C; therefore, density of water = 997 kg/m³

$$\text{TDH} = \frac{196(1000)}{997(9.81)} + \frac{1.36^2}{2(9.81)} = 20.13 \text{ m of water} \textbf{Ans}$$

4.4 PUMP SCALING LAWS

When designing a pumping station or specifying sizes of pumps, the engineer refers to a pump characteristic curve that defines the performance of a pump. Several different sizes of pumps are used, so theoretically, there should also be a number of these curves to correspond to each pump. In practice, however, this is not done. The characteristic performance of any other pump can be obtained from the curves of any one pump by the use of pump scaling laws, provided the pumps are similar. The word *similar* will become clear later.

The following dependent variables are produced as a result of independent variables either applied to a pump or are characteristics of the pump, itself: the pressure developed ΔP (corresponding to TDH), the power given to the fluid $\mathcal{P}$, and the efficiency η. The independent variables applied to the pump are the discharge Q, the viscosity of the fluid μ, and the mass density of the fluid ρ. These are variables applied, since they came from outside of and are introduced (applied) to the pump. The independent variables that are characteristics of the pump are the diameter of impeller or length of stroke D, the rotational speed or stroking speed ω, some roughness ϵ of the chamber, and some characteristic length ℓ of the chamber space. There may still be other independent variables, but experience has shown that the forgoing items are the major ones. Although they are considered major, however, some of them may still be considered redundant and can be eliminated as will be shown in the succeeding analysis.

For ΔP the functional relationship may be written as

$$\Delta P = \phi(\rho, \omega, D, Q, \mu, \epsilon, \ell) \tag{4.21}$$

At large Reynolds numbers the effect of viscosity μ is constant. For example, consider the Moody diagram. At large Reynolds numbers, the plot of the friction factor f and the Reynolds number, with f as the ordinate, is horizontal. Both μ and f are measures of resistance to flow; thus, they are directly related. Because the effect of f at large Reynolds numbers is constant, the effect of μ at large Reynolds numbers must also be constant. The rotation of the impeller or the movement of the stroke occurs at an extremely rapid rate; consequently, the flow conditions inside the pump casing are turbulent or are at large Reynolds numbers. Hence, since μ is constant at high Reynolds numbers, it does not have any functional relationship with ΔP and may be removed from Equation (4.21). ℓ as a measure of the pump chamber space is already included in D. It may also be dropped. Lastly, since the casing is too short, the effect of roughness ϵ is too small compared to the other causes of the ΔP. It may therefore be also dropped. Equation (4.21) now takes the form

$$\Delta P = \phi(\rho, \omega, D, Q) \tag{4.22}$$

Applying dimensional analysis, let [x] be read "the dimensions of x." Thus, $[\Delta P] = F/L^2$, $[\rho] = Ft^2/L^4$, $[\omega] = 1/t$, $[D] = L$, and $[Q] = L^3/t$. By inspection of these dimensions, the number of reference dimensions is three. Because five variables are used, by the pi theorem, the number of Π terms is two (number of variables minus number of reference dimensions, $5 - 3 = 2$). Let the Π terms be Π_1 and Π_2, respectively, and proceed with the dimensional analysis.

To eliminate the dimension F, divide ΔP by ρ. Thus,

$$\frac{\Delta P}{\rho} \Rightarrow \left[\frac{\Delta P}{\rho}\right] = \frac{F/L^2}{Ft^2/L^4} = \frac{L^2}{t^2} \tag{4.23}$$

To eliminate t, divide by ω^2 as follows:

$$\frac{\Delta P}{\rho\omega^2} \Rightarrow \left[\frac{\Delta P}{\rho\omega^2}\right] = \frac{L^2}{t^2\left(\frac{1}{t^2}\right)^2} = L^2 \tag{4.24}$$

To completely eliminate dimensions, divide by D^2 as follows:

$$\frac{\Delta P}{\rho\omega^2 D^2} \Rightarrow \left(\frac{\Delta P}{\rho\omega^2 D^2}\right) = \frac{L^2}{L^2} \tag{4.25}$$

Therefore $$\Pi_1 = \frac{\Delta P}{\rho\omega^2 D^2} \tag{4.26}$$

To get Π_2, operate on Q to obtain

$$\Pi_2 = \left(\frac{Q}{\omega D^3}\right) \tag{4.27}$$

The final functional relationship is

$$\frac{\Delta P}{\rho\omega^2 D^2} = \Psi\left(\frac{Q}{\omega D^3}\right) \tag{4.28}$$

But, $\Delta P = \gamma H$, where H is TDH, the total developed head. Because $\gamma = \rho g$, substituting in Equation (4.28) produces

$$\frac{Hg}{\omega^2 D^2} = \Psi\left(\frac{Q}{\omega D^3}\right) \tag{4.29}$$

$Hg/(\omega^2 D^2)$ is called the *head coefficient*, C_H, while $Q/(\omega D^3)$ is called the *flow coefficient*, C_Q. Because no one pump was chosen in the derivation, the equation is general. For any number of pumps $a, b, c, \ldots, n$ and using Equations (4.28) and (4.29), the relationships next follow:

$$\frac{H_a g}{\omega_a^2 D_a^2} = \frac{H_b g}{\omega_b^2 D_b^2} = \cdots = \frac{H_n g}{\omega_n^2 D_n^2} = C_{Hn} \tag{4.30}$$

$$\frac{Q_a}{\omega_a D_a^3} = \frac{Q_b}{\omega_b D_b^3} = \cdots = \frac{Q_n}{\omega_n D_n^3} = C_{Qn} \tag{4.31}$$

Pumps that follow the above relations are called *similar* or *homologous* *pumps*. In particular, when the Π variable C_H, which involves force are equal in the series of pumps, the pumps are said to be *dynamically similar*. When the Π variable C_Q, which relates only to the motion of the fluid are equal in the series of pumps, the pumps are said to be *kinematically similar*. Finally, when corresponding parts of the pumps are proportional, the pumps are said to be *geometrically similar*. The relationships of Eqs. (4.30) and (4.31) are called *similarity*, *affinity*, or *scaling laws*.

Considering the power P and the efficiency η as the dependent variables, similar dimensional analyses yield the following similarity relations:

$$\frac{P_a}{\rho \omega_a^3 D_a^5} = \frac{P_b}{\rho \omega_b^3 D_b^5} = \cdots = \frac{P_n}{\rho \omega_n^3 d_n^5} = C_P n \qquad (4.32)$$

$$\eta_a = \eta_b = \eta_c = \cdots = \eta_n \qquad (4.33)$$

where C_P is called the *power coefficient*. Note that the efficiencies of similar pumps are equal. The similarity relations also apply to the same pump which, in this case, the subscripts a, b, c,..., and n represent different operating conditions of this same pump.

The power P is the power given to the fluid. In plots of characteristic curves such as Figure 4.8, however, the brake power is the one plotted. Because P bears a ratio to that of the brake power in the form of the efficiency η, the similarity laws that we have developed also apply to the brake power, and figures such Figure 4.8 may be used for scaling brake powers of pumps.

Equation (4.17) expresses efficiencies in terms of heads. Letting P_{brake} represent brake power, h_{brake} may be obtained from P_{brake} as follows:

$$h_{brake} = \frac{P_{brake}}{Q_\gamma} \qquad (4.34)$$

Equation (4.20) is a special case of this equation.

From the equations derived, the following simplified scaling laws for a given pump operated at different speeds, ω, are obtained:

$$H_b = \frac{\omega_b^2}{\omega_a^2} H_a \qquad (4.35)$$

$$Q_b = \frac{\omega_b}{\omega_a} Q_a \qquad (4.36)$$

$$P_b = \frac{\omega_b^3}{\omega_a^3} P_a \qquad (4.37)$$

$$P_{brakeb} = \frac{\omega_b^3}{\omega_a^3} P_{brakea} \qquad (4.38)$$

For pumps of constant rotational or stroking speed, ω, but of different diameter or stroke, D, the following simplified scaling laws are also obtained:

$$H_b = \frac{D_b^2}{D_a^2} H_a \tag{4.39}$$

$$Q_b = \frac{D_b^3}{D_a^3} Q_a \tag{4.40}$$

$$\mathcal{P}_b = \frac{D_b^5}{D_a^5} \mathcal{P}_a \tag{4.41}$$

$$\mathcal{P}_{brakeb} = \frac{D_b^5}{D_a^5} \mathcal{P}_{brakea} \tag{4.42}$$

Example 4.3 For the pump represented by Figure 4.8, determine (a) the discharge when the pump is operating at a head of 10 m and at a speed of 875 rpm, and (b) the efficiency and the brake power.

Solution:

(a) From the figure, $Q = 0.17$ m^3/s **Ans**

(b) $\eta = 63\%$ **Ans** $\mathcal{P}_{brake} = 26$ kW **Ans**

Example 4.4 If the pump in Example 4.3 is operated at 1,170 rpm, calculate the resulting H, Q, $\mathcal{P}_{brake}$, and η.

Solution:

$$\left\{\frac{Hg}{\omega^2 D^2}\right\}_b = \left\{\frac{Hg}{\omega^2 D^2}\right\}_a \Rightarrow \frac{H_b g}{\omega_b^2 D^2} = \frac{H_a g}{\omega_a^2 D^2} \Rightarrow H_b = 10\left(\frac{1170^2}{875^2}\right) = 17.88 \text{ m} \quad \textbf{Ans}$$

$$\left\{\frac{Q}{\omega D^3}\right\}_b = \left\{\frac{Q}{\omega D^3}\right\}_a \Rightarrow \frac{Q_b}{\omega_b D^3} = \frac{Q_a}{\omega_a D^3} \Rightarrow Q_b = 0.17\left(\frac{1170}{875}\right) = 0.23 \text{ m}^3/\text{s} \quad \textbf{Ans}$$

$$\left\{\frac{\mathcal{P}_{brake}}{\rho\omega^3 D^5}\right\}_b = \left\{\frac{\mathcal{P}_{brake}}{\rho\omega^3 D^5}\right\}_a \Rightarrow \frac{\mathcal{P}_{brakeb}}{\rho\omega_b^3 D^5} = \frac{\mathcal{P}_{brakea}}{\rho\omega_a^3 D^5} \Rightarrow \mathcal{P}_{brakeb}$$

$$= 26\left(\frac{1170^3}{875^3}\right) = 62.16 \text{ kW} \quad \textbf{Ans}$$

$$\eta = 63\% \quad \textbf{Ans}$$

Example 4.5 If a homologous 30-cm pump is to be used for the problem in Example 4.4, calculate the resulting H, Q, $\mathcal{P}_{brake}$, and η for the same rpm.

Solution: The diameter of the pump represented by Figure 4.8 is 375 mm.

$$\left\{\frac{Hg}{\omega^2 D^2}\right\}_b = \left\{\frac{Hg}{\omega^2 D^2}\right\}_a \Rightarrow \frac{H_b g}{\omega^2 D_b^2} = \frac{H_a g}{\omega^2 D_a^2} \Rightarrow H_b = 17.88\left(\frac{30^2}{37.5^2}\right) = 11.44 \text{ m} \quad \textbf{Ans}$$

$$\left\{\frac{Q}{\omega D^3}\right\}_b = \left\{\frac{Q}{\omega D^3}\right\}_a \Rightarrow \frac{Q_b}{\omega D_b^3} = \frac{Q_a}{\omega D_a^3} \Rightarrow Q_b = 0.23\left(\frac{30^3}{37.5^3}\right) = 0.12 \text{ m}^3/\text{s} \quad \textbf{Ans}$$

$$\left\{\frac{\mathcal{P}_{brake}}{\rho\omega^3 D^5}\right\}_b = \left\{\frac{\mathcal{P}_{brake}}{\rho\omega^3 D^5}\right\}_a \Rightarrow \frac{\mathcal{P}_{brakeb}}{\rho\omega^3 D_b^5} = \frac{\mathcal{P}_{brakeb}}{\rho\omega^3 D_a^5} \Rightarrow \mathcal{P}_{brakeb}$$

$$= 62.16\left(\frac{30^5}{37.5^5}\right) = 20.37 \text{ kW} \quad \textbf{Ans}$$

$$\eta = 63\% \quad \textbf{Ans}$$

4.5 PUMP SPECIFIC SPEED

Raising the flow coefficient $C_Q = Q/(\omega D^3)$ to the power $\frac{1}{2}$, the head coefficient $C_H = gH/(\omega^2 D^2)$ to the power $\frac{3}{4}$, and forming the ratio of the former to that of the latter, D will be eliminated. Calling this ratio as N_s produces the expression

$$N_s = \frac{\omega\sqrt{Q}}{(gH)^{3/4}} \tag{4.43}$$

If the dimensions of N_s are substituted, it will be found dimensionless and because it is dimensionless, it can be used as a general characterization for a whole variety of pumps without reference to their sizes. Thus, a certain range of the value of N_s would be a particular type of pump such as axial (no size considered), and another range would be another particular type of pump such as radial (no size considered). N_s is called *specific speed*.

By characterizing all the pumps generally like this, N_s is of great applicability in selecting the proper type of pump, whether radial or axial or any other type. For example, refer to Figure 4.9. The radial-vane pumps are in the range of $N_s = 9.6$ to 19.2; the Francis-vane pumps are in the range of 28.9 to 76.9; and so on. Therefore, if Q, ω, and H are known, N_s can be computed using Equation (4.43); thus, depending upon the value obtained, the type of pump can be specified.

Just how is the chart of specific speeds obtained? Remember that one of the characteristics curves of a pump is the plot of the efficiency. Referring to Figure 4.8, along any curve characterized by a parameter such as rpm, there are an infinite number of efficiency values. Of these infinite number of values, there is only one maximum. As mentioned previously, this maximum is the best efficiency point.

If values of ω, Q, and H are taken from the characteristics curves at the best operating efficiencies and substituted into Equation (4.43), values of specific speeds are obtained at these efficiencies. For example, from Figure 4.8 at a Q of 0.16 m³/s, the best efficiency is approximately 66% corresponding to a total developed head

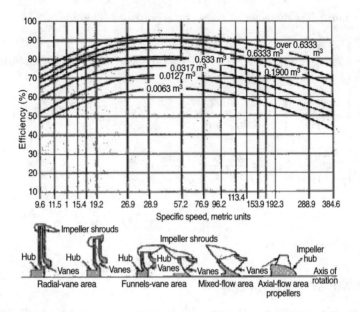

FIGURE 4.9 Specific speeds of various types of centrifugal pumps.

of about 13 meters and an rpm of 990. These values substituted into Equation (4.43), after converting the rpm of 990 rpm to radians per second, produces an N_s of 1.73. A number of calculations similar to this one need to be done on other characteristics curves in order to produce Figure 4.9. In other words, this figure has been obtained under conditions of best operating efficiencies. Therefore, specifying pumps using specific speeds as the criterion and using figures such as Figure 4.9 ensures that the pump selected operates at the best operating efficiency. From this discussion, it can be gleaned that specific speed could have gotten its name from the fact that its value is *specific* to the operating conditions at the best operating efficiency.

Example 4.6   A designer wanted to recommend the use of an axial-flow pump to move wastewater to an elevation of 30 m above a sump. Overall friction losses of the system and the velocity at the discharge side are estimated to be 20 m and 1.30 m/s, respectively. The operating drive is to be 1,200 rpm. Suction friction losses are 1.03 m; the diameter of the suction and discharge lines are 250 and 225 mm, respectively. The vertical distance from the sump pool level to the pump centerline is 2 m. Is the designer recommending the right pump? Design the pump yourself.

Solution:

$$N_s = \frac{\omega\sqrt{Q}}{(gH)^{3/4}}$$

$$-h_{lp} + h_p = \text{TDH} = \frac{P_2}{\gamma} - \frac{P_1}{\gamma} + \frac{V_2^2}{2g} - \frac{V_1^2}{2g} + h_{fs} + h_{fd} + h_{st}$$

Take the sump pool level as point 1 and the sewage discharge level as point 2. $V_1^2/2g = 0$, because the pool velocity is zero. Points 1 and 2 are both exposed to the atmosphere.

$$\frac{V_2^2}{2g} = \frac{1.30^2}{2(9.81)} = 0.086 \text{ m}$$

$$\text{TDH} = \left[\left(\frac{P_2}{\gamma} - \frac{P_1}{\gamma}\right) = 0\right] + 0.086 - 0 + 20 + 30 = 50.086 \text{ m}$$

$$\omega = 1{,}200(2\pi)/60 = 125.67 \text{ radians/s}$$

$$A_d = \text{cross-section of discharge pipe} = \frac{\pi D^2}{4} = \frac{\pi(0.225^2)}{4} = 0.040 \text{ m}^2$$

Therefore, $Q = (1.3)(0.040) = 0.052 \text{ m}^3/\text{s}$

$$N_s = \frac{125.67\sqrt{0.052}}{[(9.81)(50.086)]^{3/4}} = 0.27$$

This falls outside the range of specific speeds in Figure 4.9; however, the pump should not be an axial flow pump as recommended by the designer. **Ans**
From the figure, for a Q of 0.052 m^3/s and an N_s of 0.27, the pump would have to be of radial-vane type. **Ans**

4.6 NET POSITIVE SUCTION HEAD AND DEEP-WELL PUMPS

In order for a fluid to enter the pump, it must have sufficient energy to force itself toward the inlet. This means that a positive head (not negative head) must exist at the pump inlet. This head that must exist for pumping to be possible is termed the *net positive suction head* (NPSH). It is an absolute, not gage, positive head acting on the fluid.

Refer to the portion of Figure 4.6 where the source tank fluid level is below the center line of the impeller in either the system connected in parallel or in the system connected in series. At the surface of the wet well (point 1), the pressure acting on the liquid is equal to the atmospheric pressure P_{atm} minus the vapor pressure of the liquid P_v. This pressure is, thus, the atmospheric pressure corrected for the vapor pressure and is the pressure pushing on the liquid surface. Imagine the suction pipe devoid of liquid; if this is the case, then this pressure will push the fluid up the suction pipe. This is actually what happens as soon as the impeller starts moving and pulling the liquid up. As soon as a space is evacuated by the impeller in the suction pipe, liquid rushes up to fill the void; this is not possible, however, without a positive NPSH to push the liquid. Note that before the impeller can do its job, the fluid must, first, reach it. Thus, the need for a driving force at the inlet side.

The pushing pressure converts to available head or available energy at the suction side of the pump system. The surface of the well is below the pump, so this available energy must be subtracted by h_ℓ. The other substractions are the friction losses h_{fs}.

For the source tank fluid level above the center line of the pump impeller, h_s will be added increasing the available energy. The losses will, again, be subtracted. In symbols,

$$\text{NPSH} = \frac{P_{atm} - P_v}{\gamma} - h_\ell(\text{or} + h_s) - h_{fs} \tag{4.44}$$

It is instructive to derive Equation (4.44) by applying the energy equation between the wet well pool surface (point 1 of the lower tank) and the inlet to the pump (a or g). The equation is:

$$\frac{V_1^2}{2g} + z_1 + \frac{P_1}{\gamma} - h_{fs} = \frac{V_2^2}{2g} + z_2 + \frac{P_2}{\gamma}$$

$$0 + 0 + \frac{P_{atm} - P_v}{\gamma} - h_{fs} = \frac{V_a^2}{2g} + h_\ell + \frac{P_a}{\gamma}$$

where V_1 = velocity at the wet pool level, z_1 = elevation of the pool level with reference to a datum (the pool level, itself, in this case), P_1 = pressure at pool level, h_{fs} = friction loss from pool level to inlet of pump (the suction side friction loss), V_a = velocity at inlet to the pump, z_a = elevation of the inlet to the pump with reference to the datum (the pool level), and P_a = pressure at the inlet to the pump. P_a is the absolute pressure at the pump inlet, i.e., not a gage pressure but an absolute pressure corrected for the vapor pressure of water. This type of absolute pressure is not the same as the *normal absolute pressure* where the prevailing barometric pressure is simply added to the gage reading. This is an absolute pressure where the *vapor pressure* is first subtracted from the gage reading P_{gage} and then the result added to the prevailing atmospheric pressure. In other words, $P_a = P_{gage} - P_v + P_{atm}$. This produces the true pressure acting on the liquid at the inlet to the pump. By also considering the source tank above the center line of the impeller, the final equation after rearranging is:

$$\frac{V_a^2}{2g} + \frac{P_a}{\gamma} = \frac{V_a^2}{2g} + \frac{P_{gage} - P_v + P_{atm}}{\gamma} = \frac{P_{atm} - P_v}{\gamma} - h_\ell(\text{or} + h_s) - h_{fs} \tag{4.45}$$

Therefore, NPSH is also

$$\text{NPSH} = \frac{V_a^2}{2g} + \frac{P_{gage} - P_v + P_{atm}}{\gamma} = \frac{V_a^2}{2g} + \left(h_i + \frac{P_{atm} - P_v}{\gamma} \right) \tag{4.46}$$

Note that P_a/γ is equal to $P_{gage} - P_v + P_{atm}/\gamma = h_i + P_{atm} - P_v/\gamma$, where the vapor pressure P_v has been subtracted to obtain the true pressure acting on the fluid as mentioned.

In simple words, the NPSH is the amount of energy that the fluid possesses at the inlet to the pump. It is the inlet dynamic head that pushes the fluid into the pump impeller blades. Finally, NPSH and cavitation effects must be related. If NPSH does not exist at the suction side, cavitation will, obviously, occur.

The next point to be considered is the influence of the NPSH on deep-well pumps. It should be clear that the depth of water that can be pumped is limited by the net positive suction head. We have learned, however, that when pumps are connected in series, the heads are added. Thus, it is possible to pump groundwater from any depth, if impellers of the pump are laid out in series. This is the principle used in the design of deep-well pumps.

Refer to the deep-well pump of Figure 4.3. This pump is shown to have two stages. The water lifted by the first stage is introduced to the second stage. And, since the stages are in series, the head developed in the first stage is added to that of the second stage. Thus, this pump is capable of pumping water from deeper wells. Deep-well pumps can be designed for any number of stages, within practical limits. Because of the limitation of the NPSH, these pumps must obviously be lowered toward the bottom at a distance sufficient to have a positive NPSH on the first impeller, with an ample margin of safety. Provide a margin of safety in the neighborhood of 90% of the calculated NPSH. In other words, if the computed NPSH is in the neighborhood of 7 m, for example, assume it to be $0.9(7) \approx 6$ m.

Example 4.7 A wastewater is to be pumped to an elevation of 30 m above a sump. Overall friction losses of the system and the velocity at the discharge side are estimated to be 20 m and 1.30 m/s, respectively. The operating drive is to be 1,200 rpm. Suction friction losses are 1.03 m; the diameter of the suction and discharge lines are 250 mm and 225 mm, respectively. The vertical distance from the sump pool level to the pump centerline is 2 m. What is the NPSH?

Solution: The formula to be used is Equation 4.44.

$$\text{NPSH} = \frac{P_{atm} - P_v}{\gamma} - h_\ell - h_{fs}$$

Not all of these variables are given; therefore, some assumptions are necessary. In an actual design, this is what is actually done; the resulting design, of course, must be shown to work. Assume standard atmosphere and 20°C.

$P_{atm} = 101,325$ N/m^2 $P_v = 2340$ N/m^2 $\gamma = 997(9.81)$ N/m^3

$h_\ell = 2$ m $h_{fs} = 1.03$ m

Therefore, $\text{NPSH} = \dfrac{101,325 - 2340}{997(9.81)} - 2 - 1.03 = 7.09$ m of water   **Ans**

4.7 PUMPING STATION HEAD ANALYSIS

The *pumping station* (containing the pumps and station appurtenances) and the *system piping* constitute the *pumping system*. In this system, there are two types of characteristics: the *pump characteristics* and the *system characteristic*. The term system characteristic refers to the characteristic of the system comprising everything that contains the flow except the pump casing and the impeller inside it.

Specifically, system characteristic is the relation of discharge Q and the associated head requirements of this system which, again, does not include the pump arrangement. The pump arrangement may be called the *pump assembly*. In the design of a pumping station, both the pump characteristics and the system characteristic must be considered simultaneously.

For convenience, reproduce the formulas for TDH as follows:

$$\text{TDH} = \text{TDH0}sd = h_{st} + h_{fs} + h_{fd} \tag{a}$$

$$\text{TDH} = \text{TDH0}rm = h_{st} + h_{fr} + h_{fm} \tag{b}$$

$$\text{TDH} = \text{TDH}fullsd = \frac{P_B - P_{atm}}{\gamma} + \frac{V_B^2}{2g} + z_2 + h_{fs} + h_{fd} \tag{c}$$

$$\text{TDH} = \text{TDH}fullrm = \frac{P_B - P_{atm}}{\gamma} + \frac{V_B^2}{2g} + z_2 + h_{fr} + h_{fm} \tag{d}$$

$$\text{TDH} = \text{TDH}mano = \left(\frac{P_o}{\gamma} + \frac{V_o^2}{2g}\right) - \left(\frac{P_i}{\gamma} + \frac{V_i^2}{2g}\right) \tag{e}$$

$$\text{TDH}abso = h_{abso} + h_{vo} - (h_{absi} + h_{vi}) \tag{f}$$

$$\text{TDH} = \text{TDH}dh = \text{odh} - \text{idh} \tag{g}$$

TDH is the "total developed head" developed inside the pump casing, that is, developed by the pump assembly. This TDH is equal to any of the right-hand-side expressions of the above equations. If the TDH on the left refers to the pump assembly, then the right-hand-side expressions must refer to the system piping. By assuming different values of discharge Q, corresponding values of the right-hand-side expressions can be calculated. These values are head loss equivalents corresponding to the Q assumed. This is the relationship of the various Qs and head losses in the system characteristic mentioned above. As can be seen, these head losses are head loss requirements for the associated Q. It is head loss requirements that the TDH of the pump assembly must satisfy. Call head loss requirements TDHR. TDHR, therefore, requires the TDH of the pump.

It should be obvious that, if the TDH of the pump assembly is less than the TDHR of the system piping, no fluid will flow. To ensure that the proper size of pumps are chosen for a given desired pumping rate, the TDH of the pumps must be equal to the TDHR of the system piping. This is easily done by plotting the pump head-discharge-characteristic curve and the system-characteristic curve on the same graph. The point of intersection of the two curves is the desired operating point. The principle of series or parallel connections of pumps must be used to arrive at the proper pump combination to suit the desired system characteristic requirement. The specific speed should be checked to ensure that the pump assembly selected operates at the best operating efficiency.

The system piping is composed of the suction piping and the discharge piping system. Both the suction and the discharge piping systems would include the piping

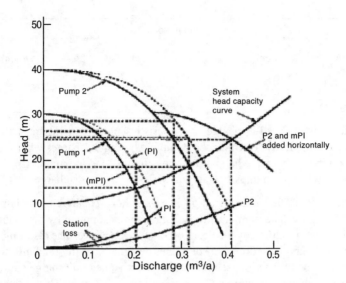

FIGURE 4.10 Use of pump and system head-discharge characteristics curves for sizing pumping stations. (From Peavy, H. S., D. R. Rowe, and G. Tchobanoglous (1985). *Environmental Engineering*. McGraw-Hill, New York, 395. With permission.)

inside the pumping station, itself. For the purpose of system head calculations, it is convenient to disregard the head losses of the pumping station piping and the suction piping. The disregarded pumping station and suction piping losses are designated as station losses and applied as corrections to the pump head-discharge curve supplied by the manufacturer. This correction produces the *effective* pump head-discharge curve. The discharge piping losses is also corrected by those portions of this loss assigned to the pumping station losses.

An illustration of sizing pumping stations is shown in Figure 4.10 and in the next example.

Example 4.8 Calculations for the system characteristic curve yield the following results:

Q (m³/s)	TDHR (m)	Q (m³/s)	TDHR (m)	Q (m³/s)	TDHR (m)
0.0	10.00	0.1	10.84	0.2	13.37
0.3	17.59	0.4	23.48	0.5	31.06

The station losses are as follows:

Pump No. 1					
Q (m³/s)	h_f (m)	Q (m³/s)	h_f (m)	Q (m³/s)	h_f (m)
0.0	0.00	0.1	0.14	0.2	0.56

Pump No. 2					
Q (m³/s)	h_f (m)	Q (m³/s)	h_f (m)	Q (m³/s)	h_f (m)
0.0	0.00	0.1	0.06	0.2	0.26
0.3	0.58	0.4	1.03		

If two pumps with head-discharge characteristics plotted in Figure 4.10 (dashed lines) are to be used, (a) determine the pumping system discharges when each pump is operated separately and, when both pumps are operated in parallel (b), determine at what head will both pumps operated in series deliver a discharge of 0.2 m³/s.

Solution: (a) The system head-discharge or head-capacity curve is plotted as shown in the figure. The pump head-discharge curves supplied by the manufacturer (dashed lines) are modified by the head losses as given above. The resulting effective head-discharge curves are drawn in solid lines designated as mP₁ and mP₂ for pumps 1 and 2, respectively. The intersection of the effective head-discharge curve and the system curve, when only pump no. 1 is operating, is 0.2 m³/s. This is the pumping system discharge. **Ans**

When only pump no. 2 is operating, the system discharge is 0.31 m³/s. **Ans**

When both are operated in parallel, the effective characteristic curve for pump no. 1 is shifted horizontally to the right until the top end of the curve coincides with a portion of the effective curve of pump no. 2, as shown. This has the effect of adding the discharges for parallel operation. As indicated, when both are operated in parallel, the system discharge is 0.404 m³/s. **Ans**

(b) For the operation in series, the TDH for pump no. 1 for a system discharge of 0.2 m²/s is 13 m. That of pump no. 2 is 32 m. Therefore, the system TDH is 32 + 13 = 45 m. **Ans**

GLOSSARY

Axial-flow pumps—Pumps that transmit the fluid pumped in the axial direction.

Best operating efficiency—Value of the efficiency that corresponds to the best operating performance of the pump.

Brake or shaft power—The power of the motor or prime mover driving the pump.

Brake efficiency—Ratio of the power given to the fluid to the brake input power (brake power) to the pump.

Cavitation—A state of flow where the pressure in the liquid becomes equal to its vapor pressure.

Centrifugal pump—A pump that conveys fluid through the momentum created by a rotating impeller.

Discharge—In a pumping system, the arrangement of elements after the pumping station.

Discharge velocity head—The velocity head at the discharge of a pumping system.

Displacement pumps—Pumps that literally pushes the fluid in order to move it.
Dynamically similar pumps—Pumps with head coefficients that are equal.
Fittings losses—Head losses in valves and fittings.
Flow coefficient C_Q—The group $Q/(\omega D^3)$.
Friction head loss—A head loss due to loss of internal energy.
Gear pump—A pump that basically operate like a lobe pump, except that instead
    of lobes, gear teeth are used to move the fluid.
Geometrically similar pumps—Pumps with corresponding parts that are
    proportional.
Head coefficient C_H—The group $Hg/(\omega^{2D} D^2)$.
Homologous pumps—Pumps that are similar. Similarities are established dyna-
    mically, kinematically, or geometrically.
Inlet dynamic head—The sum of the inlet velocity head and inlet manometric
    head of a pump.
Inlet manometric head—The manometric level at the inlet to a pump.
Kinematically similar pumps—Pumps whose flow coefficients are equal.
Lobe pump—A positive-displacement pump whose impellers are shaped like lobes.
Manifold pipe—A pipe with two or more pipes connected to it.
Manometric level—The height of liquid corresponding to the gage pressure.
Mixed-flow pump—Pump with an impeller that is designed to provide a com-
    bination of forward and radial flow.
Net positive suction head (NSPH)—The amount of energy possessed by a fluid
    at the inlet to a pump.
Non-pivot parameter—The counterpart of pivot parameter.
Outlet dynamic head—The sum of the outlet velocity head and outlet mano-
    metric head of a pump.
Outlet manometric head—The manometric level at the outlet of a pump.
Parallel connection—Mode of connection of more than one pump where the
    discharges of all the pumps are combined into one.
Positive-displacement pump—A pump that conveys fluid by directly moving it
    using a suitable mechanism such as a piston, plunger, or screw.
Power coefficient C_P—The group $\mathcal{P}/\rho\omega^3 D^5$.
Propeller pumps—The same as axial-flow pumps.
Pump assembly—The pump arrangement in a pumping station.
Pump characteristics—Set of curves that depicts the performance of a given parti-
    cular pump.
Pump loss—Head losses incurred inside the pump casing.
Pumping—A unit operation used to move fluid from one point to another.
Pumping station—A location where one or more pumps are operated to convey
    fluids.
Pumping system—The pumping station and the piping system constitute the
    pumping system.
Radial-flow pump—Pump with an impeller that directly throws the flow radially
    into the side of the chamber circumscribing it.
Scaling laws—Mathematical equations that establish the similarity of homolo-
    gous pumps.

Series connection—A mode of connecting more than one pump where the discharge of the pump ahead is introduced to the inlet of the pump following.

Similar or homologous pumps—Pumps where the head, flow, and pressure coefficients are equal.

Similarity, affinity, or scaling laws—The equations that state that the head, flow, and power coefficients of a series of pumps are equal.

Specific speed—A ratio obtained by manipulating the ratio of the flow coefficient to the head coefficient of a pump. Values obtained are values applying at the best operating efficiency.

Static discharge head—The vertical distance from the pump centerline to the elevation of the discharge liquid level.

Static suction head—The vertical distance from the elevation of the inflow liquid level above the pump centerline to the centerline of the pump.

Static suction lift—The vertical distance from the elevation of the inflow liquid level below the pump centerline to the centerline of the pump.

Suction—In a pumping system, the system of elements before the pumping station.

Suction velocity head—The velocity head at the suction side of a pumping system.

System characteristic—In a pumping system, the relationship of discharge and the associated head requirement that excludes the pump assembly.

Total dynamic head or total developed head—The head given to the pump minus pump losses.

Total developed head requirement—The equivalent head loss corresponding to a given discharge.

Total static head—The vertical distance between the elevation of the inflow liquid level and the discharge liquid level.

Transition losses—Head losses in expansions, contractions, bends, and the like.

Turbomachine—Fluid machine that turns or tends to turn about an axis.

Vane pump—A pump in which a vane pushes the water forward as it is being trapped between the vane and the side of the casing.

Volute—Casing of a centrifugal pump that is shaped into a spiral.

SYMBOLS

C_H	Head coefficient
C_Q	Flow coefficient
C_P	Power coefficient
D	Diameter of centrifugal pump or length of stroke of reciprocating pump
g	Earth's gravitational acceleration
h_{brake}	Head equivalent to brake power input to pump from a prime mover
h_f	Friction head loss
h_{fd}	Discharge side friction losses
h_{fm}	Minor head losses

h_{fr}	Head losses due to straight runs of pipes
h_{fs}	Suction side friction losses
h_i	Inlet manometric head
h_ℓ	Suction lift
h_{lp}	Pump loss
h_o	Outlet manometric head
h_p	Head given to pump
h_q	Head equivalent to heat added to system
h_s	Suction head
h_{st}	Total static head
h_{vi}	Inlet velocity head
h_{vo}	Outlet velocity head
idh	Inlet dynamic head
H	Total dynamic or developed head, TDH
N_s	Specific speed
NPSH	Net positive suction head
odh	Outlet dynamic head
P	Pressure
P_{atm}	Barometric pressure
P_{gage}	Gage reading of pressure
P_v	Vapor pressure of water
ΔP	Developed pressure, equivalent to total developed head
$\mathcal{P}$	Power to fluid.
Q	Rate of discharge
TDH	Total dynamic or developed head
TDHR	Total developed head requirement
V	Velocity
z	Elevation of fluid from a reference datum
γ	Specific weight of water
η	pump efficiency
ρ	Density of water
ω	Speed of pump in radians per second
μ	absolute or dynamic viscosity

PROBLEMS

4.1   Water at a temperature of 22°C is to be conveyed from a reservoir with a water surface elevation of 70 m to another reservoir 300 m away with a water surface elevation of 20 m. Determine the diameter of a steel pipe if the flow is 2.5 m³/s. Assume a square-edged inlet and outlet as well as two open gate valves in the system.

4.2   For the pump represented in Figure 4.8, determine the discharge when the pump is operating at a head of 20 m at a speed of 1,100 rpm.

4.3   Calculate the efficiency and power for the pump in Problem 4.2.

4.4   If the pump in Problem 4.2 is operated at 1,170 rpm, calculate the resulting H, Q, $\mathcal{P}_{brake}$, and η.

4.5   If a homologous 30-cm pump is to be used in Problem 4.2, calculate the resulting H, Q, $\mathcal{P}_{brake}$, and η for the same rpm.

4.6   The outlet manometric head at the discharge of a pump is equal to the equivalent of 50 m of water. If the discharge velocity is 2.0 m/s, what is the outlet dynamic head?

4.7   You are required to recommend the type of pump to be used to convey wastewater to an elevation of 8 m above a sump. Friction losses and the velocity at the sewage discharge level are estimated to be 3 m and 1.30 m/s, respectively. The operating drive is to be 1,200 rpm. Suction friction losses are 1.03 m; the diameter of the suction and discharge lines are 250 mm and 225 mm, respectively. The vertical distance from the sump pool level to the pump centerline is 2 m. What type of pump would you recommend?

4.8   In Problem 4.7, if the temperature is 20°C, has cavitation occurred?

4.9   Compute the inlet and outlet manometric heads in Problem 4.7.

4.10   In Problem 4.7, what are the inlet and outlet dynamic heads? From the values of idh and odh, calculate TDH.

4.11   In Problem 4.7, what is the NPSH?

4.12   In Example 4.8, calculate the percent errors in the answers if the characteristic curves of the pumps are not corrected for the station losses.

4.13   What is $-h_{lp} + h_p$ equal to? Explain why this expression is given this name.

4.14   What is the name given to the pressure that is equivalent to the pressure head?

4.15   What are inlet dynamic head and outlet dynamic head?

4.16   It is desired to pump a wastewater to an elevation of 30 m above a sump. Friction losses and velocity at the discharge side of the pump system are estimated to be 20 m and 1.30 m/s, respectively. The operating drive is to be 1,200 rpm. Suction friction losses are 1.03 m; the diameter of the suction and discharge lines are 250 mm and 225 mm, respectively. The vertical distance from the sump pool level to the pump centerline is 2 m. If the temperature is 10°C, has cavitation occurred?

4.17   What are the inlet and outlet manometric heads in Problem 4.16?

4.18   What are the inlet and outlet total dynamic heads in Problem 4.16? From the values of the idh and odh, calculate TDH.

4.19   In relation to h_{lp}, discuss the relationship of h_p and h_{brake}.

4.20   Show all the steps in the derivation of Equation (4.27).

4.21   It is desired to pump a wastewater to an elevation of 30 m above a sump. Friction losses and velocity at the discharge side of the pump system are estimated to be 20 m and 1.30 m/s, respectively. The operating drive is to be 1,200 rpm. The diameter of the suction and discharge lines are 250 mm and 225 mm, respectively. The vertical distance from the sump pool level to the pump centerline is 2 m. If the temperature is 10°C and the inlet manometric head is −3.09 m, what are the suction friction losses?

4.22   In Problem 4.21, what is the head given to the pump?

4.23 Pump characteristics curves are developed in accordance with the setup of Figure 4.7. The discharge flow is 0.15 m^3/s and the outlet diameter of the discharge pipe is 375 mm. The motor driving the pump is 50 hp. Calculate the gage pressure at the outlet of the pump.

4.24 Pump characteristics curves are developed in accordance with the setup of Figure 4.7. The pressure at the outlet of the pump is found to be 196 kN/m^2 gage. The outlet diameter of the discharge pipe is 375 mm. The motor driving the pump is 50 hp. Calculate the rate of discharge.

4.25 Pump characteristics curves are developed in accordance with the setup of Figure 4.7. The pressure at the outlet of the pump is found to be 196 kN/m^2 gage. The outlet diameter of the discharge pipe is 375 mm. If the rate of discharge is 0.15 m^3/s, calculate the input brake power to the pump.

4.26 In Problem 4.25, if the brake efficiency is 65%, calculate P_o, h_{brake}, and V_o.

4.27 If the pump in Example 4.4 has a power input of 30 kW, determine the rpm.

4.28 In Problem 4.27, calculate the resulting H.

4.29 In Problem 4.27, calculate the resulting Q and η.

4.30 If a homologous 40-cm pump is to be used for the problem in Example 4.5, calculate the resulting H, Q, $\mathcal{P}_{brake}$, and η for the same rpm.

4.31 A designer wanted to recommend the use of an axial-flow pump to move wastewater to an elevation of 50 m above a sump. Friction losses and velocity at the discharge side of the pumping system are estimated to be 20 m and 1.30 m/s, respectively. The operating drive is to be 1,200 rpm. Suction friction losses are 1.03 m; the diameter of the suction and discharge lines are 250 mm and 225 mm, respectively. The vertical distance from the sump pool level to the pump centerline is 15 m. Is the designer recommending the right pump? What is the net positive suction head? Is the pumping possible?

BIBLIOGRAPHY

Bosserman, B. and P. Behnke (1998). Selection criteria for wastewater pumps. *Water/Eng. Manage.* 145, 10.

Foster, R. S. (1998). Surge protection design for the city and county of San Francisco water transmission system. *Pipelines in the Constructed Environment, Proc. 1998 Pipeline Div. Conf.,* Aug. 23–27, San Diego, CA, 103–112, ASCE, Reston, VA.

Foster, R. S. (1998). Analysis of surge pressures in the inland feeder and eastside pipeline. *Proc. Pipelines in the Constructed Environment, Proc. 1998 Pipeline Div. Conf.,* Aug. 23–27, San Diego, CA, 88–96, ASCE, Reston, VA.

Franzini, J. B. and E. J. Finnemore (1997). *Fluid Mechanics.* McGraw-Hill, New York.

Granet, I. (1996). *Fluid Mechanics.* Prentice Hall, Englewood Cliffs, NJ.

Hammer, M. J. (1986). *Water and Wastewater Technology.* John Wiley & Sons, New York.

Kotov, A. I. (1998). Water treatment system is one of the most important elements of the ecological safety of the plant. *Tyazheloe Mashinostroenie,* 7, 58–60.

Nahm, E. S. and K. B. Woo (1998). Prediction of the amount of water supplied in wide-area waterworks. *Proc. 1998 24th Annu. Conf. IEEE Industrial Electron. Soc., IECON, Part 1,* Aug. 31–Sept. 4, Aachen, Germany, 1, 265–268. IEEE Comp. Soc., Los Alamitos, CA.

Plakhtin, V. D. (1998). Hydropneumatic system for balancing spindles of the breakdown at Mill 560. *Stal,* 3, 45–47.

Peavy, H. S., D. R. Rowe, and G. Tchobanoglous (1985). *Environmental Engineering.* McGraw-Hill, New York.

Qasim, S. R. (1985). *Wastewater Treatment Plants Planning, Design, and Operation.* Holt, Rinehart & Winston, New York.

Shafai-Bajestan, M., A. Behzadi-Poor, S. R. Abt, (1998). Control of sediment deposition at the Amir-Kabir pump station using a physical hydraulic model. *Int. Water Resour. Eng. Conf.—Proc. 1998 Int. Water Resour. Eng. Conf., Part 2,* Aug. 3–7, Memphis, TN, 2, 1553–1558, ASCE, Reston, VA.

Shames, I. (1992). *Mechanics of Fluids.* McGraw-Hill, New York.

Shi, W. (1998). Design of axial flow pump hydraulic Model ZM931 on high specific speed. *Nongye Jixie Xuebao/Trans. Chinese Soc. Agricultural Machinery,* 29, 2, 48–52.

Sincero, A. P. and G. A. Sincero (1996). *Environmental Engineering*: *A Design Approach.* Prentice Hall, Upper Saddle River, NJ.

Smolyanskij, B. G., O. V. Bakulin, and O. E. Volkov (1998). Advanced pump equipment for the acceptance and delivery of petroleum products at warehouses. *Khimicheskoe I Neftyanoe Mashinostroenie,* 6, 38–40.

Spencer, R. C., L. G. Grainger, V. G. Peacock, and J. Blois (1998). Building a 5000 hp, VFD Controlled Pump Station in 105 Days "Fact or Fantasy." *Proc. 1998 International Pipeline Conference, IPC, Part 2,* June 7–11, Calgary, Canada, 2, 1111–1117. ASME, Fairfield, NJ.

Vlasov, V. S. (1998). High pressure pump stations and their application fields. *Tyazheloe Mashinostroenie,* 7, July, 53–55.

Wang, X. T. et al. (1998). Modeling and remediation of ground water contamination at the Engelse Werk Wellfield. Ground *Water Monitoring Remediation,* 18, 3, 114–124.

Zheng, Y., et al. (1998). Hydraulic design and analysis for a water supply system modification. *Proc. 1998 Int. Water Resour. Eng. Conf., Part 2,* Aug. 3–7, Memphis, TN, 2, 1673–1678. ASCE, Reston, VA.

5 Screening, Settling, and Flotation

Screening is a unit operation that separates materials into different sizes. The unit involved is called a *screen*. As far as water and wastewater treatment is concerned, only two "sizes" of objects are involved in screening: the water or wastewater and the objects to be separated out. *Settling* is a unit operation in which solids are drawn toward a source of attraction. In gravitational settling, solids are drawn toward gravity; in centrifugal settling, solids are drawn toward the sides of cyclones as a result of the centrifugal field; and in electric-field settling, as in electrostatic precipitators, solids are drawn to charge plates. *Flotation* is a unit operation in which solids are made to float to the surface on account of their adhering to minute bubbles of gases (air) that rises to the surface. On account of the solids adhering to the rising bubbles, they are separated out from the water. This chapter discusses these three types of unit operations as applied to the physical treatment of water and wastewater.

5.1 SCREENING

Figure 5.1 shows a bar rack and a traveling screen. Bar racks (also called bar screens) are composed of larger bars spaced at 25 to 80 mm apart. The arrangement shown in the figure is normally used for shoreline intakes of water by a treatment plant. The rack is used to exclude large objects; the traveling screen following it is used to remove smaller objects such as leaves, twigs, small fish, and other materials that pass through the rack. The arrangement then protects the pumping station that lifts this water to the treatment plant. Figure 5.2 shows a bar screen installed in a detritus tank. Detritus tanks are used to remove grits and organic materials in the treatment of raw sewage. Bar screens are either hand cleaned or mechanically cleaned. The bar rack of Figure 5.1 is mechanically cleaned, as shown by the cable system hoisting the scraper; the one in Figure 5.2 is manually cleaned. Note that this screen is removable. Table 5.1 shows some design parameters and criteria for mechanically and hand-cleaned screens.

Figure 5.3 shows a microstrainer. As shown, this type of microstrainer consists of a straining material made of a very fine fabric or screen wound around a drum. The drum is about 75% submerged as it is rotated; speeds of rotation are normally about from 5 to 45 rpm. The influent is introduced from the underside of the wound fabric and exits into the outside. The materials thus strained is retained in the interior of the drum. These materials are then removed by water jets that directs the loosened strainings into a screening trough located inside the drum. In some designs, the flow is from outside to the inside.

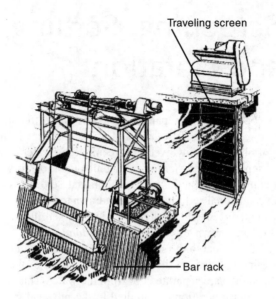

FIGURE 5.1 Bar rack and traveling screen. (Courtesy of Envirex, Inc.)

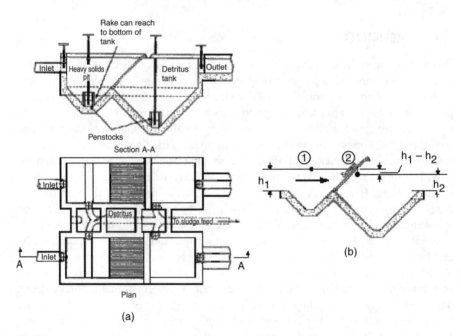

FIGURE 5.2 Bar screen in a detritus tank.

Microstrainers have been used to remove suspended solids from raw water containing high concentrations of algae. In the treatment of wastewater using oxidation ponds, a large concentration of algae normally results. Microstrainers can be used for this purpose in order to reduce the suspended solids content of the effluent

TABLE 5.1
Design Parameters and Criteria for Bar Screens

Parameter	Mechanically Cleaned	Manually Cleaned
Bar size		
Width, mm	5–20	5–20
Thickness, mm	20–80	20–80
Bars clear spacing, mm	20–50	15–80
Slope from vertical, degrees	30–45	0–30
Approach velocity, m/s	0.3–0.6	0.6–1.0

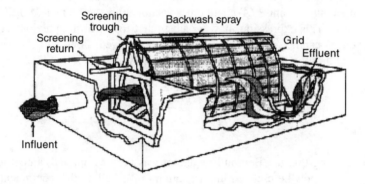

FIGURE 5.3 A Microstrainer. (Courtesy of Envirex, Inc.)

that may cause violations of the discharge permits of the plant. Microstrainers have also been used to reduce the suspended solids content of wastewaters treated by biological treatment. Openings of microstrainers are very small. They vary from 20 to 60 μm and the cloth is available in stainless steel or polyester construction.

5.1.1 HEAD LOSSES IN SCREENS AND BAR RACKS

Referring to **b** of Figure 5.2, apply the Bernoulli equation, reproduced below, between points 1 and 2.

$$\frac{P_1}{\gamma} + \frac{V_1^2}{2g} + h_1 = \frac{P_2}{\gamma} + \frac{V_2^2}{2g} + h_2 \qquad (5.1)$$

where P, V, and h are the pressure, velocity, and elevation head at indicated points; g is the acceleration due to gravity. V_1 is called the *approach velocity*; the channel in which this velocity is occurring is called the *approach channel*. To avoid sedimentation in the approach channel, the velocity of flow at this point should be maintained at the self-cleansing velocity. Self-cleansing velocities are in the neighborhood of 0.76 m/s.

Remember from fluid mechanics that the Bernoulli equation is an equation for frictionless flow along a streamline. The flow through the screen is similar to the flow through an orifice, and it is standard in the derivation of the flow through an orifice to assume that the flow is frictionless by applying the Bernoulli equation. To consider the friction that obviously is present, an orifice coefficient is simply prefix to the derived equation.

Both points 1 and 2 are at atmospheric, so the two pressure terms can be canceled out. Considering this information and rearranging the equation produces

$$V_1^2 = V_2^2 + 2g(h_2 - h_1) \qquad (5.2)$$

From the equation of continuity, V_1 may be solved in terms V_2, cross-sectional area of clear opening at point 2 (A_2), and cross-sectional area at point 1 (A_1). V_1 is then $V_1 = A_2 V_2 / A_1$. This expression may be substituted for V_1 in the previous equation, whereupon, V_2 can be solved. The value of V_2 thus solved, along with A_2, permit the discharge Q through the screen openings to be solved. This is

$$Q = A_2 V_2 = A_2 \sqrt{\frac{2g(h_1 - h_2)}{1 - \frac{A_2}{A_1}}} = A_2 \sqrt{\frac{2g\Delta h}{1 - \frac{A_2}{A_1}}} \qquad (5.3)$$

Recognizing that the Bernoulli equation was the one applied, a coefficient of discharge must now be prefixed into Equation (5.3). Calling this coefficient C_d,

$$Q = C_d A_2 \sqrt{\frac{2g\Delta h}{1 - \frac{A_2}{A_1}}} \qquad (5.4)$$

Solving for the head loss across the screen Δh,

$$\Delta h = \frac{Q^2 \left(1 - \frac{A_2}{A_1}\right)}{2g C_d^2 A_2^2} \qquad (5.5)$$

As shown in Equation (5.5), the value of the coefficient can be easily determined experimentally from an existing screen. In the absence of experimentally determined data, however, a value of 0.84 may be assumed for C_d. As the screen is clogging, the value of A_2 will progressively decrease. As gleaned from the equation, the head loss Δh will theoretically rise to infinity. At this point, the screen is, of course, no longer functioning.

The previous equations apply when an approach velocity exists. In some situations, however, this velocity does not exist. In these situations, the previous equations do not apply and another method must be developed. This method is derived in the next section on microstrainers.

5.1.2 HEAD LOSS IN MICROSTRAINERS

Referring to Figure 5.3, the flow turns a right angle as it enters the openings of the microstrainer cloth. Thus, the velocity at point 1, V_1, (refer to Figure 5.2) would be approximately zero. Therefore, for microstrainers: applying the Bernoulli equation, using the equation of continuity, and prefixing the coefficient of discharge as was done for the bar screen, produce

$$\Delta h = \frac{Q^2}{2g C_d^2 A_2^2} \qquad (5.6)$$

As in the bar screen, the value of the coefficient can be easily determined experimentally from an existing microstrainer. In the absence of experimentally determined data, a value of 0.60 may be assumed for C_d. Also, from the equation, as the microstrainer clogs, the value of A_2 will progressively decrease; thus the head loss rises to infinity, whereupon, the strainer ceases to function. Although the previous equation has been derived for microstrainers, it equally applies to ordinary screens where the approach velocity is negligible.

Example 5.1   A bar screen measuring 2 m by 5 m of surficial flow area is used to protect the pump in a shoreline intake of a water treatment plant. The plant is drawing raw water from the river at a rate of 8 m³/s. The bar width is 20 mm and the bar spacing is 70 mm. If the screen is 30% clogged, calculate the head loss through the screen. Assume $C_d = 0.60$.

Solution:

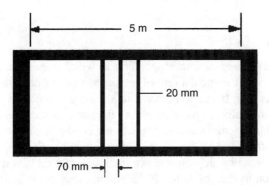

For screens used in shoreline intakes, the velocity of approach is practically zero. Thus,

$$\Delta h = \frac{Q^2}{2g C_d^2 A_2^2}$$

From the previous figure, the number of spacings is equal to one more than the number of bars. Let x = number of bars,

Therefore,

$$20x + 70(x + 1) = 5000$$
$$x = 54.77, \text{ say } 55$$
$$\text{Area of clear opening} = 70(55 + 1)(2000)$$
$$= 7,840,000 \text{ mm}^2 = 7.48 \text{ m}^2 = A_2$$

$$\Delta h = \frac{8^2}{2(9.81)(0.6)^2[7.48(0.7)]^2} = 0.33 \text{ m of water}$$

Example 5.2 In the previous example, assume that there was an approach velocity and that the approach area is 7.48 m². Calculate the head loss.

Solution:

$$\Delta h = \frac{Q^2\left(1 - \frac{A_2}{A_1}\right)}{2g C_d^2 A_2^2}$$

Assume $C_d = 0.84$

$$\Delta h = \frac{8^2\left[1 - \frac{7.48(0.7)}{5(2)}\right]}{2(9.81)(0.84^2)[7.48(0.7)]^2} = \frac{30.49}{379.54} = 0.08 \text{ m of water}$$

5.2 SETTLING

Settling has been defined as a unit operation in which solids are drawn toward a source of attraction. The particular type of settling that will be discussed in this section is gravitational settling. It should be noted that settling is different from sedimentation, although some authors consider settling the same as sedimentation.

Strictly speaking, sedimentation refers to the condition whereby the solids are already at the bottom and in the process of sedimenting. Settling is not yet sedimenting, but the particles are falling down the water column in response to gravity. Of course, as soon as the solids reach the bottom, they begin sedimenting. In the physical treatment of water and wastewater, settling is normally carried out in settling or sedimentation basins. We will use these two terms interchangeably.

Generally, two types of sedimentation basins are used: rectangular and circular. Rectangular settling basins or clarifiers, as they are also called, are basins that are rectangular in plans and cross sections. In plan, the length may vary from two to four times the width. The length may also vary from ten to 20 times the depth. The depth of the basin may vary from 2 to 6 m. The influent is introduced at one end and allowed to flow through the length of the clarifier toward the other end. The solids that settle at the bottom are continuously scraped by a sludge scraper and

FIGURE 5.4 Portion of a primary circular clarifier at the Back River Sewage Treatment Plant, Baltimore City, MD.

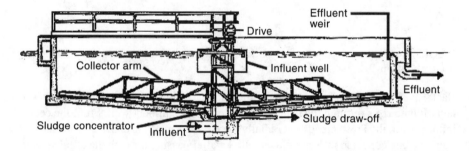

FIGURE 5.5 Elevation section of a circular radial clarifier. (Courtesy of Walker Process.)

removed. The clarified effluent flows out of the unit through a suitably designed effluent weir and launder.

Circular settling basins are circular in plan. Unlike the rectangular basin, circular basins are easily upset by wind cross currents. Because of its rectangular shape, more energy is required to cause circulation in a rectangular basin; in contrast, the contents of the circular basin is conducive to circular streamlining. This condition may cause short circuiting of the flow. For this reason, circular basins are typically designed for diameters not to exceed 30 m in diameter.

Figure 5.4 shows a portion of a circular primary sedimentation basin used at the Back River Sewage Treatment Plant in Baltimore City, MD. In this type of clarifier, the raw sewage is introduced at the center of the tank and the solids settled as the wastewater flows from the center to the rim of the clarifier. The schematic elevational section in Figure 5.5 would represent the elevational section of this clarifier at the

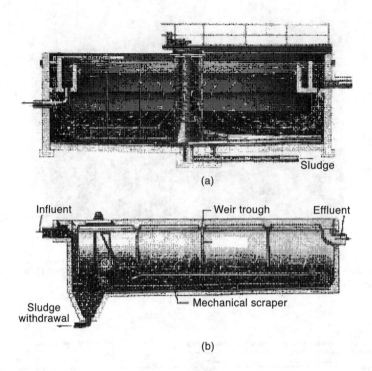

(a)

(b)

FIGURE 5.6 Elevation sections of a circular clarifier (a) and a rectangular clarifier (b). (Courtesy of Envirex, Inc.)

Back River treatment plant. As shown, the influent is introduced at the bottom of the tank. It then rises through the center riser pipe into the influent well. From the center influent well, the flow spreads out radially toward the rim of the clarifier. The clarified liquid is then collected into an effluent launder after passing through the effluent weir. The settled wastewater is then discharged as the effluent from the tank.

As the flow spreads out into the rim, the solids are deposited or settled along the way. At the bottom is shown a squeegee mounted on a collector arm. This arm is slowly rotated by a motor as indicated by the label "Drive." As the arm rotates, the squeegee collects the deposited solids or sludge into a central sump in the tank. This sludge is then bled off by a sludge draw-off mechanism.

Figure 5.6a shows a different mode of settling solids in a circular clarifier. The influent is introduced at the periphery of the tank. As indicated by the arrows, the flow drops down to the bottom, then swings toward the center of the tank, and back into the periphery, again, into the effluent launder. The solids are deposited at the bottom, where a squeegee collects them into a sump for sludge draw-off.

Figure 5.6b is an elevational section of a rectangular clarifier. In plan, this clarifier will be seen as rectangular. As shown, the influent is introduced at the left-hand side of the tank and flows toward the right. At strategic points, effluent trough (or launders) are installed that collect the settled water. On the way, the solids are then deposited at the bottom. A sludge scraper is shown at the bottom. This scraper moves the deposited sludge toward the front end sump for sludge withdrawal. Also,

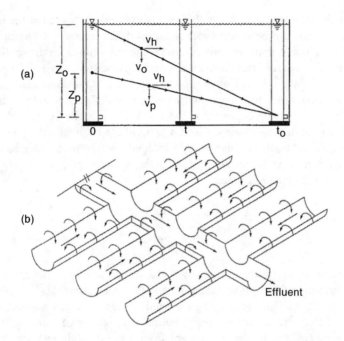

FIGURE 5.7 Removal at the settling zone (a); inboard weir design at outlet zone (b).

notice the baffles installed beneath each of the launders. These baffles would guide the flow upward, simulating a realistic upward overflow direction.

Generally, four functional zones are in a settling basin: the inlet zone, the settling zone, the sludge zone, and the outlet zone. The *inlet zone* provides a transition aimed at properly introducing the inflow into the tank. For the rectangular basin, the transition spreads the inflow uniformly across the influent vertical cross section. For one design of a circular clarifier, a baffle at the tank center turns the inflow radially toward the rim of the clarifier. On another design, the inlet zone exists at the periphery of the tank.

The *settling zone* is where the suspended solids load of the inflow is removed to be deposited into the *sludge zone* below. The *outlet zone* is where the effluent takes off into an effluent weir overflowing as a clarified liquid. Figure 5.7a and 5.7b shows the schematic of a settling zone and the schematic of an effluent weir, respectively. This effluent weir is constructed inboard. Inboard weirs are constructed when the natural side lengths or rim lengths of the basin are not enough to satisfy the weir-length requirements.

5.2.1 FLOW-THROUGH VELOCITY AND OVERFLOW RATE OF SETTLING BASINS

Figure 5.7a shows the basic principles of removal of solids in the settling zone. A settling column (to be discussed later) is shown moving with the horizontal flow of the water at velocity v_h from the entrance of the settling zone to the exit. As the column moves, visualize the solids inside it as settling; when the column reaches the end of the zone, these solids will have already been deposited at the bottom of the settling column. The behavior of the solids outside the column will be similar to that inside. Thus, a time t_o in the settling column is the same time t_o in the settling zone.

A particle possesses both downward terminal velocity v_o or v_p, and a horizontal velocity v_h (also called *flow-through velocity*). Because of the downward movement, the particles will ultimately be deposited at the bottom sludge zone to form the sludge. For the particle to remain deposited at the sludge zone, v_h should be such as not to scour it. For light flocculent suspensions, v_h should not be greater than 9.0 m/h; and for heavier, discrete-particle suspensions, it should not be more than 36 m/h. If A is the vertical cross-sectional area, Q the flow, Z_o the depth, W the width, L the length, and t_o the detention time:

$$v_h = \frac{Q}{A} = \frac{Q}{Z_o W} = \frac{L}{t_o} \qquad (5.7)$$

The *detention time* is the average time that particles of water have stayed inside the tank. Detention time is also called *retention time*. Because this time also corresponds to the time spent in removing the solids, it is also called *removal time*. For discrete particles, the detention time t_o normally ranges from 1 to 4 h, while for flocculent suspensions, it normally ranges from 4 to 6 h. Calling V the volume of the tank and L the length, t_o can be calculated in two ways: $t_o = Z_o/v_o$ and $t_o = V/Q = (WZ_oL)/Q = A_sZ_o/Q$. Also, for circular tanks with diameter D, $t_o = V/Q = (\frac{\pi D^2}{4}Z_o)/Q = A_sZ_o/Q$, also. Therefore,

$$\frac{Z_o}{v_o} = \frac{A_sZ_o}{Q} \Rightarrow v_o = \frac{Q}{A_s} = q_o \qquad (5.8)$$

where A_s is the surface area of the tank and Q/A_s is called the *overflow rate*, q_o. According to this equation, for a particle of settling velocity v_o to be removed, the overflow rate of the tank q_o must be set equal to this velocity.

Note that there is nothing here which says that the "efficiency of removal is independent of depth but depends only on the overflow rate." The statement that efficiency is independent of depth is often quoted in the environmental engineering literature; however, this statement is a fallacy. For example, assume a flow of 8 m^3/s and assert that the removal efficiency is independent of depth. With this assertion, we can then design a tank to remove the solids in this flow using any depth such as 10^{-50} meter. Assume the basin is rectangular with a width of 10^6 m. With this design, the flow-through velocity is $8/(10^{+6})(10^{-50}) = 8.0(10^{44})$ m/s. Of course, this velocity is much greater than the speed of light. The basin would be performing better if a deeper basin had been used. This example shows that the efficiency of removal is definitely not independent of depth. The notion that Equation (5.8) conveys is simply that the overflow velocity q_o must be made equal to the settling velocity v_o—nothing more. The overflow velocity multiplied by the surface area produces the *hydraulic loading rate* or *overflow rate*.

In the outlet zone, weirs are provided for the effluent to take off. Even if v_h had been properly chosen but overflow weirs were not properly sized, flows could be turbulent at the weirs; this turbulence can entrain particles causing the design to fail. Overflow weirs should therefore be loaded with the proper amount of overflow (called weir rate). Weir overflow rates normally range from 6–8 m^3/h per meter of weir length for light flocs to 14 m^3/h per meter of weir length for heavier discrete-particle suspensions. When weirs constructed along the periphery of the tank are not sufficient to meet the weir loading requirement, inboard weirs may be constructed. One such example was mentioned before and shown in Figure 5.7b. The formula to calculate weir length is as follows:

$$\text{weir Length} = \frac{Q}{\text{weir Rate}} \tag{5.9}$$

5.2.2 Discrete Settling

Generally, four types of settling occur: types 1 to 4. Type 1 settling refers to the removal of discrete particles, type 2 settling refers to the removal of flocculent particles, type 3 settling refers to the removal of particles that settle in a contiguous zone, and type 4 settling is a type 3 settling where compression or compaction of the particle mass is occurring at the same time. Type 1 settling is also called discrete settling and is the subject in this section. When particles in suspension are dilute, they tend to act independently; thus, their behaviors are therefore said to be discrete with respect to each other.

As a particle settles in a fluid, its body force f_g, the buoyant force f_b, and the drag force f_d, act on it. Applying Newton's second law in the direction of settling,

$$f_g - f_b - f_d = ma \tag{5.10}$$

where m is the mass of the particle and a its acceleration. Calling ρ_p the mass density of the particle, ρ_w the mass density of water, V_p the volume of the particle, and g the acceleration due to gravity, $f_g = \rho_p g V_p$ and $f_b = \rho_w g V_p$. The drag stress is directly proportional to the dynamic pressure, $\rho_w v^2/2$, where v is the terminal settling velocity of the particle. Thus, the drag force $f_d = C_D A_p \rho_w v^2/2$, where C_D is the coefficient of proportionality called *drag coefficient*, and A_p is the projected area of the particle normal to the direction of motion. Because the particle will ultimately settle at its terminal settling velocity, the acceleration a is equal to zero. Substituting all these into Equation (5.10) and solving for the terminal settling velocity v, produces

$$v = \sqrt{\frac{4}{3}g\frac{(\rho_p - \rho_w)d}{C_D \rho_w}} \tag{5.11}$$

assuming the particle is spherical. $A_p = \pi d^2/4$ for spherical particles, where d is the diameter.

The value of the coefficient of drag C_D varies with the flow regimes of laminar, transitional, and turbulent flows. The respective expressions are shown next.

$$C_D = \frac{24}{Re} \qquad\qquad \text{laminar flow} \qquad\qquad (5.12)$$

$$= \frac{24}{Re} + \frac{3}{\sqrt{Re}} + 0.34 \qquad \text{for transitional flow} \qquad (5.13)$$

$$= 0.4 \qquad\qquad \text{turbulent flow} \qquad\qquad (5.14)$$

where Re is the Reynolds number $= v\rho_w d/\mu$, and μ is the dynamic viscosity of water. Values of Re less than 1 indicate laminar flow, while values greater than 10^4 indicate turbulent flow. Intermediate values indicate transitional flow.

Substituting the C_D for laminar flow ($C_D = 24/Re$) in Equation (5.11), produces the Stokes equation:

$$v = \frac{g(\rho_p - \rho_w)d^2}{18\mu} \qquad\qquad (5.15)$$

To use the previous equations for non-spherical particles, the diameter d, must be the diameter of the equivalent spherical particle. The volume of the equivalent spherical particle $V_s = \frac{4}{3}\pi(\frac{d}{2})^3$, must be equal to the volume of the non-spherical particle $V_p = \beta d_p^3$, where β is a volume shape factor. Expressing the equality and solving for the equivalent spherical diameter d produces

$$d = \left(\frac{6}{\pi}\right)^{1/3} \beta^{1/3} d_p \qquad\qquad (5.16)$$

The following values of sand volumetric shape factors β have been reported: angular $= 0.64$, sharp $= 0.77$, worn $= 0.86$, and spherical $= 0.52$.

Example 5.3   Determine the terminal settling velocity of a spherical particle having a diameter of 0.6 mm and specific gravity of 2.65. Assume the settling is type 1 and the temperature of the water is 22°C.

Solution:

$$v = \sqrt{\frac{4}{3}g\frac{(\rho_p - \rho_w)d}{C_D \rho_w}}$$

$$g = 9.81 \text{ m/s}^2$$

$$\rho_{w22} = 997 \text{ kg/m}^3$$

$$\rho_p = 2.65(1000) = 2650 \text{ kg/m}^3$$

$$d = 0.6(10^{-3}) \text{ m}$$

$$\mu_{22} = 9.2(10^{-4}) \text{ kg/m-s} = 9.2(10^{-4}) \text{ N-s/m}^2$$

Therefore,

$$v = \sqrt{\frac{4}{3}(9.81)\frac{[2650 - 997][0.6(10^{-3})]}{C_D(997)}} = \frac{0.114}{\sqrt{C_D}}$$

$$C_D = \frac{24}{Re} \text{ (for } Re < 1)$$

$$C_D = \frac{24}{Re} + \frac{3}{\sqrt{Re}} + 0.34 \text{ (for } 1 \le Re \le 10^4)$$

$$C_D = 0.4 \text{ (for } Re > 10^4)$$

$$Re = \frac{dv\rho}{\mu} = \frac{dv(997)}{9.2(10^{-4})} = 1{,}083{,}695.70dv = 650.22v$$

Solve by successive iterations:

C_D	v (m/s)	Re
1.0	0.114	74
1.0	0.114	74

Therefore, $v = 0.114$ m/s **Ans**

Example 5.4 Determine the terminal settling velocity of a worn sand particle having a measured sieve diameter of 0.6 mm and specific gravity of 2.65. Assume the settling is type 1 and the temperature of the water is 22°C.

Solution:

$$v = \sqrt{\frac{4}{3}g\frac{(\rho_p - \rho_w)d}{C_D\rho_w}}$$

$$g = 9.81 \text{ m/s}^2$$

$$\rho_{w22} = 997 \text{ kg/m}^3$$

$$\rho_p = 2.65(1000) = 2650 \text{ kg/m}^3$$

$$d = 0.6(10^{-3}) \text{ m}$$

$$\mu_{22} = 9.2(10^{-4}) \text{ kg/m-s} = 9.2(10^{-4}) \text{ N-s/m}^2, \quad d = 1.24\beta^{0.333}d_p,$$
$$\text{for worn sands, } \beta = 0.86$$

Therefore,

$$d = 1.24(0.86^{0.333})(0.6)(10^{-3}) = 0.71(10^{-3}) \text{ m}$$

$$v = \sqrt{\frac{4}{3}(9.81)\frac{[2650 - 997][0.71(10^{-3})]}{C_D(997)}} = \frac{0.124}{\sqrt{C_D}}$$

$$C_D = \frac{24}{Re} \text{ (for } Re < 1)$$

$$C_D = \frac{24}{Re} + \frac{3}{\sqrt{Re}} + 0.34 \ \text{(for } 1 \le Re \le 10^4)$$

$$C_D = 0.4 \ \text{(for } Re > 10^4)$$

$$Re = \frac{dv\rho}{\mu} = \frac{dv(997)}{9.2(10^{-4})} = 1,083,695.70dv = 769.42v$$

Solve by successive iterations:

C_D	v(m/s)	R_e
1.0	0.124	95.41
0.90	0.131	100
0.88	0.132	101

Therefore, $v = 0.132$ m/s **Ans**

A raw water that comes from a river is usually turbid. In some water treatment plants, a presedimentation basin is constructed to remove some of the turbidities. These turbidity particles are composed not of a single but of a multitude of particles settling in a column of water. Since the formulas derived above apply only to a single particle, a new technique must be developed.

Consider the presedimentation basin as a *prototype*. In order to design this prototype properly, its performance is often simulated by a model. In environmental engineering, the model used is a settling column. Figure 5.8 shows a schematic of columns and the result of an analysis of a settling test.

At time equals zero, let a particle of diameter d_o be at the water surface of the column in **a**. After time t_o, let the particle be at the sampling port. Any particle that arrives at the sampling port at t_o will be considered removed. In the prototype, this removal corresponds to the particle being deposited at the bottom of the tank. t_o is the *detention time*. The corresponding settling velocity of the particle is $v_o = Z_o/t_o$, where Z_o is the depth. This Z_o corresponds to the depth of the settling zone of the prototype tank. Particles with velocities equal to v_o are removed, so particles of velocities equal or greater than v_o will all be removed. If x_o is the fraction of all

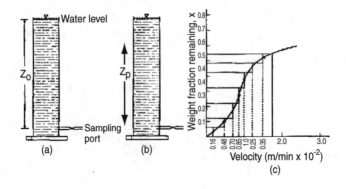

FIGURE 5.8 Settling column analysis of discrete settling.

particles having velocities less than v_o, $1 - x_o$ is the fraction of all particles having velocities equal to or greater than v_o. Therefore, the fraction of particles that are removed with certainty is $1 - x_o$.

During the interval of time t_o, some of the particles comprising x_o will be closer to the sampling port. Thus, some of them will be removed. Let dx be a differential in x_o. Assume that the average velocity of the particles in this differential is v_p. A particle is being removed because it travels toward the bottom and, the faster it travels, the more effectively it will be removed. Thus, removal is directly proportional to settling velocity. Removal is proportional to velocity, so the removal in dx is therefore $(v_p/v_o)dx$ and the total removal R comprising all of the particles with velocities equal to or greater than v_o and all particles with velocities less than v_o is then

$$R = 1 - x_o + \int_0^{x_o} \frac{v_p}{v_o} dx \qquad (5.17)$$

Note that this equation does not state that the velocity v_p must be terminal. It only states that the fractional removal R is directly proportional to the settling velocity v_p. For discrete settling, this velocity is the terminal settling velocity. For flocculent settling (to be discussed later), this velocity would be the average settling velocity of all particles at any particular instant of time.

To evaluate the integral of Equation (5.17) by numerical integration, set

$$\int_0^{x_o} \frac{v_p}{v_o} dx = \frac{1}{v_o} \sum v_p \Delta x \qquad (5.18)$$

This equation requires the plot of v_p versus x. If the original concentration in the column is $[c_o]$ and, after a time of settling t, the remaining concentration measured at the sampling port is $[c]$, the fraction of particles remaining in the water column adjacent to the port is

$$x = \frac{[c]}{[c_o]} \qquad (5.19)$$

Corresponding to this fraction remaining, the average distance traversed by the particles is $Z_p/2$, where Z_p is the depth to the sample port at time interval t from the initial location of the particles. The volume corresponding to Z_p contains all the particles that settle down toward the sampling port during the time interval t. Therefore, v_p is

$$v_p = \frac{Z_p}{2t} \qquad (5.20)$$

The values x may now be plotted against the values v_p. From the plot, the numerical integration may be carried out graphically as shown in **c** of Figure 5.8.

Example 5.5 A certain municipality in Thailand plans to use the water from the Chao Praya River as a raw water for a contemplated water treatment plant. The river is very turbid, so presedimentation is necessary. The result of a column test is as follows:

t(min)	0	60	80	100	130	200	240	420
c(mg/L)	299	190	179	169	157	110	79	28

What is the percentage removal of particles if the hydraulic loading rate is 25 $\mathrm{m^3/m^2 \cdot d}$? The column is 4-m deep.

Solution:

$$q_o = \text{hydraulic loading rate} = 25 \ \mathrm{m^3/m^2 \cdot d} = 25 \ \mathrm{m/d} = 0.0174 \ \mathrm{m/min}$$

$$R = 1 - x_o + \int_0^{x_o} \frac{v_p}{v_o} dx \qquad \int_0^{x_o} \frac{v_p}{v_o} dx = \frac{1}{v_o} \sum v_p \Delta x$$

$$v_o = 0.0174 \ \mathrm{m/min}$$

t (min)	0	60	80	100	130	200	240	420
c (mg/L)	299	190	179	169	157	110	79	28
$x = [c]/[c_o]$	1.0	0.64	0.60	0.57	0.53	0.37	0.26	0.09
$v_p = Z_p/2t = 4/2t$ (m/min)	—	0.033	0.025	0.02	0.015	0.01	0.0083	0.0048

Find the x corresponding to $v_o = 0.0174$ m/min

$$
\begin{array}{cc}
0.02 & 0.57 \\
0.0174 & x \\
0.015 & 0.53
\end{array}
\qquad
\frac{x - 0.57}{0.53 - 0.57} = \frac{0.0174 - 0.02}{0.015 - 0.02}
$$

$$x = 0.55 = x_0$$

t (min)	0	60	80	100	—	130	200	240	420	—
c (mg/L)	299	190	179	169	—	157	110	79	28	—
$x = [c]/[c_o]$	1.0	0.64	0.60	0.57	0.55	0.53	0.37	0.26	0.09	0
$v_p = Z_p/2t$ $= 4/2t$ (m/min)	—	0.033	0.025	0.02	0.0174	0.015	0.01	0.0083	0.0048	0
Δx	—	—	—	—	—	0.02[a]	0.16	0.11	0.17	0.09
v_p in Δx	—	—	—	—	—	0.0162[b]	0.0125	0.0092	0.0066	0.0024

[a] $0.02 = 0.55 - 0.53$

[b] $0.0162 = \dfrac{0.0174 + 0.015}{2}$

$$R = 1 - x_o + \frac{1}{v_o} \sum v_p \Delta x$$

$$= 1 - 0.55 + \frac{1}{0.0174}[0.0162(0.02) + 0.0125(0.16) + 0.0092(0.11)$$

$$+ 0.0066(0.17) + 0.0024(0.09)]$$

$$= 0.45 + 0.27 = 0.72 \quad \textbf{Ans}$$

5.2.3 OUTLET CONTROL OF GRIT CHANNELS

Grit channels (or chambers) are examples of units that use the concept of discrete settling in removing particles. Grit particles are hard fragments of rock, sand, stone, bone chips, seeds, coffee and tea grounds, and similar particles. In order for these particles to be successfully removed, the flow-through velocity through the units must be carefully controlled. Experience has shown that this velocity should be maintained at around 0.3 m/s. This control is normally carried out using a proportional weir or a Parshall flume. A grit channel is shown in Figure 5.9 and a proportional flow weir is shown in Figure 5.10. A proportional flow weir is just a plate with a hole shaped as shown in the figure cut through it. This plate would be installed at the effluent end of the grit channel in Figure 5.9. The Parshall flume was discussed in Chapter 3.

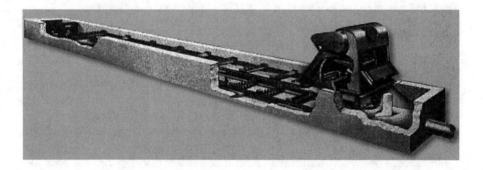

FIGURE 5.9 A grit channel. (Courtesy of Envirex, Inc.)

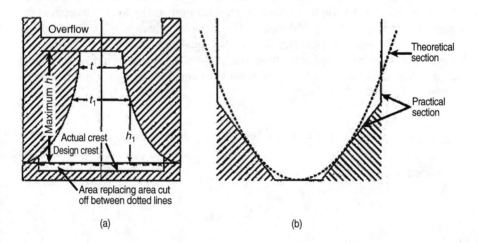

FIGURE 5.10 Velocity control of grit channels: (a) proportional-flow weir; (b) cross section of a parabolic-sectioned grit channel.

As shown in the figure, the flow area of a proportional-flow weir is an orifice. From fluid mechanics, the flow Q through an orifice is given by

$$Q = K_o \ell h^{3/2} \tag{5.21}$$

where K_o is the orifice constant, ℓ is the width of flow over the weir, and h is the head over the weir crest. There are several ways that the orifice can be cut through the plate; one way is to do it such that the flow Q will be linearly proportional to h. To fulfill this scheme, the equation is revised by letting $h^{3/2} = h^{1/2}h$. The revised equation is

$$Q = K_o \ell h^{1/2} h \tag{5.22}$$

Thus, to be linearly proportional to h, $K_o \ell h^{1/2}$ must be a constant. Or,

$$K_o \ell_1 h_1^{1/2} = K_o \ell_2 h_2^{1/2} = \text{constant} \Rightarrow \ell h^{1/2} = \text{const} \tag{5.23}$$

Equation (5.23) is the equation of the orifice opening of the proportional-flow weir in Figure 5.10. If the equation is strictly followed, however, the orifice opening will create two pointed corners that will most likely result in clogging. For this reason, for values of h less than 2.5 cm, the side curves are terminated vertically to the weir crest. The area of flow lost by terminating at this point is of no practical significance; however, if terminated at an h of greater than 2.5 cm, the area lost should be compensated for by lowering the actual crest below the design crest. This is indicated in the figure.

The general cross-sectional area of the tank may be represented by kwH, where k is a constant, w is the width at a particular level corresponding to H, the depth in the tank. Now, the flow through the tank is $Q = v_h(kwH)$, where v_h is the flow-through velocity to be made constant. This flow is also equal to the flow that passes through the control device at the end of the tank. The height of the orifice crest from the bottom of the channel is small, so h may be considered equal to the depth in the tank, H. From the equation of continuity,

$$v_h(kwH) = K_o \ell h^{1/2} h = K_o' \ell h^{1/2} H = \text{const}(H) \tag{5.24}$$

Solving for w,

$$w = \frac{\text{const}}{v_h k} = \text{constant} \tag{5.25}$$

Therefore, for grit chambers controlled by a proportional-flow weir, the width of the tank must be constant, which means that the cross-section should be rectangular.

For grit chambers controlled by other critical-flow devices, such as a Parshall flume (the proportional flow weir is also a critical-flow device), the flow through the device is also given by $Q = K_o \ell h^{3/2}$. Thus, the following equation may also be obtained:

$$v_h(kwH) = K_o \ell h^{3/2} = K_o \ell H^{3/2} \qquad (5.26)$$

Solving for H,

$$H = \text{constant}(w^2) = cw^2 = Z_o \qquad (5.27)$$

which is the equation of a parabola. Thus, for grit chambers controlled by Parshall flumes, the cross-section of flow should be shaped like a parabola. For ease in construction, the parabola is not strictly followed but approximated. This is indicated in the upper right-hand drawing of Figure 5.10. The area of the parabola is

$$A = \frac{2}{3} wH = \frac{2}{3} wZ_o \qquad (5.28)$$

Coordinates of the proportional-flow weir orifice. The opening of the weir orifice needs to be proportioned properly. To accommodate all ranges of flow during operation, the proportioning should be done for peak flows. For a given inflow peak flow to the treatment plant, not all channels may be operated at the same time. Thus, for operating conditions at peak flow, the peak flow that flows through a given grit channel will vary depending upon the number of channels put in operation. The proportioning of the orifice opening should be done on the maximum of the peak flows that flow through the channel.

Let l_{mpk} be the l of the orifice opening at the maximum peak flow through the channel. The corresponding h would be h_{mpk}. From Equation (5.23),

$$\ell h^{1/2} = \text{const} = l_{mpk} h_{mpk}^{1/2}$$

and,

$$h = \frac{l_{mpk}^2 h_{mpk}}{\ell} \qquad (5.29)$$

Let $l_{mpk} = \frac{1}{3} w$ and $h_{mpk} = Z_{ompk}$, where Z_{ompk} is the maximum depth in the channel corresponding to the maximum peak flow through the channel, Q_{mpk}. Then,

$$h = \frac{\left(\frac{1}{3}w\right)^2 Z_{ompk}}{\ell} = \frac{1}{9} \frac{w^2 Z_{ompk}}{\ell} \qquad (5.30)$$

This equation represents the coordinate of the proportional-flow weir orifice.

Coordinates of the parabolic cross section. Let A_{mpk} be the area in the parabolic section corresponding to Q_{mpk}. From Equation (5.28),

$$A_{mpk} = \frac{2}{3} w_{mpk} Z_{ompk} \tag{5.31}$$

where $\frac{1}{3} w_{mpk}$ is the top width of the parabolic section corresponding Z_{ompk}. From Equation (5.27) and the previous equation, the following equation for c can be obtained:

$$c = \frac{3 A_{mpk}}{2 w_{mpk}^3} \tag{5.32}$$

From Equation (5.27), again,

$$Z_o = \frac{3 A_{mpk}}{2 w_{mpk}^3} w^2 \tag{5.33}$$

Substituting in Equation (5.28),

$$A = \frac{2}{3} w Z_o = \frac{2}{3} w \left\{ \frac{3 A_{mpk}}{2 w_{mpk}^3} w^2 \right\} \Rightarrow w^2 = \left(\frac{w_{mpk}^3}{A_{mpk}} \right)^{2/3} A^{2/3} \tag{5.34}$$

$$w = \left(\frac{w_{mpk}^3}{A_{mpk}} \right)^{1/3} A^{1/3} \tag{5.35}$$

Thus,

$$Z_o = \frac{3 A_{mpk}}{2 w_{mpk}^3} w^2 = \frac{3 A_{mpk}}{2 w_{mpk}^3} \left(\frac{w_{mpk}^3}{A_{mpk}} \right)^{2/3} A^{2/3} = \frac{3 A_{mpk}^{1/3}}{2 w_{mpk}} A^{2/3} \tag{5.36}$$

Example 5.6 Design the cross section of a grit removal unit consisting of four identical channels to remove grit for a peak flow of 80,000 m³/d, an average flow of 50,000 m³/d and a minimum flow of 20,000 m³/d. There should be a minimum of three channels operating at any time. Assume a flow-through velocity of 0.3 m/s and that the channels are to be controlled by Parshall flumes.

Solution: Four baseline cross-sectional areas must be considered and computed as follows:

$$A_{\text{peak, three channels}} = \frac{80,000}{3(0.3)(24)(60)(60)} = 1.03 \text{ m}^2 = A_{mpk}$$

$$A_{\text{peak, four channels}} = \frac{80,000}{4(0.3)(24)(60)(60)} = 0.77 \text{ m}^2$$

$$A_{\text{ave}} = \frac{50,000}{4(0.3)(24)(60)(60)} = 0.48 \text{ m}^2$$

$$A_{\text{min}} = \frac{20,000}{4(0.3)(24)(60)(60)} = 0.19 \text{ m}^2$$

The channels are to be controlled by Parshall flumes, so the cross sections are parabolic. Thus,

$$Z_o = \frac{3 A_{mpk}^{1/3}}{2 w_{mpk}} A^{2/3}$$

and determine coordinates at corresponding areas. Let $w_{mpk} = 1.5$ m.

For $A_{peak, \text{ four chambers}} = 0.77$ m^2:

$$Z_o = \frac{3}{2} \frac{A_{mpk}^{1/3}}{w_{mpk}} A^{2/3} = \frac{3}{2} \frac{1.03^{1/3}}{1.5} (0.77^{2/3}) = 0.85 \text{ m}$$

For $A_{ave} = 0.48$ m^2:

$$Z_o = \frac{3}{2} \frac{A_{mpk}^{1/3}}{w_{mpk}} A^{2/3} = \frac{3}{2} \frac{1.03^{1/3}}{1.5} (0.48^{2/3}) = 0.62 \text{ m}$$

For $A_{min} = 0.19$ m^2:

$$Z_o = \frac{3}{2} \frac{A_{mpk}^{1/3}}{w_{mpk}} A^{2/3} = \frac{3}{2} \frac{1.03^{1/3}}{1.5} (0.19^{2/3}) = 0.33 \text{ m}$$

Area (m^2)	w(m)	Z_o(m)	Q(m^3/d)	
1.03	1.5	1.03	80,000 in three channels	
0.77	1.36	0.85	80,000 in three channels	**Ans**
0.48	1.16	0.62	50,000 in four channels	
0.19	0.85	0.33	20,000 in four channels	
0	0	0	0	

Note: In practice, these coordinates should be checked against the flow conditions of the chosen dimensions of the Parshall flume. If the flumes are shown to be submerged forcing them not to be at critical flows, other coordinates of the parabolic cross sections must be tried until the flumes show critical flow conditions or unsubmerged.

Example 5.7 Repeat previous example problem for grit channels controlled by proportional flow weirs.

Solution: For grit channels controlled by proportional weirs, the cross-section should be rectangular. Thus,

$$w = \frac{\text{constant}}{v_h k} = \text{constant; assume } w = 1.5 \text{ m}$$

Therefore, the depths, Z_o, and other parameters for various flow conditions are as follows (for a constant flow-through velocity of 0.30 m/s):

Area(m^2)	w(m)	H(m)	Q(m^3/d)
1.03	1.5	0.69	80,000 in three channels
0.77	1.5	0.51	80,000 in four channels
0.48	1.5	0.32	50,000 in four channels
0.19	1.5	0.13	20,000 in four channels
0	1.5	0	0

5.2.4 FLOCCULENT SETTLING

Particles settling in a water column may have affinity toward each other and coalesce to form flocs or aggregates. These larger flocs will now have more weight and settle faster overtaking the smaller ones, thereby, coalescing and growing still further into much larger aggregates. The small particle that starts at the surface will end up as a large particle when it hits the bottom. The velocity of the floc will therefore not be terminal, but changes as the size changes. Because the particles form into flocs, this type of settling is called *flocculent settling* or type 2 settling.

Because the velocity is terminal in the case of type 1 settling, only one sampling port was provided in performing the settling test. In an attempt to capture the changing velocity in type 2 settling, oftentimes multiple sampling ports are provided. The ports closer to the top of the column will capture the slowly moving particles, especially at the end of the settling test.

For convenience, reproduce the next equation.

$$R = 1 - x_0 + \int_0^{x_o} \frac{v_p}{v_o} \, dx \qquad (5.37)$$

As shown in this equation, the fractional removal R is a function of the settling velocity v_p. The question is that if the settling is flocculent, what would be the value of the v_p? In discrete settling, the velocity is terminal and since the velocity is terminal, the velocity substituted into the equation is the terminal settling velocity. In the case of flocculent settling, would the velocity to be substituted also be terminal?

In the derivation of Equation (5.37), however, nothing required that the velocity be terminal. If the settling is discrete, then it just happens that the velocity obtained in the settling test approximates a terminal settling velocity, and this is the velocity that is substituted into the equation. If the settling is flocculent, however, the same formula of $v_p = Z_p/2t$ is still the one used to obtain the velocity. Since removal does not require that the velocity be terminal but simply that removal is proportional to velocity, this velocity of flocculent settling can be substituted in Equation (5.37) to calculate the fractional removal, and it follows that the same formula and, thus, method can be used both in discrete settling as well as in flocculent settling.

Each of the ports in the flocculent settling test will have a corresponding Z_p. During the test each of these Z_p's will accordingly have corresponding times t and thus, will produce corresponding average velocities. These velocities and times form arrays that correspond to each other, including a corresponding array of concentration. In other words, in the flocculent settling test more test data are obtained. The method of calculating the efficiency of removal, however, is the same as in discrete settling and this is Equation (5.37).

Example 5.8 Assume Anne Arundel County wants to expand its softening plant. A sample from their existing softening tank is prepared and a settling column test is performed. The initial solids concentration in the column is 250 mg/L. The results are as follows:

	Sampling Time (min)							
Z_p/Z_o	5	10	15	20	25	30	35	40
0.1	95[a]	68	55	30	23	—	—	—
0.2	129	121	73	67	58	48	43	—
0.3	154	113	92	78	69	60	53	45

[a] Values in the table are the results of the test for the suspended solids (mg/L) concentration at the given depths.

Calculate the removal efficiency for an overflow rate of 0.16 m³/m² · min. Assume the column depth is 4 m.

Solution:

$$R = 1 - x_o + \int_0^{x_o} \frac{v_p}{v_o} dx$$

t (min)	0	5	10	15	20	25	30	35	40
				$Z_p/Z_o = 0.1$					
c (mg/L)	250	95	68	55	30	23	—	—	—
$x = [c]/[c_o]$	1.0	0.38	0.27	0.22	0.12	0.092	—	—	—
$v_p = Z_p/2t$	—	0.08[a]	0.04	0.03	0.02	0.016	—	—	—
$= 0.1(4)/2t$(m/min)									
				$Z_p/Z_o = 0.2$					
c (mg/L)	250	129	121	73	67	58	48	43	—
$x = [c]/[c_o]$	1.0	0.52	0.48	0.29	0.27	0.23	0.19	0.17	—
$v_p = Z_p/2t$	—	0.16[b]	0.08	0.05	0.04	0.032	0.027	0.023	—
$= 0.2(4)/2t$ (m/min)									
				$Z_p/Z_o = 0.3$					
c (mg/L)	250	154	113	92	78	69	60	53	45
$x = [c]/[c_o]$	1.0	0.62	0.45	0.37	0.31	0.28	0.24	0.21	0.18
$v_p = Z_p/2t$	—	0.24[c]	0.12	0.08	0.06	0.048	0.04	0.034	0.03
$= 0.3(4)/2t$ (m/min)									

[a] $0.08 = 0.1(4)/5$
[b] $0.16 = 0.2(4)/5$
[c] $0.24 = 0.3(4)/5$

v_p		0.24	0.16	0.12	0.08	0.06	0.05	0.048	0.04	0.034	0.032	0.03	0.027	0.023	0.02	0.016	0
x		0.62	0.52	0.45	0.41[a]	0.31	0.29	0.28	0.26	0.23	0.21	0.20	0.19	0.17	0.12	0.092	0
Δx		—	0.10	0.07	0.04	0.10	0.02	0.01	0.02	0.03	0.02	0.01	0.01	0.02	0.05	0.028	0.092
v_p in Δx		—	0.20[b]	0.14	0.10	0.07	0.055	0.049	0.044	0.037	0.033	0.031	0.029	0.025	0.022	0.018	0.008

[a] $0.41 = (0.37 + 0.48 + 0.38)/3$
[b] $0.20 = (0.24 + 0.16)/2$

It is not necessary to interpolate the x corresponding to $v_p = 0.16$ m/min. From the table, $x = x_o = 0.52$.
Therefore,

$$R = 1 - 0.52 + \frac{1}{0.16}[0.14(0.07) + 0.1(0.04) + 0.07(0.1) + 0.055(0.02)$$

$$+ 0.049(0.01) + 0.044(0.02) + 0.037(0.03) + 0.033(0.02) + 0.031(0.01)$$

$$+ 0.029(0.01) + 0.025(0.02) + 0.022(0.05) + 0.018(0.028) + 0.008(0.092)]$$

$$= 0.48 + 0.178 = 0.66 \Rightarrow 66\% \quad \textbf{Ans}$$

5.2.5 PRIMARY SETTLING AND WATER-TREATMENT SEDIMENTATION BASINS

The primary sedimentation tank used in the treatment of sewage and the sedimentation basin used in the treatment of raw water for drinking purposes are two of the units in the physical treatment of water and wastewater that use the concept of flocculent settling. These units are either of circular or rectangular (flow-through) design; however, they differ in one important respect: the amount of scum produced. Whereas in water treatment there is practically no scum, in wastewater treatment, a large amount of scum is produced and an elaborate scum-skimming device is used.

The primary sedimentation basins used in wastewater treatment and water treatment also differ in another respect: the length of detention time. Although longer detention times tend to effect more solids removal in water treatment, in primary sedimentation, a longer detention time can cause severe septic conditions. Septicity, because of formation of gases, makes solids rise resulting in inefficiency of the basin. Thus, in practice, there is a practical range of values of 1.5 to 2.5 h based on the average flow for primary settling detention times in wastewater treatment. This range of figures, although stated in terms of the average, really means that it takes an average of 1.5 to 2.5 h for a particle of sewage to become septic whether or not the flow is average.

Both water and wastewater treatment also need to maintain the flow-through velocity so as not to scour the sludge that has already deposited at the bottom of the settling tank. They also need properly designed overflow weirs, an example of which is shown in Figure 5.7b. The particles in both these units are flocculent, so the flow-through velocity should be maintained at no greater than 9.0 m/h and the overflow weir loading rate at no greater than 6–8 m³/h per meter of weir length, as mentioned before. Some design criteria for primary sedimentation tanks are shown in Table 5.2. Except for the detention time, the criteria values may also be used for settling tanks in water treatment.

TABLE 5.2
Design Parameters and Criteria for Primary
Sedimentation Basins

Parameter	Value	
	Range	Typical
Detention time, h	1.5–2.5	2.0
Overflow rate, m/d	30–50	40
Dimensions, m		
Rectangular		
Depth	2–6	3.5
Length	15–100	30
Width	3–30	10
Sludge scraper speed, m/min	0.5–1.5	1.0
Circular		
Depth	3–5	4.5
Diameter	3–60	30
Bottom slope, mm/m	60–160	80
Sludge scraper speed, rpm	0.02–0.05	0.03

Design flows to use. When detention time is mentioned, the flow associated with it is normally the average flow. As noted in Chapter 1, however, sewage flows are variable. In water treatment, this variation is, of course, not a problem, but it would be for sewage sedimentation basins. This situation causes a dilemma. A detention time, although customarily attributed to the average flow, may be attributed to other flow magnitudes as well. When the detention time to be used is, for example, 2.5 h in order to limit septicity, what is the corresponding flow? In this particular case, the flow would not necessarily be the average but a flow that would effect a detention time of 2.5 h. What flow then would effect this detention time?

The settling tank has an inherent capacity to damp out fluctuation in rate of inflow. In the present example, the flow is being damped out or sustained in a period of 2.5 h. Because this flow has taken effect during the detention time, it must be the flow corresponding to this detention time and, hence, must be the one adopted for design purposes. To size sedimentation basins, the flow to be used should therefore be the *sustained flow* corresponding to the detention time chosen. This detention time must, in turn, be a value that limits septicity. The method of calculating sustained flows was discussed in Chapter 1.

Suppose a sedimentation basin is to be designed for an average daily flow rate of 0.15 m^3/s, how would the sustained flow be calculated with this information on the average flow? It will be remembered that the average daily flow rate is the mean of all 24-h flow values obtained from an exhaustive length of flow record. The unit to be designed must meet this average flow requirement, but yet, the actual flow transpiring inside the tank is not always this average flow. This situation calls for a relationship between sustained flow and the average flow.

What sustained flow is equivalent to an average flow of 0.15 m^3/s? There are, actually, two sustained flows that must be determined: the sustained peak flow Q_{pks} and the sustained low flow Q_{ms}. The sustained peak flow is used to size the tank; the sustained low is used to determine the amount of recirculation desired so as to deter septicity from occurring. The occurrence of the sustained low flow will produce a much, much longer detention time. As mentioned before, detention times greater than 2.5 h (or some other value depending upon the type of waste) can cause septicity and thus inefficiency in sedimentation basins.

Using the methods developed in Chapter 1, relationships of sustained flows and average flow can be obtained. In the present example, the sustained flows analyses should be performed concurrently with that of the average flow analysis so that the aforesaid relationship can be obtained. Considering the present example, the sustained period to be used in the analysis would be the 2.5 h.

In the absence of field data and for an order-of-magnitude design, a ratio of 3:1 (peaking factor) for the sustained one-day peak flow to average may be used. This ratio was obtained from Metcalf & Eddy, Inc. (1991), which shows a lowest sustained period available of one day. Of course, this is not the correct sustained period duration since, actually, it should be 2.5 h in the present example—not one day. Now, using the ratio, the design sustained peak flow rate is $3(0.15) = 0.45$ m^3/s. This flow is the flow that must be used in conjunction with the 2.5-h detention time. Because this sustained flow rate bears a relationship to the average flow rate, using the sustained flow rate is equivalent to using the required average flow rate. In other words, the passage of the sustained flow rates (both high and low) through the tank along with the other larger and smaller flow rates, evens out to the average flow rate which, in the overall, must mean that the requirement of the average flow is met.

Using the same source of reference, the ratio of the sustained low flow to the average for a one-day sustained period is 0.3 (minimizing factor). Again, the sustained period should be 2.5 h and not one day. In the absence of field data, these values may, again, be used for an order-of-magnitude design to determine the amount of recirculation desired. Using this ratio, the sustained low flow is $0.3(0.15) = 0.045$ m^3/s.

For a detention time of 2.5 h and a sustained peak flow rate of 0.45 m^3/s, the volume of the tank V would be $0.45(60)(60)(2.5) = 4,050$ m^3. Letting Q_R be the recirculated flow, we have $Q_R(2.5) + 0.045(60)(60)(2.5) = 4,050$, and $Q_R = 1458$ m^3/hr. Thus, with a recirculated flow of 1,458 $m^3/hr = 0.405$ m^3/s and the sustained low flow of 0.045 m^3/s, the detention time is brought back to 2.5 h and septicity would be avoided when the sustained low flow occurred. This illustration demonstrates that if the calculated detention time is greater than 2.5 h, a portion of the clarified liquid must be recirculated to bring the detention time to 2.5 h, thus maintaining the efficiency of the basin. The recirculation flow is the maximum and may be used to design the recirculation pump. In operation, the rate of pumping would have to be adjusted to effect the required detention time of 1.5 to 2.5 h. The equation for recirculation flow is

$$Q_R = \frac{V - Q_{ms}t_o}{t_o} \tag{5.38}$$

Design is an iterative procedure. In the above example, the detention time was chosen as 2.5 h. It could have been chosen as 1.5 h or any value within the range allowed for no septicity and similar steps followed. No matter how the calculation is started, the ultimate goal is to have no septicity in the tank and to satisfy the overflow requirement. If the final calculations show that these goals are met, then the design is correct and the sedimentation tank should operate efficiently. Remember, however, that the 2.5 is the upper limit of the range of values. In addition, a given particular waste may have a different upper limit (or a different lower limit for that matter). If this is the case, then whatever applies in reality should be used.

One other method of design is to use equalization basins ahead of the sedimentation basins. In this case, the inflow would be the equalized flow. There should be no need for recirculation, because only one flow rate is introduced into the sedimentation basin. Design of equalization basins was discussed in Chapter 3.

Example 5.9   To meet effluent limits, it has been determined that the design of a primary sedimentation basin must remove 65% of the influent suspended solids. The average influent flow is 6,000 m^3/d; the peak daily flow is 13,000 m^3/d; and the minimum daily flow is two-thirds of the average flow. It was previously determined that 65% removal corresponds to an overflow rate of 28 m/d. Design a circular basin to meet the effluent requirement at any cost.

Solution: The data on the daily peak and daily low flow rates are not the ones that would be used in design. The peak daily flow rate is a sustained 24-h peak flow rate and the minimum daily flow rate is the sustained 24-h low flow rate. Because the detention time of the sedimentation is a mere 2.5 h, the 24-h sustained flow period is incorrect. If the maximum and minimum hourly flow values were available, however, they would be close to the sustained peak and sustained low flow values over the period of the detention time (2.5 h), respectively, and could be used. The data, as given, could not be used in design.   **Ans**

Example 5.10   The prototype detention time and overflow rate for the design of a sewage sedimentation basin were determined to be 2.5 h and 28 m/d, respectively. The peaking factor is 3.0 and the minimizing factor is 0.3. Design a rectangular settling basin for an average daily flow rate of 20,000 m^3/d.

Solution:

$$Q_{pks} = \frac{3(20,000)}{24(60)(60)} = 0.69 \ m^3 s$$

Therefore,

$$V = 0.69(2.5)(60)(60) = 6210 \ m^3$$

Assume solids are flocculent, hence v_h should not be greater than 9.0 m/h. Use 8.0 m/h = 0.0022 m/s.

Therefore,

$$\text{Area, vertical section} = \frac{0.69}{0.0022} = 313.64 \text{ m}^2$$

$$\text{Length of basin, } L = 8.0(2.5) = 20 \text{ m}$$

Considering overflow rate,

$$LW = 20W = \frac{0.69}{\frac{28}{24(60)(60)}} = 2129.14 \qquad W = 106.46 \text{ m}$$

$$\text{Depth of tank, } H = \frac{313.64}{106.46} = 2.94 \text{ m. Add freeboard of 0.3 m}$$

$$\text{Actual depth of tank} = 2.94 + 0.3 = 3.24, \text{ say 3.3 m}$$

$$\text{Volume of tank} = LWH = 20(106.46)(2.94) = 6259.8 \text{ m}^3 \simeq 6210 \text{ m}^3$$

Let there be 10 tanks; thus, width per tank = 10.65 m, say 10 m
Dimension for each tank: length = 20 m, width = 10 m, depth = 3.3 m   **Ans**

Actual detention time = 20(10)(2.94)/((0.69/10)(60)(60)) = 2.37 h. Note that the freeboard of 0.3 m does not hold water; thus, 2.94 m is used in the calculation.

Now, determine the amount of recirculation required at minimum flow.

$$Q_R = \frac{V - Q_{ms}t_o}{t_o}$$

$$V = 20(10)(2.94) = 588 \text{ m}^3, \text{ for one tank}$$

$$Q_{ms} = \frac{0.3(20,000)}{24(60)(60)(10)} = 0.0069 \text{ m}^3/\text{s}$$

Therefore,

$$Q_R = \frac{588 - 0.0069(2.37)(60)(60)}{(2.37)(60)(60)} = 0.062 \text{ m}^3/\text{s} \quad \textbf{Ans}$$

Example 5.11   Repeat the previous example for a circular settling basin.

Solution:

$$Q_{pks} = \frac{3(20,000)}{24(60)(60)} = 0.69 \text{ m}^3/\text{s}$$

Therefore,

$$V = 0.69(2.5)(60)(60) = 6210 \text{ m}^3$$

Considering overflow rate,

Overflow area of tank, $\dfrac{\pi D^2}{4} = \dfrac{0.69}{\dfrac{28}{24(60)(60)}} = 2129.14 \quad D = 52.07$ m, too large

Use two tanks, therefore $\dfrac{\pi D^2}{4} = \dfrac{2129.14}{2} \quad D = 36.82$ m, too large

Use three tanks, therefore $\dfrac{\pi D^2}{4} = \dfrac{2129.14}{3} \quad D = 30.06$ m, say 30 m

Depth of tank, $H = \dfrac{6210/3}{2129.14/3} = 2.92$ m. Add freeboard of 0.3 m

Actual depth of tank $= 2.92 + 0.3 = 3.22$, say 3.3 m

Dimension for each tank: diameter $= 30$ m, depth $= 3.3$ m   **Ans**

Actual detention time $= \dfrac{\dfrac{\pi D^2}{4}}{(Q_{pks}/3)(60)(60)} = \dfrac{\dfrac{\pi (30)^2}{4}(2.92)}{(0.69/3)(60)(60)} = 2.49$ h

Note: The freeboard of 0.3 m does not hold water; thus, 2.94 m is used in the calculation.

Now, determine the amount of recirculation required at minimum flow.

$$Q_R = \dfrac{V - Q_{ms} t_o}{t_o}$$

$$V = \dfrac{\pi (30)^2}{4}(2.92) = 2064.03 \text{ m}^3, \text{ for one tank}$$

$$Q_{ms} = \dfrac{0.3(20,000)}{24(60)(60)(3)} = 0.023 \text{ m}^3/\text{s}$$

Therefore,

$$Q_R = \dfrac{2064.03 - 0.023(2.49)(60)(60)}{(2.49)(60)(60)} = 0.207 \text{ m}^3/\text{s} \quad \textbf{Ans}$$

5.2.6 Zone Settling

The third type of settling is zone settling. This is also called type 3 settling. Figure 5.11 shows a batch analysis of zone settling showing four zones A, B, C, and D. Zone A is cleared of solids; thus, it is called a *clarification zone*. B is called uniform settling zone, since the concentration of the solids in this zone is constant. In zone C, the solids concentration increases or thickens from the value at the interface B-C to the value at the interface C-D. Thus, this zone is called a *thickening zone*. (The thickening process is a process of concentrating the solids into a sludge.) Finally, in D, the solids are compressed and the matrix is compacted and consolidated. Thus, this zone is called a *compression zone*; this is where type 4 or compression

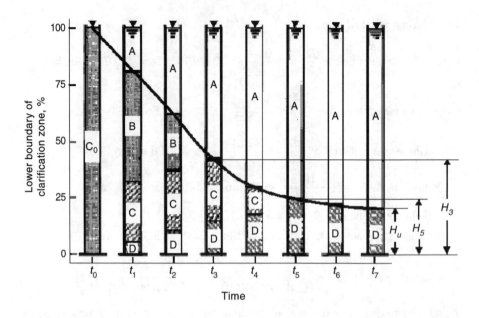

FIGURE 5.11 Batch analysis of zone settling.

settling occurs. In this zone, the solids are further thickened by the compression, compaction, and consolidation processes.

Consider that initially at time t_o, a suspension of initial concentration $[c_o]$ is poured into the graduated cylinder. Some time later at t_1, the four zones are formed. B retains the initial concentration and settles at a rate characteristic of this concentration. Since the concentration is constant at $[c_o]$ throughout this zone, B settles at a constant velocity, thus the name *uniform settling (velocity) zone*. This velocity can be determined by following the interface A-B with time.

D is formed by particles piling on top of each other producing the largest concentration of the zones. There will be a gradation of concentration from D to B. Thus, zone C will have this gradation forming a concentration gradient in the zone. This is where the thickening process first occurs. The gradient will be more or less constant. As the particles pile up on top of another, zone D lengthens. To more or less maintain the concentration gradient, the concentration immediately below interface B-C and the concentration immediately above interface C-D must remain constant for a constant length of zone C. Because D is lengthening and to maintain the length of C and thus the gradient, the interface B-C must move up at the same speed as the lengthening of D. This means that zone B is eroded both at its top and its bottom. Hence, eventually, this zone must disappear. This happens at time t_3.

After time t_3, zone C starts to diminish and totally disappear at time t_5 after which time pure compression commences. Pure compression continues until t_7 where the slope of the curve now exhibits the tendency to become horizontal. After a very long time, the slope will, indeed, be horizontal. The time midway between t_3 and t_5,

TABLE 5.3
Expected Sludge Solids Percent Concentration from Various Processes

Process	Range	Typical
Secondary settling tank		
Waste activated sludge with primary settling	0.5–1.5	0.7
Waste activated sludge without primary settling	0.8–2.5	1.3
Pure-oxygen activated sludge with primary settling	1.3–3.0	2.0
Pure-oxygen activated sludge without primary settling	1.4–4.0	3.0
Trickling filter humus sludge	1–3	2
Primary settling tank		
Primary sludge	4–10	6
Primary sludge to cyclone	0.5–3.0	2.0
Primary and waste activated sludge	3–10	4
Primary and trickling filter humus	4–10	5
Primary sludge with iron addition for phosphorus removal	5–10	8
Primary sludge with low lime addition for phosphorus removal	2–8	5
Primary sludge with high lime addition for phosphorus removal	4–16	10
Scum	3–12	10
Gravity thickener		
Primary sludge	6–12	8
Primary and waste activated sludge	3–12	5
Primary and trickling filter humus	4–10	5
Flotation thickener, waste-activated sludge	3–6	4

such as t_4, is called the *critical time* and the corresponding concentration of solids is called the *critical concentration*.

As evidenced from this batch analysis, various degrees of sludge thickening can occur. Table 5.3 shows some data that indicate the various thickening, expressed as percent solids, that can result in various units.

5.2.7 SECONDARY CLARIFICATION AND THICKENING

After the organic pollutant in raw sewage has been converted to microorganisms, the resulting solids need to be separated. This is done using the secondary clarifier. In the process of clarification, however, on account of the large concentration involved, the solids are also thickened. Because of this, the process of separating the solids in a secondary clarifier actually involves two functions: the clarification function and the thickening function. Each of these two functions will have its own clarification or thickening area requirement. The bigger of the two controls the design of the basin.

Two methods are used to thicken sludges: gravity thickening and flotation thickening. In gravitational thickening, solids are thickened as a result of solids piling

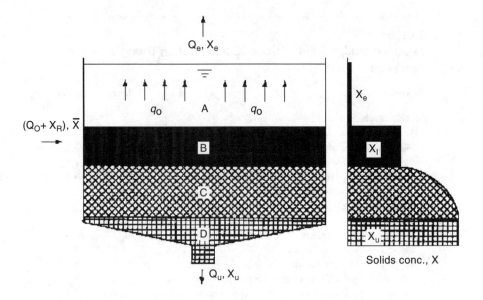

FIGURE 5.12 Gravity thickening process.

on top of each other and the action of gravity that compresses the accumulated solids. In flotation thickening, the solids stick to the rising bubbles that float toward the surface. Upon exposure to the surface, the bubbles break, leaving the solids that had stuck to the bubbles in concentrated form. This section discusses gravitational thickening as well as the clarification that occurs in the secondary clarifier.

Figure 5.12 may be considered a schematic section of either a secondary clarifier or of a gravitational thickener. This is an implementation of the batch settling test of Figure 5.11 on a continuous mode. Continuous mode means that the thickened solids are continuously removed from the bottom of the tank. As shown, there are also four zones: zone A, zone B, zone C, and zone D. Zone A is the *clarification zone*. In zone B, the concentration of the solids is constant as indicated by X_i, the biosolids coming out of the bioreactor $\bar{X}$. (X_i corresponds to the c_o of the batch settling.) In zone C, the concentration varies from low to high in going from the upper to the lower part of the zone. This is where thickening occurs. In zone D, the concentration is constant; the concentration in this zone is the *underflow concentration* $[X_u]$ and is the concentration that is withdrawn from the bottom of the tank. This zone corresponds to the compression zone in the batch experiment.

Thickening occurs in secondary clarifiers, so the sizing of the clarifier also involves the determination of the thickener area. In designing the clarifier portion of the clarifier, the subsidence velocity of zone B must be determined. This is modeled by allowing a sample to settle in a cylinder and following the movement of the interface between zones A and B over time. This subsidence velocity is equated to the overflow velocity to size the clarifier area A_c. For a pure gravity thickener, although some clarification also occurs, the thickening function is the major design parameter and the clarification function need not be investigated. The method used to size the thickener area is called the *solid flux method*.

Solids flux method. The design of the thickener area considers zone C. (*Note:* you do not consider zone D.) As indicated, the concentration of the solids in this zone is variable. Thus, the behavior of the solids is modeled by making several dilutions of the sludge to conform to the several concentrations possible in the zone. The ultimate aim of the experiment is to be able to determine the *solids loading* into the thickener. The experiment is similar to that of the clarifier design, except that several concentrations are modeled in this instance. Also, it is the relation between subsidence velocity and concentration that is sought. The reason is that if velocity is multiplied by the concentration, the result is the solids loading called the *solids flux*, the parameter that is used to size the thickener portion.

Performing a material balance at any elevation section of the thickener area (corresponding to zone C), two solid fluxes must be accounted for: the flux due to the gravitational settling of the solids and the flux due to the conveyance effect of the withdrawal of the sludge in the underflow of the tank. Calling the total flux as G_t, the gravity flux as $V_c[X_c]$ and the conveyance flux as $V_u[X_c]$, the material balance equation is

$$G_t = V_c[X_c] + V_u[X_c] \qquad (5.39)$$

where V_c is the subsidence velocity at the section of the thickening zone, $[X_c]$ is the corresponding solids concentration at the section, and V_u is the underflow velocity computed at the section as Q_u/A_t, where Q_u is the underflow rate of flow and A_t is the thickener area at the elevation section considered. Some value of this flux G_t is the one that will be used to design the thickener area and, therefore, must be determined.

Because the solids concentration in the thickening zone is variable, G_t in the zone is also variable. Of these several values of G_t, only one value would make the solids loading at the corresponding elevation section equal to the rate of withdrawal of sludge in the underflow. This particular G_t is called the *limiting flux* $G_{t\ell}$, because it is the one flux that corresponds to the underflow withdrawal rate. Because $G_{t\ell}$ corresponds to the rate of sludge withdrawal, the thickener area determined from it will be the thickener area for design, if found greater than the area determined considering the clarification function of the unit.

Equation (5.39) may be multiplied throughout by A_t producing

$$G_t A_t = V_c[X_c]A_t + V_u[X_c]A_t = Q_u[X_u] \qquad (5.40)$$

Dividing all throughout by A_t and rearranging,

$$G_t = V_u[X_u] = V_c[X_c] + V_u[X_c] \qquad (5.41)$$

The value of V_u depends on the value of Q_u. If the sludge concentration is to be withdrawn at the concentration $[X_u]$, then Q_u must be such that Equation (5.41) is satisfied. As shown, this equation contains $V_u[X_u]$. A specific value of this term determines the area A_t of the thickener. As will be shown below, this value is $G_{t\ell}$, the limiting flux.

The expression $V_c[X_c] + V_u[X_c]$ can be likened to the flow of traffic in a toll booth terminal. The cars at far distances from the terminal can travel as fast as they can, but upon reaching the terminal, they all slow down to a crawl. The number of cars that leave the terminal at any given time is a function of how fast the terminal fee is paid and processed by the teller. In other words, the terminal controls the rate of flow of traffic. A particular value of $V_u[X_u] = G_t = G_{t\ell}$ corresponds to the speed at which the fee is paid and processed by the teller, and its location in the thickener corresponds to the terminal. This location is the *critical* or *terminal section.*

A number of combinations of the values of the elements of $V_c[X_c] + V_u[X_c]$ can occur before the solids reach the critical section of the thickener. This combination corresponds to the different speeds of the cars; however, upon reaching the section, $V_c[X_c] + V_u[X_c]$ slows down to a crawl, that is, slows down to a one particular value of $V_u[X_u]$. This "crawl" value of $V_c[X_c] + V_u[X_c]$ would have to be the smallest of all the possible values it can have before reaching the section; this value would also necessarily correspond to the rate of withdrawal at the bottom of the tank. This is the limiting flux $G_{t\ell}$, mentioned previously. Thus,

$$G_{t\ell} = min\{(V_u[X_u])_i\} = [X_u]min(V_u)_i = min\{(V_c[X_c] + V_u[X_c])_i\} \quad (5.42)$$

where *min* means "minimum of" and i is an index for the several values of the parameter. This minimum value of V_u may obtained from the graph of $[X_c]$ and V_u. Note that $G_{t\ell}$ has a corresponding Q_u. If Q_u is changed, the value of $[X_u]$ is also changed and, accordingly, a new value of $G_{t\ell}$ must be determined.

Knowing $G_{t\ell}$, the thickener area A_t can now be calculated as

$$A_t = \frac{(Q_o + Q_R)[\bar{X}]}{G_{t\ell}} \quad (5.43)$$

$Q_o + Q_R = Q_i$ is the influent to the thickener (or secondary clarifier); Q_R is the recirculation flow; and Q_o is the inflow to the overall treatment plant. A_t is then compared with the clarifier area A_c; the larger of the two is the one chosen for the design. For gravity thickeners, A_t is automatically used, without comparing it to A_c, because A_c is not actually calculated in designs of thickeners.

Example 5.12 The activated sludge bioreactor facility of a certain plant is to be expanded. The results of a settling cylinder test of the existing bioreactor suspension are shown below. $Q_o + Q_R$ is 10,000 m^3/d and the influent MLSS is 3,500 mg/L. Determine the size of the clarifier that will thicken the sludge to 10,000 mg/L of underflow concentration.

MLSS = $[X_c]$(mg/L)	1,410	2,210	3,000	3,500	4,500	5,210	6,510	8,210
V_c (m/h)	2.93	1.81	1.20	0.79	0.46	0.26	0.12	0.084

Solution: First, determine area based on thickening:

$$G_{t\ell} = [X_u]min(V_u)_i = min(V_c[X_c] + V_u[X_c])$$

$$V_u = \frac{V_c[X_c]}{[X_u] - [X_c]} \qquad [X_u] = 10{,}000 \text{ mg/L}$$

$$A_t = \frac{(Q_o + Q_R)[\bar{X}]}{G_{t\ell}}$$

$[X_c]$ (mg/L)	1,410	2,210	3,000	3,500	4,500	5,210	6,510	8,210
V_c (m/h)	2.93	1.81	1.20	0.79	0.46	0.26	0.12	0.084
V_u (m/h)	0.48[a]	0.51	0.51	0.43	0.38	0.28	0.22	0.39

[a] $0.48 = 2.93(1{,}410)/(10{,}000 - 1{,}410)$.

The plot is shown next.

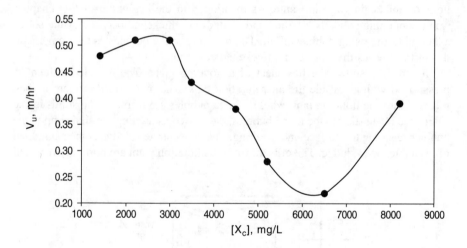

Therefore from the plot, $min(V_u) = 0.21$, and

$$G_{t\ell} = [X_u]min(V_u)_i = 10{,}000(0.21) = 2100 \ (\text{mg/L}) \cdot (\text{m/h})$$

$$A_t = \frac{\left(\frac{10{,}000}{24}\right)[3500]}{2100} = 694.4 \text{ m}^2$$

Determine area based on clarification:
For an MLSS of 3,500 mg/L, settling velocity = 0.79 m/h

Assuming the solids in the effluent are negligible, solids in the underflow = 10,000(3.5) = 35,000 kg/d = 1458.33 kg/h

$$Q_u = \frac{1458.33}{10} = 145.83 \text{ m}^3/\text{h}$$

$$Q_e = \frac{10,000}{24} - 145.83 = 270.83 \text{ m}^3/\text{h}$$

$$A_c = \frac{270.84}{0.79} = 342.83 \text{ m}^2 < 694.4 \text{ m}^2$$

Therefore, the thickening function controls and the area of the thickener is 694.4 m^2 **Ans**

5.3 FLOTATION

Flotation may be used in lieu of the normal clarification by solids-downward-flow sedimentation basins as well as thickening the sludge in lieu of the normal sludge gravity thickening. The mathematical treatments for both flotation clarification and flotation thickening are the same. As mentioned in the beginning of this chapter, water containing solids is clarified and sludges are thickened because of the solids adhering to the rising bubbles of air. The breaking of the bubbles as they emerge at the surface leaves the sludge in a thickened condition.

Figure 5.13 shows the flowsheet of a flotation plant. The recycled effluent is pressurized with air inside the air saturation tank. The pressurized effluent is then released into the flotation tank where minute bubbles are formed. The solids in the sludge feed then stick to the rising bubbles, thereby concentrating the sludge upon the bubbles reaching the surface and breaking. The concentrated sludge is then skimmed off as a thickened sludge. The effluent from the flotation plant are normally recycled

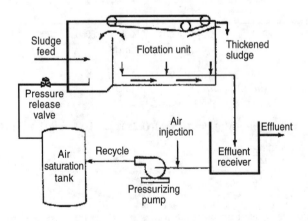

FIGURE 5.13 Schematic of a flotation plant.

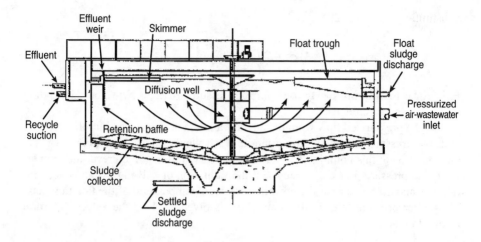

FIGURE 5.14 Elevational section of a flotation unit. (Courtesy of Enirex, Inc.)

back to the influent of the whole treatment plant for further treatment along the with the influent raw wastewater. Figure 5.14 shows an elevational section of a flotation unit. The dissolved air concentration of the wastewater in the air saturation tank $\overline{C}_{aw,t}$ is

$$\overline{C}_{aw,t} = f C_{asw,t,sp} \frac{P}{P_s} = f\beta C_{as,t,sp} \frac{P}{P_s} \qquad (5.44)$$

where f is the fraction of saturation achieved (0.5 to 0.8). Because of the small residence time allowed in the saturation tank, saturation is not achieved there but only a fraction represented by $f \cdot C_{asw,t,sp}$ is the saturation concentration of the dissolved air in the wastewater in the saturation tank at standard pressure P_s corresponding to the temperature of the wastewater; P is the pressure in the tank; and $C_{as,t,sp}$ is the saturation concentration of dissolved air in tap water at standard pressure corresponding to the temperature equal to the temperature of the wastewater. $C_{as,t,sp}$ can be obtained from the corresponding value for oxygen ($C_{os,t,sp}$) by multiplying $C_{os,t,sp}$ by $(28.84)/[0.21(32)] = 4.29$, where 28.84 is the molecular mass of air, 0.21 is the mol fraction of oxygen in air, and 32 is the molecular weight of oxygen. Thus,

$$C_{as,t,sp} = 4.29 C_{os,t,sp} \qquad (5.45)$$

The total amount of air introduced into the flotation tank comes from the air saturation tank and from the influent feed. Because the pressure in the influent feed is the same atmospheric pressure in the flotation tank and, because the temperature in the flotation tank may be assumed equal to the temperature of the influent feed, any air that happens to be with the influent does not aid in floating the solids. This source of air may therefore be neglected. Letting R be the recirculation ratio and Q_i be the influent flow, the amount of air introduced into the flotation tank A_i is

$$A_i = RQ_i f\beta C_{as,t,sp} \frac{P}{P_s} \qquad (5.46)$$

Substituting $C_{as,t,sp} = 4.29C_{os,t,sp}$,

$$A_i = 4.29RQ_i \, f\beta \frac{P}{P_s} C_{os,t,sp} \tag{5.47}$$

P_s is equal to 760 mm of Hg, one atm of pressure, 101,330 N/m², etc. depending on the system of units used.

As the pressurized flow from the air saturation tank is released into the flotation unit, the pressure reduces to atmospheric, P_a. It would be accurate to assume that the condition at this point in the flotation unit is saturation at the prevailing temperature and pressure. Let $C_{os,a,sp}$ represent the saturation dissolved oxygen concentration at standard pressure at the prevailing ambient temperature of the flotation tank. Thus, after pressure release, the remaining dissolved air A_o in the recycled portion of the flow is

$$A_o = 4.29RQ_i\beta\frac{P_a}{P_s}C_{os,a,sp} \tag{5.48}$$

Note that f is no longer in this equation, because enough time should be available for saturation to occur. The air utilized for flotation is $A_1 - A_o = A_{used}$. Or,

$$A_{used} = \frac{4.29RQ_i\beta}{P_s}(fC_{os,t,sp}P - C_{os,a,sp}P_a) \tag{5.49}$$

The solids in the influent is $Q_i[X_i]$. The air used to solids ratio A/S is then

$$A/S = \frac{\dfrac{4.29RQ_i\beta}{P_s}(fC_{os,t,sp}P - C_{os,a,sp}P_a)}{Q_i[X_i]} = \frac{4.29R\beta(fC_{os,t,sp}P - C_{os,a,sp}P_a)}{P_s[X_i]} \tag{5.50}$$

Similar derivation may be undertaken for operation without recycle. If this is done, the following equation is obtained.

$$A/S = \frac{4.29\beta(fC_{os,t,sp}P - C_{os,a,sp}P_a)}{P_s[X_i]} \tag{5.51}$$

5.3.1 LABORATORY DETERMINATION OF DESIGN PARAMETERS

To design a flotation unit, the overflow area, the pressure in the air saturation tank, and the recirculation ratio must be determined. The overflow velocity needed to estimate the overflow area may be determined in the laboratory. The right-hand side of the top drawing of Figure 5.15 shows a laboratory flotation device. A sample of the sludge is put in the pressure tank and pressurized. The tank is then shaken to ensure the sludge is saturated with the air. The pressure and temperature readings in the tank are noted. A valve leading to the flotation cylinder is then opened to allow the pressurized sludge to flow into the cylinder. The rate of rise of the sludge

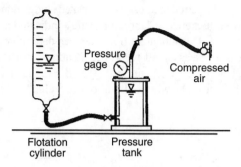

FIGURE 5.15 Laboratory flotation device.

interface in the flotation cylinder is followed with respect to time; this gives the rise velocity. This is equated to the overflow velocity to compute the overflow area.

The value of the air saturation tank pressure and the recirculation ratio may be designed depending upon the A/S ratio to be employed in the operation of the plant. This A/S ratio, in turn, is determined by the laboratory flotation experiment just described. A sample of subnatant is taken from the bottom of the flotation cylinder and the clarity or turbidity determined. Thus, by performing several runs for different values of A/S, corresponding values of clarity will be obtained, thereby producing a relationship between A/S and clarity. The A/S corresponding to the desired clarity is then chosen for the design calculations. Alternatively, the solids content of the float may be analyzed to obtain relationships between A/S and solids content. The A/S corresponding to the desired solids content may then be chosen for design.

In performing the laboratory float experiment, no recirculation is used. Thus, the formula to be used to compute the A/S is Equation 5.51. Ensuring, during the experiment, that the pressure tank is fully saturated, f can be considered unity.

Example 5.13   A laboratory experiment is performed to obtain the air-to-solids ratio A/S to be used in the design of a flotation unit. The pressure gage reads 276 kN/m^2 and the temperature of the sludge and the subnatant in the flotation cylinder is 20°C. The prevailing barometric pressure is 100.6 kN/m^2. The total solids in the sludge is 10,000 mg/L and β was originally determined to be 0.95. Determine the A/S ratio.

Solution:

$$C_{os,t,sp} \text{ at } 20°C \ = \ C_{os,a,sp} \text{ at } 20°C \ = \ 9.2 \text{ mg/L}$$

Standard barometric pressure $= 101.33$ kN/m^2

$$A/S \ = \ \frac{4.29\beta(fC_{os,t,sp}P - C_{os,a,sp}P_a)}{P_s[X_i]}$$

$$= \ \frac{4.29(0.95)[(1)(9.2)(276,000 + 100,600) - 9.2(100,600)]}{101,330(10,000)}$$

$$= 0.01 \quad \textbf{Ans}$$

Example 5.14 It is desired to thicken an activated sludge liquor from 3,000 mg/L to 4% using a flotation thickener. A laboratory study indicated an A/S ratio of 0.010 is optimal for this design. The subnatant flow rate was determined to be 8 L/m^2-min. The barometric pressure is assumed to be the standard of 101.33 kN/m^2 and the design temperature is to be 20°C. Assume $f = 0.5$; $\beta = 0.95$. The sludge flow rate is 400 m^3/d. Design the thickener with and without recycle.

Solution:
(a) Without recycle:

$$A/S = \frac{4.29\beta(fC_{os,t,sp}P - C_{os,a,sp}P_a)}{P_s[X_i]}$$

$$0.010 = \frac{4.29(0.95)[(0.5)(9.2)P - 9.2(101,330)]}{101,330(3,000)}$$

$$P = 364{,}758.82 \text{ N/m}^2 \text{ absolute} = 264{,}428.82 \text{ N/m}^2$$

$$\text{gage} = 264.43 \text{ kN/m}^2 \text{ gage} \quad \textbf{Ans}$$

Assuming solids in the subnatant is negligible, solids in the float = 3.0(400) = 1200 kg/d

$$4\% \text{ solids} = 40{,}000 \text{ mg/L} = 40 \text{ kg/m}^3$$

Therefore,

$$\text{float rate of flow, } Q_u = \frac{1{,}200}{40} = 30 \text{ m}^3/\text{d}$$

$$\text{subnatant flow, } Q_e = 400 - 30 = 370 \text{ m}^3/\text{d}$$

$$8 \text{ L/m}^2 \cdot \text{min} = 8(10^3) \text{ m}^3/\text{m}^2 \cdot \text{min}$$

Therefore,

$$\text{subnatant flow area, } A_s = \frac{370}{8(10^{-3})(60)(24)} = 32.12 \text{ m}^2 \quad \textbf{Ans}$$

(b) With recycle: Use the same operating pressure of 364,758.82 N/m^2 absolute

$$A/S = \frac{4.29R\beta(fC_{os,t,sp}P - C_{os,a,sp}P_a)}{P_s[X_i]}$$

$$0.010 = \frac{4.29(R)(0.95)[0.5(9.2)(364{,}758.82) - 9.2(101{,}330)]}{101{,}330(3{,}000)} \qquad R = 1 \quad \textbf{Ans}$$

$$\text{Subnatant flow area, } A_s = \frac{2(400) - 30}{8(10^{-3})(60)(24)} = 66.84 \text{ m}^2 \quad \textbf{Ans}$$

Note: This example shows that recycling is of no use. It increases the overflow area and it needs additional piping for the recycling.

GLOSSARY

Air-to-solids ratio—The ratio of mass of air used to the mass of solids introduced into the flotation unit.

Bar rack—Also called a bar screen, this is a device composed of large bars spaced widely far apart to separate large objects from a flowing water.

Clarification zone—In a settling process, this is the zone where the water is clarified.

Compression settling—Also called type 4 settling, this is a zone settling where compression or compaction of the particle mass is occurring at the same time.

Compression zone—In a settling or thickening process, this is the zone B where the thickened sludge from the thickening zone is further compressed, compacted, and consolidated.

Critical concentration—In a batch settling or thickening test, the concentration of solids when the uniform velocity zone disappears.

Critical section—Also called terminal section, this is the location in a thickener where the limiting flux is transpiring.

Detention time—Also called retention time and removal time, this is the average time that particles of water have stayed inside a tank.

Discrete particles—Particles that settle independent of the presence of other particles.

Discrete settling—Also called type 1 settling, this refers to the removal of discrete particles.

Flocculent particles—Particles that tend to form aggregates with other particles.

Flocculent settling—Also called type 2 settling, this is a settling of flocculent particles.

Flow-through velocity—The horizontal velocity in a rectangular clarifier.

Flotation—A unit operation in which solids are made to float to the surface on account of their adhering to minute bubbles of gases (air) that rises to the surface.

Hydraulic loading rate—Flow rate divided by the surficial area. Also called hydraulic overflow rate.

Inlet zone—In a sedimentation basin, the transition into the settling zone aimed at properly introducing the inflow into the tank.

Limiting solids flux—The solids flux that is equivalent to the rate of mass withdrawal from the bottom of the thickener.

Microstrainer—A device constructed of straining materials made of a very fine fabric or screen designed to remove minute particles from water.

Outlet zone—In a sedimentation basin, the part where the settled water is taken off into the effluent launder.

Overflow rate—Flow rate divided by the surficial area. Also called hydraulic loading rate.

Removal time—Also called detention time and retention time, this is the average time that particles of water have stayed inside a tank.

Retention time—Also called detention time and removal time, this is the average time that particles of water have stayed inside a tank.

Screening—A unit operation that separates materials into different sizes.

Settling—A unit operation in which solids are drawn toward a source of attraction.

Settling zone—In a sedimentation basin, this is the part where the suspended solids load of the inflow is removed to be deposited into the sludge zone below.

Sludge zone—In a sedimentation basin, this is the part where solids are deposited.

Solids flux—The transport of solids through a unit normal area per unit time.

Thickening zone—In a settling or thickening process, this is the zone where the sludge is concentrated.

Underflow concentration—The concentration withdrawn from the bottom of the thickener.

Uniform settling zone—In a settling or thickening process, this is the zone where an interface with the clarification zone settles at a constant rate.

Zone settling—Also called type 3 settling, this is a form of settling which refers to the removal of particles that settle in a contiguous zone.

SYMBOLS

A	Cross-sectional area of flow
A_c	Clarifier area
A_i	Amount of air introduced to the flotation tank
A_o	Amount of air remaining in the flotation tank after pressure release
A_s	Surficial area
A_t	Thickener area
A/S	Air-to-solids ratio
A_1	Cross-sectional area of flow at point 1
A_2	Cross-sectional area of flow at point 2
$[c]$	Concentration of particles
$[c_o]$	Initial concentration of particles in settling column
$C_{as,t,sp}$	Saturation air concentration of tap or distilled water corresponding to the wastewater in the air saturation tank at standard pressure
$C_{asw,t,sp}$	Saturation air concentration of the wastewater in the air saturation tank at standard pressure
$\overline{C}_{aw,t}$	Concentration of air in the air saturation tank
$C_{os,a,sp}$	Saturation dissolved oxygen concentration of the ambient wastewater at standard pressure
$C_{os,t,sp}$	Saturation dissolved oxygen concentration of the wastewater in the air saturation tank
C_d	Coefficient of discharge
C_D	Coefficient of drag
C_3	Critical concentration
d	Spherical diameter of particle
d_p	Sieve diameter of particle
g	Acceleration due to gravity

f	Fraction of saturation in the air saturation tank
G_t	Solids flux
$G_{t\ell}$	Limiting solids flux
h	Head
H_o	Initial height of sludge in graduated cylinder in the cylinder test
H_u	Height of thickened sludge corresponding to underflow concentration $[X_u]$
H_5	Height corresponding to the appearance of the critical concentration
ℓ	Top width of weir
P	Pressure in the air saturation tank
P_a	Atmospheric pressure
P_s	Standard atmospheric pressure
P_1	Pressure at point 1
P_2	Pressure at point 2
q_o	Overflow velocity
Q	Discharge flow
Q_{ms}	Minimum flow sustained
Q_i	Inflow to clarifier or thickener or any unit
Q_o	Inflow to the overall treatment plant
Q_{pks}	Peak flow sustained
Q_R	Recirculation flow
Q_u	Underflow discharge
R	Fractional removal of particles; also, recirculation ratio
Re	Reynolds number
t	Time of settling
t_o	Detention time
t_u	Time to thicken to underflow concentration
t_3	Time to appearance of critical concentration
v	Terminal settling velocity
v_h	Flow-through velocity in a rectangular clarifier
v_o	Terminal settling velocity of particle removed 100%
v_p	Terminal settling velocity of any size particle
V_c	Settling velocity at thickening zone
V_u	Underflow velocity computed at the thickening zone
V_1	Velocity at point 1
V_2	Velocity at point 2
$\mathbb{V}$	Volume of tank
w	Top width of flow in grit chambers
W	Width of settling zone of rectangular clarifier
x	Fraction of particles remaining
x_o	Fraction of particles remaining corresponding to particle size removed 100%
$[\bar{X}]$	Concentration of solids entering clarifier or thickener
$[X_c]$	Solids concentration at thickening zone
$[X_i]$	The same as $[\bar{X}]$
$[X_u]$	Underflow concentration
z_1	Elevation at point 1

z_2	Elevation at point 2
Z_o	Depth of settling zone; also, depth of settling column used to calculate terminal settling velocity of particles removed 100%
Z_p	Depth of settling column used to calculate terminal settling velocity of any particle
β	Volume shape factor; also ratio of dissolved air saturation value in waste-water to that in tap or distilled water
ρ_p	Mass density of particle
ρ_w	Mass density of water
γ	Specific weight of fluid
μ	Absolute or dynamic viscosity

PROBLEMS

5.1   A bar screen measuring 2 m by 5 m of surficial flow area is used to protect the pump in a shoreline intake of a water treatment plant. The head loss across the screen is 0.17 m. The bar width is 20 mm and the bar spacing is 70 mm. If the screen is 30% clogged, at what rate of flow is the plant drawing water from the intake? Assume $C_d = 0.84$.

5.2   A water treatment plant drawing water at rate of 8 m³/s from a shoreline intake is protected by bar screen measuring 2 m by 5 m. The head loss across the screen is 0.17 m. The bar width is 20 mm and the bar spacing is 70 mm. If the screen is 30% clogged, what is the area of the clear space opening of the bars? Assume $C_d = 0.84$.

5.3   A water treatment plant drawing water at rate of 8 m³/s from a shoreline intake is protected by bar screen. The head loss across the screen is 0.17 m. The bar width is 20 mm and the bar spacing is 70 mm. Assuming that the screen is 30% clogged and using the clear space opening in Problem 5.2, calculate the value of C_d.

5.4   A bar screen measuring 2 m × 5 m of surficial flow area is used to protect a pump. The approach area to the screen is 7.48 m² and the head loss across it is 0.17 m. The bar width is 20 mm and the bar spacing is 70 mm. If the screen is 30% clogged, at what rate of flow is the pump drawing water? Assume $C_d = 0.84$.

5.5   A water treatment plant is drawing water at rate of 8 m³/s and its intake is protected by a bar screen measuring 2 m × 5 m. The head loss across the screen is 0.17 m. The bar width is 20 mm and the bar spacing is 70 mm. If the screen is 30% clogged, what is the area of the clear space opening of the bars? Assume $C_d = 0.84$ and that the approach area is 7.48 m².

5.6   A water treatment plant drawing water at rate of 8 m³/s is protected by bar screen. The head loss across the screen is 0.17 m. The bar width is 20 mm and the bar spacing is 70 mm. Assuming that the screen is 30% clogged and using the clear space opening in Problem 5.5, calculate the value of C_d. The approach area is 7.48 m^2. Assuming the head loss across the screen is 0.2 m and using this newly found value of C_d, what is the approach area?

5.7 The terminal settling velocity of a spherical particle having a diameter of 0.6 mm is 0.11 m/s. What is the mass density of the particle? Assume the settling is type 1 and the temperature of the water is 22°C. What is the drag coefficient?

5.8 The terminal settling velocity of a spherical particle having a diameter of 0.6 mm is 0.11 m/s. Assuming the specific gravity of the particle is 2.65, at what temperature of water was the particle settling? Assume the settling is type 1 and use the value of the drag coefficient in Problem 5.7.

5.9 The terminal settling velocity of a spherical particle is 0.11 m/s. Its specific gravity is 2.65 and the temperature of the water is 22°C. What is the diameter of the particle? Assume the settling is type 1. What is the drag coefficient?

5.10 The terminal settling velocity of a worn sand particle having a sieve diameter of 0.6 mm is 0.11 m/s. What is the mass density of the particle? Assume the settling is type 1 and the temperature of the water is 22°C. What is the drag coefficient?

5.11 The terminal settling velocity of a worn sand particle having a sieve diameter of 0.6 mm is 0.11 m/s. Assuming the specific gravity of the particle is 2.65, at what temperature of water was the particle settling? Assume the settling is type 1 and use the value of the drag coefficient in Problem 5.10.

5.12 The terminal settling velocity of a worn sand particle is 0.11 m/s. Its specific gravity is 2.65 and the temperature of the water is 22°C. What is the sieve diameter of the particle? Assume the settling is type 1. What is the drag coefficient?

5.13 A certain municipality in Thailand plans to use the water from the Chao Praya River as raw water for a contemplated water treatment plant. The river is very turbid, so presedimentation is necessary. The result of a column test is as follows:

t (min)	0	60	80	100	130	200	240	420
c (mg/L)	299	190	179	169	157	110	79	28

If the percentage removal of particles is 71.58, what is the hydraulic loading rate? The column is 4 m deep.

5.14 In Problem 5.13, what is the fraction not completely removed x_o?

5.15 In Problem 5.13, what fraction in x_o is removed?

5.16 A grit removal unit consists of four identical channels to remove grit for a peak flow of 80,000 m^3/d, an average flow of 50,000 m^3/d, and a minimum flow of 20,000 m^3/d. It is to be controlled using a Parshall flume. There should be a minimum of three channels operating at any time.

Assume a flow-through velocity of 0.3 m/s. The top width of the flow at maximum flow using three channels is 1.5 m. What is depth of flow corresponding to this condition?

5.17   In Problem 5.16, what is the top width during average flow for conditions operating at 4 channels?

5.18   In Problem 5.16, what is the depth during average flow for conditions operating at 4 channels? What is the value of the constant?

5.19   Assume Anne Arundel County wants to expand its softening plant. A sample from their existing softening tank is prepared and a settling column test is performed. The initial solids concentration in the column is 140 mg/L. The results are as follows:

	Sampling Time (min)							
Z_p/Z_o	5	10	15	20	25	30	35	40
0.1	95[a]	68	55	30	23			
0.2	129	121	73	67	58	48	43	
0.3	154	113	92	78	69	60	53	45

[a] Values in the table are the results of the test for the suspended solids (mg/L) concentration at the given depths.

If the percentage removal of particles is 71.58, what is the hydraulic loading rate? The settling column has a depth of 4 m.

5.20   In Problem 5.19, what is the fraction not completely removed x_o?

5.21   In Problem 5.19, what fraction in x_o is removed?

5.22   The prototype detention time and overflow rate were calculated to be 1.5 h and 28 m/d, respectively. The peaking factor is 3.0 and the minimizing factor is 0.3. Calculate the volume of the tank. The average daily flow rate is 20,000 m^3/d. Use rectangular basin.

5.23   The prototype detention time and overflow rate were calculated to be 1.5 h and 28 m/d, respectively. The peaking factor is 3.0 and the minimizing factor is 0.3. Calculate the overflow area. The average daily flow rate is 20,000 m^3/d. Use rectangular basin.

5.24   The prototype detention time and overflow rate were calculated to be 1.5 h and 28 m/d, respectively. The peaking factor is 3.0 and the minimizing factor is 0.3. Assuming the particles are flocculent, calculate the width of the rectangular clarifier. The average daily flow rate is 20,000 m^3/d.

5.25   The prototype detention time and overflow rate were calculated to be 1.5 h and 28 m/d, respectively. The peaking factor is 3.0 and the minimizing factor is 0.3. Assuming the particles are flocculent, calculate the recirculated flow. The average daily flow rate is 20,000 m^3/d. Use rectangular basin.

5.26   The prototype detention time and overflow rate were calculated to be 1.5 h and 28 m/d, respectively. The peaking factor is 3.0 and the minimizing factor is 0.3. Calculate the volume of the tank. The average daily flow rate is 20,000 m^3/d. Use circular basin.

5.27   The prototype detention time and overflow rate were calculated to be 1.5 h and 28 m/d, respectively. The peaking factor is 3.0 and the minimizing factor is 0.3. Calculate the overflow area. The average daily flow rate is 20,000 m^3/d. Use circular basin.

5.28   The prototype detention time and overflow rate were calculated to be 1.5 h and 28 m/d, respectively. The peaking factor is 3.0 and the minimizing factor is 0.3. Assuming the particles are flocculent, calculate the recirculated flow. The average daily flow rate is 20,000 m^3/d. Use circular basin.

5.29   The activated sludge bioreactor facility of a certain plant is to be expanded. The results of a settling cylinder test of the existing bioreactor suspension are shown below. $Q_o + Q_R$ is 10,000 m^3/d and the influent MLSS is 3,500 mg/L. If the sludge is to be thickened to an underflow concentration of 10,000 mg/L, what is the limiting flux?

MLSS (mg/L)	1410	2210	3000	3500	4500	5210	6510	8210
V_c (m/h)	2.93	1.81	1.20	0.79	0.46	0.26	0.12	0.084

5.30   The activated sludge bioreactor facility of a certain plant is to be expanded. The results of a settling cylinder test of the existing bioreactor suspension are shown below. $Q_o + Q_R$ is 10,000 m^3/d and the influent MLSS is 3,500 mg/L. If the limiting flux is 2200 (m/h) · (mg/L), what is the underflow concentration?

MLSS (mg/L)	1410	2210	3000	3500	4500	5210	6510	8210
V_c (m/h)	2.93	1.81	1.20	0.79	0.46	0.26	0.12	0.084

5.31 $Q_o + Q_R$ into a secondary clarifier is 10,000 m^3/d with $R = 1$. The influent MLSS is 3,500 mg/L. If the thickener area is 651.5 m^2, what is the limiting flux?

5.32 $Q_o + Q_R$ into a secondary clarifier is 10,000 m^3/d with $R = 1$. The influent MLSS is 3,500 mg/L. If the limiting flux is 2,200 (m/h) · (mg/L), what is the thickener area?

5.33   The desired underflow concentration from a secondary clarifier is 10,500 mg/L. The influent comes from an activated sludge process operated at 3500 mg/L of MLSS. The inflow to the clarifier is 10,000 m^3/d; the thickener area is 158.7 m^2; and t_u is 7.8 min. What volume at the bottom of the thickener must be provided to hold the thickened sludge.

5.34   The A/S obtained in an experiment is 0.01. The pressure gage reads 276 kN/m^2 and the temperature of the sludge and the subnatant in the flotation cylinder is 20°C. The prevailing barometric pressure is 100.6 kN/m^2. β was originally determined to be 0.95. What is the total solids in the sludge?

5.35   The A/S obtained in an experiment is 0.01. The pressure gage reads 276 kN/m^2 and the temperature of the sludge and the subnatant in the flotation cylinder is 20°C. β was originally determined to be 0.95 and the solids are 10,000 mg/L. What is the prevailing barometric pressure?

5.36   The A/S obtained in an experiment is 0.01. The temperature of the sludge and the subnatant in the flotation cylinder is 20°C. The prevailing barometric pressure is 100.6 kN/m². β was originally determined to be 0.95. Total solids is 10,000 mg/L and the prevailing barometric pressure is 100.6 kN/m². What is the pressure in the air saturation tank?

5.37   The A/S obtained in an experiment is 0.01. The temperature of the sludge and the subnatant in the flotation cylinder is 20°C. The prevailing barometric pressure is 100.6 kN/m². β was originally determined to be 0.95. Total solids is 10,000 mg/L and the prevailing barometric pressure is 100.6 kN/m². The pressure gage of the air saturation tank reads 276 kN/m². What is the value of f ?

5.38   It is desired to thicken an activated sludge liquor from 3,000 mg/L to 4% using a flotation thickener. A laboratory study indicated an A/S ratio of 0.011 is optimal for this design. The subnatant flow rate was determined to be 8 L/m²-min. The barometric pressure is assumed to be the standard of 101.33 kN/m² and the design temperature is to be 20°C. Assume $f = 0.5$; $\beta = 0.90$. The sludge flow rate is 400 m³/d. Design the thickener with and without recycle.

BIBLIOGRAPHY

Bliss, T. (1998). Screening in the stock preparation system. *Proc. 1998 TAPPI Stock Preparation Short Course,* Apr. 29–May 1, Atlanta, GA, 151–174. TAPPI Press, Norcross, GA.

Buerger, R. and F. Concha (1998). Mathematical model and numerical simulation of the settling of flocculated suspensions. *Int. J. Multiphase Flow.* 24, 6, 1005–1023.

Chancelier, J. P., G. Chebbo, and E. Lucas-Aiguier (1998). Estimation of settling velocities. *Water Res.* 32, 11, 3461–3471.

Cheremisinoff, P. N. Treating wastewater. *Pollution Eng.* 22, 9, 60–65.

Christoulas, D. G., P. H. Yannakopoulos, and A. D. Andreadakis (1998). Empirical model for primary sedimentation of sewage. *Environment Int.* 24, 8, 925–934.

Diehl, S. and U. Jeppsson (1998). Model of the settler coupled to the biological reactor. *Water Res.* 32, 2, 331–342.

Droste, R. L. (1997). *Theory and Practice of Water and Wastewater Treatment.* John Wiley & Sons, New York.

Fernandes, L., M. A. Warith, and R. Droste (1991). Integrated treatment system for waste from food production industries. *Int. Conf. Environ. Pollut. Proc. Int. Conf. Environ. Pollut.— ICEP-1,* Apr. 1991, 671–679. Inderscience Enterprises Ltd., Geneva, Switzerland.

Hasselblad, S., B. Bjorlenius, and B. Carlsson (1997). Use of dynamic models to study secondary clarifier performance. *Water Science Technol. Proc. 7th Int. Workshop on Instrumentation, Control and Automation of Water and Wastewater Treatment and Transport Syst.,* July 6–9, Brighton, England, 37, 12, 207–212. Elsevier Science Ltd., Exeter, England.

Jefferies, C., C. L. Allinson, and J. McKeown (1997). Performance of a novel combined sewer overflow with perforated conical screen. *Water Science Technol. Proc. 1997 2nd IAWQ*

Int. Conf. on the Sewer as a Physical, Chemical and Biological Reactor, May 25–28, Aalborg, Denmark, 37, 1, 243–250. Elsevier Science Ltd., Exeter, England.

McCaffery, S., J. L. Elliott, and D. B. Ingham (1998). Two-dimensional enhanced sedimentation in inclined fracture channels. *Mathematical Eng. Industry.* 7, 1, 97–125.

Metcalf & Eddy, Inc. (1991). *Wastewater Engineering: Treatment, Disposal, and Reuse.* McGraw-Hill, New York, 37.

Renko, E. K. (1998). Modelling hindered batch settling part II: A model for computing solids profile of calcium carbonate slurry. *Water S.A.* 24, 4, 331–336.

Robinson, D. G. (1997). Rader bar screen performance at Howe Sound Pulp & Paper Ltd. *Pulp Paper Canada.* 98, 4, 21–24.

Rubio, J. and H. Hoberg (1993). Process of separation of fine mineral particles by flotation with hydrophobic polymeric carrier. *Int. J. Mineral Process.* 37, 1–2, 109–122.

Sincero, A. P. and G. A. Sincero (1996). *Environmental Engineering: A Design Approach.* Prentice Hall, Upper Saddle River, NJ.

Vanderhasselt, A. and W. Verstraete (1999). Short-term effects of additives on sludge sedimentation characteristics. *Water Res.* 33, 2, 381–390.

Wu, J. and R. Manasseh (1998). Dynamics of dual-particles settling under gravity. *Int. J. Multiphase Flow.* 24, 8, 1343–1358.

Zhang, Z. (1998). Numerical analysis of removal efficiencies in sedimentation tank. *Qinghua Daxue Xuebao/J. Tsinghua Univ.* 38, 1, 96–99.

6 Mixing and Flocculation

Mixing is a unit operation that distributes the components of two or more materials among the materials producing in the end a single blend of the components. This mixing is accomplished by agitating the materials. For example, ethyl alcohol and water can be mixed by agitating these materials using some form of an impeller. Sand, gravel, and cement used in the pouring of concrete can be mixed by putting them in a concrete batch mixer, the rotation of the mixer providing the agitation.

Generally, three types of mixers are used in the physical–chemical treatment of water and wastewater: rotational, pneumatic, and hydraulic mixers. *Rotational mixers* are mixers that use a rotating element to effect the agitation; *pneumatic mixers* are mixers that use gas or air bubbles to induce the agitation; and *hydraulic mixers* are mixers that utilize for the mixing process the agitation that results in the flowing of the water.

Flocculation, on the other hand, is a unit operation aimed at enlarging small particles through a very slow agitation of the water suspending the particles. The agitation provided is mild, just enough for the particles to stick together and agglomerate and not rebound as they hit each other in the course of the agitation. Flocculation is effected through the use of large paddles such as the one in flocculators used in the coagulation treatment of water.

6.1 ROTATIONAL MIXERS

Figure 6.1 is an example of a rotational mixer. This type of setup is used to determine the optimum doses of chemicals. Varying amounts of chemicals are put into each of the six containers. The paddles inside each of the containers are then rotated at a predetermined speed by means of the motor sitting on top of the unit. This rotation agitates the water and mixes the chemicals with it. The paddles used in this setup are, in general, called impellers. A variety of impellers are used in practice.

6.1.1 TYPES OF IMPELLERS

Figure 6.2 shows the various types of impellers used in practice: propellers (a), paddles (b), and turbines (c). *Propellers* are impellers in which the direction of the driven fluid is along the axis of rotation. These impellers are similar to the impellers used in propeller pumps treated in a previous chapter. Small propellers turn at around 1,150 to 1,750 rpm; larger ones turn at around 400 to 800 rpm. If no slippage occurs between water and propeller, the fluid would move a fixed distance axially. The ratio of this distance to the diameter of the impeller is called the *pitch*. A *square pitch* is one in which the axial distance traversed is equal to the diameter of the propeller. The pitching is obtained by twisting the impeller blade; the correct degree of twisting induces the axial motion.

FIGURE 6.1 An example of a rotational mixer. (Courtesy of Phipps & Bird, Richmond, VA. © 2002 Phipps & Bird.)

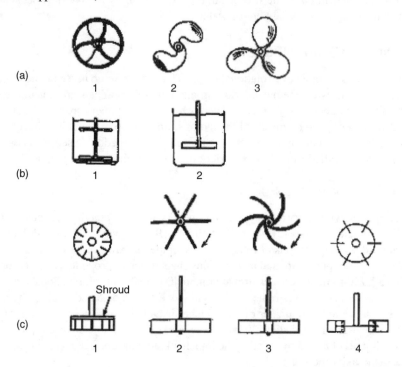

FIGURE 6.2 Types of impellers. (a) Propellers: (1) guarded; (2) weedless; and (3) standard three-blade. (b) Paddles: (1) pitched and (2) flat paddle. (c) Turbines: (1) shrouded blade with diffuser ring; (2) straight blade; (3) curved blade; and (4) vaned-disk.

Figure 6.2(a)1 is a guarded propeller, so called because there is a circular plate ring encircling the impeller. The ring guides the fluid into the impeller by constraining the flow to enter on one side and out of the other. Thus, the ring positions the flow for an axial travel. Figure 6.2(a)2 is a weedless propeller, called weedless, possibly because it originally claims no "weed" will tangle the impeller because of its two-blade design. Figure 6.2(a)3 is the standard three-blade design; this normally is square pitched.

Figure 6.2(b)1 is a paddle impeller with the two paddles pitched with respect to the other. Pitching in this case is locating the paddles at distances apart. Three or four paddles may be pitched on a single shaft; two and four-pitched paddles being more common. The paddles are not twisted as are the propellers. *Paddles* are so called if their lengths are equal to 50 to 80% of the inside diameter of the vessel in which the mixing is taking place. They generally rotate at slow to moderate speeds of from 20 to 150 rpm. Figure 6.2(b)2 shows a single-paddle agitator.

Impellers are similar to paddles but are shorter and are called *turbines*. They turn at high speeds and their lengths are about only 30 to 50% of the inside diameter of the vessel in which the mixing is taking place. Figure 6.2(c)1 shows a shrouded turbine. A shroud is a plate added to the bottom or top planes of the blades. Figures 6.2(c)2 and 6.2(c)3 are straight and curve-bladed turbines. They both have six blades. The turbine in Figure 6.2(c)4 is a disk with six blades attached to its periphery.

Paddle and turbine agitators push the fluid both radially and tangentially. For agitators mounted concentric with the horizontal cross section of the vessel in which the mixing is occurring, the current generated by the tangential push travels in a swirling motion around a circumference; the current generated by the radial push travels toward the wall of the vessel, whereupon it turns upward and downward. The swirling motion does not contribute to any mixing at all and should be avoided. The currents that bounce upon the wall and deflected up and down will eventually return to the impeller and be thrown away again in the radial and tangential direction. The frequency of this return of the fluid in agitators is called the *circulation rate*. This rate must be of such magnitude as to sweep all portions of the vessel in a reasonable amount of time.

Figure 6.3 shows a vaned-disk turbine. As shown in the elevation view on the left, the blades throw the fluid radially toward the wall thereby deflecting it up and down.

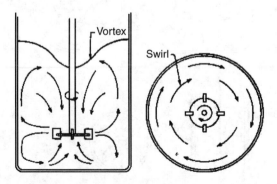

FIGURE 6.3 Flow patterns in rotational mixers.

The arrows also indicate the flow eventually returning back to the agitator blades—the circulation rate. On the right, the swirling motion is shown. The motion will simply move in a circumference unless it is broken. As the tangential velocity is increased, the mass of the swirling fluid tends to pile up on the wall of the vessel due to the increased centrifugal force. This is the reason for the formation of vortices. As shown on the left, the vortex causes the level of water to rise along the vessel wall and to dip at the center of rotation.

6.1.2 Prevention of Swirling Flow

Generally, three methods are used to prevent the formation of swirls and vortices: putting the agitator eccentric to the vessel, using a side entrance to the vessel, and putting baffles along the vessel wall. Figure 6.4 shows these three methods of prevention. The left side of Figure 6.4a shows the agitator to the right of the vessel center and in an inclined position; the right side shows the agitator to the left and in a vertical position. Both locations are no longer concentric with the vessel but eccentric to it, so the circumferential path needed to form the swirl would no longer exist, thus avoiding the formation of both the swirl and the vortex.

Figure 6.4b is an example of a side-entering configuration. It should be clear that swirls and vortices would also be avoided in this kind of configuration. Figure 6.4c shows the agitator mounted at the center of the vessel with four baffles installed on the vessel wall. The swirl may initially form close to the center. As this swirl

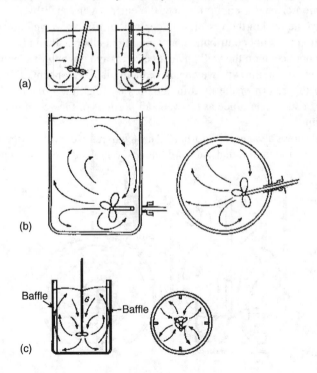

FIGURE 6.4 Methods of swirling flow prevention.

propagates toward the wall, its outer rim will be broken by the baffles, however, preventing its eventual formation.

6.1.3 Power Dissipation in Rotational Mixers

A very important parameter in the design of mixers is the power needed to drive it. This power can be known if the power given to the fluid by the mixing process is determined. The product of force and velocity is power. Given a parcel of water administered a push (force) by the blade, the parcel will move and hence attain a velocity, thus producing power. The force exists as long as the push exists; however, the water will not always be in contact with the blade; hence, the pushing force will cease. The power that the parcel had acquired will therefore simply be dissipated as it overcomes the friction imposed by surrounding parcels of water. *Power dissipation* is power lost due to frictional resistance and is equal to the power given to it by the agitator.

Let us derive this power dissipation by dimensional analysis. Recall that in dimensional analysis pi groups are to be found that are dimensionless. The power given to the fluid should be dependent on the various geometric measurements of the vessel. These measurements can be conveniently normalized against the diameter of the impeller D_a to make them into dimensionless ratios. Thus, as far as the geometric measurements are concerned, they have now been rendered dimensionless. These dimensionless ratios are called *shape factors*.

Refer to Figure 6.5. As shown, there are seven geometric measurements: W, the width of the paddle; L, the length of the paddle; J, the width of the baffle; H, the depth in the vessel; D_t, the diameter of the vessel; E, the distance of the impeller to the bottom of the vessel; and D_a, the diameter of the impeller. The corresponding shape factors are then $S_1 = W/D_a$, $S_2 = L/D_a$, $S_3 = J/D_a$, $S_4 = H/D_a$, $S_5 = D_t/D_a$, and $S_6 = E/D_a$. In general, if there are n geometric measurements, there are $n - 1$ shape factors.

The power given to the fluid should also be dependent on viscosity μ, density ρ, and rotational speed N. The higher the viscosity, the harder it is to push the fluid, increasing the power required. A similar argument holds for the density: the denser

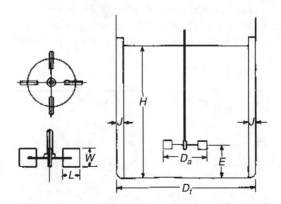

FIGURE 6.5 Normalization of geometric measurements into dimensionless ratios.

the fluid, the harder it is to push it, thus requiring more power. In addition, the power requirement must also increase as the speed of rotation is increased. Note that N is expressed in radians per second.

As shown in Figure 6.3, a vortex is being formed, raising the level of water higher on the wall and lower at the center. This rising of the level at one end and lowering at the other is has to do with the weight of the water. Because the weight of any substance is a function of gravity, the gravity g must enter into the functionality of the power given to the fluid. The shape factors have already been nondimensionalized, so we will ignore them for the time being and consider only the diameter D_a of the impeller as their representative in the functional expression for the power. Letting the power be $\mathcal{P}$,

$$\mathcal{P} = \psi(N, D_a, g, \mu, \rho) \tag{6.1}$$

Now, to continue with our dimensional analysis, let us break down the variables of the previous equation into their respective dimensions using the force-length-time (FLT) system as follows:

Variable	Dimensions, FLT
$\mathcal{P}$	FL/T
N	$1/T$ only if N is in radians per unit time; a radian by definition is dimensionless
D_a	L
g	L/T^2
ρ	$M/L^3 = FT^2/L^4$, M is the dimension of mass in the MLT system
μ	$M/LT = FT/L^2$

By inspection, the number of reference dimensions is 3; thus, the number of *pi* variables is $6 - 3 = 3$ (the number of variables minus the number of reference dimensions). Reference dimension is the smallest number of groupings obtained from grouping the basic dimensions of the variables in a given physical problem. Call the *pi* variables Π_1, Π_2, and Π_3, respectively. Letting Π_1 contain $\mathcal{P}$, write $[\mathcal{P}/N] = (FL/T)/(1/T) = FL$ to eliminate T. [ ] is read as "the dimensions of." To eliminate L, write $[\mathcal{P}/ND_a] = FL/L = F$. To eliminate F, write $[\mathcal{P}/ND_a(1/\rho N^2 D_a^4)] = F\{1/(FT^2/L^4)(1/T)^2 L^4\} = 1$. Therefore,

$$\Pi_1 = \frac{\mathcal{P}}{N^3 \rho D_a^5} = \mathcal{P}_o \quad \text{called the } power\ number \tag{6.2}$$

To solve for Π_2, write $[D_a^2 \mu] = L^2(FT/L^2) = FT$. To eliminate FT, write $[D_a^2 \mu(1/\rho ND_a^4)] = FT\{1/(FT^2/L^4)(1/T)L^4\} = 1$. Thus,

$$\Pi_2 = \frac{\mu}{\rho ND_a^2} \Rightarrow \Pi_2 = \frac{\rho ND_a^2}{\mu} = Re \quad \text{called the } Reynolds\ number \tag{6.3}$$

To solve for Π_3, write $[D_a/g] = (L)/(L/T^2) = T^2$. To eliminate T^2, write $[(D_a/g)N^2] = T^2\{1/T^2\} = 1$. Thus,

$$\Pi_3 = \frac{D_a N^2}{g} = Fr \quad \text{called the } Froude\ number \tag{6.4}$$

Including the shape factors and assuming there are n geometric measurements in the vessel, the functional relationship Equation (6.1) becomes

$$\mathcal{P} = N^3 \rho D_a^5 \phi\left(\frac{\rho N D_a^2}{\mu}, \frac{D_a N^2}{g}, S_1, S_2, ..., S_{n-1}\right) \tag{6.5}$$

$$\Rightarrow \frac{\mathcal{P}}{N^3 \rho D_a^5} = \mathcal{P}_o = \phi\left(\frac{\rho N D_a^2}{\mu}, \frac{D_a N^2}{g}, S_1, S_2, ..., S_{n-1}\right)$$

$$= \phi(Re, Fr, S_1, S_2, ..., S_{n-1}) \tag{6.6}$$

For any given vessel, the values of the shape factors will be constant. Under this condition, $\mathcal{P}_o = \mathcal{P}/N^3 \rho D_a^5$ will simply be a function of $Re = \rho N D_a^2/\mu$ and a function of $Fr = D_a N^2/g$. The effect of the Froude number Fr is manifested in the rising and lowering of the water when the vortex is formed. Thus, if vortex formation is prevented, Fr will not affect the power number $\mathcal{P}_o$ and $\mathcal{P}_o$ will only be a function of Re.

As mentioned before, the power given to the fluid is actually equal to the power dissipated as friction. In any friction loss relationships with Re, such as the Moody diagram, the friction factor has an inverse linear relationship with Re in the laminar range ($Re \leq 10$). The power number is actually a friction factor in mixing. Thus, this inverse relationship for $\mathcal{P}_o$ and Re, is

$$\mathcal{P}_o = \frac{K_L}{Re} \tag{6.7}$$

K_L is the proportionality constant of the inverse relationship. Substituting the expression for $\mathcal{P}_o$ and Re,

$$\mathcal{P} = K_L N^2 D_a^3 \mu \tag{6.8}$$

At high Reynolds numbers, friction losses become practically constant. If the Moody diagram for flow in pipes is inspected, this statement will be found to be true. Agitators are not an exception. If vortices and swirls are prevented, at high Reynolds numbers greater than or equal to 10,000, power dissipation is independent of Re and the relationship simply becomes

$$\mathcal{P}_o = K_T \tag{6.9}$$

K_T is the constant. Flows at high Reynolds numbers are characterized by turbulent conditions. Substituting the expression for $\mathcal{P}_o$,

$$\mathcal{P} = K_T N^3 D_a^5 \rho \tag{6.10}$$

TABLE 6.1
Values of Power Coefficients

Type of Impeller	K_L	K_T
Propeller (square pitch, three blades)	1.0	0.001
Propeller (pitch of 2, three blades)	1.1	0.004
Turbine (six flat blades)	1.8	0.025
Turbine (six curved blades)	1.8	0.019
Shrouded turbine (six curved blades)	2.4	0.004
Shrouded turbine (two curved blades)	2.4	0.004
Flat paddles (two blades, $D_t/W = 6$)	0.9	0.006
Flat paddles (two blades, $D_t/W = 8$)	0.8	0.005
Flat paddles (four blades, $D_t/W = 6$)	1.2	0.011
Flat paddles (six blades, $D_t/W = 6$)	1.8	0.015

Note: For vessels with four baffles at wall and $J = 0.1 \, D_t$

From W. L. McCabe and J. C. Smith (1967). *Unit Operations of Chemical Engineering*. McGraw-Hill, New York, 262.

K_L and K_T are collectively called *power coefficients* of which some values are found in Table 6.1.

For Reynolds number in the transition range ($10 < Re < 10{,}000$), the power may be taken as the average of Eqs. (6.8) and (6.10). Thus,

$$\mathcal{P} = \frac{K_L N^2 D_a^3 \mu + K_T N^3 D_a^5 \rho}{2} = \frac{1}{2} N^2 D_a^3 (K_L \mu + K_T N D_a^2 \rho) \qquad (6.11)$$

Example 6.1 A turbine with six blades is installed centrally in a baffled vessel. The vessel is 2.0 m in diameter. The turbine, 61 cm in diameter, is positioned 60 cm from the bottom of the vessel. The tank is filled to a depth of 2.0 m and is mixing alum with raw water in a water treatment plant. The water is at a temperature of 25°C and the turbine is running at 100 rpm. What horsepower will be required to operate the mixer?

Solution:

$$Re = \frac{\rho N D_a^2}{\mu} \qquad \rho = 997 \text{ kg/m}^3 \qquad N = \frac{100(2\pi)}{60} = 10.47 \text{ rad/s}$$

$$D_a = 0.61 \text{ m} \qquad \mu = 8.5(10^{-4}) \text{ kg/m} \cdot \text{s}$$

Therefore,

$$Re = \frac{997(10.47)(0.61)^2}{8.5(10^{-4})} = 4.57(10^6) \text{ turbulent}$$

Therefore,

$$P = K_T N^3 D_a^5 \rho \qquad \text{From the table, } K_T = 0.025$$

$$P = 0.025(10.47)^3(0.61)^5(997) = 2416.15 \,\text{N} \cdot \text{m/s} = \frac{2416.15}{746} = 3.24 \text{ hp } \textbf{Ans}$$

6.2 CRITERIA FOR EFFECTIVE MIXING

As the impeller pushes a parcel of fluid, this fluid is propelled forward. Because of the inherent force of attraction between molecules, this parcel drags neighboring parcels along. This is the reason why fluids away from the impeller flows even if they were not actually hit by the impeller. This force of attraction gives rise to the property of fluids called viscosity.

Visualize the filament of fluid on the left of Figure 6.6 composed of several parcels strung together end to end. The motion induced on this filament as a result of the action of the impeller may or may not be uniform. In the more general case, the motion is not uniform. As a result, some parcels will move faster than others. Because of this difference in velocities, the filament rotates. This rotation produces a torque, which, coupled with the rate of rotation produces power. This power is actually the power dissipated that was addressed before. Out of this power dissipation, the criteria are derived for effective mixing.

Refer to the right-hand side of Figure 6.6. This is a parcel removed from the filament at the left. Because of the nonuniform motion, the velocity at the bottom of the parcel is different from that at the top. Thus, a gradient of velocity will exist. Designate this as G_z. From fluid mechanics, $G_z = lim_{\Delta y \to 0} \frac{\Delta u}{\Delta y} = \partial u / \partial y$, where u is the fluid velocity in the x direction. As noted, this gradient is at a point, since Δy has been shrunk to zero. If the dimension of G_z is taken, it will be found to have per unit time as the dimension. Thus, G_z is really a rate of rotation or angular velocity. Designate this as ω_z. If Ψ_z is the torque of the rotating fluid, then in the x direction, the power P_x is

$$P_x = \Psi_z \omega_z = \Psi_z G_z = \Psi_z \partial u / \partial y \tag{6.12}$$

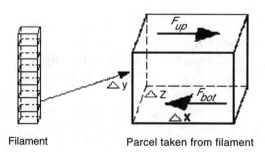

Filament   Parcel taken from filament

FIGURE 6.6 A parcel of fluid acted upon by shear forces in the x direction.

The torque Ψ_z is equal to a force times a moment arm. The force at the bottom face F_{bot} or the force at the upper face F_{up} in the parcel represents this force. This force is a force of shear. These two forces are not necessarily equal. If they were, then a couple would be formed; however, to produce an equivalent couple, each of these forces may be replaced by their average: $(F_{bot} + F_{up})/2 = \bar{F}_x$. Thus, the couple in the x direction is $\bar{F}_x \Delta y$. This is the torque Ψ_z.

The flow regime in a vessel under mixing may be laminar or turbulent. Under laminar conditions, $\bar{F}_x$ may be expressed in terms of the stress obtained from Newton's law of viscosity and the area of shear, $A_{shx} = \Delta x \Delta z$. Under turbulent conditions, the stress relationships are more complex. Simply for the development of a criterion of effective mixing, however, the conditions may be assumed laminar and base the criterion on these conditions. If this criterion is used in a consistent manner, since it is only employed as a benchmark parameter, the result of its use should be accurate.

From Newton's law of viscosity, the shear stress $\tau_x = \mu(\partial u/\partial y)$, where μ is the absolute viscosity. Substituting, Equation (6.12) becomes

$$\mathcal{P}_x = \Psi_z G_z = \bar{F}_x \Delta y \, G_z = \tau_x \Delta x \, \Delta z \, \Delta y \, G_z = \mu(\partial u/\partial y)\Delta x \, \Delta z \, \Delta y \, G_z = \mu \Delta V G_z^2$$

(6.13)

where $V = \Delta x \, \Delta z \, \Delta y$, the volume of the fluid parcel element.

Although Equation (6.13) has been derived for the fluid element power, it may be used as a model for the power dissipation for the whole vessel of volume V. In this case, the value of G_x to be used must be the average over the vessel contents. Also, considering all three component directions x, y, and z, the power is $\mathcal{P}$; the velocity gradient would be the resultant gradient of the three component gradients G_z, G_x, and G_y. Consider this gradient as $\bar{G}$, remembering that this $\bar{G}$ is the average velocity gradient over the whole vessel contents. $\mathcal{P}$ may then simply be expressed as $\mathcal{P} = \mu \Delta V \bar{G}$, whereupon solving for $\bar{G}$

$$\bar{G} = \sqrt{\frac{\mathcal{P}}{\mu V}}$$

(6.14)

Various values of this $\bar{G}$ are the ones used as criteria for effective mixing. Table 6.2 shows some criteria values that have been found to work in practice using the

TABLE 6.2
$\bar{G}$ Criteria Values for Effective Mixing

t_o, seconds	$\bar{G}$, s^{-1}
<10	4000–1500
10–20	1500–950
20–30	950–850
30–40	850–750
40–130	750–700

parameters t_o and $\overline{G} \cdot t_o$ is the detention time of the vessel. Thus, as shown in this table, the detention time to be allowed for in mixing is a function G.

The power dissipation treated above is power given to the fluid. To get the prime mover power $\mathcal{P}_b$ (the brake power), $\mathcal{P}$ must be divided by the brake efficiency η; and to get the input power from electricity $\mathcal{P}_i$, $\mathcal{P}_b$ must be divided by the motor efficiency $\mathcal{M}$.

6.3 PNEUMATIC MIXERS

Diffused aerators may also be used to provide mixing. The difference in density between the air bubbles and water causes the bubbles to rise and to quickly attain terminal rising velocities. As they rise, these bubbles push the surrounding water just as the impeller in rotational mixers push the surrounding water creating a pushing force. This force along with the rising velocity creates the power of mixing. It is evident that pneumatic mixing power is a function of the number of bubbles formed. Thus, to predict this power, it is first necessary to develop an equation to predict the number of bubbles formed.

Figure 6.7 shows designs of diffusers used to produce bubbles. In **a**, air is forced through a ceramic tube. Because of the fine opening in the ceramic mass, this design produces fine bubbles. In **d**, holes are simply pierced into the pipe creating perforations. The sizes of the bubbles would depend on how small or large the holes are. A photograph of coarse bubbles is shown in **b** while a photograph of fine bubbles is shown is **e**. The figure in **c** is simply an open pipe where air is allowed to escape; this produces large bubbles. The figure in **f** is a saran wrapped tube; this produces fine bubbles.

6.3.1 PREDICTION OF NUMBER OF BUBBLES AND RISE VELOCITY

The number of bubbles formed is equal to the volume of air in the vessel divided by the average volume of a single bubble. The volume of air in the vessel is equal to the rate of inflow of air Q_a times its detention time t_o. The detention time, in turn, is equal to the depth of submergence h (see Figure 6.7g) divided by the average total rise velocity of the bubbles. If $\bar{v}_b$ is the average rise velocity of the bubbles

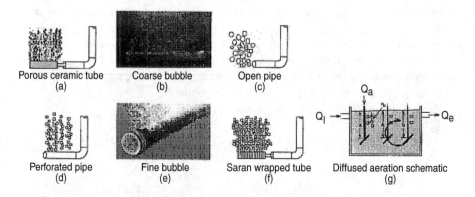

Porous ceramic tube (a) Coarse bubble (b) Open pipe (c) Perforated pipe (d) Fine bubble (e) Saran wrapped tube (f) Diffused aeration schematic (g)

FIGURE 6.7 Bubble diffuser designs.

and $\bar{v}_l$ is the net average upward velocity of the water, then the average total rise velocity of the bubbles is $\bar{v}_b + \bar{v}_l$ and $t_o = h/(\bar{v}_b + \bar{v}_l)$.

Let the average volume of a single bubble at the surface of the vessel be $\bar{V}_{bo}$ and let the influent *absolute* pressure of the air in Q_i be P_i. In order to accurately compute the number of bubbles, Q_i should be corrected so its value would correspond to $\bar{V}_{bo}$ at the surface when the pressure becomes the atmospheric pressure P_a. Since pressure and volume are in inverse ratio to each other, the rate of inflow of air corrected to its value at the surface of the vessel is then $(P_i/P_a)Q_i$. Thus, the number of bubbles n formed from a rate of inflow of air Q_i is

$$n = \frac{\left(\frac{P_i}{P_a}Q_i\right)t_o}{\bar{V}_{bo}} = \frac{\left(\frac{P_i}{P_a}Q_i\right)\frac{h}{\bar{v}_b + \bar{v}_l}}{\bar{V}_{bo}} \tag{6.15}$$

The average upward velocity of the liquid $\bar{v}_l$ is small in comparison with the rise velocity of the bubbles $\bar{v}_b$ and may be neglected. The equation then reduces to

$$n = \frac{\left(\frac{P_i}{P_a}Q_i\right)h}{\bar{V}_{bo}\bar{v}_b} = \frac{P_i Q_i h}{P_a \bar{V}_{bo}\bar{v}_b} \tag{6.16}$$

The rise velocities of bubbles were derived by Peebles and Garber (1953). Using the techniques of dimensional analysis as was used in the derivation of the power dissipation for rotational mixers, they discovered that the functionality of the rise velocities of bubbles can be described in terms of three dimensionless quantities: $G_1 = g\mu^4/\rho_l\sigma^3$, $G_2 = g(\bar{r})^4(\bar{v}_b)^4\rho_l^3/\sigma^3$, and $Re = 2\rho_l\bar{v}_b\bar{r}/\mu$. Re is a Reynolds number; g is the acceleration due to gravity; μ is the absolute viscosity of fluid; ρ_l is the mass density of fluid; σ is the surface tension of fluid; and $\bar{r}$ is the average radius of the bubbles. To give G_1 and G_2 some names, call G_1 the *Peebles number* and G_2 the *Garber number*.

We may want to perform the dimensional analysis ourselves, but the procedure is similar to the one done before. In other words, $\bar{v}_b$ is first to be expressed as a function of the variables affecting its value: $\bar{v}_b = f(g, \mu, \rho_l, \rho_g, \sigma, \bar{r})$. ρ_g is the mass density of the gas phase (air). Each of the variables in this function is then broken down into its fundamental dimensions to find the number of reference dimensions. Once the number of reference dimensions have been found, the number of pi dimensionless variables can then be determined. These dimensionless variables are then found by successive eliminations of the dimensions of the physical variables until the number of pi dimensionless ratios are obtained.

The final equations are as follows:

$$\bar{v}_b = \frac{2(\bar{r})^2(\rho_l - \rho_g)}{9\mu} = \frac{2(\bar{r})^2\rho_l}{9\mu} \qquad Re < 2 \tag{6.17}$$

$$\bar{v}_b = 0.33g^{0.76}\left(\frac{\rho_l}{\mu}\right)^{0.52}(\bar{r})^{1.28} \qquad 2 < Re < 4.02G_1^{-2.214} \tag{6.18}$$

$$\bar{v}_b = 1.35\left(\frac{\sigma}{\rho_l \bar{r}}\right)^{0.50} \qquad 4.02G_1^{-2.214} < Re < 3.10G_1^{-0.25} \qquad (6.19)$$

$$\bar{v}_b = 1.53\left(\frac{g\sigma}{\rho_l}\right)^{0.25} \qquad 3.10G_1^{-0.25} < Re < G_2 \qquad (6.20)$$

The mass density of air has been eliminated in Equation (6.17), because it is negligible.

6.3.2 POWER DISSIPATION IN PNEUMATIC MIXERS

The left-hand side of Figure 6.8 shows the forces acting on the bubble having a velocity of v_b going upward. F_B is the buoyant force acting on the bubble as a result of the volume of water it displaces; F_g is the weight of the bubble. As the bubble moves upward, it is resisted by a drag force exerted by the surrounding mass of water; this force is the drag force F_D. As the bubbles emerge from the diffusers, they quickly attain their terminal rise velocities. Thus, the bubble is not accelerated and application of Newton's second law of motion to the bubble simply results in $F_B - F_g - F_D = 0$ and $F_D = F_B - F_g$.

The right side of Figure 6.8 shows the action of the bubble upon the surrounding water as a result of Newton's third law of motion: *For every action, there is an equal and opposite reaction* $\Rightarrow$ *For every force there is an equal and opposite reactive force.* The F_D on the right is the reactive force to the F_D on the left. This force has the same action on the surrounding water as the impeller has on the water in the case of the rotational mixer. It pushes the opposing surrounding water with a force F_D traveling at an average speed of $\bar{v}_b$. The product of this force and the velocity gives the power of dissipation. Calling $\bar{V}_b$ the average volume the bubble attains as

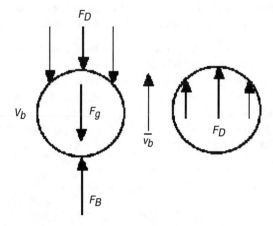

FIGURE 6.8 Bubble free-body diagrams to illustrate power dissipation.

it rises in the water column, γ_l the specific weight of the water, and γ_g the specific weight of the air, the power dissipated for n bubbles [Equation (6.16)] or simply the power dissipation in the vessel is

$$P = F_D \bar{v}_b n = (F_B - F_g) \bar{v}_b n = \bar{V}_b (\gamma_l - \gamma_g) \bar{v}_b \frac{P_i Q_i h}{P_a \bar{V}_{bo} \bar{v}_b} \qquad (6.21)$$

The value of the volume of a single bubble V_b varies as the bubble rises in the water column. By the inverse relationship of pressure and volume, V_b at any depth may be expressed in terms of $\bar{V}_{bo}$, the average volumes of the bubbles at the surface. The pressure upon V_b at depth h is $P_a + h\gamma_l$. Thus, $V_b = [P_a/(P_a + h\gamma_l)]\bar{V}_{bo}$. The value of $\bar{V}_b$ may then be derived by integrating V_b over the depth of the vessel as follows:

$$\bar{V}_b = \frac{1}{h}\int_0^h V_b\, dh = \frac{P_a \bar{V}_{bo}}{h}\int_0^h \frac{dh}{P_a + h\gamma_l} = \frac{P_a \bar{V}_{bo}}{h\gamma_l} \ln\left(\frac{P_a + h\gamma_l}{P_a}\right) \qquad (6.22)$$

And,

$$P = \bar{V}_b(\gamma_l - \gamma_g)\bar{v}_b \frac{P_i Q_i h}{P_a \bar{V}_{bo}\bar{v}_b} = \frac{P_a \bar{V}_{bo}}{h\gamma_l}\ln\left(\frac{P_a + h\gamma_l}{P_a}\right)(\gamma_l - \gamma_g)\bar{v}_b\frac{P_i Q_i h}{P_a \bar{V}_{bo}\bar{v}_b}$$

$$= P_i Q_i\left(\frac{\gamma_l - \gamma_g}{\gamma_l}\right)\ln\left(\frac{P_a + h\gamma_l}{P_a}\right) = P_i Q_i\, \ln\left(\frac{P_a + h\gamma_l}{P_a}\right) \qquad (6.23)$$

Because the specific weight of air γ_g is very much smaller than that of water γ_l, it has been neglected.

The power dissipation must be such that it causes the correct velocity gradient G. The literature have shown the criteria values for effective mixing in the case of rotational mixers. Values of G need to be determined for pneumatic mixers. As an ad hoc measure, however, the values for rotational mixers (Table 6.2) may be used.

Example 6.2 By considering the criterion for effective mixing, the volume of a rapid-mix tank used to rapidly mix an alum coagulant in a water treatment plant was found to be 6.28 m^3 with a power dissipation of 3.24 hp. Assume air is being provided at a rate of 1.12 m^3 per m^3 of water treated and that the detention time of the tank is 2.2 min. Calculate the pressure at which air is forced into the diffuser. Assume barometric pressure as 101,300 N/m^2, the depth from the surface to the diffuser as 2 m, and the temperature of water as 25°C.

Solution:

$$P = P_i Q_i \ln\left(\frac{P_a + h\gamma_l}{P_a}\right)$$

$6.28 = Q_o(2.2) \qquad Q_o = 2.85\ \text{m}^3/\text{min} = 0.047\ \text{m}^3/\text{sec of water inflow}$

therefore,

$$Q_i = 1.12(0.047) = 0.053 \text{ m}^3/\text{s}$$

$$3.24(746) = P_i(0.053) ln\left\{\frac{101,300 + 2(997)(9.81)}{101,300}\right\} P_i = 258,300.8 \text{ N/m}^2 \text{ abs}$$

Ans

6.4 HYDRAULIC MIXERS

Hydraulic mixers are mixers that use the energy of a flowing fluid to create the power dissipation required for mixing. This fluid must have already been given the energy before reaching the point in which the mixing is occurring. What needs to be done at the point of mixing is simply to dissipate this energy in such a way that the correct value of G for effective mixing is attained. The hydraulic mixers to be discussed in this chapter are the hydraulic-jump mixer and the weir mixer.

Figure 6.9 shows a hydraulic jump and its schematic. By some suitable design, the chemicals to be mixed may be introduced at the point indicated by "1" in the figure. Hydraulic-jump mixers are designed as rectangular in cross section.

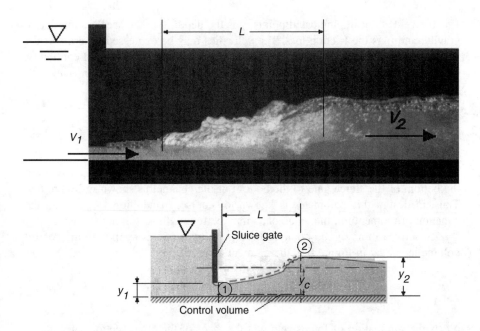

FIGURE 6.9 Hydraulic-jump mixer.

6.4.1 Power Dissipation in Hydraulic Mixers

The power of mixing is simply power dissipation. In any hydraulic process, power or energy is dissipated through friction. Thus, the power of mixing in any hydraulic mixer can be determined if the fluid friction h_f can be calculated. The product of rate of flow Q and specific weight γ is weight (force) per unit time. If this product is multiplied by h_f the result is power. Thus,

$$\mathcal{P} = Q\gamma h_f \tag{6.24}$$

The determination of the mixing power of hydraulic mixers is therefore reduced to the determination of the friction loss.

For mixing to be effective, the power derived from this loss must be such that the G falls within the realm of effective mixing. As in pneumatic mixers, G values for hydraulic-jump mixers discussed in this section need to be established. As an ad hoc measure, the values for rotational mixers (Table 6.2) may be used.

6.4.2 Mixing Power for Hydraulic Jumps

Refer to the hydraulic jump schematic of Figure 6.9. The general energy equation may be applied between points 1 and 2 producing

$$\frac{V_1^2}{2g} + y_1 - h_f = \frac{V_2^2}{2g} + y_2 \tag{6.25}$$

V is the velocity at the indicated points; y is the depth; g is the acceleration due to gravity; and h_f is the friction loss. The velocities may be expressed in terms of the flow q per unit width of the channel and the depth using the equation of continuity. Thus, $V_1 = g/y_1$ and $V_2 = g/y_2$. Substituting this in Equation (6.25), simplifying, and solving for h_f,

$$h_f = \frac{(y_2 - y_1)[q^2(y_2 + y_1) - 2gy_1^2 y_2^2]}{2gy_1^2 y_2^2} \tag{6.26}$$

For all practical purposes, the depth y_1 may be made equal to the distance from the bottom of the sluice gate to the bottom of the channel as shown in Figure 6.9. Thus, in design this parameter is known, of course, in addition to q. Using the equation of momentum, the value of y_2 may be found, thus, solving h_f.

As derived in any good book on fluid mechanics and as applied to the control volume indicated in Figure 6.9, the momentum equation is

$$\sum \vec{F} = \frac{\partial}{\partial t}\int_{CV} \vec{v}\rho\, d\mathcal{V} + \int_A \vec{v}\rho(\vec{v}\cdot\hat{n})\, dA \tag{6.27}$$

$\sum\vec{F}$ is the summation of forces acting at the faces of the cross sections at points 1 and 2; t is the time; $\vec{v}$ is the velocity vector; ρ is the mass density of water; $\mathcal{V}$ is the volume of the domain of integration; $\hat{n}$ is the unit normal vector to surface area A bounding the domain of integration. CV refers to the control volume.

Considering only the x direction in our analysis, $\Sigma \vec{F} = P_1 A_1 \hat{i} - P_2 A_2 \hat{i}$, where P is the pressure at the respective points; A is the area normal to the pressure; and $\hat{i}$ is the unit vector in the x direction. The P's and the A's may be expressed in terms of the respective depths y and specific weight γ. Thus, $\Sigma \vec{F}$ becomes $\frac{1}{2} y_1 \gamma y_1 \hat{i} - \frac{1}{2} y_2 \gamma y_2 \hat{i}$.

During operation, the mixer is at steady state; hence, $\frac{\partial}{\partial t} \int_{CM} \vec{v} \rho d V = 0$. $\oint \vec{v} \rho (\vec{v} \cdot \hat{n}) dA = -q\rho v_1 \hat{i} + q\rho v_2 \hat{i}$. Substituting all these into Equation (6.27), noting that only the x direction is to be considered, and simplifying,

$$y_2 = \frac{y_1}{2} \left(\sqrt{1 + 8 Fr_1^2} - 1 \right)$$

$$y_2 = \frac{y_1}{2} \left(\sqrt{1 + 8 \frac{v_1^2}{g y_1}} - 1 \right)$$

(6.28)

Fr_1 is the Froude number at point $1 = v_1 / \sqrt{g y_1}$.

Equation (6.26) may now be substituted into the general equation for mixing power Equation (6.24). The mixing power for hydraulic jumps is then

$$\mathcal{P}_{hydJump} = Q\gamma \frac{(y_2 - y_1)[q^2(y_2 + y_1) - 2g y_1^2 y_2^2]}{2g y_1^2 y_2^2}$$

$$= \frac{Q\gamma(y_2 - y_1)\left[\left(\frac{Q}{W}\right)^2 (y_2 + y_1) - 2g y_1^2 y_2^2 \right]}{2g y_1^2 y_2^2}$$

(6.29)

W is the width of the channel. As mentioned before, $\mathcal{P}_{hydJump}$ must have a value that corresponds to the value of G that is correct for effective mixing.

6.4.3 Volume and Detention Times of Hydraulic-Jump Mixers

Referring to the bottom of Figure 6.9, let L be the length of the hydraulic jump. Then the volume V_{jump} of the hydraulic jump is simply the volume of the trapezoidal prism of volume. Thus,

$$V_{jump} = \frac{1}{2}(y_1 + y_2)LW$$

(6.30)

The detention time t_o is then

$$t_o = \frac{V_{jump}}{Q} = \frac{\frac{1}{2}(y_1 + y_2)LW}{Q} = \frac{1}{2Q}(y_1 + y_2)LW$$

(6.31)

The length L of the hydraulic jump is measured from the front face of the jump to a point on the surface of the flow immediately after the roller as shown in Figure 6.9.

Experiments have shown that $L = 6y_2$ for $4 < Fr_1 < 20$. For Froude numbers outside this range, L is somewhat less than $6y_2$. For practical purposes, L may be taken as equal to $6y_2$. Equations (6.30) and (6.31) now become, respectively,

$$V_{jump} = \frac{1}{2}(y_1 + y_2)LW = 3y_2W(y_1 + y_2) \tag{6.32}$$

$$t_o = \frac{V_{jump}}{Q} = \frac{1}{2Q}(y_1 + y_2)LW = \frac{3y_2W}{Q}(y_1 + y_2) \tag{6.33}$$

Example 6.3 A hydraulic-jump mixer similar to the one shown in Figure 6.9 is used to mix alum in a water treatment plant. The height of the bottom of the sluice gate to the bottom of the channel is 5 cm. The rate of flow into the mixer is 0.048 m^3/s and the width of the channel is 10 cm. Calculate the mixing power developed in the jump.

Solution:

$$P_{hyd\,Jump} = \frac{Q\gamma(y_2 - y_1)\left[\left(\frac{Q}{W}\right)^2(y_2 + y_1) - 2gy_1^2y_2^2\right]}{2gy_1^2y_2^2} \qquad y_2 = \frac{y_1}{2}\left(\sqrt{1 + 8\frac{v_1^2}{gy_1}} - 1\right)$$

$$v_1 = \frac{0.048}{0.05(0.1)} = 9.6 \text{ m/s}$$

Therefore,

$$y_2 = \frac{0.05}{2}\left(\sqrt{1 + 8\left\{\frac{9.6^2}{9.81(0.05)}\right\}} - 1\right) = 0.94 \text{ m}$$

Assume temperature of water $= 25°C$; then $\rho_w = 997 \text{ kg/m}^3$; therefore,

$$P_{hyd\,Jump} = \frac{0.048(997)(9.81)(0.94 - 0.05)\left[\left(\frac{0.048}{0.10}\right)^2(0.94 + 0.05) - 2(9.81)(0.05)^2(0.94)^2\right]}{2(9.81)(0.05)^2(0.94)^2}$$

$$= 1781.14 = 2.39\text{hp} \quad \textbf{Ans}$$

Example 6.4 For the problem in Example 6.3, determine if the power dissipated conforms to the requirement of effective mixing.

Solution: To determine if the power dissipated conforms to the criterion of effective mixing, the G value would be calculated and compared with the values of Table 6.2.

$$\bar{G} = \sqrt{\frac{P}{\mu V}} \qquad \mu = 8.8(10^{-4})\text{kg/m·s}$$

$$V_{jump} = 3y_2W(y_1 + y_2) = 3(0.94)(0.1)(0.05 + 0.94) = 0.279 \text{ m}^3$$

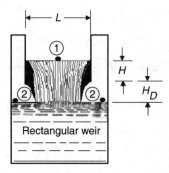

FIGURE 6.10 Power dissipation in weir mixers.

$$\bar{G} = \sqrt{\frac{1781.14}{8.8(10^{-4})(0.279)}} = 2693.43 \text{ per sec}$$

$$t_o = \frac{3y_2 W}{Q}(y_1 + y_2) = \frac{3(0.94)(0.1)}{0.048}(0.05 + 0.94) = 5.81 \text{ sec}$$

From Table 6.2, at $t_o = 5.81$ sec, $\bar{G}$ should range from 1,500 per sec to 4,000 per sec. 2693.43 per sec is within this range and the design conforms to the criteria for effective mixing. **Ans**

6.4.4 MIXING POWER FOR WEIR MIXERS

The power dissipation in weir mixers is a result of the conversion of the energy that the water possesses as it drops from the top of the weir to the bottom of the weir. To obtain the dissipation, apply the energy equation between points 1 and 2 in Figure 6.10. This will result in

$$h_f = H + H_D \tag{6.34}$$

H is the head over the weir crest and H_D is the drop provided from the weir crest to the surface of the water below. At the points directly below the falling water there is turbulence. As the particles of water reach point 2, however, turbulence ceases and the velocity becomes zero. In other words, the energy at the point of turbulence has been dissipated before reaching point 2. This dissipation is the power dissipation of mixing. Having obtained the friction loss, the power dissipation P is simply

$$P = Q\gamma h_f = Q\gamma(H + H_D) \tag{6.35}$$

Example 6.5 A suppressed rectangular is used to mix chemical in a wastewater treatment unit. The flow is 0.3 m³/s. The length of the weir L and the height P were measured and found to be 2 m and 1 m, respectively. Calculate the power dissipation in the mixer if $H_D = 2.5$ m.

Solution:

$$\mathcal{P} = Q\gamma(H + H_D)$$

$$Q = K\sqrt{2g}L\sqrt{H^3} \qquad K = 0.40 + 0.05\frac{H}{P}$$

$$Q = \left(0.40 + 0.05\frac{H}{P}\right)\sqrt{2g}L\sqrt{H^3}$$

$$0.3 = \left(0.40 + 0.05\left\{\frac{H}{1.0}\right\}\right)\sqrt{2(9.81)}(2)\sqrt{H^3}$$

By trial and error, $H = 0.19$ m
Assume temperature $= 25°C$; hence, $\rho = 997$ kg/m^3
   Therefore,

$$\mathcal{P} = Q\gamma(H + H_D) = 0.3(997)(9.81)(0.19 + 2.5) = 7892.92 \text{ N·m/sec}$$

$$= \frac{7892.92}{746} = 10.58 \text{ hp}$$

Example 6.6   For the problem in Example 6.5, determine if the power dissipated conforms to the requirement of effective mixing. Assume that mixing occurred in a volume of a rectangular parallelepiped of length equal to the length of the weir, width equal to 0.2 of the length of the weir, and depth equal to 0.5 of H_D.

Solution: To determine if the power dissipated conforms to the criterion of effective mixing, the G value would be calculated and compared with the values of Table 6.2.

$$\bar{G} = \sqrt{\frac{\mathcal{P}}{\mu\Psi}} \qquad \mu = 8.8(10^{-4})\text{kg/m·s (assuming temperature} = 25°C)$$

$$\Psi = 2\{0.2(2)\}\{0.5(2.5)\} = 1.0\,\text{m}^3$$

$$\bar{G} = \sqrt{\frac{7892.92}{8.8(10^{-4})(1)}} = 2994.87 \text{ per sec}$$

$$t_o = \frac{1.0}{0.3} = 3.33 \text{ sec}$$

From Table 6.2, at $t_o = 3.33$ sec, $\bar{G}$ should range from 1,500 per sec to 4,000 per sec. 2,994.87 per sec is within this range and the design conforms to the criteria for effective mixing.   **Ans**

6.5 FLOCCULATORS

Unlike mixing, agitation in flocculators involves gentle motion of the fluid to induce agglomeration of the smaller particles into larger flocs. This is especially used in coagulant-treated water and wastewater such as the one using alum and ferric chloride. As flocculation proceeds, smaller flocs build into larger sizes until a point is reached

TABLE 6.3
Criteria Values for Effective Flocculation

Type of Raw Water	$\bar{G}$, s^{-1}	$\bar{G}\,t_o$ (dimensionless)
Low turbidity and colored	20–70	50,000–250,000
High turbidity	70–150	80,000–190,000

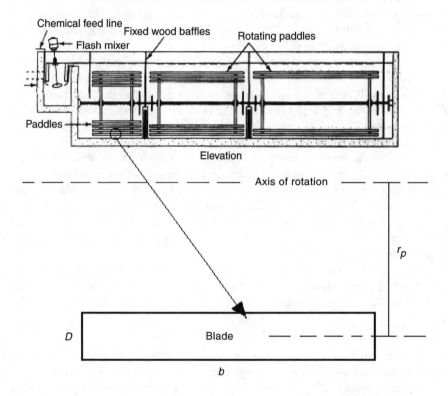

FIGURE 6.11 Longitudinal section of a flocculator (top) and a blade (bottom).

where the size cannot go on increasing. This limiting size is the critical size and depends upon the detention time and velocity gradient. Generally, the larger velocity gradients produce smaller critical sizes and the larger detention times produce larger critical sizes. Practice has it that this "mix" of detention time and velocity gradient can be combined into a product of $\bar{G}$ and t_o. Thus, the criteria values for effective flocculation are expressed in terms of two parameters: $\bar{G}t_o$ and $\bar{G}$. Table 6.3 shows some criteria values in terms of these parameters.

Figure 6.11 shows a longitudinal section of a flocculator. This flocculator belongs to the category of rotational mixers, only that it should be rotating at a very much slower speed. Notice that the compartments vary in size from the smallest one at the head end to the largest one at the exit end. This is an inverse relationship to the variation of the value of $\bar{G}$; $\bar{G}$ decreases in value, instead. As the water or wastewater is introduced into the flocculation tank, the particles are smaller at the head end and then builds up until reaching the maximum size at the exit end of the tank. At the exit, the rotation of the paddle must be made much, much slower to avoid breaking up the flocs into pieces.

The paddle configuration in each compartment is called a *paddle wheel*. A paddle wheel is composed of *paddle arms*. In the figure, it is difficult to ascertain how many paddle arms there are in each paddle wheel, but what is obvious is that at least there are two arms per wheel. In each arm are then mounted the *blades* or *slats*. Assuming there are two paddle arms per paddle wheel, the total number of blades or slats per compartment is eight in the first compartment, six in the second compartment, and four in the third compartment.

The design of the flocculator of Figure 6.11 may be made by determining the power coefficients for laminar, transitional, and turbulent regime of flow field. We will, however, discuss its design in terms of the fundamental definition of power. Consider F_D as the drag by the water on the blade; F_D is also the push of the blade upon the water. This push causes the water to move at a velocity v_p equal to the velocity of the blade.

The water is actually "touching" the blade, so the velocity that it attains on contact must be equal to that of the blade. Of course, as it departs, its velocity will be different, but this is not the critical point of power transfer. The blade transfers power to the water while still in contact. Upon detachment, the water parcel that got the power being transferred will then commence expending the power to overcome fluid friction imposed upon it by neighboring parcels; this process produces the velocity gradient required for flocculation to occur.

The paddle is rotational, therefore, v_p is a tangential velocity referred to the axis of rotation at a radial distance r_p as shown in the figure. Tangential velocity is equal to the radius times angular rotation ω in radians per unit time. Thus,

$$v_p = r_p \omega \tag{6.36}$$

The drag stress in a fluid is proportional to the dynamic pressure $\rho_l v_p^2/2$, where ρ_l is the mass density of water. Thus, the force on a single blade, $F_D = C_D A_p \rho_l v_p^2/2$. C_D is the coefficient of drag and A_p is the projected area of the blade in the direction of its motion. Power is force times velocity, so the power dissipation per blade is therefore

$$\mathcal{P}_{blade} = C_D A_p \frac{\rho_l v_p^2}{2} v_p = C_D A_p \frac{\rho_l v_p^3}{2} \tag{6.37}$$

The total power in the flocculator compartment is then the sum of the powers in each blade, thus,

$$\mathcal{P} = \sum \mathcal{P}_{blade} = \sum C_D A_p \frac{\rho_l v_p^3}{2} = C_D A_{pt} \frac{\rho_l (av_{pt})^3}{2} \tag{6.38}$$

Consider this equation as the *flocculation equation*. A_{pt} is the sum of the projected areas of the blades. v_p is the tangential velocity corresponding to r_p; however, the location of the blades requires that there be several v_p's. To use only one v_p, the paddle tip velocity v_{pt} is used multiplied by a factor a, as indicated in the equation. The product of v_{pt} and a is the equivalent of the combined effects of the v_p's of the several blades; it represents the conglomerate velocity for all the blades. The value of a is normally taken as equal to 0.75. With the expression for $\mathcal{P}$ found, the values of $\overline{G}$ and $\overline{G}t_o$ may be checked to see if the flocculator performs at conditions of effective flocculation.

As mentioned before, higher values of $\overline{G}$ produce smaller flocs, while low values of $\overline{G}$ produce larger flocs. Although larger flocs are desirable, there is a time limit as measured by $\overline{G}t_o$ to which they are allowed to form. If they are allowed to form much longer than some critical time, they will reach a critical size that, due to shearing forces, will simply break and crush to pieces.

Also, excessive velocity gradients can simply break the flocs to pieces. To prevent excessive velocity gradients between paddle tips, a minimum distance of 0.3 m should be provided between them. Also, a minimum clearance of 0.3 m should be provided between paddles and any structure inside the flocculator. Paddle tip velocity should be less than 1.0 m/s (Peavy et al., 1985).

Important parameters to be determined in design of flocculators are the dimensions of the blade. Thus, the flocculation equation, Equation (6.38), may be solved for A_{pt}; once this is known, the dimensions of the blade may be determined. A_{pt} is, of course, the sum of all the projected blade areas. The dimensions of each blade may be arbitrarily chosen based on this A_{pt}. As shown in the figure, blades are normally rectangular in shape with length b and width D. Solving for A_{pt},

$$A_{pt} = \frac{2\mathcal{P}}{C_D \rho_l (a v_{pt})^3} \tag{6.39}$$

The value of the coefficient of drag is a function of the Reynolds number $Re = Dv_p/v$; v is the kinematic viscosity. Assuming a blade of $D = 0.25$ m, the corresponding Re at $v_p = 1.0$ m/s is $0.25(1)/10^{-6} = 2.5 \times 10^5$ at 20°C. The kinematic viscosity of water at 20°C is 10^{-6} m/s (Peavy et al., 1985). At an Re equals 10^5, the formula for C_D has been determined empirically as (Munson et al., 1994).

$$C_D = 0.008\frac{b}{D} + 1.3 \tag{6.40}$$

The C_D predicted by this equation applies only to a single blade. Work needs to be done to determine the value of C_D for multiple blades.

Determination of electrical power input. To determine the electrical power input to the flocculator tank, $\mathcal{P}$ is divided by η, the brake efficiency, to obtain the *brake* or *shaft power*. The brake or shaft power is then divided by the $\mathcal{M}$, the motor efficiency, to obtain the input electrical power. The electrical power input, of course, determines how much money is spent to operate the flocculator.

Example 6.7 A flocculator tank has three compartments, with each compartment having one paddle wheel. The ratio of the length of the paddle blades of the longest compartment to that of the length of the paddle blades of the middle compartment is 2.6:2, and the ratio of the length of the paddle blades of the middle compartment to that of the length of the paddle blades of the shortest compartment is 2:1. The flocculator is to flocculate an alum treated raw water of 50,000 m^3/d at an average temperature of 20°C. Design for (a) the dimensions of the flocculator and flocculator compartments, (b) the power requirements assuming motor efficiency M of 90% and brake efficiency η of 75%, (c) the appropriate dimensions of the paddle slats, and (d) the rpm of the paddle wheels. Assume two flocculators in parallel and four paddle blades per paddle wheel, attached as two per arm.

Solution: (a) Assume an average G of $30s^-$ and a Gt_o of 80,000. Therefore, $t_o = 44.44$ min; vol. of flocculator $= (50,000/2)(44.44)\left[\frac{1}{60(24)}\right] = 771.53$ m^3. To produce uniform velocity gradient, depth must be equal to width. Thus, assuming a depth of 5 m and letting L represent the length of tank,

$$5(5)L = 771.53; \quad\quad L = 30.86 \text{ m} \quad \textbf{Ans}$$

$$\text{length of compartment no.1, the shortest} = \frac{1}{1+2+2.6}(30.86)$$

$$= \frac{1}{5.6}(30.86) = 5.51 \text{ m} \quad \textbf{Ans}$$

Width and depth of compartment no.1 = width and depth of compartment no. 2 = width and depth of compartment no. 3 = 5 m, as assumed **Ans**
Length of compartment no. 2, the middle = 2/5.6(30.86) = 11.02 m **Ans**
Length of compartment no. 3, the longest = 30.86 − 5.51 − 11.02 = 14.33 m **Ans**
Width of flocculator = depth of flocculator = 5 m, as assumed **Ans**

(b) $$\bar{G} = \sqrt{\frac{P}{\mu \Psi}} \quad\quad P = \mu \Psi\, G^2$$

Assume the following distribution of G:

Compartment no. 1, $G = 40$ s^-
Compartment no. 2, $G = 30$ s^-
Compartment no. 3, $G = 20$ s^-

$$\mu = 10(10^{-4})\,\text{kg/m}\cdot\text{s}; \quad\quad \Psi_1 = (5)(5)(5.51) = 137.75 \text{ m}^3$$

$$\text{Therefore, } P_i \text{ for compartment no. 1} = \frac{10(10^{-4})(137.75)(40^2)}{0.75(0.9)}$$

$$= 326.52 \text{ N·m/s} = 0.44 \text{ hp} \quad \textbf{Ans}$$

$$\Psi_2 = 25(11.02) = 275.50 \text{ m}^3$$

$$\text{Therefore, } P_i \text{ for compartment no. 2} = \frac{10(10^{-4})(275.50)(30^2)}{0.75(0.90)} = 0.44 \text{ hp} \quad \textbf{Ans}$$

$$\Psi_3 = 25(14.33) = 358.25 \text{ m}^3$$

Therefore, $\mathcal{P}_i$ for compartment no. 3 $= \dfrac{10(10^{-4})(358.25)(20^2)}{0.75(0.90)} = 0.28$ hp **Ans**

(c) $\mathcal{P} = C_D A_{pt} \dfrac{\rho_l(av_{pt})^3}{2}$ $\quad C_D = 0.008\dfrac{b}{D} + 1.3$

$= \left(0.008\dfrac{b}{D} + 1.3\right)(nbD)\dfrac{(997)(0.75)^3 v_{pt}^3}{2} = \left(0.008\dfrac{b}{D} + 1.3\right)(nbD)(210.3)v_{pt}^3$

n = number of paddle blades per paddle wheel

$$Re = 10^5 = \dfrac{Dv_p}{v} = \dfrac{Dav_{pt}}{v}; \qquad v_{pt} = \dfrac{10^5(v)}{0.75D}$$

Therefore,

$$\mathcal{P} = \left(0.008\dfrac{b}{D} + 1.3\right)(nbD)(210.3)\left\{\dfrac{10^5(v)}{0.75D}\right\}^3$$

For compartment no. 1:

$$b = 5.51 - 2(0.3) = 4.91 \text{ m}$$
$$v = \mu/\rho = 10(10^{-4})/998 = 10^{-6}$$

$$326.52(0.75)(0.9) = 220.05 = \left(0.008\dfrac{4.91}{D} + 1.3\right)\{4(4.91)D\}(210.3)\left\{\dfrac{10^5(10^{-6})}{0.75D}\right\}^3$$

$$= \dfrac{0.16}{D^3} + \dfrac{5.36}{D^2}$$

Let $0.16/D^3 + 5.36/D^2 = Y$ and solve by trial and error.

D	Y
0.22	125.77
0.15	285.63

$$\dfrac{y - 0.22}{0.15 - 0.22} = \dfrac{220.05 - 25.77}{285.63 - 125.77}$$

0.22	125.77	$\dfrac{y - 0.22}{0.15 - 0.22} = \dfrac{220.05 - 25.77}{285.63 - 125.77}$
y	220.05	$y = 0.18 = D$ m **Ans**
0.15	285.63	

For compartment no. 2:

$$b = 11.02 - 0.6 = 10.42 \text{ m}$$

$$327(0.75)(0.9) = 220.73 = \left(0.008\frac{10.42}{D} + 1.3\right)\{4(10.42)D)(210.3)\left\{\frac{10^5(10^{-6})}{0.75D}\right\}^3$$

$$= \frac{0.73}{D^3} + \frac{11.38}{D^2}$$

Again, by trial and error, $D = 0.25$ m   **Ans**
For compartment no. 3:
   By similar procedure, $D = 0.36$ m
   (d) For compartment no. 1:

$$v_{pt} = r\omega \ 10^5 = \frac{Dv_p}{v} = \frac{Dav_{pt}}{v} \qquad v_{pt} = 10^5(v)/D = \frac{10^{-1}}{D} = \frac{10^{-1}}{0.18} = 0.56 \text{ m/s}$$

$$r = \frac{5 - 0.6}{2} = 2.2 \text{ m}$$

$$\omega = \frac{0.56}{2.2} = 0.25 \text{ rad/s} = 15.17 \text{ rad/min} = 2.43 \text{ rpm} \textbf{Ans}$$

For compartment no. 2:

$$v_{pt} = \frac{10^{-1}}{0.25} = 0.40 \text{ m/s}$$

$$\omega = 1.74 \text{ rps}   \textbf{Ans}$$

For compartment no. 3:

$$v_{pt} = \frac{10^{-1}}{0.36} = 0.28 \text{ m/s}$$

$$\omega = 1.22 \text{ rpm}   \textbf{Ans}$$

GLOSSARY

Blade—The impeller element in a paddle wheel.
Brake or shaft power—The power given to the shaft by a prime mover.
Flocculation—A unit operation aimed at enlarging small particles through a very
   slow agitation of the water suspending the particles.
Fluid filament—A volume of fluid composed of several parcels.
Fluid parcel—A volume of fluid composed of several molecules.

Gradient of velocity—The rate of change of velocity with respect to displacement.

Hydraulic mixers—Mixers that utilize, for the mixing process, the agitation that results in the flowing of the water.

Mixing—A unit operation that distributes the components of two or more materials among the materials producing in the end a single blend of the components.

Paddles—Impellers in which the lengths are equal to 50 to 80% of the inside diameter of the vessel in which the mixing is taking place.

Paddle arm—The element extending from the axis of rotation in a paddle wheel.

Paddle wheel—The paddle configuration in a flocculator compartment.

Pitch—Assuming no slippage, the ratio of the distance traveled by the water to the diameter of the impeller.

Pneumatic mixers—Mixers that uses gas or air bubbles to induce the agitation.

Power dissipation—In fluid motion, power lost due to frictional resistance and is equal to the power given by the agitator.

Propellers—Impellers in which the direction of the driven fluid is along the axis of rotation.

Reference dimension—The smallest number of groups of the groups formed from all the possible groupings of the basic dimensions of the physical variables in a given problem.

Rotational mixers—Mixers that use a rotating element to effect the agitation.

Shape factors—The normalized distance measurements in a body to one arbitrarily chosen reference distance in the body.

Slat—Same as paddle blade.

Square pitch—A pitch in which the axial distance traversed is equal to the diameter of the propeller.

Turbines—Impellers shorter than paddles and are only about 30 to 50% of the inside diameter of the vessel in which the mixing is taking place.

SYMBOLS

a	A constant to convert paddle tip velocity to a conglomerate velocity representing all blades in paddle
A_p	Projected area of paddle blade
A_{pt}	Sum of projected areas of paddle blades
b	Length of paddle blade
C_D	Coefficient of drag
D	Width of paddle blade
D_a	The diameter of the impeller
D_t	Diameter of vessel
E	Distance of impeller to the bottom of the vessel
F_D	Drag force on bubble
F_r	Froude number
g	Acceleration due to gravity
G_x	Velocity gradient in the x direction

G_y	Velocity gradient in the y direction
G_z	Velocity gradient in the z direction
$\bar{G}$	Average G in tank
G_1	Peebles number
h	Depth of submergence of air diffuser
h_f	Friction loss
H	Depth of fluid in the vessel; head over rectangular weir
H_D	Drop allowed below crest of rectangular weir to water surface below
J	Width of the baffle
K_L	Power coefficient for mixing in laminar flow regime
K_T	Power coefficient for mixing in turbulent flow regime
L	Length of the paddle; length of hydraulic jump; length of rectangular weir
$\mathcal{M}$	Motor efficiency
n	Number of bubbles
N	Rotational speed
$\mathcal{P}$	Power dissipated
P_a	Atmospheric pressure
$\mathcal{P}_b$	Brake power
P_i	Input pressure to unit
$\mathcal{P}_i$	Input power
$\mathcal{P}_o$	Power number
$\mathcal{P}_x$	Power dissipation in the x direction in a parcel of fluid
q	Discharge per unit width of channel
Q	Discharge flow
Q_i	Input flow to unit
$\bar{r}$	Average radius of bubbles
r_p	Radius of paddle wheel
Re	Reynolds number
t_o	Detention time
u	Velocity in the x direction
$\bar{v}_b$	Average rise velocity of bubbles
$\bar{v}_l$	Average rise velocity of water
v_p	Tangential velocity of paddle
v_{pt}	Paddle tip velocity
$\mathcal{V}$	Volume of tank
V_b	Volume of bubble
$\bar{V}_b$	Average volume of bubbles in tank
$\bar{V}_{bo}$	Average volume of bubble at surface
$\mathcal{V}_{jump}$	Volume of hydraulic jump
W	Width of the paddle; width of channel
y	The y coordinate
y_1	Depth at upstream end of hydraulic jump
y_2	Depth at downstream end of hydraulic jump
η	Brake efficiency
γ	specific weight
γ_g	Specific weight of air

γ_l	Specific weight of water
μ	Absolute viscosity
ρ	Mass density
ρ_g	Mass density of air
ρ_l	Mass density of water
σ	Surface tension of water
Π	One of the π variables in dimensional analysis
Ψ_z	Fluid torque in the z direction
ω	Angular velocity of paddle wheel
ω_z	Angular velocity in the z direction

PROBLEMS

6.1   A turbine with six curved blades is installed centrally in a baffled vessel. The vessel is 2.0 m in diameter. The turbine 61 cm in diameter is positioned 60 cm from the bottom of the vessel. The tank is filled to a depth of 2.0 m and is mixing alum with raw water in a water treatment plant. The water is at a temperature of 25°C and the turbine is running at 100 rpm. What horsepower will be required to operate the mixer.

6.2   Repeat Problem 6.1 using a shrouded turbine with six curved blades.

6.3   Repeat Problem 6.1 using a shrouded turbine with two curved blades.

6.4   Repeat Problem 6.1 using a propeller with three blades and a square pitch.

6.5   Repeat Problem 6.1 using a paddle with two blades.

6.6   Repeat Problem 6.1 using a paddle with six blades.

6.7   A shrouded turbine with six curved blades is installed centrally in a baffled vessel. The vessel is 2.0 m in diameter. The turbine 61 cm in diameter is positioned 60 cm from the bottom of the vessel. The tank is filled to a depth of 2.0 m and is mixing alum with raw water in a water treatment plant. The water is at a temperature of 25°C and the turbine is running at 3.24 hp. Calculate the rpm of the motor driving the agitator.

6.8   A propeller with three blades and a square pitch is installed centrally in a baffled vessel. The vessel is 2.0 m in diameter. The propeller is positioned 60 cm from the bottom of the vessel. The tank is filled to a depth of 2.0 m and is mixing alum with raw water in a water treatment plant. The water is at a temperature of 25°C and the turbine is running at 3.24 hp at an rpm of 100. What is the diameter of the impeller?

6.9   A propeller with three blades and a square pitch is installed centrally in a baffled vessel. The vessel is 2.0 m in diameter. The propeller is positioned 60 cm from the bottom of the vessel. The tank is filled to a depth of 2.0 m and is mixing alum with raw water in a water treatment plant. The turbine is running at 3.24 hp at an rpm of 100. The propeller is 61 cm in diameter. At what temperature is the mixing being done?

6.10   For the vessel in Problem 6.1, at what detention time should it be operated for it to function effectively? At what rate of inflow should it be operated?

6.11   For the vessel in Problem 6.2, at what detention time should it be operated for it to function effectively? At what rate of inflow should it be operated?

6.12   For the vessel in Problem 6.3, at what detention time should it be operated
        for it to function effectively? At what rate of inflow should it be operated?

6.13   For the vessel in Problem 6.4, at what detention time should it be operated
        for it to function effectively? At what rate of inflow should it be operated?

6.14   For the vessel in Problem 6.5, at what detention time should it be operated
        for it to function effectively? At what rate of inflow should it be operated?

6.15   A rapid-mix tank, 2 m in diameter is filled to a depth of 2 m with water
        and alum introduced continuously in the influent. The turbine is running
        at 3.24 hp and the average velocity gradient in the tank is 950 s^{-1}. At what
        temperature is the alum being mixed with water? Determine the influent
        rate of flow to the mixer corresponding to an effective mixing.

6.16   A rapid-mix tank, 2 m in diameter is filled to a depth of 2 m with water
        and alum introduced continuously in the influent. The average velocity
        gradient in the tank is 950 s^{-1}. The temperature of mixing is 25°C.
        Calculate the power dissipation of the mixer. Determine the influent rate
        of flow to the mixer corresponding to an effective mixing.

6.17   By considering the criterion for effective mixing, the volume of a rapid-
        mix tank used to rapidly mix an alum coagulant in a water treatment plant
        was found to be 6.28 m^3. Assume air is being provided at a rate of 1.12 m^3
        per m^3 of water treated and that the detention time of the tank is 2.2 min.
        The pressure P_i at which air is forced into the diffuser is 256,920.02 N/m^2.
        Assume barometric pressure as 101,300 N/m^2, the depth from the surface
        to the diffuser as 2 m, and the temperature of water as 25°C. Calculate
        the power dissipation in the mixer.

6.18   The volume of a rapid-mix tank used to rapidly mix an alum coagulant in
        a water treatment plant is 6.28 m^3. The detention time of the tank is 2.2 min.
        The pressure P_i at which air is forced into the diffuser is 256,920.02 N/m^2.
        This developed a power dissipation for mixing of 3.24 hp. Assume baro-
        metric pressure as 101,300 N/m^2, the depth from the surface to the diffuser
        as 2 m, and the temperature of water as 25°C. Calculate the rate at which
        air is introduced into the mixer.

6.19   The volume of a rapid-mix tank used to rapidly mix an alum coagulant in a
        water treatment plant was found to be 6.28 m^3. Assume air is being provided
        at a rate of 1.12 m^3 per m^3 of water treated and that the detention time of
        the tank is 2.2 min. The power dissipation of the mixer is 3.24 hp. The
        pressure P_i at which air is forced into the diffuser is 256,920.02 N/m^2. The
        depth from the surface to the diffuser as 2 m and the temperature of water
        as 25°C. What is the barometric pressure at which the mixer is operated?

6.20   The volume of a rapid-mix tank used to rapidly mix an alum coagulant
        in a water treatment plant was found to be 6.28 m^3. Assume air is being
        provided at a rate of 1.12 m^3 per m^3 of water treated and that the detention
        time of the tank is 2.2 min. The power dissipation of the mixer is 3.24 hp.
        The pressure P_i at which air is forced into the diffuser is 256,920.02 N/m^2.
        The temperature of the water is 25°C and the barometric pressure is
        101,300 N/m^2. At what depth is the diffuser being placed at the bottom
        of the tank?

6.21   The volume of a rapid-mix tank used to rapidly mix an alum coagulant in a water treatment plant was found to be 6.28 m^3. Assume air is being provided at a rate of 1.12 m^3 per m^3 of water treated and that the detention time of the tank is 2.2 min. The power dissipation of the mixer is 3.24 hp. The pressure P_i at which air is forced into the diffuser is 256,920.02 N/m^2. The depth at which the diffuser is placed at the bottom of the tank is 2 m and the barometric pressure is 101,300 N/m^2. What is the temperature of the water?

6.22   A hydraulic-jump mixer similar to the one shown in Figure 6.9 is used to mix alum in a water treatment plant. The height of the bottom of the sluice gate to the bottom of the channel is 5 cm. The mixing power developed is 1657.2 N·m/s and the width of the channel is 10 cm. Calculate the rate of inflow into the jump. The temperature of the water is 25°C and the downstream depth of the jump is 0.94 m.

6.23   A hydraulic-jump mixer similar to the one shown in Figure 6.9 is used to mix alum in a water treatment plant. The height of the bottom of the sluice gate to the bottom of the channel is 5 cm. The mixing power developed is 1657.2 N·m/s and the width of the channel is 10 cm. The rate of inflow into the jump is 0.048 m^3/s. At what temperature does the mixing occur?

6.24   A hydraulic-jump mixer similar to the one shown in Figure 6.9 is used to mix alum in a water treatment plant. The mixing power developed is 1657.2 N·m/s and the width of the channel is 10 cm. The rate of inflow into the jump is 0.048 m^3/s. The downstream depth of the jump is 0.94 m and the temperature of mixing is 25°C. Calculate the upstream depth of the jump.

6.25   A hydraulic-jump mixer similar to the one shown in Figure 6.9 is used to mix alum in a water treatment plant. The mixing power developed is 1657.2 N·m/s and the width of the channel is 10 cm. The rate of inflow into the jump is 0.048 m^3/s. The upstream depth of the jump is 5 cm and the temperature of mixing is 25°C. Calculate the downstream depth of the jump.

6.26   A hydraulic-jump mixer similar to the one shown in Figure 6.9 is used to mix alum in a water treatment plant. The mixing power developed is 1657.2 N·m/s and the downstream depth of the jump is 0.94 m. The rate of inflow into the jump is 0.048 m^3/s. The upstream depth of the jump is 5 cm and the temperature of mixing is 25°C. What is the channel width?

6.27   At a motor efficiency of 90% and a brake efficiency of 75%, the power input into a mixer was found to be 16,459 N·m/s. The average velocity gradient in the mixer tank is 1500s^{-1} and the temperature of mixing is 25°C. Find the volume of the mixer.

6.28   A suppressed rectangular is used to mix chemical in a wastewater treatment unit. The power dissipation developed is 7423.5 N·m/s. The length of the weir L and the height P were measured and found to be 2 m and 1 m, respectively. The temperature is 25°C. Determine the rate of flow through the weir, if $H_D = 2.5$ m.

6.29   A suppressed rectangular is used to mix chemical in a wastewater treatment unit. The power dissipation developed is 7423.5 N·m/s. The length of the weir L and the height P were measured and found to be 2 m and 1 m, respectively. The rate of flow through the weir is 0.3 m^3/s and $H_D =$ 2.5 m. What is the temperature of the water?

6.30   A suppressed rectangular is used to mix chemical in a wastewater treatment unit. The power dissipation developed is 7423.5 N·m/s. The temperature is 25°C. The rate of flow through the weir is 0.3 m^3/s and $H_D =$ 2.5 m. Calculate the head over the weir.

6.31   A suppressed rectangular is used to mix chemical in a wastewater treatment unit. The power dissipation developed is 7423.5 N·m/s. The length of the weir L and the height P were measured and found to be 2 m and 1 m, respectively. The rate of flow through the weir is 0.3 m^3/s and $H_D =$ 2.5 m. The temperature is 25°C. Calculate the drop allowed H_D.

6.32   For the problem in Example 6.5, determine if the power dissipated conforms to the requirement of effective mixing. Assume that mixing occurred in a volume of a rectangular parallelepiped of length, width, and depth equal to the length of the weir.

6.33   For the mixing in Problem 6.32, check if the process conforms to the criteria of effective mixing.

6.34   Redesign the mixer in Problem 6.33 if the operation does not conform to the criteria of effective mixing.

6.35   The power input into a compartment of a flocculator is 0.33 kW. The motor efficiency is 0.90% and the brake efficiency is 75%. The coefficient of drag is 1.54 and the paddle tip velocity is 0.55 m/s. The temperature in the flocculation tank is 20°C. Calculate the total combined projected area of the paddles. If the dimension of a blade is 0.18 m by 5.4 m, how many paddle blades are there in the compartment?

6.36   The power input into a compartment of a flocculator is 0.37 kW. The motor efficiency is 0.90% and the brake efficiency is 75%. The paddle tip velocity is 0.41 m/s. The temperature in the flocculation tank is 20°C. The total combined projected area of the paddles is 10.1 m^2. Determine the coefficient of drag.

6.37   The power input into a compartment of a flocculator is 0.21 kW. The motor efficiency is 0.90% and the brake efficiency is 75%. The paddle tip velocity is 0.41 m/s. The total combined projected area of the paddles is 19.1 m^2 and the coefficient of drag is 1.6. What is the temperature at which the flocculator is operated?

6.38   The power input into a compartment of a flocculator is 0.37 kW. The motor efficiency is 0.90% and the brake efficiency is 75%. The paddle tip velocity is 0.41 m/s. The temperature in the flocculation tank is 20°C. The total combined projected area of the paddles is 10.1 m^2 and the coefficient of drag is 1.64. What is the value of the constant a?

6.39   The power input into a compartment of a flocculator is 0.37 kW. The motor efficiency is 0.90% and the brake efficiency is 75%. The temperature in the flocculation tank is 20°C. The total combined projected area of the paddles is 10.1 m^2 and the coefficient of drag is 1.64. What is the paddle tip velocity?

6.40 Repeat Example 6.7 assuming a C_D value of 1.8.

BIBLIOGRAPHY

Alfano, J. C., P. W. Carter, and A. Gerli (1998). Characterization of the flocculation dynamics in a papermaking system by non-imaging reflectance scanning laser microscopy (SLM). *Nordic Pulp Paper Res. J.* 13, 2, 159–165.

Amirtharajah, A., M. M. Clark, and R. R. Trussell (1991). *Mixing in Coagulation and Flocculation.* American Water Works Assoc., Denver, 2143–2157.

Chen, L. A., G. A. Serad, and R. G. Carbonell (1998). Effect of mixing conditions on flocculation kinetics of wastewaters containing proteins and other biological compounds using fibrous materials and polyelectrolytes. *Brazilian J. Chemical Eng.* 15, 4, 358–368.

Droste, R. L. (1997). *Theory and Practice of Water and Wastwater Treatment.* John Wiley, & Sons, New York.

Gao, S., et al. (1998). Experimental investigation of bentonite flocculation in a batch oscillatory baffled column. *Separation Science Technol.* 33, 14, 2143–2157.

Hodgson, A. T., et al. (1998). Effect of tertiary coagulation and flocculation treatment on effluent quality from a bleached kraft mill. *TAPPI J.* 81, 2, 166–172.

Liu, J., et al. (1997). Study on As removal from basic wastewater using combined aeration and flocculation method. *Zhongguo Huanjing Kexue/China Environ. Science.* 17, 2, 184–187.

Luan, Z., et al. (1997). Neutralization and flocculation of acidic is mine drainage with alkaline wastewater. *Zhongguo Huanjing Kexue/China Environ. Science.* 17, 1, 87–92.

McCabe, W. L. and J. C. Smith (1967). *Unit Operations of Chemical Engineering.* McGraw-Hill, New York, 262.

Metcalf & Eddy, Inc. (1991). *Wastewater Engineering: Treatment, Disposal, and Reuse.* McGraw-Hill, New York, 37.

Moudgil, B. M. and B. J. Scheiner (1990). Flocculation and Dewatering. *Proc. Eng. Foundation Conf.,* Palm Coast, FL, January 10–15, 1988. Am. Inst. of Chemical Eng., New York.

Munson, B. R., D. F. Young, and T. H. Okiishi (1994). *Fundamentals of Fluid Mechanics.* John Wiley & Sons, New York.

Peavy, H. S., D. R. Rowe, and G. Tchobanoglous (1985). *Environmental Engineering.* McGraw-Hill, New York, 147–151.

Peebles, F. N. and H. J. Garber (1953). The motion of gas bubbles in liquids. *Chemical Eng. Prog.,* 49, 88–97.

Roberson, J. A., J. J. Cassidy, and M. H. Chaudry (1988). *Hydraulic Engineering.* Houghton Mifflin, Boston.

Sincero, A. P. and G. A. Sincero (1996). *Environmental Engineering: A Design Approach.* Prentice Hall, Upper Saddle River, NJ.

Wanner, J., et al. (1997). National survey of activated sludge separation problems in the Czech Republic: Filaments, floc characteristics and activated sludge metabolic properties. *Water Science Technol. Proc. 1997 2nd Int. Conf. Microorganisms in Activated Sludge and Biofilm Processes,* July 21–23, Berkeley, CA, 37, 4–5, 271–279. Elsevier Science Ltd., Exeter, England.

Watanabe, Y., Sh. Kasahara, and Y. Iwasaki (1997). Enhanced flocculation/sedimentation process by a jet mixed separator. *Water Science Technol. Proc. 1997 Workshop on IAWQ-IWSA Joint Group on Particle Separation,* July 1–2, Sapporo, Japan, 37, 10, 55–67. Elsevier Science Ltd., Exeter, England.

Wistrom, A. and J. Farrell (1998). Simulation and system identification of dynamic models for flocculation control. *Water Science Technol. Proc. 7th Int. Workshop on Instrumentation, Control and Automation of Water and Wastewater Treatment and Transport Syst.,* July 6–9, 1997, Brighton, England, 37, 12, 181–192. Elsevier Science Ltd., Exeter, England.

7 Conventional Filtration

Filtration is a unit operation of separating solids from fluids. Screening is defined as a unit operation that separates materials into different sizes. Filtration also separates materials into different "sizes," so it is a form of screening, but filtration strictly pertains to the separation of solids or particles and fluids such as in water. The microstrainer discussed in Chapter 5 is a filter. In addition to the microstrainer, other examples of this unit operation of filtration used in practice include the filtration of water to produce drinking water in municipal and industrial water treatment plants, filtration of secondary treated water to meet more stringent discharge requirements in wastewater treatment plants, and dewatering of sludges to reduce their volume.

To differentiate it from Chapter 8, this chapter discusses only conventional filtration. Chapter 8 uses membranes as the medium for filtration; thus, it is titled advanced filtration.

Mathematical treatments involving the application of linear momentum to filtration are discussed. Generally, these treatments center on two types of filters called granular and cake-forming filters. These filters are explained in this chapter.

7.1 TYPES OF FILTERS

Figures 7.1 to 7.8 show examples of the various types of filters used in practice. Filters may be classified as gravity, pressure, or vacuum filters. *Gravity filters* are filters that rely on the pull of gravity to create a pressure differential to force the water through the filter. On the other hand, *pressure* and *vacuum filters* are filters that rely on applying some mechanical means to create the pressure differential necessary to force the water through the filter.

The filtration medium may be made of perforated plates, septum of woven materials, or of granular materials such as sand. Thus, according to the medium used, filters may also be classified as *perforated plate, woven septum*, or *granular filters*. The filtration medium of the microstrainer mentioned above is of perforated plate. The filter media used in plate-and-frame presses and vacuum filters are of woven materials. These units are discussed later.

Figures 7.1 and 7.2 show examples of gravity filters. The media for these filters are granular. In both figures, the influents are introduced at the top, thereby utilizing gravity to pull the water through the filter. Figure 7.1a is composed of two granular filter media anthrafilt and silica sand; thus, it is called a *dual-media gravity filter*. Figure 7.1a is a triple-media gravity filter, because it is composed of three media: anthrafilt, silica sand, and garnet sand.

Generally, two types of granular gravity filters are used: slow-sand and rapid-sand filters. In the main, these filters are differentiated by their rates of filtration.

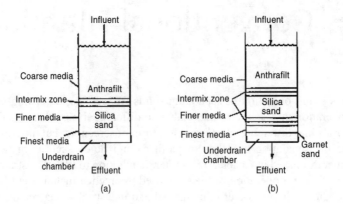

FIGURE 7.1 (a) Dual-media filter; (b) triple-media filter.

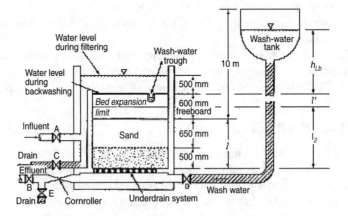

FIGURE 7.2 A typical gravity filter.

Slow-sand filters normally operate at a rate of 1.0 to 10 $m^3/d \cdot m^2$, while *rapid-sand filters* normally operate at a rate of 100 to 200 $m^3/d \cdot m^2$. A section of a typical gravity filter is shown in Figure 7.2.

The operation of a gravity filter is as follows. Referring to Figure 7.2, drain valves C and E are closed and influent value A and effluent valve B are opened. This allows the influent water to pass through valve A, into the filter and out of the filter through valve B, after passing though the filter bed.

For effective operation of the filter, the voids between filter grains should serve as tiny sedimentation basins. Thus, the water is not just allowed to swiftly pass through the filter. For this to happen, the effluent valve is slightly closed so that the level of water in the filter rises to the point indicated, enabling the formation of tiny sedimentation basins in the pores of the filter. As this level is reached, influent and effluent flows are balanced. It is also this level that causes a pressure differential pushing the water through the bed. The filter operates at this pressure differential until it is clogged and ready to be backwashed (in the case of the rapid-sand filter). Backwashing will be discussed later in this chapter. In the case of the slow-sand filter,

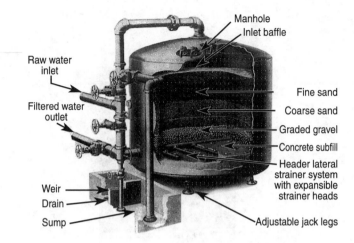

Raw water inlet

Filtered water outlet

Weir

Drain

Sump

Manhole

Inlet baffle

Fine sand

Coarse sand

Graded gravel

Concrete subfill

Header lateral strainer system with expansible strainer heads

Adjustable jack legs

FIGURE 7.3 Cutaway view of a pressure sand filter. (Courtesy of Permutit Co.)

it is not backwashed once it is clogged. Instead, the layer of dirt that collects on top of the filter (called *smutzdecke*) is scraped for cleaning.

As shown, the construction of the bed is such that the layers are supported by an underdrain mechanism. This support may simply be a perforated plate or septum. The perforations allow the filtered water to pass through. The support may also be made of blocks equipped with holes. The condition of the bed is such that the coarser heavy grains are at the bottom. Thus, the size of these holes and the size of the perforations of the septum must not allow the largest grains of the bed to pass through.

Figure 7.3 shows a cutaway view of a pressure filter. The construction of this filter is very similar to that of the gravity filter. Take note of the underdrain construction in that the filtered water is passed through perforated pipes into the filtered water outlet. As opposed to that of the gravity filter above, the filtered water does not fall through a bottom and into the underdrain, because it had already been collected by the perforated pipes. The coarse sand and graded gravel rest on the concrete subfill.

Using a pump or any means of increasing pressure, the raw water is introduced to the unit through the raw water inlet. It passes through the bed and out into the outlet. The unit is operated under pressure, so the filter media must be enclosed in a shell. As the filter becomes clogged, it is cleaned by backwashing.

Thickened and digested sludges may be further reduced in volume by dewatering. Various dewatering operations are used including vacuum filtration, centrifugation, pressure filtration, belt filters, and bed drying. In all these units, cakes are formed. We therefore call these types of filtration *cake-forming filtration* or simply *cake filtration*. Figure 7.4a shows a sectional drawing of a plate-and-frame press. In *pressure filtration*, which operates in a cycle, the sludge is pumped through the unit, forcing its way into *filter plates*. These plates are wrapped in *filter cloths*. With the filter cloths wrapped over them, the plates are held in place by *filter frames* in alternate plates-then-frames arrangement. This arrangement creates a cavity in the frame between two adjacent plates.

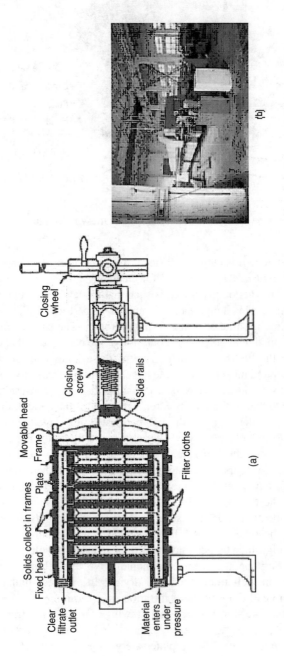

FIGURE 7.4 (a) Sectional drawing of a plate-and-frame press (from T. Shriver and Co.); (b) an installation of a plate-and-frame press (courtesy of Xingyuan Filtration Products, China).

Two channels are provided at the bottom and top of the assembly. The bottom channel serves as a conduit for the introduction of the sludge into the press, while the top channel serves as the conduit for collecting the filtrate. The bottom channel has connections to the cavity formed between adjacent plates in the frame. The top channel also has connections to small drainage paths provided in each of the plates. These paths are where the filtrate passing through cloth are collected.

As the sludge is forced through the unit at the bottom part of the assembly (at a pressure of 270 to 1,000 kPa), the filtrate passes through the filter cloth into the drainage paths, leaving the solids on the cloth to accumulate in the cavities of the frames. As determined by the cycle, the press is opened to remove the accumulated and dewatered sludge. Figure 7.4b shows an installation of a plate-and-frame press unit.

Figures 7.5 to Figure 7.7 pertain to the use of rotary vacuum filters in vacuum filtration. In *vacuum filtration*, a drum wrapped in filter cloth rotates slowly while the lower portion is submerged in a sludge tank (Figure 7.7a). A vacuum applied in the underside of the drum sucks the sludge onto the filter cloth, separating the filtrate and, thus, dewatering the sludge.

A rotary vacuum filter is actually a drum over which the filtration medium is wrapped. This medium is made of a woven material such as canvas. This medium is also called a *filter cloth*. The drum is made of an outer shell and an inner shell. These two shells form an annulus. The annulus is then divided into segments, which are normally 30 cm in width and length extending across the entire length of the drum. Figure 7.7a shows that there are twelve segments in this vacuum filter. The outer shell has perforations or slots in it, as shown in the cutaway view of Figure 7.6. Thus, each segment has a direct connection to the filter cloth. The purpose of the segments is to provide the means for sucking the sludge through the cloth while it is still submerged in the tank.

Each of the segments are connected to the rotary valve through individual pipings. As shown in Figure 7.7a, segments 1 to 5 are immersed in the sludge, while

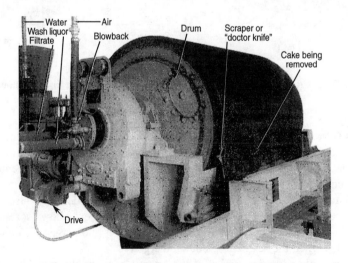

FIGURE 7.5 A rotary vacuum filter in operation. (Courtesy of Oliver United Filters.)

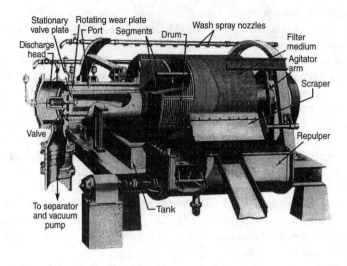

FIGURE 7.6 Cutaway view of a rotary vacuum filter. (Courtesy of Swenson Evaporator Co.)

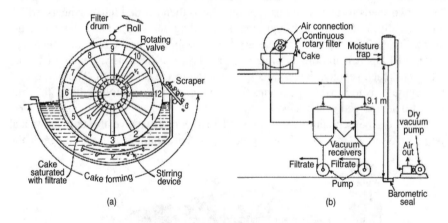

FIGURE 7.7 (a) Cross section of a rotary vacuum filter; (b) flow sheet for continuous vacuum filtration.

segments 6 to 12 are not. Pipes V_1 and V_2 of the rotary valve are connected to an external vacuum pump, as indicated in Figure 7.7b. The design of the rotary valve is such that when segments are submerged in the sludge such as segments 1 to 5, they are connected to pipe V_1 through their individual connecting pipes. When segments are not submerged such as segments 6 to 12, the design is also such that these segments are connected to pipe V_2. This arrangement allows for suction of sludge into the filter cloth over the segment when it is submerged (V_1) and drying of the sludge when the segment is not submerged (V_2).

We can finalize the description of the operation of the vacuum filter this way. As the segments that had been sucking sludge while they were still submerged in the tank emerge from the surface, their connections are immediately switched from V_1 to V_2. The connection V_2 completes the removal of removable water from the sludges, whereupon the suction switches to sucking air into the segments promoting the drying of the sludges. The "dry" sludge then goes to the scraper (also called *doctor blade*) and the sludge removed for further processing or disposal.

Figure 7.7b shows the fate of the filtrate as it is sucked from the filter cloth. Two tanks called *vacuum receivers* are provided for the two types of filtrates: the filtrate removed while the segments are still submerged in the tank and the residual filtrate removed when the segments are already out of the tank. Vacuum receivers are provided to trap the filtrate so that the filtrate will not flood the vacuum pump. Also note the barometric seal. As shown, this is in parallel connection with the suction vacuum of the filter. The vacuum pressure is normally set up to a value of 66 cm Hg below atmospheric. Any vacuum set for the filter will correspondingly exert an equal vacuum to the barometric seal, on account of the parallel connection. Hence, the length of this seal should be set equivalent to the maximum vacuum expected to be utilized in the operation of the filter. If, for example, the filter is to be operated at 51 cm,

$$51(13.6) = (1)\Delta h_{H_2O}; \qquad \Delta h_{H_2O} = 6.94 \, \text{m}$$

where 13.6 is the mass density of mercury in gm/cc, and 1 is the density of water also in gm/cc. Thus, from this result, the length of the barometric seal should be 6.94 m if the operational vacuum is 51 cm Hg. The design in Figure 7.7b shows the length as 9.1 m.

Figure 7.8a shows another type of filter that operates similar to a rotary vacuum filter in that it uses a vacuum pressure to suck sludge into the filter medium. This type of filter is called a leaf filter. A *leaf filter* is a filter that operates by immersing a component called a leaf into a bath of sludge or slurry and using a vacuum to suck the sludge onto the leaf. An example of a leaf filter is shown in Figure 7.8b. As indicated, it consists of two perforated plates parallel to each other, with a separator screen providing the spacing between them. A filter is wrapped over the plate assembly, just like in the plate-and-frame press. Each of the leaves are then attached into a hub through a clamping ring. The hub has a drainage space that connects into the central pipe through a small opening. As indicated in the cutaway view on the right of Figure 7.8a, several of these leaves are attached to the central pipe. Each of the leaves then has a connection to the central pipe through the small opening from the drainage space. The central pipe collects all the filtrates coming from each of the leaves.

In operation, a vacuum pressure is applied to each of the leaves. The feed sludge is then introduced at the feed inlet as indicated in the drawing. The sludge creates a slurry pool inside the unit immersing the leaves. Through the action of the vacuum, the sludge is sucked into the filter cloth. As the name implies, this is a rotary leaf filter. The leaves are actually in the form of a disk. The disks are rotated, immersing part of it in the slurry, just as part of the drum is immersed in the case of the rotary vacuum filter. As the immersed part of the disks emerge from the slurry pool into the air, the filtrate are continuously sucked by the vacuum resulting in a dry cake.

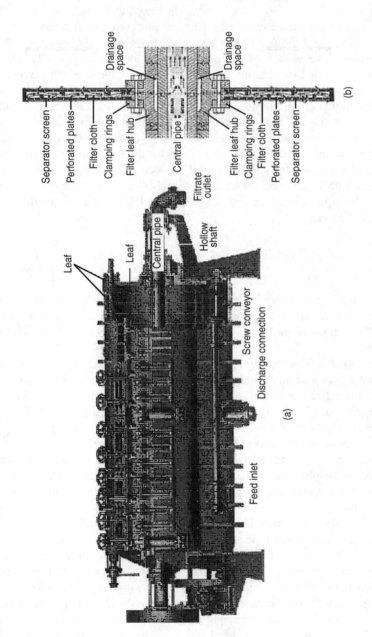

FIGURE 7.8 A rotary leaf filter showing cutaway view at right end (courtesy of Swenson Evaporator Co.); (b) section of a leaf.

A mechanism is provided for the cake to drop into a screw conveyor below for continuous removal. This mechanism does not require opening of the case for removal of the cake.

It may be noticed that some of the filters discussed are operated continuously and some are not. For example, the rapid sand filter, the slow sand filter, the pressure filter, and the rotary vacuum filter are all operated continuously. The plate-and-frame press is operated as a batch. Thus, filters may also be classified as *continuous* and *discontinuous*. Only the plate-and-frame press is discussed in this chapter as a representation of the discontinuous type, but others are used, such as the shell-and-leaf filters and the cartridge filters. The first operates in a mode that a leaf assembly is inserted into a shell while operating and retracted out from the shell when it is time to remove the cake. The second looks like a "cartridge" in outward appearance with the filter medium inside it. The medium could be thin circular plates or disks stacked on top of each other. The clearance between disks serves to filter out the solids.

7.2 MEDIUM SPECIFICATION FOR GRANULAR FILTERS

The most important component of a granular filter is the medium. This medium must be of the appropriate size. Small grain sizes tend to have higher head losses, while large grain sizes, although producing comparatively smaller head losses, are not as effective in filtering. The actual grain sizes are determined from what experience has found to be most effective. The actual medium is never uniform, so the grain sizes are specified in terms of effective size and uniformity coefficient. *Effective size* is defined as the size of sieve opening that passes the 10% finer of the medium sample. The effective size is said to be the *10th percentile size P_{10}*. The *uniformity coefficient* is defined as the ratio of the size of the sieve opening that passes the 60% finer of the medium sample (P_{60}) to the size of the sieve opening that passes the 10% finer of the medium sample. In other words, the uniformity coefficient is the ratio of the P_{60} to the P_{10}. For slow-sand filters, the effective size ranges from 0.25 mm to 0.35 mm with uniformity coefficient ranging from 2 to 3. For rapid-sand filters, the effective size ranges from 0.45 mm and higher with uniformity coefficient ranging from 1.5 and lower.

Plot of a sieve analysis of a sample of run-of-bank sand is shown in Figure 7.9 by the segmented line labeled "stock sand ...." This sample may or may not meet the required effective size and uniformity coefficient specifications. In order to transform this sand into a usable sand, it must be given some treatment. The figure shows the cumulative percentages (represented by the "normal probability scale" on the ordinate) as a function of the increasing size of the sand (represented by the "size of separation" on the abscissa).

Let p_1 be the percentage of the sample stock sand that is smaller than or equal to the desired P_{10} of the final filter sand, and p_2 be the percentage of the sample stock sand that is smaller than or equal to the desired P_{60} of the final filter sand. Since the percentage difference of the P_{60} and P_{10} represents half of the final filter sand, $p_2 - p_1$ must represent half of the stock sand that is transformed into the final

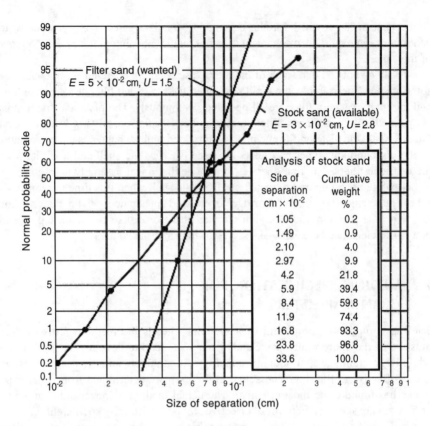

FIGURE 7.9 Sieve analysis of run-of-bank sand.

filter sand. Letting p_3 be the percentage of the stock sand that is transformed into the final filter sand,

$$p_3 = 2(p_2 - p_1) \tag{7.1}$$

Of this p_3, by definition, 10% must be the P_{10} of the final sand. Therefore, if p_4 is the percentage of the stock sand that is too fine to be usable,

$$p_4 = p_1 - 0.1p_3 = p_1 - 0.1(2)(p_2 - p_1) \tag{7.2}$$

The plot in the figure shows an increasing percentage as the size of separation increases, so the sum of p_4 and p_3 must represent the percentage of the sample stock sand above which the sand is too coarse to be usable. Letting p_5 be this percentage,

$$p_5 = p_4 + p_3 \tag{7.3}$$

Now, to convert a run-of-bank stock sand into a usable sand, an experimental curve such as Figure 7.9 is entered to determine the size of separation corresponding to p_4 and p_5. Having determined these sizes, the stock sand is washed in a sand washer that rejects the unwanted sand. The washer is essentially an upflow settling tank. By varying the upflow velocity of the water in the washer, the sand particles introduced into the tank are separated by virtue of the difference of their settling velocities. The lighter ones are carried into the effluent while the heavier ones remain. The straight line in the figure represents the size distribution in the final filter sand when the p_4 and fractions greater than p_5 have been removed.

Example 7.1 If the effective size and uniformity coefficient of a proposed filter is to be $5(10^{-2})$ cm and 1.5, respectively, perform a sieve analysis to transform the run-of-bank sand of Figure 7.9 into a usable sand.

Solution: From Figure 7.9, for a size of separation of $5(10^{-2})$ cm, the percent p_1 is 30. Also, the P_{60} size is $5(10^{-2})(1.5) = 7.5(10^{-2})$ cm. From the figure, the percent p_2 corresponding to the P_{60} size of the final sand is 53.

$$p_3 = 2(p_2 - p_1) = 2(53 - 30) = 46\%$$

$$p_4 = p_1 - 0.1(p_3) = 30 - 0.1(46)$$

$$= 25.4\%; \text{ from the figure, the corresponding size of separation}$$

$$= 4.5(10^{-2}) \text{ cm}$$

$$p_5 = 46 + 25.4 = 71.4\%; \text{ corresponding size of separation } = 1.1(10^{-1}) \text{ cm}$$

Therefore, the sand washer must be operated so that the p_4 sizes of $4.5(10^{-2})$ cm and smaller and the p_5 sizes of $1.1(10^{-1})$ cm and greater are rejected. **Ans**

7.3 LINEAR MOMENTUM EQUATION APPLIED TO FILTERS

The motion of water through a filter bed is just like the motion of water through parallel pipes. While the motion through the pipes is straightforward, however, the motion in the filter bed is tortuous. Figure 7.10 shows a cylinder or pipe of fluid and bed material. Inside this pipe is an element composed of fluid and bed material being isolated with length dl and interstitial area A and subjected to forces as shown. (We use the term interstitial area here because the bed is actually composed of grains. The fluid is in the interstitial spaces between grains.) The equation of linear momentum may be applied on the water flow in the downward direction of this element, thus,

$$\sum F_z = pA - (p + dp)A + F_g - F_{sh} = \rho \eta d\bar{V} a_z = \rho \eta d\bar{V} \frac{dv}{dt} = \rho k A_s dl \frac{dv}{dt}$$

$$\sum F_z = -dpA + F_g - F_{sh} = \rho k A_s dl \frac{dv}{dt}$$

(7.4)

where $\sum F_z$ is the net unbalanced force in the downward z direction; p is the hydrostatic pressure; A is the interstitial cross-sectional area of the cylindrical element of fluid; F_g is the weight of the fluid in the element; F_{sh} is the shear force acting on the fluid along the surface areas of the grains; $d\bar{V}$ is the volume of element of space; dl is the differential length of the element, l being any distance from some origin; A_s is the surface area of all the grains; k is a factor that converts A_s into an area such that $k A_s dl = \eta d\bar{V}$; η is the porosity of the bed; a_z is the acceleration of the fluid

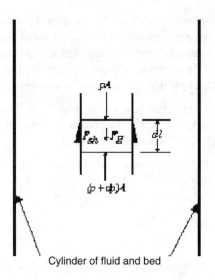

Cylinder of fluid and bed

FIGURE 7.10 Free-body diagram of a cylinder (pipe) of fluid and bed material.

element in the downward z direction; ρ is the fluid mass density; v is the component fluid element velocity in the z direction; and t is the time. Since the fluid is in the interstitial spaces, $d\Psi$ needs to be multiplied by the porosity to get the fluid volume.

The law of inertia states that a body at rest will remain at rest and a body in uniform motion will remain in this uniform motion unless acted upon by an unbalanced force. $\Sigma F_z = \rho k A_s dl(dv/dt)$ is this unbalanced force that breaks this inertia; thus, it is called the inertia force. By the chain rule, $(dv/dt) = (dv/dl)(dl/dt) = v(dv/dl)$. Thus, $\rho k A_s dl(dv/dt) = \rho k A_s dl v(dv/dl) = \rho k A_s v dv$. Let $\bar{V}$ be some characteristic average pipe velocity.

The velocity through the pipe could vary from the entrance to the exit. $\bar{V}$ represents the average of these varying values; hence, it is a constant. Note that all the velocities referred to here are interstitial velocities, the true velocities of the fluid as it travels through the pores.

Now, let $v^* = v/\bar{V}$. Hence, $dv^* = dv/\bar{V}$, and $\rho k A_s v dv = \rho k A_s(\bar{V} v^*)\bar{V} dv^* = \rho k A_s \bar{V}^2 v^* dv^*$. Thus, the inertial force $\Sigma F_z = \rho k A_s v dv$, in effect, is proportional to $\rho A_s \bar{V}^2$; the presence of $v^* dv^*$ does not nullify this fact. Calling the proportionality constant as K_i,

$$\sum F_z = K_i \rho A_s \bar{V}^2 \tag{7.5}$$

In light of this new information, Equation (7.4) may now be solved for $-dpA$, which is $-\Delta pA$ when applied to the whole length of pipe, producing

$$-\Delta pA = K_i \rho A_s \bar{V}^2 + F_{sh} - F_g \tag{7.6}$$

To address F_{sh}, let us recall the Hagen–Poiseuille equation from fluid mechanics.

This is written as

$$-\Delta p_s = \frac{32\mu l \overline{V}}{D^2} \tag{7.7}$$

$-\Delta p_s$ is the pressure drop due to shear forces; μ is the absolute viscosity of the fluid; l is the length of pipe; and D is the diameter of pipe. In a bed of grains, the cross-sectional area of flow is so small that the boundary layer created as the flow passes around one grain overlaps with the boundary layer formed in a neighboring grain. Because boundary layer flow is, by nature, laminar, flows through beds of grain is laminar and Equation (7.7) applies.

F_{sh} is a shear force acting on the fluid along the surface areas of the grains. The shear stress is this $-\Delta p_s$; thus, $F_{sh} = -\Delta p_s A_s$. From this relationship and the equation for Δp_s, we glean that the F_{sh} is directly proportional to μ, $\overline{V}$, and l and inversely proportional to the square of D. The granular filter is not really a pipe, so we replace D by the hydraulic radius r_H. With D expressed in terms of r_H, the equation becomes general. In other words, it can be used for any shape, because r_H is simply defined as the area of flow divided by the wetted perimeter. The proportionality relation is $F_{sh} \infty \mu l \overline{V} A_s / r_H^2$. Dimensionally, l and r_H^2 may be canceled leaving only r_H in the denominator. Calling the overall proportionality constant as K_s, $F_{sh} = K_s(\mu \overline{V} A_s / r_H)$; and Equation (7.6) becomes

$$-\Delta pA = K_i \rho A_s \overline{V}^2 + K_s \frac{\mu \overline{V} A_s}{r_H} - F_g \tag{7.8}$$

To address F_g, it must be noted that for a given filter installation it is a constant. It is a constant, so its effect when the variables are varied is also a constant. This effect will be subsumed into the values of K_i and K_s. We can therefore safely remove F_g from the equation and write simply

$$-\Delta pA = K_i \rho A_s \overline{V}^2 + K_s \frac{\mu \overline{V} A_s}{r_H} = A_s\left(K_i \rho \overline{V}^2 + K_s \frac{\mu \overline{V}}{r_H}\right) \tag{7.9}$$

This is the general linear momentum equation as applied to any filter.

7.4 HEAD LOSSES IN GRAIN FILTERS

Head losses in granular filters may be divided into two categories: head loss in clean filters and head loss due to the deposited materials. We will now discuss the first category.

7.4.1 CLEAN-FILTER HEAD LOSS

To derive the clean-filter head loss, we continue with Equation (7.9) by expressing A_s, r_H, and $\overline{V}$ in terms of their equivalent expressions. Let s_p be the surface area of a particle and let N be the number of grains in the bed. Thus, $A_s = N s_p$. Let S_o be the empty bed or superficial area of the bed. With the porosity η and length l, the volume of the bed grains is $S_o l(1 - \eta)$. Letting v_p represent the volume of a grain, N is also calculated as $S_o l(1 - \eta)/v_p$. Thus, A_s is

$$A_s = \frac{s_p}{v_p} S_o l(1 - \eta) \qquad (7.10)$$

Consider first that the grains are spherical. For spherical particles, $v_p = \pi d^3/6$ and $s_p = \pi d^2$, where d is the diameter of the particle. Therefore, $s_p/v_p = 6/d$. Substituting in Equations (7.10),

$$A_s = \frac{6}{d} S_o l(1 - \eta) \qquad (7.11)$$

The expression *area of flow divided by the wetted perimeter* for the hydraulic radius is the same as *volume of flow divided by the wetted area*, that is,

$$r_H = \frac{\text{volume of flow}}{\text{wetted area}} \qquad (7.12)$$

With N as the number of medium grains in the filter each of volume v_p, the volume of the filter is $Nv_p/(1 - \eta)$. Therefore, the volume of flow is $\eta[Nv_p/(1 - \eta)]$ and the hydraulic radius becomes

$$r_H = \frac{\eta \frac{Nv_p}{(1 - \eta)}}{N s_p} = \left(\frac{\eta}{1 - \eta}\right)\left(\frac{v_p}{s_p}\right) = \left(\frac{\eta}{1 - \eta}\right)\frac{d}{6} \qquad (7.13)$$

The velocity $\bar{V}$ is the interstitial velocity of the fluid through the pores of the bed. Compared to the superficial velocity $\bar{V}_s$, it is faster due to the effect of the porosity η. (The cross-sectional area of flow is much constricted, hence, the velocity is much faster.) In terms of η and $\bar{V}_s$,

$$\bar{V} = \frac{\bar{V}_s}{\eta} \qquad (7.14)$$

Substituting Eqs. (7.11), (7.13), and (7.14) in Equation (7.9) and simplifying produces

$$-\Delta pA = \frac{S_o l(1 - \eta)\bar{V}_s^2 \rho}{d\eta^2}\left[6K_i + \frac{36K_s(1 - \eta)}{d\bar{V}_s\rho/\mu}\right] \qquad (7.15)$$

$$-\Delta pA = \frac{S_o l(1 - \eta)\bar{V}_s^2 \rho}{d\eta^2}\left[6K_i + \frac{36K_s(1 - \eta)}{Re}\right] \qquad (7.16)$$

where Re is the Reynolds number defined as $d\bar{V}_s\rho/\mu$. Ergun (1952) correlating a mass of experimental data showed that $6K_i = 1.75$ and $36K_s = 150$. Thus,

$$-\Delta pA = \frac{S_o l(1 - \eta)\bar{V}_s^2 \rho}{d\eta^2}\left[1.75 + \frac{150(1 - \eta)}{Re}\right] = \frac{S_o l(1 - \eta)\bar{V}_s^2 \rho}{d\eta^2} f_p \qquad (7.17)$$

where

$$f_p = 1.75 + \frac{150(1-\eta)}{Re}, \text{ a form of friction factor.} \tag{7.18}$$

A is equal to $S_o\eta$. Substituting in Equation (7.17), the pressure drop across the filter is finally given as

$$-\Delta p = f_p \frac{2\gamma l(1-n)}{n^3 d} \frac{\overline{V}_s^2}{2g} \tag{7.19}$$

γ is the specific weight of the fluid. In terms of the equivalent height of fluid, we use the relationship $-\Delta p = \gamma h_L$, where h_L is the head loss across the filter. Thus,

$$h_L = f_p \frac{2l(1-n)}{n^3 d} \frac{\overline{V}_s^2}{2g} \tag{7.20}$$

The equations treated above all refer to the diameter of spherical particles; yet in practice, not all particles are spherical. To use the above equations for these situations, the sieve diameter must be converted to its equivalent spherical diameter. In Chapter 5, the relationship was given as $d = (6/\pi)^{1/3}\beta^{1/3} d_p$, where β is the shape factor and d_p is the sieve diameter.

Equation (7.20) has been derived for a bed of uniform grain sizes. In some types of filtration plants such as those using rapid sand filters, however, the bed is back-washed every so often. This means that after backwashing, the grain particles are allowed to settle; the grain deposits on the bed, layer by layer, are of different sieve diameters. The bed is said to be stratified. To find the head loss across a stratified medium, Equation (7.20) is applied layer by layer, each layer being converted to one single, average diameter. The head losses across each layer are then summed to produce the head loss across the bed as shown next.

$$h_L = \sum f_{pi} \frac{2l_i(1-n)}{n^3 d_i} \frac{\overline{V}_s^2}{2g} \tag{7.21}$$

where the index i refers to the ith layer. If x_i is the fraction of the d_i particles in the ith layer, then l_i equals $x_i l$. Assuming the porosity η is the same throughout the bed, the equation becomes

$$h_L = \sum f_{pi} \frac{2x_i l(1-n)}{n^3 d_i} \frac{\overline{V}_s^2}{2g} = \frac{2l(1-n)}{n^3} \frac{\overline{V}_s^2}{2g} \sum f_{pi} \frac{x_i}{d_i} \tag{7.22}$$

Note, again, that d_i is for a spherical particle. For nonspherical particles, the sieve diameter d_p must be converted into its equivalent spherical particle by the equation mentioned in a previous paragraph.

Example 7.2   A sharp filter sand has the sieve analysis shown below. The porosity of the unstratified bed is 0.39, and that of the stratified bed is 0.42. The lowest temperature anticipated of the water to be filtered is 4°C. Find the head loss if the sand is to be used in (a) a slow-sand filter 76 cm deep operated at 9.33 $m^3/m^2 \cdot d$ and (b) a rapid-sand filter 76 cm deep operated at 117 $m^3/m^2 \cdot d$.

Sieve No.	Average Size (mm)	x_i
14–20	1.0	0.01
20–28	0.70	0.05
28–32	0.54	0.15
32–35	0.46	0.18
35–42	0.38	0.18
42–48	0.32	0.20
48–60	0.27	0.15
60–65	0.23	0.07
65–100	0.18	0.01

Solution:

$$\textbf{(a)}\ h_L = f_p \frac{2l(1-n)}{n^3 d} \frac{\bar{V}_s^2}{2g}$$

$$d_{pave} = 1(0.01) + 0.7(0.05) + 0.54(0.15) + 0.46(0.18) + 0.38(0.18)$$
$$+ 0.32(0.20) + 0.27(0.15) + 0.23(0.07) + 0.18(0.01)$$
$$= 0.40\ \text{mm}$$

$$9.33\,m^3/m^2 \cdot d = 1.08(10^{-4})\ \text{m/s}$$

$$f_p = 1.75 + \frac{150(1-\eta)}{Re}$$

$$Re = d\bar{V}_s \rho/\mu$$

$$d = 1.24\beta^{0.333} d_p \qquad \beta = 0.77$$

Therefore,

$$d = 1.24(0.77^{0.333})(0.40)(10^{-3}) = 0.45(10^{-3})\ \text{m}$$

$$\mu = 15(10^{-4})\ \text{kg/m} \cdot \text{s} \quad \text{at}\ 20°\text{C}$$

$$Re = \frac{0.45(10^{-3})(1.08)(10^{-4})(1000)}{15(10^{-4})} = 0.0324$$

$$f_p = 1.75 + 150\frac{(1-0.39)}{0.0324} = 2,825.82$$

$$h_L = 2,825.82\left[\frac{2(0.76)(1-0.39)[1.08(10^{-4})]^2}{0.39^3(0.45)(10^{-3})(2)(9.81)}\right]$$

$$= \frac{3.056(10^{-5})}{5.24(10^{-4})} = 0.058\ \text{m} \quad \textbf{Ans}$$

(b) 117 m^3/m$^2 \cdot$ d = 0.00135 m/s

$$h_L = \frac{2l(1-n)\bar{V}_s^2}{n^3 \; 2g} \sum f_{pi} \frac{x_i}{d_i}$$

$$Re_i = \frac{d_i(0.00135)(1000)}{15(10^{-4})}$$

$$= \frac{1.24(0.77)^{0.333} d_{pi}(0.00135)(1000)}{15(10^{-4})} = 1022.98 d_{pi}$$

$$f_p = 1.75 + 150\frac{(1-0.42)}{1022.98 d_{pi}} = \frac{0.85}{d_{pi}} + 1.75$$

Sieve No.	Avg. Size (mm)	x_i	f_i	$f_{pi}\dfrac{x_i}{d_i} = \dfrac{f_p x_i}{1.24(0.77^{0.333})d_{pi}}$
14–20	1.0	0.01	86.75	763.28
20–28	0.70	0.05	123.18	7741.44
28–32	0.54	0.15	159.16	38,899.02
32–35	0.46	0.18	186.53	64,221.10
35–42	0.38	0.18	225.43	93,955.69
42–48	0.32	0.20	267.34	147,033.13
48–60	0.27	0.15	316.56	154,740.68
60–65	0.23	0.07	371.32	99,432.26
65–100	0.18	0.01	473.97	23,168.33
				$\Sigma = 629,977.93$

$$h_L = \frac{2l(1-n)\bar{V}_s^2}{n^3 \; 2g} \sum f_{pi} \frac{x_i}{d_i} = \frac{2(0.76)(1-0.42)(0.00135^2)}{0.42^3(2)(9.81)}(629,977.93)$$

= 0.696 m   **Ans**

7.4.2 HEAD LOSSES DUE TO DEPOSITED MATERIALS

The head loss expressions derived above pertain to head losses of clean filter beds. In actual operations, however, head loss is also a function of the amount of materials deposited in the pores of the filter. Letting q represent the deposited materials per unit volume of bed, the corresponding head loss h_d as a result of this deposited materials may be modeled as

$$h_d = a(q)^b \tag{7.23}$$

where a and b are constants. Taking logarithms,

$$\ln h_d = \ln a + b\ln q \tag{7.24}$$

This equation shows that plotting $\ln h_d$ against $\ln q$ will produce a straight line. By performing experiments, Tchobanoglous and Eliassen (1970) showed this statement to be correct.

Letting h_{Lo} represent the clean-bed head loss, the total head loss of the filter bed h_L is then

$$h_L = h_{Lo} + h_d = h_{Lo} + \Sigma a(q_i)^b \tag{7.25}$$

where h_d is the head loss over the several layers of grains in the bed due to the deposited materials.

Now, let us determine the expression for q. This expression can be readily derived from a material balance using the Reynolds transport theorem. This theorem is derived in any good book on fluid mechanics and will not be derived here. The derivation is, however, discussed in the chapter titled "Background Chemistry and Fluid Mechanics." It is important that the reader acquire a good grasp of this theorem as it is very fundamental in understanding the differential form of the material balance equation. This theorem states that the total derivative of a dependent variable is equal to the partial derivative of the variable plus its convective derivative. In terms of the deposition of the material q onto the filter bed, the total derivative is

$$\text{total derivative} = \frac{d}{dt}\left(\int_V q\, d\mathcal{V}\right) \tag{7.26}$$

$\mathcal{V}$ is the volume of the control volume. The partial derivative (also called local derivative) is

$$\text{partial derivative} = \frac{\partial}{\partial t}\left(\int_V q\, d\mathcal{V}\right) \tag{7.27}$$

and the convective derivative is

$$\text{convective derivative} = \oint_A \vec{cv} \cdot \hat{n}\eta\, dA \tag{7.28}$$

The symbol $\oint_A$ means that the integration is to be done around the surface area of the volume. The vector $\hat{n}$ is a unit vector on the surface and normal to it and the vector $\vec{v}$ is the velocity vector through the surface for the flow into the filter. Now, substituting these equations into the statement of the Reynolds theorem, we obtain

$$\frac{d}{dt}\left(\int_V q\, d\mathcal{V}\right) = \frac{\partial}{\partial t}\left(\int_V q\, d\mathcal{V}\right) + \oint_A \vec{cv} \cdot \hat{n}\eta\, dA \tag{7.29}$$

In the previous equation, the total derivative is also called *Lagrangian derivative, material derivative, substantive derivative*, or *comoving derivative*. The combination of the partial derivative and the convective derivative is also called the *Eulerian derivative*. Again, it is very important that this equation be thoroughly understood. It is to be noted that in the environmental engineering literature, many authors confuse the difference between the total derivative and the partial derivative. Some authors use the partial derivative instead of the total derivative and vise versa. As shown by the previous equation, there is a big difference between the total derivative and the partial derivative. If this difference is not carefully observed, any equation written that uses one derivative instead of the other is conceptually wrong; this confusion can be seen very often in the environmental engineering literature. Thus, caution in reading the literarture should be exercised.

Over a differential length dl and cross-sectional area A of bed,

$$\oint_A \vec{cv} \cdot \hat{n}\eta \, dA = -\dot{m} + \left(\dot{m} + \frac{\partial \dot{m}}{\partial l}dl\right) = \frac{\partial \dot{m}}{\partial l}dl \qquad (7.30)$$

where $\dot{m} = cQ$ and Q = flow, a constant through the bed. (The minus sign of $\dot{m}$ results from the fact that at the inlet the unit normal vector $\hat{n}$ is in opposite direction to that of the velocity vector $\vec{v}$.) Therefore,

$$\oint_A \vec{cv} \cdot \hat{n}\eta \, dA = \frac{\partial \dot{m}}{\partial l}dl = Q\frac{\partial c}{\partial l}dl = V_s S_o \frac{\partial c}{\partial l}dl \qquad (7.31)$$

V_s is the superficial velocity of flow in the bed.

Also, over the same differential length dl and cross-sectional area S_o of bed,

$$\frac{\partial \int_v q \, d\forall}{\partial t} = \frac{\partial (q \, d\forall)}{\partial t} = \frac{\partial q}{\partial t}d\forall = \frac{\partial q}{\partial t}S_o dl \qquad (7.32)$$

$d\forall$ in the second term has been taken out of the parentheses, because it is arbitrary and therefore independent of t.

Substituting Eqs. (7.31) and (7.32) in Equation (7.29),

$$\frac{\partial q}{\partial t}S_o dl + V_s S_o \frac{\partial c}{\partial l}dl = 0 \qquad (7.33)$$

(Since the solids are conservative substances, the total derivative is equal to zero.)

Dividing out $S_o dl$ and rearranging,

$$\frac{\partial q}{\partial t} = -V_s\frac{\partial c}{\partial l} \qquad (7.34)$$

The numerical counterpart of Equation (7.34), using n as the index for time and m as the index for distance, is

$$\frac{q_{n+1,m} - q_{n,m}}{\Delta t} = -V_s\frac{c_{n,m+1} - c_{n,m}}{\Delta l} \qquad (7.35)$$

for the first time-step. Solving for q_{n+1},

$$q_{n+1,m} = q_{n,m} - \frac{\Delta t V_s}{\Delta l}(c_{n,m-1} - c_{n,m}) = q_{n,m} - \Delta t V_s\frac{\Delta c_n}{\Delta l} \qquad (7.36)$$

The equation for the second time-step is

$$q_{n+2,m} = q_{n+1,m} - \Delta t V_s\frac{\Delta c_{n+1}}{\Delta l} = q_{n,m} - \Delta t V_s\frac{\Delta c_n}{\Delta l} - \Delta t V_s\frac{\Delta c_{n+1}}{\Delta l} \qquad (7.37)$$

Over a length Δl, the gradients $\Delta c/\Delta l$ varies negligibly from time step to time step. Thus, $\Delta c_n/\Delta l = \Delta c_{n+1}/\Delta l = \Delta c_{n+2}/\Delta l$ and so on. Equation (7.37) may then be written as

$$q_{n+2,m} = q_{n,m} - 2\Delta t V_s \frac{\Delta c}{\Delta l} \tag{7.38}$$

For the kth time step, the numerical equation becomes

$$q_{n+k,m} = q_{n,m} - k\Delta t V_s \frac{\Delta c}{\Delta l} \tag{7.39}$$

But the number of time steps k is $t/\Delta t$. Therefore, the numerical equation is simply

$$q_{n+k,m} = q_{n,m} - t V_s \frac{\Delta c}{\Delta l} \tag{7.40}$$

$q_{n+k,m}$ is actually simply q at the end of filtration. Also, $q_{n,m}$ at the beginning of the filtration run is zero. Because the effect of the deposited materials has to start from a clean filter bed, $q_{n,m}$ may be removed from the equation. The final equation becomes

$$q = -t V_s \frac{\Delta c}{\Delta l} \tag{7.41}$$

The nature and sizes of particles introduced to the filter will vary depending upon how they are produced. Thus, the influent to a filter coming from a softening plant is different than the influent coming from a coagulated surface water. Also, the influent coming from a secondary-treated effluent is different than that of any of the drinking water treatment influent. For these reasons, in order to use the previous equations for determining head losses, a pilot plant study should be conducted for a given type of influent.

Determination of constants a and b. As noted before, h_d plots a straight line in logarithmic form with q. This means that only two data points are needed in order to determine the constants a and b. Let the data points be (h_{d1}, q_1) and (h_{d2}, q_2). The two equations for h that can be used to solve for the constants are then

$$h_{d1} = a(q_1)^b \tag{7.42}$$

$$h_{d2} = a(q_2)^b \tag{7.43}$$

These equations may be solved simultaneously for a and b producing

$$b = \frac{\ln \frac{h_{d2}}{h_{d1}}}{\ln \frac{q_2}{q_1}} \qquad (7.44)$$

$$a = \frac{h_{d1}}{(q_1)^{\left(\ln \frac{h_{d2}}{h_{d1}}\right)/\left(\ln \frac{q_2}{q_1}\right)}} \qquad (7.45)$$

If the number of data points is more than two, the data may be grouped to form two data points. Using i as the index for the first group and j as the index for the second,

$$h_{d1} = \frac{\sum^{s_i} h_{di}}{s_i}; \qquad q_1 = \frac{\sum^{s_i} q_i}{s_i} \qquad (7.46)$$

$$h_{d2} = \frac{\sum^{s_j} h_{dj}}{s_j}; \qquad q_2 = \frac{\sum^{s_j} q_j}{s_j} \qquad (7.47)$$

s_i and s_j refer to the total number of member elements in the respective groups. The previous equations are actually calculating the means of the two groups.

Example 7.3 In order to determine the values of a and b of Equation (7.23), experiments were performed on uniform sands and anthracite media yielding the following results below. Calculate the a's and b's corresponding to the respective diameters.

Dia., (mm)	q (mg/cm^3)	h_d (m)
	Uniform Sand	
0.5	1.08	0.06
0.5	7.2	4.0
0.7	1.6	0.06
0.7	11.0	4.0
1.0	2.6	0.06
1.0	17.0	4.0
	Uniform Anthracite	
1.0	6.2	0.06
1.0	43	4.0
1.6	8.1	0.06
1.6	58	4.0
2.0	9.2	0.06
2.0	68	4.0

Solution: Uniform sand, diameter $= 0.5$ mm:

$$a = \frac{h_{d1}}{(q_1)^{\left(\ln\frac{h_{d2}}{h_{d1}}\right)\big/\left(\ln\frac{q_2}{q_1}\right)}} = \frac{0.06}{(1.08)^{\left(\ln\frac{4.0}{0.06}\right)\big/\left(\ln\frac{7.2}{1.08}\right)}} = 0.051 \quad \textbf{Ans}$$

$$b = \frac{\ln\frac{h_{d2}}{h_{d1}}}{\ln\frac{q_2}{q_1}} = \frac{\ln\frac{4.0}{0.06}}{\ln\frac{7.2}{1.08}} = 2.21 \quad \textbf{Ans}$$

Uniform sand, diameter $= 0.7$ mm:

$$a = \frac{h_{d1}}{(q_1)^{\left(\ln\frac{h_{d2}}{h_{d1}}\right)\big/\left(\ln\frac{q_2}{q_1}\right)}} = \frac{0.06}{(1.60)^{\left(\ln\frac{4.0}{0.06}\right)\big/\left(\ln\frac{11}{1.60}\right)}} = 0.0216 \quad \textbf{Ans}$$

$$b = \frac{\ln\frac{h_{d2}}{h_{d1}}}{\ln\frac{q_2}{q_1}} = \frac{\ln\frac{4.0}{0.06}}{\ln\frac{11}{1.60}} = 2.18 \quad \textbf{Ans}$$

Similar procedures are applied to the rest of the data. The following is the final tabulation of the answers:

	Diameter (mm)	a	b
Sand	0.5	0.051	2.24
	0.7	0.0216	2.18
	1.0	0.0133	2.24
Anthracite	1.0	0.00114	2.17
	1.6	0.00069	2.13
	2.0	0.00057	2.01

From the results of this example, the equations below with the values of b averaged are obtained.

$$h_d = 0.051(q)^{2.22} \text{ for uniform sand, 0.5 mm diameter}$$
$$h_d = 0.022(q)^{2.22} \text{ for uniform sand, 0.7 mm diameter}$$
$$h_d = 0.0133(q)^{2.22} \text{ for uniform sand, 1.0 mm diameter}$$
$$h_d = 0.00114(q)^{2.10} \text{ for uniform anthracite, 1.0 mm diameter}$$
$$h_d = 0.00069(q)^{2.10} \text{ for uniform anthracite, 1.6 mm diameter}$$
$$h_d = 0.00057(q)^{2.10} \text{ for uniform anthracite, 2.0 mm diameter}$$

The h's and q's in the previous equations are in meters and mg/cm^3, respectively. Also, the head losses for diameters not included in the equations may be interpolated or extrapolated from the results obtained from the equations.

Example 7.4 The amount of suspended solids removed in a uniform anthracite medium, 1.8 mm in diameter is 40 mg/cm^3. Determine the head loss due to suspended solids.

Solution:

$$h_d = 0.00069(q)^{2.10} \text{ for uniform anthracite, 1.6 mm diameter meter}$$

$$= 0.00069(40)^{2.10} = 1.59 \text{ m}$$

$$h_d = 0.00057(q)^{2.10} \text{ for uniform anthracite, 2.0 mm diameter}$$

$$= 0.00057(40)^{2.10} = 1.32 \text{ m}$$

Interpolating, let x be the head loss corresponding to the 1.8-mm-diameter media.

$$\frac{1.59 - x}{1.59 - 1.32} = \frac{1.6 - 1.8}{1.6 - 2.0}; \quad x = 1.46 \text{ m} \quad \textbf{Ans}$$

Example 7.5 According to a modeling evaluation of the Pine Hill Run sewage discharge permit, a secondary-treated effluent of 20 mg/L can no longer be allowed into Pine Hill Run, an estuary tributary to the Cheseapeake Bay. To meet the new, more stringent discharge requirement, the town decided to investigate filtering the effluent. A pilot study was conducted using a dual-media filter composed of anthracite as the upper 30-cm part and sand as the next lower 30-cm part of the filter. The results are shown in the following table, where c_o is the concentration of solids at the influent and c is the concentration of solids in the water in the pores of the filter.

If the respective average sizes of the anthracite and sand layers are 1.6 mm and 0.5 mm, what is the length of the filter run to a terminal head loss of 3 m at a filtration rate of 200 L/m$^2 \cdot$ min? Assume the clean water head loss is 0.793 m. Note: Terminal head loss is the loss when the filter is about to be cleaned.

Q (L/m$^2 \cdot$ min)	Depth, cm	c/c_o	Medium	Q (L/m$^2 \cdot$ min)	Depth, (cm)	c/c_o	Medium
80	0	1.0	Anthracite	80	8	0.32	Anthracite
80	16	0.27	Anthracite	80	24	0.24	Anthracite
80	30	0.23	Anthracite	80	40	0.20	Sand
80	48	0.20	Sand	80	56	0.20	Sand
160	0	1.0	Anthracite	160	8	0.46	Anthracite
160	16	0.30	Anthracite	160	24	0.27	Anthracite
160	30	0.25	Anthracite	160	40	0.22	Sand
160	48	0.22	Sand	160	56	0.22	Sand
240	0	1.0	Anthracite	240	8	0.59	Anthracite
240	16	0.39	Anthracite	240	24	0.30	Anthracite
240	30	0.27	Anthracite	240	40	0.23	Sand
240	48	0.23	Sand	240	56	0.23	Sand

Solution: Because 200 L/m^2 · min is between 160 and 240 L/m^2 · min, only the data for these two flows will be analyzed. The length of the filter run for the 200 L/m^2 · min flow will be interpolated between the lengths of filter run of the 160 and the 240 flows.

| Depth (cm) | \multicolumn{3}{c}{Filtration Rate (L/cm^2 · min)} | | | | |

	\multicolumn{3}{c}{160}			\multicolumn{3}{c}{240}	
Depth (cm)	**(1)** c/c_o	**(2)** c (mg/L)	**(3)** Δc_i	**(4)** c/c_o	**(5)** c (mg/L)	**(6)** Δc_i
0	1	20.	—	1	20	—
8	0.46	9.2	−10.8	0.59	11.8	−8.2
16	0.30	6.0	−3.2	0.39	7.8	−4.0
24	0.27	5.4	−0.6	0.3	6.0	−1.8
30	0.25	5.0	−0.4	0.27	5.4	−0.6
40	0.22	4.4	−0.6	0.23	4.6	−0.8
48	0.22	4.4	0	0.23	4.6	0
56	0.22	4.4	0	0.23	4.6	0

Column 1 = given
Column 2 = 20 (column 1)
Column 3 = $c_{i,m+1} - c_{i,m}$; for example, −10.8 = 9.2 − 20

$$q = -tV_s \frac{\Delta c}{\Delta l}$$

$$160 \text{ L/m}^2 \cdot \text{min} = \frac{160(1000) \text{ cm}^3}{100^2 \text{ cm}^2 \cdot \text{min}} = 16 \text{ cm/min}$$

$$240 \text{ L/m}^2 \cdot \text{min} = \frac{240(1000) \text{ cm}^3}{100^2 \text{ cm}^2 \cdot \text{min}} = 24 \text{ cm/min}$$

$$1 \text{ mg/L} = \frac{1 \text{ mg}}{1000 \text{ cm}^3} = 10^{-3} \text{ mg/cm}^3$$

Therefore,

$$\Delta c_i \text{ mg/L} = \Delta c_i(10^{-3}) \text{ mg/cm}^3$$

$$q = -tV_s \frac{\Delta c}{\Delta l} = -tV_s(10^{-3})\frac{\Delta c_{i,n}}{\Delta \ell_i} \text{ mg/cm}^3$$

$$h_d = 0.00069(q)^{2.10} \text{ for uniform anthracite, 1.6 mm diameter}$$

$$h_d = 0.051(q)^{2.22} \text{ for uniform sand, 0.5 mm diameter}$$

	For the 160 L/cm²·min			Run Length (min)					
				600		900		1200	
Depth (cm)	Δc_i (mg/L)	Δl_i (cm)	$\Delta c_i/\Delta \ell_i$	q_i (mg/cm³)	h_{di} (m³)	q_i (mg/cm³)	h_{di} (m³)	q_i (mg/cm³)	h_{di} (m³)
0	—			—	—	—	—	—	—
8	−10.8	8	−1.35	12.96	0.15	19.44	0.35	25.92	0.64
16	−3.2	8	−0.4	3.84	0.01	5.76	0.027	7.68	0.05
24	−0.6	8	−0.075	0.72	0.00	1.08	0.00	1.44	0.001
30	−0.4	6	−0.07	0.64	0.00	0.96	0.00	1.28	0.001
40	−0.6	10	−0.06	0.576	0.015	0.864	0.037	1.15	0.07
48	0	—	—	—	—	—	—	—	—
				$\Sigma h_i = 0.175$		$\Sigma h_i = 0.414$		$\Sigma h_i = 0.762$	

The terminal head loss of 3 m is composed of the clean water head loss of 0.793 m and the head loss due to deposited solids of h_s

$$h_d = 3 - 0.793 = 2.21 \text{ m}$$

900	0.414
1200	0.762
x	2.21

$$\frac{x - 1200}{1200 - 900} = \frac{2.21 - 0.762}{0.762 - 0.414}$$

$$x = 2448 \text{ min}$$

x	2.21
1500	2.26
2100	4.64

$$\frac{x - 1500}{1500 - 2100} = \frac{2.21 - 2.26}{2.26 - 4.64}$$

$$x = 1487$$

160	2448
200	y
240	1487

$$\frac{y - 2448}{1487 - 2448} = \frac{200 - 160}{240 - 160}$$

$$y = 1968 \text{ min} \quad \textbf{Ans}$$

	For the 240 L/cm²·min			Run Length (min)			
				1500		2100	
Depth (cm)	Δc_i (mg/L)	Δl_i (cm)	$\Delta c_i/\Delta l_i$	q_i (mg/cm³)	h_{di} (m³)	q_i (mg/cm³)	h_{di} (m³)
0	—	—	—	—	—	—	—
8	−8.2	8	−1.03	37.08	1.36	51.91	2.76
16	−4.0	8	−0.5	18	0.3	25.2	0.61
24	−1.8	8	−0.23	8.28	0.06	11.59	0.12
30	−0.6	6	−0.1	3.6	0.01	5.04	0.02
40	−0.8	10	−0.1	2.88	0.53	4.03	1.13
48	0	—	—	—	—	—	—
				$\Sigma h_i = 2.26$		$\Sigma h_i = 4.64$	

7.5 BACKWASHING HEAD LOSS
IN GRANULAR FILTERS

In the early development of filters, units that had been clogged were renewed by scraping the topmost layers of sand. The scraped sands were then cleaned by sand washers. In the nineteenth century, studies to clean the sand in place, rather than taking out of the unit, led to the development of the rapid-sand filter. The method of cleaning is called *backwashing*.

In backwashing, clean water is introduced at the filter underdrains at such a velocity as to expand the bed. The expansion frees the sand from clogging materials by causing the grains to rub against each other dislodging any material that have been clinging onto their surfaces. The dislodged materials are then discharged into washwater troughs.

To expand the bed, a force must be applied at the bottom. Per unit area of the filter, this backwashing force is simply equal to the pressure at the bottom of the filter acting upward upon a column of water with the grains suspended. Atmospheric pressure will also act on this column from its top. If the pressure used for measurement at the bottom is pressure above atmospheric, the pressure acting on the top is considered zero. The backwashing force pressure is then simply equal to the specific weight of water γ_w times all of the heights $h_{Lb} + l' + l_e$ or $\gamma_w(h_{Lb} + l' + l_e)$, where l_e is the expanded depth of the bed having a depth of l and l' is the difference between the level at the trough and the limit of bed expansion l_e (see Figure 7.2). h_{Lb} is called the *backwashing head loss*; h_{Lb} does not include friction losses through the washwater pipe, contraction loss from the washwater tank into the wash water pipe, valves, bends, and loss at the underdrain system. Thus, in design, these losses must be accounted for and added to h_{Lb}.

The weight of the suspended grains is $l_e(1 - \eta_e)\gamma_p$, where η_e is the expanded porosity of the bed and γ_p is the specific weight of the grain particles. The weight of the column of water is $(l_e\eta_e + l')\gamma_w$. These two weights are acting downward against the backwashing force. During backwashing, the whole column will be at steady state; thus, applying Newton's second law will produce $(h_{Lb} + l' + l_e)\gamma_w - l_e(1 - \eta_e)\gamma_p - (l_e\eta_e - l')\gamma_w = 0$. Solving this equation for the backwashing head loss produces

$$h_{Lb} = \frac{l_e(1 - \eta_e)(\gamma_p - \gamma_w)}{\gamma_w} \tag{7.48}$$

Calling V_b the backwashing velocity and v the settling velocity of the grains, the commonly used expression for η_e is the following empirical equation

$$\eta_e = \left(\frac{V_b}{v}\right)^{0.22} \tag{7.49}$$

For stratified beds, Equation (7.48) must be applied layer by layer. The backwashing head losses in each layer are then summed to produce the total head loss of the bed. Thus,

$$h_{Lb} = \sum h_{Lb,i} = \sum \frac{l_{e,i}(1 - \eta_{e,i})(\gamma_p - \gamma_w)}{\gamma_w} \tag{7.50}$$

where the index i refers to the ith layer. If x_i is the fraction of grain particles in layer i then $l_{e,i}$ equals $x_i l_e$. Thus, h_{Lb} becomes

$$h_{Lb} = \sum h_{Lb,i} = \frac{l_e(\gamma_p - \gamma_w)}{\gamma_w} \sum x_i(1 - \eta_{e,i}) \tag{7.51}$$

The fractional expansion of the bed may be determined by considering that the total mass of expanded and unexpanded grains are equal. Per unit surface area of the filter and assuming the porosity of the unexpanded bed η is constant throughout the depth, the material balance for each layer i (l_i for the unexpanded layer and $l_{e,i}$ for the expanded layer) is

$$l_i(1 - \eta) \gamma_p = l_{e,i}(1 - \eta_{e,i}) \gamma_p \Rightarrow l(1 - \eta) \gamma_p = l_e(1 - \eta_e) \gamma_p \tag{7.52}$$

Solving for $l_{e,i}$

$$l_{e,i} = \frac{l_i(1 - \eta)}{(1 - \eta_{e,i})} \tag{7.53}$$

Summing the expansion per layer over the depth of the bed, noting that $l_i = x_i l$, the factional bed expansion is obtained as

$$\frac{l_e}{l} = (1 - \eta) \sum \frac{x_i}{1 - \eta_{e,i}} = \frac{1 - \eta}{1 - \eta_e} \tag{7.54}$$

Example 7.6 A sharp filter sand has the sieve analysis shown in the following table. The average porosity of the stratified bed is 0.42. The lowest temperature anticipated of the water to be filtered is 4°C. Determine the percentage bed expansion and the backwashing head loss if the filter is to be backwashed at a rate of $0.4(10^{-2})$ m/s $\cdot$ $\rho_p = 2{,}650$ kg/m^3. The bed depth is 0.7 m.

Solution:

$$\frac{l_e}{l} = (1 - \eta) \sum \frac{x_i}{1 - \eta_{e,i}} = (1 - 0.42) \sum \frac{x_i}{1 - \eta_{e,i}} = (0.58) \sum \frac{x_i}{1 - \eta_{e,i}}$$

$$h_{Lb} = \sum h_{Lb,i} = \frac{l_e(\gamma_p - \gamma_w)}{\gamma_w} \sum x_i(1 - \eta_{e,i}) = \frac{l_e(9.81)(2{,}650 - 1000)}{9.81(1000)} \sum x_i(1 - \eta_{e,i})$$

$$= 1.65 l_e \sum x_i(1 - \eta_{e,i})$$

$$\eta_e = \left(\frac{V_B}{V}\right)^{0.22} = \left(\frac{0.4(10^{-2})}{V}\right)^{0.22} = \frac{0.30}{v^{0.22}}$$

$$v = \sqrt{\frac{4}{3} g \frac{(\rho_p - \rho_w)d}{C_d \rho_w}} = \sqrt{\frac{4}{3} 9.81 \frac{(2{,}650 - 1000)d}{C_D(1000)}} = 4.65 \sqrt{\frac{d}{C_D}}$$

$$\beta = 0.77$$

Sieve No.	Average Size (mm)	x_i
14–20	1.0	0.01
20–28	0.7	0.05
28–32	0.54	0.15
32–35	0.46	0.18
35–42	0.38	0.18
42–48	0.32	0.20
48–60	0.27	0.15
60–65	0.23	0.07
65–100	0.18	0.01

$$d = 1.24\beta^{0.333}d_p = 1.24(0.77)^{0.333}d_p = 1.14d_p$$

Therefore,

$$v = 4.65\sqrt{\frac{1.14d_p}{C_D}} = 4.96\sqrt{\frac{d_p}{C_D}}$$

$$C_D = \frac{24}{Re} \ (\text{for } Re < 1)$$

$$C_D = \frac{24}{Re} + \frac{3}{\sqrt{Re}} + 0.34 \ (\text{for } 1 \le Re \le 10^4)$$

$$C_D = 0.4 \ (\text{for } Re > 10^4)$$

$$Re = \frac{dv\rho}{\mu} = \frac{1.14d_p v(1000)}{15(10^{-4})} = 760{,}000 \ vd_p$$

Determine the settling velocities of the various fractions:

Sieve No.	d_{pi} (mm)	x_i	C_{Di}	v_i (m/s)	Re_i	$\eta_{e,i}$	$1 - \eta_{e,i}$	$x_i/(1 - \eta_{e,i})$	$\dfrac{x_i}{(1 - \eta_{e,i})}$
14–20	1.0	0.01	0.4	0.248	188				
—			0.69	0.189	143				
—			0.76	0.180	136				
—			0.77	0.178	135				
—			0.76	0.180	136	0.44	0.56	0.020	0.0056
20–28	0.70	0.05	1.05	0.13	69	0.47	0.53	0.09	0.0265
28–32	0.54	0.15	1.39	0.10	41	0.50	0.50	0.30	0.075
32–35	0.46	0.18	1.72	0.08	29	0.52	0.48	0.38	0.0864
35–42	0.38	0.18	2.21	0.07	20	0.54	0.46	0.39	0.0828
42–48	0.32	0.20	3.02	0.05	13	0.58	0.42	0.48	0.084
48–60	0.27	0.15	4.94	0.03	7	0.65	0.35	0.43	0.0525
60–65	0.23	0.07	6.48	0.03	5	0.65	0.35	0.20	0.0245
65–100	0.18	0.01	7.8	0.02	4	0.71	0.29	0.03	0.0029
							Σ	2.32	0.44

Therefore, percentage bed expansion $= (0.58)(\Sigma x_i/1 - \eta_{e,i})(100) = 0.58(2.32) \times (100) = 134\%$ **Ans**

Note: In practice, the bed is normally expanded to 150%.

$$h_{Lb} = 1.651 l_e \sum x_i(1 - \eta_{e,i}) = 1.65(1.34)(0.7)(0.44) = 0.68 \text{ m} \quad \textbf{Ans}$$

7.6 CAKE FILTRATION

Sludges may be reduced in volume by using the unit operation of dewatering to remove excess water. In dewatering using a vacuum (such as using a vacuum filter) or the plate-and-frame press, a cake is formed on the surface of the filter cloth. Thus, these processes of dewatering may be called *cake filtration*. In cake filtration, the flow of the filtrate in its most elementary form, may be considered as through a tightly packed bank of small crooked tubes across the cake. The rightmost drawing of Figure 7.11 is a schematic section of a filter cake and filter cloth. The general equation of momentum across a filter, Equation (7.15), may be applied across the cake; for convenience, the equation is reproduced below.

$$-\Delta p A = \frac{S_o l(1 - \eta)\overline{V}_s^2 \rho}{d\eta^2}\left[6K_i + \frac{36K_s(1 - \eta)}{d\overline{V}_s \rho/\mu}\right] \tag{7.15}$$

Because of pressure, the solids are tightly packed. The first term on the right-hand side of the above equation is the inertial force. Because the cake is tightly packed, this initial force must disappear upon entrance of flow into the cake and the flow is not accelerated; the first term is therefore equal to zero. Thus, the only valid term of the right-hand side of Equation (7.15) when applied to cake filtration is the term $36K_s(1 - \eta)/(d\overline{V}_s \rho/\mu)$ and the equation reduces to

$$-\Delta p A = \frac{S_o l(1 - \eta)\overline{V}_s^2 \rho}{d\eta^2}\left[\frac{36K_s(1 - \eta)}{d\overline{V}_s \rho/\mu}\right] = \frac{S_o l(1 - \eta)\overline{V}_s^2 \rho}{d\eta^2}\left[\frac{150(1 - \eta)}{d\overline{V}_s \rho/\mu}\right] \tag{7.55}$$

Because $A = S_o \eta$,

$$-\Delta p = \frac{l(1 - \eta)\overline{V}_s^2 \rho}{d\eta^3}\left[\frac{150(1 - \eta)}{d\overline{V}_s \rho/\mu}\right] \Rightarrow \frac{-\Delta p}{l} = \frac{(1 - \eta)\overline{V}_s^2 \rho}{d\eta^3}\left[\frac{150(1 - \eta)}{d\overline{V}_s \rho/\mu}\right].$$

In the derivation of Equation (7.15), the equation was derived by applying the equation of momentum in the direction of fluid flow. In the present case, however, the equation is applied in the direction opposite to the fluid flow. Hence, the $-\Delta p$ becomes Δp. Also, in terms of differentials, $\Delta p/l = dp/dl$; therefore,

$$\frac{dp}{dl} = \frac{(1 - \eta)\overline{V}_s^2 \rho}{d\eta^3}\left[\frac{150(1 - \eta)}{d\overline{V}_s \rho/\mu}\right] \Rightarrow dp = \frac{(1 - \eta)\rho}{d\eta^3}\left[\frac{150(1 - \eta)}{d\rho}\right]\mu\overline{V}_s \, dl \tag{7.56}$$

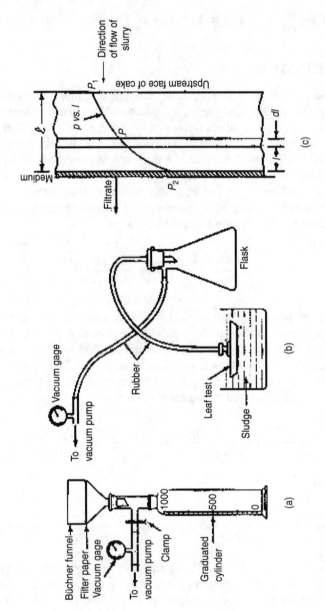

FIGURE 7.11 (a) Büchner funnel filtration assembly; (b) leaf filter assembly; and (c) mechanics of cake filtration.

If ρ_p is the density of solids, the mass dm in the differential thickness of cake dl is $dm = S_o dl(1 - \eta)\rho_p$. Solving this for dl and substituting, Equation (7.56) becomes

$$dp = \frac{(1-\eta)\rho}{d\eta^3}\left[\frac{150(1-\eta)}{d\rho}\right]\mu\bar{V}_s dl = \frac{(1-\eta)\rho}{d\eta^3}\left[\frac{150(1-\eta)}{d\rho}\right]\mu\bar{V}_s\frac{dm}{S_o(1-\eta)\rho_p}$$

(7.57)

In this equation, μ and $\bar{V}_s$ influence directly the flow characteristics of the filtering water. Since S_o determines the value of $\bar{V}_s$, it also influences directly the flow characteristics of the fluid. All the other factors in the equation are inherent characteristics of the cake. All these cake characteristics may be lumped up into a single term. Call this term as the *specific cake resistance* α. Hence, in terms of the specific cake resistance, dp is

$$dp = \alpha\frac{\mu\bar{V}_s}{S_o}dm$$

(7.58)

Examining the various parameters constituting α, as far as p and m as the variables of integration are concerned, they are all constant. Hence, α is constant over these variables of integration. In addition, $\bar{V}_s$ is also a constant over these variables of integration. Integrating from P_2 to P_1,

$$(P_1 - P_2) = -\Delta P = \alpha\frac{\mu\bar{V}_s}{S_o}m_c$$

(7.59)

m_c is the total mass of solid collected on the filter cloth.

Note that the resistance of the filter cloth has heretofore been neglected. Calling this resistance R_m and including it, the total pressure drop may be written as

$$-\Delta P = \mu\bar{V}_s\left(\frac{\alpha m_c}{S_o} + R_m\right)$$

(7.60)

7.6.1 Determination of $\bar{\alpha}$

Because V is the volume of the filtrate collected at any time t, $\bar{V}_s$ is $(d\bar{V}/dt)/S_o$; also, expressing m_c as cV (where c is the mass of cake collected per unit volume of filtrate), substituting in Equation (7.60), and integrating under the assumption of constant pressure differential operation yields

$$\frac{t}{V} = \frac{\mu c\bar{\alpha}}{2(-\Delta P)S_o^2}V + \frac{\mu R_m}{(-\Delta P)S_o}$$

(7.61)

Vacuum filtration is normally conducted under constant pressure differential; hence, $-\Delta P$ has been made constant. In addition, α may vary over time; hence, its average value $\bar{\alpha}$ has been used.

Equation (7.61) is an equation of a straight line between t/V and V, whose slope m is given by $\mu c \bar{\alpha}/[2(-\Delta P)S_o^2]$. By determining this slope from experimental data, the value of $\bar{\alpha}$ can be determined. The slope may be determined by

$$m = \frac{\frac{1}{n_2}\sum_1^{n_2}\left(\frac{t}{V}\right)_i - \frac{1}{n_1}\sum_1^{n_1}\left(\frac{t}{V}\right)_i}{\frac{1}{n_2}\sum_1^{n_2}(V)_i - \frac{1}{n_1}\sum_1^{n_1}(V)_i} \tag{7.62}$$

Thus,

$$\bar{\alpha} = \frac{2m(-\Delta P)S_o^2}{\mu c} \tag{7.63}$$

n_1 and n_2 and the number of elements in the respective group. Remember that to fit a straight line into experimental data, the data must be divided into two groups: n_1 is the number of elements in the first group and n_2 and is the number of elements in the second group.

The laboratory experiment involves using a Buchner funnel by adopting the setup shown in Figure 7.11. Operating the funnel at a constant pressure difference $-\Delta P$, the amount of filtrate collected is recorded with time. The data collected gives the relationship between t/V and V as called for by Equation (7.61). The cake collected is also weighed to determine c; μ is determined from the temperature of the filtrate.

Example 7.7 A Buchner funnel experiment to determine the specific cake resistance of a certain sludge is performed. The results are as shown in the following table. $-\Delta P = 51$ cm Hg, filter area $= 550$ cm^2, $\mu = 15(10^{-4})$ kg/m $\cdot$ s, and $c = 0.25$ g/cm^3. Determine $\bar{\alpha}$.

Volume of Filtrate (mL)	Time (sec)	t/V (10^6 s/m^3)
25	48	1.92
50	150	3.0
75	308	4.12
100	520	5.2

Solution:

$$m = \frac{\frac{1}{n_2}\sum_1^{n_2}\left(\frac{t}{V}\right)_i - \frac{1}{n_1}\sum_1^{n_1}\left(\frac{t}{V}\right)_i}{\frac{1}{n_2}\sum_1^{n_2}(V)_i - \frac{1}{n_1}\sum_1^{n_1}(V)_i}$$

$$\bar{\alpha} = \frac{2m(-\Delta P)S_o^2}{\mu c}$$

$$m = \frac{(1.92 + 3.0)/2 - (4.12 + 5.2)/2}{(25 + 50)/2 - (75 + 100)/2}(10^6)$$

$$= 0.044(10^6)\frac{s}{m^6 \cdot mL} = 4.4(10^{10})\frac{s}{m^6}$$

$$-\Delta P = \frac{51}{76}(101,330) = 67,998\frac{N}{m^2}$$

$$\text{Filter area} = 550(10^{-2})^2 = 0.55 \text{ m}^2$$

$$c = 0.25\left(\frac{10^{-3}}{(10^{-2})^3}\right) = 250\frac{kg}{m^3}$$

$$\bar{\alpha} = \frac{2m(-\Delta P)S_o^2}{\mu c} = \frac{2(4.4(10^{10})(67,998)(0.055)^2}{15(10^{-4})(250)}$$

$$\bar{\alpha} = 4.8(10^{13})\frac{1}{kg \cdot m} \quad \textbf{Ans}$$

7.6.2 DESIGN CAKE FILTRATION EQUATION

In an actual filtration installation, may it be a vacuum filter or a plate-and-frame press, the resistance of the filter medium is practically negligible; it may, therefore, be neglected. Considering this fact, Equation (7.61) may be written as

$$\frac{t}{V} = \frac{\mu c \bar{\alpha}}{2(-\Delta P)S_o^2}V \tag{7.64}$$

Since it is important to be able to calculate the amount of dewatered sludge that is finally to be disposed of, for convenience Equation (7.64) may be expressed in a form that will give this amount directly without considering the filtrate. This is done by utilizing the specific loading rate, also called *filter yield* L_f defined as

$$L_f = \frac{cV}{S_o t} \tag{7.65}$$

From this definition, L_f is the amount of cake formed per unit area of filter cloth per unit of time. In vacuum filtration, the only time that the cake is formed is when the drum is submerged in the tank. In pressure filtration, the only time that the cake is also formed is when the sludge is pumped into the plates. Thus, t in the previous equation is called the form time t_f. Also, the filters operate on a cycle. Calling the cycle time as t_c, t_f may be expressed as a fraction f of t_c. Thus, $t = t_f = ft_c$ may be substituted in Equation (7.65) and, if this equation is substituted in Equation (7.64) and the result simplified and rearranged, we have

$$L_f = \sqrt{\frac{2(-\Delta P)c}{\mu \bar{\alpha} f t_c}} \qquad (7.66)$$

For incompressible cakes, Equation (7.66) is the design cake filtration equation. In vacuum filtration, f is equal to the fraction of submergence of the drum. Also, for the pressure filter, f is the fraction of the total cycle time that the sludge is pumped into the unit.

For compressible cakes such those of sewage sludges, $\bar{\alpha}$ is not constant. Equation (7.66) must therefore be modified for the expression of the specific cake resistance. The usual form used is

$$\bar{\alpha} = \bar{\alpha}_o(-\Delta P)^s \qquad (7.67)$$

where s is a measure of cake compressibility. If s is zero, the cake is incompressible and $\bar{\alpha}$ equals $\bar{\alpha}_o$, the constant of proportionality. Substituting in Equation (7.66),

$$L_f = \sqrt{\frac{2(-\Delta P)^{1-s}c}{\mu \bar{\alpha}_o f t_c}} \qquad (7.68)$$

7.6.3 DETERMINATION OF CAKE FILTRATION PARAMETERS

In design, the parameters L_f, $-\Delta P$, f, and t_c may be specified. μ is specified from the temperature of filtration. The value of c may be determined by performing an experiment of the particular type of sludge. To determine $\bar{\alpha}_o$, take the logarithms of both sides of Equation (7.67),

$$ln\,\bar{\alpha} = ln\,\bar{\alpha}_o + s\,ln(-\Delta P) \qquad (7.69)$$

In this equation, s is the slope of the straight-line equation between $ln\,\bar{\alpha}$ as the dependent variable and $ln(-\Delta P)$ as the independent variable. By analogy with Equation Equation (7.62),

$$s = \frac{\frac{1}{n_2}\sum_1^{n_2}(ln\,\bar{\alpha})_i - \frac{1}{n_1}\sum_1^{n_1}(ln\,\bar{\alpha})_i}{\frac{1}{n_2}\sum_1^{n_2}(ln(-\Delta P))_i - \frac{1}{n_1}\sum_1^{n_1}(ln(-\Delta P))_i} \qquad (7.70)$$

And,

$$\bar{\alpha}_o = e^{\frac{1}{n_1}\sum_1^{n_1}(ln\bar{\alpha})_i - s\frac{1}{n_1}\sum_1^{n_1}[ln(-\Delta P)]_i} \qquad (7.71)$$

The experimental procedure to gather data used to obtain s and $\bar{\alpha}_o$ may be performed using the leaf filter assembly of Figure 7.11, although the Buchner experiment may also be used. The leaf filter is immersed in the sludge and a vacuum is applied to suck the filtrate during the duration of the form time t_f. For a given $-\Delta P$, the volume of filtrate over the time t_f is collected. This will give one value of $\bar{\alpha}$. To satisfy the requirement of Equation (7.69), at least two runs are made at two different pressure drops. From these pairs, the parameters $\bar{\alpha}_o$ and s may be calculated.

Example 7.8 A leaf-filter experiment is run to determined $\bar{\alpha}_o$ and s for a CaCO$_3$ slurry in water producing the results below. $\mu = 8.9(10^{-4})$ kg/m·s. The filter area is 440 cm^2, the mass of solid per unit volume of filtrate is 23.5 g/L, and the temperature is 25°C. Calculate $\bar{\alpha}_o$ and s.

$-\Delta P$ (kN/m^2)	46.18	111.67
V (L)	Time (sec)	
0.5	17.3	6.8
1.0	41.3	19.0
1.5	72.0	34.6
2.0	108.3	53.4
2.5	152.1	76.0
3.0	210.7	102.0

Solution: For $-\Delta P = 46.18$ kN/m^2:

V (L)	t (sec)	t/V
0.5	17.3	34.6
1.0	41.3	41.3
1.5	72.0	48.0
2.0	108.3	54.15
2.5	152.1	60.84
3.0	201.7	67.33

$$
m = \frac{\dfrac{1}{n_2}\sum_1^{n_2}\left(\dfrac{t}{V}\right)_i - \dfrac{1}{n_1}\sum_1^{n_1}\left(\dfrac{t}{V}\right)_i}{\dfrac{1}{n_2}\sum_1^{n_2}(V)_i - \dfrac{1}{n_1}\sum_1^{n_1}(V)_i}
$$

$$
= \frac{(34.6 + 41.3 + 48.0)/3 - (54.15 + 60.84 + 67.33)/3}{(0.5 + 1.0 + 1.5)/3 - (2.0 + 2.5 + 3.0)/3}
$$

$$
= \frac{-19.47}{-1.5} = 12.98\frac{s}{L^2} = 1.30(10^7)\frac{s}{m^6}
$$

$$\bar{\alpha} = \frac{2m(-\Delta P)S_o^2}{\mu c}$$

$$-\Delta P = 46,180 \frac{N}{m^2}$$

Filter area $= 440(10^{-2})^2 = 0.044 \text{ m}^2 = S_o$

$$c = 23.5 \left(\frac{10^{-3}}{10^{-3}} \right) = 23.5 \frac{kg}{m^3}; \qquad \mu = 8.9(10^{-4})kg/m \cdot s$$

$$\bar{\alpha} = \frac{2m(-\Delta P)S_o^2}{\mu c} = \frac{2(1.30)(10^7)(46,180)(0.044)^2}{8.9(10^{-4})(23.5)} = 1.11(10^{11}) \frac{1}{kg \cdot m}$$

For $-\Delta P = 111.67 \text{ kN/m}^2$:

V (L)	t (sec)	t/V
0.5	6.8	13.6
1.0	19.0	19.0
1.5	34.6	23.07
2.0	53.4	26.7
2.5	76.0	30.4
3.0	102.0	34

$$m = \frac{(13.6 + 19.0 + 23.07)/3 - (26.7 + 30.4 + 34)/3}{(0.5 + 1.0 + 1.5)/3 - (2.0 + 2.5 + 3.0)/3}$$

$$= 7.87 \frac{s}{L^2} = 7.87(10^6) \frac{s}{m^6}$$

$$-\Delta P = 111,670 \frac{N}{m^2}$$

$$\bar{\alpha} = \frac{2m(-\Delta P)S_o^2}{\mu c} = \frac{2(7.87)(10^6)(111,670)(0.044)^2}{8.9(10^{-4})(23.5)} = 1.63(10^{11}) \frac{1}{kg \cdot m}$$

$$s = \frac{\frac{1}{n_2}\sum_1^{n_2}(\ln \bar{\alpha})_i - \frac{1}{n_1}\sum_1^{n_1}(\ln \bar{\alpha})_i}{\frac{1}{n_2}\sum_1^{n_2}(\ln(-\Delta P))_i - \frac{1}{n_1}\sum_1^{n_1}(\ln(-\Delta P))_i} = \frac{\frac{1}{1}\ln[1.11(10^{11})] - \frac{1}{1}\ln[1.63(10^{11})]}{\frac{1}{1}\ln[46,180] - \frac{1}{1}\ln[111,670]}$$

$$= \frac{25.43 - 25.82}{10.74 - 11.62} = 0.443 \quad \textbf{Ans}$$

$$\bar{\alpha}_o = e^{\frac{1}{n_1}\sum_1^{n_1}(\ln \bar{\alpha})_i - s\frac{1}{n_1}\sum_1^{n_1}[\ln(-\Delta P)]_i}$$

$$= e^{\ln 1.11(10^{11}) - \ln 46,180} = e^{25.43 - 10.74} = 2.40(10^6) \text{ in MKS units} \quad \textbf{Ans}$$

Example 7.9 A $CaCO_3$ sludge with cake filtration parameters determined in the previous example is to be dewatered in a vacuum filter under a vacuum of 630 mm Hg. The mass of filtered solids per unit volume of filtrate is to be 60 kg/m³. The filtration temperature is determined to be 25°C and the cycle time, half of which is form time, is five minutes. Calculate the filter yield.

Solution:

$$L_f = \sqrt{\frac{2(-\Delta P)^{1-s}c}{\mu \bar{\alpha} f t_c}}$$

$$-\Delta P = \frac{630}{760}(101{,}330) = 83{,}997.24 \text{ N/m}^2; \qquad c = 60 \text{ kg/m}^3; \qquad s = 0.443$$

$$\mu = 8.9(10^{-4}) \text{ kg/m} \cdot \text{s}; \qquad \bar{\alpha}_0 = 2.40(10^6) \text{ in MKS};$$

$$f = 0.5; \qquad t_c = 5 \text{ min} = 5(60) = 300 \text{ sec}$$

$$L_f = \sqrt{\frac{2(83{,}997.24)^{1-0.443}(60)}{8.9(10^{-4})(2.40)(10^6)(0.5)(300)}} = 0.46 \text{ kg/m}^2 \cdot \text{sec} \quad \textbf{Ans}$$

GLOSSARY

Backwashing—A method of cleaning filters where a clean treated water is made to flow in reverse through the filter for the purpose of suspending the grains to effect cleaning.

Cake-forming or cake filtration—Filtration in which cakes are formed in the process.

Dual-media gravity filter—A gravity filter composed of two filtering media.

Effective size—The size of sieve opening that passes the 10% finer of a medium sample.

Filter yield—The amount of cake formed during cake filtration.

Filtration—A unit operation of separating solids from fluids.

Granular filters—Filters where media are composed of granular materials.

Gravity filters—Filters that rely on the pull of gravity to create a pressure differential to force the water through the filter.

Hydraulic radius—Area of flow divided by the wetted perimeter or volume of flow divided by the wetted area.

Leaf filter—Filter that operates by immersing a component called a leaf into a bath of sludge or slurry and using a vacuum to suck the sludge onto the leaf.

Perforated filter—Filter whose filtering medium is made of perforated plates.

Plate-and-frame press—A pressure filter in which the mechanics of filtration is through cloths wrapped on a plates alternated and held in place by frames.

Pressure filtration—A method of filtration in which the pressure differential used to drive the filtration is induced by impressing a high pressure on the inlet side of the filter medium.

Pressure and vacuum filters—Filters that rely on applying some mechanical means to create the pressure differential necessary to force the water through the filter.

Rapid-sand filter—Gravity filter normally operated at a rate of 100 to 200 $m^3/d \cdot m^2$.

Slow-sand filter—Gravity filter normally operated at a rate of 1.0 to 10 $m^3/d \cdot m^2$.

Smutzdecke—Layer of dirt that collects on top of slow-sand filters.

Specific cake resistance—A measure for the resistance of a cake to filtration.

Uniformity coefficient—The ratio of the size of the sieve opening that passes the 60% finer of the medium sample (P_{60}) to the size of the sieve opening that passes the 10% finer of the medium sample.

Vacuum filtration—A method of filtration in which the pressure differential used to drive the filtration is induced by a vacuum suction at the discharge side of the filter medium.

Woven septum filter—Filter with a filtering medium that is made of woven materials.

SYMBOLS

a	Coefficient in head loss equation for deposited materials
A	Cross-sectional area of flow (i.e., superficial area times the porosity; bounding surface area of control mass and control volume)
A_s	Surface area of grains in filter
b	Exponent in head loss equation for deposited materials
c	Concentration of solids introduced into bed
d	Diameter of spherical particle
d_p	Sieve diameter
f	Fraction of total cycle time t_c spent in actual filtration
f_p	Friction factor
g	Acceleration due to gravity
h_d	Head loss due to deposited materials
h_L	Head loss across filter
h_{Lb}	Backwashing head loss
h_{Lo}	Head loss across clean bed
K_i	Proportionality constant for the inertial force
K_s	Proportionality constant for the shear force
l	Length of bed
l_e	Expanded bed depth
L_f	Filter yield
$\dot{m}$	Rate of mass inflow into bed
m_c	Mass of cake per unit volume of filtrate
M	Control mass
$\hat{n}$	Unit normal vector

η_e	Expanded bed porosity
$\eta_{e,i}$	Expanded porosity of layer i
P_1	Pressure at point 1
P_2	Pressure at point 2
Δp	Pressure drop across filter medium
p_1	Percentage of the sample stock sand that is smaller than or equal to the desired P_{10} of the final filter sand
p_2	Percentage of the sample stock sand that is smaller than or equal to the desired P_{60} of the final filter sand
p_3	Percentage of the stock sand that is transformed into the final usable sand
p_4	Percentage of the stock sand that is too fine to be usable
p_5	Represent the percentage of the sample stock sand above which the sand is too coarse to be usable
q	Materials deposited per unit volume of filter bed
Q	Rate of flow into bed
r_H	Hydraulic radius
Re	Reynolds number
R_m	Resistance of filter medium
s	Exponent in compressible cake equation
s_p	Surface area of particles
S_o	Superficial area of bed
t_c	Cycle time of filtration
v	Settling velocity of particles
$\vec{v}$	Velocity vector in flows through bed
v_p	Volume of particles
$\overline{V}$	Characteristic average velocity in pipe; control volume
V_b	Backwashing velocity
$\overline{V}_s$	Superficial velocity
x_i	Fraction of the d_i particles in the ith layer l_i
α	Specific cake resistance
$\overline{\alpha}$	Average specific cake resistance
$\overline{\alpha}_o$	Proportionality constant in compressible cake filtration
β	Shape factor
ρ	Mass density of water
γ	Specific weight of water
γ_p	Specific weight of particle
γ_w	Specific weight of water
μ	Absolute viscosity of water
η	Porosity

PROBLEMS

7.1   A consultant decided to recommend using an effective size of 0.55 mm and a uniformity coefficient of 1.65 for a proposed filter bed. Perform a sieve analysis to transform a run-of-bank sand you provide into a usable sand.

7.2   For the usable sand percentage distribution of Figure 7.9, what is the 50th percentile size?

7.3   The percentage p_3 of a sample of stock sand is 46% and the percentage p_2 is 53%. What is the percentage of stock corresponding to desired P_{10} size? If the percentage p_1 were 30%, what would be p_2?

7.4   The percentage of stock sand corresponding to the desired P_{10} is 30% and the percentage stock sand finer than or equal to the P_{60} is 53%. What is the percentage stock sand that is too fine to be usable?

7.5   The percentage stock sand too fine to be usable is 25.4% and the percentage corresponding to the P_{60} is 53%. What is the percentage stock sand corresponding to the desired P_{10}?

7.6   The percentage stock sand corresponding to P_{10} is 30% and the percentage stock sand too fine to be usable is 25.4%. What is p_2?

7.7   The percentage stock sand too fine to be usable is 25.4% and the percentage stock sand converted to the final sand is 46%. Calculate p_5.

7.8   The percentage stock sand too coarse to be usable is 71.4% and the percentage stock sand converted to final sand is 46%. Calculate the percentage stock sand too fine to be usable.

7.9   The percentage stock sand too coarse to be usable is 71.4% and the percentage stock sand too fine to be usable is 25.4%. Calculate the percentage stock sand converted to final sand.

7.10   For the sharp filter sand of sieve analysis below, find the head loss if the sand is to be used in (a) a slow-sand filter 76 cm deep operated at 9.33 $m^3/m^2 \cdot d$ and (b) a rapid-sand filter 76 cm deep operated at 120 $m^3/m^2 \cdot d$. The porosity of the unstratified bed is 0.39, and that of the stratified bed is 0.42. The temperature of the water to be filtered is 20°C.

Sieve No.	Average Size (mm)	x_i
14–20	1.0	0.01
20–28	0.70	0.05
28–32	0.54	0.15
32–35	0.46	0.18
35–42	0.38	0.18
42–48	0.32	0.20
48–60	0.27	0.15
60–65	0.23	0.07
65–100	0.18	0.01

7.11   For the sharp filter sand of sieve analysis of Problem 7.10, calculate the depth of the bed if the sand is to be used in (a) a slow-sand filter operated at 9.33 $m^3/m^2 \cdot d$ and (b) a rapid-sand filter operated at 120 $m^3/m^2 \cdot d$. The porosity of the unstratified bed is 0.39, and that of the stratified bed is 0.42. The temperature of the water to be filtered is 20°C. The head losses in the slow-sand and rapid-sand filters are 0.06 m and 0.71 m, respectively.

7.12 For the sharp filter sand of sieve analysis of Problem 7.10, determine the average porosity of the bed if the sand is to be used in (a) a slow-sand filter operated at 9.33 $m^3/m^2 \cdot d$ and (b) a rapid-sand filter operated at 120 $m^3/m^2 \cdot d$. The temperature of the water to be filtered is 20°C. The head losses in the slow-sand and rapid-sand filters are 0.06 m and 0.71 m, respectively, and the bed depths are 76 cm, respectively.

7.13 For the sharp filter sand of sieve analysis of Problem 7.10, determine the flow rate applied to the filter in $m^3/m^2 \cdot d$ if the sand is to be used in (a) a slow-sand filter of average porosity of 0.39 and (b) a rapid-sand filter of average porosity of 0.42. The temperature of the water to be filtered is 20°C. The head losses in the slow-sand and rapid-sand filters are 0.06 m and 0.71 m, respectively and the bed depths are 76 cm, respectively.

7.14 The sharp filter sand of sieve analysis of Problem 7.10 is used to construct (a) a slow-sand filter of average porosity of 0.39 operated at 9.33 $m^3/m^2 \cdot d$ and (b) a rapid-sand filter of average porosity of 0.42 operated at 120 $m^3/m^2 \cdot d$. The head losses in the slow-sand and rapid-sand filters are 0.06 m and 0.71 m, respectively, and the bed depths are 76 cm, respectively. Determine the temperature anticipated of the operation.

7.15 A sand filter is operated at a rate of 160 $L/cm^2 \cdot min$. The total solids collected in a particular layer 8 cm deep is 12.96 mg/cm^3. The change in concentration between the lower and the upper portions of the layer is -10.8 mg/L $\cdot$ cm. How long did it take to collect this much of the solids?

7.16 The total solids collected in a particular layer of 8 cm deep of a sand filter is 12.96 mg/cm^3. The change in concentration between the lower and the upper portions of the layer is -10.8 cm $\cdot$ mg/L. It took 600 min to collect the solids. At what rate was the filter operated?

7.17 The total solids collected in a particular layer 8 cm deep of a sand filter is 12.96 mg/cm^3. It took 600 min to collect the solids when operating the filter at 160 $L/cm^2 \cdot min$. Determine the change in concentration between the lower and the upper portions of the layer.

7.18 The total solids collected in a particular layer of a sand filter is 12.96 mg/cm^3. It took 600 min to collect the solids when operating the filter at 160 $L/cm^2 \cdot min$. The change in concentration between the lower and the upper portions of the layer is -10.8 cm $\cdot$ mg/L. Determine the depth of the layer.

7.19 The amount of suspended solids removed in a uniform sand 0.8 mm in diameter is 2.0 mg/cm^3. Determine the head loss due to suspended solids.

7.20 The amount of suspended solids removed in a uniform sand 0.8 mm in diameter is 2.0 mg/cm^3. Determine the total head loss due to suspended solids removed and the bed if the clean bed head loss is 0.793 m.

7.21 Experiments were performed on uniform sands and anthracite media yielding the following results below. Calculate the a's and b's of Equation (7.23) corresponding to the respective diameters.

Dia. (mm)	$\dfrac{g}{(mg/cm^3)}$	h_L (m)
	Uniform Sand	
0.5	1.08	0.06
0.5	7.2	4.0
0.7	1.6	0.06
0.7	11.0	4.0
1.0	2.6	0.06
1.0	17.0	4.0
	Uniform Anthracite	
1.0	6.2	0.06
1.0	43	4.0
1.6	8.1	0.06
1.6	58	4.0
2.0	9.2	0.06
2.0	68	4.0

7.22 A 0.65-m deep filter bed has a uniformly sized sand with diameter of 0.45 mm, specific gravity of 2.65, and a volumetric shape factor of 0.87. If a hydrostatic head of 2.3 m is maintained over the bed, determine the backwash flow rate at 20°C. Assume the porosity is 0.4.

7.23 Determine the backwash rate at which the bed in Problem 7.22 will just begin to fluidize, assuming the bed is expanded to 150%.

7.24 A sand filter is 0.65 meter deep. It has a uniformly sized sand with diameter of 0.45 mm, specific gravity of 2.65, and a volumetric shape factor of 0.87. The bed porosity is 0.4. Determine the head required to maintain a flow rate of 10 m/h at 15°C.

7.25 A sharp filter sand has the sieve analysis shown below. The average porosity of the stratified bed is 0.42. The lowest temperature anticipated of the water to be filtered is 4°C. The backwashing head loss was originally determined as $0.67\,\text{m}$. The filter is backwashed at a rate of $0.4(10^{-2})$ m/s · $\rho_p = 2{,}650$ kg/m^3. The bed depth is 0.7 m. Calculate the expanded depth of the bed.

Sieve No.	Average Size (mm)	x_i
14–20	1.0	0.01
20–28	0.7	0.05
28–32	0.54	0.15
32–35	0.46	0.18
35–42	0.38	0.18
42–48	0.32	0.20
48–60	0.27	0.15
60–65	0.23	0.07
65–100	0.18	0.01

7.26   Use the data of Problem 7.25. The average porosity of the stratified bed is 0.42. The lowest temperature anticipated of the water to be filtered is 4°C. The backwashing head loss was originally determined as 0.67 m. The filter is backwashed at a rate of $0.4(10^{-2})$ m/s. The bed depth is 0.7 m. The ratio $l_e/l = 1.32$. Calculate the mass density of the particles.

7.27   Use the data of Problem 7.25. The average porosity of the stratified bed is 0.42. The backwashing head loss was originally determined as 0.67 m. The filter is backwashed at a rate of $0.4(10^{-2})$ m/s. The bed depth is 0.7 m. The ratio $l_e/l = 1.32$. $\rho_p = 2{,}650$ kg/m^3. Determine the temperature at which the filter is operated.

7.28   A sand filter bed with an average $d_p = 0.06$ cm is backwashed at a rate of $0.4(10^{-2})$ m/s. Calculate the resulting expanded porosity.

7.29   A sand filter bed with an average $d_p = 0.06$ cm is backwashed to an expanded porosity of 0.47. Determine the rate used in backwashing.

7.30   A sand filter bed is backwashed to an expanded porosity of 0.47. The backwashing rate is $0.4(10^{-2})$ m/s. Determine the average settling velocity of the particles.

7.31   Use the data of Problem 7.25. The backwashing rate used to clean the filter is $0.4(10^{-2})$ m/s. If the bed is expanded to 150%, calculate the unexpanded porosity of the bed.

7.32   The backwashing rate used to clean the filter is $0.4(10^{-2})$ m/s. If the bed is expanded to 150% from an unexpanded depth of 0.8 m and an unexpanded porosity of 0.4, calculate the resulting expanded porosity of the bed.

7.33   The backwashing rate used to clean the filter is $0.4(10^{-2})$ m/s. If the bed is expanded to 150% from an unexpanded depth of 0.8 m resulting in an expanded porosity of 0.7, calculate the unexpanded porosity of the bed.

7.34   Pilot plant analysis on a mixed-media filter shows that a filtration rate of 15 m^3/m^2-h is acceptable. If a surface configuration of 5 m × 8 m is appropriate, how many filter units will be required to process 100,000 m^3/d of raw water?

7.35   The results of a Buchner funnel experiment to determine the specific cake resistance of a certain sludge are as follows:

Volume of Filtrate (mL)	Time (sec)	t/V
25	48	1.92
100	520	5.2

$-\Delta P = 51$ cm Hg, filter area = 550 cm^2, $\mu = 15(10^{-4})$ kg/m · s, and $c = 0.25$ g/cm^3. Determine $\bar{\alpha}$.

7.36   A pilot study was conducted using a dual-media filter composed of anthracite as the upper 30-cm part and sand as the next lower 30-cm part of the filter. The results are shown in the table below, where c_o (= 20 mg/L) is the concentration of solids at the influent. If the respective sizes of the anthracite and sand layers are 1.6 mm and 0.5 mm, what is the length of the filter run to a terminal head loss of 3 m at a filtration rate of 120 $1/m^2$-min? Assume the clean water head loss is 0.793 m.

Q ($L/m^2 \cdot$ min)	Depth (cm)	c/c_o	Medium	Q ($L/m^2 \cdot$ min)	Depth cm	c/c_o	Medium
80	0	1.0	Anthracite	80	8	0.32	Anthracite
80	16	0.27	Anthracite	80	24	0.24	Anthracite
80	30	0.23	Anthracite	80	40	0.20	Sand
80	48	0.20	Sand	80	56	0.20	Sand
160	0	1.0	Anthracite	160	8	0.46	Anthracite
160	16	0.30	Anthracite	160	24	0.27	Anthracite
160	30	0.25	Anthracite	160	40	0.22	Sand
160	48	0.22	Sand	160	56	0.22	Sand
240	0	1.0	Anthracite	240	8	0.59	Anthracite
240	16	0.39	Anthracite	240	24	0.30	Anthracite
240	30	0.27	Anthracite	240	40	0.23	Sand
240	48	0.23	Sand	240	56	0.23	Sand

7.37   What is the resistance of the filter medium, R_m, in Problem 7.35?

7.38   In Problem 7.35, what is the volume of filtrate expected in 308 seconds?

7.39   In cake filtration, how does viscosity affect the volume of filtrate?

7.40   For the results of a leaf-filter experiment on a $CaCO_3$ slurry tabulated below, calculate $\bar{\alpha}_o$ and s. $\mu = 8.9(10^{-4})$ kg/m $\cdot$ s. The filter area is 440 cm^2, the mass of solid per unit volume of filtrate is 23.5 g/L, and the temperature is 25°C.

$-\Delta P$ (kN/m^2)	22.69	54.86
V (L)	Time (sec)	
1.0	41.3	19.0
1.5	72.0	34.6
2.5	152.1	76.0
3.0	210.7	102.0

7.41   Calculate the filter yield for a $CaCO_3$ sludge with cake filtration parameters determined in Problem 7.40. The cake is to be dewatered in a vacuum filter under a vacuum of 630 mm Hg. The mass of filtered solids per unit volume of filtrate is to be 60 kg/m^3. The filtration temperature is determined to be 25°C and the cycle time, one-fourth of which is form time, is five minutes.

7.42   What is the resistance of the filter medium R_m, in Problem 7.40?

7.43   In Problem 7.40, what is the volume of filtrate expected in 308 seconds for the 22.69-kN/m^2 test?

BIBLIOGRAPHY

Amini, F. and H. V. Truong (1998). Effect of filter media particle size distribution on filtration efficiency. *Water Quality Res. J. Canada*. 33, 4, 589–594.

American Water Works Association (1990). *Water Quality and Treatment: A Handbook of Community Water Supplies*. McGraw-Hill, New York.

American Water Works Association (1992). *Operational Control of Coagulation and Filtration Processes*. Am. Water Works Assoc., Denver.

Collins, R. and N. Graham (Eds.) (1996). *Advances in Slow Sand and Biological Filtration*. John Wiley & Sons, New York.

Cornwell, D. A., et al. (1992). *Full-Scale Evaluation of Declining and Constant Rate Filtration*. Am. Water Works Assoc., Denver.

Davies, W. J., M. S. Le, and C. R. Heath (1998). Intensified activated sludge process with submerged membrane microfiltration. *Water Science and Technology Wastewater: Industrial Wastewater Treatment, Proc. 1998 19th Biennial Conf. Int. Assoc. on Water Quality. Part 4,* June 21–26, Vancouver, Canada, 38, 4–5, 421–428. Elsevier Science Ltd., Exeter, England.

David J. Hiltebrand, Inc. (1990). *Guidance Manual for Compliance with the Filtration and Disinfection Requirements for Public Water Systems Using Surface Water Sources*. Am. Water Works Assoc., Denver.

Droste, R. L. (1997). *Theory and Practice of Water and Wastewater Treatment*. John Wiley & Sons, New York.

Eighmy, T. T., M. R. Collins, and J. P., Malley Jr. (1993). *Biologically Enhanced Slow Sand Filtration for Removal of Natural Organic Matter*. Am. Water Works Assoc., Denver.

Hall, D. and C. S. B. Fitzpatrick (1997). Mathematical filter backwash model. *Water Science and Technology Proc. 7th Int. Workshop on Instrumentation, Control and Automation of Water and Wastewater Treatment and Transport Syst.,* July 6–9, Brighton, England, 37, 12, 371–379. Elsevier Science Ltd., Exeter, England.

Hendricks, D. W. (1991). *Manual of Design for Slow Sand Filtration*. Am. Water Works Assoc., Denver.

Hendricks, D. W. (1988). *Filtration of Giardia Cysts and Other Particles under Treatment Plant Conditions* (Research Report/AWWA Research Foundation). Am. Water Works Assoc., Denver.

Holdich, R. G. (1990). Rotary vacuum filter scale-up calculations—and the use of computer spreadsheets. *Filtration and Separation.* 27, 435–439.

Holdich, R. G. (1994). Simulation of compressible cake filtration. *Filtration and Separation.* 31, 825–829.

Hornet Foundation Inc. (1992). *Water Treatment Plant Operation: A Field Study Training Guide*. California State University, Sacramento, CA.

Kimura, K., Y. Watanabe, and N. Ohkuma (1998). Filtration resistance induced by ammonia oxidizers accumulating on the rotating membrane disk. *Water Science Techno. Wastewater: Industrial Wastewater Treatment., Proc. 1998 19th Biennial Conf. Int. Assoc. on Water Quality. Part 4,* June 21–26, Vancouver, Canada, 38, 4–5, 443–452. Elsevier Science Ltd., Exeter, England.

Letterman, R. D. (1991). *Filtration Strategies to Meet the Surface Water Treatment Rule*. Am. Water Works Assoc., Denver.

Metcalf & Eddy, Inc. (1991). *Wastewater Engineering: Treatment, Disposal, and Reuse*. McGraw-Hill, New York, 37.

McCabe, W. L. and J. C. Smith (1967). *Unit Operations of Chemical Engineering*. McGraw-Hill, New York.

Peavy, H. S., D. R. Rowe, and G. Tchobanoglous (1985). *Environmental Engineering.* McGraw-Hill, New York.

Rushton, A., A. S. Ward, and R.G. Holdich (1996). *Solid-Liquid Filtration and Separation Technology.* VCH, Weinheim, Germany.

Salvato, J. A. (1982). *Environmental Engineering and Sanitation.* John Wiley & Sons, New York.

Sincero, A. P. and G. A. Sincero (1996). *Environmental Engineering: A Design Approach.* Prentice Hall, Upper Saddle River, NJ.

Tchobanoglous, G. and E. D. Schroeder (1985). *Water Quality.* Addison-Wesley, Reading, MA.

Vera, L., et al. (1998). Can microfiltration of treated wastewater produce suitable water for irrigation? *Water Science and Technology Wastewater: Industrial Wastewater Treatment, Proc. 1998 19th Biennial Conf. Int. Assoc. on Water Quality. Part 4,* June 21–26, Vancouver, Canada, 38, 4–5, 395–403. Elsevier Science Ltd., Exeter, England.

Viessman, W., Jr. and M. J. Hammer (1993). *Water Supply and Pollution Control.* Harper & Row, New York.

WEF (1994). *Preliminary Treatment for Wastewater Facilities* (Manual of Practice, Om-2). Water Environment Federation.

Williams, C. J. and R. G. J. Edyvean (1998). Investigation of the biological fouling in the filtration of seawater. *Water Science and Technology, Wastewater: Biological Processes, Proc. 1998 19th Biennial Conf. Int. Assoc. on Water Quality. Part 7,* June 21–26, Vancouver, BC, 38, 8–9, 309–316. Elsevier Science Ltd., Exeter, England.

8 Advanced Filtration and Carbon Adsorption

This chapter continues the discussion on filtration started in Chapter 7, except that it deals with advanced filtration. We have defined *filtration* as a unit operation of separating solids or particles from fluids. A unit operation of filtration carried out using membranes as filter media is advanced filtration. This chapter discusses advanced filtration using electrodialysis membranes and pressure membranes. Filtration using pressure membranes include reverse osmosis, nanofiltration, microfiltration, and ultrafiltration.

In addition to advanced filtration, this chapter also discusses carbon *adsorption*. This is a unit operation that uses the active sites in powdered, granular, and fibrous activated carbon to remove impurities from water and wastewater. Carbon adsorption and filtration share some similar characteristics. For example, head loss calculations and backwashing calculations are the same. Carbon adsorption will be discussed as the last part of this chapter.

8.1 ELECTRODIALYSIS MEMBRANES

Figure 8.1a shows a cut section of an electrodialysis filtering membrane. The filtering membranes are sheet-like barriers made out of high-capacity, highly cross-linked ion exchange resins that allow passage of ions but not of water. Two types are used: *cation membranes*, which allow only cations to pass, and *anion membranes*, which allow only anions to pass. The cut section in the figure is a cation membrane composed of an *insoluble matrix* with *water* in the pore spaces. *Negative charges* are fixed onto the insoluble matrix, and *mobile cations* reside in the pore spaces occupied by water. It is the residence of these mobile cations that gives the membrane the property of allowing cations to pass through it. These cations will go out of the structure if they are replaced by other cations that enter the structure. If the entering cations came from water external to the membrane, then, the cations are removed from the water, thus filtering them out. In anion membranes, the mechanics just described are reversed. The mobile ions in the pore spaces are the *anions*; the ions fixed to the insoluble matrix are the *cations*. The entering and replacing ions are anions from the water external to the membrane. In this case, the anions are filtered out from the water.

Figure 8.1b portrays the process of filtering out the ions in solution. Inside the tank, cation and anion membranes are installed alternate to each other. Two electrodes are put on each side of the tank. By impressing electricity on these electrodes, the positive anode attracts negative ions and the negative cathode attracts positive ions. This impression of electricity is the reason why the respective ions replace their like

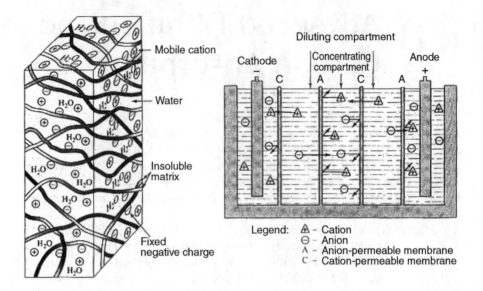

FIGURE 8.1 Cation filtering membrane (a); the electrodialysis process (b).

ions in the membranes. As shown in the figure, two compartments become "cleaned" of ions and one compartment (the middle) becomes "dirty" of ions. The two compartments are diluting compartments; the middle compartment is a concentrating compartment. The water in the diluting compartments is withdrawn as the product water, and is the filtered water. The concentrated solution in the concentrating compartment is discharged to waste.

8.1.1 POWER REQUIREMENT OF ELECTRODIALYSIS UNITS

The filtering membranes in Figure 8.1b are arranged as *CACA* from left to right, where *C* stands for cation and *A* stands for anion. In compartments *CA*, the water is deionized, while in compartment *AC*, the water is not deionized. The number of deionizing compartments is equal to two. Also, note that the membranes are always arranged in pairs (i.e., cation membrane *C* is always paired with anion membrane *A*). Thus, the number of membranes in a unit is always even. If the number of membranes is increased from four to six, the number of deionizing compartments will increase from two to three; if increased from six to eight, the number of deionizing membranes will increase from three to four; and so on. Thus, if *m* is the number of membranes in a unit, the number of deionizing compartments is equal to *m*/2.

As shown in the figure, a deionizing compartment pairs with a concentrating compartment in both directions; this pairing forms a *cell*. For example, deionizing compartment *CA* pairs with concentrating compartment *AC* in the left direction and with the concentrating compartment *AC* in the right direction of *CA*. In this paring (in both directions), however, only one cell is formed equal to the one deionizing compartment. Thus, the number of cells formed in an electrodialysis unit can be determined by counting only the number of deionizing compartments. The number of deionizing compartments in a unit is m/2, so the number of cells in a unit is also equal to m/2.

Because one equivalent of a substance is equal to one equivalent of electricity, in electrodialysis calculations, concentrations are conveniently expressed in terms of equivalents per unit volume. Let the flow to the electrodialysis unit be Q_o. The flow per deionizing compartment or cell is then equal to $Q_o/(m/2)$. If the influent ion concentration (positive or negative) is $[C_o]$ equivalents per unit volume, the total rate of inflow of ions is $[C_o]Q_o/(m/2)$ equivalents per unit time per cell. One equivalent is also equal to one Faraday. Because a Faraday or equivalent is equal to 96,494 coulombs, assuming a coulomb efficiency of η, the amount of electricity needed to remove the ions in one cell is equal to $96,494[C_o]Q_o\eta/(m/2)$ coulombs per unit time. Coulomb efficiency is the fraction of the input number of equivalents of an ionized substance that is actually acted upon by an input of electricity.

If time is expressed in seconds, coulomb per second is amperes. Therefore, for time in seconds, $96,494[C_o]Q_o\eta/(m/2)$ amperes of current must be impressed upon the membranes of the cell to effect the removal of the ions. The cells are connected in series, so the same current must pass through all of the cells in the electrodialysis unit, and the same $96,494[C_o]Q_o\eta/(m/2)$ amperes of current would be responsible for removing the ions in the whole unit. To repeat, not only is the amperage impressed in one cell but in all of the cells in the unit.

In electrodialysis calculations, a term called current density (CD) is often used. Current density is the current in milliamperes that flows through a square centimeter of membrane perpendicular to the current direction: $CD = mA/A_{cm}$, where mA is the milliamperes of electricity and A_{cm} is the square centimeters of perpendicular area. A ratio called current density to normality (CD/N) is also used, where N is the normality. A high value of this ratio means that there is insufficient charge to carry the current away. When this occurs, a localized deficiency of ions on the membrane surfaces may occur. This occurrence is called polarization. In commercial electrodialysis units CD/N of up to 1,000 are utilized.

The electric current I that is impressed at the electrodes is not necessarily the same current that passes through the cells or deionizing compartments. The actual current that successfully passes through is a function of the current efficiency which varies with the nature of the electrolyte, its concentration in solution, and the membrane system. Call $\mathcal{M}$ the current efficiency. The amperes passing through the solution is equal to the amperes required to remove the ions. Thus,

$$I\mathcal{M} = 96,494[C_o]Q_o\eta/(m/2)$$
$$I = \frac{96,494[C_o]Q_o\eta/(m/2)}{\mathcal{M}} \tag{8.1}$$

The emf E across the electrodes is given by Ohm's law as shown below, where R is the resistance across the unit.

$$E = IR \tag{8.2}$$

If I is in amperes and R is in ohms, then E is in volts.

From basic electricity, the power P is $EI = I^2R$. Thus,

$$P = EI = I^2R = 3.72(10^{10})\left(\frac{[C_o]Q_o\eta}{\mathrm{m}\mathcal{M}}\right)^2 R \qquad (8.3)$$

If I is in amperes, E is in volts, and R is in ohms, P in is in watts. Of course, the combined units of N and Q_o must be in corresponding consistent units.

Example 8.1   A brackish water of 378.51 m³/day containing 4,000 mg/L of ions expressed as Nacl is to be de-ionized using an electrodialysis unit. There are 400 membranes in the unit each measuring 45.72 cm by 50.8 cm. Resistance across the unit is 6 ohms and the current efficiency is 90%. CD/N to avoid polarization is 700. Estimate the impressed current and voltage, the coulomb efficiency, and the power requirement.

Solution:

$$[C_o] = [\mathrm{Nacl}] = \frac{4.0}{\mathrm{NaCl}} = \frac{4}{23+35.45} = 0.068\ \frac{\mathrm{geq}}{\mathrm{L}} = 68\ \frac{\mathrm{geq}}{\mathrm{m}^3};\ N = 0.068$$

Therefore,

$$CD = 700(0.068) = 47.6\ \mathrm{mA/cm}^2$$

$$I = \frac{47.6(45.72)(50.8)}{1000(0.9)} = 122.84\ \mathrm{A}\quad\textbf{Ans}$$

$$E = IR = 122.84(6) = 737\ \mathrm{V}\quad\textbf{Ans}$$

$$I = \frac{96,494[C_o]Q_o\eta/(\mathrm{m}/2)}{\mathcal{M}} = \frac{96,494(68)(\frac{378.51}{24(60)(60)})(\eta)/(400/2)}{0.90}$$

$$\eta = 0.77\quad\textbf{Ans}$$

$$P = EI = I^2R = 122.84^2(6) = 90,538\ \mathrm{W}\quad\textbf{Ans}$$

8.2 PRESSURE MEMBRANES

Pressure membranes are membranes that are used to separate materials from a fluid by the application of high pressure on the membrane. Thus, pressure membrane filtration is a high pressure filtration. This contrasts with electrodialysis membranes in which the separation is effected by the impression of electricity across electrodes. Filtration is carried out by impressing electricity, therefore, electrodialysis membrane filtration may be called *electrical filtration*.

According to Jacangelo (1989), three allied pressure-membrane processes are used: ultrafiltration (UF), nanofiltration (NF), and reverse osmosis (RO). He states that UF removes particles ranging in sizes from 0.001 to 10 μm, while RO can remove particles ranging in sizes from 0.0001 to 0.001 μm. As far as size removals are concerned, NF stays between UF and RO, being able to remove particles in the size range of the order of 0.001 μm. UF is normally operated in the range of 100 to 500 kPag (kilopascal gage); NF, in the range of 500 to 1,400 kPag; and RO, in the range of 1,400 to 8,300 kPag. Microfiltration (MF) may added to this list. MF retains larger particles than UF and operates at a lesser pressure (70 kPag).

Whereas the nature of membrane retention of particles in UF is molecular screening, the nature of membrane retention in MF is that of molecular-aggregate screening. On the other hand, comparing RO and UF, RO presents a diffusive transport barrier. Diffusive transport refers to the diffusion of solute across the membrane. Due to the nature of its membrane, RO creates a barrier to this diffusion. Figures 8.2 through 8.4 present example installations of reverse osmosis units.

FIGURE 8.2 Bank of modules at the Sanibel–Captiva reverse osmosis plant, Florida.

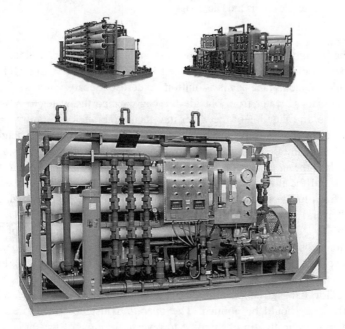

FIGURE 8.3 Installation modules of various reverse osmosis units. (Courtesy of Specific Equipment Company, Houston, TX.)

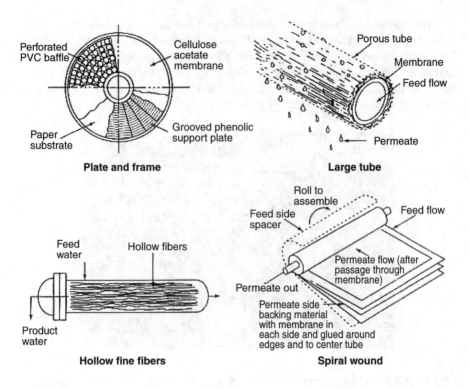

FIGURE 8.4 Reverse osmosis module designs.

The basics of a normal osmosis process are shown in Figure 8.5a. A bag of semipermeable membrane is shown placed inside a bigger container full of pure water. Inside the membrane bag is a solution of sucrose. Because sucrose has osmotic pressure, it "sucks" water from outside the bag causing the water to pass through the membrane. Introduction of the water into the membrane bag, in turn, causes the solution level to rise as indicated by the height π in the figure. The height π is a measure of the osmotic pressure. It follows that if sufficient pressure is applied to the tip of the tube in excess of that of the osmotic pressure, the height π will be suppressed and the flow of water through the membrane will be reversed (i.e., it would be from inside the bag toward the outside into the bigger container); thus, the term "reverse osmosis."

Sucrose in a concentration of 1,000 mg/L has an osmotic pressure of 7.24 kNa (kiloNewtons absolute). Thus, the reverse pressure to be applied must be, theoretically, in excess of 7.24 kNa for a sucrose concentration of 1,000 mg/L. For NaCl, its osmotic pressure in a concentration of 35,000 mg/L is 2744.07 kNa. Hence, to reverse the flow in a NaCl concentration of 35,000 mg/L, a reverse pressure in excess of 2744.07 kNa should be applied. The operation just described (i.e., applying sufficient pressure to the tip of the tube to reverse the flow of water) is the fundamental description of the *basic reverse osmosis process*.

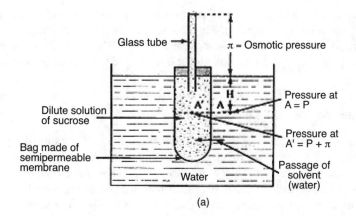

(a)

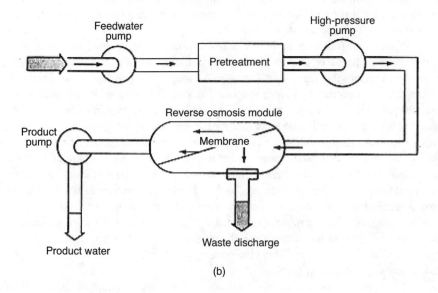

(b)

FIGURE 8.5 (a) Osmosis process; (b) reverse osmosis system.

UF, NF, MF, and RO and are all reverse osmosis filtration processes; however, when the term reverse osmosis or RO is used without qualification, it is the process operated at the highest pressure range to which it is normally referred. Figure 8.5b is a schematic of an RO plant. Figure 8.2 is a photograph of a bank of modules in the Sanibel–Captiva RO Plant in Florida. This plant treats water for drinking purposes.

Take careful note of the pretreatment requirement indicated in Figure 8.5b. As mentioned before, the RO process is an advanced mode of filtration and its purpose is to remove the very minute particles of molecules, ions, and dissolved solids. The influent to a RO plant is already "clean," only that it contains the ions, molecules, and molecular aggregates that need to be removed.

After pretreatment the high-pressure pump forces the flow into the membrane module where the solutes are rejected. The flow splits into two, one producing the product water and the other producing the waste discharge. The waste discharge has one drawback in the use of RO filtration in that it may need to be treated separately before discharge.

8.2.1 Membrane Module Designs

Over the course of development of the membrane technology, RO module designs, as shown in Figure 8.4, evolved. They are tubular, plate-and-frame, spiral wound, and hollow fine-fiber modules. In the *tubular* design, the membrane is lined inside the tube which is made of ordinary tubular material. Water is allowed to pass through the inside of the tube under excess pressure causing the water to permeate through the membrane and to collect at the outside of the tube as the *product* or *permeate*. The portion of the influent that did not permeate becomes concentrated. This is called the *concentrate* or the *reject*.

The *plate-and-frame* design is similar to the plate-and-frame filter press discussed in the previous chapter on conventional filtration. In the case of RO, the semipermeable membrane replaces the filter cloth. The *spiral-wound* design consists of two flat sheets of membranes separated by porous spacers. The two sheets are sealed on three sides; the fourth side is attached to a central collector pipe; and the whole sealed sheets are rolled around the central collector pipe. As the sheets are rolled around the pipe, a second spacer, called *influent spacer*, is provided between the sealed sheets. In the final configuration, the spiral-wound sealed membrane looks like a cylinder. Water is introduced into the influent spacer, thereby allowing it to permeate through the membrane into the spacer between the sealed membrane. The permeate, now inside the sealed membrane, flows toward the central pipe and exits through the fourth unsealed side into the pipe. The permeate is collected as the *product water*. The concentrate or the reject continues to flow along the influent spacer and is discharged as the effluent reject or effluent concentrate. This concentrate, which may contain hazardous molecules, poses a problem for disposal.

In the *hollow fine-fiber* design, the hollow fibers are a bundle of thousands of parallel, self-supporting, hair-like fibers enclosed in a fiberglass or epoxy-coated steel vessel. Water is introduced into the hollow bores of the fibers under pressure. The permeate water exits through one or more module ports. The concentrate also exits in a separate one or more module ports, depending on the design. All these module designs may be combined into banks of modules and may be connected in parallel or in series.

8.2.2 Factors Affecting Solute Rejection and Breakthrough

The reason why the product or the permeate contains solute (that ought to be removed) is that the solute has broken through the membrane surface along with the product water. It may be said that as long as the solute stays away from the membrane surface, only water will pass through into the product side and the permeate will be solute-free; However, it is not possible to exclude the solute from contacting the membrane surface; hence, it is always liable to break through. The efficiency at which solute is rejected is therefore a function of the interaction of the solute and the membrane surface. As far as solute rejection and breakthrough are concerned, a review of literature revealed the following conclusions (Sincero, 1989):

- Percentage removal is a function of functional groups present in the membrane.
- Percentage removal is a function of the nature of the membrane surface. For example, solute and membrane may have the tendency to bond by hydrogen bonding. Thus, the solute would easily permeate to the product side if the nature of the surface is such that it contains large amounts of hydrogen bonding sites.
- In a homologous series of compounds, percentage removal increases with molecular weight of solute.
- Percentage removal is a function of the size of the solute molecule.
- Percentage removal increases as the percent dissociation of the solute molecule increases. The degree of dissociation of a molecule is a function of pH, so percentage removal is also a function of pH.

This review also found that the percentage removal of a solute is affected by the presence of other solutes. For example, methyl formate experienced a drastic change in percentage removal when mixed with ethyl formate, methyl propionate, and ethyl propionate. When alone, it was removed by only 14% but when mixed with the others, the removal increased to 66%. Therefore, design of RO processes should be done by obtaining design criteria utilizing laboratory or pilot plant testing on the given influent.

8.2.3   SOLUTE–WATER SEPARATION THEORY

The sole purpose of using the membrane is to separate the solute from the water molecules. Whereas MF, UF, and NF may be viewed as similar to conventional filtration, only done in high-pressure modes, the RO process is thought to proceed in a somewhat different way. In addition to operating similar to conventional filtration, some other mechanisms operate during the process. Several theories have been advanced as to how the separation in RO is effected. Of these theories, the one suggested by Sourirajan with schematics shown in Figures 8.6a and 8.6b is the most plausible.

Sourirajan's theory is called the *preferential-sorption, capillary-flow theory*. This theory asserts that there is a competition between the solute and the water molecules for the surface of the membrane. Because the membrane is an organic substance, several hydrogen bonding sites exist on its surface which preferentially bond water molecules to them. (The hydrogen end of water molecules bonds by hydrogen bonding to other molecules.) As shown in Figure 8.6a, H_2O molecules are shown layering over the membrane surface (preferential sorption), to the exclusion of the solute ions of Na^+ and Cl^-. Thus, this exclusion brings about an initial separation. In Figure 8.6b, a pore through the membrane is postulated, accommodating two

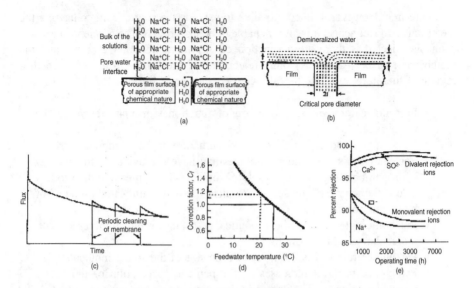

FIGURE 8.6 (a) schematic representation of preferential sorption-capillary flow theory; (b) critical pore diameter for separation; (c) flux decline with time; (d) correction factor for surface area of cellulose acetate; and (e) solute rejection as a function of operating time.

diameters of water molecules. This pore size designated as $2t$, where t is the diameter of the water molecule, is called the *critical pore diameter*. With this configuration, the final separation of the water molecules and the solutes materializes by applying pressure, pushing H_2O through the pores (capillary flow).

As the process progresses, solutes build and line up near the membrane surface creating a concentration boundary layer. This layer concentration is much larger than in the bulk solution and, also, much larger, of course, than the concentration in the permeate side. This concentration difference creates a pressure for diffusive transport. The membrane, however, creates a barrier to this diffusion, thus, retaining the solute and not allowing it to pass through easily. Eventually, however, the solute will diffuse out and leak to the permeate side.

8.2.4 Types of Membranes

The first RO membrane put to practical use was the cellulose acetate membrane (CA membrane). The technique of preparation was developed by Sourirajan and Loeb and consisted of casting step, evaporation step, gelation step, and shrinkage step. The *casting step* involves casting a solution of cellulose acetate in acetone containing an additive into flat or tubular surfaces. The additive (such as magnesium perchlorate) must be soluble in water so that it will easily leach out in the gelation step creating a porous film. After casting, the solvent acetone is *evaporated*. The material is then subjected to the *gelation step* where it is immersed in cold water. The film material sets to a gel and the additive leaches out. Finally, the film is subjected to the *shrinkage step* that determines the size of the pores, depending upon the temperature used in shrinking. High temperatures create smaller pores.

After this first development of the CA membrane, different types of membranes followed: CAB, CTA, PBIL, and PA membranes. CAB is membrane of *cellulose acetate butyrate*; CTA is *cellulose triacetate*. The PBIL membrane is a *polybenzimidazolone* polymer and PA are polyamide membranes. The structure of a PBIL unit is as follows:

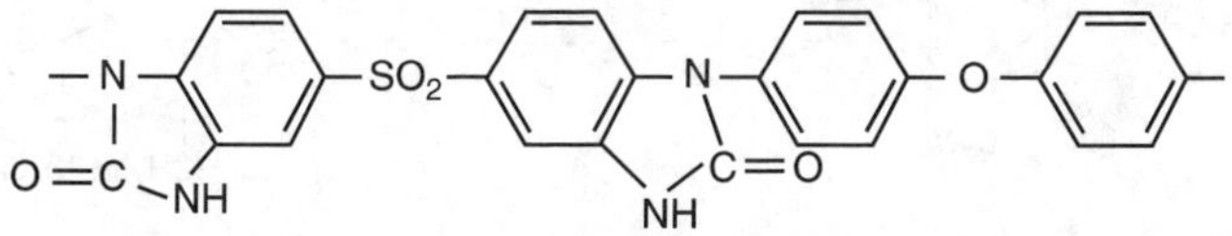

Polybenzimidazolone unit

Polyethylene amine reacted with tolylene diisocyanate produces the *NS-100 membrane* (NS stands for nonpolysaccharide). The reaction is carried out as follows:

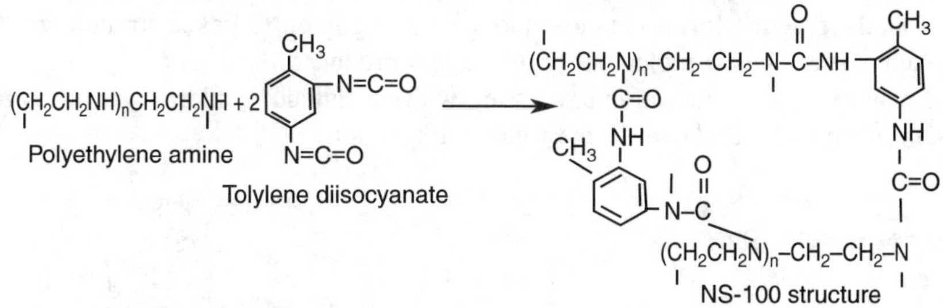

NS-100 structure

In this reaction, the H bonded to the N of the n repeating units of polyethylene amine $[(-CH_2CH_2NH)_n-]$ moves to the N of tolylene diisocyanate $[-N=C=O]$ destroying the double bond between N and C. The C of the carboxyl group $[=C=O]$ of tolylene diisocyanate then bonds with the N of the amine. The reaction above simply shows two of the tolylene molecules participating in the reaction, but in reality, there will be millions of them performing the reaction of H moving and the C of the carboxyl group of the tolylene bonding with the N of the amine and so on. The final structure is a mesh of cross-linked assembly, thus creating molecular pores.

As indicated in the NS-100 product, a closed loop structure is formed. The ethylene repeating units $[-CH_2CH_2-]$ form the backbone of the membrane, and the benzene rings ⬡ form the cross-linking mechanism that tie together the ethylene backbones forming the closed loop. The ethylene units and the benzene rings are nonpolar regions, while the peptide bonds $-\overset{\overset{O}{\|}}{C}-NH-$ and the amines $[-NH_2]$ are polar regions. In the NS-100, nonpolar regions exceed the polar regions; hence, this membrane is said to be *apolar*.

The CA membrane contains the OH^- and the acetyl $[CH_3CO_2^-]$ groups. The OH^- region exceeds the acetyl region in the membrane. OH^- is polar, while the acetyl group is nonpolar regions. Since the OH^- region exceeds the acetyl region, CA membranes are polar. The polarity or apolarity of any membrane is very important in characterizing its property to reject solutes.

Polyethylene amine reacted with isophthaloyl chloride produces the *PA-100 membrane* as shown in the following reaction:

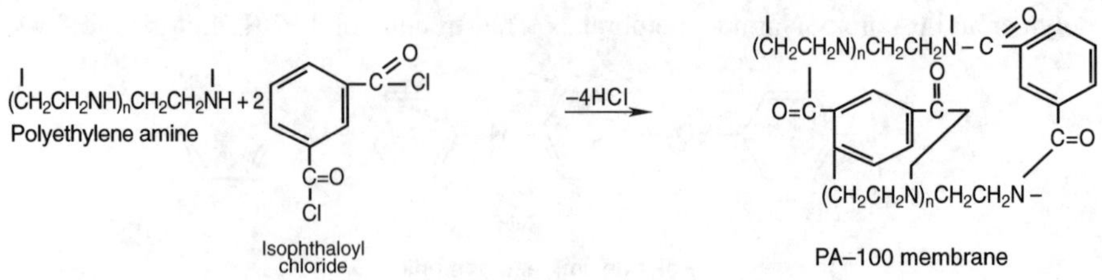

The chloride atom in isophthaloyl chloride is attached to the carboxyl group. In this reaction, the H in the amine group of polyethylene amine reacts with the chloride in the carboxyl group producing HCl, as shown by the product over the arrow, and the PA-100 membrane to the right of the reaction equation. As in the case of the NS-100, the reaction forms the closed loop resulting in cross-linked structure of the PA-membrane. Thus, a molecular pore is again produced.

Epiamine, a polyether amine, reacting with isophthaloyl chloride produces the *PA-300 membrane* as shown in the following reaction:

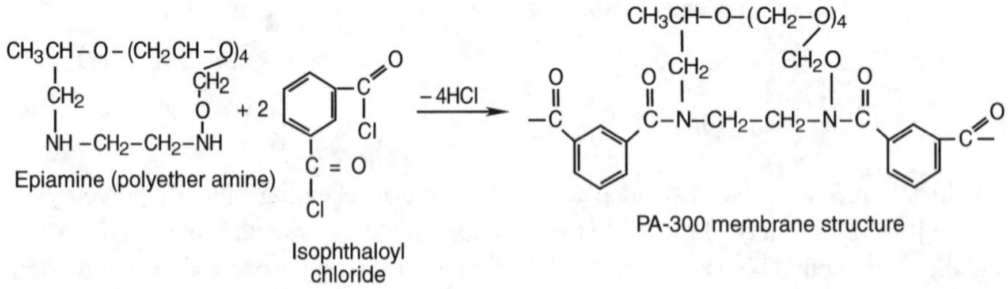

Again, in this reaction, the H in the amine group of the amine, epiamine, reacts with the chloride of isophthaloyl chloride forming HCl. The N, in turn, of the amine group, from which the H that reacts with the C chloride were taken, bonds with the C of the carboxyl group of isophthaloyl chloride producing the PA-300 membrane. As shown in its structure, this membrane is also a cross-linked membrane.

meta-Phenylene diamine reacting with trimesoyl chloride produces the *FilmTec FT-30* membrane according to the following reaction:

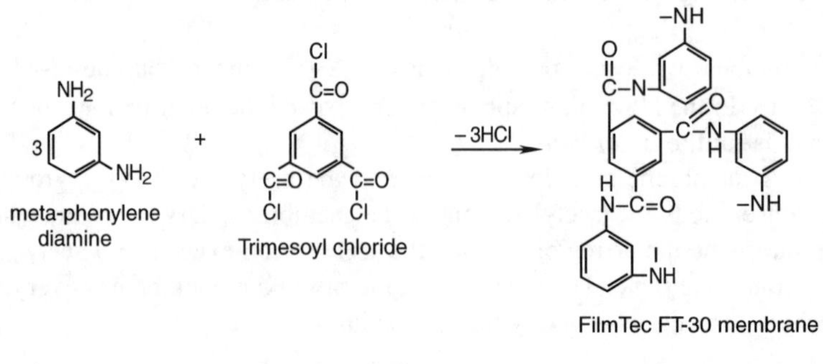

2-Hydroxy-methyl furan when dehydrated using H_2SO_4 produces the *NS-200* according to the following reaction:

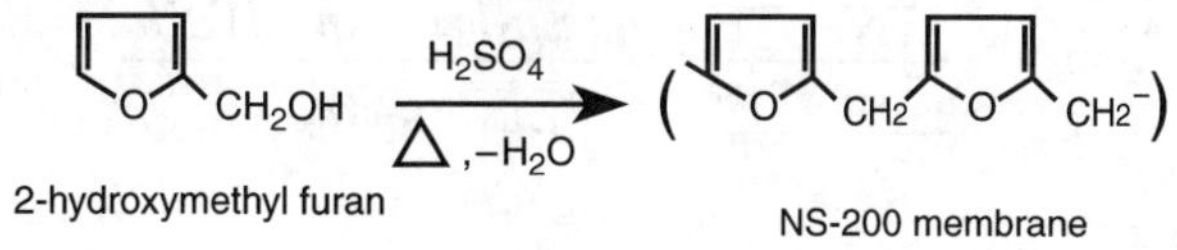

$$\underset{\text{2-hydroxymethyl furan}}{\boxed{}\text{O}\text{CH}_2\text{OH}} \quad \xrightarrow[\Delta,-H_2O]{H_2SO_4} \quad \left(\underset{\text{NS-200 membrane}}{\text{O}\text{CH}2\text{O}\text{CH}2^-}\right)$$

Another membrane formed from 2-hydroxy-methyl furan is *Toray PEC-1000*.

The PAs in the prefixes for the naming of the membranes stand for polyamide. Thus, the membranes referred to are *polyamide membranes*. PA membranes contain the amide group $[NH_2^-]$; it is for this component that they are called polyamide membranes. The formulas for the NS-100, NS-300, and the FT-30 membranes contain the amide group, thus, they are polyamide membranes. The NS-200 is not a polyamide membrane.

In the early applications of the reverse osmosis technique, the membranes available were the polysaccharide membranes such as the CA membrane. As new membranes were developed, they were differentiated from the saccharide membranes by calling them nonpolysaccharide membranes. Thus, the NS-200 is a nonpolysaccharide membrane.

8.2.5 Membrane Performance Characterization

The performance of a given membrane may be characterized according to its product flux and purity of product. Flux, which is a rate of flow per unit area of membrane, is a function of membrane thickness, chemical composition of feed, membrane porosity, time of operation, pressure across membrane, and feedwater temperature. Product purity, in turn, is a function of the rejection ability of the particular membrane.

Flux decline. Figure 8.6c shows the decline of the flux with time of operation. This curve applies to a given membrane and membrane pressure differential. The lower solid curve is the actual decline without the effect of cleaning. The saw-toothed configuration is the effect of periodic cleaning. As shown, right after cleaning, the flux rate shows an improvement, but then, it begins to decline again with time. From experience, the general trend of the curve plots a straight line in a log–log paper. Thus, empirically, the following equation fits the curve:

$$\ln F = m \ln t + \ln K \Rightarrow F = K t^m \tag{8.4}$$

where F is the flux, t is the time, m is the slope of the line, and K is a constant.

The previous equation is a straight-line equation between $\ln t$ and $\ln F$. The equation is that of a straight line, so only two data points for $\ln t$ and $\ln F$ are required to calculate the constants m and K. Using the techniques of analytic geometry as applied to straight-line equations, the following equations are produced from Equation (8.4).

$$m = \frac{\frac{\sum_{l+1}^{n} \ln F}{n-l} - \frac{\sum_{1}^{l} \ln F}{l}}{\frac{\sum_{l+1}^{n} \ln t}{n-l} - \frac{\sum_{1}^{l} \ln t}{l}} = \frac{l\sum_{l+1}^{n} \ln F - (n-l)\sum_{1}^{l} \ln F}{l\sum_{l+1}^{n} \ln t - (n-l)\sum_{1}^{l} \ln t} \tag{8.5}$$

$$K = \exp\left(\frac{\sum_{1}^{l} \ln F - m\sum_{1}^{l} \ln t}{l}\right) \tag{8.6}$$

As noted in the previous equations, the summation sign Σ is used. Although only two points are needed to determine the constants, the actual experimentation may be conducted to produce several points. Since the equation only needs two points, the data are then grouped into two groups. Thus, l is the number of experimental data in the first group in a total of n experimental data points. The second group would consist of $n - l$ data points.

Equation (8.4) may be used to estimate the ultimate flux of a given membrane at the end of its life (one to two years).

Example 8.2   A long term experiment for a CA membrane module operated at 2757.89 kPag using a feed of 2,000 mg/L of NaCl at 25°C produces the results below. What is the expected flux at the end of one year of operation? What is the expected flux at the end of two years? How long does it take for the flux to decrease to 0.37 m³/m² · day?

Time (h)	1	10,000	25,000
Flux (m³/m² · day)	0.652	0.490	0.45

Solution:

$$m = \frac{l\sum_{l+1}^{n} \ln F - (n-l)\sum_{1}^{l} \ln F}{l\sum_{l+1}^{n} \ln t - (n-l)\sum_{1}^{l} \ln t}$$

$$K = \exp\left(\frac{\sum_{1}^{l} \ln F - m(\sum_{1}^{l} \ln t)}{l}\right)$$

t (h)	1	10,000	25,000
F (m³/m² · day)	0.652	0.49	0.45
ln t	0	9.21	10.13
ln F	−0.43	−0.71	−0.799

Let $l = 1$

$$m = \frac{1(-0.71 - 0.799) - 2(-0.43)}{1(9.21 + 10.13) - 2(0)} = \frac{-1.51 + 0.86}{19.34} = -0.034$$

$$K = \exp\left(\frac{-0.43 - (-0.34)(0)}{1}\right) = 0.65$$

$$F = Kt^m$$

$$F_{1\,year} = 0.65[365(24)]^{-0.034} = 0.48 \left(\frac{m^3}{m^2 \cdot day}\right) \quad \textbf{Ans}$$

$$F_{2\,year} = 0.65[365(2)(24)]^{-0.034} = 0.47 \left(\frac{m^3}{m^2 \cdot day}\right) \quad \textbf{Ans}$$

$$0.37 = 0.65t^{-0.034} \quad t = 16,002,176h = 1,827 \text{ years}$$

that is, if the membrane has not broken before this time! **Ans**

Flux through membrane as a function of pressure drop. The flow of permeate through a membrane may be considered as a "microscopic" form of cake filtration, where the solute that polarizes at the feed side of the membrane may be considered as the cake. In cake filtration (see Chapter 7), the volume of filtrate V that passes through the cake in time t can be solved from

$$\frac{t}{V} = \frac{\mu c \bar{\alpha}}{2(-\Delta P)S_o^2}V + \frac{\mu R_m}{(-\Delta P)S_o} \tag{8.7}$$

where μ is the absolute viscosity of filtrate; c, the mass of cake per unit volume of filtrate collected; $\bar{\alpha}$, the specific cake resistance; $-\Delta P$, the pressure drop across the cake and filter; S_o, the filter area; and R_m, the filter resistance. In RO, c is the solute collected on the membrane (in the concentration boundary layer) per unit volume of permeate; and R_m, the resistance of the membrane. All the other parameters have similar meanings as explained earlier in Chapter 7.

The volume flux F is V/tS_o. Using this and solving the above equation for $V/tS_o = F$,

$$\frac{V}{tS_o} = F = \frac{2}{\mu c \,\bar{\alpha}V + 2S_o\mu R_m}(-\Delta P)S_o \tag{8.8}$$

Initially neglecting the resistance of the solute in the concentration boundary layer, $\mu c \,\bar{\alpha}V$ in the denominator of the first factor on the right-side of the equation may be set to zero, producing

$$\frac{V}{tS_o} = F = \frac{1}{\mu R_m}(-\Delta P) \tag{8.9}$$

Now, considering the resistance of the solute, designate the combined effect of compressibility, membrane resistance R_m, and solute resistance as $\bar{\alpha}_m$. Analogous to cake filtration, call $\bar{\alpha}_m$ as specific membrane resistance. Hence,

$$\frac{\mathcal{V}}{tS_o} = F = \frac{1}{\mu\bar{\alpha}_m}(-\Delta P) \qquad (8.10)$$

$$\bar{\alpha}_m = \bar{\alpha}_{mo}(-\Delta P)^s \qquad (8.11)$$

where s is an index of membrane and boundary layer compressibility. When s is equal to zero, $\bar{\alpha}_m$ is equal to $\bar{\alpha}_{mo}$, the constant of proportionality of the equation.

Calling the pressure in the feed side as P_f, the net pressure P_{fn} acting on the membrane in the feed side is

$$P_{fn} = P_f - \pi_f \qquad (8.12)$$

where π_f is the osmotic pressure in the feed side. Also, calling the pressure in the permeate side as P_p, the net pressure P_{pn} acting on the membrane in the permeate side is

$$P_{pn} = P_p - \pi_p \qquad (8.13)$$

where π_p is the osmotic pressure in the permeate side. Thus,

$$-\Delta P = P_{fn} - P_{pn} = (P_f - \pi_f) - (P_p - \pi_p) = (P_f - P_p) - (\pi_f - \pi_p) \qquad (8.14)$$

and the flux F is

$$F = \frac{1}{\mu\bar{\alpha}_m}[(P_f - P_p) - (\pi_f - \pi_p)]$$

$$= \frac{1}{\mu\bar{\alpha}_{mo}(-\Delta P)^s}[(P_f - P_p) - (\pi_f - \pi_p)] = \frac{1}{\mu\bar{\alpha}_{mo}}(-\Delta P)^{1-s} \qquad (8.15)$$

Table 8.1 shows osmotic pressure values of various solutes. Some generalizations may be made from this table. For example, comparing the osmotic pressures of 1,000 mg/L of NaCl and 1,000 mg/L of Na_2SO_4, the former has about 1.8 times that of the osmotic pressure of the latter. In solution for the same masses, NaCl yields about 1.6 times more particles than Na_2SO_4. From this it may be concluded that osmotic pressure is a function of the number of particles in solution. Comparing the 1,000 mg/L concentrations of Na_2SO_4 and $MgSO_4$, the osmotic pressure of the former is about to 1.4 times that of the latter. In solution Na_2SO_4 yields about 1.3 more particles than $MgSO_4$. The same conclusions will be drawn if other comparisons are made; therefore, osmotic pressure depends on the number of particles in solution. From this finding, osmotic pressure is, therefore, additive.

Determination of $\bar{\alpha}_{mo}$ and s. The straight-line form of Equation (8.15) is

$$ln(\mu F) = (1 - s)ln(-\Delta P) - ln\bar{\alpha}_{mo} \qquad (8.16)$$

TABLE 8.1
Osmotic Pressures at 25°C

Compound	Concentration (mg/L)	Osmotic Pressure (kPa)
NaCl	35,000	2758
NaCl	1,000	76
NaHCO$_3$	1,000	90
Na$_2$SO$_4$	1,000	41
MgSO$_4$	1,000	28
MgCl$_2$	1,000	69
CaCl$_2$	1,000	55
Sucrose	1,000	7
Dextrose	1,000	14

This equation needs only two data points of $ln(\mu F)$ and $ln(-\Delta P)$ to determine the constants $\bar{\alpha}_{mo}$ and s. Assuming there are a total of n experimental data points and using the first l data points for the first equation, the last $n - l$ data points for the second equation, and using the techniques of analytic geometry, the following equations are produced:

$$s = 1 - \frac{l\sum_{l+1}^{n} ln(\mu F) - (n - l)\sum_{1}^{l} ln(\mu F)}{l\sum_{l+1}^{n} ln(-\Delta P) - (n - l)\sum_{1}^{l} ln(-\Delta P)} \tag{8.17}$$

$$\bar{\alpha}_{mo} = \exp\left(\frac{(1 - s)\sum_{1}^{l} ln(-\Delta P) - \sum_{1}^{l} ln(\mu F)}{l}\right) \tag{8.18}$$

Example 8.3   The feedwater to an RO unit contains 3,000 mg/L of NaCl, 300 mg/L of CaCl$_2$, and 400 mg/L of MgSO$_4$. The membrane used is cellulose acetate and the results of a certain study are shown below. What will the flux be if the pressure applied is increased to 4826.31 kPag? Assume that for the given concentrations the osmotic pressures are NaCl = 235.80 kPa, CaCl$_2$ = 17.17 kPa, and MgSO$_4$ = 9.93 kPa. Also, assume the temperature during the experiment is 25°C.

Applied Pressure (kPag)	Flux (m^3/m$^2 \cdot$ day)
1723.68	0.123
4136.84	0.187

Solution:

$$F = \frac{1}{\mu\bar{\alpha}_{mo}(-\Delta P)^s}[(P_f - P_p) - (\pi_f - \pi_p)] = \frac{1}{\mu\bar{\alpha}_{mo}}(-\Delta P)^{1-s}$$

$$s = 1 - \frac{l\sum_{l+1}^{n}\ln(\mu F) - (n-l)\sum_{1}^{l}\ln(\mu F)}{l\sum_{l+1}^{n}\ln(-\Delta P) - (n-l)\sum_{1}^{l}\ln(-\Delta P)}$$

$$\bar{\alpha}_{mo} = \exp\left(\frac{(1-s)\sum_{1}^{l}\ln(-\Delta P) - \sum_{1}^{l}\ln(\mu F)}{l}\right)$$

P_1 (kPag)	$-\Delta P$ (N/m^2)	$\ln(-\Delta P)$	Flux (m^3/m$^2\cdot$ day)	μF	$\ln(\mu F)$
1723.68	1,460,780[a]	14.19	0.123	9.56	2.26
4136.84	3,873,940	15.17	0.187	14.54	2.68

[a] $(P_f - P_p) - (\pi_f - \pi_p) = -\Delta P = [(1,723,680 + 101,325) - 101,325]$

$\mu = 9.0(10^{-4})$ kg/m $\cdot$ s $= 77.76$ kg/m $\cdot$ d

$\pi_f = 235.80 + 17.17 + 9.93 = 262.9$ kN/m$^2 = 262,900$ N/m^2

$P_p = 101,325$ N/m^2; $\pi_p = 0$, assumed.

$$-[262,900 - 0] = 1,460,780 \text{ N/m}^2$$

$$s = 1 - \frac{2.68 - 2.26}{15.17 - 14.19} = 1 - 0.43 = 0.57$$

$$\bar{\alpha}_{mo} = \exp\left(\frac{(1-0.57)14.19 - 2.26}{1}\right) = 46.6$$

$$F = \frac{1}{\mu\bar{\alpha}_{mo}}(-\Delta P)^{1-s} = \frac{1}{\mu\bar{\alpha}_{mo}}(-\Delta P)^{1-s}$$

$-\Delta P$ at 4826.31 kPag $= [(P_f - P_p) - (\pi_f - \pi_p)]$

$= [(4,826,310 + 101,325) - 101,325] - [262,900 - 0] = 4,563,410$ N/m^2

Therefore,

$$F = \frac{1}{77.76(46.6)}(4,563,410)^{1-0.57} = 0.20 \frac{\text{m}^3}{\text{m}^2\cdot\text{day}} \quad \textbf{Ans}$$

Effect of temperature on permeation rate. As shown in Equation (8.15), the flux is a function of the dynamic viscosity μ. Because μ is a function of temperature, the flux or permeation rate is therefore also a function of temperature. As temperature increases, the viscosity of water decreases. Thus, from the equation, the flux is expected to increase with increase in temperature. Correspondingly, it is also expected that the flux would decrease as the temperature decreases. Figure 8.6d shows the correction factor C_f for membrane surface area (for CA membranes) as a function of temperature relative to 25°C. As shown, lower temperatures have larger correction factors. This is due to the increase of μ as the temperature decreases. The opposite is true for the higher temperatures. These correction factors are applied to the membrane surface area to produce the same flux relative to 25°C.

Percent solute rejection or removal. The other parameter important in the design and operation of RO units is the percent rejection or removal of solutes. Let Q_o be the feed inflow, $[C_o]$ be the feed concentration of solutes, Q_p be the permeate outflow, $[C_p]$ be the permeate concentration of solutes, Q_c be the concentrate outflow, and $[C_c]$ be the concentrate concentration of solutes. By mass balance of solutes, the percent rejection R is

$$R = \frac{\sum Q_o[C_{oi}] - \sum Q_p[C_{pi}]}{\sum Q_o[C_{oi}]}(100) = \frac{\sum Q_c[C_{ci}]}{\sum Q_o[C_{oi}]}(100) \qquad (8.19)$$

The index i refers to the solute species i.

Figure 8.6e shows the effect of operating time on percent rejection. As shown, this particular membrane rejects divalent ions better than it does the monovalent ions. Generally, percent rejection increases with the value of the ionic charge.

Example 8.4 A laboratory RO unit 152.4 cm in length and 30.48 cm in diameter has an active surface area of 102.18 m^2. It is used to treat a feedwater with the following composition: NaCl = 3,000 mg/L, CaCl$_2$ = 300 mg/L, and MgSO$_4$ = 400 mg/L. The product flow is 0.61 m^3/m$^2 \cdot$ day and contains 90 mg/L NaCl, 6 mg/L CaCl$_2$, and 8 mg/L MgSO$_4$. The feedwater inflow is 104.9 m^3/day. **(a)** What is the percent rejection of NaCl? **(b)** What is the over-all percent rejection of ions?

Solution:

$$\text{(a) } R = \frac{\sum Q_o[C_{oi}] - \sum Q_p[C_{pi}]}{\sum Q_o[C_{oi}]}(100) = \frac{104.9(3,000) - 0.61(102.18)(90)}{104.9(3,000)}(100)$$

$$= 98.2\% \quad \textbf{Ans}$$

$$\text{(b) } R = \frac{104.9(3,000 + 300 + 400) - 0.61(102.18)(90 + 6 + 8)}{104.9(3,000 + 300 + 400)}(100)$$

$$= \frac{388,130 - 6,482.30}{388,130}(100) = 98.3\% \quad \textbf{Ans}$$

8.3 CARBON ADSORPTION

Solids are formed because of the attraction of the component atoms within the solid toward each other. In the interior of a solid, attractive forces are balanced among the various atoms making up the lattice. At the surface, however, the atoms are subjected to unbalanced forces—the ones toward the interior are attracted, but the ones at the

FIGURE 8.7 Raw carbon material on the left transforms to the carbon on the right after activation.

surface are not. Because of this unbalanced nature, any particle that lands on the surface may be attracted by the solid. This is the phenomenon of *adsorption*, which is the process of concentrating solute at the surface of a solid by virtue of this attraction.

Adsorption may be *physical* or *chemical*. Physical adsorption is also called *van der Waals adsorption*, and chemical adsorption is also called *chemisorption*. In the former, the attraction on the surface is weak, being brought about by weak van der Waals forces. In the latter, the attraction is stronger as a result of some chemical bonding that occurs. Adsorption is a surface-active phenomenon which means larger surface areas exposed to the solutes result in higher adsorption. The solute is called the *adsorbate*; the solid that adsorbs the solute is called the *adsorbent*. The adsorbate is said to be sorbed onto the adsorbent when it is adsorbed, and it is said to be desorbed when it passes into solution.

Adsorption capacity is enhanced by activating the surfaces. In the process using steam, activation is accomplished by subjecting a prepared char of carbon material such as coal to an oxidizing steam at high temperatures resulting in the water gas reaction: $C + H_2O \rightarrow H_2 + CO$. The gases released leave behind in the char a very porous structure.

The high porosity that results from activation increases the area for adsorption. One gram of char can produce about 1000 m^2 of adsorption area. After activation, the char is further processed into three types of finished product: powdered form called *powdered activated carbon* (PAC), the granular form called *granular activated carbon* (GAC), and *activated carbon fiber* (ACF). PAC is normally less than 200 mesh; GAC is normally greater than 0.1 mm in diameter. ACF is a fibrous form of activated carbon. Figure 8.7 shows a schematic of the transformation of raw carbon to activated carbon, indicating the increase in surface area.

8.3.1 ACTIVATION TECHNIQUES

Activation is the process of enhancing a particular characteristic. Carbon whose adsorption characteristic is enhanced is called *activated carbon*. The activation techniques used in the manufacture of activated carbons are dependent on the nature and type of raw material available. The activation techniques that are principally used by commercial production operations are *chemical activation* and *steam activation*. As the name suggests, chemical activation uses chemicals in the process and is generally used for the activation of peat- and wood-based raw materials. The raw material is impregnated with a strong dehydrating agent, typically phosphoric pentoxide (P_2O_5) or zinc chloride ($ZnCl_2$) mixed into a paste and then heated to temperatures of 500–800°C to activate the carbon. The resultant activated carbon is washed, dried, and ground to desired size. Activated carbons produced by chemical activation generally exhibit a very open pore structure, ideal for the adsorption of large molecules.

Steam activation is generally used for the activation of coal and coconut shell raw materials. Activation is carried out at temperatures of 800–1100°C in the presence of superheated steam. Gasification occurs as a result of the water–gas reaction:

$$C + H_2O \rightarrow H_2 + CO + 175,440 \text{ kJ/kgmol} \tag{8.20}$$

This reaction is endothermic but the temperature is maintained by partial burning of the CO and H_2 produced:

$$2CO + O_2 \rightarrow 2CO_2 - 393,790 \text{ kJ/kgmol} \tag{8.21}$$

$$2H_2 + O_2 \rightarrow 2H_2O - 396,650 \text{ kJ/kgmol} \tag{8.22}$$

The activated carbon produced is graded, screened, and de-dusted. Activated carbons produced by steam activation generally exhibit a fine pore structure, ideal for the adsorption of compounds from both the liquid and vapor phases.

8.3.2   ADSORPTION CAPACITY

The adsorption capacity of activated carbon may be determined by the use of an adsorption isotherm. The *adsorption isotherm* is an equation relating the amount of solute adsorbed onto the solid and the equilibrium concentration of the solute in solution at a given temperature. The following are isotherms that have been developed: Freundlich; Langmuir; and Brunauer, Emmet, and Teller (BET). The most commonly used isotherm for the application of activated carbon in water and wastewater treatment are the Freundlich and Langmuir isotherms. The Freundlich isotherm is an empirical equation; the Langmuir isotherm has a rational basis as will be shown below. The respective isotherms are:

$$\frac{X}{M} = k[C]^{1/n} \tag{8.23}$$

$$\frac{X}{M} = \frac{ab[C]}{1 + b[C]} \tag{8.24}$$

X is the mass of adsorbate adsorbed onto the mass of adsorbent M; $[C]$ is the concentration of adsorbate in solution in equilibrium with the adsorbate adsorbed; n, k, a, and b are constants.

The Langmuir equation may be derived as follows. Imagine a particular experiment in which a quantity of carbon adsorbent is added to a beaker of sample containing pollutant. Immediately, the solute will be sorbed onto the adsorbent until equilibrium is reached. One factor determining the amount of the sorbed materials has to be the number of adsorption sites in the carbon. The number of these sites may be quantified by the ratio X/M. By the nature of equilibrium processes, some of the solutes adsorbed will be desorbed back into solution. While these solutes are desorbing, some solutes will also be, again, adsorbed. This process continues on, like a seesaw; this "seesaw behavior" is a characteristic of systems in equilibrium.

The rate of adsorption r_s is proportional to the concentration in solution, $[C]$, (at equilibrium in this case) and the amount of adsorption sites left vacant by the desorbing solutes. Now, let us determine these vacant adsorption sites. On a given trial of the experiment, the number of adsorption sites filled by the solute may be quantified by the ratio X/M, as mentioned previously. The greater the concentration of the solute in solution, the greater this ratio will be. For a given type of solute and type of carbon adsorbent, there will be a characteristic one maximum value for this ratio. Call this $(X/M)_{ult}$. Now, we have two ratios: X/M, which is the ratio at any time and $(X/M)_{ult}$, which is the greatest possible ratio. The difference of these two ratios is proportional to the number of adsorption sites left vacant; consequently, the rate of adsorption r_s is therefore equal to $k_s[C][(X/M)_{ult} - (X/M)]$, where k_s is a proportionality constant.

For the desorption process, as the ratio (X/M) forms on the adsorbent, it must become a driving force for desorption. Thus, letting k_d be the desorption proportionality constant, $r_d = k_d(X/M)$, where r_d is the rate of desorption. The process is in equilibrium, so the rate of adsorption is equal to the rate of desorption. Therefore,

$$k_s[C]\left[\left(\frac{X}{M}\right)_{ult} - \left(\frac{X}{M}\right)\right] = k_d\left(\frac{X}{M}\right) \tag{8.25}$$

Solving for X/M produces Equation (8.24), where $a = (X/M)_{ult}$ and $b = k_s/k_d$. Note that in the derivation no mention is made of how many layers of molecules are sorbed onto or desorbed from the activated carbon. It is simply that solutes are sorbed and desorbed, irrespective of the counts of the layers of molecules; however, it is conceivable that as the molecules are deposited and removed, the process occurs layer by layer.

The straight-line forms of the Freundlich and Langmuir isotherms are, respectively,

$$ln\frac{X}{M} = lnk + \frac{1}{n}ln[C] \quad \text{straight-line form} \tag{8.26}$$

$$\frac{[C]}{X/M} = \frac{1}{ab} + \frac{1}{a}[C] \quad \text{straight-line form} \tag{8.27}$$

Because the equations are for straight lines, only two pairs of values of the respective parameters are required to solve the constants. For the Freundlich isotherm, the required pairs of values are the parameters $ln(X/M)$ and $ln[C]$; for the Langmuir isotherm, the required pairs are the parameters $[C]/(X/M)$ and $[C]$.

To use the isotherms, constants are empirically determined by running an experiment. This is done by adding increasing amounts of the adsorbent to a sample of adsorbate solution in a container. For each amount of adsorbent added, M_i, the equilibrium concentration $[C_i]$ is determined. The pairs of experiment trial values can then be used to obtain the desired parameter values from which the constants are determined. Once the constants are determined, the resulting model is used to determine M, the amount of adsorbent (activated carbon) that is needed. From the derivation, the adsorption capacity of activated carbon is $a = (X/M)_{ult}$. From this ratio, the absorption capacity of activated carbon is shown as the maximum value of the X/M ratios. This ratio corresponds to a concentration equal to the maximum possible solute equilibrium concentration.

The value of X is obtained as follows: Let $[C_o]$ be the concentration of solute in a sample of volume $¥$ before adsorption onto a mass of adsorbent M. Then

$$X = ([C_o] - [C])¥ \qquad (8.28)$$

8.3.3 DETERMINATION OF THE FREUNDLICH CONSTANTS

Using the techniques of analytic geometry, let us derive the Freundlich constants in a little more detail than used in the derivation of the constants in the discussion of reverse osmosis treated previously. As mentioned, the straight-line form of the equation requires only two experimental data points; however, experiments are normally conducted to produce not just two pair of values but more. Thus, the experimental results must be reduced to just the two pairs of values required for the determination of the parameters; therefore, assuming there are m pairs of values, these m pairs must be reduced to just two pairs. Once the reduction to two pairs has been done, the isotherm equation may be then be written to just the two pairs of derived values as follows:

$$\sum_{1}^{l} ln\frac{X}{M} = \sum_{1}^{l} lnk + \frac{1}{n}\sum_{1}^{l} ln[C] = llnk + \frac{1}{n}\sum_{1}^{l} ln[C] \quad \text{for the first pair} \qquad (8.29)$$

$$\sum_{l+1}^{m} ln\frac{X}{M} = \sum_{l+1}^{m} lnk + \frac{1}{n}\sum_{l+1}^{m} ln[C] = (m-l)lnk + \frac{1}{n}\sum_{l+1}^{m} ln[C] \quad \text{for the second pair}$$

$$(8.30)$$

The index l is the number of data points for the first group. These equations may then be solved for n and k producing

$$n = \frac{l\sum_{l+1}^{m} ln[C] - (m-l)\sum_{m}^{l} ln[C]}{l(\sum_{l+1}^{m} ln\frac{X}{M} - (m-l)\sum_{1}^{l} ln\frac{X}{M})} \qquad (8.31)$$

$$k = exp\left(\frac{\sum_{1}^{l} ln\frac{X}{M} - \frac{1}{n}\sum_{1}^{l} ln[C]}{l}\right) \qquad (8.32)$$

8.3.4 Determination of the Langmuir Constants

As was done with the Freundlich constants, the Langmuir equation may be manipulated in order to solve the Langmuir constants. From the m pairs of experimental data

$$\sum_1^l \frac{[C]}{X/M} = l\left(\frac{1}{ab}\right) + \frac{1}{a}\sum_1^l [C] \tag{8.33}$$

$$\sum_{l+1}^m \frac{[C]}{X/M}(m-l)\left(\frac{1}{ab}\right) + \frac{1}{a}\sum_{l+1}^m [C] \tag{8.34}$$

Solving for the constants a and b produces

$$a = \left\{ \frac{l\sum_{l+1}^m [C] - (m-l)\sum_1^l [C]}{l\sum_{l+1}^m \frac{[C]}{X/M} - (m-l)\sum_1^l \frac{[C]}{X/M}} \right\} \tag{8.35}$$

$$b = \frac{l}{a\sum_1^l \frac{[C]}{X/M} - \sum_1^l [C]} \tag{8.36}$$

Example 8.5 A wastewater containing $[C_o] = 25$ mg/L of phenol is to be treated using PAC to produce an effluent concentration $[C]_{eff} =$ of 0.10 mg/L. The PAC is simply added to the stream and the mixture subsequently settled in the following sedimentation tank. The constants of the Langmuir equation are determined by running a jar test producing the results below. The volume of waste subjected to each test is one liter. If a flow rate of Q_o of 0.11 m³/s is to be treated, calculate the quantity of PAC needed for the operation. What is the adsorption capacity of the PAC? Calculate the quantity of PAC needed to treat the influent phenol to the ultimate residual concentration.

Test	PAC Added M (g)	$[C]$ (mg/L)
1	0.25	6.0
2	0.32	1.0
3	0.5	0.25
4	1.0	0.09
5	1.5	0.06
6	2.0	0.06
7	2.6	0.06

Solution:

$$a = \left\{ \frac{l\sum_{l+1}^m [C] - (m-l)\sum_1^l [C]}{l\sum_{l+1}^m \frac{[C]}{X/M} - (m-l)\sum_1^l \frac{[C]}{X/M}} \right\}$$

$$b = \frac{l}{a\sum_1^l \frac{[C]}{X/M} - \sum_1^l [C]}$$

Neglect tests 6 and 7, because the additional values of 0.06's would not conform to the Langmuir equation.

Test No.	PAC Added M (g)	[C] (mg/L)	X/M	[C]/X/M
1	0.25	6.0	0.076	78.95
2	0.32	1.0	0.075	13.33
3	0.5	0.25	0.0495	5.05
4	1.0	0.09	0.0249	3.61
5	1.5	0.06	0.0166	3.61

Let $l = 3$; $m = 5$

$$a = \frac{3(0.09 + 0.06) - 2(6.0 + 1.0 + 0.25)}{3(3.61 + 3.61) - 2(78.95 + 13.33 + 5.05)} = \frac{0.45 - 14.5}{21.66 - 194.66} = 0.081$$

$$b = \frac{l}{a\sum_1^l \frac{[C]}{X/M} - \sum_1^l[C]} = \frac{3}{0.081(78.95 + 13.33 + 5.05) - (6.0 + 1.0 + 0.25)}$$

$$= \frac{3}{7.88 - 7.25} = 4.76$$

$$\frac{X}{M} = \frac{ab[C]}{1 + b[C]} = \frac{0.081(4.76)(0.10)}{1 + (4.76)(0.10)} = 0.026 \; \frac{\text{kg phenol to be removed}}{\text{kg}\,C}$$

Total phenol to be removed $= 0.11(0.025 - 0.0001) = 0.00274$ kg/s.
Therefore,

$$PAC \text{ required} = \frac{0.00274}{0.026}(60)(60)(24) = 9,105 \text{ kg/d} \quad \textbf{Ans}$$

$$\left(\frac{X}{M}\right)_{ult} = a = 0.081 \; \frac{\text{kg phenol to removed}}{\text{kg}\,C} \quad \textbf{Ans}$$

The lowest concentration of phenol is 0.06 mg/L; at $[C] = 0.06$ mg/L,

$$\frac{X}{M} = \frac{0.081(4.76)(0.06)}{1 + (4.76)(0.06)} = 0.018 \; \frac{\text{kg phenol}}{\text{kg}\,C}$$

Total phenol to be removed to the ultimate residual concentration $= 0.11(0.025 - 0.00006) = 0.00274$ kg/s.
Therefore,

$$PAC \text{ required} = \frac{0.00274}{0.018}(60)(60)(24) = 13,152 \text{ kg/s} \quad \textbf{Ans}$$

Example 8.6 Solve the previous example using the Freundlich isotherm.

Solution:

$$n = \frac{l\sum_{l+1}^{m} ln[C] - (m-l)\sum_{1}^{l} ln[C]}{l\left(\sum_{l+1}^{m} ln\frac{X}{M} - (m-l)\sum_{1}^{l} ln\frac{X}{M}\right)}$$

$$k = \exp\left(\frac{\sum_{1}^{l} ln\frac{X}{M} - \frac{1}{n}\sum_{1}^{l} ln[C]}{l}\right)$$

Neglect tests 6 and 7, because the additional values of 0.06's would not conform to the Langmuir equation.

Test No.	PAC Added M (g)	[C] (mg/L)	In [C]	X/M	In X/M
1	0.25	6.0	1.79	0.076	−2.58
2	0.32	1.0	0	0.075	−2.59
3	0.5	0.25	−1.39	0.049	−3.02
4	1.0	0.09	−2.41	0.025	−3.69
5	1.5	0.06	−2.81	0.017	−4.07

[a] $0.076 = (25 - 6)(1)/0.25(1000)$; other values in the column are computed similarly.

Let $l = 3$; $m = 5M$

$$n = \frac{l\sum_{l+1}^{m} ln[C] - (m-l)\sum_{1}^{l} ln[C]}{l\left(\sum_{l+1}^{m} ln\frac{X}{M} - (m-l)\sum_{1}^{l} ln\frac{X}{M}\right)} = \frac{3(0.09 + 0.06) - 2(6.0 + 1.0 + 0.25)}{3(-3.69 - 4.07) - 2(-2.58 - 2.59 - 3.02)}$$

$$= \frac{0.45 - 14.5}{-23.28 + 16.38} = 2.04$$

$$k = \exp\left(\frac{\sum_{1}^{l} ln\left(\frac{X}{M} - \frac{1}{n}\right)\sum_{1}^{l} ln[C]}{l}\right)$$

$$= \exp\left[\frac{(-2.58 - 2.59 - 3.02) - \frac{1}{2.04}(1.79 + 0 - 1.39)}{3}\right]$$

$$= \exp\left[\frac{-8.19 - 0.196}{3}\right] = 0.061$$

$$\frac{X}{M} = k[C]^{1/n} = 0.061(0.1)^{1/2.04} = 0.020 \ \frac{kg \ phenol \ to \ be \ removed}{kg \, C}$$

Total phenol to be removed $= 0.11(0.025 - 0.0001) = 0.00274$ kg/s.

Therefore,

$$PAC \text{ required} = \frac{0.00274}{0.020}(60)(60)(24) = 11{,}837 \text{ kg/d} \quad \textbf{Ans}$$

From $X/M = k[C]^{1/n}$, as $[C]$ increases, X/M increases, such that, theoretically, X/M could become infinite. Practically, this means that equilibrium concentrations of C must be determined experimentally until found. The data do not reflect this. Therefore, the adsorption capacity would have to be solved using the Langmuir isotherm; thus,

$$\left(\frac{X}{M}\right)_{ult} = a = 0.081 \ \frac{\text{kg phenol to be removed}}{\text{kg}\,C} \quad \textbf{Ans}$$

The lowest concentration of phenol is 0.06 mg/L; at $[C] = 0.06$ mg/L,

$$\frac{X}{M} = 0.061(0.06)^{1/2.04} = 0.015 \ \frac{\text{kg phenol}}{\text{kg}\,C}$$

Total phenol to be removed to the ultimate residual concentration = $0.11(0.025 - 0.00006) = 0.00274$ kg/s.
Therefore,

$$PAC \text{ required} = \frac{0.00274}{0.015}(60)(60)(24) = 15{,}782 \text{ kg/s} \quad \textbf{Ans}$$

8.3.5 BED ADSORPTION AND ACTIVE ZONE

In bed adsorption, the water to be treated is passed through a bed of activated carbon, in which the form of carbon is normally GAC. The method of introduction of the influent may be made similar to sand filtration. In addition, the bed may be moving countercurrent or co-current to the flow of influent. Figure 8.8 shows a cutaway view of a carbon-bed adsorption unit. By a suitable modification of valve arrangements, this unit may be operated countercurrently or co-currently, in addition to being operated with the carbon bed stationary—the mode of operation depicted in the figure.

Figure 8.9a shows a schematic of an adsorption bed of depth L. Although the influent feed shown is introduced at the top, the following analysis applies to other modes of introduction as well, including moving beds. Hence, the figure should be viewed as a relative motion of the feed and the bed. The curved lines represent the configuration of the variation of the concentration of the adsorbate at various times as the adsorbate passes through the column. Thus, the curve labeled t_1 is the configuration at time t_1. The lower end of the curve is indicated by a zero concentration (or any concentration that represents the limit of removal) and the upper end is indicated by the influent concentration $[C_o]$. Thus, the volume of bed above curve t_1 is an

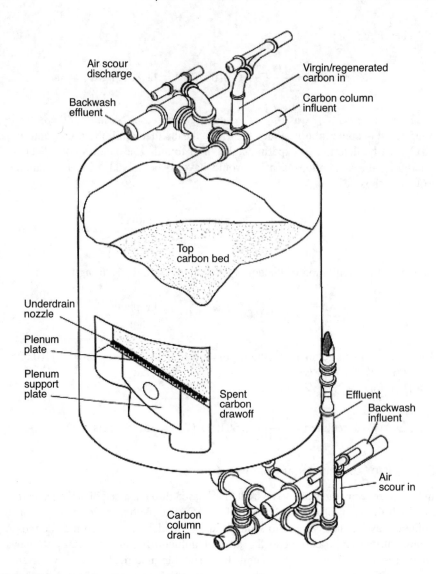

FIGURE 8.8 A bed carbon adsorption unit.

exhausted bed. Below the curve, the bed is clean (i.e., no adsorption is taking place, because all adsorbables had already been removed by the portion of the bed at curve t_1 and above). The zone of bed encompassed by the curve represents a bed partially exhausted.

The curve t_1 then advances to form the curve t_2 at time t_2. The curve shows a breakthrough of concentration of some fraction of $[C_o]$. This breakthrough concentration appears in the effluent. Finally t_2 advances to t_3 at time t_3. At this time, the bed is almost exhausted. The profile concentration represented by curve t_1 is called an *active zone*. This zone keeps on advancing until the whole bed becomes totally exhausted.

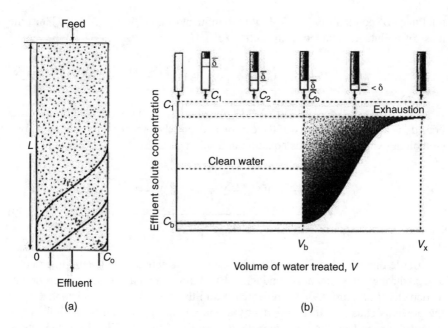

FIGURE 8.9 Active zones at various times during adsorption and the breakthrough curve.

It is to be noted that by the nature of the process, the concentration at the bottom of the active zone as it moves through the column has to be at the limiting residual concentration. Thus, as long as δ has not reached the bottom of the column, the effluent is the limiting residual concentration that a particular batch of carbon is capable of removing.

Figure 8.9b shows the movement of the active zone represented by the length δ as it advances through the bed at various times. At the beginning of breakthrough, at which the lower end of δ barely touches the bottom of the column, the total volume of treated water is represented by V_b. The shaded portion in the curve represents the total breakthrough mass of adsorbate before exhaustion. The total volume of wastewater treated at exhaustion is designated by V_x.

Determination of the length of the active zone. The length of the active zone represents a portion of bed that is no longer totally usable, since this is the particular length when breakthrough occurs. It is therefore important that this length be allowed for in design. To determine this length, perform a mass balance on the active zone during breakthrough. The total mass of pollutant that escaped removal from the beginning to the completion of breakthrough is

$$\sum (V_{n+1} - V_n)\left(\frac{[C_{n+1}] + [C_n]}{2}\right) \tag{8.37}$$

where the indices n and $n + 1$ refer to the volume or concentration that broke through δ at a time step of Δt from t_n to t_{n+1}. Also, at the completion of breakthrough, the total mass of pollutant introduced into the influent of δ is $(V_x - V_b)[C_o]$. Thus, the mass of pollutants retained in the active zone M_r is

$$M_r = (V_x - V_b)[C_o] - \sum (V_{n+1} - V_n)\left(\frac{[C_{n+1}] + [C_n]}{2}\right) \qquad (8.38)$$

M_r is also equal to $\frac{1}{2}\{A_s \delta \rho_p (X/M)_{ult}\}$, where A_s is the superficial area of the bed, ρ_p is the packed density of the carbon bed, and the others are as defined before. Combining with the previous equation and solving for δ,

$$\delta = \frac{2\left\{(V_x - V_b)[C_o] - \sum(V_{n+1} - V_n)\left(\frac{[C_{n+1}][C_n]}{2}\right)\right\}}{A_s \rho_p \left(\frac{X}{M}\right)_{ult}} \qquad (8.39)$$

To use this equation for the determination of the active zone length, in practice, a breakthrough experiment is conducted. Breakthrough volumes and concentrations as depicted in Figure 8.9b are recorded. From the data, δ may be calculated using the previous equation. It is important that the rate of application of the influent be recorded, as the length of δ is dependent upon this flow rate. The faster the rate of application, the longer the length of δ, and the slower the rate of application, the shorter the length of δ.

Example 8.7 A breakthrough experiment is conducted for phenol producing the results in the following table. Determine the length δ of the active zone. The diameter of the column used is 2.54 cm, and the packed density of the bed is 721.58 kg/m^3. $[C_o]$ is equal to 25 mg/L. $X/M = 0.020$ kg/kg.

C (mg/L)	V (L)
0.06	1.0
1.0	1.24
6.0	1.31
10	1.43
15	1.48
18	1.58
20	1.72
23	1.83
25	2.00

Solution:

$$\delta = \frac{2\left\{(V_x - V_b)[C_o] - \sum(V_{n+1} - V_n)\left(\frac{[C_{n+1}] + [C_n]}{2}\right)\right\}}{A_s \rho_p \left(\frac{X}{M}\right)_{ult}}$$

C (mg/L)	V (L)	$(V_{n+1} - V_n)/1000$ (m³)	$([C_{n+1}] + [C_n])/$ $2(1000)$ (kg/m³)	$((V_{n+1} - V_n)/1000)$ $([C_{n+1}] + [C_n])/$ $2(1000)$ (kg)
0.06	1.0	—	—	—
1.0	1.24	0.00024	0.00053	0.00000013
6.0	1.31	0.00007	0.0035	0.00000025
10	1.43	0.00012	0.008	0.00000096
15	1.48	0.00005	0.0125	0.00000063
18	1.58	0.0001	0.0165	0.0000017
20	1.72	0.00014	0.019	0.0000027
23	1.83	0.00011	0.0215	0.0000024
25	2.00	0.00017	0.024	0.0000041
				$\Sigma = 0.0000128$

$$A_s = \frac{\pi[0.0254]^2}{4} = 0.00051 \text{ m}^2$$

Therefore,

$$\delta = \frac{2\left\{(V_x - V_b)[C_o] - \Sigma(V_{n+1} - V_n)\left(\frac{[C_{n+1}] + [C_n]}{2}\right)\right\}}{A_s \rho_p \left(\frac{X}{M}\right)_{ult}}$$

$$= \frac{2(0.002 - 0.001)(0.025) - 0.0000128}{0.00051(721.58)(0.020)}$$

$$= 0.0051 \text{ m} = 5.1 \text{ mm} \quad \textbf{Ans}$$

Example 8.8 A wastewater containing 25 mg/L of phenol and having the characteristic breakthrough of the previous example is to be treated by adsorption onto an activated carbon bed. The flow rate during the breakthrough experiment is 0.11 m³/s; this is equivalent to a surficial velocity of 0.0088 m/s. The X/M ratio of the bed for the desired effluent of 0.06 mg/L is 0.02 kg solute per kg carbon. If the flow rate for design is also 0.11 m³/s, design the absorption column. Assume the influent is introduced at the top of the bed. The packed density of the carbon bed is 721.58 kg/m³.

Solution: The design will include the determination of the amount of activated carbon needed, the dimensions of the column, and the interval of activated carbon replacement.

Amount of phenol to be removed = 0.11(0.025 − 0.00006)(60)(60)(24) = 237 kg/d
Amount of activated carbon needed = 237/0.02 = 11,850 kg/d **Ans**

From the previous example, adsorbate retained in δ length of test column at exhaustion = 2(0.002 − 0.001)(0.025) − 0.0000128 = 0.000037 kg. For A_s of the test column = 0.00051 m and δ = 0.0051 m, adsorbate retained per unit volume in δ = 14.23 kg.

Assume diameter of actual column 4 m; $A_s = \pi (4^2)/4 = 12.57$ m^2
Using 2 columns, adsorbate retained in δ length of the columns
   $= 2(12.57)(0.0051)(14.23) = 1.82$ kg

Assuming carbon replacement is to be done every week, adsorbate retained in column length of length $L - \delta = 237(7) - 1.82 = 1657$ kg
   Carbon required for the 1657 kg of adsorbate $= 1657/0.02 = 82,850$ kg $= 41,425$ kg/column
Therefore,

$$L = \frac{\frac{41.425}{721.58}}{12.57} + 0.0051$$

   $= 4.75$, say 4.8 m including freeboard and other allowances.   **Ans**

Diameter $= 4$ m   **Ans**
Interval of carbon replacement $=$ once every week   **Ans**

   As soon as the bed is exhausted, as determined by the breakthrough of concentrations, the carbon may be replaced. The replaced carbon may be reactivated again for reuse. Up to 30 or more reactivations may be made without appreciable loss of adsorptive power of the reactivated carbon.

8.3.6 RELATIVE VELOCITIES IN BED ADSORPTION

As mentioned before, the unit operation of bed adsorption may be carried out in a moving-bed mode, either co-currently or countercurrently. When the breakthrough experiment is carried out, the superficial velocity should also be recorded. The reason is that adsorption is a function of the time of contact between the liquid phase containing the solute to be adsorbed and solid-phase carbon bed. Thus, for the breakthrough data to be applicable to an actual prototype adsorption column, the relative velocities that transpired during the test must be maintained in the actual column. When the relative velocities between the flowing water and the carbon bed are maintained, it is immaterial whether or not the bed is moving.
   Consider first the co-current operation. Let V_s be the superficial velocity of the flowing water relative to the stationary earth and V_b be the velocity of the bed also relative to the stationary earth. Thus, the relative velocity of the flowing water $V_{s/b}$ relative to the bed is

$$V_{s/b} = V_s - V_b$$
$$V_s = V_{s/b} + V_b \tag{8.40}$$

For the countercurrent operation, the formula is

$$V_{s/b} = V_s - (-V_b)$$
$$V_s = V_{s/b} + (-V_b) = V_{s/b} - V_b \tag{8.41}$$

In a breakthrough experiment, the superficial velocity may be obtained by dividing the volume V of water collected in t time by the superficial area of the experimental column. Breakthrough experiments are invariably conducted in stationary beds. Thus, from the previous equations this superficial velocity is actually the relative velocity of the flowing water with respect to the bed, with V_b equal to zero. This relative velocity must be maintained in the actual column design, if the data collected in the breakthrough experiment are to be applicable.

Example 8.9   Design the column of the previous example if the feed is introduced at the bottom. The carbon is continuously removed at the bottom and continuously added at the top. Due to the countercurrent operation, assume the bed expands by 40%.

Solution: The design will include, in addition to those in the previous example, the determination of the carbon removal and addition rates at the bottom and top of column, respectively. These rates are determined from the length of the packed carbon in the column and the interval of replacement as stated in the previous example.

$$\text{Removal rate} = \text{addition rate} = \frac{4.57}{7(24)}\,(12.57)(721.58) = 246.73 \text{ kg/h} \quad \textbf{Ans}$$

The superficial velocity, from the previous example = 0.0088 m/s, velocity relative to the stationary bed. In the countercurrent operation, this relative velocity must be maintained if the breakthrough curve is to be applicable. Because of the expansion, V_b is increased by 40%. Thus, V_s in present design considering the expansion:

$$V_s = V_{s/b} - V_b = 0.0088 - \frac{4.57(1.4)}{7(24)(60)(60)} = 0.0088 - 0.00001 = 0.0088 \text{ m/s}$$

Therefore,

$$A_s = \frac{0.11}{0.0088} = 12.5 \text{ m}^2 \Rightarrow \text{diameter} = 4 \text{ m} \quad \textbf{Ans}$$

$L = 1.4(4.57) = 6.4$ m say 6.7 m, including freeboard and other allowances   **Ans**

8.3.7   HEAD LOSSES IN BED ADSORPTION

The operation of carbon bed adsorption units is similar to that of filtration units. In fact, bed adsorption is partly filtration. Countercurrent flow operation is analogous to filter backwashing, and co-current flow operation is analogous to normal downflow or gravity filtration.

Head loss calculations for bed adsorption are therefore the same as those with filtration. Since head loss formulas through beds of solids have already been discussed under filtration, they will not be pursued here. The important point to remember is that for filtration formulas to apply under moving-bed adsorption operations, the superficial velocity should now be considered relative velocity.

GLOSSARY

Anion membrane—An electrodialysis membrane that allows passage of anions.

Apolar membrane—An *RO* membrane in which the nonpolar regions exceed the polar regions.

Activation—The process of enhancing a particular characteristic.

Active zone—A small length of column where adsorption is taking place.

Activated carbon—Carbon with enhanced adsorption characteristic.

Adsorbate—The solute adsorbed onto the surface of a solid.

Adsorbent—The solid that adsorbs the adsorbate.

Adsorption—The process of concentrating solute at the surface of a solid as a result of the unbalanced attraction of atoms at the surface of the solid.

Absorption capacity—The *X/M* that produces the lowest possible residual concentration of the adsorbate.

Adsorption isotherm—An equation relating the amount of solute adsorbed onto the solid and the equilibrium concentration of the solute in solution at a given temperature.

Benzene ring—

Cation membrane—An electrodialysis membrane that allows passage of cations.

Chemical activation—The process of activating carbon using chemicals.

Chemical adsorption or chemisorption—A type of adsorption aided by chemical bonding on the surfaces.

Coulomb efficiency—The fraction of the input number of equivalents of an ionized substance that is actually acted upon by an input of electricity.

Current density—The current in milliamperes that flows through a square centimeter of membrane perpendicular to the current direction.

Electrodialysis membrane—Sheet-like barriers made out of high-capacity, highly cross-linked ion exchange resins that allow passage of ions but not of water.

Flux—The rate of flow or mass flow per unit area per unit time.

Microfiltration—A reverse osmosis process that removes particles larger than those removed by ultrafiltration and operating at the lowest operating range of the reverse osmosis processes in the neighborhood of 70 kPag.

Nanofiltration—A reverse osmosis process that removes particles in the size range of the order of 0.001 μm at an operating pressure range of 500 to 1,400 kPag.

Peptide bond— $-\overset{\text{O}}{\overset{\|}{\text{C}}}-\text{NH}-$

Physical adsorption—A type of adsorption brought about by weak van der Waals forces.

Polarization—A localized deficiency of ions in the vicinity of the membrane surfaces.

Polar membrane—An *RO* membrane in which the polar regions exceed the nonpolar regions.

Pressure membrane—A membrane that is used to separate materials from a fluid by the application of high pressure on the membrane.

Product or permeate—The portion of the influent that passes (permeates) through the membrane.

Reverse osmosis—A process of removing minute particles such as ions, molecules, and molecular aggregates. The process operating at the highest pressure range of 1,400 to 8,300 kPag removes particles in the size range of 0.0001 to 0.001 μm.

Steam activation—The process of activating carbon using steam.

Ultrafiltration—A reverse osmosis process that removes particles ranging in sizes from 0.001 to 10 μm at an operating range of 100 to 500 kPag.

Water gas reaction—$C + H_2O \rightarrow H_2 + CO$.

SYMBOLS

a	A constant in the Langmuir isotherm
A_s	Cross-sectional area of carbon column
b	A constant in the Langmuir isotherm
c	Mass of cake per unit volume of filtrate
k	A constant in the Freundlich isotherm
$[C]$	Equilibrium concentration of adsorbate in solution
$[C_{ci}]$	Concentration of species i in concentrate of RO units
$[C_o]$	Influent concentration to an active zone, column, or electrodialysis RO units
$[C_{oi}]$	Influent concentration of species i to an RO unit
$[C_{pi}]$	Concentration of species i in permeate of RO units
E	Electromotive force across electrodes
F	Product flux of RO units
I	Electric current impressed on the electrodes of an electrodialysis unit
K	A constant in the flux decline equation of RO units
m	Number of membranes in an electrodialysis unit; a constant in the flux decline equation of RO units
M	Amount of adsorbent that adsorbs X amount of adsorbate
$\mathcal{M}$	Current efficiency
n	A constant in the Freundlich isotherm
P	Power to electrodialysis unit
P_f	Applied pressure on feed side of RO units
P_p	Pressure on permeate side of RO units
P_{fn}	Net applied pressure on feed side of RO units
P_{pn}	Net pressure on permeate side of RO units
ΔP	Pressure drop
Q_o	Influent flow
Q_c	Concentrate flow
Q_p	Permeate flow
R	Electric resistance across electrodialysis unit; percent rejection of RO units
R_m	Resistance of filter or RO membrane

s	Index of membrane and boundary layer compressibility
S_o	Surficial area
t	Time of operation
V_b	Absolute velocity of bed
V_s	Absolute superficial velocity of water
$V_{s/b}$	Relative superficial velocity of flowing water relative to bed
$\mathcal{V}$	Volume of sample; volume of permeate
$\mathcal{V}_b$	Volume of effluent in column at breakthrough
$\mathcal{V}_x$	Volume of effluent in column at exhaustion
X	Amount of adsorbate adsorbed to M amount of adsorbent
$(X/M)_{ult}$	Adsorption capacity
$\bar{\alpha}$	Specific cake resistance
$\bar{\alpha}_m$	Combined effect of compressibility, membrane resistance, and solute resistance in RO
$\bar{\alpha}_{mo}$	A constant of proportionality in $\bar{\alpha}_m = \bar{\alpha}_{mo} \, (-\Delta P)^s$
δ	Length of an active zone
η	Coulomb efficiency
μ	Absolute viscosity of water
ρ_p	Bulk density of carbon
π_f	Osmotic pressure on feed side of RO units
π_p	Osmotic pressure on permeate side of RO units

PROBLEMS

8.1   A wastewater containing a $[C_o] = 25$ mg/L of phenol is to be treated using PAC to produce an effluent concentration $[C]_{eff} = 0.10$ mg/L. The PAC is simply added to the stream and the mixture subsequently settled in the following sedimentation tank. The constants of the Langmuir equation are determined by running a jar test producing the results below. The volume of waste subjected to each test is one liter. If the flow rate Q_o is 0.11 m³/s, calculate the quantity of PAC needed for the operation. What is the adsorption capacity of the PAC? Calculate the quantity of PAC needed to treat the influent phenol to the ultimate residual concentration. Use the Langmuir isotherm.

Test	PAC Added X (g)	[C] (mg/L)
1	0.25	6.0
2	0.5	0.25
3	1.0	0.09
4	2.6	0.06

8.2   Solve Problem 8.1 using the Freundlich isotherm.

8.3   A wastewater containing a $[C_o] = 25$ mg/L of phenol is to be treated using PAC to produce an effluent concentration $[C]_{eff} = 0.10$ mg/L. The PAC is simply added to the stream and the mixture subsequently settled in the following sedimentation tank. The constants of the Langmuir equation is determined by running a jar test producing the results below. The volume of waste subjected to each test is one liter. If the flow rate Q_o is 0.11 m³/s, calculate the quantity of PAC needed for the operation. What is the adsorption capacity of the PAC? Calculate the quantity of PAC needed to treat the influent phenol to the ultimate residual concentration. Use the Langmuir isotherm.

Test	PAC Added X (g)	[C] (mg/L)
1	0.25	6.0
2	2.6	0.06

8.4   Solve Problem 8.3 using the Freundlich isotherm.

8.5   A wastewater containing a $[C_o]$ = 25 mg/L of phenol is to be treated using PAC to produce an effluent concentration $[C]_{eff}$ = 0.10 mg/L. The PAC is simply added to the stream and the mixture subsequently settled in the following sedimentation tank. The constants of the Langmuir equation are determined by running a jar test producing the results below. The volume of waste subjected to each test is one liter. If the flow rate Q_o is 0.11 m³/s, calculate the quantity of PAC needed for the operation. What is the adsorption capacity of the PAC? Calculate the quantity of PAC needed to treat the influent phenol to the ultimate residual concentration. Use the Langmuir isotherm.

Test	PAC Added X (g)	[C] (mg/L)
1	0.25	6.0

8.6   Solve Problem 8.5 using the Freundlich isotherm.

8.7   A breakthrough experiment is conducted for phenol producing the results below. The length of the active zone is calculated to be 5.1 mm. The diameter of the column used is 2.5 cm, and the packed density of the bed is 721.58 kg/m³ $(X/M)_{ult}$ = 0.020 kg/kg. Calculate the influent phenol concentration.

C (mg/L)	V (L)
0.06	1.0
1.0	1.24
6.0	1.31
10	1.43
15	1.48
18	1.58
20	1.72
23	1.83
25	2.00

8.8   A breakthrough experiment is conducted for phenol producing the results shown in Problem 8.7. The length of the active zone is calculated to be

5.1 mm. The influent phenol concentration is 25 mg/L. The packed density of the bed is 721.58 kg/m$^3 \cdot (X/M)_{ult} = 0.020$ kg/kg. Calculate the diameter of the column used in the experiment.

8.9   A breakthrough experiment is conducted for phenol producing the results shown in Problem 8.7. The length of the active zone is calculated to be 5.1 mm. The influent phenol concentration is 25 mg/L. $(X/M)_{ult} = 0.020$ kg/kg. The diameter of the column used in the experiment is 2.5 cm. Calculate the packed density of the carbon used.

8.10   A breakthrough experiment is conducted for phenol producing the results shown in Problem 8.7. The length of the active zone is calculated to be 5.1 mm. The influent phenol concentration is 25 mg/L. The diameter of the column used in the experiment is 2.5 cm and the packed density of the bed is 721.58 kg/m^3. Calculate the adsorption capacity of the bed.

8.11   A breakthrough experiment is conducted for phenol producing the results below. Determine the length δ of the active zone. The diameter of the column used is 2.5 cm, and the packed density of the bed is 721.58 kg/m^3. $[C_o]$ is equal to 25 mg/L. $(X/M)_{ult} = 0.020$ kg/kg.

C (mg/L)	V (L)
0.06	1.0
1.0	1.24
6.0	1.31
10	1.43
20	1.72
25	2.00

8.12   A breakthrough experiment is conducted for phenol producing the results below. Determine the length δ of the active zone. The diameter of the column used is 2.5 cm, and the packed density of the bed is 721.58 kg/m^3. $[C_o]$ is equal to 25 mg/L. $(X/M)_{ult} = 0.020$ kg/kg.

C (mg/L)	V (L)
0.06	1.0
1.0	1.24
6.0	1.31
25	2.00

8.13   A breakthrough experiment is conducted for phenol producing the results below. Determine the length δ of the active zone. The diameter of the column used is 2.5 cm, and the packed density of the bed is 721.58 kg/m^3.

$[C_o]$ is equal to 25 mg/L. $(X/M)_{ult} = 0.020$ kg/kg.

C (mg/L)	V (L)
0.06	1.0
25	2.00

8.14 A wastewater containing 25 mg/L of phenol and having the characteristic breakthrough of Problem 8.11 is to be treated by adsorption onto an activated carbon bed. Assume that the flow rate during the breakthrough experiment is 0.11 m³/s. The X/M ratio of the bed for the desired effluent of 0.1 mg/L is 0.029 kg solute per kg carbon. If the flow rate for design is also 0.11 m³/s, design the absorption column. Assume the influent is introduced at the top of the bed. The packed density of the carbon bed is 721.58 kg/m³.

8.15 A wastewater containing 25 mg/L of phenol and having the characteristic breakthrough of Problem 8.12 is to be treated by adsorption onto an activated carbon bed. Assume that the flow rate during the breakthrough experiment is 0.11 m³/s. The X/M ratio of the bed for the desired effluent of 0.1 mg/L is 0.029 kg solute per kg carbon. If the flow rate for design is also 0.11 m³/s, design the absorption column. Assume the influent is introduced at the top of the bed. The packed density of the carbon bed is 721.58 kg/m³.

8.16 A wastewater containing 25 mg/L of phenol and having the characteristic breakthrough of Problem 8.13 is to be treated by adsorption onto an activated carbon bed. Assume that the flow rate during the breakthrough experiment is 0.11 m³/s. The X/M ratio of the bed for the desired effluent of 0.1 mg/L is 0.029 kg solute per kg carbon. If the flow rate for design is also 0.11 m³/s, design the absorption column. Assume the influent is introduced at the top of the bed. The packed density of the carbon bed is 721.58 kg/m³.

8.17 A wastewater containing 25 mg/L of phenol and having the characteristic breakthrough of Problem 8.12 is to be treated by adsorption onto an activated carbon bed. Assume that the flow rate during the breakthrough experiment is 0.11 m³/s. The X/M ratio of the bed for the desired effluent of 0.1 mg/L is 0.020 kg/kg. If the flow rate for design is also 0.11 m³/s, design the absorption column. Assume the influent is introduced at the top of the bed. The packed density of the carbon bed is 721.58 kg/m³.

8.18 A wastewater containing 25 mg/L of phenol and having the characteristic breakthrough of Problem 8.13 is to be treated by adsorption onto an activated carbon bed. Assume that the flow rate during the breakthrough experiment is 0.11 m³/s. The X/M ratio of the bed for the desired effluent of 0.1 mg/L is 0.020 kg/kg. If the flow rate for design is also 0.11 m³/s, design the absorption column. Assume the influent is introduced at the top of the bed. The packed density of the carbon bed is 721.58 kg/m³.

8.19   Design the column of Problem 8.14 if the feed is introduced at the bottom. The carbon is continuously removed at the bottom and continuously added at the top. Due to the countercurrent operation, assume the bed expands by 40%.

8.20   Design the column of Problem 8.15 if the feed is introduced at the bottom. The carbon is continuously removed at the bottom and continuously added at the top. Due to the countercurrent operation, assume the bed expands by 40%.

8.21   Design the column of Problem 8.16 if the feed is introduced at the bottom. The carbon is continuously removed at the bottom and continuously added at the top. Due to the countercurrent operation, assume the bed expands by 40%.

8.22   Design the column of Problem 8.17 if the feed is introduced at the bottom. The carbon is continuously removed at the bottom and continuously added at the top. Due to the countercurrent operation, assume the bed expands by 40%.

8.23   Design the column of Problem 8.18 if the feed is introduced at the bottom. The carbon is continuously removed at the bottom and continuously added at the top. Due to the countercurrent operation, assume the bed expands by 40%.

8.24   A brackish water of 379 m^3/day containing 4000 mg/L of ions expressed as NaCl is to be deionized using an electrodialysis unit. The coulomb efficiency is 0.78 and there are 400 membranes in the unit each measuring 51 cm × 46 cm. Resistance across the unit is 6 ohms and the current efficiency is 90%. Estimate the power requirement.

8.25   A brackish water of 379 m^3/day is to be deionized using an electrodialysis unit. The coulomb efficiency is 0.78, and 400 membranes are in the unit each measuring 51 cm × 46 cm. Resistance across the unit is 6 ohms and the current efficiency is 90%. The input power required to run the unit is 93.3 kW. Estimate the concentration of ions to be removed expressed as NaCl.

8.26   A brackish water containing 4000 mg/L of ions expressed as NaCl is to be deionized using an electrodialysis unit. The coulomb efficiency is 0.78, and 400 membranes are in the unit each measuring 51 cm × 46 cm. Resistance across the unit is 6 ohms and the current efficiency is 90%. The input power required to run the unit is 93.3 kW. Estimate the input flow to the unit.

8.27   A brackish water containing 4,000 mg/L of ions expressed as NaCl is to be deionized using an electrodialysis unit. There are 400 membranes in the unit each measuring 51 cm × 46 cm. Resistance across the unit is 6 ohms and the current efficiency is 90%. The input power required to run the unit is 93.3 kW. If the inflow to the unit is 379 m^3/day, calculate the coulomb efficiency.

8.28   A brackish water containing 4,000 mg/L of ions expressed as NaCl is to be deionized using an electrodialysis unit. Resistance across the unit is 6 ohms and the current efficiency is 90%. The input power required to run the unit is 93.3 kW. If the inflow to the unit is 379 m^3/day and the coulomb efficiency is 0.78, estimate the number of membranes in the unit.

8.29   A brackish water containing 4,000 mg/L of ions expressed as NaCl is to be deionized using an electrodialysis unit. Resistance across the unit is 6 ohms. The input power required to run the unit is 93.3 kW. The inflow to the unit is 379 m^3/day, the coulomb efficiency is 0.78, and 400 membranes are in the unit each measuring 51 cm × 46 cm. Calculate the current efficiency.

8.30   A brackish water containing 4,000 mg/L of ions expressed as NaCl is to be deionized using an electrodialysis unit. The input power required to run the unit is 93.3 kW. The inflow to the unit is 379 m^3/day, the coulomb efficiency is 0.78, and 400 membranes are in the unit each measuring 51 cm × 46 cm. The current efficiency is 90%. What is the electric resistance across the unit?

8.31   A brackish water of 379 m^3/day containing 4,000 mg/L of ions expressed as NaCl is to be deionized using an electrodialysis unit. There are 400 membranes in the unit each measuring 51 cm × 46 cm inches. Resistance across the unit is 6 ohms, the current and coulomb efficiencies are, respectively, 90% and 78%. Estimate the impressed current.

8.32   A brackish water of 379 m^3/day is to be deionized using an electrodialysis unit. There are 400 membranes in the unit each measuring 51 cm × 46 cm inches. Resistance across the unit is 6 ohms, the current and coulomb efficiencies are, respectively, 90% and 78%. If the impressed current is 124.76 amperes, what is the concentration of the ions in the raw water expressed as NaCl?

8.33   A brackish water containing 4,000 mg/L of ions expressed as NaCl is to be deionized using an electrodialysis unit. There are 400 membranes in the unit each measuring 51 cm × 46 cm inches. Resistance across the unit is 6 ohms, the current and coulomb efficiencies are, respectively, 90% and 78%. If the impressed current is 124.76 amperes, what is the influent flow to the unit?

8.34   A brackish water containing 4,000 mg/L of ions expressed as NaCl is to be deionized using an electrodialysis unit. There are 400 membranes in the unit each measuring 51 cm × 46 cm inches. Resistance across the unit is 6 ohms; the current efficiency is 90%. If the impressed current is 124.76 amperes and the influent flow to the unit is 379 m^3/day, what is the coulomb efficiency?

8.35   A brackish water containing 4,000 mg/L of ions expressed as NaCl is to be deionized using an electrodialysis unit. Resistance across the unit is 6 ohms, the current and coulomb efficiencies are, respectively, 90% and 78%. If the impressed current is 124.76 amperes and the influent flow to the unit is 379 m^3/day, calculate the number of membranes in the unit.

8.36   A brackish water containing 4,000 mg/L of ions expressed as NaCl is to be deionized using an electrodialysis unit. Resistance across the unit is 6 ohms; the coulomb efficiency is 78%. If the impressed current is

124.76 amperes and the influent flow to the unit is 379 m^3/day, calculate the current efficiency.

8.37   A long term experiment for a *CA* membrane module operated at 2758 kPag using a feed of 2,000 mg/L of NaCl at 25°C produces the results below. What is the expected flux at the end of one year of operation? What is the expected flux at the end of two years? How long does it take for the flux to decrease to 0.37 m^3/m^2 · day?

Time (h)	1	25,000
Flux (m^3/ m^2 · day)	0.66	0.45

8.38   The feedwater to an *RO* unit contains 3,000 mg/L of NaCl, 300 mg/L of CaCl$_2$, and 400 mg/L of MgSO$_4$. The membrane used is cellulose acetate. Applying a pressure of 4826 kPag, the flux is found to be 0.203 m^3/m^2· day. If *s* and $\bar{\alpha}_{mo}$ are, respectively, 0.5597 and 54.72 in the MKS system of units (meter-kilogram-second), at what temperature is the unit being operated?

8.39   The feedwater to an *RO* unit contains 3,000 mg/L of NaCl, 300 mg/L of CaCl$_2$, and 400 mg/L of MgSO$_4$. The membrane used is cellulose acetate. Applying a pressure of 4,826 kPag, the flux is found to be 0.203 m^3/m^2 · day. If *s* is equal to 0.5597 in the MKS system of units (meter-kilogram-second) and the temperature of operation is 25°C, what is the value of $\bar{\alpha}_{mo}$?

8.40   The feedwater to an *RO* unit contains 3,000 mg/L of NaCl, 300 mg/L of CaCl$_2$, and 400 mg/L of MgSO$_4$. The membrane used is cellulose acetate. The flux is 0.203 m^3/m^2 · day. If *s* and $\bar{\alpha}_{mo}$ are, respectively, 0.5597 and 54.72 in the MKS system of units (meter-kilogram-second) and the temperature is 25°C, what is the pressure applied to the membrane?

8.41   The feedwater to an *RO* unit contains 3,000 mg/L of NaCl, 300 mg/L of CaCl$_2$, and 400 mg/L of MgSO$_4$. The membrane used is cellulose acetate. Applying a pressure of 4,826 kPag, the flux is found to be 0.203 m^3/m^2 · day. If $\bar{\alpha}_{mo}$ is 54.72 in the MKS system of units (meter-kilogram-second) and the temperature is 25°C, what is the value of *s*?

BIBLIOGRAPHY

Abdel-Jawad, M., et al. (1997). Pretreatment of the municipal wastewater feed for reverse osmosis plants, *Desalination*. 109, 2, 211–223.

Abuzaid, N. S. and G. Nakhla (1997). Predictability of the homogeneous surface diffusion model for activated carbon adsorption kinetics; formulation of a new mathematical model, *J. Environment. Science Health, Part A: Environment. Eng. Toxic Hazardous Substance Control*. 32, 7, 1945–1961.

Alawadhi, A. A. (1997). Pretreatment plant design—Key to a successful reverse osmosis desalination plant, *Desalination Int. Symp. Pretreatment of Feedwater for Reverse Osmosis Desalination Plants*, March 31–April 2, 110, 1–2, 1–10. Elsevier Science B.V., Amsterdam, Netherlands.

Al-Mutaz, I. S., M. A. Soliman, and A. E. S. Al-Zahrani (1997). Modeling and simulation of hollow fine fiber modules with radial dispersion—A parametric sensitivity study, *Desalination.* 110, 3, 239–250.

Baudin, I., et al. (1997). L'Apie and Vigneux case studies: First months of operation, *Desalination, Proc. 1997 Workshop on Membranes in Drinking Water Production,* June 1–4, 113, 2–3, 273–275. Elsevier Science B.V., Amsterdam, Netherlands.

Bornhardt, C., J. E. Drewes, and M. Jekel (1997). Removal of organic halogens (AOX) from municipal wastewater by powdered activated arbon (PAC)/activated sludge (AS) treatment, *Water Science Technol., Proc. 1996 IAWQ Int. Conf. Advanced Wastewater Treatment: Nutrient Removal and Anaerobic Processes,* Sept. 23–25, 1996, Amsterdam, Netherlands, 35, 10, 147–153. Elsevier Science Ltd., Oxford England.

Bou-Hamad, S., et al. (1997). Performance evaluation of three different pretreatment systems for seawater reverse osmosis technique, *Desalination Int. Symp. Pretreatment of Feedwater for Reverse Osmosis Desalination Plants,* March 31–April 2, 110, 1–2, 85–92. Elsevier Science B.V., Amsterdam, Netherlands.

Brasquet, C., E. Subrenat, and P. Le Cloirec (1997). Selective adsorption on fibrous activated carbon of organics from aqueous solution: Correlation between adsorption and molecular structure, *Water Science Technol., Proc. 1996 1st Int. Specialized Conf. on Adsorption in the Water Environment and Treatment Processes,* Nov. 5–8 1996, Shirahama, Japan 35, 7, 251–259. Elsevier Science Ltd., Oxford, England.

Chakravorty, B. and A. Layson (1997). Ideal feed pretreatment for reverse osmosis by continuous microfiltration, *Desalination Int. Symp. Pretreatment of Feedwater for Reverse Osmosis Desalination Plants,* March 31–April 2, 110, 1–2, 143–150. Elsevier Science B.V., Amsterdam, Netherlands.

Chang, C. and Y. Ku (1997). Adsorption of EDTA-chelated copper ion in aqueous solution by an activated carbon adsorption column, *J. Chinese Institute of Engineers, Trans. of the Chinese Institute of Engineers, Series A/Chung-kuo Kung Ch'eng Hsuch K'an* 20, 6, 651–659.

Cote, P., et al. (1997). Immersed membrane activated sludge for the reuse of municipal wastewater, *Desalination Proc. 1997 Workshop on Membranes in Drinking Water Production,* June 1–4, L'Aquila, Italy, 113, 2–3, 189–196. Elsevier Science B.V., Amsterdam, Netherlands.

Ghayeni, S. B., et al. (1997). Adhesion of waste water bacteria to reverse osmosis membranes, *J. Membrane Science.* 138, 1, 29–42.

de Witte, J. (1997). New development in nanofiltration and reverse osmosis membrane manufacturing, *Desalination Proc. 1997 Workshop on Membranes in Drinking Water Production,* June 1–4, L'Aquila, Italy, 113, 2–3, 153–156. Elsevier Science B.V., Amsterdam, Netherlands.

Dudley, L. Y. and E. G. Darton (1997). Pretreatment procedures to control biogrowth and scale formation in membrane systems, *Desalination Int. Symp. on Pretreatment of Feedwater for Reverse Osmosis Desalination Plants,* March 31–April 2, 110, 1–2, 11–20. Elsevier Science B.V., Amsterdam Netherlands.

El-Sayed, E., et al. (1997). Prediction of RO membrane performance for Arabian Gulf seawater, *Desalination.* 113, 1, 39–50.

Flemming, H. C., et al. (1997). Biofouling—The achilles heel of membrane processes, *Desalination, Proc. 1997 Workshop on Membranes in Drinking Water Production,* June 1–4, L'Aquila, Italy, 113, 2–3, 215–225. Elsevier Science B.V., Amsterdam, Netherlands.

Flemming, H. C. (1997). Reverse osmosis membrane biofouling, *Experimental Thermal Fluid Science.* 14, 4, 382–391.

Hand, D. W., et al. (1997). Predicting the performance of fixed-bed granular activated carbon adsorbers, *Water Science Technol., Proc. 1996 1st Int. Specialized Conf. on Adsorption in the Water Environment and Treatment Processes Shirahama*, Japan, Nov. 5–8, 1996, 35, 7, 235–241. Elsevier Science Ltd., Oxford, England.

Higgins, B. W. (1997). Managing VOC compliance at wastewater and hazardous waste facilities, *Environment. Technol*. 7, 4, 28, 30, 32–35.

Hofman, J. A. M. H., et al. (1997). Removal of pesticides and other micropollutants with cellulose-acetate, polyamide and ultra-low pressure reverse osmosis membranes, *Desalination, Proc. 1997 Workshop on Membranes in Drinking Water Production*, June 1–4, L'Aquila, Italy, 113, 2–3, 209–214. Elsevier Science B.V., Amsterdam, Netherlands.

Ishii, C., et al. (1997). Structural characterization of heat-treated activated carbon fibers, *J. Porous Materials*. 4, 3, 181–186.

Jacangelo, J. C. (1989). Membranes in water filtration, *Civil Eng*. 59, 50, 68–71.

Jacangelo, J. G., R. R. Trussell, and M. Watson (1997). Role of membrane technology in drinking water treatment in the United States, *Desalination, Proc. 1997 Workshop on Membranes in Drinking Water Production*, June 1–4, L'Aquila, Italy, 113, 2–3, 119–127. Elsevier Science B.V., Amsterdam, Netherlands.

Kastelan-Kunst, L., et al. (1997). FT30 membranes of characterized porosities in the reverse osmosis organics removal from aqueous solutions, *Water Res.* 31, 11, 2878–2884.

Kim, W. H., et al. (1997). Pilot plant study on ozonation and biological activated carbon process for drinking water treatment, *Water Science Technol., Proc. 1995 5th IAWQ Asian Regional Conf. on Water Quality and Pollut. Control*, Feb. 7–9, 1995, Manila, Philippines, 35, 8, 21–28. Elsevier Science Ltd., Oxford, England.

Kim, Y., et al. (1997). Treatment of taste and odor causing substances in drinking water, *Water Science Technol., Proc. 1995 5th IAWQ Asian Regional Conf. on Water Quality and Pollution Control*, Feb. 7–9, 1995, Manila, Philippines, 35, 8, 29–36. Elsevier Science Ltd., Oxford, England.

Kiranoudis, C. T., N. G. Voros, and Z. B. Maroulis (1997). Wind energy exploitation for reverse osmosis desalination plants, *Desalination*. 109, 2, 195–209.

Knappe, D. R. U., et al. (1997). Effect of preloading on rapid small-scale column test predictions of atrazine removal by GAC adsorbers, *Water Res.* 31, 11, 2899–2909.

Lin, S. H. and C. M. Lin (1997). Adsorption characteristics of humic acids by granular activated Carbon, *Adsorption Science Technol*. 15, 7, 507–516.

Martinez, A. A., et al. (1997). Microporous texture of activated carbon fibers prepared from aramid fiber pulp, *Microporous Mater.* 11, 5–6, 303–311.

Matatov, M. Y. and M. Sheintuch (1997). Abatement of pollutants by adsorption and oxidative catalytic regeneration, *Industrial Eng. Chem. Res.* 36, 10, 4374–4380.

McCormick, M. E. and Y. C. Kim, (1997). Ocean wave-powered desalination. *IAHR Energy and Water: Sustainable Development, Proc. 1997 27th Cong. Int. Assoc. Hydraulic Res., IAHR, Part D*, Aug. 10–15, San Francisco, CA, D, 577–582. Sponsored by: ASCE, New York.

Moreno, C., C., F. M. Carrasco, and A. Mueden (1997). Creation of acid carbon surfaces by treatment with $NH_4S_2O_8$. *Carbon*. 35 10–11, 1619–1626.

Moritz, E. J., C. R. Hoffman, and T. R. Craig (1997). Particle count monitoring of reverse osmosis water treatment for removal of low-level radionuclides. *Proc. 1995 ASME/JSME Fluids Eng. Laser Anemometry Conf. Exhibition*, Aug. 13–18, 1995, Hilton Head, SC, 221, 151–156. Sponsored by: ASME, New York.

Nishijima, W., et al. (1997). Effects of adsorbed substances on bioactivity of attached bacteria on granular activated carbon. *Water Science Technol., Proc. 1995 5th IAWQ Asian Regional Conf. on Water Quality and Pollut. Control*, Feb. 7–9, 1995, Manila, Philippines, 35, 8, 203–208. Elsevier Science Ltd., Oxford, England.

Ozoh, P. T. E. (1997). Adsorption of cotton fabric dyestuff waste water on Nigeria agricultural semi-activated carbon. *Environ. Monitoring Assessment.* 46, 3, 255–265.

Quinn, L. (1997). Reverse osmosis systems in military or emergency operations. *Desalination, Proc. 1997 Workshop on Membranes in Drinking Water Production,* June 1–4 L'Aquila, Italy, 113, 2–3, 297–301. Elsevier Science B.V., Amsterdam, Netherlands.

Ramakrishna, K. R. and T. Viraraghavan (1997). Dye removal using low cost adsorbents. *Water Science Technol., Proc. 1996 2nd IAWQ Int. Conf. Pretreatment of Industrial Wastewaters,* Oct. 16–18, 36, 2–3, 189–196. Elsevier Science Ltd., Oxford, England.

Redondo, J. A. and F. Lanari (1997). Membrane selection and design considerations for meeting European potable water requirements based on different feedwater conditions. *Desalination, Proc. 1997 Workshop on Membranes in Drinking Water Production,* June 1–4, L'Aquila, Italy, 113, 2–3, 309–323. Elsevier Science B.V., Amsterdam, Netherlands.

Rengaraj, S., B. Arabindoo, and V. Murugesan (1997). Recovery of useful material from agricultural wastes. *Proc. 1997 13th Int. Conf. Solid Waste Technol. Manage., Part 2,* Nov. 16–19, 2, 8. Widener Univ. School of England, Chester, PA.

Rosberg, R. (1997). Ultrafiltration (new technology), a viable cost-saving pretreatment for reverse osmosis and nanofiltration—A new approach to reduce costs. *Desalination, Int. Symp. Pretreatment of Feedwater for Reverse Osmosis Desalination Plants,* March 31–Apr 2, 110, 1–2, 107–114. Elsevier Science B.V., Amsterdam, Netherlands.

Sen Gupta, S. K. (1997). Evaluation of spiral wound reverse osmosis for four radioactive waste processing applications. *Proc. 1997 Canadian Nuclear Soc. Conf., Part 1,* June 8–11, Toronto, Canada, 1, 17p. Sponsored by: CNS 2 Canadian Nuclear Assoc., Toronto, Ontauo, Canada.

Shmidt, J. L., et al. (1997). Kinetics of adsorption with granular, powdered, and fibrous activated carbon. *Separation Science Technol.* 32, 13, 2105–2114.

Sincero, A. P. (1989). Reverse osmosis removal of organic compounds—A preliminary review of literature. Contract No. DAALO#-86-D-0001. U.S. Army Biomedical Research and Development Laboratory, Fort Detrick, Frederick, MD, 6–40.

Sincero, A. P. and G. A. Sincero (1996). *Environmental Engineering: A Design Approach.* Prentice Hall, Upper Saddle River, NJ.

Sirkar, K. K. (1996). Membrane separation technologies: Current developments. *Chem. Eng. Commun.* 157, 145–184.

Sun, J., et al. (1997). Activated carbon produced from an Illinois basin coal. *Carbon.* 35, 3, 341–352.

Taniguchi, Y. (1997). Overview of pretreatment technology for reverse osmosis desalination plants in Japan. *Desalination, Int. Symp. Pretreatment of Feedwater for Reverse Osmosis Desalination Plants,* March 31–Apr 2, 110, 1–2, 21–36. Elsevier Science B.V., Amsterdam, Netherlands.

Tchobanoglous, G. and E. D. Schroeder (1985). *Water Quality.* Addison-Wesley, Reading, MA.

Thacker, N. P., et al. (1997). Removal technology for pesticide contaminants in potable water. *J. Environ. Science Health, Part B: Pesticides, Food Contaminants, and Agricultural Wastes.* 32, 4, 483–496.

Torregrosa, M. R., J. M. Martin-Martinez, and M. C. Mittelmeijer-Hazeleger (1997). Porous texture of activated-carbons modified with carbohydrates. *Carbon.* 35, 4, 447–453.

Van Gauwbergen, D. and J. Baeyens (1997). Macroscopic fluid flow conditions in spiral-wound membrane elements. *Desalination.* 110, 3, 287–299.

Viessman, Jr., W. and M. J. Hammer (1993). *Water Supply and Pollution Control.* Harper & Row, New York.

Wang, C. K. and S. E. Lee (1997). Evaluation of granular activated carbon adsorber design criteria for removal of organics based on pilot and small-scale studies. *Water Science Technol., Proc. 1996 1st Int. Specialized Conf. Adsorption in the Water Environment and Treatment Processes,* Nov. 5–8, 1996, Shirahama, Japan, 35, 7, 227–234. Elsevier Science Ltd., Oxford, England.

Wang, X. L., et al. (1997). Electrostatic and Steric-Hindrance Model for the transport of charged solutes through nanofiltration membranes. *J. Membrane Science.* 135, 1, 19–32.

Wild, P. M., G. W. Vickers, and N. Djilali (1997). Fundamental principles and design considerations for the implementation of centrifugal reverse osmosis. *Proc. Inst. Mechanical Engineers, Part E: J. Process Mechanical Eng.* 211, E2, 67–81. MEP, London.

Wolf, T. and H. Bittermann (1997). Determination of adsorption parameters of active carbon filters measuring only the beginning of the breakthrough curve. *SAE Special Publications Automotive Climate Control Design Elements, Proc. 1997 Int. Cong. Exposition* Feb. 24–27, Detroit, MI, 1239, 51–54. SAE, Warrendale, PA.

Wu, S., et al. (1997). Plasma modification of aromatic polyamide reverse osmosis composite membrane surface. *J. Applied Polymer Science.* 64, 10, 1923–1926.

Zhu, X. and M. Elimelech (1997). Colloidal fouling of reverse osmosis membranes: Measurements and fouling mechanisms. *Environ. Science Technol.* 31, 12, 3654–3662.

9 Aeration, Absorption, and Stripping

Aeration, absorption, and stripping are unit operations that rely on flow of masses between phases. When a difference in concentration exists between two points in a body of mass, a flow of mass occurs between the points. When the flow occurs between two phases of masses, a transfer of mass between the phases is said to occur. This transfer of mass between phases is called *mass transfer*. Examples of unit operations that embody the concept of mass transfer are distillation, absorption, dehumidification, liquid extraction, leaching, and crystallization.

Distillation is a unit operation that separates by vaporization liquid mixtures of miscible and volatile substances into individual components or groups of components. The separation of water and alcohol into the respective components of liquid air into nitrogen, oxygen, and argon; and the separation of crude petroleum into gasoline, oil, and kerosene are examples of the distillation unit operation.

Absorption is a unit operation that removes a solute mass or masses from a gas phase into a liquid phase. Aeration of water dissolves air into it; thus, aeration is absorption. Another example of absorption is the "washing" of ammonia from an ammonia-polluted air. In this operation, ammonia is removed from the air by its dissolution into the water.

The reverse flow of masses from the liquid phase into the gas phase is called *stripping*. In stripping, the solute molecule is removed from its solution with the liquid into the gas phase.

Dehumidification is the removal of a solute liquid vapor from a gas phase by the solute condensing into its liquid phase. The removal of water vapor in air by condensation on a cold surface is dehumidification. The reverse of dehumidification is *humidification*. In this unit operation, the flow of the solute is from the liquid phase evaporating into the gas phase. The end result of this movement is saturation of the gas. For example, during heavy rains, the atmosphere may become saturated with water vapor, the degree of this saturation being measured by the *relative humidity*. *Liquid extraction* is the removal of a solute component from a liquid mixture called the *raffinate* using a liquid solvent. In this operation, the solvent preferentially dissolves the solute molecule to be extracted.

Leaching is a unit operation where a solute molecule is removed from a solid using a fluid extractor. This is similar to liquid extraction, except that the solute to be removed comes from a solid rather than from a liquid as in the case of liquid extraction. Also, the fluid extractor may be a fluid or a gas. For example, pollutants can be leached out from solid wastes in a landfill as rain percolates down the heap.

Crystallization is a unit operation where solute mass flows toward a point of concentration forming crystals. The driving force for the transfer of mass from liquid into the solid phase is the affinity of the solute to form into a solid. An example of this operation is the making of ice from liquid water. The formation of snow from water vapor in the atmosphere is also a process of crystallization.

Of all the unit operations of mass transfer, this chapter will only discuss aeration, absorption, and stripping. As mentioned, aeration is a form of absorption. Because this operation plays a very important and significant role in water and wastewater treatment, however, we will give it a separate heading and call it specifically aeration.

9.1 MASS TRANSFER UNITS

The major purpose of dissolving air is to provide oxygen to be used by microorganism in the process of wastewater treatment. This is exemplified by the aeration employed in the activated sludge process. Aeration may also be employed for the removal of iron and manganese from groundwaters. In the removal of hardness, the presence of high concentrations of carbon dioxide may result in high cost for lime, as CO_2 reacts with lime. Thus, excess concentrations of this gas may be removed from the water by stripping or spraying the water into the air. H_2S is another compound that may be removed by stripping as benzene, carbon tetrachloride, p-dicholorobenzene, vinyl chloride, and trichloroethylene may also be removed by stripping. The discussions that follow address the units or method used in aeration, absorption, and stripping.

Figure 9.1 illustrates how a pollutant may be stripped by spraying the water into the air. As the water is sprayed, droplets are formed. This creates the condition for the pollutant to transfer from the droplet phase to the air phase, in addition to the direct liberation of the pollutant as the bulk mass of water breaks up into the smaller size droplets. Figures 9.2a through 9.2d show the various types of nozzles that may be used in sprays. Figure 9.2e is an inclined apron which may be studded with riffle plates. At the air–water interface at the surface of the flowing water, the air transfers between the water and air phases. The studding creates turbulence which, in aeration, transports the water exposed at the surface to the main body or bulk of the flowing water. The whole mass of water is aerated this way, because the mass of water transported to the main body carries with it any air that was dissolved when it was exposed at the surface. In stripping, the turbulent flowing water exposes the solute at the surface triggering the process of stripping. The rate of aeration or stripping depends upon how fast the surface is renewed (i.e., how fast the water mass from the main body is transported to the surface for exposure to the air).

Figure 9.2f is a stack of perforated plates. The water is introduced at the top and allowed to trickle down the plates; the trickling water is met by a countercurrent flow of air. The process creates a droplet phase and the air-gas phase inducing a mass transfer between the droplets and the air. Figure 9.2g is a spray tower. The water is sprayed using spray nozzles at the top of the tower forming droplets. These droplets are then met by a countercurrent flow of air creating the two phase for mass transfer as in the case of the perforated plates. Figure 9.2h is a cascade aerator or deaerator as the case may be. The cascade operates on the same principle as the inclined apron, only that this is more effective because of the steps.

FIGURE 9.1 Water spray.

Figures 9.3 and 9.4 show various types of aeration devices used in wastewater treatment plants. Figure 9.3a is a turbine aerator with an air sparger at the bottom. As the air emerges from the sparger, the larger bubbles that are formed are sheared into small pieces by the turbine blade above. Figure 9.3b is a porous ceramic diffuser. Because of the small openings through which the air passes, this type of diffuser creates tiny bubbles. Tiny bubbles are more effective for mass transfer, since the many bubbles produced create a large sum total areas for transfer. Figure 9.3c is a surface aerator. Water is drawn from the bottom of the aerator and sprayed into the air creating droplets, thus, aerating the water. Figure 9.4 shows a dome-type bubble diffuser. The dome is porous, can have a diameter of 18 cm, and may be constructed of

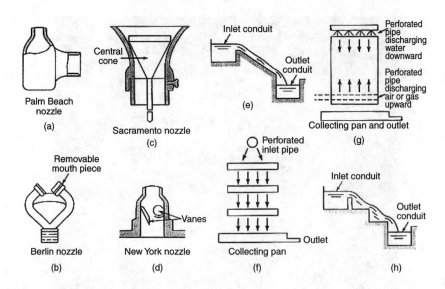

FIGURE 9.2 Nozzles for sprays and units for aeration or stripping: (a–d) nozzle types; (e) inclined apron that may be studded with riffle plates; (f) perforated plates; (g) spray tower; and (h) cascade.

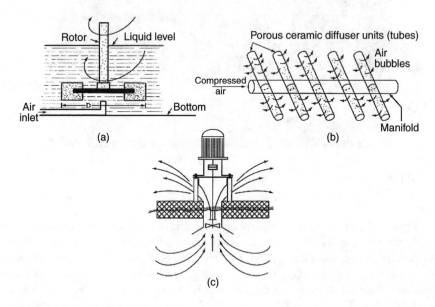

FIGURE 9.3 Aeration units: (a) turbine aerator with an air sparger; (b) porous ceramic diffuser; and (c) surface aerator.

FIGURE 9.4 Dome-type diffusers. (Courtesy of Aerocor Co.)

FIGURE 9.5 An activated sludge aeration tank at Back River wastewater treatment plant, Baltimore, MD.

aluminum oxide. This diffuser produces very tiny bubbles. As indicated, the diffusers are mounted on rows of pipes. Figure 9.5 shows an actual aeration in action. This happens to be one of the activated sludge process tanks at the Back River Wastewater Treatment Plant, Baltimore, MD.

A common device used in gas absorption and stripping is the packed tower, the elevational section of which is shown in Figure 9.6i. The device consists of a column or tower equipped with a gas inlet and distributor at the bottom and a liquid inlet and distributor at the top. It also consists of a liquid outlet at the bottom and a gas outlet at the top and a supported mass of solid shapes called *tower packing* or *filling*.

The liquid trickles down through the packing while the gas goes up the packing. The packing causes a thin film of liquid to be created on the surfaces which are contacted by the gases flowing by. Two phases and an interface between liquid and gas are therefore created inducing mass transfer.

Figures 9.6a to Figure 9.6h show the various shapes of packings used in practice. Packings are either dumped randomly into the tower or are stacked manually. Dumped packings consists of units that are 0.6 cm to 5 cm in major dimensions; they are mostly utilized in small towers. Stacked packings are 5 cm to 20 cm in major dimensions and are used in large towers. The spiral partition rings single, double, and triple are stacked. The Berl and Intalox saddles, the Raschig, Lessing, and cross-partition rings are normally dumped packings. Large Raschig rings 5 to 7 cm in diameter are often stacked.

9.2 INTERFACE FOR MASS TRANSFER, AND GAS AND LIQUID BOUNDARY LAYERS

Figure 9.7a shows the formation of boundary layers in absorption operations, and Figure 9.7b shows the formation of boundary layers in stripping operations. The right-hand side in each of these figures represents the liquid phase as in the liquid phase of a droplet and the left-hand side represents the gas phase as in the gas phase of the air.

Consider the absorption operation. Imagine the two phases being far apart initially. As the phases approach each other, a point of "touching" will eventually be reached. This point then determines a surface; being a surface, its thickness is equal to zero. This surface is identified as the interface in the figure. This figure shows the section cut across of the interface surface. The line representing the interface must have a zero thickness.

From fluid mechanics, when a fluid flows parallel to a plate, a boundary layer is formed closed to the plate surface. At the surface itself, the velocity is zero relative to the plate, because of the no-slip condition. Considering the two phases mentioned previously, either the liquid or the gas may be considered as analogous to the plate. Taking the liquid phase as analogous to the plate, the gas phase would be the fluid. The interface would then represent the surface of the plate. Because of the no-slip condition, the relative velocity of the phases parallel to the interface at the interface is equal to zero. In other words, the two phases are "glued" together at the interface and they do not move relative to each other.

As in the case of the fluid flow over a plate, a boundary layer is also formed. With the liquid phase considered as analogous to the plate, a gas phase boundary layer is formed; with the gas phase considered as analogous to the plate, the liquid phase boundary layer is formed (see figures). These boundary layers are also called films.

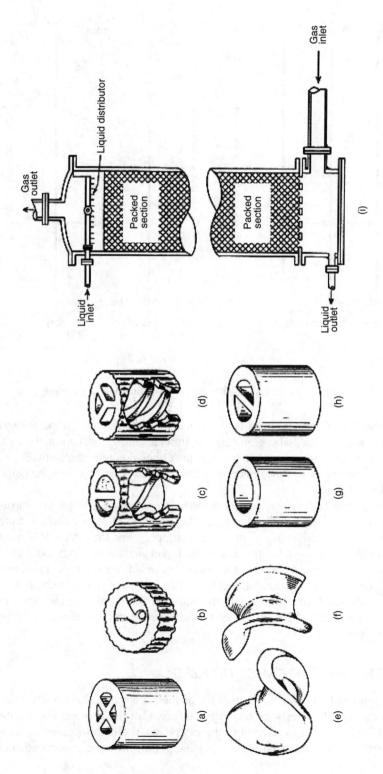

FIGURE 9.6 Various types of packings (left) and an elevational cross section of a packing tower (right): (a) cross-partition ring; (b) single-spiral ring; (c) double-spiral ring; (d) triple-spiral ring; (e) Berl saddles; (f) Intalox saddles; (g) Raschig ring; (h) Lessing ring.

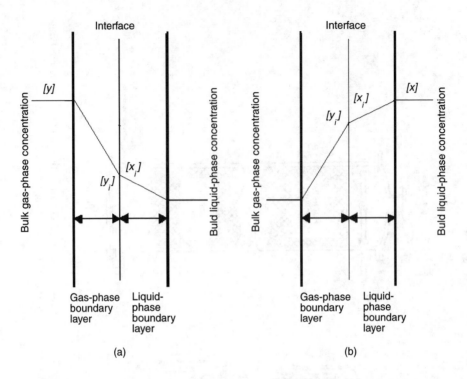

FIGURE 9.7 Formation of interfacial boundary layers: (a) absorption; (b) stripping.

Thus, the transfer of mass between the two phases must pass through the gas and liquid films. In environmental engineering literature, the term *film* is normally used. The same discussions would apply to the stripping operation represented by Figure 9.7b, with the difference that the direction of flow of mass transfer is from the liquid phase to the gas phase.

In absorption operations, the concentration in the gas phase is larger in comparison to the concentration in the liquid phase. Thus, the flow of mass transfer is from gas to liquid. The reverse is true in the case of stripping, and the direction of mass transfer is from liquid to gas. In other words, the liquid phase is said to be "stripped" of its solute component, decreasing the concentration of the solute in the liquid phase and increasing the concentration of the solute in the gas phase. In absorption, the solute is absorbed from the gas into the liquid, increasing the concentration of the solute in the liquid phase and, of course, decreasing the concentration of the solute in the gas phase.

9.3 MATHEMATICS OF MASS TRANSFER

Between liquid and gas phases, the transfer of mass from one phase to the other must pass through the interfacial boundary surface. Call the concentration of the solute at this surface as $[y_i]$ referred to the gas phase. The corresponding concentration referred to the liquid phase is $[x_i]$. $[x_i]$ and $[y_i]$ are the same concentration of the solute only that they are referred to different basis; in effect, they are equal. Because they are equal and because the thickness of the interface is zero, x_i and y_i must be in equilibrium with respect to each other.

Consider the process of absorption. If $[y]$ is the concentration in the bulk gas phase, the driving force toward the interfacial boundary is $[y] - [y_i]$ and the rate of mass transfer is $k_y([y] - [y_i])$, where k_y is the gas film coefficient of mass transfer. For this rate of mass transfer to exist, it must be balanced by an equal rate of mass transfer at the liquid film. The liquid phase mass transfer rate is $k_x([x_i] - [x])$, where k_x is the liquid film coefficient of mass transfer and $[x]$ is the bulk concentration of the solute in the liquid phase. Thus,

$$k_y([y] - [y_i]) = k_x([x_i] - [x]) \qquad (9.1)$$

What is known is that $[x_i]$ and $[y_i]$ are in equilibrium. But, determining these values experimentally would be very difficult. Thus, instead of using them, use $[x^*]$ and $[y^*]$, respectively. $[x^*]$ is the concentration that $[x]$ would attain if it were to reach equilibrium value. By parallel deduction, $[y^*]$ is also the concentration that $[y]$ would attain if it were to reach equilibrium value. The corresponding driving forces are now $[y] - [y^*]$ and $[x^*] - [x]$, respectively. To comprehend the physical meaning of the driving forces, refer to Figure 9.8.

As shown, $[x_i]$, $[y_i]$ is on the equilibrium curve. The *equilibrium curve* is the relationship between the concentration $[x]$ in the liquid phase and the concentration $[y]$ in the gas phase when there is no net mass transfer between the phases. For any given pair of values $[x]$ and $[y]$ in the liquid and gas phase, respectively, the point $([x_i], [y_i])$ represents the "distance" that $([x], [y])$ will have to "move" to attain their equilibrium values concurrently at the equilibrium curve. This distance, represented by the line segment $([x], [y] \rightarrow [x_i], [y_i])$, is the actual driving force for mass transfer;

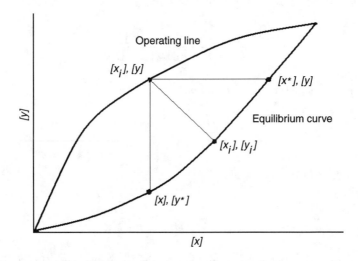

FIGURE 9.8 Relationship among the various mole fractions.

however, as mentioned before, ($[x_i]$, $[y_i]$) is impossible to determine experimentally. Thus, locate the point ($[x]$, $[y^*]$) to view the surrogate driving force in the gas film and locate point ($[x^*]$, $[y]$) to view the surrogate driving force in the liquid film.

As seen, $[y] - [y^*]$ is greater than $[y] - [y_i]$; however, it is not the true driving force for transfer. Also $[x^*] - [x]$ is greater than $[x_i] - [x]$, but, again, it is not the true driving force for transfer. When the transfer equation is written, however, it is prefixed with a proportionality constant. This situation is therefore taken advantage of by using a different proportionality constant for the case of the surrogate driving forces. Thus, using K_y as the proportionality constant for the gas-side mass transfer equation in the surrogate situation,

$$k_y(y - y_i) = K_y(y - y^*) \tag{9.2}$$

On the liquid side, using K_x as the proportionality constant,

$$k_x([x_i] - [x]) = K_x([x^*] - [x]) \tag{9.3}$$

K_y and K_x are called *overall mass transfer coefficients* for the gas and liquid sides, respectively. To differentiate, k_y and k_x are called *individual mass transfer coefficients* for the respective sides.

It is instructive to determine the equation relating the overall and the individual mass transfer coefficients. Equation (9.2) may be rearranged to obtain

$$\frac{1}{K_y} = \frac{[y] - [y^*]}{k_y([y] - [y_i])} = \frac{([y] - [y_i]) + ([y_i] - [y^*])}{k_y([y] - [y_i])} = \frac{1}{k_y} + \frac{[y_i] - [y^*]}{k_y([y] - [y_i])} \tag{9.4}$$

Replacing $k_y([y] - [y_i])$ using Equation (9.1),

$$\frac{1}{K_y} = \frac{1}{k_y} + \frac{[y_i] - [y^*]}{k_x([x_i] - [x])} \tag{9.5}$$

Letting $([y_i] - [y^*])/([x_i] - [x])$ equal m,

$$\frac{1}{K_y} = \frac{1}{k_y} + \frac{m}{k_x} \tag{9.6}$$

A parallel derivation for K_x yields

$$\frac{1}{K_x} = \frac{1}{k_x} + \frac{1}{mk_y} = \frac{1}{mK_y} \tag{9.7}$$

The coordinates $[x_i]$ and $[y_i]$ are the coordinates of the point ($[x_i]$, $[y_i]$) on the equilibrium curve. On the other hand, the coordinates $[x]$ and $[y]$ are coordinates of point ([x], [y]) representing the concentration [x] in the liquid phase and the concentration [y] in the gas. Because the phases are not in equilibrium, ([x], [y]) is not on the equilibrium curve. The various values of the pair ([x], [y]) in the liquid and gas phases can be plotted; this plot is called an *operating line* (see figure). Any ([x], [y]) pair is called an *operating point*. For the operating point ([x], [y]) the corresponding points on the equilibrium curve are ([x], [y*]) and ([x*], [y]), based on the surrogate equation. Point ([x*], [y*]) can only exist if ([x], [y]) is on the equilibrium curve.

The slope between ([x], [y*]) and ([x_i], [y_i]) is ([y_i] − [y*])/([x_i] − [x]); this is equal to m. Thus, m is the slope of the equilibrium curve if it is a straight line.

Parallel derivations may be performed for the stripping operation; the results are similar. The only difference is that there will be interchange of subscripts and superscripts. Thus, the following analogous equations will be obtained: $k_x([x] − [x_i]) = k_y([y_i] − [y])$, $k_x([x] − [x_i]) = K_x([x] − [x*])$ and $k_y([y_i] − [y]) = K_y([y*] − [y])$. Refer to the figure to visualize that the mass flow is from the liquid phase to the gas phase.

9.4 DIMENSIONS OF THE OVERALL MASS TRANSFER COEFFICIENTS

$K_x([x*] − [x])$ is the mass of solute passing across the interfacial area per unit time per unit cross-sectional area of the interface. The interfacial area of mass transfer is largely indeterminate. To make sense of the expression, first define the term destination medium. *Destination medium* is the overall entity inside the control volume from or into which the gas and liquid phases flow. For example, in an aeration basin, the destination media are the bulk contents of the tank; in a packed tower, the destination media are the packings inside the tower; and in a trickling filter, the destination media are the rocks that make up the filter.

Now, invent a parameter a that defines the area of interfacial area per unit bulk volume of the mass transfer destination medium and form the expression $K_x a([x*] − [x])$. The dimension of $K_x a([x*] − [x])$ is now mass per unit time per unit volume or $M/t \cdot L^3$, where M is the dimension of mass, t is the dimension of time, and L is the dimension of length. From this expression, the dimension of $K_x a$ is per unit time or $1/t$. It will be shown later in this section on absorption towers, however, that the dimensions of $K_x a$ will not be $1/t$ if mole fraction units are used for the concentrations. Both K_x and $K_x a$ are also called overall mass transfer coefficients based on the liquid side. The corresponding overall mass transfer coefficients based on the gas side are K_y and $K_y a$, respectively.

9.5 MECHANICS OF AERATION

Oxygen is a necessary nutrient. In suspended-growth processes, such as the activated sludge process, air must be literally forced into the liquid. The air, thus, dissolved provides the necessary oxygen nutrient for the microorganism stabilizing the wastewater.

The basic process for oxygen mass transfer from air to water is absorption. Call the equilibrium concentration of oxygen in water at a particular temperature and pressure as $[C_{os}]$. This equilibrium concentration is also called the *saturation concentration* of the dissolved oxygen (DO) and this corresponds to $[x^*]$. Let the concentration in the water at any given moment be $[\bar{C}]$; this corresponds to x. The driving force for mass transfer is then $[C_{os}] - [\bar{C}]$. The rate at which the concentration of oxygen will increase is $(d[\bar{C}]/dt)$. In aeration, $K_x a$ is normally written as $K_L a$. Thus,

$$\frac{d[\bar{C}]}{dt} = K_L a([C_{os}] - [\bar{C}])$$ (9.8)

9.5.1 Equipment Specification

The rating of aeration equipment is reported at standard conditions defined as 20°C, one atmosphere pressure, and 0 mg/L of dissolved oxygen concentration in tap water (or distilled water). Under these conditions, Equation (9.8) becomes

$$\frac{d[\bar{C}]}{dt} = (K_L a)_{20}[C_{os,20,sp}]$$ (9.9)

$(K_L a)_{20}$ = the $K_L a$ at standard conditions and $[C_{os,20,sp}]$ = the saturation DO at 20°C and standard pressure. This equation is the *standard oxygen rate* (SOR). Equipment is specified in terms of SOR. Testing is not normally done at standard conditions, so $(K_L a)_{20}$ must be obtained from the $K_L a$ obtained at the condition of testing using the Arrhenius temperature relation,

$$K_L a = (K_L a)_{20}\theta^{T-20}$$ (9.10)

where θ is the temperature correction factor, and T is the temperature in degrees Celsius at testing conditions. This equation assumes that the effect of pressure on $K_L a$ is negligible. In wastewater treatment, the value of θ is usually taken as 1.024.

The ability of aeration equipment to transfer oxygen at field conditions is, using Equation (9.8),

$$\frac{d[\bar{C}]}{dt} = (K_L a)_w([C_{os,w}] - [\bar{C}])$$ (9.11)

where $(K_L a)_w$ and $[C_{os,w}]$ are the $K_L a$ and the $[C_{os}]$ of the wastewater at field conditions, respectively. This equation represents the *actual oxygenation rate* (AOR). $(K_L a)_w$ may also be expressed in terms of its value at 20°C, $(K_L a)_{w,20}$, by the Arrhenius temperature relation

$$(K_L a)_w = (K_L a)_{w,20}\theta^{T-20}$$ (9.12)

From Eqs. (9.9), (9.11), and (9.12),

$$\text{SOR} = \frac{\text{AOR}}{\alpha \dfrac{(\beta[C_{os}] - [\bar{C}])}{[C_{os,20,sp}]} \theta^{T-20}} \qquad (9.13)$$

where

$$\alpha = \frac{(K_L a)_{w,20}}{(K_L a)_{20}} \quad \text{and} \quad \beta = \frac{[C_{os,w}]}{[C_{os}]}.$$

As mentioned before, specification of aeration equipment requires the determination of SOR. From Equation (9.13), this involves finding the values of the aeration parameters AOR, β, and α. The parameter α, in turn, requires the determination of the $K_L a$ values. Each wastewater is unique in its characteristics, so these parameters should be determined experimentally. The literature reports values of $(K_L a)_{w,20}$ in the neighborhood of 2.5 per hour, α in the range of 0.7 to 0.9, and β in the range of 0.9 to 1.0. AOR in the neighborhood of 1.40 kg/m$^3 \cdot$ day has also been obtained. The determination of these parameters will addressed in the succeeding discussions.

Example 9.1 The value of $(K_L a)_{w,20}$ for a certain industrial waste is 2.46 per hour. What is the value of $(K_L a)_w$ at 25°C?

Solution:

$$(K_L a)_w = (K_L a)_{w,20} \theta^{T-20}$$

Therefore,

$$(K_L a)_{w,25} = 2.46(1.024)^{25-20} = 2.77 \text{ per hour} \quad \textbf{Ans}$$

Example 9.2 A chemical engineer proposes to purchase an aerator for an activated sludge reactor. In order to do so, she performs a series of experiments obtaining the following results: AOR $= 1.30$ kg/m$^3 \cdot$ day, $\alpha = 0.91$, and $\beta = 0.94$. If the aeration is to be maintained to effect a dissolved oxygen concentration of 1.0 mg/L at 25°C in the reactor, what SOR should be specified to the manufacturer of the aerator?

Solution:

$$\text{SOR} = \frac{\text{AOR}}{\alpha \dfrac{(\beta[C_{os}] - [\bar{C}])}{[C_{os,20,sp}]} \theta^{T-20}}$$

$$[\bar{C}] \text{ at } 25°C = 8.38 \text{ mg/L}$$

$$[C_{os,20,sp}] = 9.17 \text{ mg/L}$$

Therefore,

$$\text{SOR} = \frac{1.30}{0.91 \left(\dfrac{0.94(8.38) - 1.0}{9.17} \right)(1.024)^{25-20}} = 0.768 \text{ kg/m}^3 \quad \textbf{Ans}$$

Example 9.3   A civil engineer performs an experiment for the purpose of determining the value of α of a particular wastewater. $(K_La)_{20}$ and $(K_La)_{w,20}$ were found, respectively, to be 2.46 per hour and 2.25 per hour. Calculate α.

Solution:

$$\alpha = \frac{(K_La)_{w,20}}{(K_La)_{20}} = \frac{2.25}{2.46} = 0.91$$

9.5.2 Determination of Aeration Parameters

The environmental engineer specifies an aeration equipment based upon standard laboratory tests performed by equipment manufacturers. In other words, although the engineer can easily determine the AOR, AOR must still be converted to SOR to match with the standard manufacturers value. To perform this conversion, the α and β parameters must be determined.

Determination of β.   The determination of β is very simple. Take a liter jar and fill it half with the sample. The jar is then vigorously shaken to saturate the sample with air or oxygen and the dissolved oxygen concentration measured. Table 9.1 shows the saturated concentrations of dissolved oxygen in clean water exposed to one atmosphere barometric pressure at various temperatures. From this table, at the temperature corresponding to the temperature of the experiment, the saturation DO for clean water at one atmosphere barometric pressure can be obtained. This concentration is $[C_{os}]$. From this, along with the saturation DO of the sample determined in the experiment $[C_{os,w}]$, β may be calculated as

$$\beta = \frac{[C_{os,w}]}{[C_{os}]} \tag{9.14}$$

TABLE 9.1
Saturation DO in Distilled Water under One Atmosphere
of Atmospheric Pressure

Temperature (°C)	DO (mg/L)	Temperature (°C)	DO (mg/L)	Temperature (°C)	DO (mg/L)
0	14.6	1	14.2	2	13.8
3	13.5	4	13.1	5	12.8
6	12.5	7	12.2	8	11.9
9	11.6	10	11.3	11	11.1
12	10.8	13	10.6	14	10.4
15	10.2	16	10.0	17	9.7
18	9.5	19	9.4	20	9.2
21	9.0	22	8.8	23	8.7
24	8.5	25	8.4	26	8.2
27	8.1	28	7.9	29	7.8
30	7.6				

If the test is being conducted at elevation other than zero or one atmosphere of barometric pressure, the $[C_{os}]$ obtained from Table 9.1 must be corrected for the pressure of the test. Thus, during the performance of the experiment, the barometric pressure at the location should be recorded. Also, the barometric pressure at the test location may be obtained using the *barometric equation*. This equation relates the barometric pressure P_b with altitude z in the troposphere, and from fluid mechanics, this equation is

$$P_b = P_{bo}\left(1 - \frac{Bz}{T_o}\right)^{g/RB} = 101325\left(1 - \frac{0.0065z}{T_o}\right)^{9.81/[(286.9)(0.0065)]}$$

$$= 101325\left(1 - \frac{0.0065z}{T_o}\right)^{5.26} \tag{9.15}$$

where P_{bo} = barometric pressure at $z = 0$ (= 101325 N/m^2); B (the temperature lapse rate) = 0.0065° K/m for the standard atmosphere; T_o = temperature in °K at $z = 0$; $g = 9.81$ m/s^2; and R (gas constant for air) = 286.9 N · m/kg · °K. As written, P_b has the unit of N/m^2. The equation assumes that the temperature varies linearly with altitude. Therefore, if the elevation where the test is conducted and the temperature at mean sea level are known, the corresponding barometric pressure can be found. Because $[C_{os}]$ varies directly as the pressure and if the condition of test is at barometric pressure, the $[C_{os}]$ at test conditions will be

$$[C_{os}] = \frac{P_b}{P_s}[C_{os,sp}] \tag{9.16}$$

where $[C_{os,sp}]$ is the $[C_{os}]$ from Table 9.1 at the standard pressure of $P_s = P_{bo} = 760$ mm Hg.

Pressures corresponding to $[C_{os}]$. In sizing aerators, what pressure to use to determine $[C_{os}]$ depends upon the type of aeration device used. For apron, cascade and surface aerators and for spray and plate towers, the corresponding pressure should be taken as barometric. For aerators that are submerged below the surface of water such as the bubble-diffusion and turbine type aerators, since the point of release of the air is submerged, the pressure must correspond to the average depth of submergence. If the submergence depth is Z_d, the corresponding average pressure P is

$$P = P_b + \frac{Z_d}{2}\gamma \tag{9.17}$$

where γ is the specific weight of the water or the mixed liquor, in the case of the activated sludge process tank. The equation for $[C_{os}]$ is then revised to

$$[C_{os}] = \frac{P}{P_s}[C_{os,sp}] \tag{9.18}$$

Example 9.4 An environmental engineer performs an experiment for the purpose of determining the value of β of a particular wastewater. The $[C_{os,w}]$ of the wastewater after shaking the jar thoroughly is 7.5 mg/L. The temperature of the wastewater is 25°C. Calculate β.

Solution:

$$\beta = \frac{[C_{os,w}]}{[C_{os}]} \qquad [C_{os}] \text{ at } 25°C = 8.4 \text{ mg/L}$$

Therefore,

$$\beta = \frac{7.5}{8.4} = 0.89 \quad \textbf{Ans}$$

Example 9.5 An environmental engineer performs an experiment for the purpose of determining the value of β of a particular wastewater in the mountains of Allegheny County, MD. On a normal day, what is the prevailing barometric pressure, if the temperature of the air on the Chesapeake Bay is 30°C?

Solution:

$$P_b = 101325\left(1 - \frac{0.0065z}{T_o}\right)^{5.26}$$

In Allegheny, $z = 756$ m above the Chesapeake Bay, consulting the county topographical map

$$P_b = 101325\left(1 - \frac{0.0065(756)}{30 + 273}\right)^{5.26} = 92,975 \text{ N/m}^2 \quad \textbf{Ans}$$

Example 9.6 An activated sludge reactor is aerated using a turbine aerator located 5.5 m below the surface of the mixed liquor. What is the pressurizing pressure if the prevailing barometric pressure is 761 mm Hg and the water temperature 20°C?

Solution:

$$P = P_b + \frac{Z_d}{2}\gamma \qquad \gamma = 997(1000)$$

$$P = 0.761(13.6)(1000)(9.81) + \left(\frac{5.5}{2}\right)(997)(9.81) = 128,426.14 \text{ N/m}^2 \quad \textbf{Ans}$$

Example 9.7 An activated sludge reactor is aerated using a turbine aerator located 5.5 m below the surface of the mixed liquor. What is $[C_{os}]$ if the prevailing barometric pressure is 761 mm Hg and the water temperature 20°C?

Solution: From the previous example, $P = 128,426.14$ N/m². From Table 9.1, $[C_{os}] = 9.2$ mg/L at 20°C and 1 atm pressure.

Therefore,

$$[C_{os}] \text{ in reactor} = \frac{128,426.14}{101,325}(9.2) = 11.66 \text{ mg/L} \quad \textbf{Ans}$$

Determination of α. Integrating Equation (9.8) from $t = 0$ and $\bar{C} = 0$ to $t = t$ and $\bar{C} = \bar{C}$ produces

$$ln\frac{[C_{os}] - [\bar{C}]}{[C_{os}]} = -K_L at \tag{9.19}$$

In this equation, because $[C_{os}]$ is a constant, only one pair of values of t and $\bar{C}$ is needed to determine $K_L a$. In practice, however, there can be several pair of these values obtained from an experiment. One way to determine $K_L a$ is to plot the straight line of the relationship between t and $ln([C_{os}] - [\bar{C}])/[C_{os}]$. The slope of this line determines $K_L a$. A more practical and easy method, however, is to average the t's and the $(ln([C_{os}] - [\bar{C}])/[C_{os}])$'s to obtain a single pair of value. This pair is then used to solve for $K_L a$.

The averaging for the t's and the $(ln[C_{os}] - [\bar{C}])/[C_{os}])$'s have the same number of addends, their sums may be simply equated. This is shown below.

$$\sum_{m=1}^{m=n}\left(ln\frac{[C_{os}] - [\bar{C}]}{[C_{os}]}\right)_m = -K_L a\sum_{m=1}^{m=n}t_m \tag{9.20}$$

Solving for From $K_L a$,

$$K_L a = -\frac{\sum_{m=1}^{m=n}\left(ln\frac{[C_{os}] - [\bar{C}]}{[C_s]}\right)_m}{\sum_{m=1}^{m=n}t_m} \tag{9.21}$$

From this equation, $K_L a$ may be calculated, which can then be corrected to obtain $(K_L a)_{20}$ using the temperature relation. $(K_L a)_{20}$ is one of the factors needed to calculate α.

The organisms in wastewater respire, so oxygen utilization must be incorporated. Calling the respiration rate by $\bar{r}$, Equation (9.11) is modified to

$$\frac{d[\bar{C}]}{dt} = (K_L a)_w([C_{os,w}] - [\bar{C}]) - \bar{r} \tag{9.22}$$

This equation may be rewritten as

$$\frac{d[\bar{C}]}{dt} = (K_L a)_w([C_{os,w}] - [\bar{C}]) - \bar{r} = (K'_L a)_w([C_{os,w}] - [\bar{C}]) \tag{9.23}$$

where $(K'_L a)_w$ is an apparent overall mass transfer coefficient. It encompasses both $(K_L a)_w$, the true overall mass transfer coefficient, and $\bar{r}$.

The second part of Equation (9.23), $[d(\bar{C})/dt] = (K'_L a)_w([C_{os,w}] - [\bar{C}])$, is similar in form to Equation (9.8). It can therefore be manipulated to obtain an equation similar to Equation (9.21). This is shown below.

$$\sum_{m=1}^{m=n}\left(ln\frac{[C_{os,w}] - [\bar{C}]}{[C_{os,w}]}\right)_m = \sum_{m=1}^{m=n}\left(ln\frac{\beta[C_{os}] - [\bar{C}]}{\beta[C_{os}]}\right)_m = -(K'_L a)_w\sum_{m=1}^{m=n}t_m \tag{9.24}$$

$(K'_L a)_w$ may now be solved as

$$(K'_L a)_w = - \frac{\sum_{m=1}^{m=n} \left(ln \, \frac{\beta[C_{os}] - [\bar{C}]}{\beta[C_{os}]} \right)_m}{\sum_{m=1}^{m=n} t_m} \tag{9.25}$$

Having solved $(K'_L a)_w$ and for n values of $[\bar{C}]$, Equation (9.23) may be written as

$$(K_L a)_w \sum_{m=1}^{m=n} (\beta[C_{os}] - [\bar{C}])_m - \sum_{m=1}^{m=n} \bar{r}_m = (K'_L a)_w \sum_{m=1}^{m=n} (\beta[C_{os}] - [\bar{C}])_m \tag{9.26}$$

from which $(K_L a)_w$ may be solved once $\bar{r}$ is determined in a separate test. For a constant $\bar{r}$, $\sum_{m=1}^{m=n} \bar{r}_m = n\bar{r}$.

Thus,

$$(K_L a)_w = \frac{(K'_L a)_w \sum_{m=1}^{m=n} (\beta[C_{os}] - [\bar{C}])_m + n\bar{r}}{\sum_{m=1}^{m=n} (\beta[C_{os}] - [\bar{C}])_m} \tag{9.27}$$

$(K_L a)_w$ may now be corrected to obtain $(K_L a)_{w,20}$ using the Arrhenius temperature relation. Once $(K_L a)_{w,20}$ and $(K_L a)_{20}$ are known, α can be computed.

The actual laboratory experimentation involves deaerating the sample, first. This is done by consuming the dissolved oxygen using sodium sulfate (Na_2SO_3) with cobalt chloride ($CoCl_2$) added as a catalyst. The sulfite converts to sulfate when reacted with the dissolved oxygen. From the stoichiometry of the reaction, 7.9 mg/L of the sulfate is needed per mg/L of the dissolved oxygen. Ten to 20% excess is normally used. Cobalt chloride has been used in concentration of 1.5 mg/L to act as catalyst. As soon as the sample is completely deoxygenated, reaeration is allowed to take place using the type of aeration system to be employed in the prototype such as bubble-diffusion, turbine, surface-aeration, cascade, perforated plate, and spray tower. The increase in dissolved oxygen concentration with respect to time is monitored. The data obtained are then used to calculate the aeration parameters. In the cases of cascades, perforated plate towers, and spray towers, the time may be taken as the time it takes the mass of water or droplets to fall through the height. The concentration at the top of the cascade or tower would have to be zero; that at the bottom would have to be whatever is measured.

Example 9.8   A settling column 4 m in height is used to determine the α of a wastewater. The wastewater is to be aerated using a fine-bubble diffuser in the prototype aeration tank. The laboratory diffuser releases air at the bottom of the tank. The result of the unsteady state aeration test is shown below. Assume $\beta = 0.926$, $\bar{r} = 1.0$ mg/L · h and the plant is 304.79 m above mean sea level. For practical purposes, assume mass density of water = 1000 kg/m^3. Assume an ambient temperature of 25°C. Calculate α.

Tap Water at 6.5°C		Wastewater at 25°C	
Time (min)	$[\bar{C}]$ (mg/L)	Time (min)	$[\bar{C}]$ (mg/L)
3	0.6	3	0.9
6	1.6	6	1.8
9	3.1	9	2.4
12	4.3	12	3.3
15	5.4	15	4.0
18	6.0	18	4.7
21	7.0	21	5.3

Solution:

$$\alpha = \frac{(K_L a)_{w,20}}{(K_L a)_{20}} \qquad K_L a = (K_L a)_{20} \theta^{T-20} \qquad (K_L a)_w = (K_L a)_{w,20} \theta^{T-20}$$

$$K_L a = - \frac{\sum_{m=1}^{m=n}\left(\ln \frac{[C_{os}]-[\bar{C}]}{[C_s]} \right)_m}{\sum_{m=1}^{m=n} t_m} \qquad (K_L a)_w = \frac{(K'_L a)_w \sum_{m=1}^{m=n}(\beta[C_{os}]-[\bar{C}])_m + n\bar{r}}{\sum_{m=1}^{m=n}(\beta[C_{os}]-[\bar{C}])_m}$$

$$(K'_L a)_w = - \frac{\sum_{m=1}^{m=n}\left(\ln \frac{\beta[C_{os}]-[\bar{C}]}{\beta[C_{os}]} \right)_m}{\sum_{m=1}^{m=n} t_m}$$

$$P_b = 101325\left(1 - \frac{0.0065z}{T_o} \right)^{5.26} \qquad P = P_b + \frac{Z_d}{2}\gamma$$

Temperature at mean sea level $= 25 + 0.0065(304.79) = 26.981°C \Rightarrow 299.98°K$

$$P_b = 101325\left(1 - \frac{0.0065(304.79)}{299.98} \right)^{5.26} = 97{,}854.31 \text{ N/m}^2$$

Therefore,

$$P = 97{,}854.31 + \frac{4}{2}(1000)(9.81) = 117{,}474.31 \text{ N/m}^2$$

Tap water at 6.5°C:

$[C_{os}]$ at 6.5°C and 1 atm $(= 101{,}325 \text{ N/m}^2) = 12.30$ mg/L

$[C_{os}]$ at 6.5°C and 117,474.31 N/m$^2 = 12.30\left(\dfrac{117{,}474.31}{101{,}325} \right) = 14.26$ mg/L

Time (min)	$[\bar{C}]$ (mg/L)	$\ln [C_{os}] - [C]/[C_s]$
3	0.5	−0.0357
6	1.7	−0.1269
9	3.1	−0.2451
12	4.4	−0.3690
15	5.5	−0.4872
18	6.1	−0.5582
21	7.1	−0.6889
$\Sigma = 84$		$\Sigma = -2.511$

Therefore,

$$K_L a = -\frac{\sum_{m=1}^{m=n}\left(\ln\frac{[C_{os}] - [\bar{C}]}{[C_s]}\right)_m}{\sum_{m=1}^{m=n} t_m} = -\frac{-2.511}{84} = 0.0298 \text{ per min}$$

$$0.0298 = (K_L a)_{20}(1.024^{6.5-20}) \qquad (K_L a)_{20} = 0.041 \text{ per min} = 2.46 \text{ per hr}$$

Wastewater at 25°C:

$$[C_{os}] \text{ at } 25°C \text{ and } 1 \text{ atm } (= 101325 \text{ N/m}^2) = 8.4 \text{ mg/L}$$

$$[C_{os}] \text{ at } 25°C \text{ and } 117,474.31 \text{ N/m}^2 = 8.4\left(\frac{117,474.31}{101,325}\right) = 9.74 \text{ mg/L}$$

Therefore,

$$[\beta C_{os}] = 0.926(9.74) = 6.35 \text{ mg/L}$$

Time (min)	$[\bar{C}]$ (mg/L)	$\beta[C_{os}] - [\bar{C}]$	$\ln \beta[C_{os}] - [C]/\beta[C_{os}]$
3	0.9	8.12	−0.1049
6	1.8	7.22	−0.2224
9	2.4	6.62	−0.3091
12	3.3	5.72	−0.4553
15	4.0	5.02	−0.5858
18	4.7	4.32	−0.7360
21	5.3	3.72	−0.8855
$\Sigma = 84$		$\Sigma = 40.74$	$\Sigma = -3.299$

Therefore,

$$(K_L'a)_w = -\frac{\sum_{m=1}^{m=n}\left(\ln\frac{\beta[C_{os}] - [\bar{C}]}{\beta[C_{os}]}\right)_m}{\sum_{m=1}^{m=n} t_m} = -\frac{-3.299}{84} = 0.393 \text{ per min} = 2.36 \text{ per hr}$$

$$(K_L a)_w = \frac{(K_L'a)_w \sum_{m=1}^{m=n}(\beta[C_{os}] - [\bar{C}])_m + n\bar{r}}{\sum_{m=1}^{m=n}(\beta[C_{os}] - [\bar{C}])_m}$$

$$= \frac{2.36(40.74) + (7)(1.0)}{40.74} = 2.53 \text{ per hr}$$

$$2.53 = (K_L a)_{w,20}(1.024^{25-20})(K_L a)_{w,20} = 2.23 \text{ per hr}$$

$$\alpha = \frac{2.23}{2.46} = 0.91 \quad \textbf{Ans}$$

9.5.3 CALCULATION OF ACTUAL OXYGEN REQUIREMENT, THE AOR

Consider the contents inside an activated sludge reactor laden with dissolved oxygen and pollutants subjected to aeration. Also, consider the sedimentation basin that follows the reactor and all the associated pipings. For the purpose of performing the material balance, let the boundaries of these items encompass the control volume. According to the Reynolds transport theorem, the total rate of increase of the concentration of dissolved oxygen is equal to its partial (or local) rate of increase plus its convective rate of increase.

The convective rate of increase is equal to $-Q_o[\bar{C}_o] + (Q_o - Q_w)[\bar{C}_e] + Q_w[\bar{C}_w]$, where Q_o is the inflow to the reactor; $[\bar{C}_o]$ is the concentration of dissolved oxygen in Q_o; Q_w is the outflow of wasted sludge; $[\bar{C}_e]$ is the concentration of dissolved oxygen in the effluent of the secondary sedimentation basin; and $[\bar{C}_w]$ is the concentration of dissolved oxygen in the wasted sludge. $[\bar{C}_w]$ is equal to zero. The local rate of increase is simply given by $(\partial[\bar{C}]/\partial t)\mathcal{V}$.

The total rate of increase is $(d[\bar{C}]/dt)\mathcal{V}$, where $\mathcal{V}$ is the volume of the system which, as mentioned, is composed of the reactor, secondary basin, and the associated pipings. From Equation (9.22), $d[\bar{C}]/dt$ is given by the rate of aeration, $(K_L a)_w([C_{os,w}] - [\bar{C}]) = \text{AOR}$, and the rate of respiration by the organisms, $\bar{r}$. Thus, the total rate of increase is

$$\frac{d[\bar{C}]}{dt}\mathcal{V} = [(K_L a)_w([C_{os,w}] - [\bar{C}]) - \bar{r}]\mathcal{V} = [\text{AOR} - \bar{r}]\mathcal{V} \qquad (9.28)$$

The total rate of increase is equal to the local rate of increase plus the convective rate of increase, so the following equation is obtained:

$$\frac{d[\bar{C}]}{dt}\mathcal{V} = V[\text{AOR} - \bar{r}]\mathcal{V} = \frac{\partial[\bar{C}]}{\partial t}\mathcal{V} - Q_o[\bar{C}_o] + (Q_o - Q_w)[\bar{C}_e] \qquad (9.29)$$

Note that $[\bar{C}_w]$ is equal to zero; thus, it is not appearing in Equation (9.29). It will be recalled that the left-hand side of the equation, $(d[\bar{C}]/dt)\mathcal{V}$, is called the *Lagrangian derivative* and the right-hand side exp-ressions, $(\partial[\bar{C}]/\partial t)\mathcal{V} - Q_o[\bar{C}_o] + (Q_o - Q_w)[\bar{C}_e]$, are collectively called the *Eulerian derivative*.*

---

* As mentioned in the chapter, "Background Chemistry and Fluid Mechanics," the Reynolds transport theorem distinguishes the difference between the full derivative and the partial derivative. As stated in that chapter, the environmental engineering literature is very confusing with respect to the use of these derivatives. Some authors use the full derivative and some use the partial derivative to express the same meaning.

At steady state, the local derivative, $(\partial[\bar{C}]/\partial t)\Psi$, of the Eulerian derivative is zero. Also, as the wastewater enters the reactor, its dissolved oxygen content is practically zero; thus, $[\bar{C}_o]$ is equal to zero. $[\bar{C}_e]$, as it goes out of the secondary basin must also be equal to zero. Thus,

$$\text{AOR} = \bar{r} \tag{9.30}$$

The respiration rate $\bar{r}$ is due to the consumption of substrate BOD which is composed of CBOD and NBOD. Let $(\bar{r})_s$ be the respiration due to CBOD and $(\bar{r})_n$ be the respiration due to NBOD. $\bar{r}$ is then

$$\bar{r} = (\bar{r})_s + (\bar{r})_n \tag{9.31}$$

Now, apply the Reynolds transport theorem to the fate of the CBOD. As a counterpart to $[\bar{C}]$ in the case of dissolved oxygen, let $[S]$ represent CBOD. The Lagrangian rate of decrease (opposite to increase, thus, will have a negative sign) of CBOD is $-(d[S]/dt)\Psi$. This decrease represents the consumption of CBOD substrates to produce energy by respiration, $(\bar{r})_s$, and the consumption of the substrates for synthesis or to replace dead cells, $(syn)_s$. Thus,

$$-\left(\frac{d[S]}{dt}\right)\Psi = \{(\bar{r})_s + (syn)_s\}\Psi \tag{9.32}$$

For the Eulerian derivative, the local derivative is $-(\partial[S]/\partial t)\Psi$ and the convective derivative is $-(-Q_o[S_o] + Q_o[S])$, where $[S_o]$ is the influent CBOD concentration and $[S]$ is the outgoing CBOD concentration from the control volume. The outgoing concentrations are those coming out from the effluent of the secondary basin and the wasted sludge. Note that the convective derivative is preceded by a negative sign. The negative sign is used to precede it, since this derivative is a convective rate of decrease, as distinguished from the convective rate of increase which has a positive sign preceding it.

At steady state, the local derivative is equal to zero. Thus, from the Reynolds transport theorem, (Lagrangian derivative = to the Eulerian derivative):

$$-\left(\frac{d[S]}{dt}\right)\Psi = \{(\bar{r})_s + (syn)_s\}\Psi = -(-Q_o[S_o] + Q_o[S]) \tag{9.33}$$

$$\Psi(\bar{r})_s = Q_o([S_o] - [S]) - \Psi(syn)_s \tag{9.34}$$

$\Psi(syn)_s$ is the equivalent amount of CBOD in the body of organisms in the waste sludge. If $[\bar{X}_u]$ is the concentration of organisms in the waste sludge, $\Psi(syn)_s = f_s Q_w[\bar{X}_u]$, where f_s is the factor that converts the mass of microorganisms in the waste sludge to the equivalent oxygen concentration $\Psi(syn)_s$. Therefore,

$$\Psi(\bar{r})_s = Q_o([S_o] - [S]) - f_s Q_w[\bar{X}_u] \tag{9.35}$$

The factor f_s converts $Q_w[\overline{X}_u]$ to its equivalent oxygen value. The value of f_s is normally taken as 1.42; however, to do an accurate job for a given specific waste, it should be determined experimentally.

By analogy with Equation (9.35), the respiration rate due to NBOD, $(\bar{r})_n$, may be obtained from

$$\mathcal{V}(\bar{r})_n = f_N Q_o([N_o] - [N]) - f_s Q_w[\overline{X}_u] \qquad (9.36)$$

where f_N is the factor for converting nitrogen concentrations to oxygen equivalent, $[N_o]$ and $[N]$ are the nitrogen concentrations in the influent and effluent, respectively, and f_n is the factor for converting the nitrogen in the wasted sludge $(Q_w[\overline{X}_u])$ to the oxygen equivalent.

Having determined the expressions for $(\bar{r})_s$ and $(\bar{r})_n$, the equation for AOR is finally

$$\mathcal{V}(AOR) = \mathcal{V}r = \mathcal{V}\{(\bar{r})_s + (\bar{r})_n\}$$

$$\mathcal{V}(AOR) = Q_o([S_o] - [S]) - f_s Q_w[\overline{X}_u] + f_N Q_o([N_o] - [N]) - f_n Q_w[\overline{X}_u] \qquad (9.37)$$

In the derivations of the equations above, $\mathcal{V}$ was the volume of the control volume which is composed of the volume of the reactor, secondary clarifier, and the associated pipings. As the effluent from the reactor is introduced into the secondary clarifier, it is true that microorganisms continue to respire. In the absence of aeration in the basin, however, this respiration is but for a few moments and consumption of substrates ceases. It is also true that there will be no respiration and consumption of substrates in the associated pipings. Thus, the only volume of the control volume applicable to the material balance is the volume of the reactor. Therefore, $\mathcal{V}$ may be considered simply as the volume of the reactor.

Determination of f_s, f_N, and f_n. The formula for microorganisms has been given as $C_5H_7NO_2$ (Mandt and Bell, 1982). To find the oxygen equivalent of the mass synthesized, f_s, react this "molecule" with oxygen as follows:

$$C_5H_7NO_2 + 5O_2 \rightarrow 5CO_2 + 2H_2O + NH_3 \qquad (9.38)$$

From this equation, the oxygen equivalent of the mass synthesized is 1.42 mg O_2 per mg $C_5H_7NO_2$. Therefore, $f_s = 1.42$.

The reduction in the concentration of nitrogen is also brought about by reaction with oxygen for energy and for the requirement for synthesis. NBOD is actually in the form of NH_3. Reacting with O_2,

$$NH_3 + 2O_2 \rightarrow HNO_3 + H_2O \qquad (9.39)$$

From this reaction, the oxygen equivalent per mg $NH_3 - N$ is 4.57 mg. Thus, f_N is equal to 4.57.

The nitrogen for synthesis goes with the sludge wasted, which can be expressed in terms of the total Kjeldahl nitrogen (TKN). Bacteria (volatile solids, VS) contain approximately 14 nitrogen. (Protein contains approximately 16% nitrogen.) Thus, the equivalent oxygen of the nitrogen in the sludge wasted is $4.57(0.14)Q_w[X_u] = 0.64\ Q_w[X_u]$ and the value of f_n is 0.64.

The total actual oxygen requirement, AOR, is therefore

$$\text{AOR} = \frac{1}{V}\{Q_o([S_o]-[S])-1.42Q_w[\bar{X}_u]+4.57Q_o([N_o]-[N])-0.64Q_w[\bar{X}_u]\}$$

(9.40)

This AOR is used to find SOR in order to size the aerator needed.

Example 9.9 The influent of 10,000 m³/day to a secondary reactor has a BOD₅ of 150 mg/L. It is desired to have an effluent BOD₅ of 5 mg/L, an MLVSS (mixed liquor volatile suspended solids) of 3000 mg/L, and an underflow concentration of 10,000 mg/L. The effluent suspended solids concentration is 7 mg/L at 71% volatile suspended solids content. The volume of the reactor is 1611 m³ and the sludge is wasted at the rate of 43.3 m³/day. Calculate the SOR. Assume the aerator to be of the fine-bubble diffuser type with an $\alpha = 0.55$; depth of submergence equals 2.44 m. Assume β of liquor is 0.90. The influent TKN is 25 mg/L and the desired effluent $NH_3 - N$ concentration is 5.0 mg/L $\cdot f = 1.43$. The average temperature of the reactor is 25°C and is operated at an average of 1.0 mg/L of dissolved oxygen.

Solution:

$$\text{AOR} = \frac{1}{V}\{Q_o([S_o]-[S])-1.42Q_w[\bar{X}_u]+4.57Q_o([N_o]-[N])-0.64Q_w[\bar{X}_u]\}$$

$$Q_w[\bar{X}_u] = 43.3[10,000(0.001)] = 433.0\,\text{kg/d}$$

$$\text{AOR} = \frac{1}{1611}\{10,000(0.15-0.005)(1.43)-1.42(433.3)+4.57(10,000)$$

$$\times\,(0.025-0.005)-0.64(433.3)\} = 2092.8\text{ kg/d};\quad \text{AOR} = 1.30\frac{\text{kg}}{\text{m}^3\cdot\text{d}}$$

$$\text{SOR} = \frac{\text{AOR}}{\alpha\dfrac{(\beta[C_{os}]-[C])}{[C_{os,20,sp}]}\theta^{T-20}}$$

At 25°C, $[C_{os,sp}] = 8.4$ mg/L; $[C_{os,20,sp}] = 9.20$ mg/L

$$P = P_b + \frac{Z_d}{2}\gamma;\ \text{assume plant is located at sea level}$$

Therefore,

$$P = 101,325 + \frac{2.44}{2}[997(9.81)] = 113,257.3\text{ N/m}^2$$

$$[C_{os}] = \frac{P}{P_s}[C_{os,sp}] = \frac{113,257.3}{101,325}(8.4) = 9.39\text{ mg/L}$$

$$\text{SOR} = \frac{1.30}{0.55\left(\frac{[0.9(9.39)-1.0]}{9.20}\right)(1.024^{25-20})} = 2.62\text{ kg/d}\cdot\text{m}^3\quad\textbf{Ans}$$

9.5.4 TIME OF CONTACT

Having determined the overall coefficient of mass transfer $(K_L a)_w$, the differential equation

$$\frac{d[\bar{C}]}{dt} = (K_L a)_w([C_{os,w}] - [\bar{C}]) - \bar{r} \tag{9.41}$$

may be integrated. Take note that the only variables in this equation are t and $\bar{C}$, all the others are constant. Integrating from $t = 0$ to $t = t$ and from $[\bar{C}] = 0$ to $[\bar{C}] = [\bar{C}]$,

$$t = -\frac{1}{(K_L a)_w} ln \frac{(K_L a)_w([C_{os,w}] - [\bar{C}]) - \bar{r}}{(K_L a)_w[C_{os,w}] - \bar{r}} = -\frac{1}{(K_L a)_w} ln \frac{(K_L a)_w(\beta[C_{os}] - [\bar{C}]) - \bar{r}}{(K_L a)_w \beta[C_{os}] - \bar{r}}$$

$$(9.42)$$

t is the time of contact for aeration or mass transfer. The apparatus must provide this time if the liquid phase is to have the concentration of $[\bar{C}]$. For example, for a spray tower, this time must be the time it takes for the droplets to fall from the top of the tower to the bottom. For a cascade aerator, this time is the time for the water to fall through the cascade from the top to the bottom.

9.5.5 SIZING OF AERATION BASINS AND RELATIONSHIP TO CONTACT TIME

The sizing of aeration basins (for the activated sludge reactor, for example) are determined using the parameters of hydraulic detention time and mean cell retention time. *Hydraulic detention time* is the average time that the particles of water in an inflow to a basin are retained in the basin before outflow. *Mean cell retention time*, on the other hand, is the average time that cells of organisms (not water) are retained in the basin.

A third parameter that is used not to size aeration basins but to design the aerators used in the basin is the contact time (as derived previously). The size of the basin must be such that it provides this time to effect the required length of time of contact between the gas phase and liquid phase. There are situations where contact time for aeration is equal to the hydraulic detention time. For example, in the case of the trickling filter, as the water flows down the bed, a volume of this water occupies the interstices of the bed. The total volume of these interstices multiplied by a factor to account for the partial filling of the interstices divided by the inflow Q_o is the hydraulic detention time. Because contact between the water phase and the air also occurs during the time that the water occupies the interstices, however, detention time is equal to contact time.

In other situations, contact time is not equal to detention time. For example, in the case of the activated sludge reactor, contact time is the rise time of the bubbles to the surface, whereas detention time is the average time of travel of the water between the inlet to the outlet of the reactor. Contact time refers solely to the bubbles contacting the water during the time they were rising from the diffusers. Hydraulic detention time, in this case, has no relation to the contact time.

Example 9.10 An experiment was performed on a trickling filter for the purpose of determining the overall mass transfer coefficient. Raw sewage with zero dissolved oxygen was introduced at the top of the filter and allowed to flow down the bed. A tracer was also introduced at the top to determine how long it takes for the sewage to trickle down the bed. At the bottom, the resulting DO was then measured. From this experiment, the overall mass transfer coefficient $(K_La)_w$ was found to be 2.53 per hour. This information is then used to design another trickling filter. The β of the waste = 0.9 and the respiration rate $\bar{r}$ = 1.0 mg/L · hr. Assume the average temperature in the filter is 25°C and the effluent should have a DO = 1.0 mg/L. What should be the detention time of the sewage in the filter to effect this DO at the effluent? What void volume should be provided in the interstices of the bed to effect this detention time? The inflow rate is 10,000 m³/d. Assume the voids are 0.70 filled with water.

Solution:

$$t = -\frac{1}{(K_La)_w}\ln\frac{(K_La)_w(\beta[C_{os}]-[\bar{C}])-\bar{r}}{(K_La)_w\beta[C_{os}]-\bar{r}} \quad \text{At } 25°C, [C_{os,sp}] = 8.4 \text{ mg/L} = [C_{os}]$$

Therefore,

$$t = -\frac{1}{2.53}\ln\frac{2.53(0.9\times8.4-1.0)-1.\bar{0}}{2.53(0.9\times8.4)-1.\bar{0}} = 0.059 \text{ hr} \quad \textbf{Ans}$$

$$\text{Volume occupied by water} = 10,000\left(\frac{0.059}{24}\right) = 24.58 \text{ m}^3$$

Therefore,

$$\text{Volume of interstices} = \frac{24.58}{0.7} = 35.1 \text{m}^3 \quad \textbf{Ans}$$

9.5.6 CONTACT FOR BUBBLE AERATORS

Bubble aerators were discussed in a previous Chapter 6 (Mixing and Flocculation). In that chapter, the rise velocities of bubbles $\bar{v}_b$ as derived by Peebles and Garber were presented. The formulas are:

$$\bar{v}_b = \frac{2(\bar{r})^2(\rho_l-\rho_g)}{9\mu} = \frac{2(\bar{r})^2\rho_l}{9\mu} \quad Re = \frac{2\rho_l\bar{v}_b\bar{r}}{\mu} < 2$$

$$\bar{v}_b = 0.33g^{0.76}\left(\frac{\rho_l}{\mu}\right)^{0.52}(\bar{r})^{1.28} \quad 2 < Re < 4.02G_1^{-2.214}$$

$$\bar{v}_b = 1.35\left(\frac{\sigma}{\rho_l\bar{r}}\right)^{0.50} \quad 4.02G_1^{-2.214} < Re < 3.10G_1^{-0.25}$$

$$\bar{v}_b = 1.53\left(\frac{g\sigma}{\rho_l}\right)^{0.25} \quad 3.10G_1^{-0.25} < Re < G_2$$

$$G_1 = \frac{g\mu^4}{\rho_l\sigma^3} \qquad G_2 = \frac{g(\bar{r})^4(\bar{v}_b)^4\rho_l^3}{\sigma^3}$$

Re is a Reynolds number; *g* is the acceleration due to gravity; μ is the absolute viscosity of fluid; ρ_l is the mass density of fluid; σ is the surface tension of fluid; ρ_g is the mass density of the gas phase (air); and $\bar{r}$ is the average radius of the bubbles. To give G_1 a name, we called it the *Peebles number*. If the depth of submergence of the bubble diffuser and the rise velocity computed from one of the above equations are known, the time of contact between the gas phase in the bubbles and the surrounding water can be determined. This is illustrated in the next example.

Example 9.11 An aerator in an activated sludge reactor is located 3.5 m below the surface. The temperature inside the reactor is 30°C. What are the rise velocity of the bubbles, contact time for aeration, *Re*, G_1 and G_2? The approximate average diameter of the bubbles at mid-depth is 0.25 cm.

Solution:

$$\text{For } \bar{v}_b = \frac{2(\bar{r})^2\rho_l}{9\mu} \qquad Re = \frac{2\rho_l\bar{v}_b\bar{r}}{\mu} < 2:$$

$$\rho_l = 996 \text{ kg/m}^3 \quad \mu = 8(10^{-4}) \text{ kg/m} \cdot \text{s} \quad \text{all at 30°C}$$

Therefore,

$$\bar{v}_b = \frac{2(0.0025/2)^2(996)}{9(8)(10^{-4})} = 0.43 \text{ m/s}$$

$$Re = \frac{2\rho_l\bar{v}_b\bar{r}}{\mu} = \frac{2(996)(0.43)(0.0025/2)}{8(10^{-4})} = 1338 \gg 2$$

Thus, formula is not applicable.

$$\text{For } \bar{v}_b = 0.33g^{0.76}\left(\frac{\rho_l}{\mu}\right)^{0.52}(\bar{r})^{1.28} \quad 2 < Re < 4.02G_1^{-2.214}:$$

$$\bar{v}_b = 0.33(9.81)^{0.76}\left(\frac{996}{8(10^{-4})}\right)^{0.52}(0.0025/2)^{1.28}$$

$$= 1.87(1477.37)(0.00019) = 0.52 \text{ m/s}$$

$$Re = \frac{2(996)(0.52)(0.0025/2)}{8(10^{-4})} = 1618.5 \qquad G_1 = \frac{g\mu^4}{\rho_l\sigma^3} \qquad \sigma = 0.0712 \text{ N/m}$$

$$G_1 = \frac{9.81[8(10^{-4})]^4}{996(0.0712)^3} = 7.958(10^{-13})$$

$$4.02G_1^{-2.214} = 4.02[(7.958(10^{-13})]^{-2.214} = 2.46(10^{27})$$

Therefore,

$2 < Re < 4.02 G_1^{-2.214}$ satisfied and $\bar{v}_b = 0.52$ m/s **Ans**

$t = \dfrac{3.5}{0.52} = 6.73$ sec **Ans** $Re = 1618.5$ **Ans** $G_1 = 7.958(10^{-13})$ **Ans**

$$G_2 = \frac{g(\bar{r})^4 (\bar{v}_b)^4 \rho_l^3}{\sigma^3} = \frac{9.81(0.0025/2)^4 (0.52)^4 (996)^3}{8(10^{-4})^3} = 21.71 \quad \textbf{Ans}$$

9.6 ABSORPTION AND STRIPPING

As mentioned before, aeration is absorption. Thus, the discussions that follow apply equally to aeration (and air stripping). More specifically, the following discussions address the sizing of absorption and stripping towers.

9.6.1 SIZING OF ABSORPTION AND STRIPPING TOWERS

Absorption and stripping are reverse processes to each. Thus, discussing one is the same as discussing the other. Two design parameters required for the design of absorption towers are the cross section and the height. The cross section is a function of the mass velocity through the section. If the mass or volume flow rate is known, the cross section can be found using the equation of continuity $\rho_1 A_1 \bar{V}_1 = \rho_2 A_2 \bar{V}_2$ or $A_1 \bar{V}_1 = A_2 V_2$, if the mass density ρ is constant. Indices 1 and 2 refer to points in the elevation of the tower; A is the superficial area of the tower; and $\bar{V}$ is the average superficial velocity through the tower.

9.6.2 OPERATING LINE

The plot between the concentration of the solute in the liquid phase and that in the gas phase is called the *operating line*. Consider an absorption operation in a tower and let G be the mole flow rate of solute-free gas phase (*carrier gas*) carrying solute at a concentration $[Y]$ mole units per unit mole of the gas phase solute-free carrier gas. The corresponding quantities for the liquid phase are L and $[X]$, where L is the mole flow rate of solute-free liquid phase (*carrier liquid*) and $[X]$ is the mole of solute per unit mole of the solute-free liquid carrier.

The mole flow rate of solute out of the gas phase must be equal to the mole flow rate of the solute into the liquid phase. Thus, if $d[Y]$ and $d[X]$ are the respective differential changes of the concentrations $[Y]$ and $[X]$,

$$G d[Y] = L d[X] \tag{9.43}$$

$[Y]$ and $[X]$ are also respectively equal to $[Y] = [y_f]/1-[y_f]$ and $[X] = [x_f]/1-[x_f]$, where $[y_f]$ is the mole fraction of solute in the gas phase and $[x_f]$ is the mole fraction of the solute in the liquid phase. Thus, $d[Y] = d[y_f]/(1-[y_f])^2$ and $d[X] = d[x_f]/(1-[x_f])^2$. Substituting all into the equation and integrating from $[x_f] = [x_{f1}]$ to $[x_f] = [x_f]$ and from $[y_f] = [y_{f1}]$ to $[y_f] = [y_f]$ yield

$$G\left(\frac{1}{1-[y_f]} - \frac{1}{1-[y_{f1}]}\right) = L\left(\frac{1}{1-[x_f]} - \frac{1}{1-[x_{f1}]}\right) \quad (9.44)$$

$[x_{f1}]$ and $[y_{f1}]$ are the concentrations at the bottom of the tower. Note that G and L are constants. Equation (9.44) yields the concentration at any elevation in the tower; thus, it is called the equation of the *operating line*. Solving for $[y_f]$,

$$[y_f] = 1 - \frac{G(1-[y_{f1}])(1-[x_{f1}])(1-[x_f])}{G(1-[x_{f1}])(1-[x_f]) + L\{1-[y_{f1}])(1-[x_{f1}]) - (1-[y_{f1}])(1-[x_f])\}} \quad (9.45)$$

When plots of the operating line and the equilibrium line are far apart, the driving force will be large and the rate of mass transfer will also be large. When the two lines are close to each other, the driving force will be small, and accordingly, the rate of mass transfer will also be small. The limiting condition is reached when the two lines touch or intersect each other. The liquid flow rate (or the gas flow rate) corresponding to this condition is the *minimum flow rate*. In absorption, a multiplying factor of 1.5 to 2 is applied to the minimum liquid flow rate to get the actual liquid flow rate in the column; in stripping, this range is applied to the minimum gas flow rate to get the actual gas flow rate in the column.

9.6.3 TOWER HEIGHT

Let us first derive the units of the overall mass transfer coefficients when the concentration units used are in mole fractions. Let the overall mass transfer coefficient for the gas side be K_{yf} and that for the liquid side be K_{xf}. $Gd[Y]$ is mole of solute flowing per unit time. Mass transfer is a process where mass crosses an area perpendicular to the direction of motion of the solute particles. This area is the contact area for mass transfer. Let the differential area be designated as dA. Thus, in terms of mass transfer, $Gd[Y]$ is equal to $K_{yf}([y_f] - [y_f^*])dA$. From this expression, the dimensions of K_{yf} are mole per unit time per unit mole fraction per unit square area or $M/t/mole\ fraction \cdot L^2$. In an analogous manner, $Ld[X]$ is equal to $K_{xf}([x_f^*] - [x_f])dA$; the dimensions of K_{xf} are similarly $M/t/mole\ fraction \cdot L^2$. The asterisks used as "exponent" denotes equilibrium concentrations.

Because the objective is to design the height of the tower, express dA in terms of height, along with other parameters. Call the superficial area as S, the height as Z, and the interfacial contact area per unit bulk volume of tower (the destination medium) as a. Thus, $dA = aSdZ$. Also, calling G' as the gas flow rate (mixture of carrier gas and solute), G is equal to $G'(1 - [y_f])$; calling L' as the liquid flow rate (mixture of carrier liquid and solute), L is equal to $L'(1 - [x_f])$. As found before $[Y] = [y_f]/1 - [x_f]$ and $[X] = [x_f]/1 - [x_f]$. Performing the necessary substitutions, the mass transfer expressions become, for the gas side and liquid side, respectively,

$$K_{yf}a([y_f] - [y_f^*])dZ = \frac{G'}{S}\frac{d[y_f]}{1-[y_f]} = V_{My}\frac{d[y_f]}{1-[y_f]} \quad (9.46)$$

$$K_{yf}a([x_f^*] - [x_f])dZ = \frac{L'}{S}\frac{d[x_f]}{1-[x_f]} = V_{Mx}\frac{d[x]}{1-[x_f]} \tag{9.47}$$

where V_{My} is called the gas side *molar mass velocity* and V_{Mx} is called the *liquid side molar velocity*. Integrating the equations and simplifying produce the formulas for tower height based on the gas and liquid sides, respectively:

$$\int_{[y_{f1}]}^{[y_{f2}]} \frac{d[y_f]}{(1-[y_f])([y_f]-[y_f^*])} = \left(\overline{\frac{K_{yf}a}{V_{My}}}\right)Z_T \tag{9.48}$$

$$\int_{[x_{f1}]}^{[x_{f2}]} \frac{d[x_f]}{(1-[x_f])([x_f^*]-[x_f])} = \left(\overline{\frac{K_{xf}a}{V_{Mx}}}\right)Z_T \tag{9.49}$$

Z_T is the tower height and the expressions with overbars are average values between elevations 1 and 2, respectively.

If the Z_T's of the previous equations are solved, they will be expressed in terms of the product of the reciprocal of the overbarred factors by the respective integrals. Thus, the tower height may be expressed as the product of two factors. Consider the first as the *height of a mass transfer unit, H*, and the second as the *number of mass transfer units, N_t*. Therefore,

$$Z_T = HN_t \tag{9.50}$$

H based on the gas side will be designated as H_y and H based on the liquid side will be designated as H_x. The corresponding designations for N_t are N_{ty} and N_{tx}, respectively. Thus,

$$H_y = \frac{1}{\left(\overline{\frac{K_{yf}a}{V_{My}}}\right)} \tag{9.51}$$

$$H_x = \frac{1}{\left(\overline{\frac{K_{xf}a}{V_{Mx}}}\right)} \tag{9.52}$$

$$N_{ty} = \int_{[y_{f1}]}^{[y_{f2}]} \frac{d[y_f]}{(1-[y_f])([y_f]-[y_f^*])} \tag{9.53}$$

$$N_{tx} = \int_{[x_{f1}]}^{[x_{f2}]} \frac{d[x_f]}{(1-[x_f])([x_f^*]-[x_f])} \tag{9.54}$$

Example 9.12 A packed absorption tower is designed to removed SO_2 from a coke oven stack. The stack gas flow rate measured at one atmosphere and 30°C is 10 m³/s, and the SO_2 content is 3.0%. Using an initially pure water, 90% removal is desired. The equilibrium curve of SO_2 in water may be approximated by $y_i = 30x_i$. Determine the water requirement if 150% of the minimum flow rate is deemed adequate. Calculate the height of the tower. Assume $K_{yf}a = 0.024$ kgmols/m³·s·mol fraction. Assume the total cross-sectional area of tower equals 11.0 m².

Solution:

$$G\left(\frac{1}{1-[y_f]}-\frac{1}{1-[y_{f1}]}\right) = L\left(\frac{1}{1-[x_f]}-\frac{1}{1-[x_{f1}]}\right)$$

$$y_{f1} = 0.03$$

$$y_{f2} = \frac{0.1(0.03)(1)}{(1-0.03)(1)+0.1(0.03)(1)} = 0.0031$$

Therefore,

$$x_{f2}, y_{f2} = 0.0, 0.0031$$

$$x_{f1}, y_{f1} = x_{f1}, 0.03$$

To get the minimum flow rate, assume that at the bottom, the operating line is at equilibrium point; thus,

$$y_i = 30x_i; \quad 0.03 = 30x_{f1}; \quad x_{f1} = 0.001$$

Therefore,

$$G\left(\frac{1}{1-0.0031}-\frac{1}{1-0.03}\right) = L\left(\frac{1}{1-0}-\frac{1}{1-0.001}\right)$$

$$L/G = 28$$

$$G = 10\left(\frac{273}{30+273}\right)(10^3)\left(\frac{1}{22.4}\right)(10^{-3})(1-0.03) = 0.39 \text{ kg·mol/s}$$

Therefore,

$$L = (1.5)(28)(0.39)(18) = 294.84 \text{ kg/s} = \frac{294.84}{\rho_{\text{water}}} = \frac{294.84}{981}$$

$$= 0.30 \text{ m}^3/\text{s} \quad \textbf{Ans}$$

$$Z_T = HN_t$$

$$H_y = \frac{1}{\left(\dfrac{K_{yf}a}{V_{My}}\right)}$$

$$N_{ty} = \int_{[y_{f1}]}^{[y_{f2}]} \frac{d[y_f]}{(1-[y_f])([y_f]-[y_f^*])}$$

Average $y_f = (0.03+0.0031)/2 = 0.017$

$$G' = \frac{0.39}{1-0.017} = 0.40 \text{ kg·mol/s}$$

$$V_{My} = \frac{0.40}{11} = 0.036 \text{ kg·mol/m}^2\cdot\text{s}$$

$$H_y = \frac{1}{0.024/0.036} = 1.5 \text{ per transfer unit}$$

$$G\left(\frac{1}{1-[y_f]} - \frac{1}{1-0.03}\right) = (1.5)(28)(0.39)\left(\frac{1}{1-[x_f]} - \frac{1}{1-0.001}\right)$$

$$x_f = \frac{0.39 y_f}{16.38 - 15.99 y_f}$$

From $y_i = 30 x_i$, $y_f^* = 30 x_f$

y_f	x_f	y_f^*	$1/(y_f - y_f^*)$	$1/(1 - y_f)$	Δy_f	$\Delta y_f/(1-y_f)(y_f-y_f^*)$
0.0031	0.0000	0.0000	—	—	—	—
0.0050	0.00012	0.0036	714.3	1.0050	0.002	1.44
0.0080	0.00019	0.0057	429.19	1.0081	0.003	1.38
0.0110	0.00026	0.0078	309.9	1.0111	0.003	0.94
0.0130	0.00031	0.0092	261	1.0132	0.002	0.53
0.0160	0.00038	0.011	210.7	1.0163	0.003	0.64
0.0190	0.00044	0.013	172	1.0194	0.003	0.53
0.0210	0.00049	0.0147	159	1.0215	0.002	0.32
0.0240	0.00056	0.0168	138	1.0245	0.003	0.28
0.0270	0.00063	0.0188	122	1.0277	0.003	0.38
0.03	0.00060	0.021	109	1.0309	0.003	0.34
						$\Sigma = 6.78$

$$Z_T = 1.5(6.78) = 10.17 \text{ m} \quad \textbf{Ans}$$

9.6.4 AMMONIA STRIPPING (OR ABSORPTION)

The reaction of NH_3 with water can be represented by

$$NH_3 + H_2O \rightleftharpoons NH_4^+ + OH^- \tag{9.55}$$

From this reaction, raising the pH (as represented by the OH^-) will drive the reaction to the left, increasing the concentration of NH_3. This makes ammonia more easily removed by stripping. In practice, the pH is raised to about 10 to 11 using lime. Stripping is done by introducing the wastewater at the top of the column and allowing it to flow down countercurrent to the flow of air introduced at the bottom. The stripping medium inside the column may be composed of packings or fillings, such as Raschig rings and Berl saddles, or sieve trays and bubble caps. The liquid flows in thin sheets around the medium, thereby, allowing more intimate contact between liquid and the stripping air.

Let G be the moles per unit time of ammonia-free air, L be the moles per unit time of ammonia-free water, $[Y_1]$ be the moles of ammonia per mole of ammonia-free air at the bottom, $[Y_2]$ be the moles of ammonia per mole of ammonia-free air at the top, $[X_1]$ be the moles of ammonia per mole ammonia-free water at the bottom, and $[X_2]$ be the moles of ammonia per mole of ammonia-free water at the top. At any section between the bottom and the top of the column, the $[X]$s and $[y]$s are simply $[X]$ and $[Y]$. By mass balance, the ammonia stripped from the wastewater is equal to the ammonia absorbed by the air. Thus,

$$G([Y_2] - [Y_1]) = L([X_2] - [X_1]) \tag{9.56}$$

Let the equilibrium concentrations corresponding to $[X]$ and $[Y]$ be $[X^*]$ and $[Y^*]$, respectively. From the equilibrium between ammonia in water and ammonia in air, the following equations for the relationships between $[X^*]$ and $[Y^*]$ at various temperatures and one atmosphere of total pressure may be derived (Metcalf and Eddy, 1991):

at 10°C: $[Y^*] = 0.469[X^*]$
at 20°C: $[Y^*] = 0.781[X^*]$
at 30°C: $[Y^*] = 1.25[X^*]$
at 40°C: $[Y^*] = 2.059[X^*]$
at 50°C: $[Y^*] = 2.692[X^*]$

The coefficients of $[X^*]$ in these equations plot a straight line with the temperatures. Letting these coefficients be K produces the regression equation,

$$K = 0.0585T - 0.338 \quad \text{with } T \text{ in } °C$$
$$K = 0.0585T_k - 16.31 \quad \text{with } T_k \text{ in degrees Kelvin}$$

Therefore, in general, the relationship between $[X^*]$ and $[Y^*]$ using the temperature in degrees Kelvin, is

$$[Y^*] = (0.0585T_k - 16.31)[X^*] \tag{9.57}$$

This is the equation of the equilibrium line for ammonia stripping or absorption. It may be expressed in terms of the respective mole fractions $[y_f^*]$ and $[x_f^*]$. Performing the operation and solving for $[y_f^*]$ produces

$$[y_f^*] = \frac{(0.0585T_k - 16.31)\frac{[x_f^*]}{1 - [x_f^*]}}{1 + (0.0585T_k - 16.31)\frac{[x_f^*]}{1 - [x_f^*]}} \tag{9.58}$$

L, $[X_1]$, $[X_2]$, and $[Y_1]$ are known values. Assuming the ammonia in the air leaving the top of the tower is in equilibrium with the ammonia content of the entering wastewater, $[Y_2]$ may be written as

$$[Y_2] = [Y_2^*] = (0.0585T_k - 16.31)[X_2] \tag{9.59}$$

Thus, the ratio G/L, the moles of air needed for stripping per mole of water all in solute-free basis is

$$\frac{G}{L} = \frac{([X_2] - [X_1])}{(0.0585T_k - 16.31)[X_2] - [Y_1]} \tag{9.60}$$

In terms of mole fractions,

$$\frac{G}{L} = \frac{(1-[y_{f1}])[(1-[x_{f1}])[x_{f2}]-(1-[x_{f2}])[x_{f1}]]}{(1-[x_{f1}])[(0.0585T_k-16.31)(1-[y_{f1}])[x_{f2}]-(1-[x_{f2}])[y_{f1}]]} \quad (9.61)$$

G is the theoretical amount of air needed to strip ammonia in the wastewater from a concentration of $[X_2]$ down to a concentration of $[X_1]$. In practice, the minimum theoretical G is often multiplied by a factor of 1.5 to 2.0. Once G has been established, the cross-sectional area of the tower may be computed using the equation of continuity. The superficial velocity through the tower should be less than the velocity that will cause flooding or boiling up of the incoming wastewater. The method for estimating tower height was previously discussed.

Operating line for ammonia. The equation $G([Y] - [Y_1]) = L([X] - [X_1])$ is the equation for the operating line for ammonia. Expressing in terms of the equivalent mole fractions $[y_f]$, $[y_{f1}]$, $[x_f]$, and $[x_{f1}]$, respectively, and solving the resulting expression for $[y_f]$,

$$[y_f] = \frac{\frac{G[y_{f1}](1-[x_{f1}])(1-[x_f])+L[[x_f](1-[y_{f1}])(1-[x_{f1}])-[x_{f1}](1-[y_{f1}])(1-[x_f])]}{(1-[y_{f1}])(1-[x_{f1}])(1-[x_f])}}{G+\frac{G[y_{f1}](1-[x_{f1}])(1-[x_f])+L[[x_f](1-[y_{f1}])(1-[x_{f1}])-[x_{f1}](1-[y_{f1}])(1-[x_f])]}{(1-[y_{f1}])(1-[x_{f1}])(1-[x_f])}}$$

$$(9.62)$$

Example 9.13   A wastewater containing 25 mg/L of NH_3–N is to be stripped. The temperature of operation is 15°C and the flow is 0.013 m³/s. Determine the amount of air needed assuming a multiplying factor of 1.5 for the minimum theoretical G. Calculate the cross-sectional area of the tower. Assume a superficial velocity of 0.30 m/s.

Solution: To get the minimum G, assume the operation at the top of the tower is at equilibrium. Thus,

$$\frac{G}{L} = \frac{([X_2]-[X_1])}{(0.0585T_k-16.31)[X_2]-[Y_1]} \quad 1L \text{ of water } = 1000 \text{ g} = 1,000,000 \text{ mg}$$

Therefore,

$$[X_2] \simeq \frac{25/N}{(1,000,000-25)/H_2O} = \frac{25/14}{(1,000,000-25)/18}$$

$$= 3.2(10^{-5})\frac{\text{mol}(\ NH_3-N)}{\text{mol } H_2O \text{ (ammonia free)}}$$

$$[Y_1] = 0. \quad T_k = 15+273 = 288°K$$

$$\frac{G}{L} = \frac{3.2(10^{-5})-0}{[.0585(288)-16.31](3.2)(10^{-5})-0} = 1.86$$

$$L \simeq 0.013 \text{ m}^3/\text{s} = 0.013\left(\frac{1000}{18}\right) = 0.722 \text{ kgmol/s}$$

Therefore,

$$G = 1.5(1.86)(0.722) = 2.01 \text{ kgmol/s} = 1.74(10^8) \text{ gmol/d}$$

$$= 1.74(10^8)(22.4)\left(\frac{288}{273}\right)\left(\frac{1}{1000}\right) = 4.11(10^6) \text{ m}^3/\text{d} \quad \textbf{Ans}$$

$$\text{Overall cross-sectional area of towers} = \frac{4.11(10^6)}{0.3(60)(60)(24)} = 158.56 \text{ m}^3$$

$$\text{Using three towers, cross-sectional area} = \frac{158.56}{3} = 52.85 \text{ m}^2 \quad \textbf{Ans}$$

GLOSSARY

Absorption—A unit operation that removes a solute mass or masses from a gas phase into a liquid phase.

Actual oxygenation rate—The rate of transfer of oxygen to water at actual operating conditions.

Aeration—The unit operation of absorption where the gas involved is air.

Carrier gas—The gas phase that contains the solute.

Carrier liquid—The liquid phase that contains the solute.

Contact time—The time allowed for the process to take place.

Control volume—The volume in space that contains the materials addressed in any particular problem setting.

Crystallization—A unit operation where solute mass flows toward a point of concentration forming crystals.

Dehumidification—The removal of a solute liquid vapor from a gas phase by the solute condensing into its liquid phase.

Distillation—A unit operation that separates by vaporization liquid mixtures of miscible and volatile substances into individual components or groups of components.

Destination medium—The overall entity inside the control volume from, through, and into which the gas and liquid phases flow.

Equilibrium curve—The relationship between the concentration x in the liquid phase and the concentration y in the gas phase when there is no net mass transfer between phases.

Eulerian derivative—The total rate or rate of change of a quantity that reflects the effect of convection.

Humidification—The reverse of dehumidification where the solute flow from the liquid phase into the gas phase as in stripping, although in humidification nothing is stripped, because the liquid phase is composed of the same molecule as the solute molecule.

Hydraulic detention time—The average time that the particles of water in an inflow to a basin are retained in the basin before outflow.

Kjeldahl nitrogen—The sum of organic and ammonia nitrogens.

Leaching—A unit operation where the solute molecule is removed from a solid using a fluid extractor.

Lagrangian derivative—The total rate of change of a quantity as if convection is absent in the process.

Liquid extraction—The removal of a solute component from a liquid mixture called the raffinate using a liquid solvent.

Mass velocity—Mass flow rate divided by cross-sectional area.

Mass transfer—The transfer of mass between phases.

Mean cell retention time—The average time that cells of organisms are retained in the basin.

Minimum flow rate—The liquid flow rate corresponding to the condition when the operating line and the equilibrium line intersect each other.

Molar mass velocity—Mass velocity expressed in terms of mole unit.

Operating line—The plot between the concentration in the liquid phase and the concentration in gas phase.

Packing or filling—A supported mass of solid shapes in towers and similar structures.

Standard oxygen rate, SOR—The rate of transfer of oxygen from air into tap water or distilled at a pressure of one atmosphere and a dissolved oxygen concentration of 0 mg/L

Stripping—The flow of masses from the liquid phase into the gas phase.

SYMBOLS

a	Interfacial contact area per unit bulk volume of tower
AOR	Actual oxygenation rate
B	Temperature lapse rate
BOD	Biochemical oxygen demand
BOD_5	Five-day biochemical oxygen demand
$[\bar{C}]$	Concentration of dissolved oxygen in water at any given moment at a corresponding temperature and pressure
CBOD	Carbonaceous oxygen demand
$[\bar{C}_e]$	Concentration of dissolved oxygen in the effluent of the secondary sedimentation basin
$[\bar{C}_o]$	Concentration of dissolved oxygen in Q_o
$[C_{os}]$	Equilibrium concentration of oxygen in water at a particular temperature and pressure. This equilibrium concentration is also called the saturation concentration of the dissolved oxygen (DO)
$[C_{os,sp}]$	$[C_{os}]$ at standard pressure
$[C_{os,w}]$	$[C_{os}]$ of the wastewater at field conditions
$[C_{os,20,sp}]$	Saturation DO at 20°C and standard pressure
$[\bar{C}_w]$	Concentration of dissolved oxygen in the wasted sludge
DO	Saturation dissolved oxygen
f	Factor to convert BOD_5 concentrations to equivalent CBOD concentrations in terms of S

f_n	Factor for converting the nitrogen in the wasted sludge to oxygen equivalent
f_N	Factor for converting nitrogen concentrations to oxygen equivalent
f_s	Factor that converts the mass of microorganisms in the waste sludge to the equivalent oxygen concentration
g	Acceleration due to gravity
G	Mole flow rate of solute-free gas phase (*carrier gas*); the moles of ammonia per unit time of ammonia-free air
G'	Gas flow rate as a mixture of carrier gas and solute
G_1	Peebles number
H	Height of transfer unit
H_x	Height of transfer unit based on the liquid side
H_y	Height of transfer unit based on the gas side
$K_L a$	Dissolved oxygen overall mass transfer coefficient of tap or distilled water
$(K_L a)_{20}$	Dissolved oxygen overall mass transfer coefficient of tap or distilled water at standard conditions
$(K_L a)_w$	Dissolved oxygen overall mass transfer coefficient of wastewater at field conditions
$(K'_L a)_w$	Apparent overall mass transfer coefficient of wastewater
$(K_L a)_{w,20}$	Dissolved oxygen overall mass transfer coefficient of wastewater at standard conditions
k_x	Individual liquid film coefficient of mass transfer
K_x	Overall mass transfer coefficient for the liquid side
$K_x a$	Overall mass transfer coefficient for the liquid side
k_y	Individual gas film coefficient of mass transfer
K_y	Overall mass transfer coefficient for the gas side
$K_y a$	Overall mass transfer coefficient for the gas side
L	Mole flow rate of solute-free liquid phase (*carrier liquid*); the moles ammonia per unit time of ammonia-free water
L'	Liquid flow rate as mixture of carrier liquid and solute
m	Slope of operating line if it were a straight line
N_t	Number of transfer units
N_{ty}	Number of transfer units based on the gas side
N_{tx}	Number of transfer units based on the liquid side
$[N]$	Nitrogen concentration in the effluent
$[N_o]$	The nitrogen concentration in the influent
NBOD	Nitrogenous oxygen demand
P	Pressurization pressure
P_b	Barometric at a given elevation
P_s	Standard pressure
P_{bo}	Barometric pressure at elevation $z = 0$
Q_o	Inflow to reactor
Q_w	Outflow of wasted sludge
$\bar{r}$	Respiration rate; average radius of bubbles
$(\bar{r})_n$	Respiration due to NBOD
$(\bar{r})_s$	Respiration due to CBOD

R	Gas constant for air ($= 286.9$ N·m/kg·°K)
Re	Reynolds number
$(syn)_s$	Portion of S consumed for synthesis for growth or to replace dead cells
$[S]$	Concentration of CBOD; superficial area
$[S_o]$	Influent CBOD concentration
SOR	Standard oxygen rate
t	Time of contact for aeration or mass transfer
T	Temperature in degree Celsius or Kelvin
T_o	Temperature in °K at $z = 0 =$ ambient temperature corresponding P_b
$\bar{v}_b$	Bubble rise velocity
$\forall$	Volume of the control volume or volume of reactor
V_{My}	Gas side molar mass velocity
V_{Mx}	Liquid side molar mass velocity
$[x_i]$	Concentration of solute at the interface surface referred to the liquid phase
$[x^*]$	The concentration that x would attain if it were to reach equilibrium value
$[X]$	Mole of solute per unit mole of the solute-free liquid
$[X_1]$	Moles of ammonia per mole ammonia-free water at the bottom
$[X_2]$	Moles of ammonia per mole of ammonia-free water at the top.
$[\bar{X}_u]$	Concentration of organisms in the waste sludge
$[y]$	Concentration of solute in the bulk gas phase
$[y_i]$	Concentration of solute at the interface surface referred to the gas phase
$[y^*]$	The concentration that y would attain if it were to reach equilibrium value
$[Y]$	Mole units of solute per unit mole of solute-free gas
$[Y_1]$	Moles of ammonia per mole of ammonia-free air at the bottom
$[y_2]$	Moles of ammonia per mole of ammonia-free air at the top
Z	Tower height
Z_T	Tower height
α	$(K_La)_{w,20}/(K_La)_{20}$
β	$[C_{os,w}]/[C_{os}]$
θ	Temperature correction factor
μ	Absolute viscosity of fluid
ρ_g	Mass density of the gas phase (air)
ρ_l	Mass density of fluid
σ	Surface tension of fluid

PROBLEMS

9.1   The value of $(K_La)_{20}$ for a certain industrial waste is 2.46 per hour and the value of $(K_La)_w$ at 25°C = 2.77 per hour. What is the value of the temperature correction factor?

9.2   The value of $(K_La)_w$ at 25°C = 2.77 per hour. If the value of the temperature correction factor is 1.024, what is the value of $(K_La)_{20}$?

9.3   The value of $(K_La)_w = 2.77$ per hour. If the value of the temperature correction factor is 1.024 and the value of $(K_La)_{20} = 2.46$ per hour, at what temperature is the value of $(K_La)_w$ for?

9.4   A chemical engineer proposes to purchase an aerator for an activated sludge reactor. In order to do so, he performs a series of experiments obtaining the following results: AOR $= 1.30$ kg/m^3 · day, and $\beta = 0.94$. The aeration is to be maintained to effect a dissolved oxygen concentration of 1.0 mg/L at 25°C in the reactor. The SOR specified to the manufacturer of the aerator is 2.05 kg/m^3 · day. Determine the value of α.

9.5   A chemical engineer proposes to purchase an aerator for an activated sludge reactor. In order to do so, he performs a series of experiments obtaining the following results: $\alpha = 0.91$, and $\beta = 0.94$. The aeration is to be maintained to effect a dissolved oxygen concentration of 1.0 mg/L at 25°C in the reactor. If the SOR specified to the manufacturer of the aerator is 2.05 kg/m^3 · day, what was the AOR?

9.6   A chemical engineer proposes to purchase an aerator for an activated sludge reactor. In order to do so, he performs a series of experiments obtaining the following results: $\alpha = 0.91$, and AOR $= 1.30$ kg/m^3·day. The aeration is to be maintained to effect a dissolved oxygen concentration of 1.0 mg/L at 25°C in the reactor. The SOR specified to the manufacturer of the aerator is 2.05 kg/m^3 · day. Determine the value of β.

9.7   A chemical engineer proposes to purchase an aerator for an activated sludge reactor. In order to do so, he performs a series of experiments obtaining the following results: $\alpha = 0.91$, $\beta = 0.94$, and AOR $= 1.30$ kg/m^3·day. The aeration is to be maintained to effect a dissolved oxygen concentration of 1.0 mg/L at 25°C in the reactor. The SOR specified to the manufacturer of the aerator is 2.05 kg/m^3 · day and the concentration of dissolved oxygen at 20°C and one atmosphere pressure of tap water is 9.2 mg/L. The Arrhenius temperature correction factor is 1.024. From these data, calculate the saturation dissolved oxygen concentration of tap water corresponding to the temperature of the proposed reactor.

9.8   A chemical engineer plans to purchase an aerator for a proposed activated sludge reactor. In order to do so, he performs a series of experiments obtaining the following results: $\alpha = 0.91$, $\beta = 0.94$, and AOR $= 1.30$ kg/m^3·day. The average temperature of operation of the reactor is expected to be 25°C. The SOR specified to the manufacturer of the aerator is 2.05 kg/m^3·day and the concentration of dissolved oxygen at 20°C and one atmosphere pressure of tap water is 9.2 mg/L. The Arrhenius temperature correction factor is 1.024. The saturation dissolved oxygen concentration of tap water corresponding to the temperature of the proposed reactor at 25°C is 8.4 mg/L. Using these data, calculate the dissolved oxygen concentration at which the reactor can be maintained.

9.9   A chemical engineer plans to purchase an aerator for a proposed activated sludge reactor. In order to do so, he performs a series of experiments obtaining the following results: $\alpha = 0.91$, $\beta = 0.94$, and AOR $= 1.30$ kg/m^3· day. The average temperature of operation of the reactor is expected to be 25°C.

The SOR specified to the manufacturer of the aerator is 2.05 kg/m³ · day. The Arrhenius temperature correction factor is 1.024. The saturation dissolved oxygen concentration of tap water corresponding to the temperature of the proposed reactor at 25°C is 8.4 mg/L. The proposed reactor is to be maintained at a dissolved oxygen concentration of 1.0 mg/L. Using these data, calculate the dissolved oxygen at 20°C and one atmosphere pressure of tap water.

9.10   A chemical engineer plans to purchase an aerator for a proposed activated sludge reactor. In order to do so, he performs a series of experiments obtaining the following results: $\alpha = 0.91$, $\beta = 0.94$, and AOR = 1.30 kg/m³ · day. The average temperature of operation of the reactor is expected to be 25°C. The SOR specified to the manufacturer of the aerator is 2.05 kg/m³ · day. The saturation dissolved oxygen concentration of tap water corresponding to the temperature of the proposed reactor at 25°C is 8.4 mg/L. The proposed reactor is to be maintained at a dissolved oxygen concentration of 1.0 mg/L. The dissolved oxygen at 20°C and one atmosphere pressure of tap water is 9.2 mg/L. Using these data, calculate the Arrhenius temperature correction factor θ.

9.11   A chemical engineer plans to purchase an aerator for a proposed activated sludge reactor. In order to do so, he performs a series of experiments obtaining the following results: $\alpha = 0.91$, $\beta = 0.94$, and AOR = 1.30 kg/m³ · day. The SOR specified to the manufacturer of the aerator is 2.05 kg/m³ · day. The saturation dissolved oxygen concentration of tap water corresponding to the temperature of the proposed reactor is 8.4 mg/L. The proposed reactor is to be maintained at a dissolved oxygen concentration of 1.0 mg/L. The dissolved oxygen at 20°C and one atmosphere pressure of tap water is 9.2 mg/L. The Arrhenius temperature correction factor θ is 1.024. Using these data, calculate the average temperature that the proposed reactor will be operated.

9.12   A civil engineer performs an experiment for the purpose of determining the values of some aeration parameters of a particular wastewater. $(K_L a)_{20}$ and α were found, respectively, to be 2.46 per hr and 0.91. Calculate $(K_L a)_{w,20}$.

9.13   A civil engineer performs an experiment for the purpose of determining the values of some aeration parameters of a particular wastewater. $(K_L a)_{w,20}$ and α were found, respectively, to be 2.25 per hr and 0.91. Calculate $(K_L a)_{20}$.

9.14   An environmental engineer performs an experiment for the purpose of determining the value of β of a particular wastewater. If β is found to be equal to 0.89 and the temperature of the wastewater is 25°C, calculate $[C_{os,w}]$.

9.15   An environmental engineer performs an experiment for the purpose of determining the value of β of a particular wastewater. If β is found to be equal to 0.89 and the $[C_{os,w}]$ of the wastewater after shaking the jar thoroughly is 7.5 mg/L , calculate $[C_{os}]$.

9.16   An environmental engineer performs an experiment for the purpose of determining the value of β of a particular wastewater in the mountains of Allegheny County, MD. The prevailing barometric pressure is 92,974 N/m² at an average atmospheric temperature of 30°C. At what elevation in the county is the experiment conducted?

9.17 An environmental engineer performs an experiment for the purpose of determining the value of β of a particular wastewater in the mountains of Allegheny County, MD. The prevailing barometric pressure is 92,974 N/m^2 and the elevation where the experiment is conducted is 756 m above mean sea level. What is the average temperature of the atmosphere?

9.18 An activated sludge reactor is aerated using a turbine aerator located 5.5 m below the surface of the mixed liquor. The pressurizing pressure is calculated to be 101,436 N/m^2. If the water temperature is 20°C, calculate the prevailing barometric pressure.

9.19 An activated sludge reactor is aerated using a turbine aerator. The pressurizing pressure is calculated to be 101,436 N/m^2 at a prevailing barometric pressure of 760 mm Hg. If the water temperature is 20°C, determine the depth of submergence of the aerator.

9.20 An activated sludge reactor is aerated using a turbine aerator. The pressurizing pressure is calculated to be 101,436 N/m^2 at a prevailing barometric pressure of 760 mm Hg. The depth of submergence of the aerator is 5.5 m. What is the specific weight of the water in the reactor?

9.21 An activated sludge reactor is aerated using a turbine aerator. The $[C_{os}]$ corresponding to the dissolution pressure is 9.2 mg/L and the prevailing barometric pressure is 761 mm Hg. The water temperature is 20°C. Calculate the depth of submergence.

9.22 An activated sludge reactor is aerated using a turbine aerator. The $[C_{os}]$ corresponding to the dissolution pressure is 9.2 mg/L and the prevailing barometric pressure is 761 mm Hg. The water temperature is 20°C and the depth of submergence is 5.5 m. Using the calculated value of the pressuring pressure, calculate $[C_{os,sp}]$.

9.23 An experiment to determine the overall mass transfer coefficient of a tap water is performed using a settling column 4 m in height. The result of the unsteady state aeration test is shown below. The experiment is performed in Allegheny County, MD. For practical purposes, assume mass density of water = 1000 kg/m^3. Assume an ambient temperature of 28°C. Calculate $K_L a$.

Tap Water at 5.5°C	
Time (min)	$[\bar{C}]$(mg/L)
3	0.5
6	1.7
9	3.1
12	4.4
15	5.5
18	6.1
21	7.1

9.24 An activated sludge reactor is aerated using a turbine aerator. The $[C_{os}]$ corresponding to the dissolution pressure is 9.2 mg/L and the prevailing barometric pressure is 761 mm Hg. The water temperature is 20°C and the depth of submergence is 5.5 m. Using the calculated value of the pressuring pressure and using a value of 8.6 for $[C_{os,sp}]$, calculate the standard pressure P_s.

9.25 An experiment to determine the overall mass transfer coefficient of a tap water is performed using a settling column 4 m in height. The result of the unsteady state aeration test is shown below. The experiment is performed in Allegheny County, MD. For practical purposes, assume mass density of water = 1000 kg/m^3. Assume an ambient temperature of 28°C. Calculate $(K_La)_{20}$.

Tap Water at 5.5°C	
Time (min)	$[\overline{C}]$,(mg/L)
3	0.5
6	1.7
9	3.1
12	4.4
15	5.5
18	6.1
21	7.1

9.26 A settling column 4 m in height is used to determine the overall mass transfer coefficient of a wastewater. The wastewater is to be aerated using a fine-bubble diffuser in the prototype aeration tank. The laboratory diffuser releases air at the bottom of the tank. The result of the unsteady state aeration test is shown below. Assume $\beta = 0.94$, $\overline{r} = 1.0$ mg/L · hr and the plant is in the mountains of Allegheny County, MD. For practical purposes, assume mass density of water = 1000 kg/m^3. Assume an ambient temperature of 28°C. Calculate the apparent overall mass transfer coefficient, overall mass transfer coefficient, and the overall mass transfer coefficient corrected to 20°C.

Wastewater at 25°C	
Time (min)	$[\overline{C}]$(mg/L)
3	0.9
6	1.8
9	2.4
12	3.3
15	4.0
18	4.7
21	5.3

9.27 The influent of 10,000 m^3/day to a secondary reactor has a BOD_5 of 150 mg/L. It is desired to have an effluent BOD_5 of 5 mg/L, an MLVSS (mixed liquor volatile suspended solids) of 3000 mg/L, and an underflow concentration of 10,000 mg/L. The effluent suspended solids concentration is 7 mgL at 71% volatile suspended solids content. The sludge is wasted at the rate of 43.3 m^3/day. Assume the aerator to be of the fine-bubble diffuser type with an $\alpha = 0.55$; depth of submergence equals 2.44 m. Assume β of liquor is 0.90. The influent TKN is 25 mg/L and the desired effluent $NH_3 - N$ concentration is 5.0 mg/L $\cdot f = 1.47$. The average temperature of the reactor is 25°C and is operated at an average of 1.0 mg/L of dissolved oxygen. The AOR is calculated to be 3.36 kg/d $\cdot m^3$. Calculate the volume of the reactor.

9.28 The volume of a secondary reactor is 1611 m^3. It receives a primary-treated wastewater containing a BOD_5 of 150 mg/L. It is desired to have an effluent BOD_5 of 5 mg/L, an MLVSS (mixed liquor volatile suspended solids) of 3000 mg/L, and an underflow concentration of 10,000 mg/L. The effluent suspended solids concentration is 7 mgL at 71% volatile suspended solids content. The sludge is wasted at the rate of 43.3.m^3/day. Assume the aerator to be of the fine-bubble diffuser type with an $\alpha = 0.55$; depth of submergence equals 2.44 m. Assume β of liquor is 0.90. The influent TKN is 25 mg/L and the desired effluent $NH_3 - N$ concentration is 5.0 mg/L $\cdot f = 1.47$. The average temperature of the reactor is 25°C and is operated at an average of 1.0 mg/L of dissolved oxygen. The AOR is calculated to be 3.36 kg/d $\cdot m^3$. Determine the volume of inflow to the reactor.

9.29 The volume of a secondary reactor is 1611 m^3. It receives an inflow of 10,000 m^3/day from primary-treated wastewater. It is desired to have an effluent BOD_5 of 5 mg/L, an MLVSS (mixed liquor volatile suspended solids) of 3000 mg/L, and an underflow concentration of 10,000 mg/L. The effluent suspended solids concentration is 7 mg/L at 71% volatile suspended solids content. The sludge is wasted at the rate of 43.3 m^3/day. Assume the aerator to be of the fine-bubble diffuser type with an $\alpha = 0.55$; depth of submergence equals 2.44 m. Assume β of liquor is 0.90. The influent TKN is 25 mg/L and the desired effluent $NH_3 - N$ concentration is 5.0 mg/L. $f = 1.47$. The average temperature of the reactor is 25°C and is operated at an average of 1.0 mg/L of dissolved oxygen. The AOR is calculated to be 3.36 kg/d $\cdot m^3$. What is the influent BOD_5?

9.30 Using the data in Problem 9.29, calculate the effluent BOD_5 from the secondary reactor if the influent BOD_5 is 150 mg/L.

9.31 Using the data in Problem 9.29, calculate the rate of sludge wasting from the process if the influent BOD_5 is 150 mg/L.

9.32 Using the data in Problem 9.29, calculate the underflow concentration from the secondary clarifier if the influent BOD_5 is 150 mg/L.

9.33 An experiment was performed on a trickling filter for the purpose of determining the overall mass transfer coefficient. Raw sewage with zero dissolved oxygen was introduced at the top of the filter and allowed to flow down the bed. A tracer was also introduced at the top to determine how long it takes for the sewage to trickle down the bed. At the bottom, the resulting DO was then measured. From this experiment, the overall mass transfer coefficient $(K_La)_w$ was found to be 2.53 per hour. This information is then used to design another trickling filter. The β of the waste $= 0.9$ and the respiration rate $\bar{r} = 1.0$ mg/L $\cdot$ hr. Assume the average temperature in the filter is 25°C and the effluent should have a DO $= 1.0$ mg/L. What void volume should be provided in the interstices of the bed to effect this detention time? The inflow rate is 10,000 m^3/d. The detention time of the sewage in the proposed filter to effect this DO at the effluent of 1.0 mg/L is calculated to be 0.06 hour. Calculate the overall mass transfer coefficient of the wastewater.

9.34   Using the data in Problem 9.33, calculate β if the overall mass transfer of the wastewater is 2.53 per hour.

9.35   Using the data in Problem 9.33, calculate the saturation dissolved oxygen concentration at the dissolution pressure if the overall mass transfer of the wastewater is 2.53 per hour.

9.36   Using the data in Problem 9.33, calculate the effluent DO if the overall mass transfer of the wastewater is 2.53 per hour.

9.37   Using the data in Problem 9.33, calculate the respiration rate if the overall mass transfer of the wastewater is 2.53 per hour.

9.38   An aerator in an activated sludge reactor is located below the surface. The temperature inside the reactor is 29°C. The rise velocity of the bubbles is determined to be 0.26 m/s at a time of rise of 13.46 sec. The approximate average diameter of the bubbles at mid-depth is 0.25 cm. Calculate the depth of submergence.

9.39   An aerator in an activated sludge reactor is located 3.5 m below the surface. The temperature inside the reactor is 29°C. The time of rise of the bubbles to the surface is 13.46 sec. The approximate average diameter of the bubbles at mid-depth is 0.25 cm. Calculate the rise velocity.

9.40   A packed absorption tower is designed to removed SO_2 from a coke oven stack. The water flow rate $L = 16.64$ kgmols/s and the gas flow rate $G = 0.39$ kgmols/s. The mole fraction of SO_2 in the air at the bottom of the tower is 0.03 and its mole fraction in the absorption water at the bottom of the tower is 0.0007. If the mole fraction of SO_2 in the air at a point in the tower 7 meters from the bottom is 0.016, what is the corresponding mole fraction of the SO_2 in the downward flowing scrubbing water.

9.41   Using the data in Problem 9.40, calculate the corresponding mole fraction of SO_2 in air if its mole fraction in water at the 7-m point is 0.0004.

9.42   Using the data in Problem 9.40, calculate the gas flow rate G if the mole fraction of SO_2 in water at the 7-m point is 0.0004.

9.43   Using the data in Problem 9.40, calculate the water flow rate L if the mole fraction of SO_2 in water at the 7-m point is 0.0004.

9.44   Using the data in Problem 9.40, calculate the mole fraction of SO_2 in the air at the bottom of the tower if the mole fraction of SO_2 in water at the 7-m point is 0.0004.

9.45 Using the data in Problem 9.40, calculate the mole fraction of SO_2 in the water at the bottom of the tower if the mole fraction of SO_2 in water at the 7-m point is 0.0004.

BIBLIOGRAPHY

Benyahia, F., et al. (1996). Mass transfer studies in pneumatic reactors. *Chem. Eng. Technol.,* 19, 5, 425–43.

Boswell, S. T. and D. A. Vaccari (1994). Plate and frame membrane air stripping. *Proc 21st Annu. Conf. on Water Policy and Manage.: Solving the Problems,* May 23–26, 1994, Denver, CO, 726–729. ASCE. New York.

Chern, J. and C. Yu (1997). Oxygen transfer modeling of diffused aeration systems. *Industrial Eng. Chem.,* 36, 5447–5453.

Droste, R. L. (1997). *Theory and Practice of Water and Wastewater Treatment.* John Wiley & Sons, New York.

Gupta, H. S. and T. K. Marshall (1997). Biopulsing: An *in situ* aeration remediation strategy. *Geotechnical Special Publication* In Situ *Remediation of the Geoenvironment Proc. 1997 ASCE Annu. Fall Natl. Convention,* Oct 5–8 1997, Minneapolis, MN, 516–531, ASCE, New York.

He, M. and X. Zhang (1997). Study on the volatile property of organics in coke-plant wastewater under the aerated stripping conditions. *Huanjing Kexue/Environmental Science,* 18, 5, 34–36.

Jiang, Z., H. Bo, and Y. Li (1995). Study of aeration mechanism of free-jet flotation column. *J. China Coal Soc.* 20, 432–436.

Kouda, T., et al. (1997). Effects of oxygen and carbon dioxide pressures on bacterial cellulose production by acetobacter in aerated and agitated culture. *J. Fermentation Bioengineering,* 84, 2, 124–127.

Leu, H.G., C. F. Ouyang, and T. Y. Pai (1997). Effects of flow velocity and depth on the rates of reaeration and BOD removal in a shallow open channel. *Water Science Technol. Proc. 1995 5th IAWQ Asian Regional Conf. on Water Quality and Pollut. Control,* Feb. 7–9 1995, Manila, Philippines, 35, 8, 57–67. Elsevier Science Ltd. Oxford. England.

Mandt, M. G. and B. A. Bell (1982). *Oxidation Ditches in Wastewater Treatment.* Ann Arbor Science Publishers, Ann Arbor, MI, 48.

Marx, J. and M. G. Rieth (1997). QA for effective fine pore aeration diffusers in wastewater treatment systems. *ASCE Annu. Convention on Innovative Civil Eng. for Sustainable Development Quality Assurance—A National Commitment Proc. 1997 ASCE Annu. Convention on Innovative Civil Eng. for Sustainable Development,* Oct. 5–8, Minneapolis, MN, 132–141. Sponsored by ASCE, New York.

Metcalf & Eddy, Inc. (1991). *Wastewater Engineering: Treatment, Disposal, and Reuse.* McGraw-Hill, New York, 37.

McCabe, W. L. and J. C. Smith (1967). *Unit Operations of Chemical Engineering.* McGraw-Hill, New York.

Olsson, M. P. and K. V. Lo (1997). Sequencing batch reactors for the treatment of egg processing wastewater. *Canadian Agricultural Eng.* 39, 3, 195–202.

Peavy, H. S., D. R. Rowe, and G. Tchobanoglous (1985). *Environmental Engineering.* McGraw-Hill, New York.

Ramamurthy, R., et al. (1996). Control strategies for activated sludge treatment plants. *Pulp Paper Canada.* 97, 11, 32–37.

Salvato, J. A. (1982). *Environmental Engineering and Sanitation.* John Wiley & Sons, New York.

Sincero, A. P. and G. A. Sincero (1996). *Environmental Engineering: A Design Approach.* Prentice Hall, Upper Saddle River, NJ.

APHA, AWWA, WEF (1992). *Standard Methods for the Examination of Water and Wastewater.* American Public Health Association, Washington.

Tchobanoglous, G. and E. D. Schroeder (1985). *Water Quality.* Addison-Wesley, Reading, MA.

Viessman, W., Jr. and M. J. Hammer (1993). *Water Supply and Pollution Control.* Harper & Row, New York.

Wilson, D. J. and R. D. Norris (1997). Groundwater cleanup by *in-situ* sparging, engineered bioremediation with aeration curtains. *Separation Science Technol.* 32, 16, 2569–2589.

Part III

Unit Processes of Water and Wastewater Treatment

Part III covers the various unit processes employed in water and wastewater treatment including water softening; water stabilization; coagulation; removal of iron and manganese by chemical precipitation; removal of nitrogen by nitrification–denitrification; removal of phosphorus by chemical precipitation; ion exchange; and disinfection.

10 Water Softening

Softening is the term given to the process of removing ions that interfere with the use of soap. These ions are called *hardness ions* due to the presence of multivalent cations, mostly calcium and magnesium. In natural waters, other ions that may be present to cause hardness but not in significant amounts are iron (Fe^{2+}), manganese (Mn^{2+}), strontium (Sr^{2+}), and aluminum (Al^{3+}).

In the process of cleansing using soap, lather is formed causing the surface tension of water to decrease. This decrease in surface tension makes water molecules partially lose their mutual attraction toward each other, allowing them to wet "foreign" solids, thereby, suspending the solids in water. As the water is rinsed out, the solids are removed from the soiled material. In the presence of hardness ions, however, soap does not form the lather immediately but reacts with the ions, preventing the formation of lather and forming scum. Lather will only form when all the hardness ions are consumed. This means that hard waters are hard to lather. *Hard waters* are those waters that contain these hardness ions in excessive amounts. Softening using chemicals is discussed in this chapter. Other topics related to softening are discussed as necessary.

10.1 HARD WATERS

The following lists the general classification of hard waters:

Soft	<50 mg/L as $CaCO_3$
Moderately hard	50–150 mg/L as $CaCO_3$
Hard	150–300 mg/L as $CaCO_3$
Very hard	>300 mg/L as $CaCO_3$

A very soft water has a slimy feel. For example, rainwater, which is exceedingly soft, is slimy when used with soap. For this reason, hardness in water used for domestic purposes is not completely removed. Hardness is normally removed to the level of 75 to 120 mg/L as $CaCO_3$.

10.2 TYPES OF HARDNESS

Two basic types of hardness are associated with the ions causing hardness: *carbonate* and *noncarbonate hardness*. When the hardness ions are associated with the HCO_3^- ions in water, the type of hardness is called *carbonate hardness*; otherwise, the type of hardness is called *noncarbonate hardness*. An example of carbonate hardness is $Ca(HCO_3)_2$, and an example of noncarbonate hardness is $MgCl_2$.

In practice, when one addresses hardness removal, it means the removal of the calcium and magnesium ions associated with the two types of hardness. Our discussion of hardness removal therefore is divided into the categories of calcium and magnesium. Corresponding chemical reactions are developed. If other specific hardness ions are also present and to be removed, such as iron, manganese, strontium, and aluminum, corresponding specific reactions have to be developed for them. The removal of iron and manganese will be discussed in a separate chapter.

In general, water is softened in three ways: chemical precipitation, ion exchange, and reverse osmosis. Only the chemical precipitation method is discussed in this chapter.

10.3 PLANT TYPES FOR HARDNESS REMOVAL

In practice, two types of plants are generally used for chemical precipitation hardness removal: One type uses a sludge blanket contact mechanism to facilitate the precipitation reaction. The second type consists of a flash mix, a flocculation basin, and a sedimentation basin. The former is called a *solids-contact* clarifier. The latter arrangement of flash mix, flocculation, and sedimentation were discussed in previous chapters on unit operations.

A solids-contact clarifier is shown in Figure 10.1. The chemicals are introduced into the primary mixing and reaction zone. Here, the fresh reactants are mixed by the swirling action generated by the rotor impeller and also mixed with a return sludge that are introduced under the hood from the clarification zone. The purpose of the return sludge is to provide nuclei that are important for the initiation of the chemical reaction. The mixture then flows up through the sludge blanket where secondary reaction and mixing occur. The reaction products then overflow into the clarification zone, where the clarified water is separated out by sedimentation of the reaction product solids. The clarified water finally overflows into the effluent discharge. The settled sludge from the clarification is drawn off through the sludge discharge pipe.

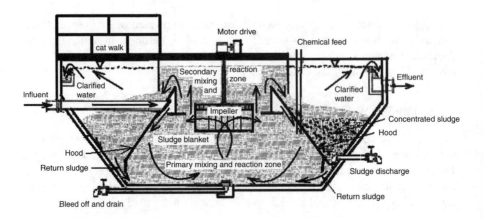

FIGURE 10.1 Solids-contact clarifier. (Courtesy of Infilco Degremont, Inc.)

10.4 THE EQUIVALENT CaCO₃ CONCENTRATION

In the literature, hardness is frequently expressed in terms of $CaCO_3$. Expressing concentrations in terms of $CaCO_3$ can be confusing. For example, when the hardness of water is 60 mg/L as $CaCO_3$, what does this really mean? Related to this question is a second question: Given the concentration of a hardness substance, how is this converted to the equivalent $CaCO_3$ hardness concentration?

To obtain the mass of any substance, the number of equivalents of that substance is multiplied by its equivalent mass. Because the number of equivalents of all substances participating in a given chemical reaction are equal, what differentiates the various species in this chemical reaction are their respective equivalent masses. Thus, to obtain the concentration of 60 mg/L as $CaCO_3$, this equal number of equivalents must have been multiplied by the equivalent mass of calcium carbonate. This section will determine this equivalent mass. This is actually 50 and converting the concentration of any hardness substance to the equivalent $CaCO_3$ hardness concentration is obtained by multiplying the number of equivalents of the substance by 50.

Let $[C_{T,hard}]_{eq}$ represent the total concentration of hardness in equivalents, where the symbol [ ] is read as "the concentration of" and the subscript eq means that the concentration is expressed in terms of equivalents. If the only hardness ions present are calcium and magnesium,

$$[C_{T,hard}]_{eq} = [Ca^{2+}]_{eq} + [Mg^{2+}]_{eq} \tag{10.1}$$

The number of equivalents of one substance in a given chemical reaction is equal to the same number of equivalents of any other substance in this reaction, so it is entirely correct to arbitrarily express the total concentration of hardness in terms of only one of the ions that participates in the chemical reaction. The concentrations of the other ions must then be subsumed in the concentration of this one ion being chosen. For example, if the total hardness is to be expressed in terms of the magnesium ion only, the previous equation will become

$$[C_{T,hard}]_{eq} = [Mg^{2+}]'_{eq} \tag{10.2}$$

In this equation, the part of the total hardness contained in calcium is subsumed in magnesium, $[Mg^{2+}]'_{eq}$. Note the prime. The term $[Mg^{2+}]'_{eq}$ means that it is a concentration in terms of magnesium but that the calcium ion concentration is subsumed in it and expressed in terms of magnesium equivalents. Moreover, since an equivalent of one is equal to the equivalent of another, it is really immaterial under what substance the total equivalents of hardness is expressed. Then, in a similar manner, the total concentration of hardness may also be expressed in terms of the calcium ion alone as follows

$$[C_{T,hard}]_{eq} = [Ca^{2+}]'_{eq} \tag{10.3}$$

In this equation, the part of the total hardness contained in magnesium is now subsumed in calcium, $[Ca^{2+}]'_{eq}$. Again, note the prime. Similar to the term $[Mg^{2+}]'_{eq}$, $[Ca^{2+}]'_{eq}$ means that it is a concentration in terms of calcium but the magnesium ion is subsumed in it and expressed in terms of calcium equivalents. If other hardness ions are present as well, then the concentrations of these ions will also be subsumed in the one particular ion chosen to express the hardness.

When any ion participates in a chemical reaction, it will react to the satisfaction of its ionic charge. For example, when the calcium and magnesium ions participate in a softening reaction, they will react to the satisfaction of their ionic charges of two. Thus, if the chemical reaction is written out, the number of reference species for Ca^{2+} and Mg^{2+} will be found to be two and the respective equivalent masses are then Ca/2 and Mg/2. In terms of molar concentrations, $[Ca^{2+}]_{eq}$ is then equal to $[Ca^{2+}](Ca/Ca/2) = 2[Ca^{2+}]$. By analogy, $[Ca^{2+}]'_{eq} = 2[Ca^{2+}]'$, where $[Ca^{2+}]'$ is now an equivalent molar concentration in terms of the calcium ion that subsumes all concentrations. Also, $[Mg^{2+}]_{eq} = [Mg^{2+}](Mg/Mg/2) = 2[Mg^{2+}]$ and, again, by analogy, $[Mg^{2+}]'_{eq} = 2[Mg^{2+}]'$. $[Mg^{2+}]'$ is the equivalent molar concentration in terms of the magnesium ion that also subsumes all concentrations. Considering Eqs. (10.2) and (10.3) simultaneously and expressing the molar concentrations of the calcium and magnesium hardness in terms of calcium only,

$$[C_{T,hard}]_{eq} = [Mg^{2+}]'_{eq} = [Ca^{2+}]'_{eq} = 2[Ca^{2+}]' \tag{10.4}$$

The term $2[Ca^{2+}]'$ can be converted to $[CaCO_3]'$. This is done as follows: In $CaCO_3$, one mole of $[Ca^{2+}]$ is equal to one mole of $[CaCO_3]$. Hence, $2[Ca^{2+}]' = 2[CaCO_3]'$ and Equation (10.4) becomes

$$[C_{T,hard}]_{eq} = [Mg^{2+}]'_{eq} = [Ca^{2+}]'_{eq} = 2[Ca^{2+}]' = 2[CaCo_3]' \quad \text{or} \tag{10.5}$$

$$[CaCO_3]' = \frac{[C_{T,hard}]_{eq}}{2} = \frac{[Mg^{2+}]'_{eq}}{2} = \frac{[Ca^{2+}]'_{eq}}{2} \tag{10.6}$$

$[CaCO_3]'$ is an equivalent molar concentration expressed in terms of moles of $CaCO_3$. Thus, to get the equivalent calcium carbonate mass concentration, it must be multiplied by $CaCO_3 = 100$. Therefore,

$$[CaCO_3]'(100) = \left(\frac{[C_{T,hard}]_{eq}}{2}\right)(100) = 50[C_{T,hard}]_{eq}$$

$$= \left(\frac{[Mg^{2+}]'_{eq}}{2}\right)(100) = 50[Mg^{2+}]'_{eq}$$

$$= \left(\frac{[Ca^{2+}]'_{eq}}{2}\right)(100) = 50[Ca^{2+}]'_{eq}$$

Or,

$$[C_{T,hard}]_{asCaCO_3} = 50[C_{T,hard}]_{eq} = 50[Mg^{2+}]'_{eq} = 50[Ca^{2+}]'_{eq} \qquad (10.7)$$

Thus, the concentration of hardness expressed in terms of the mass of $CaCO_3$ is equal to the number of equivalents of hardness ($[C_{T,hard}]_{eq}$, $[Mg^{2+}]'_{eq}$, or $[Ca^{2+}]'_{eq}$) divided 2 times the molecular mass of calcium carbonate. Or, simply put, the concentration of hardness expressed in terms of the mass of $CaCO_3$ is equal to the number of equivalents of hardness ($[C_{T,hard}]_{eq}$, $[Mg^{2+}]'_{eq}$, or $[Ca^{2+}]'_{eq}$) times 50.

Note that Equation (10.7) merely states the concentration of hardness in terms of the mass of $CaCO_3$. Although the symbol for calcium carbonate is being used, it does not state anything about the actual concentration of the $CaCO_3$ species present; it is even possible that no species of calcium carbonate exists, but $MgCO_3$ or any other species where the concentrations are simply being expressed in terms of $CaCO_3$.

To apply Equation (10.7), suppose that the concentration of Mg^{2+} in a sample of water is given as 60 mg/L as $CaCO_3$. The meq/L of Mg^{2+} is then equal to $[Mg^{2+}]_{eq} = [Mg^{2+}]'_{eq} = 60/50 = 1.2$ and the concentration of magnesium in mg/L is $1.2(Mg/2) = 1.2(24.3/2) = 14.58$ mg/L.

Take the following reaction and the example of the magnesium ion:

$$Mg(HCO_3)_2 + 2Ca(OH)_2 \rightarrow Mg(OH)_2\downarrow + 2CaCO_3\downarrow + 2HOH$$

In this reaction, the equivalent mass of $CaCO_3$ is $2(CaCO_3)/2 = 100$. Again, in any given chemical reaction, the number of equivalents of all the participating species are equal. Thus, if the number of equivalents of the magnesium ion is 1.2 meq/L, the number of equivalents of $CaCO_3$ also must be 1.2 meq/L. Thus, from this reaction, the calcium carbonate concentration corresponding to the 1.2 meq/L of Mg^{2+} would be $1.2(100) = 120$ mg/L as $CaCO_3$. Or, because the species is really calcium carbonate, it is simply 120 mg/L $CaCO_3$—no more "as."

How is the concentration of 120 mg/L $CaCO_3$ related to the concentration of 60 mg/L as $CaCO_3$ for the magnesium ion? The 60 mg/L is not a concentration of calcium carbonate but a concentration of the magnesium expressed as $CaCO_3$. These two are very different. In the "120," there is really calcium carbonate present, while in the "60," there is none.

Again, expressing the magnesium concentration as 60 mg/L $CaCO_3$ does not mean that there are 60 mg/L of the $CaCO_3$ but that there are 60 mg/L of the ion of magnesium expressed as $CaCO_3$. Expressing the concentration of one substance in terms of another is a normalization. This is analogous to expressing other currencies in terms of the dollar. Go to the Philippines and you can make purchases with the dollar, because there is a normalization (conversion) between the dollar and the peso.

Example 10.1 The concentration of total hardness in a given raw water is found to be 300 mg/L as $CaCO_3$. Calculate the concentration in milligram equivalents per liter.

Solution:

$$[C_{T,hard}]_{meq} = \frac{300}{50} = 6.0 \text{ meq/L} \quad \textbf{Ans}$$

10.5 SOFTENING OF CALCIUM HARDNESS

Calcium hardness may either be carbonate or noncarbonate. The solubility product constant of $CaCO_3$ is $K_{sp} = [Ca^{2+}][CO_3^{2-}] = 5(10^{-9})$ at 25°C. A low value of the K_{sp} means that the substance has a low solubility; a value of $5(10^{-9})$ is very low. Because of this very low solubility, calcium hardness is removed through precipitation of $CaCO_3$. Because there are two types of calcium hardness, there corresponds two general methods of removing it. When calcium is associated with the bicarbonate ion, the hardness metal ion can be easily removed by providing the hydroxide radical. The H^+ of the bicarbonate becomes neutralized by the OH^- provided forming water and the CO_3^{2-} ion necessary to precipitate calcium carbonate. The softening reaction is as follows:

$$Ca(HCO_3)_2 + 2OH^- \rightarrow CaCO_3\downarrow + CO_3^{2-} + 2HOH \quad (10.8)$$

The precipitate, $CaCO_3$, is indicated by a downward pointing arrow, $\downarrow$. As shown, the two bicarbonate species in the reactant side are converted to the two carbonate species in the product side. The carbonate ion shown unpaired will pair with whatever cation the OH^- was with in the reactant side of the reaction.

Once all the bicarbonate ions have been destroyed by the provision of OH^-, they convert to the carbonate ion, as indicated. If the OH^- source contains the calcium ion, the carbonates all precipitate as $CaCO_3$ according to the following softening reaction:

$$Ca^{2+} + CO_3^{2-} \rightarrow CaCO_3\downarrow \quad (10.9)$$

In this case, no carbonate ion is left unpaired, because the calcium ion is present to cause precipitation.

The other method of calcium removal involves the case when the hardness is in the form of the noncarbonate, such as in the form of $CaCl_2$. In this case, a carbonate ion must be provided, whereupon Equation (10.9) will apply, precipitating calcium as calcium carbonate. The usual source of the carbonate ion is soda ash.

10.6 SOFTENING OF MAGNESIUM HARDNESS

As in the case of calcium hardness, magnesium can also be present in the form of carbonate and noncarbonate hardness. The K_{sp} of $Mg(OH)_2$ is a low value of $9(10^{-12})$. Thus, the hardness is removed in the form of $Mg(OH)_2$. To remove the carbonate hardness of magnesium, a source of the OH^- ion is therefore added to precipitate the $Mg(OH)_2$ as shown in the following softening chemical reaction:

$$Mg(HCO_3)_2 + 4OH^- \rightarrow Mg(OH)_2\downarrow + 2CO_3^{2-} + 2HOH \quad (10.10)$$

The carbonate ions in the product side will pair with whatever cation the OH⁻ was with in the reactant side of the reaction. If this cation is calcium, in the form of $Ca(OH)_2$, then the product will again be the calcium carbonate precipitate.

In natural waters, the form of noncarbonate hardness normally encountered is the one associated with the sulfate anion; although, occasionally, large quantities of the chloride and nitrate anions may also be found. The softening reactions for the removal of the noncarbonate hardness of magnesium associated with the possible anions are as follows:

$$MgSO_4 + 2OH^- \rightarrow Mg(OH)_2\downarrow + SO_4^{2-} \qquad (10.11)$$

$$MgCl_2 + 2OH^- \rightarrow Mg(OH)_2\downarrow + 2Cl \qquad (10.12)$$

$$Mg(NO_3)_2 + 2OH^- \rightarrow Mg(OH)_2\downarrow + 2NO_3 \qquad (10.13)$$

As shown in all the previous reactions, the removal of the magnesium hardness, both carbonate and noncarbonate, can use just one chemical. This chemical is normally lime, CaO.

10.7 LIME–SODA PROCESS

As shown by the various chemical reactions above, the chemicals soda ash and lime may be used for the removal of hardness caused by calcium and magnesium. Thus, the *lime–soda process* is used. This process, as mentioned, uses lime (CaO) and soda ash (Na_2CO_3). As the name of the process implies, two possible sets of chemical reactions are involved: the reactions of lime and the reactions of soda ash. To understand more fully what really is happening in the process, it is important to discuss these chemical reactions. Let us begin by discussing the lime reactions.

CaO first reacts with water to form slaked lime, before reacting with the bicarbonate. The slaking reaction is

$$CaO + HOH \rightarrow Ca(OH)_2 \qquad (10.14)$$

After slaking, the bicarbonates are neutralized according to the following reactions:

$$Ca(HCO_3)_2 + Ca(OH)_2 \rightarrow 2CaCO_3\downarrow + 2HOH \qquad (10.15)$$

$$Mg(HCO_3)_2 + 2Ca(OH)_2 \rightarrow Mg(OH)_2\downarrow + 2CaCO_3\downarrow + 2HOH \qquad (10.16)$$

Note that in Equation (10.16) two types of solids are produced: $Mg(OH)_2$ and $CaCO_3$ and that the added calcium ion from the lime that would have produced an added hardness to the water has been removed as $CaCO_3$. Although the hardness ions have been precipitated out, the resulting solids, however, pose a problem of disposal in water softening plants.

Magnesium, whether in the form of the carbonate or noncarbonate hardness, is always removed in the form of the hydroxide. Thus, to remove the total magnesium hardness, more lime is added to satisfy the overall stoichiometric requirements for both the carbonates and noncarbonates. (Later, we will also discuss the requirement of adding more lime to raise the pH.) The pertinent softening reactions for the removal of the noncarbonate hardness of magnesium follow.

$$MgSO_4 + Ca(OH_2) \rightarrow Mg(OH)_2\downarrow + CaSO_4 \qquad (10.17)$$

$$MgCl_2 + Ca(OH)_2 \rightarrow Mg(OH)_2\downarrow + CaCl_2 \qquad (10.18)$$

$$Mg(NO_3)_2 + Ca(OH)_2 \rightarrow Mg(OH)_2\downarrow + Ca(NO_3)_2 \qquad (10.19)$$

As shown from the previous reactions, there is really no net mole removal of hardness that results from the addition of lime. For every mole of magnesium hardness removed [$MgSO_4$, $MgCl_2$, or $Mg(NO_3)_2$], there is a corresponding mole of by-product noncarbonate calcium hardness produced [$CaSO_4$, $CaCl_2$, or $Ca(NO_3)_2$]. Despite the fact that these reactions are useless, they still must be considered, because they will always transpire if the noncarbonate hardness is removed using lime. For this reason, the implementation of the lime-soda process should be such that as much magnesium as possible is left unremoved and rely only on the removal of calcium to meet the desired treated water hardness. If the desired hardness level is not met by the removal of only the calcium ions, then removal of the magnesium may be initiated. This will entail the use of lime followed by the possible addition of soda ash to remove the resulting noncarbonate hardness of calcium. As shown by Eqs. (10.17), (10.18), and (10.19), soda ash is needed to remove the by-product calcium hardness formed from the use of the lime.

As noted before, the calcium ion is removed in the form of $CaCO_3$. This is the reason for the use of the second chemical known as soda ash for the removal of the noncarbonate hardness of calcium. Thus, another set of chemical reactions involving soda ash and calcium would have to written. The pertinent softening reactions are as follows:

$$CaSO_4 + Na_2CO_3 \rightarrow CaCO_3\downarrow + Na_2SO_4 \qquad (10.20)$$

$$CaCl_2 + Na_2CO_3 \rightarrow CaCO_3\downarrow + 2NaCl \qquad (10.21)$$

$$Ca(NO_3)_2 + Na_2CO_3 \rightarrow CaCO_3\downarrow + 2NaNO_3 \qquad (10.22)$$

Some of the calcium hardness of these reactions would be coming from the by-product noncarbonate hardness of calcium that results if lime were added to remove the noncarbonate hardness of magnesium.

It is worth repeating that soda ash is used for two purposes only: to remove the original calcium noncarbonate hardness in the raw water and to remove the by-product calcium noncarbonate hardness that results from the precipitation of the noncarbonate hardness of magnesium. It is also important to remember that by using lime the carbonate hardness of magnesium does not produce any calcium noncarbonate hardness. It is only the noncarbonate magnesium that is capable of producing the by-product calcium noncarbonate hardness when lime is used.

10.7.1 CALCULATION OF STOICHIOMETRIC LIME REQUIRED IN THE LIME–SODA PROCESS

Because the raw water is exposed to the atmosphere, there will always be some CO_2 dissolved in it. As will be shown later, carbon dioxide also consumes lime. We will, however, ignore this requirement for the moment, and discuss it in the latter part of this chapter. Ignoring carbon dioxide, the amount of lime needed comes from the requirement to remove the carbonate hardness of calcium and the requirements to remove both the carbonate and the noncarbonate hardness of magnesium. We will first calculate the amount of lime required for the removal of the carbonate hardness of calcium.

Let M_{CaHCO_3} be the mass of $Ca(HCO_3)_2$ in the carbonate hardness of calcium to be removed. From Equation (10.15), taking the positive oxidation state, the number of reference species is 2 moles of positive charges. Thus, the equivalent mass of calcium bicarbonate is $Ca(HCO_3)_2/2$ and the number of equivalents of the M_{CaHCO_3} mass of calcium bicarbonate is $M_{CaHCO_3}/(Ca(HCO_3)_2/2)$. This is also the same number of equivalents of CaO (or slaked lime) used to neutralize the bicarbonate.

From Eqs. (10.14) and (10.15), the equivalent mass of CaO is CaO/2. Thus, the mass of CaO, $M_{CaOCaHCO_3}$, needed to neutralize the calcium bicarbonate is $M_{CaHCO_3}/(Ca(HCO_3)_2/2)(CaO/2)$. Let M_{TCaHCO_3} be the total calcium bicarbonate in the raw water and f_{CaHCO_3} be its fractional removal. Thus, $M_{CaHCO_3} = f_{CaHCO_3} M_{TCaHCO_3}$ and

$$M_{CaOCaHCO_3} = \frac{f_{CaHCO_3} M_{TCaHCO_3}}{Ca(HCO_3)_2/2}(CaO/2) = 0.35 f_{CaHCO_3} M_{TCaHCO_3} \quad (10.23)$$

Now, calculate the amount of lime needed to remove $Mg(HCO_3)_2$. Let M_{MgHCO_3} be the mass of $Mg(HCO_3)_2$ in the carbonate hardness of magnesium to be removed. From Equation (10.16), again taking the positive oxidation state, the number of reference species is 2 moles of positive charges. Thus, the equivalent mass of magnesium bicarbonate is $Mg(HCO_3)_2/2$ and the number of equivalents of the M_{MgHCO_3} mass of magnesium bicarbonate is $M_{MgHCO_3}/(Mg(HCO_3)_2/2)$. This is also the same number of equivalents of CaO (or slaked lime) used to neutralize the magnesium bicarbonate.

From Eqs. (10.14) and (10.16), the equivalent mass of CaO is 2CaO/2. Thus, the mass of CaO, $M_{CaOMgHCO_3}$, needed to neutralize the magnesium bicarbonate is $M_{MgHCO_3}/(Mg(HCO_3)_2/2)CaO$. Let M_{TMgHCO_3} be the total magnesium bicarbonate in the raw water and f_{MgHCO_3} be its fractional removal. Thus, $M_{MgHCO_3} = f_{MgHCO_3} M_{TMgHCO_3}$ and

$$M_{CaOMgHCO_3} = \frac{f_{MgHCO_3} M_{TMgHCO_3}}{Mg(HCO_3)_2/2} CaO = 0.77(f_{MgHCO_3} M_{TMgHCO_3}) \quad (10.24)$$

The removal of the noncarbonate hardness of magnesium using lime as the precipitant produces a corresponding amount of noncarbonate hardness of calcium as a by-product. This by-product requires the use of soda ash for its removal. Thus, the amount of lime needed for the removal of this magnesium noncarbonate hardness will be discussed in conjunction with the determination of the stoichiometric soda ash required to remove the by-product calcium noncarbonate hardness that results.

10.7.2 Key to Understanding Subscripts

So far, we have used mass variables such as M_{TCaHCO_3}, M_{CaHCO_3}, $M_{CaOMgHCO_3}$ and M_{MgHCO_3} and fractional variables such as f_{MgHCO_3} and f_{CaHCO_3}. Later on, we will also be using other mass variables such as $M_{solidsNonCarb}$, and M_{CaCO_3MgCa} and fractional variables such as f_{MgCa} and f_{Ca}. As we continue to develop equations, it would be found that ascertaining the meaning of the subscripts of the variables would become difficult. Thus, a technique must be developed to aid in remembering.

First, for the mass variables, M refers to the mass of the type of species involved. The subscript of M is a string of words. In this string, T may or may not be present. If the M of a given species can be both a partial mass and a total mass, the T would be present as the first one in the subscipt to represent the total mass. Then if it is not present, the mass is a partial mass. Partial here means a part of the total. For example, consider M_{TCaHCO_3} and M_{CaHCO_3}. As shown, the M's in these symbols can be both a partial mass and a total mass. Thus, the first one refers to the total mass and the second refers to the partial mass of the type of species involved.

The next part of the string subscript is $CaHCO_3$. This refers to the type of species involved. This is not really the correct formula for calcium bicarbonate, but because it is used as a subscript, it can be easily taken to mean calcium bicarbonate. Thus, the symbols M_{TCaHCO_3} and M_{CaHCO_3} refer to the total mass of calcium bicarbonate and the partial mass of calcium bicarbonate, respectively. Calcium bicarbonate, $Ca(HCO_3)_2$, is said to be the type of species involved.

If M can only have a total mass, the T is not used. For example, the total solids, M_{solids}, can come from the solids produced from carbonate hardness, $M_{solidsCarb}$, from the solids produced from noncarbonate hardness, $M_{solidsNonCarb}$, and from the calcium carbonate solids produced from carbon dioxide, $M_{CaCO_3CO_2}$. In this case, since the designation for the components of total solids use different subscripts, the M of total solids can only have the total mass and therefore M_{solids} is used instead of $M_{Tsolids}$. No confusion occurs, however, if $M_{Tsolids}$ is used, because this would definitely refer to the total.

Not counting T, a second portion of the string subscript (the third portion, if T is present) may also be present. This will then refer to the reason for the existence of the type of mass involved. For example, consider $M_{CaCO_3CO_2}$, mentioned previously. In this symbol, CO_2 is the reason for the existence of the mass of $CaCO_3$. Thus, $M_{CaCO_3CO_2}$ is the mass of calcium carbonate produced from carbon dioxide.

Another example where the second portion of a string subscript is the reason for existence of the type of mass involved is $M_{CaOCaHCO_3}$. In this symbol, $Ca(HCO_3)_2$ is the reason for the existence of CaO. In the pertinent chemical reaction, calcium oxide is not produced from calcium bicarbonate, but, rather, calcium oxide is used because of the bicarbonate. Nevertheless, the bicarbonate is still the reason for the existence of calcium oxide.

The portions of a string subscript can be easily identified, since they conspicuously stand out. For example, portions $CaCO_3$ and CO_2 conspicuously stand out in $M_{CaCO_3CO_2}$. $CaCO_3$ and CO_2 are said to be portions of the string subscript of $M_{CaCO_3CO_2}$.

In some situations, a string subscript may appear to have several portions conspicuously standing out. For example, consider $M_{solidsNonCarb}$. This symbol seems to have three portions standing out; however, *solidsNonCarb* does not refer to three different entities but to only one and this is the noncarbonate solids. Thus, $M_{solidsNonCarb}$ does not have three subscript portions.

As far as the masses are concerned, consider this last example, M_{CaCO_3MgCa}. This symbol indicates that the reason for the existence of the $CaCO_3$ solid is Mg, however, a next symbol Ca also appears. The appearance of Ca simply indicates that Mg, as the reason for the existence of $CaCO_3$, bears a relationship to calcium. This relationship can only happen when the noncarbonate hardness of magnesium is involved, for this will produce the corresponding noncarbonate hardness of calcium—the relationship. An important thing to remember, as far as our method of subscripting is concerned, is that whenever Mg and Ca are used as subscripts, they refer to the *noncarbonate* form of their respective hardness. For example, M_{TCa} refers to the total noncarbonate calcium hardness and M_{Ca} refers to a partial noncarbonate calcium hardness. As another example, consider M_{CaCO_3MgCa}. In this symbol, noncarbonate Mg is the reason for the existence of the calcium carbonate. This magnesium, of course, produced the noncarbonate Ca (the relationship) for the eventual production of the calcium carbonate. In other words, to repeat, whenever Ca and Mg are used as subscripts, they are intended to refer only to their respective noncarbonate form of hardness.

Now, consider the fractional variables. For example, consider f_{MgCa} and f_{Ca}. As in the mass variables, the first subscript of the fractional variable refers to the type of fraction of the mass. Thus, in f_{MgCa}, the type of fraction is the fraction of Mg; and, considering the fact that when Mg and Ca are used as subscripts, they refer to the noncarbonate forms of hardness, f_{MgCa} stands for the fraction of the noncarbonate form of magnesium hardness, with Ca reminding about the relationship. f_{MgCa} could just simply be written as f_{Mg} and they would be the same, but Ca is there, again, just as a reminder that it is also involved in the chemistry of the reaction. By the convention we have just discussed, f_{Ca} stands for the fraction of the noncarbonate form of calcium hardness. As a further example, to what does f_{CaHCO_3} refer? The is the fraction of the calcium carbonate (bicarbonate) hardness.

10.7.3 Calculation of Stoichiometric Soda Ash Required

The amount of soda ash needed comes from the requirement to remove the noncarbonate hardness of calcium. In addition, if the noncarbonate hardness of magnesium was precipitated using lime, additional soda ash will also be required to remove the by-product noncarbonate hardness of calcium that results from the magnesium precipitation (if additional hardness removal is desired).

Let M_{Ca} be the mass of the calcium species to be removed from the noncarbonate hardness of calcium. From Eqs. (10.20), (10.21), or (10.22), taking the positive oxidation state, the number of reference species is 2 moles of positive charges for each of these reactions. Thus, the equivalent mass of calcium is Ca/2 and the number of equivalents of the M_{Ca} mass of calcium is $M_{Ca}/(Ca/2)$. This is also the same number of equivalents of Na_2CO_3 needed to precipitate the noncarbonate hardness of calcium to $CaCO_3$. Again, from Eqs. (10.20), (10.21), or (10.22), the equivalent mass of Na_2CO_3 is $Na_2CO_3/2$. Thus, the mass of soda ash, $M_{sodAshCa}$, needed to precipitate the calcium species from the noncarbonate hardness of calcium is $M_{Ca}/(Ca/2)(Na_2CO_3/2)$. Let M_{TCa} be the total calcium in the noncarbonate hardness of calcium in the raw water and f_{Ca} be its fractional removal. Thus, $M_{Ca} = f_{Ca}M_{TCa}$ and

$$M_{sodAshCa} = \frac{f_{Ca}M_{TCa}}{Ca/2}(Na_2CO_3/2) = 2.65 f_{Ca}M_{TCa} \qquad (10.25)$$

Note that in $M_{sodAshCa}$, Ca is the reason for the use of soda ash.

To calculate the soda ash requirement for the calcium noncarbonate hardness by-product, let M_{MgCa} represent the mass of the magnesium species that precipitates and results in the production of the additional calcium hardness cation, where the Ca, again, is written as a reminder. Refer to Eqs. (10.17), (10.18), and (10.19) to see how the calcium hardness is produced from the precipitation of noncarbonate magnesium. The number of equivalents of the calcium hardness produced is equal to the number of equivalents of the noncarbonate hardness of magnesium precipitated [which is equal to $M_{MgCa}/(Mg/2)$, where Mg/2 is the equivalent mass of Mg as obtained from Eqs. (10.17), (10.18), and (10.19)]. Because this is calcium hardness, we already have the method of determining the amount of soda ash needed to remove it as shown in Equation (10.25). Letting this amount be $M_{sodAshMgCa}$, $M_{sodAshMgCa}$ is equal to $M_{MgCa}/(Mg/2)(Na_2CO_3/2)$. Let M_{TMg} be the total magnesium in the noncarbonate hardness of magnesium and f_{MgCa} be its fractional removal. Thus, $M_{MgCa} = f_{MgCa}M_{TMg}$, with this and Equation (10.25),

$$M_{sodAshMgCa} = \frac{f_{MgCa}M_{TMg}}{Mg/2}(Na_2CO_3/2) = 4.36 f_{MgCa}M_{TMg} \qquad (10.26)$$

The mass of magnesium species from the noncarbonate hardness of magnesium that precipitates also needs additional lime for its precipitation as magnesium hydroxide. Note that this is the same mass of magnesium that produced the additional by-product calcium hardness. As mentioned above, this mass is M_{MgCa}. Again, from Eqs. (10.17), (10.18), and (10.19), the number of equivalents of the M_{MgCa} mass of magnesium is $M_{MgCa}/(Mg/2)$. This is the same number of equivalents of the lime needed to precipitate $Mg(OH)_2$. From the preceding equations, the equivalent mass of CaO is CaO/2. Let the mass of lime needed to precipitate the $Mg(OH)_2$ resulting from M_{MgCa} mass of magnesium be $M_{CaOMgCa}$. Thus, $M_{CaOMgCa}$ is $M_{MgCa}/(Mg/2)(CaO/2)$, but because $M_{MgCa} = f_{MgCa}M_{TMg}$,

$$M_{CaOMgCa} = \frac{f_{MgCa}M_{TMg}}{Mg/2}(CaO/2) = 2.30 f_{MgCa}M_{TMg} \qquad (10.27)$$

10.7.4 CALCULATION OF SOLIDS PRODUCED

The solids produced in water softening plants, if not put to use, pose a disposal problem. Conceptually, because of their basic nature, they can be used in absorption towers that use alkaline solutions to scrub acidic gas effluents. These solids, as far as the lime-soda process is concerned, come from the solids produced in the removal of (1) the carbonate hardness and (2) the noncarbonate hardness.

Let us first derived the equations for solids production from the removal of the carbonate hardness. The sources of these solids are the $CaCO_3$ produced from the removal of the carbonate hardness of calcium [Equation (10.15)], the $CaCO_3$ produced from the removal of the carbonate hardness of magnesium [Equation (10.16)], and the $Mg(OH)_2$ produced also from the removal of the carbonate hardness of magnesium [Equation (10.16)].

The amount of calcium bicarbonate removed is $M_{CaHCO_3} = f_{CaHCO_3}M_{TCaHCO_3}$. Thus, using Equation (10.15), the corresponding production of $CaCO_3$ is

$$M_{CaCO_3CaHCO_3} = \frac{2CaCO_3}{Ca(HCO_3)_2}f_{CaHCO_3}M_{TCaHCO_3} = 1.23 f_{CaHCO_3}M_{TCaHCO_3} \qquad (10.28)$$

$M_{CaCO_3CaHCO_3}$ is the calcium carbonate produced. Note that the second subscript, $CaHCO_3$, is the reason for the existence of the carbonate.

Now, the amount of magnesium bicarbonate removed is $M_{MgHCO_3} = f_{MgHCO_3}M_{TMgHCO_3}$. From Equation (10.16), the corresponding production of $CaCO_3$ is

$$M_{CaCO_3MgHCO_3} = \frac{2CaCO_3}{Mg(HCO_3)_2}f_{MgHCO_3}M_{TMgHCO_3} = 1.37 f_{MgHCO_3}M_{TMgHCO_3} \qquad (10.29)$$

$M_{CaCO_3MgHCO_3}$ is the calcium carbonate produced. Note that, similarly with $M_{CaCO_3CaHCO_3}$, the second subscript, $MgHCO_3$, is the reason for the existence of the carbonate. Also, from the same Equation (10.16), the amount of $Mg(OH)_2$ solids produced is

$$M_{MgOHMgHCO_3} = \frac{Mg(OH)_2}{Mg(HCO_3)_2}f_{MgHCO_3}M_{TMgHCO_3} = 0.40 f_{MgHCO_3}M_{TMgHCO_3} \qquad (10.30)$$

$M_{MgOHMgHCO_3}$ is the amount of the $Mg(OH)_2$ solids produced. Also, note the reason for its existence, which is $MgHCO_3$.

Combining Eqs. (10.28), (10.29), and (10.30), the total mass of solids, $M_{soildsCarb}$, produced from the removal of the carbonate hardness using the lime–soda ash process is therefore

$$M_{soildsCarb} = M_{CaCO_3CaHCO_3} + M_{CaCO_3MgHCO_3} + M_{MgOHMgHCO_3} \qquad (10.31)$$

Now, let us derive the equations for solids production from the removal of the noncarbonate hardness. The sources of these solids are the $Mg(OH)_2$ produced from the removal of the M_{MgCa} mass of magnesium [Eqs. (10.17), (10.18), and (10.19)]; the $CaCO_3$ produced from the removal of the M_{Ca} mass of calcium [Eqs. (10.20), (10.21), and (10.22)]; and the $CaCO_3$ produced from the M_{MgCa} mass of magnesium [Eqs. (10.17), (10.18), and (10.19)] and [Eqs. (10.20), (10.21), and (10.22)].

Equations (10.17), (10.18), and (10.19) show the production of the $Mg(OH)_2$ solids. From these equations, the amount of hydroxide solids, $M_{MgOHMgCa}$, produced from the M_{MgCa} mass of magnesium, is

$$M_{MgOHMgCa} = \frac{Mg(OH)_2}{Mg}(f_{MgCa}M_{TMg}) = 2.40f_{MgCa}M_{TMg} \qquad (10.32)$$

The amount of noncarbonate calcium removed is $M_{Ca}=f_{Ca}M_{TCa}$. Thus, Eqs. (10.20), (10.21), and (10.22), the corresponding production of $CaCO_3$ is

$$M_{CaCO_3Ca} = \frac{CaCO_3}{Ca}f_{Ca}M_{TCa} = 2.49f_{Ca}M_{TCa} \qquad (10.33)$$

M_{CaCO_3Ca} is the amount of the $CaCO_3$ solids produced. Note that Ca is the reason for its existence.

The mass of $CaCO_3$ solids, M_{CaCO_3MgCa}, produced from the removal of the $M_{MgCa}=f_{MgCa}M_{TMg}$ mass of magnesium is obtained as follows: From Eqs. (10.17), (10.18), and (10.19), one mole of Mg produces one mole of Ca. From Eqs. (10.20), (10.21), and (10.22), one mole of Ca also produces the same one mole of $CaCO_3$. Thus,

$$M_{CaCO_3MgCa} = \frac{CaCO_3}{Mg}f_{MgCa}M_{TMg} = 4.11f_{MgCa}M_{TMg} \qquad (10.34)$$

From Eqs. (10.32), (10.33), and (10.34), the total mass of solids, $M_{soildsNonCarb}$, produced from the removal of the noncarbonate hardness is

$$M_{solidsNonCarb} = M_{MgOHMgCa} + M_{CaCO_3Ca} + M_{CaCO_3MgCa} \qquad (10.35)$$

10.8 ORDER OF REMOVAL

Suppose that there are four gram-equivalents of Ca^{2+}, one gram-equivalent of Mg^{2+}, and 2.5 gram-equivalents of HCO_3^-. Suppose further that the decision has been made to use lime. The question is which reaction takes precedence, Equation (10.15) or Equation (10.16)? Assuming the former takes precedence, the bicarbonate species present will be composed of 2.5 gram-equivalent of $Ca(HCO_3)_2$ and none of the $Mg(HCO_3)_2$. The amount of lime needed to remove this bicarbonate is 2.5(CaO/2) = 70 g. Assuming the latter takes precedence, the bicarbonate species will now be composed of one gram-equivalent of $Mg(HCO_3)_2$ and 1.5 gram-equivalent of $Ca(HCO_3)_2$. The corresponding amount of lime needed to remove the bicarbonates is 1.0(2CaO/ 2) + 1.5(CaO/2) = 98 g. These two simple calculations produce two very different results, so knowledge of the order of precedence of the reactions is very important. The order of precedence can be judged from the solubility of the precipitates, namely: $CaCO_3$ and $Mg(OH)_2$.

The K_{sp} for $CaCO_3$ is $5(10^{-9})$ at 25°C, while that for $Mg(OH)_2$ is $9(10^{-12})$, also at 25°C. Consider calcium carbonate, first. Before the solid is formed, the product of the concentrations of the calcium ions and the carbonate ions must be, at least, equal to $5(10^{-9})$. Let x be the concentration of the calcium ion; thus, x will also be the concentration of the carbonate ion. Therefore,

$$x^2 = 5(10^{-9}) \Rightarrow x = 0.000071 \text{ gmole/L}$$

Now, consider the magnesium hydroxide. Again, before the solid is formed, the product of the concentrations of the magnesium ions and the hydroxide ions must be, at least, equal to $9(10^{-12})$. In magnesium hydroxide, there are two hydroxide ions to one of the magnesium ion. Thus, letting y be the concentration of the magnesium ions, the concentration of the hydroxide ions is $2y$. Therefore,

$$y(y^2) = 9(10^{-12}) \Rightarrow y = 0.00021 \text{ gmole/L}$$

Now, comparing the two answers, $CaCO_3$ is less soluble than $Mg(OH)_2$, because it only needs 0.000071 gmole/L to precipitate it compared to 0.00021 gmole/L for the case of $Mg(OH)_2$. Therefore, calcium carbonate will precipitate first and Equation (10.15) takes precedence over Equation (10.16). In this example, the answer would be 70 g instead of 98 g.

10.9 ROLE OF CO₂ IN REMOVAL

For the removal of the hardness of calcium and magnesium, the effect of carbon dioxide on the amounts of sludge produced and chemicals used were not considered; but because the type of raw water treated is exposed to the atmosphere, CO_2 always dissolves in it. Therefore, in addition to the chemical reactions that have been portrayed, the precipitant chemicals must also react with carbon dioxide. This means that more lime than indicated in previous reactions would be needed. The reaction that portrays the consumption of lime by carbon dioxide is as follows:

$$CO_2 + Ca(OH)_2 \rightarrow CaCO_3\downarrow + H_2O \qquad (10.36)$$

As shown in Equation (10.36), solids of $CaCO_3$ are produced. Let M_{CO_2} be the mass of carbon dioxide in the raw water. Therefore, letting $M_{CaCO_3CO_2}$ be the carbonate solids produced from the dissolved carbon dioxide,

$$M_{CaCO_3CO_2} = \frac{CaCO_3}{CO_2}(M_{CO_2}) = 2.27(M_{CO_2}) \qquad (10.37)$$

Also, from Equation (10.36), the lime needed, M_{CaOCO_2}, to react with the carbon dioxide in the raw water is

$$M_{CaOCO_2} = \frac{CaO}{CO_2}(M_{CO_2}) = 1.27(M_{CO_2}) \qquad (10.38)$$

10.10 EXCESS LIME TREATMENT AND OPTIMUM OPERATIONAL pH

Refer to Figure 10.2. As shown, the lowest solubility of Ca^{2+} in equilibrium with $CaCO_3$ is at a pH of approximately 9.3. This represents the optimum pH for adding lime or soda ash to precipitate calcium as $CaCO_3$. Raising the pH to about 11 practically dissolves all the $CaCO_3$. This is illustrated in the reaction below. The anion species responsible for raising the pH to 11 is the OH^-. Thus, the chemical reaction is

$$CaCO_3 + OH^- \rightarrow Ca(OH)_2 + CO_3^{2-}$$

$$Ca^{2+} + OH^- \rightleftharpoons CaOH^+ \qquad K_{sp} = \frac{\{Ca^{2+}\}\{OH^-\}}{CaOH^+} = 10^{-1.49} \qquad (10.39)$$

This means that, in order to precipitate the calcium carbonate optimally, the pH should be maintained at the vicinity of 9.3 and that it should not be raised to pH 11. As shown in the chemical reaction, the calcium ion is converted into the complex $CaOH^+$, which has a relatively high K_{sp} of $10^{-1.49}$, indicating the calcium carbonate is practically dissolved when the pH is raised to 11.

Also, from the figure, the lowest solubility of Mg^{2+} in equilibrium with $Mg(OH)_2$ occurs at about pH 10.4; raising the pH above this value does not affect the solubility, since the solubility is already zero at pH 10.4. At this value of pH, the solubility of Ca^{2+} in equilibrium with $CaCO_3$ is not affected very much. Also, lowering the pH to below 9.8 practically dissolves the $Mg(OH)_2$. Thus, these two limiting conditions leave only a very small window of pH 9.8 to 10.4 for an optimal concurrent precipitation of Ca^{2+} and Mg^{2+}. Above pH 10.4, the precipitation of $CaCO_3$ suffers and below pH 9.8, the precipitation of $Mg(OH)$ suffers.

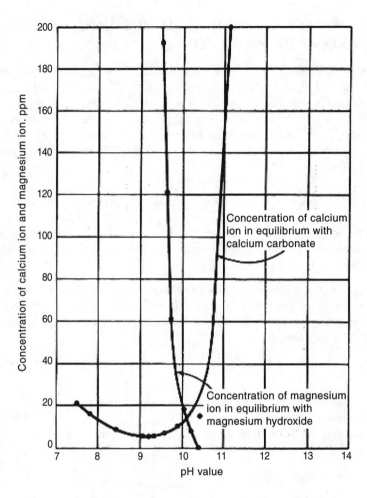

FIGURE 10.2 Concentrations of calcium and magnesium ions in equilibrium with the calcium carbonate and magnesium hydroxide solids, respectively. (From Powell, S. T. (1954). *Water Conditioning for Industry.* McGraw-Hill, New York. With permission.)

It has been found experimentally that adding lime to satisfy the stoichiometric amount for precipitating $Mg(OH)_2$ does not raise the pH value to 10.4. Lime dissolves in water according to the following reaction: $CaO + HOH \rightarrow Ca(OH)_2 \rightleftharpoons Ca^{2+} + 2OH^-$. Empirically, based on this reaction (equivalent mass = $CaO/2$), 1.0 milligram-equivalent per liter of water of excess alkalinity over the computed stoichiometric amount is needed to raise the pH to 10.4. This means that this amount must be added if the pH is to be maintained at this level. Letting $M_{CaOExcess}$ be the total amount of excess lime and V be the volume of water treated in cubic meters,

$$M_{CaOExcess} = \frac{1.0(CaO/2)(1000)V}{1000(1000)} = 0.028V \text{ kg} \qquad (10.40)$$

10.11 SUMMARY OF CHEMICAL REQUIREMENTS AND SOLIDS PRODUCED

To summarize, let M_{CaO} be the total lime requirement; M_{sodAsh} be the total soda ash requirement; and M_{solids} be the total solids produced. M_{CaO} is equal to the amount of lime used for the removals of the carbonate hardness of calcium ($M_{CaOCaHCO_3}$), the carbonate hardness of magnesium ($M_{CaOMgHCO_3}$), the noncarbonate hardness of magnesium ($M_{CaOMgCa}$), the requirement to neutralize the dissolved carbon dioxide (M_{CaOCO_2}), and the requirement to raise the pH to 10.4 for the precipitation of $Mg(OH)_2$. Thus,

$$M_{CaO} = M_{CaOCaHCO_3} + M_{CaOMgHCO_3} + M_{CaOMgCa} + M_{CaOCO_2} + M_{CaOExcess} \quad (10.41)$$

Note: The unit of $M_{CaOExcess}$ is kilograms. Thus, in order for the terms of this equation to be dimensionally equivalent to $M_{CaOExcess}$, they must all be expressed in terms of kilograms.

M_{sodAsh} is equal to the amount of soda ash needed to precipitate the noncarbonate hardness of calcium and the amount of soda ash needed to precipitate the noncarbonate hardness of calcium produced from the precipitation of the noncarbonate hardness of magnesium. Thus,

$$M_{sodAsh} = M_{sodAshCa} + M_{sodAshMgCa} \quad (10.42)$$

M_{solids} is equal to the solids produced from the removals of the carbonate and the noncarbonate hardness ($M_{solidsCarb}$ and $M_{solidsNonCarb}$, respectively) and the solids produced from the neutralization of the dissolved carbon dioxide using lime ($M_{CaCO_3CO_2}$). Thus,

$$M_{solids} = M_{solidsCarb} + M_{solidsNonCarb} + M_{CaCO_3CO_2} \quad (10.43)$$

10.12 SLUDGE VOLUME PRODUCTION

The mass of solids precipitated enmeshes in an enormous amount of water. As these solids are flocculated and settled, they retain extraordinary amounts of water resulting in a huge volume of sludge to be disposed of. The amount of the solids is negligible compared to the total amount of sludge. In fact, sludges contains approximately 99% water. Let f_{solids} be the fraction of solids in the sludge. Then the mass of sludge M_{sludge} is M_{solids}/f_{solids}. If S_{sl} is the specific gravity of the sludge, its volume V_{sl} in cubic meters is

$$V_{sl} = \frac{M_{solids}}{f_{solids}S_{sl}(1000)} = \frac{M_{sludge}}{S_{sl}(1000)} \quad (10.44)$$

The 1,000 in the denominator is the mass density of water in kg/m^3 at the temperature of 4°C used as the reference temperature in the definition of specific gravity. All units in this equation must be in the mks (meter-kilogram-second) system.

10.13 CHEMICAL SPECIES IN THE TREATED WATER

In addition to the H^+ and OH^- species, a great majority of ions present in the treated water are the cations Ca^{2+}, Mg^{2+}, and Na^+ and the anions SO_4^{2-} and HCO_3^-. Occasionally, the anions Cl^- and NO_3^- may also be present when the hardness cations in natural waters are associated with these anions.

The effluent water from the hardness removal reactor is basic and tends to deposit scales in distribution pipes. For this reason, this water should be stabilized. Stabilization is normally done using carbon dioxide, a process called *recarbonation*. Stabilization using carbon dioxide affects the concentration of the bicarbonate ion in the treated water. The concentrations of the SO_4^{2-}, Cl^- and NO_3^- ions are not affected, however, because they do not react with carbon dioxide. Their concentrations remain the same as when they were in the influent to the treatment plant. The original cation Na^+ from the influent raw water is also not affected for the same reason that it does not react with carbon dioxide. Na^+ is, however, introduced with the soda ash.

Let $[SO_4^{2-}]_{meq,inf}$, $[Cl^-]_{meq,inf}$, and $[NO_3^-]_{meq,inf}$ be the concentrations of the indicated ions in milligram equivalents per liter in the influent to the reactor. Also, let $[SO_4^{2-}]_{meq}$, $[Cl^-]_{meq}$, and $[NO_3^-]_{meq}$ be the concentrations of the ions in the treated water. Thus,

$$[SO_4^{2-}]_{meq} = [SO_4^{2-}]_{meq,inf}$$

$$[Cl^-]_{meq} = [Cl^-]_{meq,inf}$$

$$[NO_3^-]_{meq} = [NO_3^-]_{meq,inf}$$

10.13.1 LIMITS OF TECHNOLOGY

The precipitation of $CaCO_3$ and $Mg(OH)_2$ is never complete. At $0°C$, the solubility of $CaCO_3$ is 15 mg/L and at $25°C$, it is 14 mg/L. These solubility values imply that no matter how much precipitant is applied to the water, there will always remain some calcium and carbonate ions which, in the aggregate, amounts to 14 to 15 mg/L of $CaCO_3$ dissolved in the treated water.

For $Mg(OH)_2$, its solubility at $0°C$ is 17 mg/L as $CaCO_3$; at $18°C$ it is 15.5 mg/L as $CaCO_3$. As in the case of $CaCO_3$, some ions of magnesium and, of course, the hydroxide will always remain dissolved in the treated effluent no matter how much precipitant is employed. Operationally, let us adopt the following figures as the limit of technology: $CaCO_3 = 15$ mg/L; $Mg(OH)_2 = 16$ mg/L as $CaCO_3$. This brings a total to $15 + 16 = 31$ mg/L as $CaCO_3$.

10.13.2 CONCENTRATION OF Ca^{2+}

The concentration of the calcium ion in the treated water comes from the calcium bicarbonate not precipitated, the calcium from the noncarbonate hardness of calcium not precipitated, and the calcium that results from the precipitation of the noncarbonate hardness of magnesium. Note that, as the treated water is recarbonated, the 15 mg/L of $CaCO_3$ from the limit of technology converts to the bicarbonate form,

$$CaCO_3 + H_2CO_3(CO_2 + HOH \rightarrow H_2CO_3) \rightarrow Ca(HCO_3)_2 \qquad (10.45)$$

From this reaction, the number of milligram equivalents per liter of Ca^{2+} and HCO_3^- are $15/(CaCO_3/2) = 0.3$, respectively.

Recall that $M_{CaHCO_3}(= f_{CaHCO_3}M_{TCaHCO_3})$ mass of calcium bicarbonate from the carbonate hardness of calcium was precipitated. Let this mass be measured in kilograms. Also, let $[Ca(HCO_3)_2]_{mgnot}$, in mg/L, be the concentration of calcium bicarbonate not precipitated. Thus,

$$[Ca(HCO_3)_2]_{mgnot} = \frac{M_{TCaHCO_3} - f_{CaHCO_3}M_{TCaHCO_3}}{V}\left[\frac{1000(1000)}{1000}\right]$$

$$= 1000\frac{M_{TCaHCO_3}(1 - f_{CaHCO_3})}{V} \qquad (10.46)$$

V is the volume of water treated in cubic meters.

$Ca(HCO_3)_2$ ionizes to $Ca(HCO_3)_2 \rightarrow Ca^{2+} + 2HCO_3^-$; thus, the equivalent mass of calcium bicarbonate is $Ca(HCO_3)_2/2$. The calcium bicarbonate not precipitated therefore produces $[Ca(HCO_3)_2]_{mgnot}/(Ca(HCO_3)_2/2) = 0.0123([Ca(HCO_3)_2]_{mgnot}) = 0.0123[1000M_{TCaHCO_3}(1 - f_{CaHCO_3})/V] = 12.3[M_{TCaHCO_3}(1 - f_{CaHCO_3})/V]$ milligram equivalents per liter of Ca^{2+} and HCO_3^-, respectively.

Now, also recall that $M_{Ca} (= f_{Ca}M_{TCa})$ mass of calcium from the noncarbonate hardness of calcium precipitated. As before, let this mass be measured in kilograms. Also, let $[Ca^{2+}]_{mgnot}$, in mg/L, be the concentration of calcium not precipitated. Thus,

$$[Ca^{2+}]_{mgnot} = \frac{M_{TCa} - f_{Ca}M_{TCa}}{V}\left[\frac{1000(1000)}{1000}\right] = 1000\left[\frac{M_{TCa}(1 - f_{Ca})}{V}\right] \qquad (10.47)$$

Consider $CaSO_4$ as representing the noncarbonate hardness of calcium species. This ionizes as $CaSO_4 \rightarrow Ca^{2+} + SO_4^{2-}$; thus, the equivalent mass of calcium in the noncarbonate hardness of calcium is $Ca/2$. The calcium not precipitated therefore produces $[Ca^{2+}]_{mgnot}/(Ca/2) = 0.050[Ca^{2+}]_{mgnot} = 50.0 [M_{TCa} (1 - f_{Ca})/V]$ milligram equivalents of Ca^{2+}.

Summing all the calcium ion concentrations, the total concentration $[Ca^{2+}]_{meq}$ is

$$[Ca^{2+}]_{meq} = 12.3\left[\frac{M_{TCaHCO_3}(1 - f_{CaHCO_3})}{V}\right] + 50.0\left[\frac{M_{TCa}(1 - f_{Ca}) + f_{MgCa}M_{TMg}}{V}\right]$$

$$(10.48)$$

Again, take note that M_{TCaHCO_3} and M_{TCa} must be in kilograms and V must be in cubic meters. $[Ca^{2+}]_{meq}$ is then in milligram equivalents per liter.

10.13.3 CONCENTRATION OF Mg^{2+}

The concentration of the magnesium ion in the treated water comes from the magnesium from the magnesium bicarbonate not precipitated, and the magnesium from the noncarbonate hardness of magnesium not precipitated.

As the treated water is recarbonated, the 16 mg/L of Mg(OH)$_2$ as CaCO$_3$ from the limit of technology converts to the bicarbonate form,

$$Mg(OH)_2 + 2H_2CO_3 \rightarrow Mg(HCO_3)_2 + 2HOH \qquad (10.49)$$

Equation (10.7) has been written as

$$[C_{T,hard}]_{asCaCO_3} = \left(\frac{[Mg^{2+}]'_{eq}}{2}\right)(100) = 50[Mg^{2+}]'_{eq} \qquad (10.50)$$

Thus, the 16 mg/L of Mg(OH)$_2$ as CaCO$_3$ is equivalent to 16/50 = 0.32 meq/L of Mg^{2+} = 0.32 meq/L of HCO$_3^-$.

The possibility exists that the 16 mg/L of Mg(OH)$_2$ as CaCO$_3$ will react in the presence of excess bicarbonates to form magnesium carbonate; however, the K_{sp} of magnesium carbonate is around 10^{-5} compared to that of the K_{sp} of magnesium hydroxide, which is $9(10^{-12})$. Thus, Mg(OH)$_2$ will stay as Mg(OH)$_2$ until recarbonation.

Recall that M_{MgHCO_3} $(= f_{MgHCO_3}M_{TMgHCO_3})$ mass of magnesium bicarbonate from the carbonate hardness of magnesium was precipitated. Again, let all this mass be measured in kilograms. Also, let [Mg(HCO$_3$)$_2$]$_{mgnot}$, in mg/L, be the concentration of magnesium bicarbonate not precipitated. Thus,

$$[Mg(HCO_3)_2]_{mgnot} = \frac{M_{TMgHCO_3} - f_{MgHCO_3}M_{TMgHCO_3}}{\Psi}\left[\frac{1000(1000)}{1000}\right]$$

$$= \frac{M_{TMgHCO_3}(1 - f_{MgHCO_3})}{\Psi}[1000] \qquad (10.51)$$

Ψ is the volume of water treated in cubic meters.

Mg(HCO$_3$)$_2$ ionizes to Mg(HCO$_3$)$_2 \rightarrow$ Mg^{2+} + 2HCO$_3^-$; thus, the equivalent mass of magnesium bicarbonate is Mg(HCO$_3$)$_2$/2. The magnesium bicarbonate not precipitated therefore produces:

$$\frac{[Mg(HCO_3)_2]_{mgnot}}{Mg(HCO_3)_2/2} = 0.0137([Mg(HCO_3)_2]_{mgnot})$$

$$= 13.7\left[\frac{M_{TMgHCO_3}(1 - f_{MgHCO_3})}{\Psi}\right]$$

milligram equivalents of Mg^{2+} and HCO$_3^-$, respectively.

Now, also recall that M_{MgCa} $(= f_{MgCa}M_{TMg})$ mass of magnesium from the noncarbonate hardness of magnesium precipitated. As before, let this mass be measured in kilograms. Also, let $[Mg^{2+}]_{mgnot}$, in mg/L, be the concentration of magnesium not precipitated. Thus,

$$[Mg^{2+}]_{mgnot} = \frac{M_{TMg} - f_{MgCa}M_{TMg}}{V}\left[\frac{1000(1000)}{1000}\right] = \frac{M_{TMg}(1 - f_{MgCa})}{V}[1000]$$

(10.52)

Consider $MgSO_4$ as representing the noncarbonate hardness of calcium species. This ionizes as $MgSO_4 \rightarrow Mg^{2+} + SO_4^{2-}$; thus, the equivalent mass of magnesium in the noncarbonate hardness of magnesium is $Mg/2$. The magnesium not precipitated therefore produces:

$$\frac{[Mg^{2+}]_{mgnot}}{Mg/2} = 0.0823[Mg^{2+}]_{mgnot} = 82.3\left(\frac{M_{TMg}(1 - f_{MgCa})}{V}\right)$$

milligram equivalents of Mg^{2+}.

Summing all the magnesium ion concentrations, the total concentration $[Mg^{2+}]_{meq}$ is

$$[Mg^{2+}]_{meq} = 13.7\left[\frac{M_{TMgHCO_3}(1 - f_{MgHCO_3})}{V}\right] + 82.3\left[\frac{M_{TMg}(1 - f_{MgCa})}{V}\right]$$

(10.53)

Again, take note that M_{TMgHCO_3} and M_{TMg} must be in kilograms and V must be in cubic meters. $[Mg^{2+}]_{meq}$ is then in milligram equivalents per liter.

10.13.4 CONCENTRATION OF HCO$_3^-$

Two sources of the bicarbonate ion in the treated water are those coming from the bicarbonates of the carbonate hardness of calcium and magnesium not removed $[12.3(M_{TCaHCO_3} - M_{CaHCO_3})/V + 13.7(M_{TMgHCO_3} - M_{MgHCO_3})/V]$. In addition, bicarbonates may also result from the recarbonation of the magnesium hydroxide from the limits of technology. Call this concentration $[HCO_3]_{OHmeq}$. If no carbonate hardness of calcium is present but noncarbonate, bicarbonates may also result from the recarbonation of the calcium carbonate that precipitates when soda ash is used to remove the noncarbonate hardness of calcium. Call this concentration $[HCO_3]_{CO_3meq}$ and let $[HCO_3]_{OHCO_3meq}$ equal $[HCO_3]_{OHmeq}$ plus $[HCO_3]_{CO_3meq}$. Letting the total concentration of bicarbonate be designated as $[HCO_3^-]_{meqtent}$, in milligram equivalents per liter, we obtain

$$[HCO_3^-]_{meqtent} = [HCO_3]_{OHCO_3meq} + 12.3\left[\frac{M_{TCaHCO_3}(1 - f_{CaHCO_3})}{V}\right]$$

$$+ 13.7\left[\frac{M_{TMgHCO_3}(1 - f_{MgHCO_3})}{V}\right]$$

(10.54)

One of the subscripts is *tent*. This stands for tentative, because the concentration will be modified as discussed below. $[HCO_3]_{OHmeq}$ is equal to 0.32 meq/L, if magnesium hydroxide is recarbonated, and it is equal to zero, if no magnesium hydroxide is recarbonated. $[HCO_3]_{CO_3meq}$ is equal to zero, if the carbonate hardness of calcium is present, and it is equal to 3.0 meq/L, if no carbonate hardness of calcium is present but noncarbonate and the noncarbonate hardness of calcium is removed using soda ash.

Because of equilibrium, however, HCO_3^- dissociates into H^+ and CO_3^{2-} as follows:

$$HCO_3^- \rightleftharpoons H^+ + CO_3^{2-} \quad K_{sp,HCO_3} = \frac{\{H^+\}\{CO_3^{2-}\}}{\{HCO_3^-\}} \tag{10.55}$$

where the activities must be in gram moles per liter and K_{sp,HCO_3} is the K_{sp} of the bicarbonate. Thus, the concentration of bicarbonate at the treated effluent will be less than that portrayed by Equation (10.54). In addition, some CO_3^{2-} is also present in the effluent.

The geq/L of HCO_3^- is equal to the gmol/L of HCO_3^-. Now, let us find the gram mole per liter of the bicarbonate ion at equilibrium that results from an original one gram mole per liter of the ion. Let x be the concentration of CO_3^{2-} and H^+, respectively, produced from the dissociation. Thus, at equilibrium, $\{HCO_3^-\} = 1 - x$. Substituting into the K_{sp} equation,

$$K_{sp,HCO_3} = \frac{\{H^+\}\{CO_3^{2-}\}}{\{HCO_3^-\}} = \frac{x^2}{1-x} \tag{10.56}$$

Solving for x,

$$x = \frac{-K_{sp,HCO_3} + \sqrt{K_{sp,HCO_3}^2 + 4K_{sp,HCO_3}}}{2} \quad \text{gmol/L(or geq/L)} \tag{10.57}$$

Therefore, one milligram equivalent per liter of original bicarbonate produces $1 - (-K_{sp,HCO_3} + \sqrt{K_{sp,HCO_3}^2 + 4K_{sp,HCO_3}})/2$ mgeq/L of bicarbonate at equilibrium, and thus, the $[HCO_3^-]_{meqtent}$ of original bicarbonate produces $\{[HCO_3^-]_{meqtent}\}\{1 - (-K_{sp,HCO_3} + \sqrt{K_{sp,HCO_3}^2 + 4K_{sp,HCO_3}})/2\}$ mgeq/L. From Equation (10.54) and letting $[HCO_3^-]_{meq}$ be the milligram equivalent per liter of the bicarbonate in the effluent,

$$[HCO_3^-]_{meq} = \left([HCO_3]_{OHCO_3meq} + 12.3 \left[\frac{M_{TCaHCO_3}(1 - f_{CaHCO_3})}{\Psi} \right] \right.$$

$$\left. + 13.7 \left[\frac{M_{TMgHCO_3}(1 - f_{MgHCO_3})}{\Psi} \right] \right)$$

$$\left(1 - \frac{-K_{sp,HCO_3} + \sqrt{K_{sp,HCO_3}^2 + 4K_{sp,HCO_3}}}{2} \right) \tag{10.58}$$

10.13.5 Concentration of CO_3^{2-}

The one milligram equivalent per liter of original bicarbonate also produces $(-K_{sp,HCO_3} + \sqrt{K_{sp,HCO_3}^2 + 4K_{sp,HCO_3}})/2$ gram moles per liter of the carbonate ion. The equivalent mass of carbonate is $CO_3/2$. Thus, the milligram equivalent per liter of the carbonate, $[CO_3^-]_{meq}$, appearing in the effluent is

$$[CO_3^-]_{meq} = 2\left(\frac{-K_{sp,HCO_3} + \sqrt{K_{sp,HCO_3}^2 + 4K_{sp,HCO_3}}}{2}\right) \qquad (10.59)$$

10.13.6 Concentration of Na^+

The sources of the sodium ion are the original sodium from the influent and the sodium added with the soda ash. Recall that there were M_{sodAsh} mass of Na_2CO_3 used. Let this be measured in terms of kilograms. Its reaction with the Ca^{2+} is $Na_2CO_3 + Ca^{2+} \rightarrow CaCO_3 + 2Na^+$. Thus, the equivalent mass of Na_2CO_3 is $Na_2CO_3/2$ and the number of kilogram equivalents of the M_{sodAsh} mass of Na_2CO_3 is $M_{sodAsh}/(Na_2CO_3/2) = 0.01887$ M_{sodAsh}. For the V cubic meters of water treated, the equivalent concentration in meq/L is $0.01887 M_{sodAsh}/V$ [1000(1000)/1000] = $18.9(M_{sodAsh}/V)$. Let the concentration in the influent be $[Na^+]_{meq,inf}$ milligram equivalents per liter. Then, if $[Na^+]_{meq}$ is the concentration of the sodium ion in the treated water,

$$[Na^+]_{meq} = [Na^+]_{meq,inf} + 18.9\left(\frac{M_{sodAsh}}{V}\right) \qquad (10.60)$$

10.14 RELATIONSHIPS OF THE FRACTIONAL REMOVALS

The fractional removals f_{CaHCO_3}, f_{MgHCO_3}, f_{Ca}, and f_{MgCa} have been used in the foregoing derivations. Let us formalize the definitions of these parameters in this heading as follows:

$$f_{CaHCO_3} = \frac{M_{CaHCO_3}}{M_{TCaHCO_3}} \qquad (10.61)$$

$$f_{MgHCO_3} = \frac{M_{MgHCO_3}}{M_{TMgHCO_3}} \qquad (10.62)$$

$$f_{Ca} = \frac{M_{Ca}}{M_{TCa}} \qquad (10.63)$$

$$f_{MgCa} = \frac{M_{MgCa}}{M_{TMg}} \qquad (10.64)$$

In addition, define f_1 as the overall removal of the total calcium hardness, f_2 as the overall removal of the total magnesium hardness, and f as the overall removal of total hardness. Recall that V is the total volume of water treated and that the concentrations of the calcium and magnesium ions in the effluent are $[Ca^{2+}]_{meq}$ and $[Mg^{2+}]_{meq}$, respectively. From this information, the total mass of calcium in the effluent is $[Ca^{2+}]_{meq}[(Ca/2)/1000(1000)](1000)V = 0.02[Ca^{2+}]_{meq}\,V$ kg. Similarly, the total mass of magnesium in the effluent is $[Mg^{2+}]_{meq}[(Mg/2)/1000(1000)](1000)V = 0.012\,[Mg^{2+}]_{meq}\,V$ kg. Therefore,

$$f_1 = 1 - \frac{0.02[Ca^{2+}]_{meq}V}{M_{TCa} + \dfrac{Ca}{Ca(HCO_3)_2}M_{TCaHCO_3}}$$

$$= 1 - \frac{V}{M_{TCa} + \dfrac{Ca}{Ca(HCO_3)_2}M_{TCaHCO_3}}\left\{0.246\left[\frac{M_{TCaHCO_3}(1 - f_{CaHCO_3})}{V}\right]\right.$$

$$\left. + \left[\frac{M_{TCa}(1 - f_{Ca})}{V}\right]\right\} \tag{10.65}$$

$$f_2 = 1 - \frac{0.012[Mg^{2+}]_{meq}V}{M_{TMg} + \dfrac{Mg}{Mg(HCO_3)_2}M_{TMgHCO_3}}$$

$$= 1 - \frac{V}{M_{TMg} + \dfrac{Mg}{Mg(HCO_3)_2}M_{TMgHCO_3}}\left\{0.16\left[\frac{M_{TMgHCO_3}(1 - f_{MgHCO_3})}{V}\right]\right.$$

$$\left. + 0.99\left[\frac{M_{TMg}(1 - f_{MgCa})}{V}\right]\right\} \tag{10.66}$$

To derive the equation for f, it is important that the calcium and magnesium be expressed in terms of equivalents. The total equivalents of calcium, $TECa$, and the total equivalents of magnesium, $TEMg$, are, respectively, equal to

$$TECa = \frac{M_{TCaHCO_3}}{Ca(HCO_3)_2/2} + \frac{M_{TCa}}{Ca/2} \tag{10.67}$$

$$TEMg = \frac{M_{TMgHCO_3}}{Mg(HCO_3)_2/2} + \frac{M_{TMg}}{Mg/2} \tag{10.68}$$

Thus, f equals

$$f = \frac{f_1\left(\dfrac{M_{TCaHCO_3}}{Ca(HCO_3)_2/2} + \dfrac{M_{TCa}}{Ca/2}\right) + f_2\left(\dfrac{M_{TMgHCO_3}}{Mg(HCO_3)_2/2} + \dfrac{M_{TMg}}{Mg/2}\right)}{\left(\dfrac{M_{TCaHCO_3}}{Ca(HCO_3)_2/2} + \dfrac{M_{TCa}}{Ca/2}\right) + \left(\dfrac{M_{TMgHCO_3}}{Mg(HCO_3)_2/2} + \dfrac{M_{TMg}}{Mg/2}\right)}$$

$$= \frac{f_1(0.0123M_{TCaHCO_3} + 0.05M_{TCa}) + f_2(0.0137M_{TMgHCO_3} + 0.082M_{TMg})}{(0.0123M_{TCaHCO_3} + 0.05M_{TCa}) + (0.0137M_{TMgHCO_3} + 0.082M_{TMg})}$$

$$\tag{10.69}$$

In the design of the water softening process, the various removal fractions need to be assumed. Normally, the desired effluent quality is known, thus knowing the value of f. It depends upon the designer how to apportion the respective removal fractions for the calcium and magnesium ions. Magnesium in excess of 40 mg/L as $CaCO_3$ deposits as scales on heat exchange elements. In addition, $CaSO_4$ tends to deposit at high temperatures. These two constraints should be considered in making assumptions regarding the various fractional removals.

10.15 NOTES ON EQUIVALENT MASSES

If all the softening reactions were reviewed, it would be observed that the equivalent masses of the hardness "molecules" and their associated ionic species are obtained by dividing the molecular masses by the respective total number of valences of the positive or negative ions of the associated species, irrespective of the coefficients in the reaction. For example, in Equation (10.17), the equivalent mass of the "molecule" $MgSO_4$ is obtained by dividing the molecular mass $MgSO_4$ by 2, the valence of the associated ionic species Mg^{2+} or SO_4^{2-}. The equivalent masses of Mg^{2+} and SO_4^{2-} would be Mg/2 and SO_4/2, respectively. Also, in Equation (10.16), the equivalent mass of $Mg(HCO_3)_2$ is obtained by dividing the molecular mass $Mg(HCO_3)_2$ by 2, the valence of Mg^{2+} or the total number of valences of the associated species HCO_3^-; similar calculations will hold for the associated species. The equivalent mass of the hardness and associated species in solution is equal to the molecular mass divided by the total number of valences of the positive or negative ions of the associated ionic species, irrespective of the coefficient.

The previous findings, however, cannot be generalized to the precipitant species or species other than the hardness and its associated ionic species. For example, in Equation (10.16), if the above findings were applied to the precipitant $Ca(OH)_2$, its equivalent mass would be $Ca(OH)_2$/2; however, this is not correct—the equivalent mass of $Ca(OH)_2$ in this equation is $2Ca(OH)_2$/2. To conclude, the equivalent mass of a precipitant species or species other than the hardness and its associated species cannot be generalized as molecular mass divided by the total number of valences of the species but must be deduced from the chemical reaction.

For emphasis, we make the following summary: For the hardness ions or associated species, the equivalent mass is obtained by dividing the molecular mass by the total number of valences of the positive or negative charges of the species, irrespective of the coefficient; for all other species, the equivalent mass must be deduced from the balanced chemical reaction.

For convenience, the equations for chemical requirements, solids productions, fractional removals, and effluent quality are summarized in Tables 10.1 to 10.4.

Example 10.2 A raw water to be treated by the lime-soda process to the minimum hardness possible has the following characteristics: CO_2 = 22.0 mg/L, Ca^{2+} = 80 mg/L, Mg^{2+} = 12.0 mg/L, Na^+ = 46.0 mg/L, HCO_3^- = 152.5 mg/L, and

---

TABLE 10.1
Chemical Requirement

Equation

$$M_{CaOCaHCO_3} = 0.35 f_{CaHCO_3} M_{TCaHCO_3}$$

$$M_{CaOMgHCO_3} = 0.77 f_{MgHCO_3} M_{TMgHCO_3}$$

$$M_{CaOMgCa} = 2.30 f_{MgCa} M_{TMg}$$

$$M_{CaOCO_2} = 1.27 (M_{CO_2})$$

$$M_{CaOExcess} = 0.028 V \text{ kg}$$

$$M_{CaO} = M_{CaOCaHCO_3} + M_{CaOMgHCO_3} + M_{CaOMgCa} + A$$

$$A = M_{CaOCO_2} + M_{CaOExcess}$$

$$M_{sodAshCa} = 2.65 f_{Ca} M_{TCa}$$

$$M_{sodAshMgCa} = 4.36 f_{MgCa} M_{TMg}$$

$$M_{sodAsh} = M_{sodAshCa} + M_{sodAshCaMg}$$

TABLE 10.2
Solids Production

Equation

$$M_{CaCO_3CaHCO_3} = 1.23 f_{CaHCO_3} M_{TCaHCO_3}$$

$$M_{MgOHMgHCO_3} = 0.40 f_{MgHCO_3} M_{TMgHCO_3}$$

$$M_{CaCO_3MgHCO_3} = 1.37 f_{MgHCO_3} M_{TMgHCO_3}$$

$$M_{solidsCarb} = M_{CaCO_3CaHCO_3} + M_{MgOHMgHCO_3} + M_{CaCO_3MgHCO_3}$$

$$M_{MgOHMgCa} = 2.40 f_{MgCa} M_{TMg}$$

$$M_{CaCO_3Ca} = 2.49 f_{Ca} M_{TCa}$$

$$M_{CaCO_3MgCa} = 4.11 f_{MgCa} M_{TMg}$$

$$M_{solidsNonCarb} = M_{MgOHMgCa} + M_{CaCO_3Ca} + M_{CaCO_3MgCa}$$

$$M_{CaCO_3CO_2} = 2.27 (M_{CO_2})$$

$$M_{solids} = M_{solidsCarb} + M_{solidsNonCarb} + M_{CaCO_3CO_2}$$

$SO_4^{2-} = 216.0$ mg/L. (**a**) Check if the number of equivalents of positive and negative ions are balanced. For a flow of 25,000 m³/d and a complete removal of hardness, calculate (**b**) the lime requirement, (**c**) the soda ash requirement, and (**d**) the mass of solids and volume of sludge produced. Assume that the lime used is 90% pure and the soda ash used is 85% pure. Also, assume that the specific gravity of the sludge is 1.04.

TABLE 10.3
Fractional Removals

Equation

$$f_1 = 1 - \frac{\forall}{M_{TCa} + \frac{Ca}{Ca(HCO_3)_2}M_{TCaHCO_3}} \left\{ 0.246 \left[\frac{M_{TCaHCO_3}(1 - f_{CaHCO_3})}{\forall} \right] \right.$$

$$\left. + \left[\frac{M_{TCa}(1 - f_{Ca})}{\forall} \right] \right\}$$

$$f_2 = 1 - \frac{\forall}{M_{TMg} + \frac{Mg}{Mg(HCO_3)_2}M_{TMgHCO_3}} \left\{ 0.16 \left[\frac{M_{TMgHCO_3}(1 - f_{MgHCO_3})}{\forall} \right] \right.$$

$$\left. + 0.99 \left[\frac{M_{TMg}(1 - f_{MgCa})}{\forall} \right] \right\}$$

$$f = \frac{f_1(0.0123 M_{TCaHCO_3} + 0.05 M_{TCa}) + f_2(0.0137 M_{TMgHCO_3} + 0.082 M_{TMg})}{(0.0123 M_{TCaHCO_3} + 0.05 M_{TCa}) + (0.0137 M_{TMgHCO_3} + 0.082 M_{TMg})}$$

TABLE 10.4
Effluent Quality after Recarbonation

Equation

$$[Ca^{2+}]_{meq} = 12.3 \left[\frac{M_{TCaHCO_3}(1 - f_{CaHCO_3})}{\forall} \right] + 50.0 \left[\frac{M_{TCa}(1 - f_{Ca})}{\forall} \right]$$

$$[Mg^{2+}]_{meq} = 13.7 \left[\frac{M_{TMgHCO_3}(1 - f_{MgHCO_3})}{\forall} \right] + 82.3 \left[\frac{M_{TMg}(1 - f_{MgCa})}{\forall} \right]$$

$$[Na^+]_{meq} = [Na^+]_{meq,inf} + 18.9 \left(\frac{M_{sodAsh}}{\forall} \right)$$

$$[HCO_3^-]_{meq} = \left([HCO_3]_{OHCO_3 meq} + 12.3 \left[\frac{M_{TCaHCO_3}(1 - f_{CaHCO_3})}{\forall} \right] \right.$$

$$\left. + 13.7 \left[\frac{M_{TMgHCO_3}(1 - f_{MgHCO_3})}{\forall} \right] \right) \{A\}$$

$$A = \left(1 - \frac{-K_{sp,HCO_3} + \sqrt{K_{sp,HCO_3}^2 + 4K_{sp,HCO_3}}}{2} \right)$$

$$[CO_3^-]_{meq} = 2 \left(\frac{-K_{sp,HCO_3} + \sqrt{K_{sp,HCO_3}^2 + 4K_{sp,HCO_3}}}{2} \right)$$

$$[SO_4^{2-}]_{meq} = [SO_4^{2-}]_{meq,inf}$$

$$[Cl^-]_{meq} = [Cl^-]_{meq,inf}$$

$$[NO_3^-]_{meq} = [NO_3^-]_{meq,inf}$$

Solution:

(a)

Ions	Conc (mg/L)	Equiv. mass	No. of Equiv. (meq/L)
Ca^{2+}	80	20.05	3.99
Mg^{2+}	12.0	12.15	1.0
Na^+	46.0	23	2
			$\Sigma = 6.99$
HCO_3^-	152.5	61[a]	2.5
SO_4^{2-}	216.0	48.05[b]	4.49
			$\Sigma = 6.99$

$^a \dfrac{1.008+12.0+3(16)}{1} = 61$

$^b \dfrac{32.1+4(16)}{2} = 48.05$

Therefore, ions are balanced. **Ans**

(b) $M_{CaO} = M_{CaOCaHCO_3} + M_{CaOMgHCO_3} + M_{CaOMgCa} + M_{CaOCO_2} + M_{CaO\,Excess}$

$M_{CaOCaHCO_3} = 0.35 f_{CaHCO_3} M_{TCaHCO_3}$:

Removal of calcium bicarbonate takes precedence. Therefore, meq/L of $Ca(HCO_3)_2 = 2.5$

$f_{CaHCO_3} = \dfrac{2.5 - 0.3}{2.5} = 0.88$; 0.3 meq/L is the limit of technology of calcium carbonate.

$M_{TCaHCO_3} = 2.5\left(\dfrac{1}{1000}\right)(25,000) = 62.5 \text{ kgeq/d}$

$= 62.5(20.05)\dfrac{Ca(HCO_3)_2}{Ca} = 62.5(20.05)\left(\dfrac{40.1+2(61)}{40.1}\right) = 5065.63 \text{ kg/d}$

Therefore, $M_{CaOCaHCO_3} = 0.35(0.88)(5065.63) = 1560.21 \text{ kg/d}$

$M_{CaOMgHCO_3} = 0.77 f_{MgHCO_3} M_{TMgHCO_3} = 0$

$M_{CaOMgCa} = 2.30 f_{MgCa} M_{TMg}$:

$f_{MgCa} = \dfrac{1 - 0.32}{1} = 0.68$; 0.32 meq/L is the limit of technology of magnesium hydroxide

$M_{TMg} = 1.0\left(\dfrac{1}{1000}\right)(25,000) = 25 \text{ kgeq/d} = 25(12.15) = 303.75 \text{ kg/d}$

Therefore, $M_{CaOMgCa} = 2.30(0.68)(303.75) = 475.07 \text{ kg/d}$

$M_{CaOCO_2} = 1.27(M_{CO_2}) = 1.27(22.0)\left(\dfrac{1}{1000}\right)(25,000) = 698.5 \text{ kg/d}$

$M_{CaO\,Excess} = 0.028\,\forall \text{ kg} = 0.028(25,000) = 700 \text{ kg/d}$

Therefore, $M_{\text{CaO}} = 1560.21 + 0 + 475.07 + 698.5 + 700 =$ kg/d of pure lime

$$= \frac{3433.78}{0.90} = 3815.31 \text{ kg/d of lime} \quad \textbf{Ans}$$

(c) $M_{sodAsh} = M_{sodAshCa} + M_{sodAshCaMg}$

$M_{sodAshCa} = 2.65 \, f_{\text{Ca}} M_{T\text{Ca}}$:

$f_{\text{Ca}} = 1 \; M_{T\text{Ca}} = (3.99 - 2.5)(20.05)\left(\frac{1}{1000}\right)(25,000) = 745.75 \text{ kg/d}$

Therefore, $M_{sodAshCa} = 2.65(1.0)(745.75) = 1976.24 \text{ kg/d}$

$M_{sodAshCaMg} = 4.36 \, f_{\text{MgCa}} M_{T\text{Mg}}$:

$$= 4.36(1.0)(303.75) = 1324.35 \text{ kg/d}$$

Therefore, $M_{sodAsh} = 1976.24 + 1324.35 = 3300.59 \text{ kg/d of pure soda ash}$

$$= \frac{3300.59}{0.85} = 3883.05 \text{ kg/d of soda ash} \quad \textbf{Ans}$$

(d) $M_{solids} = M_{solidsCarb} + M_{solidsNonCarb} + M_{\text{CaCO}_3\text{CO}_2}$

$M_{solidsCarb} = M_{\text{CaCO}_3\text{CaHCO}_3} + M_{\text{MgOHMgHCO}_3} + M_{\text{CaCO}_3\text{MgHCO}_3}$:

$\quad M_{\text{CaCO}_3\text{CaHCO}_3} = 1.23 \, f_{\text{CaHCO}_3} M_{T\text{CaHCO}_3}$:

$\quad f_{\text{CaHCO}_3} = 0.88 \quad M_{T\text{CaHCO}_3} = 5065.63 \text{ kg/d}$

$\quad$ Therefore, $M_{\text{CaCO}_3\text{CaHCO}_3} = 1.23(0.88)(5065.63) = 5483.04 \text{ kg/d}$

$M_{\text{MgOHMgHCO}_3} = 0.40 \, f_{\text{MgHCO}_3} M_{T\text{MgHCO}_3} = 0$

$M_{\text{CaCO}_3\text{MgHCO}_3} = 1.37 \, f_{\text{MgHCO}_3} M_{T\text{MgHCO}_3} = 0$

Therefore, $M_{solidsCarb} = 5483.04 + 0 + 0 = 5483.04 \text{ kg/d}$

$M_{solidsNonCarb} = M_{\text{MgOHMgCa}} + M_{\text{CaCO}_3\text{Ca}} + M_{\text{CaCO}_3\text{MgCa}}$:

$M_{\text{MgOHMgCa}} = 2.40 \, f_{\text{MgCa}} M_{T\text{Mg}} = 2.40(0.68)(303.75) = 495.72 \text{ kg/d}$

$M_{\text{CaCO}_3\text{Ca}} = 2.49 \, f_{\text{Ca}} M_{T\text{Ca}} = 2.49(1)(745.75) = 1856.92 \text{ kg/d}$

$M_{\text{CaCO}_3\text{MgCa}} = 4.11 \, f_{\text{MgCa}} M_{T\text{Mg}} = 4.11(0.68)(303.75) = 848.92 \text{ kg/d}$

Therefore, $M_{solidsNonCarb} = 495.72 + 1856.92 + 848.92 = \textbf{3201.56 kg/d}$

$M_{\text{CaCO}_3\text{CO}_2} = 2.27(M_{\text{CO}_2}) = 2.27(22.0)\left(\frac{1}{1000}\right)(25,000) = 1248.5 \text{ kg/d}$

Therefore, M_{solids} = 5483.04 + **3201.56** + 1248.5 = 9933.1 kg/d **Ans**

$$V_{sl} = \frac{M_{solids}}{f_{solids}S_{sl}(1000)}; \text{ assume } f_{solids} = 0.99$$

Therefore, $V_{sl} = \dfrac{9933.1}{0.99(1.04)(1000)} = 9.65 \text{ m}^3/\text{d}$

Example 10.3 In the previous example, **(a)** determine the ionic composition of the finished water and **(b)** show that the cations and anions are balanced. Assume the temperature is 25°C such that $K_{sp,HCO_3} = 10^{-10.33}$. Also, assume that the effluent is recarbonated.

Solution:

(a) $\qquad [Ca^{2+}]_{meq} = 12.3\left[\dfrac{M_{T CaHCO_3}(1 - f_{CaHCO_3})}{V}\right] + 50.0\left[\dfrac{M_{T Ca}(1 - f_{Ca})}{V}\right]$

Recarbonation will transform the 15.0 mg/L as $CaCO_3$ of carbonate ions from the limit of technology to 15.0 mg/L as $CaCO_3$ of bicarbonate ions. From the order of removal, these carbonate ions will be coming solely from the removal of the raw calcium bicarbonate. The carbonate ions coming from the use of soda ash to form calcium carbonate will not be accounted for in this limit of technology consideration, since by the time they are formed, the water is already saturated with the carbonate coming from the removal of the raw calcium bicarbonate.

Therefore,

$$f_{CaHCO_3} = 1 - \frac{15}{152.5} = 0.90 \quad f_{Ca} = 1 \quad M_{T CaHCO_3} = 5065.63 \text{ kg/d}$$

$$M_{T Ca} = 745.75 \text{ kg/d} \qquad V = 25{,}000 \text{ m}^3/\text{d}$$

Therefore,

$$[Ca^{2+}]_{meq} = 12.3\left[\frac{5065.63(1 - 0.90)}{25{,}000}\right] + 50.0\left[\frac{745.75(1 - 1)}{25{,}000}\right] = 0.25 \text{ meq/L} \textbf{ Ans}$$

$$[Mg^{2+}]_{meq} = 13.7\left[\frac{M_{T MgHCO_3}(1 - f_{MgHCO_3})}{V}\right] + 82.3\left[\frac{M_{T Mg}(1 - f_{MgCa})}{V}\right]:$$

From the limit of technology, $Mg(OH)_2$ = 16 mg/L as $CaCO_3$ will dissolve. The given concentration of 1.0 meq/L of magnesium is equal to 50 mg/L as $CaCO_3$.

Therefore,

$$f_{MgCa} = 1 - \frac{16}{50} = 0.68; \quad M_{T Mg} = 303.75 \text{ kg/d}; \quad M_{T MgHCO_3} = 0$$

Therefore,

$$[Mg^{2+}]_{meq} = 13.7\left[\frac{0(1 - f_{MgHCO_3})}{25,000}\right] + 82.3\left[\frac{303.75(1 - 0.68)}{25,000}\right] = 0.32 \text{ meq/L} \quad \textbf{Ans}$$

$$[Na^+]_{meq} = [Na^+]_{meq,inf} + 18.9\frac{M_{sodAsh}}{V}:$$

$[Na^+]_{meq,inf} = 2$ meq/L $M_{sodAsh} = 3300.59$ kg/d

Therefore,

$$[Na^+]_{meq} = 2 + 18.9\left(\frac{3300.59}{25,000}\right) = 4.50 \text{ meq/L} \quad \textbf{Ans}$$

$$[HCO_3^-]_{meq} = \left([HCO_3]_{OHCO_3,meq} + 12.3\left[\frac{M_{TCaHCO_3}(1 - f_{CaHCO_3})}{V}\right]\right.$$

$$\left. + 13.7\left[\frac{M_{TMgHCO_3}(1 - f_{MgHCO_3})}{V}\right]\right)\{A\}$$

$$A = \left(1 - \frac{-K_{sp,HCO_3} + \sqrt{K_{sp,HCO_3}^2 + 4K_{sp,HCO_3}}}{2}\right)$$

$$= \left(1 - \frac{-10^{-10.33} + \sqrt{(10^{-10.33})^2 + 4(10^{-10.33})}}{2}\right) \approx 1.0$$

$[HCO_3]_{OHmeq} = 0.32$ meq/L; $[HCO_3]_{CO_3meq} = 0$

Therefore,

$$[HCO_3^-]_{meq} = \left(0.32 + 12.3\left[\frac{5065.63(1 - 0.90)}{25,000}\right]\right.$$

$$\left. + 13.7\left[\frac{0(1 - f_{MgHCO_3})}{25,000}\right]\right)\{1\} = 0.57 \text{ meq/L} \quad \textbf{Ans}$$

$$[CO_3^-]_{meq} = 2\left(\frac{-K_{sp,HCO_3} + \sqrt{K_{sp,HCO_3}^2 + 4K_{sp,HCO_3}}}{2}\right)$$

$$= 2\left(\frac{-10^{-10.33} + \sqrt{(10^{-10.33})^2 + 4(10^{-10.33})}}{2}\right) \approx 0$$

$$[SO_4^{2-}]_{meq} = [SO_4^{2-}]_{meq,inf} = 4.49 \text{ meq/L} \quad \textbf{Ans}$$

$$[Cl^-]_{meq} = [Cl^-]_{meq,inf} = 0$$

$$[NO_3^-]_{meq} = [NO_3^-]_{meq,inf} = 0$$

(b)

Ions	No. of Equiv. (meq/L)
Ca^{2+}	0.25
Mg^{2+}	0.32
Na^+	4.50
	$\Sigma = 5.07$
HCO_3^-	0.57
SO_4^{2-}	4.49
	$\Sigma = 5.06$

As shown in the table, the ions in the finished water are balanced.   **Ans**

10.16 TYPICAL DESIGN PARAMETERS AND CRITERIA

Table 10.5 shows typical design criteria for softening systems.

10.17 SPLIT TREATMENT

Water with a high concentration of magnesium is often softened by a process called *split treatment*. Water softening may be done in either a single- or two-stage treatment. In a single-stage treatment, all the chemicals are added in just one basin, whereas, in a two-stage treatment, chemicals are added in two stages. In split treatment, the operation is in two stages. Part of the raw water is bypassed from the first stage (split). Excess lime to facilitate the precipitation of magnesium hydroxide to the limit of technology is added in the first stage but, instead, of neutralizing this excess, it is used in the second stage to react with the calcium hardness of the bypassed flow that is introduced into the second stage.

Referring to Figure 10.3, let Q be the rate of flow of water treated, Mgr be the concentration of Mg in the raw water, and Mgf be the concentration of magnesium

TABLE 10.5
Design Parameters and Criteria for Softening Systems

Parameter	Mixer	Flocculator	Settling Basin	Solids-Contact Basin
Detention time	5 min	30–60 min	2–4 h	2–4 h
Velocity gradient, 1/sec	800	10–90	—	—
Flow-through velocity, m/s	—	0.15–0.5	0.15–0.5	—
Overflow rate, m³/min · m²	—	—	0.8–1.8	4.0

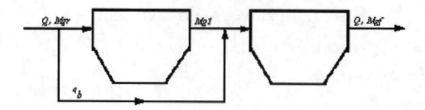

FIGURE 10.3 Schematic of the split-treatment mode of softening.

in the finished water. The amount of magnesium removed is then equal to $Q\text{Mg}r - Q\text{Mg}f$. If q_b is the bypassed flow, the amount of magnesium introduced to the second stage from the bypassed flow is $q_b\text{Mg}r$. The flow coming out of the first stage and flowing into the second stage is $Q - q_b$; the corresponding amount of magnesium introduced into the second stage is $(Q - q_b)\text{Mg}1$, where Mg1 is the concentration of magnesium in the effluent of the first stage. The total amount of magnesium introduced into the second stage is then the sum of those coming from the bypassed flow and those coming from the first stage and is equal to $q_b\text{Mg}r + (Q - q_b)\,\text{Mg}1$. No removal of magnesium is effected in the second stage, so this is equal to $Q\text{Mg}f$. Expressing in terms of an equation,

$$Q\text{Mg}f = q_b\text{Mg}r + (Q - q_b)\text{Mg}1 \qquad (10.70)$$

Solving for the fraction of bypassed flow f_b,

$$f_b = \frac{q_b}{Q} = \frac{\text{Mg}f - \text{Mg}1}{\text{Mg}r - \text{Mg}1} \qquad (10.71)$$

Example 10.4   For the raw water of Example 10.2, the lime–soda process in the split treatment mode is used to remove the total hardness to 120 mg/L as $CaCO_3$ containing magnesium hardness of 30 mg/L as $CaCO_3$. Calculate the chemical requirements in the first stage.

Solution:

Ions	Conc. (mg/L)	Equiv. Mass	No. of Equiv. (meq/L)	Conc. as $CaCO_3$ (mg/L)
Ca^{2+}	80	20.05	3.99	199.5
Mg^{2+}	12.0	12.15	1.0	50
Na^+	46.0	23	2	100
			$\Sigma = 6.99$	349.5
HCO_3^-	152.5	61[a]	2.5	125
SO_4^{2-}	216.0	48.05[b]	4.49	224.5
			$\Sigma = 6.99$	349.5

[a] $\dfrac{1.008 + 12.0 + 3(16)}{1} = 61$

[b] $\dfrac{32.1 + 4(16)}{2} = 48.05$

$$f_b = \frac{q_b}{Q} = \frac{Mgf - Mg1}{Mgr - Mg1}$$

Mg1 = 16 mg/L as $CaCO_3$ Mgf = 30 mg/L as $CaCO_3$ Mgr = 50 mg/L as $CaCO_3$

$$f_b = \frac{q_b}{Q} = \frac{30 - 16}{50 - 16} = 0.41$$

Therefore, bypassed flow to the second stage = 0.41(25,000) = 10,250 m^3d and flow to the first stage = 14,750 m^3d.

$$M_{CaO} = M_{CaOCaHCO_3} + M_{CaOMgHCO_3} + M_{CaOMgCa} + M_{CaOCO_2} + M_{CaO\,Excess}$$

$$M_{CaOCaHCO_3} = 0.35\,f_{CaHCO_3}\,M_{TCaHCO_3}:$$

Removal of calcium bicarbonate takes precedence. Therefore, meq/L of $Ca(HCO_3)_2 = 2.5$

$$f_{CaHCO_3} = \frac{2.5 - 0.3}{2.5} = 0.88$$

$$M_{TCaHCO_3} = 2.5\left(\frac{1}{1000}\right)(14{,}750) = 36.88 \text{ kgeq/d}$$

$$= 36.88(20.05)\frac{Ca(HCO_3)_2}{Ca} = 36.88(20.05)\left(\frac{40.1 + 2(61)}{40.1}\right)$$

$$= 2989.12 \text{ kg/d}$$

Therefore, $M_{CaOCaHCO_3} = 0.35(0.88)(2989.12) = 920.65$ kg/d

$$M_{CaOMgHCO_3} = 0.77\,f_{MgHCO_3}\,M_{TMgHCO_3} = 0$$

$$M_{CaOMgCa} = 2.30\,f_{MgCa}\,M_{TMg}:$$

$$f_{MgCa} = \frac{1 - 0.32}{1} = 0.68; \quad M_{TMg} = 1.0\left(\frac{1}{1000}\right)(14{,}750) = 14.75 \text{ kgeq/d}$$

$$= 14.75(12.15) = 179.21 \text{ kg/d}$$

Therefore,

$$M_{CaOMgCa} = 2.30(0.68)(179.21) = 280.28 \text{ kg/d}$$

$$M_{CaOCO_2} = 1.27(M_{CO_2}) = 1.27(22.0)\left(\frac{1}{1000}\right)(14{,}750) = 412.12 \text{ kg/d}$$

$$M_{CaO\,Excess} = 0.028 \; \forall \text{ kg} = 0.028(14{,}750) = 413 \text{ kg/d}$$

Therefore,

$$M_{CaO} = 920.65 + 0 + 280.28 + 412.12 + 413 = 2026.05 \text{ kg/d of pure lime}$$

$$= \frac{2026.05}{0.90} = 2251.17 \text{ kg/d of lime} \quad \textbf{Ans}$$

Soda ash will not be used in the first stage.

Example 10.5 In Example 10.4, calculate the chemical requirements in the second stage.

Solution:

$$M_{CaO} = M_{CaOCaHCO_3} + M_{CaOMgHCO_3} + M_{CaOMgCa} + M_{CaOCO_2} + M_{CaO\,Excess}$$

First, we have to find the influent hardness concentrations to the second stage. Calcium introduced to the second stage from the first stage is equal to the limit of technology of 15.0 mg/L $CaCO_3$, plus the calcium from the lime used to remove the noncarbonate hardness of magnesium, which is 50 mg/L $CaCO_3$, plus the noncarbonate calcium from the raw water, which is $199.5 - 125 = 74.5$ mg/L $CaCO_3$. This will produce a total outflow of calcium from the first stage of $50 + 15 + 74.5 = 139.5$ mg/L as $CaCO_3$ from a flow of 14,750 m^3/d. As this outflow mixes with the bypass flow, the influent calcium concentration to the second stage then becomes $(139.5(14,750) + 199.5(10,250))/25,000 = 164.1$ mg/L as $CaCO_3 = 3.28$ meq/L. Calcium to remain in the effluent is $120 - 30 = 90$ mg/L as $CaCO_3 = 1.8$ meq/L. Because of the limit of technology of 0.3 meq/L, pseudo calcium concentration to remain in effluent is $1.8 - 0.3 = 1.5$ meq/L. Thus, calcium to be removed is $3.28 - 1.5 = 1.78$ meq/L.

Considering the bypass, the influent HCO_3^- concentration to the second stage is $= (0(14,750) + 2.5(10,250))/25,000 = 1.025$ meq/L. Because we have 3.28 meq/L of influent calcium but only 1.025 meq/L of bicarbonate, by the order of removal, influent $Ca(HCO_3)_2 = 1.025$ meq/L. An amount of 1.78 meq/L of calcium is to be removed. Thus, noncarbonate calcium to be removed $= 1.78 - 1.025 = 0.755$ meq/L. Noncarbonate calcium $= 3.28 - 1.025 = 2.255$ meq/L.

The limit of technology for magnesium hydroxide is 16 mg/L as $CaCO_3 = 0.32$ meq/L. Considering the bypass flow, Mg influent concentration to the second stage $= (0.32(14,750) + 1.0(10,250))/25,000 = 0.60$ meq/L.

$$M_{CaOCaHCO_3} = 0.35 f_{CaHCO_3} M_{TCaHCO_3}$$

$$f_{CaHCO_3} = \frac{1.025 - 0.3}{1.025} = 0.71$$

$$M_{TCaHCO_3} = 1.025 \left\{ \frac{1}{(1000)} \right\} (25,000) = 25.625 \text{ keq/d}$$

$$= 25.625(20.05)\frac{Ca(HCO_3)_2}{Ca} = 25.625(20.05)\left(\frac{40.1+2(61)}{40.1}\right)$$

$$= 2076.91 \text{ kg/d}$$

Therefore,

$$M_{CaOCaHCO_3} = 0.35(0.71)(2076.91) = 514.16 \text{ kg/d}$$

$$M_{CaOMgHCO_3} = 0; \quad + M_{CaOMgCa} = 0;$$

$$M_{CaOCO_2} = 1.27(M_{CO_2}) = 1.27(22.0)\left(\frac{1}{1000}\right)(10,250) = 286.39 \text{ kg/d}$$

$$M_{CaOExcess} = -413 \text{ kg/d}$$

Therefore,

$$M_{CaO} = 514.16 + 0 + 0 + 286.39 - 413 = 387.55 \text{ kg/d of pure lime}$$

$$= \frac{387.55}{0.9} = 430.61 \text{ kg/d of lime} \quad \textbf{Ans}$$

$$M_{sodAsh} = M_{sodAshCa} + M_{sodAshCaMg}$$

$$M_{sodAshCa} = 2.65 f_{Ca} M_{TCa}:$$

$$M_{TCa} = 0.755\left\{\frac{1}{(1000)}\right\}(25,000) = 18.875 \text{ keq/d}$$

$$= 18.875(20.05) = 378.44 \text{ kg/d}$$

Therefore,

$$M_{sodAshCa} = 2.65(1)(378.44) = 1002.88 \text{ kg/d of pure soda ash}$$

$$M_{sodAshCaMg} = 0$$

Therefore,

$$M_{sodAsh} = 1002.88 + 0 = 1002.88 \text{ kg/d}$$

$$= \frac{1002.88}{0.85} = 1179.85 \text{ kg/d of soda ash} \quad \textbf{Ans}$$

Example 10.6 In Example 10.4, calculate the ionic concentrations of the finished water assuming recarbonation is done and show that the ions are balanced.

Solution: The solution will pertain only to the second stage.

$$[Ca^{2+}]_{meq} = 12.3\left[\frac{M_{TCaHCO_3}(1 - f_{CaHCO_3})}{\forall}\right] + 50.0\left[\frac{M_{TCa}(1 - f_{Ca})}{\forall}\right]:$$

$$[Ca(HCO_3)_2] = 1.025 \text{ meq/L}$$

$$M_{TCaHCO_3} = 1.025\left\{\frac{1}{(1000)}\right\}(25,000) = 25.625 \text{ keq/d}$$

$$= 25.625(20.05)\frac{Ca(HCO_3)_2}{Ca} = 25.625(20.05)\left(\frac{40.1 + 2(61)}{40.1}\right)$$

$$= 2076.91 \text{ kg/d}$$

$$f_{CaHCO_3} = \frac{1.025 - 0.3}{1.025} = 0.71 \qquad V = 25,000 \text{ m}^3/\text{d}$$

$$M_{TCa} = 2.255\left(\frac{1}{1000}\right)(25,000) = 56.375 \text{ kgeq/d}$$

$$= 56.375(20.05) = 1130.32 \text{ kg/d}$$

$$f_{Ca} = \frac{0.755}{2.255} = 0.33$$

Therefore,

$$[Ca^{2+}]_{meq} = 12.3\left[\frac{2076.91(1 - 0.71)}{25,000}\right] + 50.0\left[\frac{1130.32(1 - 0.33)}{25,000}\right]$$

$$= 0.296 + 1.51 = 1.8 \text{ meq/L } \textbf{Ans}$$

$$[Mg^{2+}]_{meq} = 13.7\left[\frac{M_{TMgHCO_3}(1 - f_{MgHCO_3})}{V}\right] + 82.3\left[\frac{M_{TMg}(1 - f_{MgCa})}{V}\right]$$

$$M_{TMgHCO_3} = 0; \quad M_{TMg} = 0.60\left(\frac{1}{1000}\right)(12.15)(25,000) = 182.25 \text{ kg/d};$$

$$f_{MgCa} = 0$$

Therefore,

$$[Mg^{2+}]_{meq} = 13.7\left[\frac{0(1 - f_{MgHCO_3})}{25,000}\right] + 82.3\left[\frac{182.25(1 - 0)}{25,000}\right] = 0.6 \text{ meq/L} \quad \textbf{Ans}$$

$$[Na^+]_{meq} = [Na^+]_{meq,inf} + 18.9\left(\frac{M_{sodAsh}}{V}\right) = 2 + 18.9\left(\frac{1002.88}{25,000}\right)$$

$$= 2.758 \text{ meq/L} \quad \textbf{Ans}$$

$$[HCO_3^-]_{meq} = \left\{ [HCO_3]_{OHCO_3meq} + 12.3\left[\frac{M_{TCaHCO_3}(1 - f_{CaHCO_3})}{V}\right] \right.$$

$$\left. + 13.7\left[\frac{M_{TMgHCO_3}(1 - f_{MgHCO_3})}{V}\right] \right\}\{A\}:$$

$$[HCO_3]_{OHmeq} = 0.32 \text{ meq/L}; \quad [HCO_3]_{CO_3meq} = 0$$

$$A = \left(1 - \frac{-K_{sp,HCO_3} + \sqrt{K_{sp,HCO_3}^2 + 4K_{sp,HCO_3}}}{2}\right)$$

Assuming temperature = 25°C, $K_{sp,HCO_3} = 10^{-10.33}$

$$A = \left(1 - \frac{-10^{-10.33} + \sqrt{(10^{-10.33})^2 + 4(10^{-10.33})}}{2}\right) \approx 1.0$$

Therefore,

$$[HCO_3^-]_{meq} = \left\{0.32 + 12.3\left[\frac{2076.91(1 - 0.71)}{25,000}\right] + 13.7\left[\frac{0(1 - 0)}{25,000}\right]\right\}\{1\}$$

$$[HCO_3^-]_{meq} = 0.32 + 0.296 + 0 = 0.62 \text{ meq/L} \quad \textbf{Ans}$$

$$[CO_3^-]_{meq} = 2\left(\frac{-K_{sp,HCO_3} + \sqrt{K_{sp,HCO_3}^2 + 4K_{sp,HCO_3}}}{2}\right)$$

$$= 2\left(\frac{-10^{-10.33} + \sqrt{(10^{-10.33})^2 + 4(10^{-10.33})}}{2}\right) \approx 0$$

$$[SO_4^{2-}]_{meq} = [SO_4^{2-}]_{meq,inf} = 4.49 \text{ meq/L} \quad \textbf{Ans}$$

Ions	No. of Equiv. (meq/L)
Ca^{2+}	1.80
Mg^{2+}	0.60
Na^+	2.758
	$\Sigma = 5.15$
HCO_3^-	0.62
SO_4^{2-}	4.49
	$\Sigma = 5.11$

The ions may be considered balanced   **Ans**

10.18   USE OF ALKALINITY IN WATER SOFTENING CALCULATIONS

It will be noted that we have avoided the use of alkalinity in any of the softening equations, yet, in the literature, this parameter is used in softening calculations. The reason is that the use of this parameter in softening is often misleading. It will be recalled that the carbonate ion and the bicarbonate ion play very different roles in the unit process reactions. They are on the opposite sides, the carbonate being produced from the bicarbonate.

How then is the literature able to use alkalinity in calculations? The answer is that the literature is conceptually wrong. Alkalinity values cannot be used in any softening calculations, unless it is assumed that the alkalinity is mostly bicarbonate. If this is assumed, then the calculation can proceed, because as we have seen, this bicarbonate will then produce the carbonate. Although this assumption is, of course, accurate, it does not make its use conceptually correct. To repeat, alkalinity values should not be used in any softening calculations, without prior knowledge of its constituent species. Its various constituents have very different roles to play in the process.

GLOSSARY

Carbonate hardness—Hardness ions associated with the HCO_3^- ions in water.

Equivalent mass—Mass of any substance participating in a reaction per unit of reference species.

Equivalents or number of equivalents—Mass of a substance divided by the equivalent mass.

Hardness ions—Multivalent cations mostly calcium and magnesium that interfere in the use of soap.

Lime–soda process—A process of removing hardness ions that uses lime and soda ash.

Noncarbonate hardness—Hardness ions are associated with anions other than the bicarbonate ion.

Recarbonation—A method of stabilizing water that uses carbon dioxide.

Reference species—A number that represents the combining or reacting capacity of the substances participating in a given chemical reaction.

Softening—Term given to the process of removing ions that interfere with the use of soap.

Solids-contact clarifier—A clarifier that uses a sludge blanket contact mechanism to facilitate the precipitation reaction.

SYMBOLS

$[Ca^{2+}]_{eq}$	Concentration of the calcium ion in equivalents per unit volume
$[Ca^{2+}]'_{eq}$	Concentration of equivalent calcium ions per unit volume
$[Ca^{2+}]_{meq}$	Milligramequivalent per liter of the calcium ion in the treated water
$[Cl^-]_{meq}$	Milligramequivalent per liter of the chloride ion in treated water
$[Cl^-]_{meq,inf}$	Milligramequivalent per liter of the chloride ion in influent to the softening plant
$[C_{T,hard}]$	Concentration of total hardness
$[C_{T,hard}]_{eq}$	Concentration of total hardness in equivalents per unit volume
f	Overall removal of total hardness.
f_{Ca}	Fraction of noncarbonate hardness of calcium removed
f_1	Overall removal of the calcium hardness
f_2	Overall removal of the magnesium hardness
f_{CaHCO_3}	Fraction of $Ca(HCO_3)_2$ removed

f_{MgCa}	Fraction of mass of magnesium noncarbonate hardness removed
f_{MgHCO_3}	Fraction of $Mg(HCO_3)_2$ removed
$[HCO_3^-]_{meq}$	Milligramequivalent per liter of the bicarbonate ion in the treated water
K	Equilibrium constant
K_{sp}	Solubility product constant
M_{Ca}	Mass of noncarbonate hardness of calcium removed
M_{CaCO_3Ca}	Mass of calcium carbonate solids produced from the precipitation of the noncarbonate hardness of calcium
$M_{CaCO_3CaHCO_3}$	Mass of calcium carbonate solid produced from the calcium bicarbonate removed
$M_{CaCO_3CO_2}$	Mass of calcium carbonate solids produced from reaction of lime with dissolved carbon dioxide
M_{CaCO_3MgCa}	Mass of calcium carbonate solids produced from the precipitation of the calcium that results from the precipitation of the noncarbonate hardness of magnesium
$M_{CaCO_3MgHCO_3}$	Mass of calcium carbonate produced from the precipitation of the magnesium bicarbonate removed
M_{CaHCO_3}	Mass of $Ca(HCO_3)_2$ removed
$M_{CaOCaHCO_3}$	Mass of lime needed to precipitate the $Ca(HCO_3)_2$ removed
M_{CaO}	Mass of lime used in the lime-soda process in kg
M_{CaOCO_2}	Mass of lime needed for the dissolved carbon dioxide
$M_{CaOExcess}$	Mass of excess lime in kg needed to raise pH to 10.4
$M_{CaOMgCa}$	Mass of lime needed to precipitate the noncarbonate hardness of magnesium removed
$M_{CaOMgHCO_3}$	Mass of lime needed to precipitate the $Mg(HCO_3)_2$ removed
M_{CO_2}	Mass of dissolve carbon dioxide
$[Mg^{2+}]_{eq}$	Concentration of the magnesium ion in equivalents per unit volume
$[Mg^{2+}]_{meq}$	Milligramequivalent per liter of the magnesium ion in the treated water
$[Mg^{2+}]'_{eq}$	Concentration of equivalent magnesium ions
M_{MgCa}	Mass of noncarbonate hardness of magnesium removed
M_{MgHCO_3}	Mass of magnesium bicarbonate removed
$M_{MgOHMgHCO_3}$	Mass of magnesium hydroxide solid produced from the magnesium bicarbonate removed
$M_{MgOHMgCa}$	Mass of magnesium solids produced from the removal of the noncarbonate hardness of magnesium
$moleMass$	Molecular mass
M_{sodAsh}	Amount of soda ash required in the lime-soda process
$M_{sodAshCa}$	Mass of soda ash needed to precipitate the noncarbonate calcium hardness removed
$M_{sodAshMgCa}$	Mass of soda ash needed to precipitate the calcium produced from the removal of the noncarbonate hardness of magnesium
$M_{solidsCarb}$	Mass of solids produced from the removal of carbonate hardness
$M_{solidsNonCarb}$	Mass of solids produced from the removal of noncarbonate hardness

M_{solids}	Mass of solids produced in the lime-soda process
M_{TCa}	Total mass of noncarbonate hardness of calcium
M_{TCaHCO_3}	Total mass of $Ca(HCO_3)_2$
M_{TMg}	Total mass of noncarbonate hardness of magnesium
M_{TMgHCO_3}	Total mass of $Mg(HCO_3)_2$
$[Na^+]_{meq}$	Milligramequivalent per liter of the sodium ion in treated water
$[NO_3^-]_{meq}$	Milligramequivalent per liter of the nitrate ion in treated water
$[NO_3^-]_{meq,inf}$	Milligramequivalent per liter of the nitrate ion in influent to softening plant
q_b	Bypass flow in the split-treatment mode
Q	Inflow to softening plant
Mg1	Concentration of magnesium in influent from first stage of split-treatment softening plant
Mgf	Concentration of magnesium in treated water
Mgr	Concentration of raw water to softening plant
$ratioChem$	Chemical ratio
$[SO_4^{2-}]_{meq}$	Milligramequivalent per liter of the sulfate ion in treated water
$[SO_4^{2-}]_{meq,inf}$	Milligramequivalent per liter of the sulfate ion in influent to softening plant
Ψ	Volume of water softened in cubic meters
Ψ_{sl}	Volume of sludge in cubic meters

PROBLEMS

10.1 Write the balanced chemical reaction in the softening of calcium bicarbonate using lime.

10.2 Write the balanced chemical reaction in the softening of calcium sulfate using soda ash.

10.3 The total mass of calcium carbonate solids produced from softening a bicarbonate of calcium using lime is 500 g. Calculate the original number of equivalents of the bicarbonate. Calculate the original number of moles of the bicarbonate.

10.4 The total mass of calcium carbonate solids produced from softening of calcium sulfate using soda ash is 150 g. Calculate the original number of equivalents of the sulfate. Calculate the original number of moles of the sulfate.

10.5 The number of moles of a particular substance is 2.5 and its mass is 500 g. What is its molecular mass?

10.6 The reaction in the softening of magnesium bicarbonate using lime is

$$Mg(HCO_3)_2 + 2Ca(OH)_2 \rightarrow Mg(OH)_2\downarrow + 2CaCO_3\downarrow + 2HOH$$

Calculate the chemical ratio between $CaCO_3$ and $Mg(HCO_3)_2$.

10.7 From the chemical reaction in Problem 10.6, if the mass of $CaCO_3$ is 100 grams, how many grams of magnesium carbonate were removed?

10.8 A raw water to be treated by the lime-soda process to the minimum hardness possible has the following characteristics: $CO_2 = 22.0$ mg/L, $Ca^{+2} = 80$ mg/L,

$Mg^{+2} = 12.0$ mg/L, $Na^+ = 46.0$ mg/L, $HCO_3^- = 152.5$ mg/L, and $SO_4^{-2} = 216.0$ mg/L. Check if the number of equivalents of positive and negative ions are balanced. The flow is 25,000 m^3/d.

10.9 In Problem 10.8 for a complete removal of hardness, calculate the lime requirement. Assume that the lime used is 90% pure.

10.10 In Problem 10.8 for a complete removal of hardness, calculate the soda ash requirement. Assume that the soda ash used is 85% pure.

10.11 In Problem 10.8 for a complete removal of hardness, calculate the mass of solids and the volume of sludge produced. Assume that the lime used is 90% pure and the soda ash used is 85% pure. Also, assume that the specific gravity of the sludge is 1.04.

10.12 Taking Problems 10.8 through 10.11 together, (a) determine the ionic composition of the finished water and (b) show that the cations and anions are balanced.

10.13 For the raw water of Example 10.8, using the lime-soda process in the split treatment mode to remove the total hardness to 130 mg/L as $CaCO_3$ containing magnesium hardness of 40 mg/L as $CaCO_3$, calculate the total lime needed.

10.14 For the raw water of Example 10.8, using the lime-soda process in the split treatment mode to remove the total hardness to 130 mg/L as $CaCO_3$ containing magnesium hardness of 40 mg/L as $CaCO_3$, calculate the total soda ash needed.

10.15 Taking Problems 10.13 and 10.14 together, calculate the ionic concentrations of the finished water and show whether or not the ions are balanced. Assume that recarbonation is done.

10.16 For the raw water of Example 10.8, using the lime-soda process in the split treatment mode to remove the total hardness to 130 mg/L as $CaCO_3$ containing magnesium hardness of 40 mg/L as $CaCO_3$, calculate the total mass of lime used to precipitate the calcium bicarbonate.

10.17 For the raw water of Example 10.8, using the lime-soda process in the split treatment mode to remove the total hardness to 130 mg/L as $CaCO_3$ containing magnesium hardness of 40 mg/L as $CaCO_3$, calculate the total mass of lime used to precipitate the magnesium bicarbonate.

10.18 For the raw water of Example 10.8, using the lime-soda process in the split treatment mode to remove the total hardness to 130 mg/L as $CaCO_3$ containing magnesium hardness of 40 mg/L as $CaCO_3$, calculate the total mass of lime needed to precipitate the magnesium in the noncarbonate hardness of magnesium removed.

10.19 For the raw water of Example 10.8, using the lime-soda process in the split treatment mode to remove the total hardness to 130 mg/L as $CaCO_3$ containing magnesium hardness of 40 mg/L as $CaCO_3$, calculate the total mass of lime needed to neutralize the dissolved carbon dioxide.

10.20 For the raw water of Example 10.8, using the lime-soda process in the split treatment mode to remove the total hardness to 130 mg/L as $CaCO_3$ containing magnesium hardness of 40 mg/L as $CaCO_3$, calculate the total mass of lime used to raise the pH to 10.4.

10.21 For the raw water of Example 10.8, using the lime-soda process in the split treatment mode to remove the total hardness to 130 mg/L as $CaCO_3$ containing magnesium hardness of 40 mg/L as $CaCO_3$, calculate the total mass of soda ash needed to precipitate the calcium in the noncarbonate hardness of calcium.

10.22 For the raw water of Example 10.8, using the lime-soda process in the split treatment mode to remove the total hardness to 130 mg/L as $CaCO_3$ containing magnesium hardness of 40 mg/L as $CaCO_3$, calculate the total mass of soda ash needed to precipitate the calcium ion produced from the precipitation of the magnesium hardness in the noncarbonate hardness of magnesium.

10.23 For the raw water of Example 10.8, using the lime-soda process in the split treatment mode to remove the total hardness to 130 mg/L as $CaCO_3$ containing magnesium hardness of 40 mg/L as $CaCO_3$, calculate the total mass of soda ash needed to precipitate the noncarbonate hardness.

10.24 For the raw water of Example 10.8, using the lime-soda process in the split treatment mode to remove the total hardness to 130 mg/L as $CaCO_3$ containing magnesium hardness of 40 mg/L as $CaCO_3$, calculate the total mass of calcium carbonate solids that precipitated from the removal of the carbonate hardness of calcium.

10.25 For the raw water of Example 10.8, using the lime-soda process in the split treatment mode to remove the total hardness to 130 mg/L as $CaCO_3$ containing magnesium hardness of 40 mg/L as $CaCO_3$, calculate the total mass of magnesium hydroxide solids that precipitated from the removal of the magnesium bicarbonate of the carbonate hardness of magnesium.

10.26 For the raw water of Example 10.8, using the lime-soda process in the split treatment mode to remove the total hardness to 130 mg/L as $CaCO_3$ containing magnesium hardness of 40 mg/L as $CaCO_3$, calculate the total mass of calcium carbonate solids precipitated from the removal of the magnesium bicarbonate of the carbonate hardness of magnesium.

10.27 For the raw water of Example 10.8, using the lime-soda process in the split treatment mode to remove the total hardness to 130 mg/L as $CaCO_3$ containing magnesium hardness of 40 mg/L as $CaCO_3$, calculate the total mass of solids precipitated from the removal of the carbonate hardness.

10.28 For the raw water of Example 10.8, using the lime-soda process in the split treatment mode to remove the total hardness to 130 mg/L as $CaCO_3$ containing magnesium hardness of 40 mg/L as $CaCO_3$, calculate the total mass of magnesium hydroxide solids produced from the precipitation of magnesium in the noncarbonate hardness of magnesium.

10.29 For the raw water of Example 10.8, using the lime-soda process in the split treatment mode to remove the total hardness to 130 mg/L as $CaCO_3$ containing magnesium hardness of 40 mg/L as $CaCO_3$, calculate the total mass of calcium carbonate solids produced from the precipitation of calcium from the noncarbonate hardness of calcium.

10.30 For the raw water of Example 10.8, using the lime-soda process in the split treatment mode to remove the total hardness to 130 mg/L as $CaCO_3$ containing magnesium hardness of 40 mg/L as $CaCO_3$, calculate the total mass of calcium carbonate solids produced from the precipitation of magnesium in the magnesium of the noncarbonate hardness of magnesium.

10.31 For the raw water of Example 10.8, using the lime-soda process in the split treatment mode to remove the total hardness to 130 mg/L as $CaCO_3$ containing magnesium hardness of 40 mg/L as $CaCO_3$, calculate the total mass of solids produced from the removal of the noncarbonate hardness.

10.32 For the raw water of Example 10.8, using the lime-soda process in the split treatment mode to remove the total hardness to 130 mg/L as $CaCO_3$ containing magnesium hardness of 40 mg/L as $CaCO_3$, calculate the total mass of calcium carbonate solids produced from neutralizing the dissolved carbon dioxide.

10.33 Calculate the total amount of solids produced from applying the lime-soda process to the raw water of Example 10.8 in the split treatment mode to remove the total hardness to 130 mg/L as $CaCO_3$ containing magnesium hardness of 40 mg/L as $CaCO_3$.

10.34 What is the anion mostly associated with the hardness ions in natural waters?

10.35 Write the chemical formula of EDTA.

10.36 What are the two principal cations that cause hardness in natural waters?

10.37 The hardness of a given sample of water is 120 mg/L as $CaCO_3$. Calculate the hardness in terms of the calcium ion and the magnesium.

10.38 How does soap aid in cleansing?

BIBLIOGRAPHY

Davis, L. M. and D. A. Cornwell (1991). *Introduction to Environmental Engineering.* MacGraw-Hill, New York.

Doran, G. F., et al. (1998). Evaluation of technologies to treat oil field produced water to drinking water or reuse quality. *Proc. SPE Annu. Western Regional Meeting,* May 11–15, Bakersfield, CA, Soc. Pet. Eng. (SPE). Richardson, TX.

Kedem, O. and G. Zalmon (1997). Compact accelerated precipitation softening (CAPS) as a pretreatment for membrane desalination I. softening by NaOH, *Desalination,* 113, 1, 65–71.

Malakhov, I. A., et al. (1995). Small-waste technology of water softening and decarbonization for heat network feeding. *Teploenergetika,* 12, 61–63.

Rich, L. G. (1963). *Unit Process of Sanitary Engineering.* John Wiley & Sons, New York.

Sincero, A. P. and G. A. Sincero (1996). *Environmental Engineering: A Design Approach.* Prentice Hall, Upper Saddle River, NJ.

Sladeckova, A. (1994). Biofilm and periphyton formation in storage tanks, *Water Supply Proc. 19th Int. Water Supply Cong. Exhibition* Oct. 2–8 1993, Budapest, Hungary, 12, 1–2, Blackwell Scientific Publishers, Oxford.

Snoeyink, V. L. (1980). *Water Chemistry.* John Wiley & Sons, New York.

Tchobanoglous, G. and E. D. Schroeder (1985). *Water Quality.* Addison-Wesley, Reading, MA.

11 Water Stabilization

As mentioned in Chapter 10 on water softening, as long as the concentrations of $CaCO_3$ and $Mg(OH)_2$ exceed their solubilities, the solids may continue to precipitate. This condition can cause scale to form, a solid that deposits due to precipitation of ions in solution. To prevent scale formation, the water must be stabilized. A water is said to be stable when it neither dissolves nor deposits precipitates. If the pH is high, stabilization may be accomplished using one of several acids or using CO_2, a process called *recarbonation*. If the pH is low, stabilization may be accomplished using lime or some other bases.

Because of the universal presence of carbon dioxide, any water body is affected by the reaction products of carbon dioxide and water. The species produced from this reaction form the carbonate system equilibria. As discussed later, the stability or instability of water can be gaged using these equilibria. Thus, this chapter discusses this concept. It also discusses criteria for stability and the recarbonation process after water softening.

11.1 CARBONATE EQUILIBRIA

The carbonate equilibria is a function of the ionic strength of water, activity coefficient, and the effective concentrations of the ionic species. The equilibrium coefficients that are calculated from the species concentrations are a function of the temperature. This functionality of the coefficients can, in turn, be calculated using the *Van't Hoff equation*, to be addressed later.

One of the major cations that can form scales as a result of the instability of water is calcium. Calcium plays an important role in the carbonate equilibria. We will therefore express the carbonate equilibria in terms of the interaction of the calcium ion and the carbonate species which are the reaction products of carbon dioxide and water. In addition, since the equilibria occur in water, the dissociation of the water molecule must also be involved. Using calcium as the cation, the equilibrium equations of the equilibria along with the respective equilibrium constants at 25°C are as follows (Rich, 1963):

$$K_w = 10^{-14} = \{H^+\}\{OH^-\} \tag{11.1}$$

$$K_1 = 10^{-6.35} = \frac{\{H^+\}\{HCO_3^-\}}{\{H_2CO_3^*\}} \tag{11.2}$$

$$K_2 = 10^{-10.33} = \frac{\{H^+\}\{CO_3^{2-}\}}{\{HCO_3^-\}} \tag{11.3}$$

$$K_{sp,CaCO_3} = 4.8(10^{-9}) = \{Ca^{2+}\}\{CO_3^{2-}\} \tag{11.4}$$

The Ks are the values of the respective equilibrium constants. $K_{sp,CaCO_3}$ is the equilibrium constant for the solubility of $CaCO_3$. The pair of braces, { }, are read as "the activity of," the meaning of which is explained in the Background Chemistry and Fluid Mechanics chapter in the Background Prerequisites section.

As shown, the equilibrium constants are calculated using the activity. In simple language, *activity* is a measure of the effectiveness of a given species in its participation in a reaction. It is proportional to concentration; it is an *effective* or *active concentration* and has units of concentrations. Because activity bears a relationship to concentration, its value may be obtained using the value of the corresponding concentration. This relationship is expressed as follows:

$$\{sp\} = \gamma[sp] \tag{11.5}$$

where sp represents any species involved in the equilibria such as Ca^{2+}, CO_3^{2-}, HCO_3^- and so on. The pair of brackets, [], is read as "the concentration of," γ is the activity coefficient.

11.1.1 Ionic Strength

As the particle ionizes, the number of particles increases. Thus, it is not a surprise that activity coefficient is a function of the number of particles in solution. The number of particles is characterized by the ionic strength μ. This parameter was devised by Lewis and Randall (1980) to describe the electric field intensity of a solution:

$$\mu = \frac{1}{2}\sum[sp_i]z_i^2 \tag{11.6}$$

i is the index for the particular species and z is its charge. The concentrations are in gmmols/L. In terms of the ionic strength, the activity coefficient is given by the *DeBye-Huckel law* as follows (Snoeyink and Jenkins, 1980; Rich, 1963):

$$-\log\gamma = \frac{0.5z_i^2(\sqrt{\mu})}{1 + 1.14(\sqrt{\mu})} \tag{11.7}$$

$$\gamma = 10^{-\frac{0.5z_i^2(\sqrt{\mu})}{1+1.14(\sqrt{\mu})}} \tag{11.8}$$

In 1936, Langelier presented an approximation to the ionic strength μ. Letting TDS in mg/L represent the total dissolved solids, his approximation is

$$\mu = 2.5(10^{-5})TDS \tag{11.9}$$

Also, in terms of the specific conductance, *sp conduc* (in mmho/cm), Russell, another researcher, presented yet another approximation as

$$\mu = 1.6(10^{-5})sp\ conduc \tag{11.10}$$

Example 11.1 The pH of a solution is 7. Calculate the hydrogen ion concentration?

Solution:

$$pH = -\log_{10}\{H^+\}$$

$$7 = -\log_{10}\{H^+\} \qquad \{H^+\} = 10^{-7}\ \text{gmols/L} \quad \textbf{Ans}$$

Example 11.2 The concentration of carbonic acid was analyzed to be 0.2 mgmol/L. If the pH of the solution is 7, what is the concentration of the bicarbonate ion if the temperature is 25°C?

Solution:

$$K_1 = 10^{-6.35} = \frac{\{H^+\}\{HCO_3^-\}}{\{H_2CO_3^*\}} = \frac{10^{-7}\{HCO_3^-\}}{0.2/1000}$$

$$\{HCO_3^-\} = 8.93(10^{-4})\ \text{gmol/L} = [HCO_3^-] \quad \textbf{Ans}$$

Example 11.3 A sample of water has the following composition: $CO_2 = 22.0$ mg/L, $Ca^{2+} = 80$ mg/L, $Mg^{2+} = 12.0$ mg/L, $Na^+ = 46.0$ mg/L, $HCO_3^- = 152.5$ mg/L, and $SO_4^{2-} = 216$ mg/L. What is the ionic strength of the sample?

Solution:

$$\mu = \frac{1}{2}\sum [sp_i]z_i^2$$

Ion	mg/L	Mol. Mass	gmols/L
Ca^{2+}	80	40.1	0.001995
Mg^{2+}	12.0	24.3	0.0004938
Na^+	46.0	23	0.002
HCO_3^-	152.5	61	0.0025
SO_4^{2-}	216	96.1	0.0022

$$\mu = \frac{1}{2}[0.001995(2^2) + 0.0004938(2^2) + 0.002(1) + 0.0025(1) + 0.0022(2^2)]$$

$$= 0.023 \quad \textbf{Ans}$$

Example 11.4 In Example 11.3, calculate the activity coefficient and the activity in mg/L of the bicarbonate ion.

Solution:

$$\gamma = 10^{-\frac{0.5z_i^2(\sqrt{\mu})}{1+1.14(\sqrt{\mu})}} = 10^{-\frac{0.5(1)(\sqrt{0.023})}{1+1.14(\sqrt{0.023})}=0.86}$$

$$\{sp\} = \gamma[sp] = 0.86(0.0025) = 0.00215 \text{ mg/L} \quad \textbf{Ans}$$

11.1.2 Equilibrium Constant as a Function of Temperature

The equilibrium constants given previously were at 25°C. To find the values of the equilibrium constants at other temperatures, the *Van't Hoff equation* is needed. According to this equation, the equilibrium constant K (K_{sp} for the solubility product constants) is related to temperature according to a derivative as follows:

$$\frac{d\ln K}{dT} = \frac{\Delta H^o}{RT^2} \tag{11.11}$$

T is the absolute temperature; ΔH^o is the standard enthalpy change, where the standard enthalpy change has been adopted as the change at 25°C at one atmosphere of pressure; and R is the universal gas constant.

The value of R depends upon the unit used for the other variables. Table 11.1 gives its various values and units, along with the units used for ΔH^o and T. By convention, the concentration units used in the calculation of K are in gmmols/L.

Enthalpy is heat released or absorbed in a chemical reaction at *constant pressure*. Table 11.2 shows values of interest in water stabilization. It is normally reported as enthalpy changes. There is no such thing as An absolute value of an enthalpy does not exist, only a change in enthalpy. Enthalpy is a heat exchange at constant pressure, so enthalpy changes are measured by allowing heat to transfer at constant pressure; the amount of heat measured during the process is the enthalpy change. Also, the table indicates enthalpy of formation. This means that the values in the table are the heat

TABLE 11.1
Values and Units of R

R Value	R Units	K Concentration Units Used	ΔH^o Units	T Units
0.08205	$\dfrac{L \text{ atm}}{gmmol.K^o}$	$\dfrac{gmmols}{L}$	—	°K
8.315	$\dfrac{J}{gmmol.K^o}$	$\dfrac{gmmols}{L}$	$\dfrac{J}{gmmol}$	°K
1.987	$\dfrac{cal}{gmmol.K^o}$	$\dfrac{gmmols}{L}$	$\dfrac{cal}{gmmol}$	°K
82.05	$\dfrac{atm.cm^3}{gmmol.K^o}$	$\dfrac{gmmols}{L}$	—	°K

From J. M. Montgomery Engineers, Pasadena, CA.

TABLE 11.2
Enthalpies of Formation of Substances of
Interest in Stabilization

Substance	ΔH^o_{298}, kcal/gmmol
$HOH_{(l)}$	−68.317
$H^+_{(aq)}$	0
$OH^-_{(aq)}$	−54.96
$H_2CO_3^*$	−167.0
$CO_{2(aq)}$	−98.69
$CO_{3(aq)}^{2-}$	−161.63
$HCO_{3(aq)}^-$	−165.18
$CaCO_3(s)$	−288.45
$Ca^{2+}_{(aq)}$	−129.77
$Ca(OH)_{2(aq)}$	−239.2
$Mg(OH)_{2(aq)}$	−221.0
Mg^{2+}	−110.41

---

measured when the particular substance was formed from its elements. For example, when calcium carbonate solid was formed from its elements calcium, carbon and oxygen, −288.45 kcal of heat per gmmol of calcium carbonate was measured. The negative sign means that the heat measured was released or liberated in the chemical reaction.

Also, the state of the substance when it was formed is also indicated in the table. For example, the state when calcium carbonate is formed liberating heat in the amount of −288.45 kcal/gmmol is solid, indicated as s. The symbol l means that the state is liquid and the symbol aq means that the substance is being formed in water solution.

Also, note the subscript and superscript. They indicate that the values in the table were obtained at standard temperature and pressure and one unit of activity for the reactants and products. The standard temperature is 25°C; thus the 298, which is the *Kelvin* equivalent of 25°C. The standard pressure is 1 atmosphere. The superscript ^{o} symbolizes unit activity of the substances. This means that the elements from which the substances are formed were all at a unit of activity and the product substances formed are also all at a unit of activity.

The enthalpy change is practically constant with temperature; thus ΔH^o may be replaced by ΔH^o_{298}. Doing this and integrating the Van't Hoff equation from K_{T1} to K_{T2} for the equilibrium constant K and from T_1 to T_2 for the temperature,

$$K_{T2} = K_{T1}\exp\left[\frac{\Delta H^o_{298}}{RT_1 T_2}(T_2 - T_1)\right] \qquad (11.12)$$

This equation expresses the equilibrium constant as a function of temperature.

11.1.3 ΔH^o_{298}'s for Pertinent Chemical Reactions of the Carbonate Equilibria

Let us now derive the values of the ΔH^o_{298} of the various pertinent chemical reactions in the carbonate equilibria as shown in Eqs. (11.1) through (11.4). According to *Hess's law*, if the chemical reaction can be written in steps, the enthalpy changes can be obtained as the sum of the steps.

Thus, consider Equation (11.1). The corresponding reaction is

$$HOH \; \rightleftharpoons \; H^+ + OH^- \tag{11.13}$$

Writing in steps to conform to Hess's law:

$$HOH_{(l)} \rightarrow H_2 + \frac{1}{2}O_2 \qquad \Delta H^o_{298} = +68.317 \text{ kcal/gmmol of } HOH_{(l)}$$

$$\frac{1}{2}H_2 \rightarrow H^+_{(aq)} \qquad \Delta H^o_{298} = 0$$

$$\frac{1}{2}H_2 + \frac{1}{2}O_2 \rightarrow OH^-_{(aq)} \qquad \Delta H^o_{298} = -54.96 \text{ kcal/gmmol of } OH^-_{(aq)}$$

$$\overline{HOH_{(l)} \; \rightleftharpoons \; H^+_{(aq)} + OH^-_{(aq)}} \qquad \Delta H^o_{298} = +13.36 \text{ kcal/gmmol of } HOH_{(l)}$$

The values of the ΔH^o_{298}'s are obtained from Table 11.2. Note that the values in the table indicate ΔH^o_{298} of formation having negative values. Thus, if the reaction is not a formation but a breakup such as $HOH_{(l)} \rightarrow H_2 + \frac{1}{2}O_2$, the sign is positive. This reaction indicates that to break the water molecule into its constituent atoms +68.317 kcal/gmmol of energy is required. The + sign indicates that the reaction is endothermic requiring energy for the reaction to occur. For the ionization of the water molecule as represented by $HOH_{(l)} \rightleftharpoons H^+_{(aq)} + OH^-_{(aq)}$ and using Hess's law as shown previously, +13.36 kcal/gmol of $HOH_{(l)}$ is required.

The Hess's law steps for the rest of Eqs. (11.1) through (11.4) are detailed as follows:

$$H_2CO^*_3 \rightarrow H_2 + C + 6O_2 \qquad \Delta H^o_{298} = +167.0 \text{ kcal/gmmol of } H_2CO^*_{3(aq)}$$

$$\frac{1}{2}H_2 \rightarrow H^+_{(aq)} \qquad \Delta H^o_{298} = 0$$

$$\frac{1}{2}H_2 + C + 6O_2 \rightarrow HCO^-_{3(aq)} \qquad \Delta H^o_{298} = -165.18 \text{ kcal/gmmol of } HCO_{3(aq)}$$

$$\overline{H_2CO^*_{3(aq)} \; \rightleftharpoons \; H^+_{(aq)} + HCO^-_{3(aq)}} \qquad \Delta H^o_{298} = +1.82 \text{ kcal/gmmol of } H_2CO_{3(aq)}$$

$$HCO_{3(aq)}^- \rightarrow \frac{1}{2}H_2 + C + 6O_2 \qquad \Delta H_{298}^o = +165.18 \text{ kcal/gmmol of } HCO_{3(aq)}$$

$$\frac{1}{2}H_2 \rightarrow H_{(aq)}^+ \quad \Delta H_{298}^o = 0$$

$$C + 6O_2 \rightarrow CO_{3(aq)}^{2-} \qquad \Delta H_{298}^o = -161.63 \text{ kcal/gmmol of } CO_{3(aq)}^{2-}$$

$$HCO_{3(aq)} \rightleftharpoons H_{(aq)}^+ + CO_{3(aq)}^{2-} \qquad \Delta H_{298}^o = +3.55 \text{ kcal/gmmol of } HCO_{3(aq)}$$

$$CaCO_{3(s)} \rightarrow Ca + C + 6O_2 \qquad \Delta H_{298}^o = +288.45 \text{ kcal/gmmol of } CaCO_{3(s)}$$

$$Ca \rightarrow Ca_{(aq)}^{2+} \qquad \Delta H_{298}^o = -129.77 \text{ kcal/gmmol of } Ca_{(aq)}^{2+}$$

$$C + 6O_2 \rightarrow CO_{3(aq)}^{2-} \qquad \Delta H_{298}^o = -161.63 \text{ kcal/gmmol of } CO_{3(aq)}^{2-}$$

$$CaCO_{3(s)} \rightarrow Ca_{(aq)}^{2+} + CO_{3(aq)} \qquad \Delta H_{298}^o = -2.95 \text{ kcal/gmmol of } CaCO_{3(s)}$$

Example 11.5 A softened municipal water supply enters a residence at 15°C and is heated to 60°C in the water heater. Compare the values of the equilibrium constants for $CaCO_3$ at these two temperatures. If the water was at equilibrium at 25°C, determine if $CaCO_3$ will deposit or not at these two temperatures.

Solution:

$$K_{T2} = K_{T1}\exp\left[\frac{\Delta H_{298}^o}{RT_1T_2}(T_2 - T_1)\right]$$

$$K_{sp,CaCO_3} = \{Ca^{2+}\}\{CO_3^{2-}\} = 4.8(10^{-9})$$

$$= K_{T1} = 4.8(10^{-9}) \text{ in gmol units at } 25°C$$

$$\Delta H_{298}^o = -2.95 \text{ kcal/gmmol of } CaCO_{3(s)} = -2,950 \text{ cal/gmmol of } CaCO_{3(s)}$$

$$R = 1.987\frac{cal}{gmmol.K°} \qquad T_1 = 25 + 273 = 298°K$$

$$T_2 = 15 + 273 = 288°K \qquad T_2 = 60 + 273 = 333°K$$

Therefore,

$$K_{T2} \text{ at } 15°K = 4.8(10^{-9})\exp\left[\frac{2,950}{1.987(298)(288)}(288 - 298)\right]$$

$$= 4.038(10^{-9}) \text{ in gmol units}$$

Therefore,

$$K_{T2} \text{ at } 60°K = 4.8(10^{-9})\exp\left[\frac{2{,}950}{1.987(298)(333)}(333 - 298)\right]$$

$$= 8.10(10^{-9}) \text{ in gmol units}$$

Thus, the value of equilibrium constant is greater at 60°C than at 15°C. **Ans**

The value of the equilibrium constant for calcium carbonate at 25°C is $4.8(10^{-9})$. At this condition, the ions Ca^{2+} and CO_3^{2-} ions are given to be in equilibrium; thus, will neither deposit nor dissolve $CaCO_3$. At the temperature of 15°C, the value of the equilibrium constant is $4.038(10^{-9})$. This value is less than $4.8(10^{-9})$ and will require less of the ionized ion; therefore at 15°C, the water is oversaturated and will deposit $CaCO_3$. **Ans**

At 60°C, the equilibrium constant is $8.10(10^{-9})$, which is greater than that at 25°C. Thus, at this temperature, the water is undersaturated and will not deposit $CaCO_3$. **Ans**

11.2 CRITERIA FOR WATER STABILITY AT NORMAL CONDITIONS

In the preceding discussions, a criterion for stability was established using the equilibrium constant called K_{sp}. At normal conditions, as especially used in the water works industry, specialized forms of water stability criteria have been developed. These are saturation pH, *Langelier index*, and the precipitation potential of a given water.

11.2.1 SATURATION pH AND THE LANGELIER INDEX

Because pH is easily determined by simply dipping a probe into a sample, determination of the saturation pH is a convenient method of determining the stability of water. The concentrations of any species at equilibrium conditions are in equilibrium with respect to each other. Also, for solids, if the condition is at equilibrium no precipitate or scale will form. One of the concentration parameters of equilibrium is the hydrogen ion concentration, which can be ascertained by the value of the pH. Thus, if the pH of a sample is determined, this can be compared with the equilibrium pH to see if the water is stable or not. Therefore, we now proceed to derive the equilibrium pH. Equilibrium pH is also called *saturation pH*.

In natural systems, the value of the pH is strongly influenced by the carbonate equilibria reactions. The CO_3^{2-} species of these reactions will pair with a cation, thus "guiding" the equilibrium reactions into a *dead end* by forming a precipitate. For example, the complete carbonate equilibria reactions are as follows:

$$HOH \rightleftharpoons H^+ + OH^- \quad K_w \tag{11.14}$$

$$H_2CO_3^* \rightleftharpoons H^+ + HCO_3^- \quad K_1 \tag{11.15}$$

$$HCO_{3(aq)}^- \rightleftharpoons H_{(aq)}^+ + CO_3^{2-} \quad K_2 \tag{11.16}$$

$$Cation_2(CO_3)_{c(s)}\downarrow \rightleftharpoons 2Cation^{c+} + cCO_3^{2-} \quad K_{sp} \tag{11.17}$$

c is the charge of the cation that pairs with CO_3^{2-} forming the precipitate $Cation_2(CO_3)_{c(s)}$. We call the formation of this precipitate as the dead end of the carbonate equilibria, since the carbonate species in solution are diminished by the precipitation.

Let us digress for a moment from our discussion of the saturation pH in order to find the dead end cation for the carbonate system equilibria. Several of these cations can possibly pair with the carbonate. The pairing will be governed by the value of the solubility product constant, K_{sp}. A small value of the K_{sp} means that only small values of the concentration of the constituent species are needed to form a product equal to the K_{sp}. This, in turn, means that solids with smaller K_{sp}'s will easily form the solids. Thus, of all the possible cations that can pair with the carbonate, the one with the smallest K_{sp} value is the one that can form a dead end for the carbonate equilibria reactions. Mg forms $MgCO_3$ with a K_{sp} of 10^{-5}. Ca forms $CaCO_3$ with a K_{sp} of $4.8(10^{-9})$. Table 11.3 shows other carbonate solids with the respective solubility product constants.

From the previous table, the smallest of the K_{sp}'s is that for Hg_2CO_3. Thus, considering all of the possible candidates that we have written, Hg_2CO_3 is the one that will form a dead end for the carbonate equilibria; however, of all the possible cations, Ca^{2+} is the one that is found in great abundance in nature compared to the rest. Thus, although all the other cations have much more smaller K_{sp}'s than calcium,

---

TABLE 11.3
Solubility Product Constants of Solid Carbonates at 25°C

Carbonate Solid	K_{sp}
$BaCO_3$	$8.1(10^{-9})$
$CdCO_3$	$2.5(10^{-14})$
$CaCO_3$	$4.8(10^{-9})$
$CoCO_3$	$1.0(10^{-12})$
$CuCO_3$	$1.37(10^{-10})$
$FeCO_3$	$2.11(10^{-11})$
$PbCO_3$	$1.5(10^{-13})$
$MgCO_3$	$1.0(10^{-5})$
$MnCO_3$	$8.8(10^{-11})$
Hg_2CO_3	$9(10^{-17})$
$NiCO_3$	$1.36(10^{-7})$
Ag_2CO_3	$8.2(10^{-12})$
$SrCO_3$	$9.42(10^{-10})$
$ZnCO_3$	$6(10^{-11})$

they are of no use as dead ends if they do not exist. The other cation that exists in abundance in natural waters is magnesium. In fact, this is the other constituent hardness ion in water. Comparing the K_{sp}'s of the carbonate of these cations, however, $CaCO_3$ is the smaller. Thus, as far as the carbonate equilibria reactions are concerned, the calcium ion is the one to be considered to form a dead end in the carbonate system equilibria. $Cation_2(CO_3)_{c(s)}$ is therefore $CaCO_{3(s)}$. For this reason, Equation (11.4) was written in terms of $CaCO_3$. (See Table 11.4).

As will be shown later, the saturation pH may conveniently be expressed in terms of total alkalinity and other parameters. The *alkalinity of water* is defined as its capacity to neutralize any acid added to it. When an acid represented by H^+ is added to a hydroxide represented by OH^-, the acid will be neutralized according to the reaction $H^+ + OH^- \rightleftharpoons HOH$. Thus, the hydroxide is an alkaline substance. When the acid is added to a carbonate, the acid is also neutralized according to the reaction $2H^+ + CO_3^{2-} \rightleftharpoons H_2CO_3$. Carbonate is therefore also an alkaline substance. By writing a similar reaction, the bicarbonate ion will also be shown to be an alkaline substance.

As we know, these species are the components of the carbonate equilibria. They also represent as components of the total alkalinity of the carbonate system equilibria. They may be added together to produce the value of the total alkalinity. To be additive, each of these component alkalinities should be expressed in terms of a common unit. A convenient common unit is the *gram equivalent*.

$[OH^-]$ is equal to $\{OH^-\}/\gamma_{OH}$, where γ_{OH} is the activity coefficient of the hydroxyl ion. $\{OH^-\}$ could be eliminated in terms of the ion product of water, $K_w = \{H^+\}\{OH^-\}$.

To establish the equivalence of the component alkalinities, they must all be referred to a common end point when the acid H^+ is added to the solution. From general chemistry, we learned that this is the methyl orange end point. As far as the OH^- ion is concerned, the end point for the reaction $H^+ + OH^- \rightleftharpoons HOH$ has already been reached well before the methyl orange end point. Thus, for the purpose of determining equivalents, the reaction for the hydroxide alkalinity is simply $H^+ + OH^- \rightleftharpoons HOH$ and the equivalent mass of the hydroxide is $OH/1$. One gram equivalent of the hydroxide is then equal to one gram mole. Therefore,

$$[OH^-]_{geq} = [OH^-] = \frac{\{OH^-\}}{\gamma_{OH}} = \frac{K_w}{\gamma_{OH}\{H^+\}} \tag{11.18}$$

$[HCO_3^-] = \{HCO_3^-\}/\gamma_{HCO_3}$, where γ_{HCO_3} is the activity coefficient of the bicarbonate ion. From Equation (11.3), $\{HCO_3^-\} = \{H^+\}\{CO_3^{2-}\}/K_2$; thus, $[HCO_3^-] = \{H^+\}\{CO_3^{2-}\}/\gamma_{HCO_3}K_2$. Reaction of the acid H^+ with the bicarbonate given by $H^+ + HCO_3^- \rightleftharpoons H_2CO_3^{2-}$ ends exactly at the methyl orange end point. From this reaction, the equivalent weight of the bicarbonate ion is $HCO_3/1$; thus, one gram equivalent is equal to one gram mole. Therefore,

$$[HCO_3^-]_{geq} = [HCO_3^-] = \frac{\{H^+\}\{CO_3^{2-}\}}{\gamma_{HCO_3}K_2} \tag{11.19}$$

$[CO_3^{2-}] = \{CO_3^{2-}\}/\gamma_{CO_3}$, where γ_{CO_3} is the activity coefficient of the carbonate ion. From Equation (11.4), $\{CO_3^{2-}\} = K_{sp,CaCO_3}/\{Ca^{2+}\}$; thus, $[CO_3^{2-}] = K_{sp,CaCO_3}/\gamma_{CO_3}\{Ca^{2+}\}$. Reaction of the acid H^+ with the carbonate ion given by $2H^+ + CO_3^{2-} \rightleftharpoons H_2CO_3$ also ends exactly at the methyl orange end point. From this reaction, the equivalent mass of the carbonate ion is $CO_3/2$; thus, one gram equivalent is equal to $1(CO_3/2)/CO_3 = \frac{1}{2}$ gram mole. Therefore,

$$[CO_3^{2-}]_{geq} = \frac{[CO_3^{2-}]}{2} = \frac{K_{sp,CaCO_3}}{2\gamma_{CO_3}\{Ca^{2+}\}} \tag{11.20}$$

Using the equation $\{CO_3^{2-}\} = K_{sp,CaCO_3}/\{Ca^{2+}\}$, Equation (11.19) becomes

$$[HCO_3^-]_{geq} = [HCO_3^-] = \frac{\{H^+\}\{CO_3^{2-}\}}{\gamma_{HCO_3}K_2} = \frac{\{H^+\}K_{sp,CaCO_3}}{\gamma_{HCO_3}K_2\{Ca^{2+}\}} \tag{11.21}$$

The alkalinity of water has been defined as its capacity to react with any acid added to it. Thus, if any hydrogen ion is present, this must be subtracted to reflect the overall alkaline capacity of the water. Letting $[A]_{geq}$ represent the total alkalinity,

$$[A]_{geq} = [OH^-]_{geq} + [HCO_3^-]_{geq} + [CO_3^{2-}]_{geq} - [H^+] \tag{11.22}$$

$$[A]_{geq} = \frac{K_w}{\gamma_{OH}\{H^+\}} + \frac{\{H^+\}K_{sp,CaCO_3}}{\gamma_{HCO_3}K_2\{Ca^{2+}\}} + \frac{K_{sp,CaCO_3}}{2\gamma_{CO_3}\{Ca^{2+}\}} - \frac{\{H^+\}}{\gamma_H} \tag{11.23}$$

Let

$$A = [A]_{geq}, \quad B = \frac{K_w}{\gamma_{OH}}, \quad C = \frac{K_{sp,CaCO_3}}{\gamma_{HCO_3}K_2\{Ca^{2+}\}} - \frac{1}{\gamma_H}, \quad \text{and} \quad D = \frac{K_{sp,CaCO_3}}{2\gamma_{CO_3}\{Ca^{2+}\}}.$$

Solving for $\{H^+\}$,

$$\{H^+\} = \frac{-(D-A) + \sqrt{(D-A)^2 - 4CB}}{2C} = \{H_s^+\} \tag{11.24}$$

Thus, the saturation pH, pH_s is

$$pH_s = -\log_{10}\{H_s^+\} \tag{11.25}$$

The *Langlier Index* (or *Saturation Index*) (LI) is the difference between the actual pH and the saturation pH of a solution, thus

$$LI = pH - pH_s \tag{11.26}$$

A positive value of the Langelier index indicates that the water is supersaturated and will deposit $CaCO_3$, whereas a negative value indicates that the water is undersaturated and will dissolve any $CaCO_3$ that happens to exist at the particular moment.

Example 11.6 Analysis of a water sample yields the following results: [TDS] = 140 mg/L, $[Ca^{2+}]$ = 0.7 mgmol/L, $[Mg^{2+}]$ = 0.6 mgmol/L, $[A]_{mgeq}$ = 0.4 mgeq/L, temperature = 20°C, and pH = 6.7. Calculate the saturation pH and determine if the water will deposit or dissolve $CaCO_3$.

Solution:

$$\{H_s^+\} = \frac{-(D-A)+\sqrt{(D-A)^2-4CB}}{2C}$$

$$A = [A]_{geq}, \quad B = \frac{K_w}{\gamma_{OH}}, \quad C = \frac{K_{sp,CaCO_3}}{\gamma_{HCO_3}K_2\{Ca^{2+}\}} - \frac{1}{\gamma_H}, \quad \text{and} \quad D = \frac{K_{sp,CaCO_3}}{2\gamma_{CO_3}\{Ca^{2+}\}}$$

$A = 0.0004$ geq/L

$$B = \frac{K_w}{\gamma_{OH}}$$

$$\mu = 2.5(10^{-5})\text{TDS} \quad \gamma = 10^{-\frac{0.5z_i^2(\sqrt{\mu})}{1+1.14(\sqrt{\mu})}}$$

$$\mu = 2.5(10^{-5})(140) = 3.5(10^{-3}) \qquad \gamma_{OH} = 10^{-\frac{0.5(1)^2\{\sqrt{3.5(10^{-3})}\}}{1+1.14\{\sqrt{3.5(10^{-3})}\}}} = 0.94$$

$$K_{T2} = K_{T1}\exp\left[\frac{\Delta H_{298}^o}{RT_1T_2}(T_2-T_1)\right] = K_w 20°C$$

$$K_w = 10^{-14} = \{H^+\}\{OH^-\} \text{ at } 25°C = K_{T1}$$

ΔH_{298}^o for K_w = +13.36 kcal/gmmol of $HOH_{(l)}$ = +13,360 cal/gmmol

$R = 1.987\dfrac{\text{cal}}{\text{gmmol.K}°} \quad T_2 = 20+273 = 293°\text{K} \quad T_1 = 25+273 = 298°\text{K}$

Therefore,

$$K_{T2} = K_w 20°C = (10^{-14})\exp\left[\frac{13,360}{1.987(298)(293)}(293-298)\right] = 6.80(10^{-15})$$

Therefore,

$$B = \frac{6.80(10^{-15})}{0.94} = 7.24(10^{-15})$$

$$C = \frac{K_{sp,CaCO_3}}{\gamma_{HCO_3} K_2 \{Ca^{2+}\}} - \frac{1}{\gamma_H}:$$

$$K_{sp,CaCO_3} = 4.8(10^{-9}) \text{ at } 25°C \quad \Delta H^o_{298} \text{ for } CaCO_3$$
$$= -2.95 \text{ kcal/gmmol of } CaCO_{3(s)}$$

Therefore,

$$K_{sp,CaCO_3} \text{ at } 20°C = 4.8(10^{-9}) \exp\left[\frac{-2,950}{1.987(298)(293)}(293 - 298)\right] = 5.23(10^{-9})$$

$$\gamma_{HCO_3} = \gamma_{OH} = 0.94 \quad K_2 = 10^{-10.33} \ 25°C$$

$$\Delta H^o_{298} \text{ for } HCO_3^- = +3.55 \text{ kcal/gmmol of } HCO_{3(aq)}$$

Therefore,

$$K_2 \text{ at } 20°C = 10^{-10.33} \exp\left[\frac{3,550}{1.987(298)(293)}(293 - 298)\right] = 4.22(10^{-11})$$

$${Ca^{2+}} = \gamma_{Ca}[Ca^{2+}] = \gamma_{Ca}(0.0007) \ \gamma_{Ca} = 10^{-\frac{0.5(2)^2\{\sqrt{3.5(10^{-3})}\}}{1+1.14\{\sqrt{3.5(10^{-3})}\}}} = 0.77$$

$${Ca^{2+}} = (0.77)(0.0007) = 0.00054 \text{ gmol/L}$$

Therefore,

$$C = \frac{5.23(10^{-9})}{0.94(4.22)(10^{-11})(0.00054)} - \frac{1}{0.94} = 2.44(10^5)$$

$$D = \frac{K_{sp,CaCO_3}}{2\gamma_{CO_3}\{Ca^{2+}\}}:$$

$$\gamma_{CO_3} = \gamma_{Ca} = 0.77$$

Therefore,

$$D = \frac{5.23(10^{-9})}{2(0.77)(0.00054)} = 6.28(10^{-6})$$

Therefore,

$$\{H_s^+\} = \frac{-[6.28(10^{-6}) - 0.0004] + \sqrt{[6.28(10^{-6}) - 0.0004]^2 - 4[2.44(10^5)][7.24(10^{-15})]}}{2[2.44(10^5)]}$$

$$= \frac{3.94(10^{-4}) + \sqrt{1.55(10^{-7}) - 7.07(10^{-9})}}{2[2.44(10^5)]}$$

$$= \frac{3.94(10^{-4}) + 3.85(10^{-4})}{2[2.44(10^5)]} = 1.60(10^{-9})$$

$$pH_s = -\log_{10}\{H_s^+\} = -\log_{10}[1.60(10^{-9})] = 8.8 \quad \textbf{Ans}$$

LI = pH − pH_s = 6.7 − 8.8 = −2.1 and the water will not deposit $CaCO_3$ but will dissolve it.   **Ans**

Example 11.7   In Example 11.6, if the pH_s were actually 8.0, what would be the concentration of the calcium ion in equilibrium with $CaCO_3$ at this condition? Assume the rest of the data holds.

Solution:

$$[A]_{geq} = \frac{K_w}{\gamma_{OH}\{H^+\}} + \frac{\{H^+\}K_{sp,CaCO_3}}{\gamma_{HCO_3}K_2\{Ca^{2+}\}} + \frac{K_{sp,CaCO_3}}{2\gamma_{CO_3}\{Ca^{2+}\}} - \frac{\{H^+\}}{\gamma_H}$$

$$0.0004 = \frac{6.80(10^{-15})}{0.94(10^{-8})} + \frac{10^{-8}[5.23(10^{-9})]}{0.94[4.22(10^{-11})]\{Ca^{2+}\}} + \frac{5.23(10^{-9})}{2(0.77)\{Ca^{2+}\}} - \frac{10^{-8}}{0.94}$$

$$0.0004 = 7.23(10^{-7}) + \frac{1.32(10^{-6})}{\{Ca^{2+}\}} + \frac{3.4(10^{-9})}{\{Ca^{2+}\}} - 1.06(10^{-8})$$

$$3.99(10^{-4})\{Ca^{2+}\} = 1.323(10^{-6})$$

$$\{Ca^{2+}\} = 3.32(10^{-3}) \text{ gmol/L} \gg 7(10^{-4}) \text{ gmol/L} \quad \textbf{Ans}$$

$$\text{or } [Ca^{2+}] = 133 \text{ mg/L} = \frac{133}{0.77} = 173 \text{ mg/L} \quad \textbf{Ans}$$

11.2.2 DETERMINATION OF $\{Ca^{2+}\}$

The activity of the calcium ion is affected by its complexation with anions. Ca^{2+} forms complexes with the carbonate species, OH^- and SO_4^{2-}. The complexation reactions are shown as follows:

$$CaCO_3^o \rightleftharpoons Ca^{2+} + CO_3^{2-} \tag{11.27}$$

$$CaHCO_3^+ \rightleftharpoons Ca^{2+} + HCO_3^- \tag{11.28}$$

TABLE 11.4
Equilibrium Constants for Various Complexes of Calcium at 25°C

Complex	Constant Symbol	Value
$CaCO_3^o$	$K_{CaCO_3 c}$	$10^{-3.22}$
$CaHCO_3^+$	$K_{CaHCO_3 c}$	$10^{-1.26}$
$CaOH^+$	$K_{CaOH c}$	$10^{-1.49}$
$CaSO_4^o$	$K_{CaSO_4 c}$	$10^{-2.31}$

$$CaOH^+ \rightleftharpoons Ca^{2+} + OH^- \qquad (11.29)$$

$$CaSO_4^o \rightleftharpoons Ca^{2+} + SO_4^{2-} \qquad (11.30)$$

These complexes are weak enough that, in the complexometric titration determination using EDTA, they break and are included in the total calcium hardness reported. Thus, the total calcium hardness is composed of the "legitimate" cation, Ca^{2+}, plus the complex ions as shown in the previous equations. This total calcium hardness must be corrected by the concentrations of the complexes in order to determine the correct activities of the calcium ions. Let the total concentration of the calcium species as determined by the EDTA titration be $[Ca_T]$. Thus, the concentration of the calcium ion $[Ca_{(aq)}^{2+}]$ is

$$[Ca_{(aq)}^{2+}] = [Ca_T] - ([CaCO_3^o] + [CaHCO_3^+] + [CaOH^+] + [CaSO_4^o]) \quad (11.31)$$

Table 11.4 shows the equilibrium constants of the previous complexes at 25°C. For other temperatures, these values must be corrected using the Van't Hoff equation. The use of this equation, however, requires the value of the standard heat of formation ΔH_{298}^o. At present, none are available for $CaCO_3^o$, $CaHCO_3^+$, $CaOH^+$, and $CaSO_4^o$. Research is therefore needed to find these values.

Determination of the Calcium Complexes. The whole thrust of the discussion regarding water stability is the determination of whether $CaCO_3$ precipitates at a given solution condition. The concentration of $CaCO_3^o$ will therefore be determined in relation to the solubility product of $CaCO_3$. Applying Hess's law,

$$CaCO_{3s} \rightleftharpoons Ca^{2+} + CO_3^{2-} \qquad K_{sp,CaCO_3}$$

$$Ca^{2+} + CO_3^{2-} \rightleftharpoons CaCO_3^o \qquad \frac{1}{K_{CaCO_3 c}}$$

$$CaCO_{3s} \rightleftharpoons CaCO_3^o \qquad K_{sp,CaCO_3}\left(\frac{1}{K_{CaCO_3 c}}\right)$$

Note: When adding the equations, the respective equilibrium constants are multiplied. The activity of a solid is equal to unity.

Thus, $\{CaCO_{3(s)}\} = 1$ and

$$\frac{\{CaCO_3^o\}}{\{CaCO_{3(s)}\}} = \frac{\{CaCO_3^o\}}{1} = K_{sp,CaCO_3}\left(\frac{1}{K_{CaCO_3c}}\right) \tag{11.32}$$

Therefore,

$$[CaCO_3^o] = \frac{\{CaCO_3^o\}}{\gamma_{CaCO_3c}} = \frac{K_{sp,CaCO_3}}{K_{CaCO_3c}} \tag{11.33}$$

The activity coefficient γ_{CaCO_3c} of $CaCO_3^o$ is unity, since it is not dissociated.

The determination of $[CaHCO_3^+]$ will use, in addition to the equilibrium constant for $CaCO_{3(s)}$, the ionization constant for the bicarbonate ion, K_2. The steps are as follows:

$$CaCO_{3(s)} \rightleftharpoons Ca^{2+} + CO_3^{2-} \qquad K_{sp,CaCO_3}$$

$$H^+ + CO_3^{2-} \rightleftharpoons HCO_3^- \qquad \frac{1}{K_2}$$

$$Ca^{2+} + HCO_3^- \rightleftharpoons CaHCO_3^+ \qquad \frac{1}{K_{CaHCO_3c}}$$

$$\overline{CaCO_{3(s)} + H^+ \rightleftharpoons CaHCO_3^+ \qquad K_{sp,CaCO_3}\left(\frac{1}{K_2}\right)\left(\frac{1}{K_{CaHCO_3c}}\right)}$$

Therefore,

$$\frac{\{CaHCO_3^+\}}{\{CaCO_{3(s)}\}\{H^+\}} = \frac{\{CaHCO_3^+\}}{\{H^+\}} = \frac{K_{sp,CaCO_3}}{K_2 K_{CaHCO_3c}} \tag{11.34}$$

$$[CaHCO_3^+] = \frac{\{CaHCO_3^+\}}{\gamma_{CaHCO_3c}} = \frac{\gamma_H K_{sp,CaCO_3}}{\gamma_{CaHCO_3c} K_2 K_{CaHCO_3c}}[H^+] \tag{11.35}$$

γ_H and γ_{CaHCO_3c} are the activity coefficients of H^+ and $CaHCO_3^+$, respectively. The determination of $[CaOH^+]$ is shown in the steps that follow:

$$HOH \rightleftharpoons H^+ + OH^- \qquad K_w$$

$$Ca^{2+} + OH^- \rightleftharpoons CaOH^+ \qquad \frac{1}{K_{CaOHc}}$$

$$\overline{HOH + Ca \rightleftharpoons H^+ + CaOH^+ \qquad (K_w)\left(\frac{1}{K_{CaOHc}}\right)}$$

Therefore,

$$\frac{\{H^+\}\{CaOH^+\}}{\{Ca\}} = \frac{K_w}{K_{CaOHc}} \tag{11.36}$$

$$[CaOH^+] = \frac{\{CaOH^+\}}{\gamma_{CaOHc}} = \frac{\gamma_{Ca}K_w[Ca]}{\gamma_{CaOHc}K_{CaOHc}\gamma_H[H^+]} \tag{11.37}$$

γ_{CaOHc}, γ_{Ca}, and γ_H are the activity coefficients of $CaOH^+$, Ca, and H^+, respectively. $[CaSO_4^o]$ is calculated as follows:

$$CaSO_4^o \rightleftharpoons Ca^{2+} + SO_4^{2-} \quad K_{CaSO_4c}$$

$$\frac{\{Ca^{2+}\}\{SO_4^{2-}\}}{\{CaSO_4^o\}} = K_{CaSO_4c} \tag{11.38}$$

$$[CaSO_4^o] = \frac{\{CaSO_4^o\}}{\gamma_{CaSO_4c}} = \frac{\gamma_{Ca}[Ca^{2+}][SO_4^{2-}]}{K_{CaSO_4c}}$$

The activity coefficient $\gamma_{CaSO_4c} = 1$, because $CaSO_4^o$ is not dissociated. γ_{SO_4} is the activity coefficient of the sulfate ion.

The previous expressions for $[CaCO_3^o]$, $[CaHCO_3^+]$, $[CaOH^+]$, and $[CaSO_4^o]$ may now be substituted into Equation (11.31) and the result solved for $[Ca^{2+}]$ to produce

$$[Ca^{2+}] = \frac{[Ca_T] - [CaCO_3^o] - [CaHCO_3^+]}{1 + \dfrac{\gamma_{Ca}K_w}{\gamma_{CaOHc}K_{CaOHc}\gamma_H[H^+]} + \dfrac{\gamma_{Ca}\gamma_{SO_4}[SO_4^{2-}]}{K_{CaSO_4c}}} \tag{11.39}$$

Example 11.8 Analysis of a water sample yields the following results: [TDS] = 140 mg/L, $[Ca_T] = 0.7$ mgmol/L, $[Mg_T] = 0.6$ mgmol/L, $[A]_{mgeq} = 0.4$ mg/L, sulfate ion = 0.3 mgmol/L, temperature = 25°C, and pH = 6.7. Calculate the concentration of the calcium ion corrected for the formation of the complex ions.

Solution:

$$[Ca^{2+}] = \frac{[Ca_T] - [CaCO_3^o] - [CaHCO_3^+]}{1 + \dfrac{\gamma_{Ca}K_w}{\gamma_{CaOHc}K_{CaOHc}\gamma_H[H^+]} + \dfrac{\gamma_{Ca}\gamma_{SO_4}[SO_4^{2-}]}{K_{CaSO_4c}}}$$

$$[Ca_T] = \frac{0.7}{1000} = 7.0(10^{-4}) \text{ gmol/L}$$

$$[CaCO_3^o] = \frac{K_{sp,CaCO_3}}{K_{CaCO_3c}} = \frac{4.8(10^{-9})}{10^{-3.22}} = 7.97(10^{-6}) \ gmol/L$$

$$[CaHCO_3^+] = \frac{\gamma_H K_{sp,CaCO_3}}{\gamma_{CaHCO_3c} K_2 K_{CaHCO_3c}}[H^+]:$$

$$\gamma_H = 0.94 = \gamma_{CaHCO_3c} = \gamma_{CaOHc} \quad K_2 = 10^{-10.33}$$

$$K_{CaHCO_3c} = 10^{-1.26}\{H^+\} = 10^{-6.7}$$

Therefore,

$$[CaHCO_3^+] = \frac{0.94[4.8(10^{-9})]}{0.94(10^{-1.26})}(10^{-6.7}) = 1.74(10^{-14}) \ gmol/L$$

$$\gamma_{Ca} = 0.77 \quad K_{CaOHc} = 10^{-1.49} \quad \gamma_{SO_4} = 0.77$$

$$[SO_4^{2-}] = \frac{0.3}{(1000)[32.1 + 4(16)]} = 3.12(10^{-6}) \ gmol/L$$

$$K_{CaSO_4c} = 10^{-2.31}$$

Therefore,

$$[Ca^{2+}] = \frac{7.0(10^{-4}) - 7.97(10^{-6}) - 1.74(10^{-14})}{1 + \dfrac{0.77(10^{-14})}{0.94(10^{-1.49})(0.94)(10^{-6.7})} + \dfrac{(0.77)(0.77)[3.12(10^{-6})]}{10^{-2.31}}}$$

$$= \frac{6.92(10^{-4})}{1 + 1.35(10^{-6}) + 3.78(10^{-4})} = 6.92(10^{-4}) \ gmol/L$$

$$= 0.69 \ mgmol/L = 27.67 \ mg/L \quad \textbf{Ans}$$

11.2.3 TOTAL ALKALINITY AS CALCIUM CARBONATE

The unit of concentration that we have used for alkalinity is equivalents per unit volume. This unit follows directly from the chemical reaction, so this method of expression is fairly easy to understand. In practice, however, alkalinity is also expressed in terms of $CaCO_3$. Expressing alkalinities in these terms is a sort of equivalence, although this practice can become very confusing. Depending upon the chemical reaction it is involved with, calcium carbonate can have more than one value for its equivalent mass, and this is the source of the confusion. Thus, to understand the underpinnings of this method of expression, one must look at the reference chemical reaction.

As shown previously, the equivalence of all forms of alkalinity was unified by using a common end point—the methyl orange end point that corresponds to a pH of about 4.5. For example, the OH⁻ alkalinity was assumed to go to completion at the methyl orange end point, although the reaction is complete long beforehand at around a pH of 10.8. The same argument held for the carbonate and the bicarbonate alkalinities, which, of course, would be accurate, since their end points are legitimately at the methyl orange end point. This was done in order to have a common equivalence point. Thus, to express alkalinities in terms of $CaCO_3$, the reaction of calcium carbonate must also be assumed to complete at the same end point. Even without assuming, this, of course, happens to be true. The reaction is

$$CaCO_3 + 2H^+ \rightleftharpoons H_2CO_3 + Ca^{2+} \tag{11.40}$$

and $CaCO_3$ has an equivalent mass of $CaCO_3/2$, because the number of reference species is 2.

To illustrate the use of this concept, assume 10^{-3} gmmol/L of the hydroxide ion and express this concentration in terms of $CaCO_3$. The pertinent reaction is $OH^- + H^+ \rightleftharpoons HOH$. From this reaction, OH^- has an equivalent mass of OH/1 and the number of equivalents per liter of the 10^{-3} gmol/L of the hydroxide is $10^{-3}(OH)/OH = 10^{-3}$. Thus, expressing the concentration in terms of $CaCO_3$

$$[OH^-] = 10^{-3} \text{ gmmol/L} = 10^{-3} \text{ geq/L} = 10^{-3}(CaCO_3/2) \text{ g/L as } CaCO_3$$
$$= 0.05 \text{ g/L as } CaCO_3 = 50 \text{ mg/L as } CaCO_3$$

11.2.4 PRECIPITATION POTENTIAL

Figure 11.1 shows a pipe that is almost completely blocked due to precipitation of $CaCO_3$. Precipitation potential is another criterion for water stability, and application of this concept can help prevent situations like the one shown in this figure. Understanding this concept requires prerequisite knowledge of the charge balance.

All solutions are electrically neutral and negative charges must balance the positive charges. Thus, the balance of charges, where concentration must be expressed in terms of equivalents, is

$$[CO_3^{2-}]_{eq} + [HCO_3^-]_{eq} + [OH]_{eq} = [H^+]_{eq} + [Ca^{2+}]_{eq} \tag{11.41}$$

Expressing in terms of moles,

$$2[CO_3^{2-}] + [HCO_3^-] + [OH] = [H^+] + 2[Ca^{2+}] \tag{11.42}$$

Now, the amount of calcium carbonate that precipitates is simply the equivalent of the calcium ion that precipitates, Ca_{ppt}. Because the number of moles of Ca_{ppt} is equal to the number of moles of the carbonate solid $CaCO_{3ppt}$ that precipitates,

$$[CaCO_{3ppt}] = [Ca_{ppt}] \tag{11.43}$$

FIGURE 11.1 A water distribution pipe almost completely blocked with precipitated calcium carbonate.

$[CaCO_{3ppt}]$ is the precipitation potential of calcium carbonate. Ca_{ppt}, in turn, can be obtained from the original calcium, Ca_{before}^{2+}, minus the calcium at equilibrium, Ca_{after}^{2+}.

$$[Ca_{ppt}] = [Ca_{before}^{2+}] - [Ca_{after}^{2+}] \qquad (11.44)$$

To use Equation (11.44), $[Ca_{before}^{2+}]$ must first be known. To determine $[Ca_{after}^{2+}]$, the charge balance equation derived previously will be used. $[Ca_{after}^{2+}]$ is the $[Ca^{2+}]$ in the charge balance equation, thus

$$2[CO_3^{2-}] + [HCO_3^-] + [OH] = [H^+] + 2[Ca_{after}^{2+}] \qquad (11.45)$$

Making the necessary substitutions of the carbonate system equilibria equations to Equation (11.45) and solving for $[Ca_{after}^{2+}]$, produces

$$2[CO_3^{2-}] + [HCO_3^-] + [OH] = [H^+] + 2[Ca_{after}^{2+}]$$

$$2\frac{K_{sp,CaCO_3}}{\gamma_{CO_3}\gamma_{Ca}[Ca_{after}^{2+}]} + \frac{[H^+]K_{sp,CaCO_3}}{\gamma_{HCO_3}\gamma_H K_2 \gamma_{Ca}[Ca_{after}^{2+}]} + \frac{K_w}{\gamma_{OH}\gamma_H[H^+]} = [H^+] + 2[Ca_{after}^{2+}]$$

$$2[H^+][Ca_{after}^{2+}]^2 + \left([H^+]^2 - \frac{K_w}{\gamma_{OH}\gamma_H}\right)[Ca_{after}^{2+}] - \left(2\frac{K_{sp,CaCO_3}}{\gamma_{CO_3}\gamma_{Ca}}[H^+] + \frac{[H^+]^2 K_{sp,CaCO_3}}{\gamma_{HCO_3}\gamma_H K_2 \gamma_{Ca}}\right) = 0$$

$$[Ca_{after}^{2+}] = \frac{-\left([H^+]^2 - \dfrac{K_w}{\gamma_{OH}\gamma_H}\right) + \sqrt{\left([H^+]^2 - \dfrac{K_w}{\gamma_{OH}\gamma_H}\right)^2 + 8[H^+]\left(2\dfrac{K_{sp,CaCO_3}}{\gamma_{CO_3}\gamma_{Ca}}[H^+] + \dfrac{[H^+]^2 K_{sp,CaCO_3}}{\gamma_{HCO_3}\gamma_H K_2 \gamma_{Ca}}\right)}}{4[H^+]}$$

$$(11.46)$$

The $[H^+]$ in the previous equations is the saturation pH of Equation (11.25). Now, finally, the precipitation potential $[CaCO_{3ppt}]$ is

$$[CaCO_{3ppt}] = [Ca_{ppt}] = [Ca_{before}^{2+}] - [Ca_{after}^{2+}] \qquad (11.47)$$

11.2.5 DETERMINATION OF PERCENT BLOCKING POTENTIAL OF PIPES

Let Vol_{pipe} be the volume of the pipe segment upon which the percent blocking potential is to be determined. The amount of volume precipitation potential in this volume after a time t is $(100([Ca_{before}^{2+}] - [Ca_{after}^{2+}])Vol_{pipe}t)/\rho_{CaCO_3}t_d$, where ρ_{CaCO_3} is the mass density of the carbonate precipitate and t_d is the detention time of the pipe segment. Letting Q_{pipe} be the rate of flow through the pipe, $t_d = Vol_{pipe}/Q_{pipe}$. Substituting this expression for t_d, the percent blocking potential P_{block} after time t is

$$P_{block} = \frac{\dfrac{100([Ca_{before}^{2+}] - [Ca_{after}^{2+}])Vol_{pipe}t}{Vol_{pipe}}}{\rho_{CaCO_3}\dfrac{Vol_{pipe}}{Q_{pipe}}}(100) \qquad (11.48)$$

$$= \frac{100([Ca_{before}^{2+}] - [Ca_{after}^{2+}])Q_{pipe}t}{\rho_{CaCO_3}Vol_{pipe}}(100) \qquad (11.49)$$

$\rho_{CaCO_3} = 2.6$ g/cc $= 2600$ g/L. Note that since the concentrations are expressed in gram moles per liter, volumes and rates should be expressed in liters and liters per unit time, respectively. ρ_{CaCO_3} should be in g/L.

Example 11.9 Water of the following composition is obtained after a softening–recarbonation process: $[Ca^{2+}] = 10^{-3}$ gmol/L, $[HCO_3^-] = 10^{-4}$ gmol/L, $[CO_3^{2-}] = 3.2(10^{-3})$ gmol/L, $[H_2CO_3^*] = 10^{-9}$ gmol/L, pH $= 8.7$, pH$_s = 10$, temperature $= 25°C$, $\mu = 5(10^{-3})$. Calculate the equilibrium calcium ion concentration precipitation potential. Express precipitation potential in gmols/L and mg/L.

Solution: The saturation pH is given as pH$_s = 10$. The actual pH is 8.7, so the system is not saturated with calcium carbonate and no carbonate will precipitate; the precipitation potential is therefore zero. At equilibrium at the Langelier saturation pH, the calcium ion concentration will remain the same at $[Ca^{2+}] = 10^{-3}$ gmol/L. **Ans**

Example 11.10 Assume that the pH of the treated water in Example 11.9 was raised to cause the precipitation of the carbonate solid. The water is distributed through a distribution main at a rate of 0.22 m³/s. Determine the length of time it takes to clog a section of the distribution main 1 km in length, if the diameter is 0.42 m.

Solution:

$$P_{block} = \frac{100([Ca^{2+}_{before}] - [Ca^{2+}_{after}])Q_{pipe}t}{\rho_{CaCO_3}Vol_{pipe}}(100)$$

$$[Ca^{2+}_{before}] = 10^{-3} \text{gmol/L} \quad [Ca^{2+}_{after}] = 0$$

$$Q_{pipe} = 0.22 \text{ m}^3/\text{s} = 220 \text{ L/s} \quad \rho_{CaCO_3} = 2600 \text{ g/L}$$

$$Vol_{pipe} = \frac{\pi(0.42)^2}{4}(1000)(1000) = 138{,}544.24 \text{ L}$$

Therefore,

$$100 = \frac{100(10^{-3} - 0)(220)t}{(2600)(138{,}544.24)}(100)t = 16{,}373{,}410.18 \ s = 189.51 \text{ days} \quad \textbf{Ans}$$

This example shows the importance of controlling the precipitation of $CaCO_3$. Of course, in this particular situation, once the pipe is constricted due to blockage, the consumers would complain and the water utilities would then solve the problem; however, the problem situation may become too severe and the distribution pipe would have to be abandoned.

11.3 RECARBONATION OF SOFTENED WATER

After the softening process, the pH is so high that reduction is necessary to prevent deposition of scales in distribution pipes. This can be accomplished inexpensively using carbon dioxide. We will therefore develop the method for determining the carbonic acid necessary to set the water to the equilibrium pH.

In recarbonation, the available calcium ion in solution is prevented from precipitation. Therefore, it remains to determine at what pH will the equilibrium condition be, given this calcium concentration. This determination is, in fact, the basis of the Langelier saturation pH. Adding carbonic acid will increase the acidity of the solution after it has neutralized any existing alkalinity.

Let the current pH be pH_{cur} and the pH to which it is to be adjusted (the destination pH) be pH_{to}. If pH_{cur} is greater than pH_{to} an acid is needed. No matter how insignificant, a natural water will always have an alkalinity in it. Alkalinities of surface water can vary from 10 to 800 mg/L (Sincero, 1968). Until it is all consumed, this alkalinity will resist the change in pH. Let the current total alkalinity be $[A_{cur}]_{geq}$ in gram equivalents per liter. Let the total acidity to be added be $[A_{cadd}]_{geq}$ in gram equivalents per liter.

TABLE 11.5
Fractional Dissociations, $\phi_{aH_2CO_3}$, of
H^+ from Carbonic Acid

Molar Concentration	$\phi_{aH_2CO_3}$
0.1	0.00422
0.5	0.00094
1.0	0.000668

The hydrogen ion concentration corresponding to pH_{cur} is $10^{-pH_{cur}}$ gram moles per liter and that corresponding to pH_{to} is $10^{-pH_{to}}$ gram moles per liter.

Note: The gram moles per liter of hydrogen ion is equal to its number of equivalents per liter.

Assuming no alkalinity present, the total acid to be added is $10^{-pH_{to}} - 10^{-pH_{cur}}$ gram moles per liter. Alkalinity is always present, however, so more acid must be added to counteract the natural alkalinity, $[A_{cur}]_{geq}$. Thus, the total acidity to be added, $[A_{cadd}]_{geq}$, is

$$[A_{cadd}]_{geq} = [A_{cur}]_{geq} + \frac{10^{-pH_{to}} - 10^{-pH_{cur}}}{\phi_a} \tag{11.50}$$

where ϕ_a is the fractional dissociation of the hydrogen ion from the acid supplied. For strong acids, ϕ_a is unity; for weak acids, it may be calculated from equilibrium constants (Table 11.5).

To determine the equivalent mass of carbon dioxide needed, write the following chemical reactions, noting that carbon dioxide must react with existing alkalinity:

$$CO_2 + H_2O \rightleftharpoons H_2CO_3$$
$$H_2CO_3 + 2OH^- \rightarrow 2H_2O + CO_3^{2-} \tag{11.51}$$

From these reactions, the equivalent mass of CO_2 is $CO_2/2$.

Let us calculate some fractional dissociations of H_2CO_3. Its dissociation reaction is as follows:

$$H_2CO_3 \rightleftharpoons H^+ + HCO_3^- \qquad K_1 = 10^{-6.35} \tag{11.52}$$

First, assume a 0.1 M concentration of carbonic acid and let x be the concentration of H^+ and HCO_3^- at equilibrium. At equilibrium, the concentration of carbonic acid will be $0.1 - x$. Thus, substituting into the equilibrium expression produces

$$\frac{x^2}{0.1 - x} = 10^{-6.35}$$

$$x = \frac{-10^{-6.35} + \sqrt{(10^{-6.35})^2 + 4(0.1)(10^{-6.35})}}{2} = 0.000211 \tag{11.53}$$

Therefore,

$$\phi_{aH_2CO_3} = \frac{0.000211}{0.1} = 0.00211$$

Note: In the previous calculation, the activity coefficients have been ignored.

Example 11.11 A water sample from the reactor of a softening plant has a total alkalinity of $[A]_{geq} = 2.74(10^{-3})$ geq/L, pH of 10, $[Ca^{2+}] = 0.7$ mgmol/L, $[Mg^{2+}] = 0.6$ mgmol/L, and $\mu = 3.7(10^{-3})$. Using the Langelier saturation pH equation, $pH_s = 8.7$. Calculate the amount of carbon dioxide necessary to lower the pH to the saturation value. Use $\phi_{H_2CO_3} = 0.00422$.

Solution:

$$[A_{cadd}]_{geq} = [A_{cur}]_{geq} + \frac{10^{-pH_{to}} - 10^{-pH_{cur}}}{\phi}$$

$$[A_{cur}]_{geq} = 2.74(10^{-3}) \text{ geq/L}$$

$$[A_{cadd}]_{geq} = 2.74(10^{-3}) + \frac{10^{-8.7} - 10^{-10}}{0.00422} = 2.74(10^{-3}) \text{ geq/L} \quad \textbf{Ans}$$

GLOSSARY

Activity—Measure of the effectiveness of a given species in its participation in a reaction.

Alkalinity—The capacity of a solution to neutralize any acid added to it.

Activity coefficient—A proportionality constant relating activity and concentration.

Dead end—For a system in equilibrium, the reaction that causes solids to precipitate.

Enthalpy—Heat released or absorbed in a chemical reaction at constant pressure.

Equilibrium constant—The product of the activities of the reactants and products (raised to appropriate powers) of a chemical reaction in equilibrium.

Ionic strength—A term devised by Lewis and Randall to describe the electric field intensity of a solution.

Langlier index—The difference between the actual pH and the saturation pH of a solution.

Precipitation potential—The amount of calcium carbonate that will precipitate when the solution is left by itself from its supersaturating condition.

Precipitation-potential pH—The pH attained at the precipitation potential condition.

Proton condition—A condition of balance between species that contain the proton and counteracting species that do not contain the proton at a particular end point such as the dead end.

Recarbonation—The process of adding carbon dioxide to water for the purpose of lowering the pH to the desired value.

Saturation index—The same as Langelier index.

Saturation pH—The pH attained at equilibrium with original calcium ion prevented from precipitation.

Solubility product constant—The term given to the equilibrium constant when products are in equilibrium with solid reactants.

SYMBOLS

aq	Subscript symbol for "aqueous"
$[Ca_T]$	Total molar concentration of species containing the calcium atom
$[Ca^{2+}_{(aq)}]$	Molar concentration of the calcium ion
$[Ca^{2+}_{after}]$	Calcium ion concentration after occurrence of precipitation potential
$[Ca^{2+}_{before}]$	Calcium ion concentration before occurrence of precipitation potential
$[Ca_{ppt}]$	Precipitation potential of the calcium ion
$[CaCO_{3ppt}]$	Precipitation potential of $CaCO_3$
$CaCO_3^o$	Carbonate complex of calcium ion
$CaHCO_3^+$	Bicarbonate complex of calcium ion
$CaOH^+$	Hydroxide complex of calcium ion
$[\Delta carb]$	Moles of carbon dioxide per unit volume of flow required to bring the pH of water to the saturation pH
$CaSO_4^o$	Sulfate complex of calcium ion
$[\Delta Decarb]$	Moles of base per unit volume of flow required to bring the pH of water to the saturation pH
$[\Delta H_2CO_3^*]$	Change in concentration of carbonic acid needed to lower actual pH to the saturation pH
$H_2CO_3^*$	Mixture of CO_2 in water, $CO_{2(aq)}$, and H_2CO_3; carbonic acid
ΔH^o	Standard enthalpy change
ΔH^o_{298}	Standard enthalpy change at 25°C and unit of activity
H_s	Saturation hydrogen ion concentration corresponding to the saturation pH
K	Equilibrium constant
K_1	Ionization constant of $H_2CO_3^*$
K_2	Ionization constant of the bicarbonate ion
K_{CaCO_3c}	Equilibrium constant of $CaCO_3^o$
K_{CaHCO_3c}	Equilibrium constant of $CaHCO_3^+$
K_{CaOHc}	Equilibrium constant of $CaOH^+$
K_{CaSO_4c}	Equilibrium constant of $CaSO_4^o$
K_{sp}	Solubility product constant
$K_{sp,CaCO_3}$	K_{sp} of $CaCO_3$
K_{T1}	Equilibrium constant at temperature T_1
K_{T2}	Equilibrium constant at temperature T_2
K_w	Ion product of water

l	Symbol for liquid state of a substance
LI	Langelier of Saturation Index
P_{block}	Percent blocking potential of a segment of pipe due to potential deposition of $CaCO_3$
pH_s	Saturation pH
Q_{pipe}	Rate of flow through pipe segment in which percent blocking is to be determined
R	Universal gas constant.
s	Symbol for solid state of a substance
sp	Any species in equilibrium
$sp\ conduc$	Specific conductance
t	Time to percent blocking potential
t_d	Detention time in pipe segment for which percent blocking potential is to be determined
T	Absolute temperature
Vol_{pipe}	Volume of pipe segment in which blocking potential is to be determined
z	Charge of ion
$\{\ \}$	Read as "activity of"
$[\ ]$	Read as "concentration of"
γ	Activity coefficient
γ_{Ca}	Activity coefficient of the calcium ion
γ_{CO_3}	Activity coefficient of the carbonate ion
γ_H	Activity coefficient of the hydrogen ion
γ_{HCO_3}	Activity coefficient of the bicarbonate ion
γ_{OH}	Activity coefficient of the hydroxyl ion
μ	Ionic strength
ρ_{CaCO_3}	Mass density of $CaCO_3$

PROBLEMS

11.1 The pH of a solution is 14. Calculate the hydrogen ion concentration.

11.2 The pH of a solution is 0. Calculate the hydrogen ion concentration.

11.3 The pH of a solution is 0. Calculate the hydroxyl ion concentration.

11.4 The pH of a solution is 14. Calculate the hydroxyl ion concentration.

11.5 The concentration of carbonic acid was analyzed to be 0.2 mgmol/L. If the concentration of the bicarbonate ion is $8.93(10^{-4})$ gmol/L, what is the pH of the solution if the temperature is 25°C?

11.6 The concentrations of carbonic acid and bicarbonate were analyzed to be 0.2 mgmol/L and $8.93(10^{-4})$ gmol/L, respectively. If the pH is equal to 7, calculate the value of K_1.

11.7 The concentration of the bicarbonate ion was analyzed to be $8.93(10^{-4})$ gmol/L. If the pH is equal to 7 and the temperature is 25°C calculate the concentration of carbonic acid.

11.8   The concentrations of carbonic acid and bicarbonate were analyzed to be
       0.2 mgmol/L and $8.93(10^{-4})$ gmol/L, respectively. If the pH is equal to 7,
       determine the temperature of the solution.

11.9   A sample of water has the following composition: $CO_2 = 22.0$ mg/L,
       $Ca^{2+} = 80$ mg/L, $Mg^{2+} = 12.0$ mg/L, $Na^+ = 46.0$ mg/L, $HCO^{3-} = 152.5$
       mg/L, and $SO_4^{2-} = 216$ mg/L. What is the value of z for carbon dioxide?

11.10 In Problem 11.9, calculate the activity coefficient and the activity in mg/L
       of the carbon dioxide.

11.11 In Problem 11.9, calculate the activity coefficient and the activity in mg/L
       of the sulfate ion.

11.12 In Problem 11.9, calculate the activity coefficient and the activity in mg/L
       of the magnesium ion.

11.13 In Problem 11.9, calculate the activity coefficient and the activity in mg/L
       of the sodium ion.

11.14 A softened municipal water supply enters a residence at 15°C and is heated
       to 60°C in the water heater. If the water is at equilibrium at 15°C, deter-
       mine if $CaCO_3$ will deposit or not.

11.15 A softened municipal water supply enters a residence at 15°C and is heated
       to 60°C in the water heater. If the water is at equilibrium at 60°C, deter-
       mine if $CaCO_3$ will deposit or not.

11.16 Analysis of a water sample yields the following results: [TDS] = 140
       mg/L, $[Ca^{2+}]$ = 0.7 mgmol/L, $[Mg^{2+}]$ = 0.6 mgmol/L, $[A]_{mgeq}$ = 0.4
       mgeq/L, temperature = 20°C, and pH = 6.7. The saturation pH was
       calculated to be equal to 8.67. Calculate the values of K_2, K_{spCaCO_3}, γ_{OH},
       γ_{CO_3}, K_w, and γ_H.

11.17 Analysis of a water sample yields the following results: [TDS] = 140
       mg/L, $[Ca^{2+}] = 0.7$ mgmol/L, $[Mg^{2+}] = 0.6$ mgmol/L, $[A]_{mgeq} = 2.74$ geq/L,
       temperature = 20°C, and pH = 6.7. The saturation pH was calculated to
       be equal to 8.67. Calculate the values of γ_{HCO_3}, K_{spCaCO_3}, γ_{OH}, γ_{CO_3}, K_w,
       and γ_H.

11.18 Analysis of a water sample yields the following results: [TDS] = 140
       mg/L, $[Ca^{2+}]$ = 0.7 mgmol/L, $[Mg^{2+}]$ = 0.6 mgmol/L, $[A]_{mgeq}$ = 0.4
       mgeq/L, temperature = 20°C, and pH = 6.7. The saturation pH was
       calculated to be equal to 8.67. Calculate the values of γ_{HCO_3}, K_2, γ_{OH}, γ_{CO_3},
       K_w, and γ_H.

11.19 Analysis of a water sample yields the following results: [TDS] = 140
       mg/L, $[Ca^{2+}] = 0.7$ mgmol/L, $[Mg^{2+}] = 0.6$ mgmol/L, $[A]_{mgeq} = 2.74$ geq/L,
       temperature = 20°C, and pH = 6.7. The saturation pH was calculated
       to be equal to 8.67. Calculate the values of γ_{HCO_3}, K_2, K_{spCaCO_3}, γ_{CO_3}, K_w,
       and γ_H.

11.20 Analysis of a water sample yields the following results: [TDS] = 140
       mg/L, $[Ca^{2+}] = 0.7$ mgmol/L, $[Mg^{2+}] = 0.6$ mgmol/L, $[A]_{mgeq} = 0.4$ mgeq/L,
       temperature = 20°C, and pH = 6.7. The saturation pH was calculated
       to be equal to 8.67. Calculate the values of γ_{HCO_3}, K_2, K_{spCaCO_3}, γ_{OH}, K_w,
       and γ_H.

11.21 Analysis of a water sample yields the following results: [TDS] = 140
       mg/L, $[Ca^{2+}] = 0.7$ mgmol/L, $[Mg^{2+}] = 0.6$ mgmol/L, $[A]_{mgeq} = 2.74$ geq/L,
       temperature = 20°C, and pH = 6.7. The saturation pH was calculated to
       be equal to 8.67. Calculate the values of γ_{HCO_3}, K_2, K_{spCaCO_3}, γ_{OH}, γ_{CO_3},
       and γ_H.

11.22 Analysis of a water sample yields the following results: [TDS] = 140
       mg/L, $[Ca^{2+}] = 0.7$ mgmol/L, $[Mg^{2+}] = 0.6$ mgmol/L, $[A]_{mgeq} = 0.4$ mgeq/L,
       temperature = 20°C, and pH = 6.7. The saturation pH was calculated
       to be equal to 8.67. Calculate the values of γ_{HCO_3}, K_2, K_{spCaCO_3}, γ_{OH}, γ_{CO_3},
       and K_w.

11.23 Analysis of a water sample yields the following results: [TDS] = 140
       mg/L, $[Ca^{2+}] = 3.2$ mgmol/L, $[Mg^{2+}] = 0.6$ mgmol/L, temperature = 20°C,
       and pH = 6.7. The saturation pH was calculated to be 8.0. Calculate the
       alkalinity.

11.24 Analysis of a water sample yields the following results: [TDS] = 140
       mg/L, $[Mg^{2+}] = 0.6$ mgmol/L, $[A]_{mgeq} = 0.4$ mgeq/L, temperature = 20°C,
       and pH = 6.7. The saturation pH was calculated to be 8.0. Calculate the
       calcium ion concentration.

11.25 Analysis of a water sample yields the following results: [TDS] = 140
       mg/L, $[Ca^{2+}] = 3.2$ mgmol/L, $[Mg^{2+}] = 0.6$ mgmol/L, $[A]_{mgeq} = 0.4$ mgeq/L,
       temperature = 20°C, and pH = 6.7. Calculate the saturation pH.

11.26 Analysis of a water sample yields the following results: [TDS] = 140
       mg/L, $[Ca^{2+}] = 3.2$ mgmol/L, $[Mg^{2+}] = 0.6$ mgmol/L, $[A]_{mgeq} = 0.4$ mgeq/L,
       temperature = 25°C, and pH = 6.7. Calculate the saturation pH.

11.27 Analysis of a water sample yields the following results: [TDS] = 140
       mg/L, $[Ca_T] = 0.7$ mgmol/L, $[Mg_T] = 0.6$ mgmol/L, $[A]_{mgeq} = 0.4$ mgeq/L,
       temperature = 25°C, and pH = 6.7. The concentration of the calcium after
       correction for the formation of the complex ion is 0.3 mgmol/L. Calculate
       the sulfate ion concentration.

11.28 Analysis of a water sample yields the following results: [TDS] = 140
       mg/L, $[Mg_T] = 0.6$ mgmol/L, $[A]_{mgeq} = 0.4$ mgeq/L, sulfate ion = 0.3
       mgmol/L, temperature = 25°C, and pH = 6.7. The concentration of the
       calcium after correction for the formation of the complex ion is 0.3
       mgmole/L. Calculate $[Ca_T]$.

11.29 Analysis of a water sample yields the following results: $[Mg_T] = 0.6$
       mgmol/L, sulfate ion = 0.3 mgmol/L, $[A]_{mgeq} = 2.74$ geq/L, [TDS] = 140
       mg/L, $[Ca_T] = 0.7$ mgmole/L, and temperature = 25°C. The concentration
       of the calcium after correction for the formation of the complex ion is 0.3
       mgmol/L. Calculate the pH.

11.30 Analysis of a water sample yields the following results: [TDS] = 140
       mg/L, $[Ca_T] = 0.7$ mgmol/L, $[Mg_T] = 0.6$ mgmol/L, $[A]_{mgeq} = 0.4$ mgeq/L,
       sulfate ion = 0.3 mgmol/L, temperature = 25°C, and pH = 6.7. The
       concentration of the calcium after correction for the formation of the
       complex ion is 0.3 mgmol/L. Calculate K_w, γ_{CaOHc}, K_{CaOHc}, γ_H, γ_{SO_4}, and
       K_{CaSO_4c}.

11.31 Analysis of a water sample yields the following results: [TDS] = 140 mg/L, $[Ca_T]$ = 0.7 mgmol/L, $[Mg_T]$ = 0.6 mgmol/L, $[A]_{mgeq}$ = 0.4 mgeq/L, sulfate ion = 0.3 mgmol/L, temperature = 25°C, and pH = 6.7. The concentration of the calcium after correction for the formation of the complex ion is 0.3 mgmol/L. Calculate γ_{Ca}, γ_{CaOHc}, K_{CaOHc}, γ_H, γ_{SO_4}, and K_{CaSO_4c}.

11.32 Analysis of a water sample yields the following results: [TDS] = 140 mg/L, $[Ca_T]$ = 0.7 mgmol/L, $[Mg_T]$ = 0.6 mgmol/L, $[A]_{mgeq}$ = 0.4 mgeq/L, sulfate ion = 0.3 mgmol/L, temperature = 25°C, and pH = 6.7. The concentration of the calcium after correction for the formation of the complex ion is 0.3 mgmol/L. Calculate γ_{Ca}, K_w, K_{CaOHc}, γ_H, γ_{SO_4}, and K_{CaSO_4c}.

11.33 Analysis of a water sample yields the following results: [TDS] = 140 mg/L, $[Ca_T]$ = 0.7 mgmol/L, $[Mg_T]$ = 0.6 mgmol/L, $[A]_{mgeq}$ = 0.4 mgeq/L, sulfate ion = 0.3 mgmol/L, temperature = 25°C, and pH = 6.7. The concentration of the calcium after correction for the formation of the complex ion is 0.3 mgmol/L. Calculate γ_{Ca}, K_w, γ_{CaOHc}, γ_H, γ_{SO_4}, and K_{CaSO_4c}.

11.34 Analysis of a water sample yields the following results: [TDS] = 140 mg/L, $[Ca_T]$ = 0.7 mgmol/L, $[Mg_T]$ = 0.6 mgmol/L, $[A]_{mgeq}$ = 2.74 geq/L, sulfate ion = 0.3 mgmol/L, temperature = 25°C, and pH = 6.7. The concentration of the calcium after correction for the formation of the complex ion is 0.3 mgmol/L. Calculate γ_{Ca}, K_w, γ_{CaOHc}, K_{CaOHc}, γ_{SO_4}, and K_{CaSO_4c}.

11.35 Analysis of a water sample yields the following results: [TDS] = 140 mg/L, $[Ca_T]$ = 0.7 mgmol/L, $[Mg_T]$ = 0.6 mgmol/L, $[A]_{mgeq}$ = 0.4 mgeq/L, sulfate ion = 0.3 mgmol/L, temperature = 25°C, and pH = 6.7. The concentration of the calcium after correction for the formation of the complex ion is 0.3 mgmol/L. Calculate γ_{Ca}, K_w, γ_{CaOHc}, K_{CaOHc}, γ_H, and K_{CaSO_4c}.

11.36 Analysis of a water sample yields the following results: [TDS] = 140 mg/L, $[Ca_T]$ = 0.7 mgmol/L, $[Mg_T]$ = 0.6 mgmol/L, $[A]_{mgeq}$ = 0.4 mgeq/L, sulfate ion = 0.3 mgmol/L, temperature = 25°C, and pH = 6.7. The concentration of the calcium after correction for the formation of the complex ion is 0.3 mgmol/L. Calculate γ_{Ca}, K_w, γ_{CaOHc}, K_{CaOHc}, γ_H, and γ_{SO_4}.

11.37 Assume that the pH of the treated water in Problem 11.10 was adjusted to cause the precipitation of the carbonate solid, if it was not precipitating already. The water is distributed through a distribution main at a rate of 0.22 m³/s. Determine the length of time it takes to clog a section of the distribution main 1 km in length, if the diameter is 0.42 m. Assume $[Ca^{2+}_{after}]$ = 1.47(10^{-4}) gmol/L.

BIBLIOGRAPHY

Al-Rqobah, H. E. and A. Al-Munayyis (1989). Recarbonation process for treatment of distilled water produced by MSF plants in Kuwait. *Desalination Vol. I—Desalination in Total Water Resource Manage.; Non-Traditional Desalination Technologies; Operational Experience—Evaporative Proc. Fourth World Congress on Desalination and Water Reuse*, Nov. 4–8, 73, 1–3, 295–312. Kuwait Foundation for the Advancement of Sciences (KFAS), Kuwait.

Coucke, D., et al. (1997). Comparison of the different methods for determining the behaviour of water to calcium carbonate. *Aqua (Oxford)*. 46, 2, 49–58.

Edwards, M., M. R. Schock, and E. T. Meyer (1996). Alkalinity, pH, and copper corrosion by-product release. *J. Am. Water Works Assoc.* 88, 3, 81–94.

Gallup, D. L. and K. Kitz (1997). Low-cost silica, calcite and metal sulfide scale control through on-site production of sulfurous acid from H₂S or elemental sulfur. *Trans.—Geothermal Resour. Council, Proc. 1997 Annu. Meeting Geothermal Resour. Council*, Oct. 12–15, 21, 399–403, Burlingame, CA. Geothermal Resources Council, Davis, CA.

Gill, J. S. (1998). Silica Scale Control. *Materials Performance*. 37, 11, 41–45.

Gnusin, N. P. and O. A. Demina (1993). Ionic equilibria in the treatment of natural waters. *Khimiya i Tekhnologiya Vody*. 15, 6, 468–474.

Gomolka, E. and B. Gomolka (1991). Application of disk aerators to recarbonation of alkaline wastewater from wet gasification of carbide. *Water Science Technol. Advanced Wastewater Treatment and Reclamation, Proc. IAWPRC Conf.*, Sep. 25–27, 1989, 24, 7, 277–284. Polish Soc. of Sanitary Engineers.

Holtzclaw Jr., H. F. and W. R. Robinson (1988). *General Chemistry*. A7–A8. D. C. Heath and Company, Lexington, MA.

Langelier, W. F. (1936). The analytical control of anti-corrosion water treatment. *J. Am. Water Works Assoc.*, 28.

Moller, N., J. P. Greenberg, and J. H. Weare (1998). Computer modeling for geothermal systems: Predicting carbonate and silica scale formation, CO₂ breakout and H₂S Exchange. *Transport in Porous Media*. 33, 1–2, 173–204.

Murphy, B., J. T. O'Connor, and T. L. O'Connor, (1997). Willmar, Minnesota battles copper corrosion: Part 1. *Public Works*. 128, 11, 65–68.

Neal, C., W. Alan House, and K. Down (1998). Assessment of excess carbon dioxide partial pressures in natural waters based on pH and alkalinity measurements. *Science Total Environment*. 210–211, 1–6, 173–185.

Pandya, M. T. (1997). Novel energy saving approach for scale and corrosion control in pipelines. *Chem. Eng. World*. 32, 2, 73–75.

Perry, R. H. and C. H. Chilton (1973). *Chemical Engineers' Handbook*. McGraw-Hill, New York, 3–138.

Price, S. and F. T. Jefferson (1997). Corrosion control strategies for changing water supplies in Tucson, Arizona. *J. New England Water Works Assoc.* 111, 3, 285–293.

Przybylinski, J. L. (1992). How pressure and CO₂ affect reservoirs and influence the selection of scale control treatments. *Proc. 39th Annu. Meeting Southwestern Petroleum Short Course Conf.* Apr. 22–23, 432–445. Southwestern Petroleum Short Course Assoc., Lubbock, TX.

Rich, L. G. (1963). *Unit Process of Sanitary Engineering*. John Wiley & Sons, New York, 90, 93.

Risser, T. M. (1997). Use of potassium carbonate for corrosion control in small systems, *J. New England Water Works Assoc.* 111, 3, 253–257.

Roque, C., M. Renard, and N. Van Quy (1994). Numerical model for predicting complex sulfate scaling during waterflooding. *Proc. Int. Symp. Formation Damage Control*, Feb. 9–10, Lafayette, LA, 429–440. Society of Petroleum Engineers (SPE), Richardson, TX.

Sincero, A. P. (1968). *Sludge production in coagulation treatment of water*. Master's thesis, Asian Institute of Technology, Bangkok, Thailand, 34.

Sincero, A. P. and G. A. Sincero (1996). *Environmental Engineering: A Design Approach*. Prentice Hall, Upper Saddle River, NJ, 42.

Snoeyink, V. L. and D. Jenkins (1980). *Water Chemistry.* John Wiley & Sons, New York, 64, 75, 77, 295–298.

Yongbo, S. and Z. Shao (1997). Study on preservative and preventing incrustation agent SPZ-1 in coding system of blast furnace. *J. Northeastern Univ.* 18, 2, 222–225.

Svardal, K. (1990). Calcium carbonate precipitation in anaerobic waste water treatment. *Water Science Technol., Proc. 15th Biennial Conf. Int. Assoc. Water Pollut. Res. Control,* July, 29–Aug. 3, 23, 7–9, 1239–1248.

Zannoni, R., et al. (1991). Limestone hardening system for brindisi's power station. First limestone recarbonation unit installed in Italy for industrial water production. *Institution of Chem. Engineers Symp. Series, Proc. Twelfth Int. Symp. Desalination and Water Re-Use,* Apr. 15–18, Malta, 2, 125, 351–357. Inst. of Chemical Engineers, Rugby, England.

12 Coagulation

Colloids are agglomerates of atoms or molecules whose sizes are so small that gravity has no effect on settling them but, instead, they stay in suspension. Because they stay in suspension, they are said to be stable. The reason for this stability is the mutual repulsion between colloid particles. They may, however, be destabilized by application of chemicals. *Coagulation* is the unit process of applying these chemicals for the purpose of destabilizing the mutual repulsion of the particles, thus causing the particles to bind together. This process is normally applied in conjunction with the unit operation of flocculation. The colloid particles are the cause of the turbidity and color that make waters objectionable, thus, should, at least, be partially removed.

This chapter applies the techniques of the unit process of coagulation to the treatment of water and wastewater for the removal of colloids that cause turbidity and color. It also discusses prerequisite topics necessary for the understanding of coagulation such as the behavior of colloids, zeta potential, and colloid stability. It then treats the coagulation process, in general, and the unit process of the use of alum and the iron salts, in particular. It also discusses chemical requirements and sludge production.

12.1 COLLOID BEHAVIOR

Much of the suspended matter in natural waters is composed of silica, or similar materials, with specific gravity of 2.65. In sizes of 0.1 to 2 mm, they settle rapidly; however, in the range of the order of 10^{-5} mm, it takes them a year, in the overall, to settle a distance of only 1 mm. And, yet, it is the particle of this size range that causes the turbidity and color of water, making the water objectionable. The removal of particles by settling is practical only if they settle rapidly in the order of several hundreds of millimeters per hour. This is where coagulation can perform its function, by destabilizing the mutual repulsions of colloidal particles causing them to bind together and grow in size for effective settling. Colloidal particles fall in the size range of 10^{-6} mm to 10^{-3} mm. They are aggregates of several hundreds of atoms or molecules, although a single molecule such as those of proteins is enough to be become a colloid. The term colloid comes from the two Greek words *kolla*, meaning glue, and *eidos*, meaning like.

A colloid system is composed of two phases: the *dispersed phase*, or the *solute*, and the *dispersion medium*, or the *solvent*. Both of these phases can have all three states of matter which are solid, liquid, and gas. For example, the dispersion medium may be a liquid and the dispersed phase may be a solid. This system is called a *liquid sol*, an example of which is the turbidity in water. The dispersion medium may be a

TABLE 12.1
Types of Colloidal Systems

Dispersion Medium	Dispersed Phase	Common Name	Example
Solid	Solid	Solid sol	Colored glass and gems, some alloys
Solid	Liquid	Solid emulsion	Jelly, gel, opal (SiO_2 and H_2O), pearl ($CaCO_3$ and H_2O)
Solid	Gas	None	Pumice, floating soap
Liquid	Solid	Liquid sol	Turbidity in water, starch suspension, ink, paint, milk of magnesia
Liquid	Liquid	Liquid emulsion	Oil in water, milk, mayonnaise, butter
Liquid	Gas	Foam	Whipped cream, beaten egg whites
Gas	Solid	Gaseous sol	Dust, smoke
Gas	Liquid	Gaseous emulsion	Mist, fog, cloud, spray
Gas	Gas	Not applicable	None

gas and dispersed phase may be solid. This system is called a *gaseous sol*, and examples are dust and smoke. Table 12.1 shows the different types of colloidal systems. Note that it is not possible to have a colloidal system of gas in a gas, because gases are completely soluble in each other. In the coagulation treatment of water and wastewater, we will be mainly interested in the solid being dispersed in water, the liquid sol or simply sol. Unless required for clarity, we will use the word "sol" to mean liquid sol.

Sols are either *lyophilic* or *lyophobic*. Lyophilic sols are those that bind the solvent, while the lyophobic sols are those that do not bind the solvent. When the solvent is water, lyophilic and lyophobic sols are, respectively, called *hydrophilic* and *hydrophobic sols*.

The affinity of the *hydrophilic sols* for water is due to polar functional groups that exist on their surfaces. These groups include such polar groups as –OH, –COOH, and –NH₂. They are, respectively, called the *hydroxyl, carboxylic*, and *amine* groups. Figure 12.1a shows the schematic of a hydrophilic colloid. As portrayed, the functional polar groups are shown sticking out from the surface of the particle. Because of the affinity of these groups for water, the water is held tight on the surface. This water is called *bound water* and is fixed on the surface and moves with the particle.

The *hydrophobic colloids* do not have affinity for water; thus, they do not contain any bound water. In general, inorganic colloids are hydrophobic, while organic colloids are hydrophilic. An example of an inorganic colloid is the clay particles that cause turbidity in natural water, and an example of an organic colloid is the colloidal particles in domestic sewage.

12.2 ZETA POTENTIAL

The repulsive property of colloid particles is due to electrical forces that they possess. The characteristic of these forces is indicated in the upper half of Figure 12.1b. At a short distance from the surface of the particle, the force is very high. It dwindles down to zero at infinite distance from the surface.

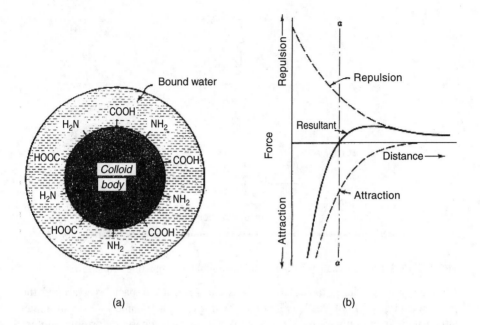

(a)   (b)

FIGURE 12.1 (a) Hydrophilic colloid encased in bound water; (b) interparticle forces as a function of interparticle distance.

The electrical forces are produced due to the charges that the particles possess at their surfaces. These charges called *primary charges* are, in turn, produced from one or both of two phenomena: the dissociation of the polar groups and preferential adsorption of ions from the dispersion medium. The primary charges on hydrophobic colloids are due to preferential adsorption of ions from the dispersion medium.

The primary charges on hydrophilic colloids are due chiefly to the polar groups such as the carboxylic and amine groups. The process by which the charges on these types of colloids are produced is indicated in Figure 12.2. The symbol R represents the colloid body. First, the colloid is represented at the top of the drawing, without the effect of pH. Then by a proper combination of the H^+ and OH^- being added to the solution, the colloid attains ionization of both carboxylic and the amine groups. At this point, both ionized groups neutralize each other and the particle is neutral. This point is called the *isoelectric point*, and the corresponding ion of the colloid is called the *zwitter ion*. Increasing the pH by adding a base cause the added OH^- to neutralize the acid end of the zwitter ion (the NH_3^+); the zwitter ion disappears, and the whole particle becomes negatively charged. The reverse is true when the pH is reduced by the addition of an acid. The added H^+ neutralizes the base end of the zwitter ion (the COO^-); the zwitter ion disappears, and the whole particle becomes positively charged. From this discussion, a hydrophilic colloid can attain a primary charge of either negative or positive depending upon the pH.

The primary charges on a colloid which, as we have seen, could either be positive or negative, attract ions of opposite charges from the solution. These opposite charges are called *counterions*. This is indicated in Figure 12.3. If the primary charges are

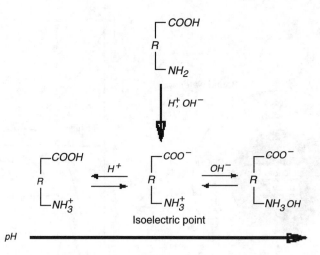

FIGURE 12.2 Primary charges of a hydrophilic colloid as a function of pH.

sufficiently large, the attracted counterions can form a compact layer around the primary charges. This layer is called the *Stern layer*. The counterions, in turn, can attract their own counterions, the *coions* of the primary charges, forming another layer. Since these coions form a continuous distribution of ions into the bulk of the solution, they tend to be diffused and form a diffused layer. The second layer is called the *Gouy layer*. Thus, the *Stern* and *Gouy* layers form an envelope of electric double layer around the primary charges.

All of the charges in the Stern layer move with the colloid; thus, this layer is a fixed layer. In the Gouy layer, part of the layer may move with the colloid particle by shearing at a *shear plane*. This layer may shear off beyond the boundary of the fixed Stern layer measured from the surface of the colloid. Thus, some of the charges in the layer move with the particle, while others do not. This plane is indicated in Figure 12.3.

The charges are electric, so they possess electrostatic potential. As indicated on the right-hand side of Figure 12.3, this potential is greatest at the surface and decreases to zero at the bulk of the solution. The potential at a distance from the surface at the location of the shear plane is called the *zeta potential*. Zeta potential meters are calibrated to read the value of this potential. The greater this potential, the greater is the force of repulsion and the more stable the colloid.

12.3 COLLOID DESTABILIZATION

Colloid stability may further be investigated by the use Figure 12.1b. This figure portrays the competition between two forces at the surface of the colloid particle: the *van der Waal's force of attraction*, represented by the lower dashed curve, and the force of repulsion, represented by the upper dashed curve. The solid curve represents the resultant of these two forces. As shown, this resultant becomes zero at $a - a'$ and becomes fully an attractive force to the left of the line. When the resultant force becomes fully attractive, two colloid particles can bind themselves together.

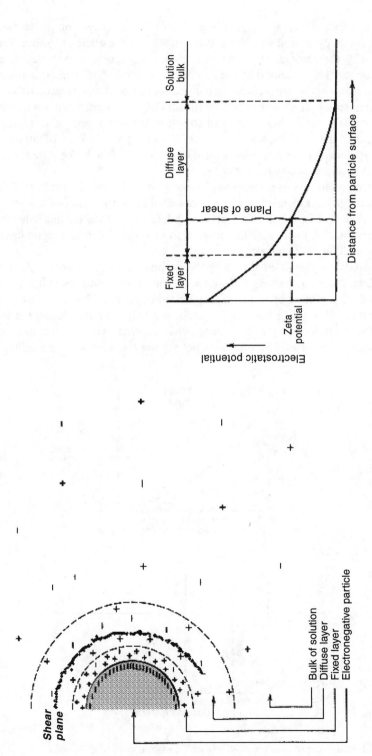

FIGURE 12.3 Charged double layer around a negatively charged colloid particle (left) and variation of electrostatic potential with distance from particle surface (right).

The force of repulsion, as we have seen, is due to the charges on the surface. Inherent in any body is a natural force that tends to bind particles together. This force is exactly the same as the force that causes adsorption of particles to an adsorbing surface. This is caused by the imbalance of atomic forces on the surface.

Whereas atoms below the surface of a particle are balanced with respect to forces of neighboring atoms, those at the surface are not. Thus, the unbalanced force at the surface becomes the van der Waal's force of attraction. By the presence of the primary charges that exert the repulsive force, however, the van der Waal's force of attraction is nullified until a certain distance designated by $a - a'$ is reached. The distance can be shortened by destabilizing the colloid particle.

The use of chemicals to reduce the distance to $a - a'$ from the surface of the colloid is portrayed in Figure 12.4. The zeta potential is the measure of the stability of colloids. To destabilize a colloid, its zeta potential must be reduced; this reduction is equivalent to the shortening of the distance to $a - a'$ and can be accomplished through the addition of chemicals.

The chemicals to be added should be the counterions of the primary charges. Upon addition, these counterions will neutralize the primary charges reducing the zeta potential. This process of reduction is indicated in Figures 12.4a and 12.4b; the potential is reduced in going from Figure 12.4a to 12.4b. Note that destabilization is simply the neutralization of the primary charges, thus reducing the force of repulsion between particles. The process is not yet *the coagulation* of the colloid.

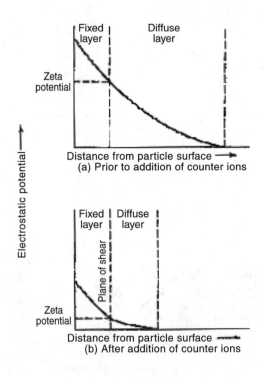

FIGURE 12.4 Reduction of zeta potential to cause destabilization of colloids.

12.4 COAGULATION PROCESS

The destabilization of colloids through the addition of counterions should be done in conjunction with the application of the complete coagulation process. Four methods are used to bring about this process: double-layer compression, charge neutralization, entrapment in a precipitate, and intraparticle bridging.

When the concentration of counterions in the dispersion medium is smaller, the thickness of the electric double layer is larger. Two approaching colloid particles cannot come close to each other because of the thicker electric double layer, therefore, the colloid is stable. Now, visualize adding more counterions. When the concentration is increased, the attracting force between the primary charges and the added counterions increases causing the double layer to shrink. The layer is then said to be compressed. As the layer is compressed sufficiently by the continued addition of more counterions, a time will come when the van der Waals force exceeds the force of repulsion and coagulation results.

The charge of a colloid can also be directly neutralized by the addition of ions of opposite charges that have the ability to directly adsorb to the colloid surface. For example, the positively charged dodecylammoniun, $C_{12}H_{25}NH_3^+$, tends to be hydrophobic and, as such, penetrates directly to the colloid surface and neutralize it. This is said to be a direct charge neutralization, since the counterion has penetrated directly into the primary charges. Another direct charge neutralization method would be the use of a colloid of opposite charge. Direct charge neutralization and the compression of the double layer may compliment each other.

A characteristic of some cations of metal salts such as Al(III) and Fe(III) is that of forming a precipitate when added to water. For this precipitation to occur, a colloidal particle may provide as the seed for a nucleation site, thus, entrapping the colloid as the precipitate forms. Moreover, if several of this particles are entrapped and are close to each other, coagulation can result by direct binding because of the proximity.

The last method of coagulation is intraparticle bridging. A bridging molecule may attach a colloid particle to one active site and a second colloid particle to another site. An *active site* is a point in the molecule where particles may attach either by chemical bonding or by mere physical attachment. If the two sites are close to each other, coagulation of the colloids may occur; or, the kinetic movement may loop the bridge assembly around causing the attached colloids to bind because for now they are hitting each other, thus bringing out coagulation.

12.4.1 COAGULANTS FOR THE COAGULATION PROCESS

Electrolytes and polyelectrolytes are used to coagulate colloids. *Electrolytes* are materials which when placed in solution cause the solution to be conductive to electricity because of charges they possess. *Polyelectrolytes* are polymers possessing more than one electrolytic site in the molecule, and *polymers* are molecules joined together to form larger molecules. Because of the charges, electrolytes and polyelectrolytes coagulate and precipitate colloids. The coagulating power of electrolytes is summed up in the *Schulze–Hardy rule* that states: the coagulation of a colloid is affected by that ion of an added electrolyte that has a charge opposite in sign to that of the colloidal particle; the effect of such an ion increases markedly with the number of charges carried. Thus, comparing the effect of $AlCl_3$ and $Al_2(SO_4)_3$ in coagulating positive colloids, the latter is 30 times more effective than the former, since sulfate has two negative charges while the chloride has only one. In coagulating negative colloids, however, the two have about the same power of coagulation.

The most important coagulants used in water and wastewater treatment are alum, copperas (ferrous sulfate), ferric sulfate, and ferric chloride. Later, we will specifically discuss the chemical reactions of these coagulants at greater lengths. Other coagulants have also been used but, owing to high cost, their use is restricted only to small installations. Examples of these are sodium aluminate, $NaAlO_2$; ammonia alum, $Al_2(SO_4)_3 \cdot (NH_4)_2 \cdot 24H_2O$; and potash alum, $Al_2(SO_4)_3 \cdot K_2SO_4 \cdot 24H_2O$. The reactions of sodium aluminate with aluminum sulfate and carbon dioxide are:

$$6NaAlO_2 + Al_2(SO_4)_3 \cdot 14.3H_2O \rightarrow 8Al(OH)_3 + 3Na_2SO_4 + 2.3H_2O \quad (12.1)$$

$$2NaAlO_2 + CO_2 + 3H_2O \rightarrow 2Al(OH)_3\downarrow + Na_2CO_3 \quad (12.2)$$

12.4.2 Coagulant Aids

Difficulties with settling often occur because of flocs that are slow-settling and are easily fragmented by the hydraulic shear in the settling basin. For these reasons, coagulant aids are normally used. Acids and alkalis are used to adjust the pH to the optimum range. Typical acids used to lower the pH are sulfuric and phosphoric acids. Typical alkalis used to raise the pH are lime and soda ash. Polyelectrolytes are also used as coagulant aids. The cationic form has been used successfully in some waters not only as a coagulant aid but also as the primary coagulant. In comparison with alum sludges that are gelatinous and voluminous, sludges produced by using cationic polyelectrolytes are dense and easy to dewater for subsequent treatment and disposal. Anionic and nonionic polyelectrolytes are often used with primary metal coagulants to provide the particle bridging for effective coagulation. Generally, the use of polyelectrolyte coagulant aids produces tougher and good settling flocs.

Activated silica and clays have also been used as coagulant aids. Activated silica is sodium silicate that has been treated with sulfuric acid, aluminum sulfate, carbon dioxide, or chlorine. When the activated silica is applied, a stable negative sol is produced. This sol unites with the positively charged primary-metal coagulant to produce tougher, denser, and faster settling flocs.

Bentonite clays have been used as coagulant aids in conjunction with iron and alum primary coagulants in treating waters containing high color, low turbidity, and low mineral content. Low turbidity waters are often hard to coagulate. Bentonite clay serves as a weighting agent that improves the settleability of the resulting flocs.

12.4.3 Rapid Mix for Complete Coagulation

Coagulation will not be as efficient if the chemicals are not dispersed rapidly throughout the mixing tank. This process of rapidly mixing the coagulant in the volume of the tank is called *rapid* or *flash mix*. Rapid mixing distributes the chemicals immediately throughout the volume of the mixing tank. Also, coagulation should be followed by flocculation to agglomerate the tiny particles formed from the coagulation process.

If the coagulant reaction is simply allowed to take place in one portion of the tank because of the absence of the rapid mix rather than being spread throughout the volume, all four mechanisms for a complete coagulation discussed above will not be utilized. For example, charge neutralization will not be utilized in all portions of the tank because, by the time the coagulant arrives at the point in question, the reaction of charge neutralization will already have taken place somewhere.

Interparticle bridging will not be as effective, since the force to loop the bridge around will not be as strong without the force of the rapid mix. Colloid particles will not effectively be utilized as seeds for nucleation sites because, without rapid mix, the coagulant may simply stay in one place. Finally, the compression of the double layer will not be as effective if unaided by the force due to the rapid mix. The force of the rapid mix helps push two colloids toward each other, thus enhancing coagulation. Hence, because of all these stated reasons, coagulation should take place in a rapidly mixed tank.

12.4.4 THE JAR TEST

In practice, irrespective of what coagulant or coagulant aid is used, the optimum dose and pH are determined by a jar test. This consists of four to six beakers (such as 1000 ml in volume) filled with the raw water into which varying amounts of dose are administered. Each beaker is provided with a variable-speed stirrer capable of operating from 0 to 100 rpm.

Upon introduction of the dose, the contents are rapidly mixed at a speed of about 60 to 80 rpm for a period of one minute and then allowed to flocculate at a speed of 30 rpm for a period of 15 minutes. After the stirring is stopped, the nature and settling characteristics of the flocs are observed and recorded qualitatively as *poor, fair, good*, or *excellent*. A hazy sample denotes poor coagulation; a properly coagulated sample is manifested by well-formed flocs that settle rapidly with clear water between flocs. The lowest dose of chemicals and pH that produce the desired flocs and clarity represents the optimum. This optimum is then used as the dose in the actual operation of the plant. See Figure 12.5 for a picture of a jar testing apparatus.

12.5 CHEMICAL REACTIONS OF ALUM

The alum used in water and wastewater treatment is $Al_2(SO_4)_3 \cdot 14H_2O$. (The "14" actually varies from 13 to 18.) For brevity, this will simply be written without the water of hydration as $Al_2(SO_4)_3$. When alum is dissolved in water, it dissociates according to the following equation (Sincero, 1968):

$$Al_2(SO_4)_3 \rightarrow 2Al^{3+} + 3SO_4^{2-} \qquad (12.3)$$

By rapid mix, the ions must be rapidly dispersed throughout the tank in order to effect the complete coagulation process.

FIGURE 12.5 A Phipps and Bird jar testing apparatus. (Courtesy of Phipps & Bird, Richmond, VA. © 2002 Phipps & Bird.)

Because the water molecule is polar, it attracts Al^{3+} forming a complex ion according to the following:

$$Al^{3+} + 6H_2O \rightarrow Al(H_2O)_6^{3+} \tag{12.4}$$

In the complex ion $Al(H_2O)_6^{3+}$, Al is called the *central atom* and the molecules of H_2O are called *ligands*. The subscript 6 is the coordination number, the number of ligands attached to the central atom; the superscript 3+ is the charge of the complex ion. The whole assembly of the complex forms what is called a *coordination sphere*.

As indicated in Equation (12.4), aluminum has a coordination number of 6 with the water molecule. This means that no more water molecules can bind with the central atom but that any interaction would not be a mere insertion into the coordination sphere. In fact, further reaction with the water molecule involves hydrolysis of the water molecule and exchanging of the resulting OH^- ion with the H_2O ligand inside the coordination sphere. This type of reaction is called *ligand exchange reaction*.

Some of the hydrolysis products of the ligand exchange reaction are mononuclear, which means that only one central atom of aluminum is in the complex; and some are polynuclear, which means that more than one central atom of aluminum exists in the complex. Because the water molecule is not charged, $Al(H_2O)_6^{3+}$ may simply be written as Al^{3+}. This is the symbol to be used in the complex reactions that follow. Without going into details, we will simply write at once all the complex ligand exchange equilibrium reactions.

$$Al^{3+} + H_2O \rightleftharpoons Al(OH)^{2+} + H^+ \qquad K_{Al(OH)c} = 10^{-5} \tag{12.5}$$

$$7Al^{3+} + 17H_2O \rightleftharpoons Al_7(OH)_{17}^{4+} + 17H^+ \qquad K_{Al_7(OH)_{17}c} = 10^{-48.8} \tag{12.6}$$

$$13Al^{3+} + 34H_2O \rightleftharpoons Al_{13}(OH)_{34}^{5+} + 34H^+ \qquad K_{Al_{13}(OH)_{34}c} = 10^{-97.4} \tag{12.7}$$

$$Al(OH)_{3(s)}(\text{fresh precipitate}) \rightleftharpoons Al^{3+} + 3OH^- \qquad K_{sp,Al(OH)_3} = 10^{-33} \qquad (12.8)$$

$$Al(OH)_{3(s)} + OH^- \rightleftharpoons Al(OH)_4^- \qquad K_{Al(OH)_4 c} = 10^{+1.3} \qquad (12.9)$$

$$2Al^{3+} + 2H_2O \rightleftharpoons Al_2(OH)_2^{4+} + 2H^+ \qquad K_{Al_2(OH)_2 c} = 10^{-6.3} \qquad (12.10)$$

The equilibrium constants apply at 25°C:

Note: From the number of aluminum atoms they contain $Al_7(OH)_{17}^{4+}$, $Al_{13}(OH)_{34}^{5+}$, and $Al_2(OH)_2^{4+}$ are polynuclear complexes.

Also, the H^+ and the OH^- are participants in these reactions. This means that the concentrations of each of these complex ions are determined by the pH of the solution.

In the application of the previous equations in coagulation treatment of water, conditions must be adjusted to allow maximum precipitation of the solid represented by $Al(OH)_{3(s)}$. To allow for this maximum precipitation, the concentrations of the complex ions must be held to a minimum.

12.5.1 DETERMINATION OF THE OPTIMUM pH

For effective removal of the colloids, as much of alum should be converted to the solid $Al(OH)_{3(s)}$. Also, as much of the concentrations of the complex ions should neutralize the primary charges of the colloids to effect their destabilization. Overall, this means that once the solids have been formed and the complex ions have neutralized the colloid charges, the concentrations of the complex ions standing in solution should be at the minimum. The pH corresponding to this condition is called the optimum pH.

Let sp_{Al} represent all the species that contain the aluminum atom standing in solution. Thus, the concentration of all the species containing the aluminum atom Al(III), is

$$[sp_{Al}] = [Al^{3+}] + [Al(OH)^{2+}] + 7[Al_7(OH)_{17}^{4+}] + 13[Al_{13}(OH)_{34}^{5+}]$$
$$+ [Al(OH)_4^-] + 2[Al_2(OH)_2^{4+}] \qquad (12.11)$$

All the concentrations in the right-hand side of the previous equation will now be expressed in terms of the hydrogen ion concentration. This will result in the expressing of $[sp_{Al}]$ in terms of the hydrogen ion. Differentiating the resulting equation of $[sp_{Al}]$ with respect to $[H^+]$ and equating the result to zero will produce the minimum concentration of sp_{Al} and, thus, the optimum pH determined. Using the equilibrium reactions, Eqs. (12.5) through (12.10), along with the ion product of water, we now proceed as follows:

$$[Al^{3+}] = \frac{\{Al^{3+}\}}{\gamma_{Al}} = \frac{K_{sp,Al(OH)_3}}{\gamma_{Al}\{OH^-\}^3} = \frac{K_{sp,Al(OH)_3}\{H^+\}^3}{\gamma_{Al}K_w^3} = \frac{K_{sp,Al(OH)_3}\gamma_H^3[H^+]^3}{\gamma_{Al}K_w^3}$$
$$(12.12)$$

$$[Al(OH)^{2+}] = \frac{\{Al(OH)^{2+}\}}{\gamma_{Al(OH)c}} = \frac{K_{Al(OH)c}\{Al^{3+}\}}{\gamma_{Al(OH)c}\{H^+\}} = \frac{K_{Al(OH)c}K_{sp,Al(OH)_3}\gamma_H^2[H^+]^2}{\gamma_{Al(OH)c}K_w^3}$$

$$(12.13)$$

$$[Al_7(OH)_{17}^{4+}] = \frac{[Al_7(OH)_{17}]}{\gamma_{Al_7(OH)_{17}c}} = \frac{K_{Al_7(OH)_{17}c}\{Al^{3+}\}^7}{\gamma_{Al_7(OH)_{17}c}\{H^+\}^{17}} = \frac{K_{Al_7(OH)_{17}c}\gamma_{Al}^7[Al^{3+}]^7}{\gamma_{Al_7(OH)_{17}c}\gamma_H^{17}[H^+]^{17}}$$

$$= \frac{K_{Al_7(OH)_{17}c}K_{sp,Al(OH)_3}^7\gamma_H^4[H^+]^4}{\gamma_{Al_7(OH)_{17}c}K_w^{21}}$$

$$(12.14)$$

$$[Al_{13}(OH)_{34}^{5+}] = \frac{\{Al_{13}(OH)_{34}^{5+}\}}{\gamma_{Al_{13}(OH)_{34}c}} = \frac{K_{Al_{13}(OH)_{34}c}\{Al^{3+}\}^{13}}{\gamma_{Al_{13}(OH)_{34}c}\{H^+\}^{34}} = \frac{K_{Al_{13}(OH)_{34}c}\gamma_{Al}^{13}[Al^{3+}]^{13}}{\gamma_{Al_{13}(OH)_{34}c}\gamma_H^{34}[H^+]^{34}}$$

$$= \frac{K_{Al_{13}(OH)_{34}c}K_{sp,Al(OH)_3}^{13}\gamma_H^5[H^+]^5}{\gamma_{Al_{13}(OH)_{34}c}K_w^{39}}$$

$$(12.15)$$

$$[Al(OH)_4^-] = \frac{\{Al(OH)_4^-\}}{\gamma_{Al(OH)_4c}} = \frac{K_{Al(OH)_4c}\{OH^-\}}{\gamma_{Al(OH)_4c}} = \frac{K_{Al(OH)_4c}K_w}{\gamma_{Al(OH)_4c}\gamma_H[H^+]} \quad (12.16)$$

$$[Al_2(OH)_2^{4+}] = \frac{\{Al_2(OH)_2^{4+}\}}{\gamma_{Al_2(OH)_2c}} = \frac{K_{Al_2(OH)_2c}\{Al^{3+}\}^2}{\gamma_{Al_2(OH)_2c}\{H^+\}^2} = \frac{K_{Al_2(OH)_2c}\gamma_{Al}^2[Al^{3+}]^2}{\gamma_{Al_2(OH)_2c}\gamma_H^2[H^+]^2}$$

$$= \frac{K_{Al_2(OH)_2c}K_{sp,Al(OH)_3}^2\gamma_H^4[H^+]^4}{\gamma_{Al_2(OH)_2c}K_w^6}$$

$$(12.17)$$

γ_{Al}, γ_H, $\gamma_{Al(OH)c}$, $\gamma_{Al_7(OH)_{17}c}$, $\gamma_{Al_{13}(OH)_{34}c}$, $\gamma_{Al(OH)_4c}$, $\gamma_{Al_2(OH)_2c}$ are, respectively, the activity coefficients of the aluminum ion and the hydrogen ion and the complexes $Al(OH)^{2+}$, $Al_7(OH)_{17}^{4+}$, $Al_{13}(OH)_{34}^{5+}$, $Al(OH)_4^-$, and $Al_2(OH)_2^{4+}$. $K_{sp,Al(OH)_3}$ is the solubility product constant of the solid $Al(OH)_{3(s)}$ and K_w is the ion product of water. $K_{Al(OH)c}$, $K_{Al_7(OH)_{17}c}$, $K_{Al_{13}(OH)_{34}c}$, $K_{Al(OH)_4c}$, and $K_{Al_2(OH)_2c}$ are, respectively, the equilibrium constants of the complexes $Al(OH)^{2+}$, $Al_7(OH)_{17}^{4+}$, $Al_{13}(OH)_{34}^{5+}$, $Al(OH)_4^-$, and $Al_2(OH)_2^{4+}$.

Remember that the equilibrium constants are a function of temperature. To obtain the corresponding values at other temperatures, the *Van't Hoff equation* should be used. The use of this equation, however, requires the value of the standard enthalpy ΔH_{298}^o. At present, none are available for the aluminum complexes. Research is therefore needed to find these values.

Equations (12.12) through (12.17) may now be substituted into Equation (12.11) to produce

$$[sp_{Al}] = \frac{K_{sp,Al(OH)_3}\gamma_H^3[H^+]^3}{\gamma_{Al}K_w^3} + \frac{K_{Al(OH)c}K_{sp,Al(OH)_3}\gamma_H^2[H^+]^2}{\gamma_{Al(OH)c}K_w^3}$$
$$+ \frac{7K_{Al_7(OH)_{17}c}K_{sp,Al(OH)_3}^7\gamma_H^4[H^+]^4}{\gamma_{Al_7(OH)_{17}c}K_w^{21}} + \frac{13K_{Al_{13}(OH)_{34}c}K_{sp,Al(OH)_3}^{13}\gamma_H^5[H^+]^5}{\gamma_{Al_{13}(OH)_{34}c}K_w^{39}}$$
$$+ \frac{K_{Al(OH)_4c}K_w}{\gamma_{Al(OH)_4c}\gamma_H[H^+]} + \frac{2K_{Al_2(OH)_2c}K_{sp,Al(OH)_3}^2\gamma_H^4[H^+]^4}{\gamma_{Al_2(OH)_2c}K_w^6} \qquad (12.18)$$

To obtain the optimum pH, differentiate $[sp_{Al}]$ of Equation (12.18) with respect to $[H^+]$ and equate the result to zero. Doing the differentiation, rearranging the resulting equation, and calling the resulting solution for $[H^+]$ as $[H_{opt}^+]$, obtain the following equation:

$$\left\{\frac{2K_{Al(OH)c}K_{sp,Al(OH)_3}\gamma_H^2}{\gamma_{Al(OH)c}K_w^3}\right\}[H_{opt}^+]^3 + \left\{\frac{3K_{sp,Al(OH)_3}\gamma_H^3}{\gamma_{Al}K_w^3}\right\}[H_{opt}^+]^4$$
$$+ \left\{\frac{28K_{Al_7(OH)_{17}c}K_{sp,Al(OH)_3}^7\gamma_H^4}{\gamma_{Al_7(OH)_{17}c}K_w^{21}} + \frac{8K_{Al_2(OH)_2c}K_{sp,Al(OH)_3}^2\gamma_H^4}{\gamma_{Al_2(OH)_2c}K_w^6}\right\}[H_{opt}^+]^5$$
$$+ \left\{\frac{65K_{Al_{13}(OH)_{34}c}K_{sp,Al(OH)_3}^{13}\gamma_H^5}{\gamma_{Al_{13}(OH)_{34}c}K_w^{39}}\right\}[H_{opt}^+]^6 = \frac{K_{Al(OH)_4c}K_w}{\gamma_{Al(OH)_4c}\gamma_H} \qquad (12.19)$$

By trial and error, Equation (15.19) may now be solved for $[H_{opt}^+]$. Thus, the optimum pH is

$$pH = -\log\{H_{opt}^+\} = \log(\gamma_H[H_{opt}^+]) \qquad (12.20)$$

Example 12.1 A raw water containing 140 mg/L of dissolved solids is subjected to coagulation treatment using alum. Calculate the optimum pH that the operation should be conducted. Assume the temperature of operation is 25°C.

Solution:

$$\left\{\frac{2K_{Al(OH)c}K_{sp,Al(OH)_3}\gamma_H^2}{\gamma_{Al(OH)c}K_w^3}\right\}[H_{opt}^+]^3 + \left\{\frac{3K_{sp,Al(OH)_3}\gamma_H^3}{\gamma_{Al}K_w^3}\right\}[H_{opt}^+]^4$$
$$+ \left\{\frac{28K_{Al_7(OH)_{17}c}K_{sp,Al(OH)_3}^7\gamma_H^4}{\gamma_{Al_7(OH)_{17}c}K_w^{21}} + \frac{8K_{Al_2(OH)_2c}K_{sp,Al(OH)_3}^2\gamma_H^4}{\gamma_{Al_2(OH)_2c}K_w^6}\right\}[H_{opt}^+]^5$$
$$+ \left\{\frac{65K_{Al_{13}(OH)_{34}c}K_{sp,Al(OH)_3}^{13}\gamma_H^5}{\gamma_{Al_{13}(OH)_{34}c}K_w^{39}}\right\}[H_{opt}^+]^6 = \frac{K_{Al(OH)_4c}K_w}{\gamma_{Al(OH)_4c}\gamma_H}$$

$$\frac{2K_{Al(OH)c}K_{sp,Al(OH)_3}\gamma_H^2}{\gamma_{Al(OH)c}K_w^3}:$$

$$K_{Al(OH)c} = 10^{-5} \quad K_{sp,Al(OH)_3} = 10^{-33} \quad \mu = 2.5(10^{-5})TDS \quad \gamma = 10^{-\frac{0.5z_i^2(\sqrt{\mu})}{1+1.14(\sqrt{\mu})}}$$

$$\mu = 2.5(10^{-5})(140) = 3.5(10^{-3}) \qquad \gamma_H = 10^{-\frac{0.5(1)^2[\sqrt{3.5(10^{-3})}]}{1+1.14[\sqrt{3.5(10^{-3})}]}} = 0.94$$

$$\gamma_{Al(OH)c} = 10^{-\frac{0.5(2)^2[\sqrt{3.5(10^{-3})}]}{1+1.14[\sqrt{3.5(10^{-3})}]}} = 0.77 \qquad K_w = 10^{-14}$$

Therefore,

$$\frac{2K_{Al(OH)c}K_{sp,Al(OH)_3}\gamma_H^2}{\gamma_{Al(OH)c}K_w^3} = \frac{2(10^{-5})(10^{-33})(0.94)^2}{(0.77)(10^{-14})^3} = 2.30(10^4)$$

$$\frac{3K_{sp,Al(OH)_3}\gamma_H^3}{\gamma_{Al}K_w^3}:$$

$$\gamma_{Al} = 10^{-\frac{0.5(3)^2[\sqrt{3.5(10^{-3})}]}{1+1.14[\sqrt{3.5(10^{-3})}]}} = 0.56$$

Therefore,

$$\frac{3K_{sp,Al(OH)_3}\gamma_H^3}{\gamma_{Al}K_w^3} = \frac{3(10^{-33})(0.94)^3}{(0.56)(10^{-14})^3} = 4.45(10^9)$$

$$\frac{28K_{Al_7(OH)_{17}c}K_{sp,Al(OH)_3}^7\gamma_H^4}{\gamma_{Al_7(OH)_{17}c}K_w^{21}} + \frac{8K_{Al_2(OH)_2c}K_{sp,Al(OH)_3}^2\gamma_H^4}{\gamma_{Al_2(OH)_2c}K_w^6}:$$

$$K_{Al_7(OH)_{17}c} = 10^{-48.8} \quad \gamma_{Al_7(OH)_{17}c} = \gamma_{Al_2(OH)_2c} = 10^{-\frac{0.5(4)^2[\sqrt{3.5(10^{-3})}]}{1+1.14[\sqrt{3.5(10^{-3})}]}} = 0.36$$

$$K_{Al_2(OH)_2c} = 10^{-6.3}$$

Therefore,

$$\frac{28K_{Al_7(OH)_{17}c}K_{sp,Al(OH)_3}^7\gamma_H^4}{\gamma_{Al_7(OH)_{17}c}K_w^{21}} + \frac{8K_{Al_2(OH)_2c}K_{sp,Al(OH)_3}^2\gamma_H^4}{\gamma_{Al_2(OH)_2c}K_w^6}$$

$$= \frac{28(10^{-48.8})(10^{-33})^7(0.94)^4}{(0.36)(10^{-14})^{21}} + \frac{8(10^{-6.3})(10^{-33})^2(0.94)^4}{(0.36)(10^{-14})^6}$$

$$= 9.62(10^{15}) + 8.70(10^{12}) = 9.63(10^{15})$$

$$\frac{65K_{Al_{13}(OH)_{34}c}K_{sp,Al(OH)_3}^{13}\gamma_H^5}{\gamma_{Al_{13}(OH)_{34}c}K_w^{39}}:$$

$$K_{Al_{13}(OH)_{34}c} = 10^{-97.4} \qquad \gamma_{Al_{13}(OH)_{34}c} = 10^{-\frac{0.5(5)^2[\sqrt{3.5(10^{-3})}]}{1+1.14[\sqrt{3.5(10^{-3})}]}} = 0.20$$

Therefore,

$$\frac{65K_{Al_{13}(OH)_{34}c}K_{sp,Al(OH)_3}^{13}\gamma_H^5}{\gamma_{Al_{13}(OH)_{34}c}K_w^{39}} = \frac{65(10^{-97.4})(10^{-33})^{13}(0.94)^5}{(0.20)(10^{-14})^{39}} = \frac{65(0.94)^5}{(0.20)(10^{-19.6})}$$

$$= 9.50(10^{21})$$

$$\frac{K_{Al(OH)_4c}K_w}{\gamma_{Al(OH)_4c}\gamma_H}:$$

$$K_{Al(OH)_4c} = 10^{+1.3} \qquad \gamma_{Al(OH)_4c} = \gamma_H = 0.94$$

Therefore,

$$\frac{K_{Al(OH)_4c}K_w}{\gamma_{Al(OH)_4c}\gamma_H} = \frac{(10^{+1.3})(10^{-14})}{(0.94)(0.94)} = 2.26(10^{-13})$$

Therefore,

$$2.30(10^4)[H_{opt}^+]^{+3} + 4.45(10^9)[H_{opt}^+]^4 + \{9.63(10^{15})[H_{opt}^+]^5 + 9.50(10^{21})[H_{opt}^+]^6$$
$$= 2.26(10^{-13})$$

Solving by trial and error, let

$$Y = 2.30(10^4)[H_{opt}^+]^3 + 4.45(10^9)[H_{opt}^+]^4 + \{9.63(10^{15})[H_{opt}^+]^5 + 9.50(10^{21})[H_{opt}^+]^6$$

$[H_{opt}^+]$	Y
10^{-6}	$4.66(10^{-14})$
10^{-8}	$2.30(10^{-20})$

$$y \qquad 2.26(10^{-13}) \qquad \frac{y-10^{-6}}{10^{-6}-10^{-8}} = \frac{2.26(10^{-13})-4.66(10)^{-14}}{4.66(10)^{-14}-2.30(10)^{-20}}$$

$$10^{-6} \qquad 4.66(10^{-14}) \qquad y = 4.81(10^{-6})$$

$$10^{-8} \qquad 2.30(10^{-20}) \qquad \text{Therefore, pH} = -\log_{10}[4.81(10^{-6})] = 5.32 \quad \textbf{Ans}$$

12.6 CHEMICAL REACTIONS OF THE FERROUS ION

The ferrous salt used as coagulant in water and wastewater treatment is copperas, $FeSO_4 \cdot 7H_2O$. For brevity, this will simply be written without the water of hydration as $FeSO_4$. When copperas dissolves in water, it dissociates according to the following equation:

$$FeSO_4 \rightarrow Fe^{2+} + SO_4^{2-} \tag{12.21}$$

As in the case of alum, the ions must be rapidly dispersed throughout the tank in order to effect the complete coagulation process. The solid precipitate $Fe(OH)_{2(s)}$ and complexes are formed and expressing in terms of equilibrium with the solid $Fe(OH)_{2(s)}$, the following reactions transpire (Snoeyink and Jenkins, 1980):

$$Fe(OH)_{2(s)} \rightleftharpoons Fe^{2+} + 2OH^- \qquad K_{sp,Fe(OH)_2} = 10^{-14.5} \tag{12.22}$$

$$Fe(OH)_{2(s)} \rightleftharpoons FeOH^+ + OH^- \qquad K_{FeOHc} = 10^{-9.4} \tag{12.23}$$

$$Fe(OH)_{2(s)} + OH^- \rightleftharpoons Fe(OH)_3^- \qquad K_{Fe(OH)_3c} = 10^{-5.1} \tag{12.24}$$

The complexes are $FeOH^+$ and $Fe(OH)_3^-$. Also note that the OH^- ion is a participant in these reactions. This means that the concentrations of each of these complex ions are determined by the pH of the solution. In the application of the above equations in an actual coagulation treatment of water, conditions must be adjusted to allow maximum precipitation of the solid represented by $Fe(OH)_{2(s)}$. To allow for this maximum precipitation, the concentrations of the complex ions must be held to the minimum. The values of the equilibrium constants given above are at 25°C.

12.6.1 DETERMINATION OF THE OPTIMUM pH

For effective removal of the colloids, as much of the copperas should be converted to the solid $Fe(OH)_{2(s)}$. Also, as much of the concentrations of the complex ions should neutralize the primary charges of the colloids to effect their destabilization. Overall, this means that once the solids have been formed and the complex ions have neutralized the colloid charges, the concentrations of the complex ions standing in solution should be at the minimum. The pH corresponding to this condition is the optimum pH for the coagulation using copperas.

Let sp_{FeII} represent all the species that contain the Fe(II) ion standing in solution. Thus, the concentration of all the species containing the ion is

$$[sp_{FeII}] = [Fe^{2+}] + [FeOH^+] + [Fe(OH)_3^-] \tag{12.25}$$

All the concentrations in the right-hand side of the previous equation will now be expressed in terms of the hydrogen ion concentration. As in the case of alum, this will result in the expressing of $[sp_{FeII}]$ in terms of the hydrogen ion. Differentiating the resulting equation of $[sp_{FeII}]$ with respect to $[H^+]$ and equating the result to zero will produce the minimum concentration of sp_{FeII} and, thus, the optimum pH determined. Using the equilibrium reactions, Eqs. (12.22) through (12.24), along with the ion product of water, we now proceed as follows:

$$[\text{Fe}^{2+}] = \frac{\{\text{Fe}^{2+}\}}{\gamma_{FeII}} = \frac{K_{sp,\text{Fe(OH)}_2}}{\gamma_{FeII}\{\text{OH}^-\}^2} = \frac{K_{sp,\text{Fe(OH)}_2}\{\text{H}^+\}^2}{\gamma_{FeII}K_w^2} = \frac{K_{sp,\text{Fe(OH)}_2}\gamma_H^2[\text{H}^+]^2}{\gamma_{FeII}K_w^2} \quad (12.26)$$

$$[\text{FeOH}^+] = \frac{\{\text{FeOH}^+\}}{\gamma_{FeOHc}} = \frac{K_{FeOHc}}{\gamma_{FeOHc}\{\text{OH}^-\}} = \frac{K_{FeOHc}\{\text{H}^+\}}{\gamma_{FeOHc}K_w} = \frac{K_{FeOHc}\gamma_H[\text{H}^+]}{\gamma_{FeOHc}K_w} \quad (12.27)$$

$$[\text{Fe(OH)}_3^-] = \frac{\{\text{Fe(OH)}_3^-\}}{\gamma_{\text{Fe(OH)}_3c}} = \frac{K_{\text{Fe(OH)}_3c}\{\text{OH}^-\}}{\gamma_{\text{Fe(OH)}_3c}} = \frac{K_{\text{Fe(OH)}_3c}K_w}{\gamma_{\text{Fe(OH)}_3c}\{\text{H}^+\}} = \frac{K_{\text{Fe(OH)}_3c}K_w}{\gamma_{\text{Fe(OH)}_3c}\gamma_H[\text{H}^+]}$$

$$\qquad\qquad (12.28)$$

γ_{FeII}, γ_{FeOHc}, $\gamma_{\text{Fe(OH)}_3c}$ are, respectively, the activity coefficients of the ferrous ion and the complexes FeOH$^+$ and $[\text{Fe(OH)}_3^-]$. $K_{sp,\text{Fe(OH)}_2}$ is the solubility product constant of the solid Fe(OH)$_{2(s)}$. K_{FeOHc} and $K_{\text{Fe(OH)}_3c}$ are, respectively, the equilibrium constants of the complexes FeOH$^+$ and Fe(OH)$_3^-$.

Equations (12.26) through (12.28) may now be substituted into Equation (12.25) to produce

$$[sp_{FeII}] = \frac{K_{sp,\text{Fe(OH)}_2}\gamma_H^2[\text{H}^+]^2}{\gamma_{FeII}K_w^2} + \frac{K_{FeOHc}\gamma_H[\text{H}^+]}{\gamma_{FeOHc}K_w} + \frac{K_{\text{Fe(OH)}_3c}K_w}{\gamma_{\text{Fe(OH)}_3c}\gamma_H[\text{H}^+]} \quad (12.29)$$

Differentiating with respect to [H$^+$], equating to zero, rearranging, and changing H$^+$ to H$^+_{opt}$, the concentration of the hydrogen ion at optimum conditions,

$$\left\{\frac{2K_{sp,\text{Fe(OH)}_2}\gamma_H^2}{\gamma_{FeII}K_w^2}\right\}[\text{H}^+_{opt}]^3 + \left\{\frac{K_{FeOHc}\gamma_H}{\gamma_{FeOHc}K_w}\right\}[\text{H}^+_{opt}]^2 = \frac{K_{\text{Fe(OH)}_3c}K_w}{\gamma_{\text{Fe(OH)}_3c}\gamma_H} \quad (12.30)$$

The value of [H$_{opt}$] may be solved by trial error.

12.7 CHEMICAL REACTIONS OF THE FERRIC ION

The ferric salts used as coagulant in water and wastewater treatment are FeCl$_3$ and Fe$_2$(SO$_4$)$_3$. They have essentially the same chemical reactions in that both form the Fe(OH)$_{3(s)}$ solid. When these coagulants are dissolved in water, they dissociate according to the following equations:

$$\text{FeCl}_3 \rightarrow \text{Fe}^{3+} + 3\text{Cl}^- \quad (12.31)$$

$$\text{Fe}_2(\text{SO}_4)_3 \rightarrow 2\text{Fe}^{3+} + 3\text{SO}_4^{2-} \quad (12.32)$$

As in any coagulation process, these ions must be rapidly dispersed throughout the tank in order to effect the complete coagulation process. The solid precipitate $Fe(OH)_{3(s)}$ and complexes are then formed. The reactions, together with the respective equilibrium constants at 25°C, are as follows (Snoeyink and Jenkins, 1980):

$$Fe(OH)_{3(s)} \rightleftharpoons Fe^{3+} + 3OH^- \qquad K_{sp, Fe(OH)_3} = 10^{-38} \qquad (12.33)$$

$$Fe(OH)_{3(s)} \rightleftharpoons FeOH^{2+} + 2OH^- \qquad K_{FeOHc} = 10^{-26.16} \qquad (12.34)$$

$$Fe(OH)_{3(s)} \rightleftharpoons Fe(OH)_2^+ + OH^- \qquad K_{Fe(OH)_2c} = 10^{-16.74} \qquad (12.35)$$

$$Fe(OH)_{3(s)} + OH^- \rightleftharpoons Fe(OH)_4^- \qquad K_{Fe(OH)_4c} = 10^{-5} \qquad (12.36)$$

$$2Fe(OH)_{3(s)} \rightleftharpoons Fe_2(OH)_2^{4+} + 4OH^- \qquad K_{Fe_2(OH)_2c} = 10^{-50.8} \qquad (12.37)$$

The complexes are $FeOH^{2+}$, $Fe(OH)_2^+$, $Fe(OH)_4^-$, and $Fe_2(OH)_2^{4+}$. Also note that the OH^- ion is a participant in these reactions. This means that the concentrations of each of these complex ions are determined by the pH of the solution.

In the application of the above equations in an actual coagulation treatment of water as in all applications of coagulants, conditions must be adjusted to allow maximum precipitation of the solid which in the present case is represented by $Fe(OH)_{3(s)}$. To allow for this maximum precipitation, the concentrations of the complex ions must be held to a minimum.

12.7.1 DETERMINATION OF THE OPTIMUM pH

For effective removal of the colloids, as much of the ferric ions should be converted to the solid $Fe(OH)_{3(s)}$. Also, as much of the concentrations of the complex ions should neutralize the primary charges of the colloids to effect their destabilization. Overall, this means that once the solids have been formed and the complex ions have neutralized the colloid charges, the concentrations of the complex ions standing in solution should be at the minimum, which corresponds to the optimum pH for the coagulation process.

Let sp_{FeIII} represent all the species that contain the Fe(III) ion standing in solution. Thus, the concentration of all the species containing the ion is

$$[sp_{FeIII}] = [Fe^{3+}] + [FeOH^{2+}] + [Fe(OH)_2^+] + [Fe(OH)_4^-] + 2[Fe_2(OH)_2^{4+}] \quad (12.38)$$

All the concentrations in the right-hand side of the above equation will now be expressed in terms of the hydrogen ion concentration. This will result in the expressing of $[sp_{FeIII}]$ in terms of the hydrogen ion. Differentiating the resulting equation of

$[sp_{\text{FeIII}}]$ with respect to $[\text{H}^+]$ and equating the result to zero will produce the minimum concentration of sp_{FeIII} and, thus, the optimum pH determined. Using the equilibrium reactions, Eqs. (12.33) through (12.37), along with the ion product of water, we now proceed as follows:

$$[\text{Fe}^{3+}] = \frac{\{\text{Fe}^{3+}\}}{\gamma_{\text{FeIII}}} = \frac{K_{sp,\text{Fe(OH)}_3}}{\gamma_{\text{FeIII}}\{\text{OH}^-\}^3} = \frac{K_{sp,\text{Fe(OH)}_3}\{\text{H}^+\}^3}{\gamma_{\text{FeIII}}K_w^3} = \frac{K_{sp,\text{Fe(OH)}_3}\gamma_{\text{H}}^3[\text{H}^+]^3}{\gamma_{\text{FeIII}}K_w^3}$$

(12.39)

$$[\text{FeOH}^{2+}] = \frac{\{\text{FeOH}^{2+}\}}{\gamma_{\text{FeOH}c}} = \frac{K_{\text{FeOH}c}}{\gamma_{\text{FeOH}c}\{\text{OH}^-\}^2} = \frac{K_{\text{FeOH}c}\{\text{H}^+\}^2}{\gamma_{\text{FeOH}c}K_w^2} = \frac{K_{\text{FeOH}c}\gamma_{\text{H}}^2[\text{H}^+]^2}{\gamma_{\text{FeOH}c}K_w^2}$$

(12.40)

$$[\text{Fe(OH)}_2^+] = \frac{\{\text{Fe(OH)}_2^+\}}{\gamma_{\text{Fe(OH)}_2c}} = \frac{K_{\text{Fe(OH)}_2c}}{\gamma_{\text{Fe(OH)}_2c}\{\text{OH}^-\}} = \frac{K_{\text{Fe(OH)}_2c}\{\text{H}^+\}}{\gamma_{\text{Fe(OH)}_2c}K_w} = \frac{K_{\text{Fe(OH)}_2c}\gamma_{\text{H}}[\text{H}^+]}{\gamma_{\text{Fe(OH)}_2c}K_w}$$

(12.41)

$$[\text{Fe(OH)}_4^-] = \frac{\{\text{Fe(OH)}_4^-\}}{\gamma_{\text{Fe(OH)}_4c}} = \frac{K_{\text{Fe(OH)}_4c}\{\text{OH}^-\}}{\gamma_{\text{Fe(OH)}_4c}} = \frac{K_{\text{Fe(OH)}_4c}K_w}{\gamma_{\text{Fe(OH)}_4c}\{\text{H}^+\}} = \frac{K_{\text{Fe(OH)}_4c}K_w}{\gamma_{\text{Fe(OH)}_4c}\gamma_{\text{H}}\{\text{H}^+\}}$$

(12.42)

$$[\text{Fe}_2(\text{OH})_2^{4+}] = \frac{\{\text{Fe}_2(\text{OH})_2^{4+}\}}{\gamma_{\text{Fe}_2(\text{OH})_2c}} = \frac{K_{\text{Fe}_2(\text{OH})_2c}}{\gamma_{\text{Fe}_2(\text{OH})_2c}\{\text{OH}^-\}^4} = \frac{K_{\text{Fe}_2(\text{OH})_2c}\gamma_{\text{H}}^4[\text{H}^+]^4}{\gamma_{\text{Fe}_2(\text{OH})_2c}K_w^4}$$

(12.43)

γ_{FeIII}, γ_{H}, $\gamma_{\text{FeOH}c}$, $\gamma_{\text{Fe(OH)}_2c}$, $\gamma_{\text{Fe(OH)}_4c}$, $\gamma_{\text{Fe}_2(\text{OH})_2c}$ are, respectively, the activity coefficients of the ferric and the hydrogen ions and the complexes FeOH^{2+}, Fe(OH)_2^+, Fe(OH)_4^-, and $\text{Fe}_2(\text{OH})_2^{4+}$. $K_{sp,\text{Fe(OH)}_3}$ is the solubility product constant of the solid $\text{Fe(OH)}_{3(s)}$ and K_w is the ion product of water. $K_{\text{FeOH}c}$, $K_{\text{Fe(OH)}_2c}$, $K_{\text{Fe(OH)}_4c}$, and $K_{\text{Fe}_2(\text{OH})_2c}$ are, respectively, the equilibrium constants of the complexes FeOH^{2+}, Fe(OH)_2^+, Fe(OH)_4^-, and $\text{Fe}_2(\text{OH})_2^{4+}$.

Equations (12.39) through (12.43) may now be substituted into Equation (12.38) to produce

$$[sp_{\text{FeIII}}] = \frac{K_{sp,\text{Fe(OH)}_3}\gamma_{\text{H}}^3[\text{H}^+]^3}{\gamma_{\text{FeIII}}K_w^3} + \frac{K_{\text{FeOH}c}\gamma_{\text{H}}^2[\text{H}^+]^2}{\gamma_{\text{FeOH}c}K_w^2} + \frac{K_{\text{Fe(OH)}_2c}\gamma_{\text{H}}[\text{H}^+]}{\gamma_{\text{Fe(OH)}_2c}K_w}$$

$$+ \frac{K_{\text{Fe(OH)}_4c}K_w}{\gamma_{\text{Fe(OH)}_4c}\{\text{H}^+\}} + \frac{2K_{\text{Fe}_2(\text{OH})_2c}\gamma_{\text{H}}^4[\text{H}^+]^4}{\gamma_{\text{Fe}_2(\text{OH})_2c}K_w^4}$$

(12.44)

Differentiating with respect to $[H^+]$, equating to zero, rearranging, and changing H^+ to H_{opt}^+, the concentration of the hydrogen ion at optimum conditions,

$$\left\{\frac{8K_{Fe_2(OH)_2}c\,\gamma_H^4}{\gamma_{Fe_2(OH)_2}cK_w^4}\right\}[H_{opt}^+]^5 + \left\{\frac{3K_{sp,Fe(OH)_3}\gamma_H^3}{\gamma_{FeIII}K_w^3}\right\}[H_{opt}^+]^4 + \left\{\frac{2K_{FeOHc}\gamma_H^2}{\gamma_{FeOHc}K_w^2}\right\}[H_{opt}^+]^3$$

$$+\left\{\frac{K_{Fe(OH)_2}c\,\gamma_H}{\gamma_{Fe(OH)_2}cK_w}\right\}[H_{opt}^+]^2 = \frac{K_{Fe(OH)_4}cK_w}{\gamma_{Fe(OH)_4}c\,\gamma_H} \qquad (12.45)$$

The value of $[H_{opt}]$ may be solved by trial error.

Example 12.2   A raw water containing 140 mg/L of dissolved solids is subjected to coagulation treatment using copperas. Calculate the optimum pH that the operation should be conducted. Assume the temperature of operation is 25°C.

Solution:

$$\left\{\frac{2K_{sp,Fe(OH)_2}\gamma_H^2}{\gamma_{FeII}K_w^2}\right\}[H_{opt}^+]^3 + \left\{\frac{K_{FeOHc}\gamma_H}{\gamma_{FeOHc}K_w}\right\}[H_{opt}^+]^2 = \frac{K_{Fe(OH)_3}cK_w}{\gamma_{Fe(OH)_3}c\,\gamma_H}$$

$$\frac{2K_{sp,Fe(OH)_2}\gamma_H^2}{\gamma_{FeII}K_w^2}:$$

$$K_{sp,Fe(OH)_2} = 10^{-14.5} \qquad \gamma_H = 0.94 \qquad \gamma_{FeII} = 10^{-\frac{0.5(2)^2[\sqrt{3.5(10^{-3})}]}{1+1.14[\sqrt{3.5(10^{-3})}]}} = 0.77$$

Therefore,

$$\frac{2K_{sp,Fe(OH)_2}\gamma_H^2}{\gamma_{FeII}K_w^2} = \frac{2(10^{-14.5})(0.94)^2}{(0.77)(10^{-14})^2} = 7.26(10^{13})$$

$$\frac{K_{FeOHc}\gamma_H}{\gamma_{FeOHc}K_w}:$$

$$K_{FeOHc} = 10^{-9.4} \qquad \gamma_{FeOHc} = \gamma_H = 0.94$$

Therefore,

$$\frac{K_{FeOHc}\gamma_H}{\gamma_{FeOHc}K_w} = \frac{(10^{-9.4})(0.94)}{(0.94)(10^{-14})} = 3.98(10^4)$$

$$\frac{K_{Fe(OH)_3}cK_w}{\gamma_{Fe(OH)_3}c\,\gamma_H}:$$

$$K_{Fe(OH)_3c} = 10^{-5.1} \qquad \gamma_{Fe(OH)_3c} = \gamma_H = 0.94$$

$$\frac{K_{Fe(OH)_3}cK_w}{\gamma_{Fe(OH)_3}c\,\gamma_H} = \frac{(10^{-5.1})(10^{-14})}{(0.94)(0.94)} = 8.99(10^{-20})$$

Therefore,

$$7.26(10^{13})[H_{opt}^+]^3 + 3.98(10^4)[H_{opt}^+] = 8.99(10^{-20})$$

Solving by trial and error, let

$$Y = 7.26(10^{13})[H_{opt}^+]^3 + 3.98(10^4)[H_{opt}^+]^2$$

$[H_{opt}^+]$	Y
10^{-11}	$4.05(10^{-18})$
10^{-12}	$3.99(10^{-20})$

10^{-11}	$4.05(10^{-18})$	$\dfrac{y - 10^{-12}}{10^{-12} - 10^{-11}} = \dfrac{8.99(10^{-20}) - 3.99(10^{-20})}{3.99(10^{-20}) - 4.05(10^{-18})}$
y	$8.99(10^{-20})$	$y = 1.11(10^{-12})$
10^{-12}	$8.99(10^{-20})$	Therefore, pH $= -\log_{10}[1.11(10^{-12})] = 11.95$ **Ans**

Example 12.3 A raw water containing 140 mg/L of dissolved solids is subjected to coagulation treatment using a ferric salt. Calculate the optimum pH that the operation should be conducted. Assume the temperature of operation is 25°C.

Solution:

$$\left\{ \frac{8K_{Fe_2(OH)_2c}\gamma_H^4}{\gamma_{Fe_2(OH)_2c}K_w^4} \right\}[H_{opt}^+]^5 + \left\{ \frac{3K_{sp,Fe(OH)_3}\gamma_H^3}{\gamma_{FeIII}K_w^3} \right\}[H_{opt}^+]^4 + \left\{ \frac{2K_{FeOHc}\gamma_H^2}{\gamma_{FeOHc}K_w^2} \right\}[H_{opt}^+]^3$$

$$+ \left\{ \frac{K_{Fe(OH)_2c}\gamma_H}{\gamma_{Fe(OH)_2c}K_w} \right\}[H_{opt}^+]^2 = \frac{K_{Fe(OH)_4c}K_w}{\gamma_{Fe(OH)_4c}\gamma_H}$$

$$\frac{8K_{Fe_2(OH)_2c}\gamma_H^4}{\gamma_{Fe_2(OH)_2c}K_w^4}:$$

$$K_{Fe_2(OH)_2c} = 10^{-50.8} \qquad \gamma_H = 0.94 \qquad \gamma_{Fe_2(OH)_2c} = 10^{\frac{0.5(4)^2[\sqrt{3.5(10^{-3})}]}{1+1.14[\sqrt{3.5(10^{-3})}]}} = 0.36$$

Therefore,

$$\frac{8K_{Fe_2(OH)_2c}\gamma_H^4}{\gamma_{Fe_2(OH)_2c}K_w^4} = \frac{8(10^{-50.8})(0.94)^4}{(0.36)(10^{-14})^4} = 2.75(10^6)$$

$$\frac{3K_{sp,Fe(OH)_3}\gamma_H^3}{\gamma_{FeIII}K_w^3}:$$

$$K_{sp,Fe(OH)_3} = 10^{-38} \qquad \gamma_{FeIII} = 10^{\frac{0.5(3)^2[\sqrt{3.5(10^{-3})}]}{1+1.14[\sqrt{3.5(10^{-3})}]}} = 0.56$$

Therefore,

$$\frac{3K_{sp,Fe(OH)_3}\gamma_H^3}{\gamma_{FeIII}K_w^2} = \frac{3(10^{-38})(0.94)^3}{0.56(10^{-14})^3} = 4.45(10^4)$$

$$\frac{2K_{FeOHc}\gamma_H^2}{\gamma_{FeOHc}K_w^2}:$$

$$K_{FeOHc} = 10^{-26.16} \qquad \gamma_{FeOHc} = 10^{-\frac{0.5(2)^2[\sqrt{3.5(10^{-3})}]}{1+1.14[\sqrt{3.5(10^{-3})}]}} = 0.77$$

Therefore,

$$\frac{2K_{FeOHc}\gamma_H^2}{\gamma_{FeOHc}K_w^2} = \frac{2(10^{-26.16})(0.94)^2}{(0.77)(10^{-14})^2} = 158.78$$

$$\frac{K_{Fe(OH)_2c}\gamma_H}{\gamma_{Fe(OH)_2c}K_w}:$$

$$K_{Fe(OH)_2c} = 10^{-16.74} \qquad \gamma_{Fe(OH)_2c} = \gamma_H = 0.94$$

Therefore,

$$\frac{K_{Fe(OH)_2c}\gamma_H}{\gamma_{Fe(OH)_2c}K_w} = \frac{10^{-16.74}(0.94)}{(0.94)(10^{-14})} = 10^{-2.74}$$

$$\frac{K_{Fe(OH)_4c}K_w}{\gamma_{Fe(OH)_4c}\gamma_H}:$$

$$K_{Fe(OH)_4c} = 10^{-5} \qquad \gamma_{Fe(OH)_4c} = \gamma_H = 0.94$$

Therefore,

$$\frac{K_{Fe(OH)_4c}K_w}{\gamma_{Fe(OH)_4c}\gamma_H} = \frac{10^{-5}(10^{-14})}{0.94(0.94)} = 1.13(10^{-19})$$

Therefore,

$$2.75(10^6)[H_{opt}^+]^5 + 4.45(10^4)[H_{opt}^+]^4 + 158.78[H_{opt}^+]^3 + 10^{-2.74}[H_{opt}^+]^2 = 1.13(10^{-19})$$
$$5(12) = 60 \qquad 4(12) = 48$$

Solving by trial and error, let

$$Y = 2.75(10^6)[H_{opt}^+]^5 + 4.45(10^4)[H_{opt}^+]^4 + 158.78[H_{opt}^+]^3 + 10^{-2.74}[H_{opt}^+]^2$$

$[H^+_{opt}]$	Y
10^{-8}	$1.82(10^{-19})$
10^{-10}	$1.82(10^{-23})$

10^{-8}	$1.82(10^{-19})$	$\dfrac{y - 10^{-8}}{10^{-8} - 10^{-10}} = \dfrac{1.13(10^{-19}) - 1.82(10^{-19})}{1.82(10^{-19}) - 1.82(10^{-23})}$
y	$1.13(10^{-19})$	$y = 0.63(10^{-8})$
10^{-10}	$1.82(10^{-23})$	Therefore, $pH = -\log_{10}[0.63(10^{-8})] = 8.2$ **Ans**

12.8 JAR TESTS FOR OPTIMUM pH DETERMINATION

We may summarize the optimum pH's of the coagulants obtained in the previous examples: alum = 5.32, ferrous = 11.95, and ferric = 8.2. The problem with these values is that they only apply at a temperature of 25°C. If the formulas for the determination of these pH's are reviewed, they will be found to be functions of equilibrium constants. By the use of the Van't Hoff equation, values at other temperatures for the equilibrium constants can be found. These, however, as mentioned before, also need the value of the standard enthalpy change, ΔH^o_{298}, as discussed in the chapter on water stabilization. For the aforementioned coagulants, no values of the enthalpy change are available. Thus, until studies are done to determine these values, optimum pH values must be determined using the jar test.

In addition, the optimum pH's of 5.32, 11.95, and 8.2 were obtained at a dissolved solids of 140 mg/L. The value of the dissolved solids predicts the values of the activity coefficients of the various ions in solution, which, in turn, determine the activities of the ions, including that of the hydrogen ion. It follows that, if the dissolved solids concentration is varied, other values of optimum pH's will also be obtained not only the respective values of 5.32, 11.95, and 8.2. This is worth repeating: the values of 5.32, 11.95, and 8.2 apply only at a dissolved solids concentration of 140 mg/L. In addition, they only apply provided the temperature is 25°C. In subsequent discussions, mention of these optimum pH values would mean values at the conditions of 25°C of temperature and a solids concentration of 140 mg/L.

12.9 CHEMICAL REQUIREMENTS

The chemical reactions written so far are not usable for determining chemical requirements. They all apply at equilibrium conditions. When chemicals are added to water, it first has to react, after which equilibrium will set in. It is this first reaction that determines the amount of chemical, not when equilibrium has set in. Chemical requirements for the three methods of coagulation treatments discussed will now be addressed.

The respective chemical reactions will first be derived. From the concept of equivalence, the number of equivalents of all the species participating in a given chemical reaction are equal. From the reaction, the equivalent masses will then be calculated. Once this is done, equations for chemical requirements can be derived.

12.9.1 CHEMICAL REQUIREMENTS IN ALUM COAGULATION TREATMENT

The water of hydration for alum varies from 13 to 18. For the purposes of calculating the chemical requirements, this range of values will be designated by x. In actual applications, the correct value of x must be obtained from the label of the container used to ship the chemical. Using x as the water of hydration, the chemical reaction for alum is

$$Al_2(SO_4)_3 \cdot xH_2O + 3Ca(HCO_3)_2 \rightarrow 2Al(OH)_3\downarrow + 6CO_2 + 3CaSO_4 + xH_2O$$

$$(12.46)$$

The rationale behind the previous reaction is explained here. The bicarbonate is known to act as a base as well as an acid. As a base, its interaction is $HCO_3^- + H_2O$ $\rightleftharpoons H_2CO_3 + OH^- \rightleftharpoons CO_2 + H_2O + OH^-$. The K_{sp} of aluminum hydroxide is approximately (10^{-33}). This means that the hydroxide is very insoluble. Thus, with the ions of the coagulant and the bicarbonate dispersed in the water, Al^{3+} "grabs" whatever OH^- there is to form the precipitate, $Al(OH)_3$, and the reaction portrayed above ensues.

A very important point must be discussed with respect to the previous coagulation reaction, in comparison with those found in the literature. The environmental engineering literature normally uses the equilibrium arrows, $\rightleftharpoons$, instead of the single forward arrow, $\rightarrow$, as written previously. Equilibrium arrows indicate that a particular reaction is in equilibrium, which would mean for the present case, that alum is produced in the backward reaction. Alum, however, is never produced by mixing aluminum hydroxide, carbon dioxide, calcium sulfate, and water, the species found on the right-hand side of the previous equation. Once $Al_2(SO_4)_3 \cdot 14H_2O$ is mixed with $Ca(HCO_3)_2$, the alum is gone forever producing the aluminum hydroxide precipitate—it cannot be recovered. After the formation of the precipitate, any backward reaction would be for the complex reactions and not for the formation of $Al_2(SO_4)_3 \cdot xH_2O$, as would be inferred if the above reaction were written with the equilibrium arrows.

In addition, coagulation is a process of expending the coagulant. In the process of expenditure, the alum must react to produce its products. This means that what must "exist" is the forward arrow and not any backward arrow. Portraying the backward arrow would mean that the alum is produced, but it is known that it is not produced but expended. During expenditure, no equilibrium must exist. To reiterate, the coagulation reaction should be represented by the forward arrow and not by the equilibrium arrows.

As shown in Equation (12.46), an alkaline substance is needed to react with the alum. The bicarbonate alkalinity is used, since it is the alkalinity that is always found in natural waters. In practice, its concentration must be determined to ascertain if enough is present to satisfy the optimum alum dose. If found deficient, then lime is normally added to satisfy the additional alkalinity requirement. As we have found, the reaction is optimum at a pH of 5.32 at 25°C when the dissolved solids concentration is 140 mg/L.

In any given application, it is uncertain whether the bicarbonate alone or in combination is needed, therefore, the most practical way of expressing the alkalinity requirement is through the use of equivalents. Using this method, the number of equivalents of the alum used is equal to the number of equivalents of the alkalinity required; in fact, it is equal to the same number of equivalents of any species participating in the chemical reaction. All that is needed, therefore, is to find the number of equivalent masses of the alum and the alkalinity needed is equal to this number of equivalent masses.

In the previous reaction, the number of references is 6. Thus, the equivalent mass of alum is $Al_2(SO_4)_3 \cdot xH_2O/6 = 57.05 + 3x$ and that of the calcium bicarbonate species is $3Ca(HCO_3)_2/6 = 81.05$. The other alkalinity sources that can be used are lime, caustic soda, and soda ash. Lime is used in the discussion that follows. Also, alkalinity requirements are usually expressed in terms of $CaCO_3$. Therefore, we also express the reactions of alum in terms of calcium carbonate. The respective chemical reactions are:

$$Al_2(SO_4)_3 \cdot xH_2O + 3Ca(OH)_2 \rightarrow 2Al(OH)_3\downarrow + 3CaSO_4 + xH_2O \qquad (12.47)$$

$$Al_2(SO_4)_3 \cdot xH_2O + 3CaCO_3 + 3HOH \rightarrow 2Al(OH)_3\downarrow + 3CaSO_4 + 3CO_2 + xH_2O$$
$$(12.48)$$

$Ca(OH)_2$ is actually slaked lime: $CaO + H_2O$. Note that in order to find the equivalent masses, the same number of molecules of alum in the balanced chemical reaction as used in Equation (12.46) should be used in Eqs. (12.47) and (12.48); otherwise, the equivalent masses obtained are equivalent to each other. From the reactions, the equivalent mass of lime (CaO) is $3CaO/6 = 28.05$ and that of calcium carbonate is $3CaCO_3/6 = 50$.

It is impossible to determine the optimum dose of alum using chemical reaction. This value must be obtained through the jar test. Let $[Alopt]_{mg}$ and $[Alopt]_{geq}$ be the milligrams per liter and gram equivalents per liter of optimum alum dose, respectively and let V be the cubic meters of water or wastewater treated. Also, let $M_{CaOkgeqAl}$ and M_{CaOAl} be the kilogram equivalents and kilogram mass of lime, respectively, used at a fractional purity of P_{CaO}. In the case of $Ca(HCO_3)_2$, the respective symbols are $M_{Ca(HCO_3)_2kgeqAl}$ and $M_{Ca(HCO_3)_2Al}$ at a fractional purity of $P_{Ca(HCO_3)_2} \cdot P_{Ca(HCO_3)_2} = 1$ in natural waters. Note the Al is one of the subscripts. This is to differentiate when ferrous and ferric are used as the coagulants.

The number of kilogram equivalents of alum needed is $([Alopt]_{geq}(1000)/1000)$ $V = [Alopt]_{geq}V = ([Alopt]_{mg}/1000(57.05 + 3x))V$. Let M_{Alkgeq} and M_{Al} be the kilogram equivalents and kilograms of alum used, respectively, at a fractional purity of P_{Al}. Thus,

$$M_{Alkgeq} = \frac{[Alopt]_{mg}}{1000(57.05 + 3x)P_{Al}}V \qquad (12.49)$$

$$M_{Al} = \frac{[Alopt]_{mg}}{1000P_{Al}}V \qquad (12.50)$$

The most general notion is that M_{Alkgeq} is reacted by all the alkalinities. This, of course, is not necessarily true in practice, but is conceptually correct. Let $f_{AlCa(HCO_3)_2}$ be the fraction of M_{Alkgeq} reacted by calcium bicarbonate and f_{AlCaO}, the fraction reacted by lime. If calcium bicarbonate and lime are the only alkalinities reacting with the alum, $f_{AlCa(HCO_3)_2} + f_{AlCaO} = 1$. We now have

$$M_{Ca(HCO_3)_2kgeqAl} = \frac{f_{AlCa(HCO_3)_2}[Alopt]_{mg}}{1000(57.05 + 3x)}V \tag{12.51}$$

$$M_{Ca(HCO_3)_2Al} = \frac{0.081 f_{AlCa(HCO_3)_2}[Alopt]_{mg}V}{(57.05 + 3x)} \tag{12.52}$$

$$M_{CaOkgeqAl} = \frac{f_{AlCaO}[Alopt]_{mg}V}{1000(57.05 + 3x)P_{CaO}} \tag{12.53}$$

$$M_{CaOAl} = \frac{0.0281 f_{AlCaO}[Alopt]_{mg}V}{(57.05 + 3x)P_{CaO}} \tag{12.54}$$

Example 12.4   A raw water containing 140 mg/L of dissolved solids is subjected to a coagulation treatment using alum. If the optimum alum dose as determined by a jar test is 40 mg/L, calculate the alkalinity of calcium bicarbonate required. If the natural alkalinity is 100 mg/L as $CaCO_3$, is there enough alkalinity to neutralize the alum dose. Note that, as will be explained later, the equivalent mass of alkalinity expressed as $CaCO_3$ is $CaCO_3/2$.

Solution:

$$M_{Ca(HCO_3)_2kgeqAl} = \frac{f_{AlCa(HCO_3)_2}[Alopt]_{mg}}{1000(57.05 + 3x)}V$$

$$f_{AlCa(HCO_3)_2} = 1 \qquad V = 1 \text{ m}^3 \qquad \text{let } x = 18$$

Therefore,

$$M_{Ca(HCO_3)_2kgeqAl} = \frac{(1)(40)}{1000[57.05 + 3(18)]}V = 3.6(10^{-4}) \text{ kgeq/m}^3 \quad \textbf{Ans}$$

$$100 \text{ mg/L as } CaCO_3 = \frac{100}{1000(50)} = 2.0(10^{-3}) \text{ kgeq/m}^3$$

Therefore, enough alkalinity is present.   **Ans**

12.9.2 KEY TO UNDERSTANDING SUBSCRIPTS

At this juncture, we will digress from our discussion and address the question of understanding the subscripts of the mass and fractional variables. Consider first, the mass variables $M_{Ca(HCO_3)_2kgeqAl}$, $M_{Ca(HCO_3)_2Al}$, and M_{CaOAl}.

As discussed in Chapter 10, if T is used as the first subscript, then it refers to the total of the type of species that follows it. In our example variables, this T is not present. For example, in $M_{Ca(HCO_3)_2kgeqAl}$, the T is not present; but the first subscript is calcium bicarbonate, which is the type of species involved, and there can be no other calcium bicarbonate. Thus, in this instance, the T is not needed. The second subscript in our example is kgeq. From previous conventions, this would be the reason for the existence of the calcium bicarbonate; however, kgeq cannot be a reason for the existence of calcium bicarbonate, since it is a unit of measurement. Therefore, if we see a unit of measurement used as a subscript, that simply indicates the unit of measurement of the type of mass. Thus, kgeq is the unit of the calcium bicarbonate. The last subscript, Al, is the one which is the reason for the existence of the calcium bicarbonate.

$M_{Ca(HCO_3)_2kgeqAl}$ then stands for the mass of calcium bicarbonate expressed in kilogram equivalents, with alum used as the coagulant. By the same token, $M_{Ca(HCO_3)_2Al}$ is the mass of calcium bicarbonate, however, with no unit specified for the type of mass. Al, again, indicates that alum is the coagulant used. The unit in this symbol cannot be ascertained, but must be established by convention, and the convention we have used must be in kilograms. Finally, then, by using this mass convention, M_{CaOAl} stands for the kilograms of lime used with alum as the coagulant. Other symbols of mass variables can be interpreted similarly.

Now, consider the fractional variables such as $f_{AlCa(HCO_3)_2}$ and f_{AlCaO}. Take $f_{AlCa(HCO_3)_2}$ first. By convention, the first subscript Al is the type of fraction of the species involved. In this instance, it would refer to the mass of alum, which could be M_{Al} or M_{Alkeq}, if the masses are expressed as absolute mass or equivalent mass, respectively. The second subscript, $Ca(HCO_3)_2$, is the reason for the existence of Al, which means that calcium bicarbonate is the one reacting with the alum in this fraction. Thus, in other words, $f_{AlCa(HCO_3)_2}$ is the fraction of the alum that is reacted by calcium bicarbonate. By the same token, f_{AlCaO} is the fraction of alum that is reacted by lime. Other symbols of fractional variables can be interpreted similarly.

12.9.3 CHEMICAL REQUIREMENTS IN FERROUS COAGULATION TREATMENT

The chemical reactions for the ferrous coagulant $FeSO_2 \cdot 7H_2O$ are:

$$FeSO_4 \cdot 7H_2O + Ca(HCO_3)_2 + 2Ca(OH)_2 \rightarrow Fe(OH)_2\downarrow + CaSO_4 + 2CaCO_3\downarrow + 9H_2O \tag{12.55a}$$

$$FeSO_4 \cdot 7H_2O + Ca(OH)_2 \rightarrow Fe(OH)_2\downarrow + CaSO_4 + 7H_2O \tag{12.55b}$$

These reactions are optimum at pH of 11.95 when the dissolved solids concentration is 140 mg/L and when the temperature is 25°C.

In the first reaction, because calcium bicarbonate is always present in natural waters, it is a participant. Only when all the natural alkalinities are expended will the second reaction proceed. The second reaction assumes that lime is the one used. In the presence of dissolved oxygen, $Fe(OH)_2$ oxidizes to $Fe(OH)_3$. Ferric hydroxide is the intended precipitate in the coagulation process using copperas, since it is a more insoluble precipitate than the ferrous form. In addition, the reaction with oxygen cannot be avoided, because oxygen will always be present in water. The oxidation reaction is

$$Fe(OH)_2 + \frac{1}{4}O_2 + \frac{1}{2}H_2O \rightarrow Fe(OH)_3\downarrow \qquad (12.56)$$

Note that to have equivalence among the species of Reactions (12.55a), (12.55b), and (12.56), $Fe(OH)_2$ is used as the "tying species" between the three reactions. To satisfy the equivalence, one mole of $Fe(OH)_2$ is used in each of the reactions. The number of reference species is then equal to 2. Thus, the equivalent masses are as follows: copperas $= FeSO_4 \cdot 7H_2O/2 = 138.95$; $Ca(HCO_3)_2 = Ca(HCO_3)_2/2 = 81.05$; lime $= 2CaO/2 = 56.1$ [Reaction (12.55a)]; lime $= CaO/2 = 28.05$ [Reaction (12.55b)]; and oxygen $= (1/4)O_2/2 = 4$.

In Equation (12.56), the equivalent masses could have been based on the number of electrons involved. The ferrous ion is oxidized to the ferric ion; thus, the number of reference species is 1 mole of electrons, for which the equivalent masses would have been divided by 1. If this were done, however, Equation (12.56) would stand unrelated to Eqs. (12.55a), (12.55b).

As in the case of alum, it is impossible to determine the optimum dose using chemical reaction. This value must be obtained through the jar test. Let $[FeIIopt]_{mg}$ and $[FeIIopt]_{geq}$ be the milligrams per liter and gram equivalents per liter of optimum copperas dose, respectively, and let V be the cubic meters of water or wastewater treated. Also, let $M_{CaOkgeqFeII}$ and $M_{CaOFeII}$ be the kilogram equivalents and kilogram mass of lime, respectively, used at a fractional purity of P_{CaO}. In the case of $Ca(HCO_3)_2$, the respective symbols are $M_{Ca(HCO_3)_2kgeqFeII}$ and $M_{Ca(HCO_3)_2FeII}$ at a fractional purity of $P_{Ca(HCO_3)_2}$. $P_{Ca(HCO_3)_2} = 1$ in natural waters; and, in the case of oxygen, the respective symbols are M_{O_2kgeq} and M_{O_2}. The fractional purity will be 1.

The number of kilogram equivalents of copperas needed is $([FeIIopt]_{geq}(1000)/1000)V = [FeIIopt]_{geq}V = ([FeIIopt]_{mg}/1000(138.95))V$. Let $M_{FeIIkeq}$ and M_{FeII} be the kilogram equivalents and kilograms of copperas used, respectively, at a fractional purity of P_{FeII}. Thus,

$$M_{FeIIkgeq} = \frac{[FeIIopt]_{mg}}{1000(138.95)P_{FeII}}V \qquad (12.57)$$

$$M_{FeII} = \frac{[FeIIopt]_{mg}}{1000P_{FeII}}V \qquad (12.58)$$

The number of kilogram equivalents of pure copperas is the same number of equivalents of pure calcium calcium bicarbonate, lime, and dissolved oxygen. Let $f_{FeIICaO}$ be the fraction of $M_{FeIIkgeq}$ reacted by lime alone without the presence of bicarbonate and $f_{FeIICa(HCO_3)_2}$ be the fraction reacted in the presence of calcium bicarbonate and lime. $f_{FeIICaO} + f_{FeIICa(HCO_3)_2} = 1$. Therefore, considering fractional purities,

$$M_{CaOakgeqFeII} = \frac{f_{FeIICa(HCO_3)_2}[FeIIopt]_{mg}}{1000(138.95)P_{CaO}}\mathcal{V} \quad [\text{Reaction (12.55a)}] \quad (12.59)$$

$$M_{CaOaFeII} = \frac{0.056 f_{FeIICa(HCO_3)_2}[FeIIopt]_{mg}}{(138.95)P_{CaO}}\mathcal{V} \quad [\text{Reaction (12.55a)}] \quad (12.60)$$

$$M_{CaObkgeqFeII} = \frac{f_{FeIICaO}[FeIIopt]_{mg}}{1000(138.95)P_{CaO}}\mathcal{V} \quad [\text{Reaction (12.55b)}] \quad (12.61)$$

$$M_{CaObFeII} = \frac{0.028 f_{FeIICaO}[FeIIopt]_{mg}}{(138.95)P_{CaO}}\mathcal{V} \quad [\text{Reaction (12.55b)}] \quad (12.62)$$

$$M_{Ca(HCO_3)_2 kgeqFeII} = \frac{f_{FeIICa(HCO_3)_2}[FeIIopt]_{mg}}{1000(138.95)}\mathcal{V} \quad (12.63)$$

$$M_{Ca(HCO_3)_2 FeII} = \frac{0.081 f_{FeIICa(HCO_3)_2}[FeIIopt]_{mg}}{(138.95)}\mathcal{V} \quad (12.64)$$

$$M_{O_2 kgeq} = \frac{[FeIIopt]_{mg}}{1000(138.95)}\mathcal{V} \quad (12.65)$$

$$M_{O_2} = \frac{0.004[FeIIopt]_{mg}}{(138.95)}\mathcal{V} \quad (12.66)$$

Example 12.5 A raw water containing 140 mg/L of dissolved solids is subjected to coagulation treatment using copperas. If the optimum dose as determined by a jar test is 50 mg/L, calculate the alkalinity of calcium bicarbonate, lime, and dissolved oxygen required. If the natural alkalinity is 100 mg/L as $CaCO_3$, is there enough alkalinity to neutralize the coagulant dose? If the dissolved oxygen is 6.0 mg/L, is this enough to satisfy the DO requirement?

Solution: First, convert 100 mg/L as $CaCO_3$ to kgeq/L. Thus,

$$100 \text{ mg/L as } CaCO_3 = \frac{100}{1000(50)} = 2.0(10^{-3}) \text{ kgeq/m}^3$$

Assume that the optimum dose is reacted in the presence of both the bicarbonate and lime. Therefore,

$$M_{Ca(HCO_3)_2 kgeqFeII} = \frac{f_{FeIICa(HCO_3)_2}[FeIIopt]_{mg}}{1000(138.95)}\mathcal{V}$$

$$f_{FeIICa(HCO_3)_2} = 1$$

Therefore,

$$M_{Ca(HCO_3)_2 kgeqFeII} = \frac{(1)(50)}{1000(138.95)}(1)$$

$$= 3.60(10^{-4}) \text{ kgeq/m}^3 \ll 2.0(10^{-3}) \text{ kgeq/m}^3.$$

Therefore, enough natural alkalinity is present and the assumption that the optimum dose is reacted in the presence of both the bicarbonate and lime is correct. **Ans**

$$M_{CaOakgeqFeII} = \frac{f_{FeIICa(HCO_3)_2}[FeIIopt]_{mg}}{1000(138.95)P_{CaO}} V \quad [\text{Reaction (12.55a)}]$$

Assume $P_{CaO} = 0.85$
   Therefore,

$$M_{CaOakgeqFeII} = \frac{(1)(50)}{1000(138.95)(0.85)}(1) = 4.23(10^{-4}) \text{ kgeq/L} \quad \textbf{Ans}$$

$$M_{O_2} = \frac{0.004[FeIIopt]_{mg}}{(138.95)} V = \frac{0.004(50)}{(138.95)}(1)$$

$$= 1.44(10^{-3}) \text{ kg/L} = 1.44 \text{ mg/L} \quad \textbf{Ans}$$

Because 6.0 mg/L is greater than 1.44 mg/L, enough dissolved oxygen is present in the water.

12.9.4 CHEMICAL REQUIREMENTS IN FERRIC COAGULATION TREATMENT

The chemical reactions for the ferric coagulants $FeCl_3$ and $Fe_2(SO_4)_2$ involve precipitations in the form of ferric hydroxide. Calcium bicarbonate is always present in natural waters, so it must first be satisfied before any external source of the hydroxide ion is provided. This hydroxide can be provided using lime, and as in the case of alum, this is the hydroxide that will be utilized in the coagulation reactions to be discussed. The respective reactions for the satisfaction of calcium bicarbonate are as follows:

$$2FeCl_3 + 3Ca(HCO_3)_2 \rightarrow 2Fe(OH)_3\downarrow + 6CO_2 + 3CaCl_2 \qquad (12.67)$$

$$Fe_2(SO_4)_3 + 3Ca(HCO_3)_2 \rightarrow 2Fe(OH)_3\downarrow + 6CO_2 + 3CaSO_4 \qquad (12.68)$$

The precipitation of $Fe(OH)_3$ is optimal at a pH of 8.2, therefore, the coagulation pH should be adjusted to this value. This is assuming the coagulation temperature is 25°C and at a solids concentration of 140 mg/L.

   For the two reactions, the number of reference species is 6. Thus, the equivalent masses are ferric chloride = $2FeCl_3/6$ = 54.1, ferric sulfate = $Fe_2(SO_4)_3/6$ = 66.65 and calcium bicarbonate = $3Ca(HCO_3)_2/6$ = 81.05.

Upon satisfaction of the natural alkalinity, other alkalinity sources may be used such as lime, caustic soda, and soda ash. Also, as in the case of alum coagulation, alkalinity requirements are usually expressed in terms of $CaCO_3$. Therefore, we also express the reactions of the ferric salts in terms of calcium carbonate. The respective chemical reactions are:

$$2FeCl_3 + 3Ca(OH)_2 \rightarrow 2Fe(OH)_3\downarrow + 3CaCl_2 \qquad (12.69)$$

$$Fe_2(SO_4)_3 + 3Ca(OH)_2 \rightarrow 2Fe(OH)_3\downarrow + 3CaSO_4 \qquad (12.70)$$

Note: In order to find the equivalent masses, the same number of atoms of iron in the balanced chemical reactions as used in Eqs. (12.67) and (12.68) should be used in Eqs. (12.69) through (12.70).

Otherwise, the equivalent masses obtained will not be equivalent to each other. From the reactions, the equivalent mass of lime = 3CaO/6 = 28.05.

As determined by a jar test, let $[FeIIIopt]_{mg}$ and $[FeIIIopt]_{geq}$ be the milligrams per liter and gram equivalents per liter of optimum ferric salt dose, respectively, and let V be the cubic meters of water or wastewater treated. Also, let $M_{CaOkgeqFeIII}$ and $M_{CaOFeIII}$ be the kilogram equivalents and kilogram mass of lime, respectively, used at a fractional purity of P_{CaO}. In the case of $Ca(HCO_3)_2$, the respective symbols are $M_{Ca(HCO_3)_2kgeqFeIII}$ and $M_{Ca(HCO_3)_2FeIII}$ at a fractional purity of $P_{Ca(HCO_3)_2}$. $P_{Ca(HCO_3)_2}$ is, of course, equal to 1 in natural waters.

The number of kilogram equivalents, $M_{FeIIIkgeq}$, of any one of the ferric salts needed is $([FeIIIopt]_{geq}(1000)/1000)V = [FeIIIopt]_{geq}V$. For $FeCl_3$, $[FeIIIopt]_{geq}V = [FeIIIopt]_{mg}V/1000(54.1)$ and that for $Fe_2(SO_4)_3$, it is $[FeIIIopt]_{geq}V = [FeIIIopt]_{mg}V/1000(66.65)$. Let $M_{FeIIIClkeq}$ and $M_{FeIIICl}$ be the kilogram equivalents and kilograms of $FeCl_3$ used, respectively, at a fractional purity of $P_{FeIIICl}$. Also, let $M_{FeIIISO_4keq}$ and $M_{FeIIISO_4}$ be the kilogram equivalents and kilograms of $Fe_2(SO_4)_3$ used, respectively, at a fractional purity of $P_{FeIIISO_4}$. Thus,

$$M_{FeIIIkeq} = [FeIIIopt]_{geq}V \quad \text{(in pure state)} \qquad (12.71)$$

$$M_{FeIIICl} = \frac{[FeIIIopt]_{mg}V}{1000P_{FeIIICl}} \qquad (12.72)$$

$$M_{FeIIISO_4} = \frac{[FeIIIopt]_{mg}V}{1000P_{FeIIISO_4}} \qquad (12.73)$$

As in the case of alum, the most general notion is that $M_{FeIIIkgeq}$ is reacted by all the alkalinities. Let $f_{FeIIICa(HCO_3)_2}$ be the fraction of $M_{FeIIIkgeq}$ reacted by calcium bicarbonate and $f_{FeIIICaO}$, the fraction reacted by lime. If calcium bicarbonate and lime are the only alkalinities reacting with the ferric salt, $f_{FeIIICa(HCO_3)_2} + f_{FeIIICaO} = 1$. We now have

$$M_{Ca(HCO_3)_2kgeqFeIIIchl} = \frac{f_{FeIIICa(HCO_3)_2}[FeIIIopt]_{mg}V}{1000(54.1)} \text{(using ferric chloride)} \qquad (12.74)$$

$$M_{Ca(HCO_3)_2FeIIIchl} = 0.0015 f_{FeIIICa(HCO_3)_2}[FeIIIopt]_{mg} V \quad \text{(using ferric chloride)}$$

$$M_{Ca(HCO_3)_2kgeqFeIIIsul} = \frac{f_{FeIIICa(HCO_3)_2}[FeIIIopt]_{mg} V}{1000(66.65)} \quad \text{(using ferric sulfate)} \quad (12.75)$$

$$M_{FeIIICa(HCO_3)_2sul} = 0.0012 f_{FeIIICa(HCO_3)_2}[FeIIIopt]_{mg} V \quad \text{(using ferric sulfate)}$$

$$M_{CaOkgeqFeIIIchl} = \frac{f_{FeIIICaO}[FeIIIopt]_{mg} V}{1000(54.1)P_{CaO}} \quad \text{(using ferric chloride)} \quad (12.76)$$

$$M_{CaOFeIIIchl} = \frac{0.00052 f_{FeIIICaO}[FeIIIopt]_{mg} V}{P_{CaO}} \quad \text{(using ferric chloride)}$$

$$M_{CaOkgeqFeIIIsul} = \frac{f_{FeIIICaO}[FeIIIopt]_{mg} V}{1000(66.65)P_{CaO}} \quad \text{(using ferric sulfate)} \quad (12.77)$$

$$M_{CaOFeIIIsul} = \frac{0.00042 f_{FeIIICaO}[FeIIIopt]_{mg} V}{P_{CaO}} \quad \text{(using ferric sulfate)}$$

The additional subscripts chl and sul refer to ferric chloride and ferric sulfate, respectively.

Example 12.6 A raw water containing 140 mg/L of dissolved solids is subjected to a coagulation treatment using $Fe_2(SO_4)_3$. If the optimum coagulant dose as determined by a jar test is 40 mg/L, calculate the alkalinity of calcium bicarbonate required. If the natural alkalinity is 100 mg/L as $CaCO_3$, is there enough alkalinity to neutralize the coagulant dose?

Solution:

$$M_{Ca(HCO_3)_2kgeqFeIIIsul} = \frac{f_{FeIIICa(HCO_3)_2}[FeIIIopt]_{mg} V}{1000(66.65)}$$

$$f_{FeIIICa(HCO_3)_2} = 1 \quad V = 1 \text{ m}^3$$

Therefore,

$$M_{Ca(HCO_3)_2kgeqFeIIIsul} = \frac{(1)(40)(1)}{1000(66.65)} = 6.0(10^{-4}) \text{ kgeq/m}^3 \quad \textbf{Ans}$$

$$100 \text{ mg/L as CaCO}_3 = \frac{100}{1000(50)} = 2.0(10^{-3}) \text{ kgeq/m}^3 \gg 6.0(10^{-4}) \text{ kgeq/m}^3.$$

Therefore, enough alkalinity is present. **Ans**

12.10 CHEMICAL REQUIREMENTS FOR pH ADJUSTMENTS

As we have seen, the pH needs to be adjusted to the optimum for optimal coagulation. In general, there are two directions in which this can happen. If the pH is high, then it needs to be adjusted in the downward direction and if it is low, it needs to be adjusted in the upward direction. To adjust in the downward direction, an acid is needed. The cheapest way to do this is to use carbon dioxide. To adjust in the upward direction, a base is needed and the cheapest way to do this is to use lime. In Chapter 11, the following formula was derived:

$$[A_{cadd}]_{geq} = [A_{cur}]_{geq} + \frac{10^{-pH_{to}} - 10^{-pH_{cur}}}{\phi_a} \tag{12.78}$$

where

$[A_{cadd}]_{geq}$ = gram equivalents per liter of acid needed
$[A_{cur}]_{geq}$ = current alkalinity
pH_{to} = destination pH
pH_{cur} = current pH
ϕ_a = fractional dissociation of the hydrogen ion from acid used

Carbon dioxide will be used as the source of acidity in the derivation below. To determine the equivalent mass of carbon dioxide, the species it is reacting with must be known. Whatever alkalinity species are present, it is expected that carbon dioxide will react with them until the end point is reached. Since the hydroxyl ion is always present no matter what stage of pH any solution is in, the OH⁻ is the very ion that will take carbon dioxide to this end point; thus, it determines the equivalent mass of carbon dioxide.

Therefore, proceed as follows:

$$CO_2 + 2OH^- \rightarrow H_2O + CO_3^{2-} \tag{12.79}$$

This reaction represents one of the alkalinity reactions that carbon dioxide must neutralize before it can provide the needed hydrogen ion concentration that lowers the pH. These hydrogen ions, it provides from the H^+ of the H_2CO_3 it will eventually become as follows:

$$CO_2 + H_2O \rightleftharpoons H_2CO_3 \rightleftharpoons H^+ + HCO_3^- \tag{12.80}$$

From Equation (12.79), the equivalent mass of carbon dioxide is $CO_2/2 = 22$. We must make this also as its equivalent mass in Equation (12.80), if the two equations are to be compatible. Let M_{CO_2pH} be the kilograms of carbon dioxide added to lower the pH from pH_{cur} to pH_{to} and V be the corresponding cubic meters of water treated. Then,

$$M_{CO_2pH} = 22\left\{[A_{cadd}]_{geq} = [A_{cur}]_{geq} + \frac{10^{-pH_{to}} - 10^{-pH_{cur}}}{\phi_a}\right\}V \tag{12.81}$$

Now, consider the situation where pH_{cur} is less than pH_{to}. In this case, a base is needed. As in the case of alkalinity, a natural water will always have acidity. Until it is all consumed, this acidity will resist the change in pH when alkalinity is added to the water. Let the current total acidity be $[A_{cur}]_{geq}$ in gram equivalents per liter. Also, let the total alkalinity be $[A_{add}]_{geq}$ in gram equivalents per liter.

Assuming no acidity present, the total base to be added is $10^{-pH_{cur}} - 10^{-pH_{to}}$ gram moles per liter of the equivalent hydrogen ions. But since there is always acidity present, the total alkalinity to be added must include the amount for neutralizing the natural acidity, $[A_{cnat}]_{geq} = [A_{ccur}]_{geq}$. Thus, the total alkalinity to be added, $[A_{add}]_{geq}$, is

$$[A_{add}]_{geq} = [A_{ccur}]_{geq} + \frac{10^{-pH_{cur}} - 10^{-pH_{to}}}{\phi_b} \qquad (12.82)$$

where ϕ_b is the fractional dissociation of the hydroxide ion from the base supplied. For strong bases, ϕ_b is unity; for weak bases, it may be calculated from equilibrium constants.

To obtain the equivalent mass of the lime, consider that it must neutralize the acidity first before raising the pH. Thus,

$$Ca(OH)_2 + 2H^+ \rightarrow Ca^{2+} + 2H_2O \qquad (12.83)$$

Thus, the equivalent mass of lime is $CaO/2 = 28.05$.

Let M_{CaOpH} be the kilograms of lime to be added to raise the pH from pH_{cur} to pH_{to} and V be the corresponding cubic meters of volume treated. Then,

$$M_{CaOph} = 28.05[A_{add}]_{geq}V = 28.05\left([A_{ccur}]_{geq} + \frac{10^{-pH_{cur}} - 10^{-pH_{to}}}{\phi_b}\right)V \qquad (12.84)$$

In Chapter 13, we will calculate some values of ϕ_a and ϕ_b.

12.11 ALKALINITY AND ACIDITY EXPRESSED AS CaCO₃

To find the equivalent concentration of alkalinity in terms of $CaCO_3$ proceed as follows:

$$CaCO_3 + 2H^+ \rightarrow H_2CO_3 + Ca^{2+} \qquad (12.85)$$

Note: We have reacted the alkalinity species ($CaCO_3$) with the hydrogen ion, since it is the property of alkalinity to react with an acid.

From this reaction, the equivalent mass of $CaCO_3$ is 50. Therefore, a mass equivalent of alkalinity is equal to 50 mass units expressed as $CaCO_3$. For example, 1 gram equivalent of alkalinity is equal to 50 grams of alkalinity expressed as $CaCO_3$.

The determination of the equivalent concentration of acidity in terms of $CaCO_3$ is a bit tricky. Calcium carbonate does not react with an OH^- which ought to determine the amount of acidity that $CaCO_3$ contains, if it does. But since no reaction can ever occur, $CaCO_3$ does not have any acidity; it has alkalinity, instead, as shown by the reaction in Equation (12.85). If $CaCO_3$ does not have any acidity, why then express acidity in terms of calcium carbonate? This is one of the biggest blunders in the environmental engineering literature, and it ought to be a big mistake; however, right or not, things can always be made arbitrary and then rationalized—this is what is done and used in practice. Arbitrarily, Equation (12.85) has been made the basis for the determination of the equivalent mass of $CaCO_3$ when expressing concentrations of acidity in terms of calcium carbonate. The rationalization is that the two H^+ in Equation (12.85) is equivalent to two OH^-. Thus, the reaction is equivalent as if $CaCO_3$ is reacting directly with OH^-. Reasonable? No, but you have to take it. This rationalization is the same as the following rationalization: Pedro and Maria are friends. Maria and Jose are also friends. Therefore, Pedro and Jose are friends. Is this correct? No, because Pedro and Jose are, in reality, irreconcilable enemies. Nonetheless, equivalent mass of $CaCO_3 = 50$, no questions asked.

12.12 SLUDGE PRODUCTION

Sludge is composed of solids and water in such a mixture that the physical appearance looks more of being composed of wet solids than being a concentrated water. Analysis of the sludge, however, will show that it is practically water containing about 99% of it.

Because of the high percentage of water that it contains, the volume of sludge produced in chemical coagulation is the biggest drawback for its use. Nevertheless, depending upon circumstances, it has to be used. Therefore, we now derive the formulas for estimating the volume of sludge produced in coagulation treatment of water.

Two sources of solids are available for the production of sludge: the suspended solids or turbidity in the raw water and the solids produced from the coagulant. The suspended solids in surface raw waters can vary from 5 to 1000 mg/L (Sincero, 1968). The aim of coagulation treatment is to produce a clear water. Thus, for practical purposes, suspended solids can be assumed to be all removed. If V is the cubic meters of water coagulated and sp_{ss} are the solids, the kilograms of solids produced from the suspended solids is $([sp_{ss}]_{mg}/1000)V$, where $[sp_{ss}]_{mg}$ is in mg/L.

For convenience, the chemical reactions for the coagulants are reproduced as follows:

$$Al_2(SO_4)_3 \cdot xH_2O + 3Ca(HCO_3)_2 \rightarrow 2Al(OH)_3\downarrow + 6CO_2 + 3CaSO_4 + xH_2O$$

$$(12.86)$$

$$Al_2(SO_4)_3 \cdot xH_2O + 3Ca(OH)_2 \rightarrow 2Al(OH)_3\downarrow + 3CaSO_4 + xH_2O \quad (12.87)$$

$$Al_2(SO_4)_3 \cdot xH_2O + 3CaCO_3 + 3HOH \rightarrow 2Al(OH)_3\downarrow + 3CaSO_4 + 3CO_2 + xH_2O$$

$$(12.88)$$

$$FeSO_4 \cdot 7H_2O + Ca(HCO_3)_2 + 2Ca(OH)_2$$
$$\rightarrow Fe(OH)_2\downarrow + CaSO_4 + 2CaSO_3\downarrow + 9H_2O \qquad (12.89)$$

$$FeSO_4 \cdot 7H_2O + Ca(OH)_2 \rightarrow Fe(OH)_2\downarrow + CaSO_4 + 7H_2O \qquad (12.90)$$

$$Fe(OH)_2 + \frac{1}{4}O_2 + \frac{1}{2}H_2O \rightarrow Fe(OH)_3\downarrow \qquad (12.91)$$

$$2FeCl_3 + 3Ca(HCO_3)_2 \rightarrow 2Fe(OH)_3\downarrow + 6CO_2 + 3CaCl_2 \qquad (12.92)$$

$$Fe_2(SO_4)_3 + 3Ca(HCO_3)_2 \rightarrow 2Fe(OH)_3\downarrow + 6CO_2 + 3CaSO_4 \qquad (12.93)$$

$$2FeCl_3 + 3Ca(OH)_2 \rightarrow 2Fe(OH)_3\downarrow + 3CaCl_2 \qquad (12.94)$$

$$Fe_2(SO_4)_3 + 3Ca(OH)_2 \rightarrow 2Fe(OH)_3\downarrow + 3CaSO_4 \qquad (12.95)$$

$$2Fe^{3+} + 3CaCO_3 + 3H_2O \rightarrow 2Fe(OH)_3\downarrow + 3Ca^{2+} + 3CO_2 \qquad (12.96)$$

The solids produced from the coagulation reactions are $Al(OH)_3$, $Fe(OH)_3$ and $CaCO_3$. Referring to the reactions of alum, the equivalent mass of aluminum hydroxide is $2Al(OH)_3/6 = 26.0$. The ferric hydroxide is produced through the use of copperas and the ferric salts. Its equivalent mass from the use of copperas [Eqs. (12.89) through (12.91)] is $Fe(OH)_3/2 = 53.4$ and from the use of the ferric salts, its equivalent mass is $2Fe(OH)_3/6 = 35.6$. Calcium carbonate is produced from the reaction of copperas, Equation (12.89). From this reaction, the equivalent mass of calcium carbonate is $2CaCO_3/2 = 100$.

Let M_{AlOH_3} be the kilogram mass of aluminum hydroxide (plus the solids coming from the suspended solids) produced from treating Ψ m^3 of water and from an optimum alum dose of $[Alopt]_{geq}$ gram equivalents per liter or, equivalently, $[Alopt]_{mg}$ milligrams per liter. Also, let M_{FeOH_3FeII} be the kilogram mass of ferric hydroxide (including the calcium carbonate plus the solids coming from the suspended solids) resulting from the use of an optimum dose of copperas of $[FeIIopt]_{geq}$ gram equivalents per liter or $[FeIIopt]_{mg}$ milligrams per liter, and let M_{FeOH_3FeIII} be the kilogram mass of ferric hydroxide (plus the solids coming from the suspended solids) resulting from the use of an optimum dose of $[FeIIIopt]_{geq}$ gram equivalents per liter or $[FeIIIopt]_{mg}$ milligrams per liter of any of the ferric salt coagulants.

$[Alopt]_{geq}\Psi = ([ALopt]_{mg}/1000(57.05 + 3x))\Psi$ and $[FeIIopt]_{geq}\Psi = ([FeIIopt]_{mg}/1000(138.95))\Psi$. $[FeIIIopt]_{geq}\Psi = [FeIIIopt]_{mg}\Psi/1000(54.1)$ for $FeCl_3$ and $[FeIIIopt]_{geq}\Psi = [FeIIIopt]_{mg}\Psi/1000(66.65)$ for $Fe_2(SO_4)_3$. Thus,

$$M_{AlOH_3} = 26[Alopt]_{geq}\Psi + \frac{[sp_{ss}]_{mg}}{1000}\Psi$$

$$= \frac{26[Alopt]_{mg}}{1000(57.05 + 3x)}\Psi + \frac{[sp_{ss}]_{mg}}{1000}\Psi$$

$$= \frac{0.026[Alopt]_{mg}}{(57.05 + 3x)}\Psi + \frac{[sp_{ss}]_{mg}}{1000}\Psi \qquad (12.97)$$

$$M_{FeOH_3FeII} = (53.4 + 100)[FeIIopt]_{geq}\mathcal{V} + \frac{[sp_{ss}]_{mg}}{1000}\mathcal{V}$$

$$= (153.4)\frac{[FeIIopt]_{mg}}{1000(138.95)}\mathcal{V} + \frac{[sp_{ss}]_{mg}}{1000}\mathcal{V}$$

$$= 0.0011[FeIIopt]_{mg}\mathcal{V} + \frac{[sp_{ss}]_{mg}}{1000}\mathcal{V} \quad\quad (12.98)$$

$$M_{FeOH_3FeIII} = 35.6[FeIIopt]_{geq}\mathcal{V} + \frac{[sp_{ss}]_{mg}}{1000}\mathcal{V} \quad\quad (12.99)$$

$$M_{FeOH_3FeIIIchl} = 35.6\frac{[FeIIopt]_{mg}\mathcal{V}}{1000(54.1)} + \frac{[sp_{ss}]_{mg}}{1000}\mathcal{V} \quad (\text{using } FeCl_3)$$

$$= 0.00066[FeIIopt]_{mg}\mathcal{V} + \frac{[sp_{ss}]_{mg}}{1000}\mathcal{V} \quad (\text{using } FeCl_3)$$

$$M_{FeOH_3FeIIIsul} = 35.6\frac{[FeIIopt]_{mg}\mathcal{V}}{1000(66.65)} + \frac{[sp_{ss}]_{mg}}{1000}\mathcal{V} \quad (\text{using } Fe_2(SO_4)_3)$$

$$= 0.00053[FeIIopt]_{mg}\mathcal{V} + \frac{[sp_{ss}]_{mg}}{1000}\mathcal{V} \quad (\text{using } Fe_2(SO_4)_3)$$

$([sp_{ss}]_{mg}/1000)\mathcal{V}$ is the kilograms of solids produced from the suspended solids of the raw water. The corresponding volumes of the sludges may be obtained by using the percent solids and the mass density of sludge.

Example 12.7 A raw water containing 100 mg/L of suspended solids is subjected to a coagulation treatment using $Fe_2(SO_4)_3$. The current acidity and pH are, respectively, 30 mg/L as $CaCO_3$ and 5.9. Calculate the amount of sludge produced if the optimum dose is 40 mg/L of $Fe_2(SO_4)_3$.

Solution:

$$M_{FeOH_3FeIIIsul} = 0.00053[FeIIIopt]_{mg}\mathcal{V} + \frac{[sp_{ss}]_{mg}}{1000}\mathcal{V}$$

$$= 0.00053(40)(1) + \frac{100}{1000}(1) = 0.12 \text{ kg/m}^3 \quad \textbf{Ans}$$

GLOSSARY

Activated silica—Sodium silicate that has been treated with sulfuric acid, aluminum sulfate, carbon dioxide, or chlorine.

Acidity—The capacity of a solution to absorb the effect of the addition of a base.

Alkalinity—The capacity of a solution to absorb the effect of the addition of an acid.

Anionic polyelectrolytes—Polyelectrolytes possessing a net negative charge in the molecules.

Cationic polyelectrolytes—Polyelectrolytes possessing a net positive charge in the molecules.

Charge neutralization—A mode of destabilizing a colloid by directly neutralizing the primary charges using counterions or colloids of opposite charges.

Central atom—In a complex ion, the atom to which several atoms are bonded.

Coagulation—The unit process of applying coagulant chemicals for the purpose of destabilizing the mutual repulsion of colloid particles, thus causing the particles to bind together.

Coagulation process—The process consisting of double-layer compression, charge neutralization, entrapment in a precipitate, and intraparticle bridging in the destabilization of a colloid.

Colloids—Agglomerates of atoms or molecules whose sizes are so small that gravity has no effect on settling them but they, instead, stay in suspension.

Copperas—Copperas is $FeSO_4 \cdot 7H_2O$.

Coordination sphere—The ligands and the central atom in a complex ion.

Counterions—Ions opposite in charge to a given ion.

Double-layer compression—A mode of destabilizing a colloid produced by "thinning out" the electric double layer.

Electric double layer—The layers surrounding a colloid body composed of the Stern and Gouy layers.

Electrolytes—Materials that, when placed in solution, cause the solution to be conductive to electricity.

Foam—A colloidal system of gases dispersed in a liquid.

Gaseous emulsion—A colloidal system of liquids dispersed in a gas.

Gaseous sol—A colloidal system of solids dispersed in a gas.

Gouy layer—A diffuse layer of coions to the primary charges surrounding the Stern layer.

Hydrophilic sols—Lyophilic sols that have water for the solvent.

Hydrophobic sols—Lyophobic sols that have water for the solvent.

Intraparticle bridging—A mode of destabilizing colloids using active sites in polymeric molecules for the colloids to become attached, thereby putting them in close association with each other for actual contact to promote agglomeration.

Isoelectric point—The condition in functional groups in a colloid attain equal positive and negative charges.

Jar test—A procedure where a number of doses are administered into a series of beakers for the purpose of determining the optimum dose.

Ligand exchange reaction—A complex reaction where the ligands are replaced by outside atoms.

Ligands—Atoms bonded to the central atom of a complex ion.

Liquid emulsion—A colloidal system of liquids dispersed in a liquid.

Liquid sol—A colloidal system of solids dispersed in a liquid.

Lyophilic sols—Liquid sols that bind the solvent.

Lyophobic sols—Liquid sols that do not bind the solvent.

Nonionic polyelectrolytes—Polyelectrolytes where the negative and positive charges that result in the ionization of the functional groups are balanced.

Optimum pH—The pH that produces optimal precipitation of the coagulant.

pH—The negative of the logarithm to the base 10 of the hydrogen ion activity.

Polyelectrolytes—Polymers possessing more than one electrolytic sites in the molecule.

Polymers—Molecules joined together by a chemical reaction to form larger molecules.

Precipitate entrapment—A mode of destabilizing colloid particles by using the colloids as nuclei for the initiation of the chemical precipitation of the coagulant.

Primary charges—Charges that colloids concentrate at their immediate surfaces.

Rapid or flash mix—Process of rapidly mixing the coagulant in the volume of the mixing tank.

Shear plane—The boundary between ions that move with the colloid body and ions that do not move with the body.

Solid emulsion—A colloidal system of liquids dispersed in a solid.

Solid sol—A colloidal system of solids dispersed in a solid.

Stern layer—Layer of counterions around the primary charges of a colloid.

van der Waal's force—A force of attraction that exists at the surface of a particle as a result of the unbalanced atomic forces at the surface of the particle.

Zeta potential—The electric potential at the shear plane.

Zwitter ion—Ion that exists at the isoelectric point.

SYMBOLS

$[Alopt]_{geq}$	Gram equivalents per liter of optimum alum dose
$Al(OH)^{2+}$	First hydroxide complex of the aluminum ion
$Al_2(OH)_2^{4+}$	Second hydroxide complex of the aluminum ion
$Al(OH)_4^-$	Fourth hydroxide complex of the aluminum ion
$Al_7(OH)_{17}^{4+}$	Seventeenth hydroxide complex of the aluminum ion
$Al_{13}(OH)_{34}^{5+}$	Thirty-fourth hydroxide complex of the aluminum ion
$[AL(OH)^{2+}]$	Concentration of the first hydroxide complex of the aluminum ion, gmols/L
$[Al_2(OH)_2^{4+}]$	Concentration of the second hydroxide complex of the aluminum ion, gmols/L
$[Al(OH)_4^-]$	Concentration of the fourth hydroxide complex of the aluminum ion, gmols/L
$[Al_7(OH)_{17}^{4+}]$	Concentration of the seventeenth hydroxide complex of the aluminum ion, gmols/L
$[Al_{13}(OH)_{34}^{5+}]$	Concentration of the thirty-fourth hydroxide complex of the aluminum ion, gmols/L
$f_{AlCa(HCO_3)_2}$	Fraction of M_{Alkgeq} reacted by calcium bicarbonate
f_{AlCaO}	Fraction of M_{Alkgeq} reacted by lime

$f_{\text{FeIICa(HCO}_3)_2}$	Fraction of M_{FeIIkgeq} reacted in the presence of calcium bicarbonate
f_{FeIICaO}	Fraction of M_{FeIIkgeq} reacted by lime alone
$f_{\text{FeIIICa(HCO}_3)_2}$	Fraction of $M_{\text{FeIIIkgeq}}$ reacted by calcium bicarbonate
f_{FeIIICaO}	Fraction of $M_{\text{FeIIIkgeq}}$ reacted by lime
$[\text{FeII}opt]_{\text{geq}}$	Gram equivalents per liter of optimum copperas dose
$[\text{FeIII}opt]_{\text{geq}}$	Gram equivalents per liter optimum dose of any of the two ferric salts
FeOH^+	First hydroxide complex of the ferrous ion
Fe(OH)_3^-	Third hydroxide complex of the ferrous ion
$[\text{FeOH}^+]$	Concentration of the first hydroxide complex of the ferrous ion, gmols/L
$[\text{Fe(OH)}_3^-]$	Concentration of the first third hydroxide complex of the ferrous ion, gmols/L
FeOH^{2+}	First hydroxide complex of the ferric ion
Fe(OH)_2^+	Second hydroxide complex of the ferric ion
$\text{Fe}_2(\text{OH})_2^{4+}$	Second hydroxide complex of the ferric double ion
Fe(OH)_4^-	Fourth hydroxide complex of the ferric ion
$[\text{FeOH}^{2+}]$	Concentration of the first hydroxide complex of the ferric ion, gmols/L
$[\text{Fe(OH)}_2^+]$	Concentration of the second hydroxide complex of the ferric ion, gmols/L
$[\text{Fe}_2(\text{OH})_2^{4+}]$	Concentration of the second hydroxide complex of the ferric double ion, gmols/L
$[\text{Fe(OH)}_4^-]$	Concentration of the fourth hydroxide complex of the ferric ion, gmols/L
$[\text{H}_{opt}^+]$	Optimum hydrogen ion concentration, gmoles/L
$K_{\text{Al(OH)}_c}$	Equilibrium constant of Al(OH)^{2+}
$K_{\text{Al}_2(\text{OH})_2c}$	Equilibrium constant of $\text{Al}_2(\text{OH})_2^{4+}$
$K_{\text{Al(OH)}_4c}$	Equilibrium constant of Al(OH)_4^-
$K_{\text{Al}_7(\text{OH})_{17}c}$	Equilibrium constant of $\text{Al}_7(\text{OH})_{17}^{4+}$
$K_{\text{Al}_{13}(\text{OH})_{34}c}$	Equilibrium constant of $\text{Al}_{13}(\text{OH})_{34}^{5+}$
$K_{\text{FeOH}c}$	Equilibrium constant of FeOH^+
$K_{\text{FeOH}c}$	Equilibrium constant of FeOH^{2+}
$K_{\text{Fe(OH)}_2c}$	Equilibrium constant of Fe(OH)_2^+
$K_{\text{Fe}_2(\text{OH})_2c}$	Equilibrium constant of $\text{Fe}_2(\text{OH})_2^{4+}$
$K_{\text{Fe(OH)}_4c}$	Equilibrium constant of Fe(OH)_4^-
$K_{\text{Fe(OH)}_3c}$	Equilibrium constant of Fe(OH)_3^-
$K_{sp,\text{Al(OH)}_3}$	Solubility product constant of $\text{Al(OH)}_{3(s)}$
$K_{sp,\text{Fe(OH)}_2}$	Solubility product constant of $\text{Fe(OH)}_{2(s)}$
$K_{sp,\text{Fe(OH)}_3}$	Solubility product constant of $\text{Fe(OH)}_{3(s)}$
K_w	Ion product of water
M_{Al}	Kilograms of alum used
M_{Alkgeq}	Kilogram equivalents of alum used
$M_{\text{Ca(HCO}_3)_2\text{Al}}$	Kilogram mass of calcium bicarbonate used with alum as coagulant

$M_{Ca(HCO_3)_2kgeqAl}$	Kilogram equivalents of calcium bicarbonate used with alum as coagulant
M_{CaOAl}	Kilogram mass of lime used with alum as coagulant
$M_{CaOkgeqAl}$	Kilogram equivalents of lime with alum as coagulant
M_{AlOH_3}	Kilograms of aluminum hydroxide plus the solids produced from the suspended solids resulting from the use of an optimum dose of $[Alopt]_{geq}$ of alum
M_{FeII}	Kilogram of copperas used
$M_{Ca(HCO_3)_2FeII}$	Kilograms of calcium bicarbonate used with copperas as coagulant
$M_{Ca(HCO_3)_2kgeqFeII}$	Kilogram equivalents of calcium bicarbonate used with copperas as coagulant
$M_{CaOFeII}$	Kilograms of lime used with copperas as coagulant
$M_{CaOkgeqFeII}$	Kilogram equivalents of lime used with copperas as coagulant
M_{FeOH_3FeII}	Kilograms of ferric hydroxide (including the calcium carbonate plus the solids from the suspended solids) resulting from the use of an optimum dose of copperas of $[FeIIopt]_{geq}$
M_{FeOH_3FeIII}	Kilograms of ferric hydroxide plus the solids produced from the raw water suspended solids resulting from the use of an optimum dose of $[FeIIIopt]_{geq}$
$M_{FeOH_3FeIIIchl}$	M_{FeOH_3FeIII} using ferric chloride
$M_{FeOH_3FeIIIsul}$	M_{FeOH_3FeIII} using ferric sulfate
$M_{FeIIkeq}$	Kilogram equivalents of copperas used
$M_{Ca(HCO_3)_2FeIII}$	Kilograms of calcium bicarbonate used with one of the ferric salts as coagulant
$M_{Ca(HCO_3)_2kgeqFeIII}$	Kilogram equivalents of calcium bicarbonate used with one of the ferric salts as coagulant
$M_{CaOFeIII}$	Kilograms of lime used with one of the ferric salts used as coagulant
$M_{CaOkgeqFeIII}$	Kilogram equivalents of lime used with one of the ferric salts as coagulant
$M_{FeIIICl}$	Kilograms of $FeCl_3$ used
$M_{FeIIIClkeq}$	Kilogram equivalents of $FeCl_3$
$M_{FeIIIClkgeq}$	Kilogram equivalents of any one of the ferric salts used
$M_{FeIIISO_4}$	Kilograms of $Fe_2(SO_4)_3$ used
$M_{FeIIISO_4keq}$	Kilogram equivalents of $Fe_2(SO_4)_3$ used
M_{O_2}	Kilograms of dissolved oxygen used with copperas as coagulant
M_{O_2kgeq}	Kilogram equivalents of dissolved oxygen used with copperas as coagulant
M_{CaOpH}	Kilograms of carbon dioxide needed to raise pH from pH_{cur} to pH_{to}
M_{CO_2pH}	Kilograms of carbon dioxide needed to lower the pH from pH_{cur} to pH_{to}
P_{Al}	Fractional purity of alum
P_{CaO}	Fractional purity of lime
P_{FeII}	Fractional purity of copperas
$P_{FeIIICl}$	Fractional purity of $FeCl_3$

P_{FeIIISO_4}	Fractional purity of $Fe_2(SO_4)_3$
pH_{cur}	Current pH value of water
pH_{to}	pH value to which water will be adjusted
$[sp_{\text{Al}}]$	Concentration of all species containing the aluminum ion, gmols/L
$[sp_{\text{FeII}}]$	Concentration of all species containing the ferrous ion, gmols/L
$[sp_{\text{FeIII}}]$	Concentration of all species containing the ferric ion, gmols/L
$\forall$	Cubic meters of water or wastewater treated
γ_{Al}	Activity coefficient of the aluminum ion
γ_{H}	Activity coefficient of the hydrogen ion
$\gamma_{\text{Al(OH)}c}$	Activity coefficient of $AL(OH)^{2+}$
$\gamma_{\text{Al}_2(\text{OH})_2 c}$	Activity coefficient $Al_2(OH)_2^{4+}$
$\gamma_{\text{Al(OH)}_4 c}$	Activity coefficient of $Al(OH)_4^-$
$\gamma_{\text{Al}_7(\text{OH})_{17} c}$	Activity coefficient of $Al_7(OH)_{17}^{4+}$
$\gamma_{\text{Al}_{13}(\text{OH})_{34} c}$	Activity coefficient of $Al_{13}(OH)_{34}^{5+}$
γ_{FeII}	Activity coefficient of the ferrous ion
$\gamma_{\text{FeOH}c}$	Activity coefficient of $FeOH^+$
$\gamma_{\text{Fe(OH)}_3 c}$	Activity coefficient of $Fe(OH)_3^-$
γ_{FeIII}	Activity coefficient of the ferric ion
$\gamma_{\text{FeOH}c}$	Activity coefficient of $FeOH^{2+}$
$\gamma_{\text{Fe(OH)}_2 c}$	Activity coefficient of $Fe(OH)_2^+$
$\gamma_{\text{Fe(OH)}_4 c}$	Activity coefficient of $Fe(OH)_4^-$
$\gamma_{\text{Fe}_2(\text{OH})_2 c}$	Activity coefficient of $Fe_2(OH)_2^{4+}$

PROBLEMS

12.1   A raw water containing 140 mg/L of dissolved solids is subjected to coagulation treatment using alum. The optimum pH was determined to be equal to 5.32 at a temperature of 25°C. Assume that all parameters to solve the problem can be obtained from the given conditions except the hydrogen ion activity coefficient. Calculate the hydrogen ion activity coefficient.

12.2   A raw water containing 140 mg/L of dissolved solids is subjected to coagulation treatment using alum. The hydrogen ion activity corresponding to the optimum pH was calculated to be equal to $4.81(10^{-6})$ gmols/L at a temperature of 25°C. Assume that all parameters to solve the problem can be obtained from the given conditions except the activity coefficient of the aluminum ion. Calculate the activity coefficient of the aluminum ion.

12.3   A raw water containing 140 mg/L of dissolved solids is subjected to coagulation treatment using alum. The optimum pH was determined to be equal to 5.32 at a temperature of 25°C. Assume that all parameters to solve the problem can be obtained from the given conditions except the equilibrium constant of $Al(OH)^{2+}$. Calculate the equilibrium constant of $Al(OH)^{2+}$.

12.4   A raw water containing 140 mg/L of dissolved solids is subjected to coagulation treatment using alum. The optimum pH was determined to be equal to 5.32 at a temperature of 25°C. Assume that all parameters to solve the problem can be obtained from the given conditions except the activity coefficient $Al(OH)^{2+}$. Calculate the activity coefficient of $Al(OH)^{2+}$.

12.5 A raw water containing 140 mg/L of dissolved solids is subjected to coagulation treatment using alum. The optimum pH was determined to be equal to 5.32 at a temperature of 25°C. Assume that all parameters to solve the problem can be obtained from the given conditions except the equilibrium constant of $Al_7(OH)_{17}^{4+}$. Calculate the equilibrium constant of $Al_7(OH)_{17}$.

12.6 A raw water containing 140 mg/L of dissolved solids is subjected to coagulation treatment using alum. The optimum pH was determined to be equal to 5.32 at a temperature of 25°C. Assume that all parameters to solve the problem can be obtained from the given conditions except the activity coefficient of $Al_7(OH)_{17}$. Calculate the activity coefficient of $Al_7(OH)_{17}$.

12.7 A raw water containing 140 mg/L of dissolved solids is subjected to coagulation treatment using copperas. The optimum pH was determined to be equal to 11.95 at a temperature of 25°C. Assume that all parameters to solve the problem can be obtained from the given conditions except the hydrogen ion activity coefficient. Calculate the hydrogen ion activity coefficient.

12.8 A raw water containing 140 mg/L of dissolved solids is subjected to coagulation treatment using copperas. The hydrogen ion activity corresponding to the optimum pH was calculated to be equal to $1.11(10^{-12})$ gmols/L at a temperature of 25°C. Assume that all parameters to solve the problem can be obtained from the given conditions except the activity coefficient of the ferrous ion. Calculate the activity coefficient of the ferrous ion.

12.9 A raw water containing 140 mg/L of dissolved solids is subjected to coagulation treatment using copperas. The optimum pH was determined to be equal to 11.95 at a temperature of 25°C. Assume that all parameters to solve the problem can be obtained from the given conditions except the equilibrium constant of $FeOH^+$. Calculate the equilibrium constant of $FeOH^+$.

12.10 A raw water containing 140 mg/L of dissolved solids is subjected to coagulation treatment using copperas. The optimum pH was determined to be equal to 11.95 at a temperature of 25°C. Assume that all parameters to solve the problem can be obtained from the given conditions except the activity coefficient $FeOH^+$. Calculate the activity coefficient of $FeOH^+$.

12.11 A raw water containing 140 mg/L of dissolved solids is subjected to coagulation treatment using copperas. The optimum pH was determined to be equal to 11.95 at a temperature of 25°C. Assume that all parameters to solve the problem can be obtained from the given conditions except the equilibrium constant of $Fe(OH)_3^-$. Calculate the equilibrium constant of $Fe(OH)_3^-$.

12.12 A raw water containing 140 mg/L of dissolved solids is subjected to coagulation treatment using copperas. The optimum pH was determined to be equal to 11.95 at a temperature of 25°C. Assume that all parameters to solve the problem can be obtained from the given conditions except the activity coefficient $Fe(OH)_3^-$. Calculate the activity coefficient of $Fe(OH)_3^-$.

12.13 A raw water containing 140 mg/L of dissolved solids is subjected to coagulation treatment using ferric chloride. The optimum pH was determined to be equal to 8.2 at a temperature of 25°C. Assume that all parameters to solve the problem can be obtained from the given conditions except the hydrogen ion activity coefficient. If the activity coefficient of the hydrogen ion can be determined, calculate the activity coefficient of the hydroxyl ion.

12.14 A raw water containing 140 mg/L of dissolved solids is subjected to coagulation treatment using ferric sulfate. The hydrogen ion activity corresponding to the optimum pH was calculated to be equal to $6.3(10^{-9})$ gmols/L at a temperature of 25°C. Assume that all parameters to solve the problem can be obtained from the given conditions except the activity coefficient of the ferric ion. Calculate the activity coefficient of the ferric ion.

12.15 A raw water containing 140 mg/L of dissolved solids is subjected to coagulation treatment using ferric chloride. The optimum pH was determined to be equal to 8.2 at a temperature of 25°C. Assume that all parameters to solve the problem can be obtained from the given conditions except the equilibrium constant of $FeOH^{2+}$. Calculate the equilibrium constant of $FeOH^{2+}$.

12.16 A raw water containing 140 mg/L of dissolved solids is subjected to coagulation treatment using ferric sulfate. The optimum pH was determined to be equal to 8.2 at a temperature of 25°C. Assume that all parameters to solve the problem can be obtained from the given conditions except the activity coefficient $FeOH^{2+}$. Calculate the activity coefficient of $FeOH^{2+}$.

12.17 A raw water containing 140 mg/L of dissolved solids is subjected to coagulation treatment using ferric sulfate. The optimum pH was determined to be equal to 8.2 at a temperature of 25°C. Assume that all parameters to solve the problem can be obtained from the given conditions except the equilibrium constant of $Fe_2(OH)_2^{4+}$. Calculate the equilibrium constant of $Fe_2(OH)_2^{4+}$.

12.18 A raw water containing 140 mg/L of dissolved solids is subjected to coagulation treatment using ferric chloride. The optimum pH was determined to be equal to 8.2 at a temperature of 25°C. Assume that all parameters to solve the problem can be obtained from the given conditions except the activity coefficient $Fe_2(OH)_2^{4+}$. Calculate the activity coefficient of $Fe_2(OH)_2^{4+}$.

12.19 A raw water containing 140 mg/L of dissolved solids is subjected to a coagulation treatment using alum. The calcium bicarbonate requirement is 20.2 mg/L as $CaCO_3$. Calculate the optimum alum dose. Make appropriate assumptions.

12.20 A raw water containing 140 mg/L of dissolved solids is subjected to a coagulation treatment using alum. The calcium bicarbonate requirement is 20.2 mg/L as $CaCO_3$ and the optimum alum dose is 44.86 mg/L. Calculate the fraction of the optimum alum dose neutralized by calcium bicarbonate. Make appropriate assumptions, except the fraction neutralized by calcium bicarbonate.

12.21 A raw water containing 140 mg/L of dissolved solids is subjected to a coagulation treatment using alum. The calcium bicarbonate requirement is 20.2 mg/L as $CaCO_3$ and the optimum alum dose is 44.86 mg/L. Calculate the cubic meters of water treated. Make appropriate assumptions, except the cubic meters of water treated.

12.22 A raw water containing 140 mg/L of dissolved solids is subjected to a coagulation treatment using alum. The calcium bicarbonate requirement is 20.2 mg/L as $CaCO_3$ and the optimum alum dose is 44.86 mg/L. Calculate the water of hydration of the alum used. Make appropriate assumptions, except the water of hydration.

12.23 A raw water containing 140 mg/L of dissolved solids is subjected to a coagulation treatment using alum. If the optimum alum dose as determined by a jar test is 44.86 mg/L, calculate the kilograms of calcium bicarbonate required.

12.24 A raw water containing 140 mg/L of dissolved solids is subjected to coagulation treatment using copperas. The alkalinity requirement is 18 mg/L as $CaCO_3$ and the natural alkalinity of the raw water is 100 mg/L as CaCO. Calculate the optimum dose of the coagulant.

12.25 A raw water containing 140 mg/L of dissolved solids is subjected to coagulation treatment using copperas. The alkalinity requirement is 18 mg/L as $CaCO_3$ and the natural alkalinity of the raw water is 100 mg/L as CaCO. Calculate the cubic meters of water coagulated, if the optimum dose of the coagulant is 50 mg/L.

12.26 A raw water containing 140 mg/L of dissolved solids is subjected to coagulation treatment using copperas. The alkalinity requirement is 18 mg/L as $CaCO_3$ and the natural alkalinity of the raw water is 100 mg/L as CaCO. The optimum dose of the coagulant is 50 mg/L. Calculate the fraction of the coagulant dose neutralized in the presence of calcium bicarbonate.

12.27 A raw water containing 140 mg/L of dissolved solids is subjected to coagulation treatment using copperas. The lime used is 22.4 mg/L at a purity of 90%. Making appropriate assumptions and calculate the optimum coagulant dose.

12.28 A raw water containing 140 mg/L of dissolved solids is subjected to coagulation treatment using copperas. The lime used is 22.4 mg/L at a purity of 90%. The optimum coagulant dose is 50 mg/L. If the fraction of the coagulant dose neutralized by calcium bicarbonate is 0.90, calculate the volume of water treated.

12.29 A raw water containing 140 mg/L of dissolved solids is subjected to coagulation treatment using copperas. The lime used is 22.4 mg/L. The optimum coagulant dose is 50 mg/L. If the fraction of the coagulant dose neutralized by calcium bicarbonate is 0.90 and the volume of water treated is 1 m^3, calculate the fractional purity of the lime.

12.30 A raw water containing 140 mg/L of dissolved solids is subjected to a coagulation treatment using $Fe_2(SO_4)_3$. The calcium bicarbonate alkalinity required is 30 mg/L as $CaCO_3$. The natural alkalinity of the raw water is 100 mg/L as $CaCO_3$. Calculate the optimum coagulant dose.

12.31 A raw water containing 140 mg/L of dissolved solids is subjected to a coagulation treatment using $Fe_2(SO_4)_3$. The calcium bicarbonate alkalinity required is 30 mg/L as $CaCO_3$ and the optimum coagulant dose is 40 mg/L. The natural alkalinity of the raw water is 100 mg/L as $CaCO_3$. Calculate the cubic meters of water treated.

12.32   A raw water containing 140 mg/L of dissolved solids is subjected to a coagulation treatment using $Fe_2(SO_4)_3$. The calcium bicarbonate alkalinity required is 30 mg/L as $CaCO_3$ and the optimum coagulant dose is 40 mg/L. Per cubic meter of water treated, calculate the fraction of the coagulant dose neutralized by calcium bicarbonate.

12.33   A raw water containing 140 mg/L of dissolved solids is subjected to a coagulation treatment using $Fe_2(SO_4)_3$. The optimum coagulant dose as determined by a jar test is 40 mg/L. The natural alkalinity is 5 mg/L as $CaCO_3$. The fraction of the coagulant dose neutralized by lime of 90% purity is 0.6. Per cubic meter of water coagulated, calculate the kilogram of lime used.

12.34   A raw water containing 140 mg/L of dissolved solids is subjected to a coagulation treatment using $Fe_2(SO_4)_3$. The kilogram of lime used is 800 at a percentage purity of 90. The optimum coagulant dose as determined by a jar test is 40 mg/L. The natural alkalinity is 5 mg/L as $CaCO_3$. If the fraction of the coagulant dose neutralized by lime is 0.6, calculate cubic meters of water coagulated.

12.35   A raw water containing 140 mg/L of dissolved solids is subjected to a coagulation treatment using $Fe_2(SO_4)_3$. The current acidity and pH are, respectively, 30 mg/L as $CaCO_3$ and 5.9. The amount of pure lime used per cubic meter of water coagulated is 0.016865135 kg. Calculate the value of the pH that the coagulation was adjusted to.

12.36   A raw water containing 140 mg/L of dissolved solids is subjected to a coagulation treatment using $Fe_2(SO_4)$. The current acidity is 30 mg/L as $CaCO_3$. The amount of pure lime used per cubic meter of water coagulated is 0.016865135 kg. The pH of coagulation was adjusted to 8.15. Calculate the value of the current pH of the water.

12.37   A raw water containing 140 mg/L of dissolved solids is subjected to a coagulation treatment using $Fe_2(SO_4)_3$. The amount of pure lime used per cubic meter of water coagulated is 0.0169 kg. The current pH of the raw water is 5.9 and is adjusted to 8.15. Calculate the current acidity.

12.38   A raw water containing 100 mg/L of suspended solids is subjected to a coagulation treatment using $FeCl_3$. The current acidity and pH are, respectively, 30 mg/L as $CaCO_3$ and 5.9. Calculate the amount of sludge produced if the optimum dose is 40 mg/L of $FeCl_3$.

12.39   A raw water is subjected to a coagulation treatment using $FeCl_3$. The current acidity and pH are, respectively, 30 mg/L as $CaCO_3$ and 5.9. The solids produced amount to 0.132 kg/m^3 of water treated. What was the suspended solids content of the raw water, if the optimum coagulant dose is 40 mg/L?

12.40   A raw water is subjected to a coagulation treatment using $FeCl_3$. The current acidity and pH are, respectively, 30 mg/L as $CaCO_3$ and 5.9. The solids produced is 132 kg and the suspended solids of the water is 100 m/L. Calculate the cubic meters of water treated, if the optimum coagulant dose is 40 mg/L.

BIBLIOGRAPHY

Abuzaid, N. S., M. H. Al-Malack, and A. H. El-Mubarak (1998). Separation of colloidal polymeric waste using a local soil, *Separation Purification Technol.* 13, 2, Apr. 15, 1998, 161–169.

Al-Malack, M. H., N. S. Abuzaid, and A. H. El-Mubarak (1999). Coagulation of polymeric wastewater discharged by a chemical factory. *Water Research.* 33, 2, 521–529.

Chen, W. J., D. P. Lin, and J. P. Hsu (1998). Contribution of electrostatic interaction to the dynamic stability coefficient for coagulation-flocculation kinetics of beta-iron oxy-hydroxides in polyelectrolyte solutions. *J. Chem. Eng. Japan.* 31, 5, 722–733.

Cho, Y., et al. (1998). Mobility and size-distribution analyses of hetero-coagulated structures of colloidally mixed mullite precursor suspensions and their sinterability. *J. Ceramic Soc. Japan.* 106, 1230, 189–193.

Chou, S., S. Lin, and C. Huang (1997). Application of optical monitor to evaluate the coagulation of pulp wastewater. *Water Science Technol., Proc. 7th Int. Workshop on Instrumentation, Control and Automation of Water and Wastewater Treatment and Transport Syst.*, July 6–9, Brighton, England, 37, 12, 111–119. Elsevier Science Ltd., Exeter, England.

Chuang, C. J. and K. Y. Li (1997). Effect of coagulant dosage and grain size on the performance of direct filtration. *Separation Purification Technol.* 12, 3, 229–241.

Dalrymple, C. W. (1997). Electrocoagulation of industrial wastewaters. *Proc. 1997 Air Waste Manage. Assoc. 90th Annu. Meet. Exhibition,* June 8–13, Toronto, Canada, Air & Waste Manage. Assoc., Pittsburgh, PA.

Gregor, J. E., C. J. Nokes, and E. Fenton (1997). Optimising natural organic matter removal from low turbidity waters by controlled pH adjustment of aluminium coagulation. *Water Res.* 31, 12, 2949–2958.

Hodgson, A. T., et al. (1998). Effect of tertiary coagulation and flocculation treatment on effluent quality from a bleached kraft mill. *TAPPI J.* 81, 2, 166–172.

Hodgson, A. T., et al. (1997). Effect of tertiary coagulation and flocculation technologies on bleached kraft mill effluent quality. *TAPPI Proc.—Environmental Conf. Exhibition, Proc. 1997 Environmental Conf. Exhibit, Part 1,* May 5–7, Minneapolis, MN, 1, 307–313. TAPPI Press, Norcross, GA.

Holtzclaw Jr., H. F. and W. R. Robinson (1988). *General Chemistry.* D. C. Heath and Company, Lexington, MA, 347.

Hsu, J. P. and B. T. Liu (1998). Critical coagulation concentration of a colloidal suspension at high particle concentrations. *J. Physical Chem. B,* 102, 2, 334–337.

Hsu, Y. C., C. H. Yen, and H. C. Huang (1998). Multistage treatment of high strength dye wastewater by coagulation and ozonation. *J. Chem. Technol. Biotechnol.* 71, 1, 71–76.

Kang, S. F. and H. M. Chang (1997). Coagulation of textile secondary effluents with fenton's reagent. *Asia–Pacific Regional Conf., Asian Waterqual '97,* May 20–23, Seoul, South Korea, 36, 12, 215–222. Elsevier Science Ltd., Oxford, England.

Lei, Z. and Y. Lu (1997). Analysis of Lignin Coagulation. *China Environmental Science.* 17, 6, 535–539.

Lichtenfeld, H., V. Shilov, and L. Knapschinsky (1998). Determination of the distance of nearest approach of coagulated particles in haematite and latex dispersions. *Colloids Surfaces A: Physicochem. Eng. Aspects.* 142, 2–3, 155–163.

Ma, J., N. Graham, and G. Li (1997). Effect of permanganate preoxidation in enhancing the coagulation of surface waters—Laboratory case studies. *Aqua (Oxford).* 46, 1, 1–10.

Malhotra, S., D. N. Kulkarni, and S. P. Pande (1997). Effectiveness of polyaluminum chloride (PAC) vis-a-vis alum in the removal of fluorides and heavy metals. *J. Environmental*

Science Health, Part A: Environmental Science Eng. Toxic Hazardous Substance Control. 32, 9–10, 2563–2574.

Muller, V. M. (1996). Theory of reversible coagulation. *Kolloidnyi Zhurnal.* 58, 5, 634–647.

Muyibi, S. A. and L. M. Evison (1996). Coagulation of turbid water and softening of hardwater with *Moringa oleifera* seeds. *Int. J. Environmental Studies A & B.* 49, 3–4, 247–259.

Otto, E., et al. (1999). Log-normal size distribution theory of Brownian aerosol coagulation for the entire particle size range: Part II—Analytical solution using Dahneke's coagulation kernel. *J. Aerosol Science.* 30, 1, 17–34.

Panswad, T. and B. Chan-narong (1998). Turbidity removal by the up-flow pelletisation process for low turbidity water. *Aqua (Oxford).* 47, 1, 36–40.

Park, S. H., et al. (1999). Log-normal size distribution theory of brownian aerosol coagulation for the entire particle size range: Part I—Analytical solution using the harmonic mean coagulation kernel. *J. Aerosol Science.* 30, 1, 3–16.

Philipps, C. A. and G. D. Boardman (1997). Physicochemical and biological treatability study of textile dye wastewater. *Proc. 1997 29th Mid-Atlantic Industrial Hazardous Waste Conf.* July 13–16, Blacksburg, VA, 493–504. Technomic Publ. Co. Inc., Lancaster, PA.

Ramakrishnan, V. and S. G. Malghan (1998). Stability of alumina-zirconia suspensions. *Colloids Surfaces A: Physicochem. Eng. Aspects.* 133, 1–2, 135–142.

Reade, W. and L. R. Collins (1998). Collision and coagulation in the infinite-stokes-number regime. *Aerosol Science Technol.* 29, 6, 493–509.

Relle, S. and B. S. Grant (1998). One-step process for particle separation by magnetic seeding, *Langmuir.* 14, 9, 2316–2328.

Rich, G. (1963). *Unit Process of Sanitary Engineering.* John Wiley & Sons, New York, 135.

Ruan, F., et al. (1998). Computer simulation for the coagulation of iron-based water clarifying agent. *Chem. Reaction Eng. Technol.* 14, 1, 66–70.

Ruehl, K. E. (1998). How to conduct a successful primary coagulant trial. *Public Works.* 129, 2.

Sincero, A. P. and G. A. Sincero (1996). *Environmental Engineering: A Design Approach.* Prentice Hall, Upper Saddle River, NJ, 260–265.

Sincero, A. P. (1968). *Sludge production in coagulation treatment of water.* Master's thesis, Asian Institute of Technology, Bangkok, Thailand, 12–13, 34.

Snoeyink, V. L. (1980). *Water Chemistry.* John Wiley & Sons, New York, 264–266, 271.

Tang, W., et al. (1997). Study on coagulate mechanism of polyaluminium chloride. *J. Nanjing Univ. Science Technol.* 21, 4, 325–328.

Thompson, P. L. and W. L. Paulson (1998). Dewaterability of alum and ferric coagulation sludges. *J. Am. Water Works Assoc.* 90, 4, 164–170.

van Leeuwen, J., et al. (1997). Comparison of coagulant doses determined using a charge titration unit with a jar test procedure for eight german surface waters. *Aqua (Oxford).* 46, 5, 261–273.

Vrijenhoek, E. M., et al. (1998). Removing particles and THM precursors by enhanced coagulation. *J. Am. Water Works Assoc.* 90, 4, 139–150.

Yaroshevskaya, N. V., V. R. Murav'ev, and T. Z. Sotskova (1997). Effect of flocculants LT27 and 573C on the quality of water purification by contact coagulation. *Khimiya i Tekhnologiya Vody.* 19, 3, 308–314.

Yoon, J., et al. (1998). Characteristics of coagulation of fenton reaction in the removal of landfill leachate organics. *Water Science Technol. Solid Wastes: Sludge Landfill Management. Proc. 1998 19th Biennial Conf. Int. Assoc. Water Quality, Part 2,* June 21–26, Vancouver, Canada, 38, 2, 209–214, Elsevier Science Ltd. Exeter, England.

13 Removal of Iron and Manganese by Chemical Precipitation

Iron concentrations as low as 0.3 mg/L and manganese concentrations as low as 0.05 mg/L can cause dirty water complaints. At these concentrations, the water may appear clear but imparts brownish colors to laundered goods. Iron also affects the taste of beverages such as tea and coffee. Manganese flavors tea and coffee with medicinal tastes.

Some types of bacteria derive their energy by utilizing soluble forms of iron and manganese. These organisms are usually found in waters that have high levels of iron and manganese in solution. The reaction changes the species from soluble forms into less soluble forms, thus causing precipitation and accumulation of black or reddish brown gelatinous slimes. Masses of mucous iron and manganese can clog plumbing and water treatment equipment. They also slough away in globs that become iron or manganese stains on laundry.

Standards for iron and manganese are based on levels that cause taste and staining problems and are set under the Environmental Protection Agency Secondary Drinking Water Standards (EPA SDWA). They are, respectively, 0.3 mg/L for iron and 0.05 mg/L for manganese. Iron and manganese are normally found in concentrations not exceeding 10 mg/L and 3 mg/L, respectively, in natural waters. Iron and manganese can be found at higher concentrations; however, these conditions are rare. Iron concentrations can go as high as 50 mg/L.

Iron and manganese may be removed by *reverse osmosis* and *ion exchange*. The unit operation of reverse osmosis was discussed in a previous chapter; the unit process of ion exchange is discussed in a later chapter. This chapter discusses the removal of iron and manganese by the unit process of chemical precipitation.

One manufacturer claimed that these elements could also be removed by biological processes. It claimed that a process called the *bioferro* process encourages the growth of naturally occurring iron assimilating bacteria, such as *Gallionella ferruginea*, thus reducing iron concentration. An experimental result shows a reduction of iron from 6.0 mg/L to less than 0.1 mg/L. They also claimed that a companion process called *bioman* could remove manganese down to 0.08 mg/L. This uses naturally occurring manganese bacteria to consume manganese. Figure 13.1 is a photograph showing growths of "bioferro" and "bioman" bacteria.

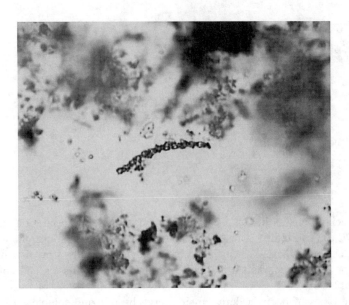

FIGURE 13.1 A photograph of "bioferro" and "bioman" bacteria.

13.1 NATURAL OCCURRENCES OF IRON AND MANGANESE

Iron and manganese have the electronic configurations of $[Ar]3d^6 4s^2$ and $[Ar]3d^5 4s^2$, and are located in Groups *VIIIB* and *VIIB* of the Periodic Table, respectively. They are both located in the fourth period. [Ar] means that these elements have the electronic configuration of the noble gas argon. The letters *d* and *s* refer to the *d* and *s* orbitals; the superscripts indicate the number of electrons that the orbitals contain. Thus, the *d* orbital of iron contains 6 electrons and that of manganese contains 5 electrons. Both elements contain 2 electrons in their *s* orbitals. This means that in their most reduced positive state, they acquire oxidation states of 2+. Also, because of the *d* orbitals, they can form a number of oxidation states. The multiplicity of oxidation states give iron and manganese the property of imparting colors such as the imparting of brownish colors to laundered goods.

Surface waters always contain dissolved oxygen in it. Thus, iron and manganese would not exist in their most reduced positive state of 2+ in these waters. The reason is that they will simply be oxidized to higher states of oxidation by the dissolved oxygen forming hydroxides and precipitate out. Groundwater is a source where these elements could come from. Groundwaters occurring deep down in the earth can become devoid of oxygen, thus, any iron or manganese present would have to be reduced. Therefore, the waters where removal of iron and manganese could be undertaken are groundwaters and the form of the elements are in the 2+ oxidation states, Fe(II) and Mn(II), respectively

---

TABLE 13.1
Solubility Product Constant of Iron and
Manganese Precipitation Products

Precipitation Product	Solubility Product, K_{sp}
$Fe(OH)_2$	$3.16(10^{-15})$
$FeCO_2$	$2.11(10^{-11})$
FeS	$8(10^{-26})$
$Fe(OH)_3$	$1.1(10^{-38})$
$Mn(OH)_2$	$4.5(10^{-14})$
$MnCO_3$	$8.8(10^{-11})$
MnS	$4.3(10^{-22})$

13.2 MODES OF REMOVAL OF IRON AND MANGANESE

The best place to investigate for determining the mode by which the elements can be removed is the table of solubility products constants as shown in Table 13.1. In general, a precipitation product that has the lowest K_{sp} means that the substance is the most insoluble. As shown in the table, for iron, the lowest K_{sp} is that of $Fe(OH)_3$, an Fe(III) iron, and has the value of $1.1(10^{-38})$. For manganese, the lowest K_{sp} is that of MnS, an Mn(II) manganese, and has the value of $1.1(10^{-22})$.

These K_{sp}'s indicate that the elements must be removed in the form of ferric hydroxide and manganese sulfide, respectively; however, from the table, manganese can also be removed as $Mn(OH)_2$ at a $K_{sp} = 4.5(10^{-14})$. Of course, lime has many uses, while sulfide has only few. Sodium sulfide is used in photographic film development; however, lime is used in water and wastewater treatment, as an industrial chemical, as well as being used in agriculture. Thus, because of its varied use, lime is much cheaper. In addition, using a sulfide to remove iron and manganese would be a new method. Its health effect when found in drinking water is not documented. On the other hand, lime has been used for years. We will therefore use lime as the precipitant for the removal of iron and manganese. The probable use of sulfide in removing iron and manganese could be a topic for investigation in applied research.

13.3 CHEMICAL REACTIONS OF THE FERROUS AND THE FERRIC IONS

The chemical reactions of the ferrous and the ferric ions were already discussed in a previous chapter. From the topic in the preceding section, iron is more efficiently removed as ferric hydroxide. The natural iron is in the form of Fe(II), so this ferrous must therefore oxidize to the ferric form in order to precipitate as the ferric hydroxide, if, in fact, the iron is to be removed in the ferric form. In Chapter 12, this was done using the dissolved oxygen that is relatively abundant in natural waters. It must be noted, however, that based on the K_{sp} values, iron can also be removed in the ferrous form as $Fe(OH)_2$. For convenience, the reactions are reproduced next.

Ferrous:

$$Fe(OH)_{2(s)} \rightleftharpoons Fe^{2+} + 2OH^- \qquad K_{sp,Fe(OH)_2} = 10^{-14.5} \qquad (13.1)$$

$$Fe(OH)_{2(s)} \rightleftharpoons FeOH^+ + OH^- \qquad K_{FeOHc} = 10^{-9.4} \qquad (13.2)$$

$$Fe(OH)_{2(s)} + OH^- \rightleftharpoons Fe(OH)_3^- \qquad K_{Fe(OH)_3c} = 10^{-5.1} \qquad (13.3)$$

$$Fe(OH)_2 + \tfrac{1}{4}O_2 + \tfrac{1}{2}H_2O \rightarrow Fe(OH)_3 \downarrow \text{ (conversion from ferrous to ferric)} \qquad (13.4)$$

Ferric:

$$Fe(OH)_{3(s)} \rightleftharpoons Fe^{3+} + 3OH^- \qquad K_{sp,Fe(OH)_3} = 10^{-38} \qquad (13.5)$$

$$Fe(OH)_{3(s)} \rightleftharpoons FeOH^{2+} + 2OH^- \qquad K_{FeOHc} = 10^{-26.16} \qquad (13.6)$$

$$Fe(OH)_{3(s)} \rightleftharpoons Fe(OH)_2^+ + OH^- \qquad K_{Fe(OH)_2c} = 10^{-16.74} \qquad (13.7)$$

$$Fe(OH)_{3(s)} + OH^- \rightleftharpoons Fe(OH)_4^- \qquad K_{Fe(OH)_4c} = 10^{-5} \qquad (13.8)$$

$$2Fe(OH)_{3(s)} \rightleftharpoons Fe_2(OH)_2^{4+} + 4OH^- \qquad K_{Fe_2(OH)_2c} = 10^{-50.8} \qquad (13.9)$$

Remember that the values of the solubility product constants, $K_{sp,Fe(OH)_2}$ and $K_{sp,Fe(OH)_3}$, and all the other equilibrium constants for the complex ions apply only at 25°C.

The presence of the complex ions increase the solubility of the iron species and therefore increase the concentration of these species in solution. For the ferrous and the ferric species, they are sp_{FeII} and sp_{FeIII} and were derived in Chapter 12, respectively, as

$$[sp_{FeII}] = [Fe^{2+}] + [FeOH^+] + [Fe(OH)_3^-]$$

$$= \frac{K_{sp,Fe(OH)_2}\gamma_H^2[H^+]^2}{\gamma_{FeII}K_w^2} + \frac{K_{FeOHc}\gamma_H[H^+]}{\gamma_{FeOHc}K_w} + \frac{K_{Fe(OH)_3c}K_w}{\gamma_{Fe(OH)_3c}\gamma_H[H^+]} \qquad (13.10)$$

$$[sp_{FeIII}] = [Fe^{3+}] + [FeOH^{2+}] + [Fe(OH)_2^+] + [Fe(OH)_4^-] + 2[Fe_2(OH)_2^{4+}]$$

$$= \frac{K_{sp,Fe(OH)_3}\gamma_H^3[H^+]^3}{\gamma_{FeIII}K_w^3} + \frac{K_{FeOHc}\gamma_H^2[H^+]^2}{\gamma_{FeOHc}K_w^2} + \frac{K_{Fe(OH)_2c}\gamma_H[H^+]}{\gamma_{Fe(OH)_2c}K_w} + \frac{K_{Fe(OH)_4c}K_w}{\gamma_{Fe(OH)_4c}[H^+]}$$

$$+ \frac{2K_{Fe_2(OH)_2c}\gamma_H^4[H^+]^4}{\gamma_{Fe_2(OH)_2c}K_w^4} \qquad (13.11)$$

Example 13.1 From the respective optimum pH's of 11.95 and 8.2 for sp_{FeII} and sp_{FeIII}, calculate the concentrations $[sp_{FeII}]$ and $[sp_{FeIII}]$, respectively. Assume the water contains 140 mg/L of dissolved solids.

Solution:

$$[sp_{FeII}] = \frac{K_{sp,Fe(OH)_2}\gamma_H^2[H^+]^2}{\gamma_{FeII}K_w^2} + \frac{K_{FeOHc}\gamma_H[H^+]}{\gamma_{FeOHc}K_w} + \frac{K_{Fe(OH)_3c}K_w}{\gamma_{Fe(OH)_3c}\gamma_H[H^+]}$$

$$K_{sp,Fe(OH)_2} = 10^{-14.5} \qquad \mu = 2.5(10^{-5})TDS \qquad \gamma = 10^{-\frac{0.5z_i^2(\sqrt{\mu})}{1+1.14(\sqrt{\mu})}}$$

$$\mu = 2.5(10^{-5})(140) = 3.5(10^{-3})$$

$$\gamma_H = \gamma_{FeOHc} = \gamma_{Fe(OH)_3c} = 10^{-\frac{0.5(1)^2[\sqrt{3.5(10^{-3})}]}{1+1.14[\sqrt{3.5(10^{-3})}]}} = 0.94$$

$$\gamma_{FeII} = 10^{-\frac{0.5(2)^2[\sqrt{3.5(10^{-3})}]}{1+1.14[\sqrt{3.5(10^{-3})}]}} = 0.77 K_w = (10^{-14})$$

$$K_{FeOHc} = 10^{-9.4} \qquad K_{Fe(OH)_3c} = 10^{-5.1}$$

Therefore,

$$[sp_{FeII}] = \frac{10^{-14.5}(0.94)^2[10^{-11.95}]^2}{(0.77)(10^{-14})^2} + \frac{(10^{-9.4})(0.94)[10^{-11.95}]}{(0.94)(10^{-14})} + \frac{(10^{-5.1})(10^{-14})}{(0.94)(0.94)[10^{-11.95}]}$$

$$= 4.57(10^{-11}) + 4.47(10^{-8}) + 8.01(10^{-8}) = 1.0(10^{-7}) \text{ gmol/L} \quad \textbf{Ans}$$

$$= 0.0056 \text{ mg/L}$$

$$[sp_{FeIII}] = \frac{K_{sp,Fe(OH)_3}\gamma_H^3[H^+]^3}{\gamma_{FeIII}K_w^3} + \frac{K_{FeOHc}\gamma_H^2[H^+]^2}{\gamma_{FeOHc}K_w^2} + \frac{K_{Fe(OH)_2c}\gamma_H[H^+]}{\gamma_{Fe(OH)_2c}K_w} + \frac{K_{Fe(OH)_4c}K_w}{\gamma_{Fe(OH)_4c}[H^+]}$$

$$+ \frac{2K_{Fe_2(OH)_2c}\gamma_H^4[H^+]^4}{\gamma_{Fe_2(OH)_2c}K_w^4}$$

$$K_{sp,Fe(OH)_3} = 10^{-38} \quad \gamma_{FeIII} = 10^{-\frac{0.5(3)^2[\sqrt{3.5(10^{-3})}]}{1+1.14[\sqrt{3.5(10^{-3})}]}} = 0.56 \qquad K_{FeOHc} = 10^{-26.16}$$

$$\gamma_{FeOHc} = 10^{-\frac{0.5(2)^2[\sqrt{3.5(10^{-3})}]}{1+1.14[\sqrt{3.5(10^{-3})}]}} = 0.77 \qquad K_{Fe(OH)_2c} = 10^{-16.74}$$

$$\gamma_{Fe(OH)_2c} = 10^{-\frac{0.5(1)^2[\sqrt{3.5(10^{-3})}]}{1+1.14[\sqrt{3.5(10^{-3})}]}} = 0.94$$

$$K_{Fe(OH)_4c} = 10^{-5} \quad \gamma_{Fe(OH)_4c} = \gamma_{Fe(OH)_2c} = 0.94 \quad K_{Fe_2(OH)_2c} = 10^{-50.8}$$

$$\gamma_{Fe_2(OH)_2c} = 10^{\frac{0.5(4)^2[\sqrt{3.5(10^{-3})}]}{1+1.14[\sqrt{3.5(10^{-3})}]}} = 0.36$$

Therefore,

$$[sp_{FeIII}] = \frac{(10^{-38})(0.94)^3[10^{-8.2}]^3}{(0.56)(10^{-14})^3} + \frac{(10^{-26.16})(0.94)^2[10^{-8.2}]^2}{(0.77)(10^{-14})^2} + \frac{(10^{-16.74})(0.94)[10^{-8.2}]}{(0.94)(10^{-14})}$$

$$+ \frac{(10^{-5})(10^{-14})}{(0.94)[10^{-8.2}]} + \frac{2(10^{-50.8})(0.94)^4[10^{-8.2}]^4}{(0.36)(10^{-14})^4}$$

$$= \frac{2.09(10^{-63})}{5.6(10^{-43})} + \frac{2.43(10^{-43})}{7.7(10^{-29})} + \frac{1.08(10^{-25})}{9.4(10^{-15})} + \frac{1.0(10^{-19})}{5.93(10^{-9})} + \frac{3.92(10^{-84})}{3.6(10^{-57})}$$

$$= 3.73(10^{-21}) + 3.16(10^{-15}) + 1.15(10^{-11}) + 1.69(10^{-11}) + 1.08(10^{-27})$$

$$= 2.84(10^{-11}) \text{ gmol/L} = 0.0000016 \text{ mg/L} \quad \textbf{Ans}$$

13.3.1 PRACTICAL OPTIMUM pH RANGE FOR THE REMOVAL OF FERROUS AND FERRIC

As shown in Chapter 12, at 25°C and at a solids concentration of 140 mg/L, the optimum pH's correspond to 11.95 and 8.2 (or around 12 and 8), respectively, for ferrous and ferric. The respective concentrations for sp_{FeII} and sp_{FeII} at these conditions as obtained in the previous example are $[sp_{FeII}] = 0.0056$ mg/L and $[sp_{FeIII}] = 0.0000016$ mg/L. A pH range exists, however, at which units used for the removal of the elements can be operated and effect good results. This range is called the *practical optimum pH range*.

Tables 13.2 through 13.5 show the respective concentrations of sp_{FeII} and sp_{FeIII} at other conditions of pH and total solids. The values for $[sp_{FeII}]$ were obtained using Equation (13.10) and the values for $[sp_{FeIII}]$ were obtained using Equation (13.11). Note that these equations require the values of the activity coefficients of the ions. The activity coefficients are needed by the equations and, since activity coefficients are functions of the dissolved solids, dissolved solids are used as parameters in the tables, in addition to pH.

The previous tables indicate that the total solids (or equivalently, the activities of the ions) do not have a significant effect on the optimum pH values, which for sp_{FeII} remain at about 12.0 and for sp_{FeIII} remain at about 8.0. For practical purposes, however, the *practical optimum pH* for sp_{FeII} ranges from 11 to 13 and for sp_{FeIII}, it ranges from 5.0 to 13.0. Note that for the range of pH for sp_{FeII}, it is assumed the element is to be removed as $Fe(OH)_2$. If it is to be removed as $Fe(OH)_3$, the pH of the solution during its oxidation by dissolved oxygen or any oxidizer need not be adjusted since the practical optimum pH for the precipitation of ferric hydroxide varies over a wide range from 5.0 to 13.0 and already includes the range for the ferrous removal.

TABLE 13.2
Concentration of sp_{FeII} as a Function of pH at 25°C

pH, Dissolved Solids = 140 mg/L	$[sp_{FeII}]$, mg/L
0	$5.0(10^{+18})$
1	$2.0(10^{+16})$
2	$2.0(10^{+14})$
3	$2.0(10^{+12})$
4	$2.0(10^{+10})$
5	$2.0(10^{+8})$
6	$2.0(10^{+6})$
7	$2.0(10^{+4})$
8	$2.2(10^{+2})$
9	$4.2(10^{+0})$
10	$2.4(10^{-1})$
11	$2.3(10^{-2})$
12	$7.3(10^{-3})$
13	$5.1(10^{-2})$
14	$5.0(10^{-1})$
15	$5.0(10^{+0})$

TABLE 13.3
Concentration of sp_{FeII} as a Function of pH at 25°C

pH, Dissolved Solids = 35,000 mg/L	$[sp_{FeII}]$, mg/L
0	$5.0(10^{+18})$
1	$5.0(10^{+16})$
2	$5.0(10^{+14})$
3	$5.0(10^{+12})$
4	$5.0(10^{+10})$
5	$5.0(10^{+8})$
6	$5.0(10^{+6})$
7	$5.0(10^{+4})$
8	$5.2(10^{+2})$
9	$7.2(10^{+0})$
10	$2.7(10^{-1})$
11	$2.4(10^{-2})$
12	$1.5(10^{-2})$
13	$1.3(10^{-1})$
14	$1.3(10^{+0})$
15	$1.3(10^{+1})$

TABLE 13.4
Concentration of sp_{FeIII} as a
Function of pH at 25°C

pH, Dissolved Solids = 140 mg/L	$[sp_{FeII}]$, mg/L
0	$1.2(10^{+10})$
1	$8.6(10^{+5})$
2	$1.3(10^{+3})$
3	$5.3(10^{+0})$
4	$5.5(10^{-2})$
5	$1.4(10^{-3})$
6	$1.1(10^{-4})$
7	$1.0(10^{-5})$
8	$1.6(10^{-6})$
9	$6.0(10^{-6})$
10	$6.0(10^{-5})$
11	$6.0(10^{-4})$
12	$6.0(10^{-3})$
13	$6.0(10^{-2})$
14	$6.0(10^{-1})$
15	$6.0(10^{+0})$

TABLE 13.5
Concentration of sp_{FeIII} as a
Function of pH at 25°C

pH, Dissolved Solids = 35,000 mg/L	$[sp_{FeIII}]$, mg/L
0	$1.2(10^{+10})$
1	$1.2(10^{+7})$
2	$1.3(10^{+4})$
3	$2.3(10^{+1})$
4	$1.2(10^{-1})$
5	$1.1(10^{-3})$
6	$1.1(10^{-5})$
7	$2.0(10^{-7})$
8	$9.4(10^{-7})$
9	$9.4(10^{-6})$
10	$9.4(10^{-5})$
11	$9.4(10^{-4})$
12	$9.4(10^{-3})$
13	$9.4(10^{-2})$
14	$9.4(10^{-1})$
15	$9.4(10^{+0})$

The previous optimum values of pH apply only at 25°C. As indicated in the formulas, equilibrium constants are being used to compute these values. The values of the equilibrium constants vary with temperature, so the optimum range at other temperatures would be different. Equilibrium constants at other temperatures can be calculated using the *Van't Hoff equation* which, however, requires the values of the standard enthalpy changes. At present, these values are unavailable making values of optimum pH range at other temperatures impossible to calculate. For this reason, the pH range found above must be modified to a conservative range. Hence, adopt the following: for ferrous removal as $Fe(OH)_2$, $11.5 \leq$ optimum pH ≤ 12.5 and for ferrous removal as $Fe(OH)_3$, $5.5 \leq$ optimum pH ≤ 12.5.

13.4 CHEMICAL REACTIONS OF THE MANGANOUS ION [Mn(II)]

Manganese can be removed as $Mn(OH)_2$ using a suitable OH^- source. Upon introduction of the hydroxide source, however, it is not only this solid that is produced. Manganese forms complex ions with the hydroxide. The complex equilibrium reactions are as follows (Snoeyink and Jenkins, 1980):

$$Mn(OH)^+ \rightleftharpoons Mn^{2+} + OH^- \qquad K_{MnOHc} = 10^{-3.4} \qquad (13.12)$$

$$Mn(OH)_{2(s)} \rightleftharpoons Mn^{2+} + 2OH^- \qquad K_{sp,Mn(OH)_2} = 4.5(10^{-14}) \qquad (13.13)$$

$$Mn(OH)_2^0 \rightleftharpoons Mn^{2+} + 2OH^- \qquad K_{Mn(OH)_2c} = 10^{-6.8} \qquad (13.14)$$

$$Mn(OH)_3^- \rightleftharpoons Mn^{2+} + 3OH^- \qquad K_{Mn(OH)_3c} = 10^{-7.8} \qquad (13.15)$$

The values of the equilibrium constants given above are at 25°C. The complexes are $Mn(OH)^+$, $Mn(OH)_2^0$, and $Mn(OH)_3^-$. Also note that the OH^- ion is a participant in these reactions. This means that the concentrations of each of these complex ions are determined by the pH of the solution. In the application of the previous equations in an actual treatment of water, conditions must be adjusted to allow maximum precipitation of the solid represented by $Mn(OH)_{2(s)}$. To allow for this maximum precipitation, the concentrations of the complex ions must be held to the minimum. The pH corresponding to this condition is the optimum pH. From the previous reactions, the equivalent mass of Mn^{2+} is $Mn/2 = 27.45$.

13.4.1 DETERMINATION OF THE OPTIMUM pH

Let sp_{Mn} represent the collection of species standing in solution containing the Mn(II) species. Thus,

$$[sp_{Mn}] = [Mn^{2+}] + [Mn(OH)^+] + [Mn(OH)_2^0] + [Mn(OH)_3^-] \qquad (13.16)$$

All the concentrations in the right-hand side of the above equation will now be expressed in terms of the hydrogen ion concentration. This will result in expressing

$[sp_{Mn}]$ in terms of the hydrogen ion. Differentiating the resulting equation of $[sp_{Mn}]$ with respect to $[H^+]$ and equating the result to zero will produce the minimum concentration of sp_{Mn} and, thus, the optimum pH determined.

Using the equations and equilibrium constants of Eqs. (13.12) through (13.15), along with the ion product of water, we proceed as follows:

$$[Mn^{2+}] = \frac{\{Mn^{2+}\}}{\gamma_{Mn}} = \frac{K_{sp,\,Mn(OH)_2}}{\gamma_{Mn}\{OH^-\}^2} = \frac{K_{sp,\,Mn(OH)_2}\{H^+\}^2}{\gamma_{Mn}K_w^2} = \frac{K_{sp,\,Mn(OH)_2}\gamma_H^2[H^+]^2}{\gamma_{Mn}K_w^2}$$

(13.17)

$$[Mn(OH)^+] = \frac{\{Mn(OH)^+\}}{\gamma_{Mn(OH)c}} = \frac{\{Mn^{2+}\}\{OH^-\}}{\gamma_{Mn(OH)c}K_{MnOHc}} = \frac{K_{sp,\,Mn(OH)_2}\gamma_H[H^+]}{\gamma_{Mn(OH)c}K_{MnOHc}K_w}$$

(13.18)

$$[Mn(OH)_2^0] = \frac{\{Mn(OH)_2^0\}}{\gamma_{Mn(OH)_2 c=1}} = \{Mn(OH)_2^0\} = \frac{\{Mn^{2+}\}\{OH^-\}^2}{K_{Mn(OH)_2 c}} = \frac{K_{sp,\,Mn(OH)_2}}{K_{Mn(OH)_2 c}}$$

(13.19)

$$[Mn(OH)_3^-] = \frac{\{Mn(OH)_3^-\}}{\gamma_{Mn(OH)_3 c}} = \frac{\{Mn^{2+}\}\{OH^-\}^3}{\gamma_{Mn(OH)_3 c}K_{Mn(OH)_3 c}} = \frac{K_{sp,\,Mn(OH)_2}K_w}{\gamma_{Mn(OH)_3 c}K_{Mn(OH)_3 c}\gamma_H[H^+]}$$

(13.20)

γ_{Mn}, γ_{MnOHc}, $\gamma_{Mn(OH)2c}$, and $\gamma_{Mn(OH)3c}$ are, respectively, the activity coefficients of Mn(II) and the complexes $Mn(OH)^+$, $Mn(OH)_2^0$, and $Mn(OH)_3^-$. $K_{sp,Mn(OH)_2}$ is the solubility product constant of the solid $Mn(OH)_{2(s)}$. K_{MnOHc}, $K_{Mn(OH)_2 c}$, and $K_{Mn(OH)_3 c}$ are, respectively, the equilibrium constants of the complexes $Mn(OH)^+$, $Mn(OH)_2^0$, and $Mn(OH)_3^-$.

Equations (13.17) through (13.20) may now be substituted into Equation (13.16) to produce

$$[sp_{Mn}] = \frac{K_{sp,\,Mn(OH)_2}\gamma_H^2[H^+]^2}{\gamma_{Mn}K_w^2} + \frac{K_{sp,\,Mn(OH)_2}\gamma_H[H^+]}{\gamma_{Mn(OH)c}K_{MnOHc}K_w} + \frac{K_{sp,\,Mn(OH)_2}}{K_{Mn(OH)_2 c}}$$

$$+ \frac{K_{sp,\,Mn(OH)_2}K_w}{\gamma_{Mn(OH)_3 c}K_{Mn(OH)_3 c}\gamma_H\{H^+\}}$$

(13.21)

Differentiating with respect to $[H^+]$, equating to zero, rearranging, and changing H^+ to H_{opt}^+, the concentration of the hydrogen ion at optimum conditions, the following expression is produced:

$$\frac{2\gamma_H^2[H_{opt}^+]^3}{\gamma_{Mn}K_w^2} + \frac{\gamma_H[H_{opt}^+]^2}{\gamma_{Mn(OH)c}K_{MnOHc}K_w} = \frac{K_w}{\gamma_{Mn(OH)_3 c}K_{Mn(OH)_3 c}\gamma_H}$$

(13.22)

The value of $[H_{opt}]$ may be solved by trial error.

Example 13.2 Calculate the optimum pH for precipitating $Mn(OH)_2$. Assume the water contains 140 mg/L of dissolved solids.

Solution:

$$\frac{2\gamma_H^2[H_{opt}^+]^3}{\gamma_{Mn}K_w^2} + \frac{\gamma_H[H_{opt}^+]^2}{\gamma_{Mn(OH)c}K_{MnOHc}K_w} = \frac{K_w}{\gamma_{Mn(OH)_3c}K_{Mn(OH)_3c}\gamma_H}$$

$$\mu = 2.5(10^{-5})(140) = 3.5(10^{-3})$$

$$\gamma_H = \gamma_{Mn(OH)c} = \gamma_{Mn(OH)_3c} = 10^{-\frac{0.5(1)^2[\sqrt{3.5(10^{-3})}]}{1+1.14[\sqrt{3.5(10^{-3})}]}} = 0.94$$

$$\gamma_{Mn} = 10^{-\frac{0.5(2)^2[\sqrt{3.5(10^{-3})}]}{1+1.14[\sqrt{3.5(10^{-3})}]}} = 0.77 \quad K_w = (10^{-14}) \quad K_{MnOHc} = 10^{-3.4}$$

$$K_{Mn(OH)_3c} = 10^{-7.8}$$

Therefore,

$$\frac{2(0.94)^2[H_{opt}^+]^3}{0.77(10^{-14})^2} + \frac{0.94[H_{opt}^+]^2}{(0.94)(10^{-3.4})(10^{-14})} = \frac{10^{-14}}{0.94(10^{-7.8})(0.94)}$$

$$2.30(10^{28})[H_{opt}^+]^3 + 2.51(10^{17})[H_{opt}^+]^2 = 7.14(10^{-7})$$

Solving by trial and error, let $Y = 2.30(10^{28})[H_{opt}^+]^3 + 2.51(10^{17})[H_{opt}^+]^2$

$[H_{opt}^+]$	$2.30(10^{28})[H_{opt}^+]^3$	$2.51(10^{17})[H_{opt}^+]^2$	Y
10^{-11}	$2.3(10^{-5})$	$2.51(10^{-5})$	$4.81(10^{-5})$
10^{-12}	$2.3(10^{-8})$	$2.51(10^{-7})$	$2.74(10^{-7})$

10^{-11} $\quad 4.81(10^{-5})$

$$\frac{y-10^{-11}}{10^{-11}-10^{-12}} = \frac{7.14(10^{-7})-4.81(10^{-5})}{4.81(10^{-5})-2.74(10^{-7})}$$

$y \quad 7.14(10^{-7})$ $\qquad y = 1.08(10^{-12})$ gmol/L $= [H_{opt}^+]$

$10^{-12} \quad 2.74(10^{-7})$ $\qquad$ Therefore, pH $= -\log_{10}1.08(10^{-12}) = 11.97$ **Ans**

Example 13.3 From the optimum pH's of 11.97, calculate the concentrations sp_{Mn} in mg/L. Assume the water contains 140 mg/L of dissolved solids.

Solution:

$$[sp_{Mn}] = \frac{K_{sp,Mn(OH)_2}\gamma_H^2[H^+]^2}{\gamma_{Mn}K_w^2} + \frac{K_{sp,Mn(OH)_2}\gamma_H[H^+]}{\gamma_{Mn(OH)c}K_{MnOHc}K_w} + \frac{K_{sp,Mn(OH)_2}}{K_{Mn(OH)_2c}}$$

$$+ \frac{K_{sp,Mn(OH)_2}K_w}{\gamma_{Mn(OH)_3c}K_{Mn(OH)_3c}\gamma_H[H^+]}$$

$$K_{sp,\,Mn(OH)_2} = 4.5(10^{-14}) \qquad \gamma_H = \gamma_{Mn(OH)c} = \gamma_{Mn(OH)_3c} = 0.94 \qquad \gamma_{Mn} = 0.77$$

$$K_{MnOHc} = 10^{-3.4}$$

$$K_{Mn(OH)_2c} = 10^{-6.8} \qquad K_{Mn(OH)_3c} = 10^{-7.8}$$

Therefore,

$$[sp_{Mn}] = \frac{(4.5)(10^{-14})(0.94)^2[10^{-11.97}]^2}{(0.77)(10^{-14})^2} + \frac{(4.5)(10^{-14})(0.94)[10^{-11.97}]}{(0.94)(10^{-3.4})(10^{-14})}$$

$$+ \frac{(4.5)(10^{-14})}{10^{-6.8}} + \frac{(4.5)(10^{-14})(10^{-14})}{(0.94)(10^{-7.8})(0.94)[10^{-11.97}]}$$

$$= \frac{4.57(10^{-38})}{7.7(10^{-29})} + \frac{4.53(10^{-26})}{3.74(10^{-18})} + 2.84(10^{-7}) + \frac{4.5(10^{-28})}{1.5(10^{-20})}$$

$$= 3.27(10^{-7}) \text{ geq/L} = 0.0179 \text{ mg/L} \quad \textbf{Ans}$$

13.4.2 PRACTICAL OPTIMUM pH RANGE
FOR THE REMOVAL OF MANGANESE

From Example 13.2, at 25°C and at a solids concentration of 140 mg/L, the optimum pH for the removal of manganese is 11.97. The corresponding concentration for sp_{Mn} is 0.0179 mg/L. As in the case for the removal of ion, there is a practical pH range at which units used for the removal of manganese can be operated and effect good results. Tables 13.6 and 13.7 show the respective concentrations of sp_{Mn} at other conditions of pH and total solids. The values for $[sp_{Mn}]$ were obtained using Equation (13.21).

Again, as in the case of the removal of iron, the tables show that total solids (or equivalently, the activities of the ions) do not have a significant effect on the optimum pH value for the removal of manganese as $Mn(OH)_2$. The optimum pH remains at around 12.0. For the practical optimum pH, we adopt the following:

manganese removal as $Mn(OH)_2$: $11.5 \leq$ optimum pH ≤ 12.5,

which is the same as that for ferrous removed as $Fe(OH)_2$.

13.5 OXIDATION OF IRON AND MANGANESE TO REDUCE PRECIPITATION pH

To summarize, the optimum pH's of precipitation are as follows: ferrous removed as $Fe(OH)_2$, $11.5 \leq$ optimum pH ≤ 12.5; ferrous removed as $Fe(OH)_3$, $5.5 \leq$ optimum pH ≤ 12.5; and Mn removed as $Mn(OH)_2$, $11.5 \leq$ optimum pH ≤ 12.5. The reason for the high pH values for the removal of iron as $Fe(OH)_2$ and for the removal of manganese as $Mn(OH)_2$ is the formation of the complex ions. These ions are $FeOH^+$ and $Fe(OH)_3^-$ for the ferrous ion and $Mn(OH)^+$, $Mn(OH)^0$, and $Mn(OH)_3^-$ for the

TABLE 13.6
Concentration of sp_{Mn} as a
Function of pH at 25°C

pH, Dissolved Solids = 140 mg/L	$[sp_{Mn}]$, mg/L
0	$1.4(10^{+19})$
1	$1.4(10^{+17})$
2	$1.4(10^{+15})$
3	$1.4(10^{+13})$
4	$1.4(10^{+11})$
5	$1.4(10^{+9})$
6	$1.4(10^{+7})$
7	$1.4(10^{+5})$
8	$1.4(10^{+3})$
9	$1.5(10^{+1})$
10	$1.8(10^{-1})$
11	$1.8(10^{-2})$
12	$9.0(10^{-3})$
13	$1.7(10^{-2})$
14	$9.5(10^{-1})$
15	$9.0(10^{-1})$

TABLE 13.7
Concentration of sp_{Mn} as a
Function of pH at 25°C

pH, Dissolved Solids = 35,000 mg/L	$[sp_{Mn}]$, mg/L
0	$3.5(10^{+19})$
1	$3.5(10^{+17})$
2	$3.5(10^{+15})$
3	$3.5(10^{+13})$
4	$3.5(10^{+11})$
5	$3.5(10^{+9})$
6	$3.5(10^{+7})$
7	$3.5(10^{+5})$
8	$3.5(10^{+3})$
9	$3.5(10^{+1})$
10	$3.9(10^{-1})$
11	$1.5(10^{-2})$
12	$1.1(10^{-2})$
13	$3.1(10^{-2})$
14	$2.3(10^{-1})$
15	$2.2(10^{+0})$

manganous ion. The values of the equilibrium constants of these complex ions are such that their concentrations diminish the hydroxide ions needed to precipitate the solid hydroxides $Fe(OH)_2$ and $Mn(OH)_2$. For these solids to precipitate out, more hydroxide ions must be added, resulting in a high pH. If these complex ions were destroyed, then the metal ions released would precipitate as the respective hydroxides, preventing the complex equilibria to occur.

To destroy the complex ions requires the use of oxidants. Chlorine, potassium permanganate, and ozone are normally used for this purpose. In the case of iron, the ferrous state is oxidized to the ferric state. This oxidation includes the oxidation of the ferrous complexes. Because $Fe(OH)_3$ precipitates at a wider optimum pH range, the precipitation pH can therefore be reduced to a much lower value, i.e., to even a pH of 5.5.

Let us tackle the situation of precipitating manganese at a lower pH range. Once the complex ions have been destroyed, the reactions left would be those for a precipitation product of a higher oxidation state than 2+. The reactions for the destruction using chlorine, potassium permanganate, and ozone are, respectively, as follows:

$$Mn^{2+} + Cl_2 + 2H_2O \rightarrow MnO_{2(s)} \downarrow + 2Cl^- + 4H^+ \qquad (13.23)$$

$$3Mn^{2+} + 2KMnO_4 + 2H_2O \rightarrow 5MnO_{2(s)} \downarrow + 2K^+ + 4H^+ \qquad (13.24)$$

$$3Mn^{2+} + O_3 + 3H_2O \rightarrow 3MnO_2(s) \downarrow + 6H^+ \qquad (13.25)$$

The previous reactions show that manganese is oxidized from Mn(II) to Mn(IV). This oxidation poses the possibility of Mn(IV) forming a complex with the hydroxides. The review of the literature, however, did not uncover any evidence for this to be so. It did reveal that Mn(IV) forms a complex with fluorine and potassium (Holtzclaw and Robinson, 1988). This complex is $K_2[MnF_6]$. The review also did not reveal any solubility product constant for MnO_2. It could very well be that it does not have any. If, in fact, MnO_2 does not have any solubility product constant, then the previous reactions can be construed to be possible at any pH. Until a study is done to show the complex formation of Mn(IV) and the accompanying solubility constant and to be compatible with the removal of ferrous as $Fe(OH)_3$, we will adopt the practical pH for removal of manganese through its oxidation by the above reactions as $5.5 \leq$ optimum pH ≤ 12.5. This range is, however, arbitrary and applies when the complexes have been destroyed. For more accurate results, a pilot plant investigation for a given raw water should be undertaken.

13.6 UNIT OPERATIONS FOR IRON AND MANGANESE REMOVAL

The unit operations involved can be divided into two general categories: removal at the pH range of 11.5 to 12.5 and removal at the pH range of 5.5 to 12.5. The former range is called the *high pH range* and the latter is called the *low pH range*. The latter is arbitrarily categorized as low, because it is possible to adjust the operation at a low *pH* value. In these ranges, both iron and manganese can be removed at the same time and at the same unit. The differences and similarities of these categories of operations will be explored.

13.6.1 HIGH pH RANGE

Under this category, iron is removed as $Fe(OH)_2$ and manganese is removed as $Mn(OH)_2$. The unit operations will then involve a unit (tank) for containing the addition of the hydroxide source (lime) to raise the pH and for mixing the chemicals and allowing completion of the precipitation of $Fe(OH)_2$ and $Mn(OH)_2$. The precipitates formed will still be in suspension; thus, it is necessary for them to flocculate by adding a flocculation tank. After the flocculation tank will be a settling tank. The settling tanks removes the flocculated solids. A filter then follows to filter out any solids that were not removed by the settling tank.

The sequence of unit operations discussed above may not all be needed in a given application. What will actually be used depends upon the concentrations of iron and manganese in the raw water. If the concentrations are small such as less than 10 mg/L, the resulting floc concentration will be small; the chemically treated water, in this instance, may simply be introduced directly into the filter, without using any flocculation or settling basin. In all cases, however, the mixing chamber for the reactions to take place and the filter are always necessary. The unit operations of mixing, flocculation, settling, and filtration were already discussed in the earlier chapters of this book.

As can be deduced from the addition of lime, the unit operations above are the same as those that would be applied in water softening. Thus, for reasons of economics, iron and manganese removal are normally incorporated into water softening and all the unit operations involved are identical. In fact, iron and manganese are hardness ions.

13.6.2 LOW pH RANGE

Under this category, iron is removed as $Fe(OH)_3$ and manganese is removed as MnO_2, which involves oxidation of the metal ions to their higher oxidation states. The unit operations under the previous category are applicable under the present category, only that under the present scheme, oxidation of the metals is involved. The unit process involves one of Eqs. (13.23) through (13.25) for the case of manganese and similar equations for iron. The equation designed into the unit operations depends upon the unit process contemplated. The unit process part, however, will only be at the mixing tank where the reactions should occur.

As in the previous category, the unit operations of flocculation and settling may not be necessary. Again, in all cases, the mixing chamber for the reaction to take place and filtration is always necessary. The reaction does not immediately produce particles capable of being filtered, therefore, sufficient detention time should be provided in the reaction chamber to allow the particles to grow into filtrable sizes. This normally ranges from 20 to 30 minutes. In any case, a pilot plant may be necessary to determine the exact detention time.

When oxidizing iron and manganese using dissolved oxygen, the process is usually carried out under catalytic reaction on some contact surfaces. To accomplish the reaction, the water is generally made to trickle over small rock surfaces such as limestone, coke, or pyrolusite (MnO_2). Pyrolusite possesses high catalytic power, springing from the iron and manganese oxides precipitated from the raw water that coat over and around the rocks. These coatings act as catalysts between the reaction of the ferrous and manganous ions in the water and the oxygen from the air as they come into intimate contact over the contact surfaces of the rocks.

The contact medium sizes vary from 3.5 to 5 cm. The accumulated flocs produced from the deposition of the hydroxides or oxides are periodically flushed out by rapid drainage. This is done by filling the tank containing the bed of rocks to capacity and quickly releasing the water.

13.7 CHEMICAL REQUIREMENTS

The chemical requirements are those for the hydroxide source, chlorine, permanganate, ozone, and oxygen. The discussion will be subdivided into requirements in the ferrous reactions and into requirements in the manganous reactions. The treatment on chemical requirements, in effect, is reduced to the determination of the equivalent masses of the pertinent chemicals.

13.7.1 REQUIREMENTS IN THE FERROUS REACTIONS

Under these requirements are the reactions at the high and low pH ranges. At the high range, the hydroxide reaction is

$$Fe^{2+} + 2OH^- \rightarrow Fe(OH)_{2(s)} \downarrow \qquad (13.26)$$

The equivalent mass of Fe^{2+} in the previous reaction is Fe/2 = 27.9 and that of the hydroxide ion is $2OH^-/2 = 17$.

The possible hydroxide sources are caustic and lime. To make these sources jointly equivalent to Equation (13.26), they must all be made equivalent to $2OH^-$. Thus,

$$2NaOH \Rightarrow 2OH^- \qquad (13.27)$$

$$Ca(OH)_2 \Rightarrow 2OH^- \qquad (13.28)$$

From the previous equivalence, the equivalent of mass of NaOH is 2NaOH/2 = 40 and the equivalent mass of CaO is CaO/2 = 28.05.

Chlorine, permanganate, and ozone, will only be used at the low pH range. There is no need for them to be used at the high pH range, because under these conditions, the ferrous ion will be forced to precipitate as the ferrous hydroxide. It is true that after precipitating, chlorine, permanganate, or ozone could then be used to oxidize ferrous hydroxide to the ferric state, but no practical reason exists for designing the unit process this way. Thus, the unit process design will be to oxidize the ferrous ion at the low pH range, converting it directly to the ferric hydroxide solid. Including dissolved oxygen, the reactions are, respectively, as follows:

$$Fe^{2+} + \frac{1}{2}Cl_2 + 3H_2O \rightarrow Fe(OH)_{3(s)} \downarrow + Cl^- + 3H^+ \tag{13.29}$$

$$Fe^{2+} + \frac{1}{3}KMnO_4 + \frac{7}{3}H_2O \rightarrow Fe(OH)_{3(s)} \downarrow + \frac{1}{3}MnO_2 \downarrow + \frac{1}{3}K^+ + \frac{5}{3}H^+ \tag{13.30}$$

$$Fe^{2+} + \frac{1}{6}O_3 + \frac{15}{6}H_2O \rightarrow Fe(OH)_{3(s)} \downarrow + 2H^+ \tag{13.31}$$

$$Fe^{2+} + \frac{5}{2}H_2O + \frac{1}{4}O_2 \rightarrow Fe(OH)_{3(s)} \downarrow + 2H^+ \tag{13.32}$$

The previous reactions are oxidation-reduction reactions, so it is appropriate to use as the reference species the number of electron moles involved, although the choice of the reference species is completely arbitrary. One atom of Fe^{2+} is being oxidized to Fe^{3+}, therefore, one electron mole is involved (thus, the number of reference species = 1) and the equivalent masses are ferrous = Fe/1 = 55.8, chlorine = $\frac{1}{2}Cl_2/1 = 35.5$, potassium permanganate = $\frac{1}{3}$ KMnO$_4$/1 = 52.67, ozone = $\frac{1}{6}O_3/1 = 8.0$, and oxygen $\frac{1}{4}O_2/1 = 8.0$.

Occasionally, the hypochlorites are used as the chlorinating–oxidizing agent. To determine the equivalent masses of the hypochlorites, they must be made to interact with electrons. On the basis of one electron mole,

$$\frac{1}{2}OCl^- + e^- + H^+ \rightarrow \frac{1}{2}Cl^- + \frac{1}{2}H_2O \tag{13.33}$$

From this reaction, the equivalent mass of sodium hypochlorite is $\frac{1}{2}$ NaOCl/1 = 37.25 and that of calcium hypochlorite, it is $\frac{1}{4}$ Ca(OCl)$_2$/1 = 35.78.

13.7.2 REQUIREMENTS IN THE MANGANOUS REACTIONS

Under these requirements are also the reactions at the high and low pH ranges. At the high range, the hydroxide reaction is

$$Mn^{2+} + 2OH^- \rightarrow Mn(OH)_{2(s)} \downarrow$$

The equivalent mass of Mn^{2+} in the previous reaction is Mn/2 = 27.45 and that of the hydroxide ion is 2OH$^-$/2 = 17. If the reactions for the caustic and lime are written, the equivalent masses will be found to be the same as those in the case of ferrous, which are 40 and 28.05, respectively.

For the same reasons as in the case of ferrous, chlorine, permanganate, and ozone will only be used at the low pH range. The unit process design is to oxidize

Mn^{2+} at the low pH range converting it directly to MnO_2 as shown in Eqs. (13.23) through (13.25). Converting these equations in terms of one atom of Mn^{2+}, along with the reaction of oxygen, the following equations are produced:

$$Mn^{2+} + Cl_2 + 2H_2O \rightarrow MnO_{2(s)} \downarrow + 2Cl^- + 4H^+ \tag{13.34}$$

$$Mn^{2+} + \frac{2}{3}KMnO_4 + \frac{2}{3}H_2O \rightarrow \frac{5}{3}MnO_{2(s)} \downarrow + \frac{2}{3}K^+ + \frac{4}{3}H^+ \tag{13.35}$$

$$Mn^{2+} + \frac{1}{3}O_3 + H_2O \rightarrow MnO_{2(s)} \downarrow + 2H^+ \tag{13.36}$$

$$Mn^{2+} + \frac{1}{2}O_2 + H_2O \rightarrow MnO_{2(s)} \downarrow + 2H^+ \tag{13.37}$$

In each of the previous reactions, the manganese atom is oxidized from 2+ to 4+, involving 2 electron moles. Thus, the equivalent masses are manganese $Mn/2 = 27.45$, chlorine $Cl_2/2 = 35.5$, potassium permanganate $= \frac{2}{3}KMnO_4/2 = 52.67$, ozone $= \frac{1}{3}O_3/2 = 8$, and oxygen $= \frac{1}{2}O_2/2 = 8$. The equivalent masses of sodium hypochlorite and calcium hypochlorite are the same as those in the case of the ferrous, which are 37.25 and 35.78, respectively.

Tables 13.8 and 13.9 summarize the equivalent masses.

TABLE 13.8
Summary of Equivalent Masses, Ferrous Reactions

	Fe^{2+}	OH^-	CaO	NaOH	Cl_2	NaClO	$Ca(ClO)_2$	$KMnO_4$	O_3	O_2
High pH range	27.9	17.0	28.05	40.0	—	—	—	—	—	—
Low pH range	55.8	—	—	—	35.5	37.25	35.78	52.67	8.0	8.0

TABLE 13.9
Summary of Equivalent Masses, Manganous Reactions

	Mn^{2+}	OH^-	CaO	NaOH	Cl_2	NaClO	$Ca(ClO)_2$	$KMnO_4$	O_3	O_2
High pH range	27.45	17.0	28.05	40.0	—	—	—	—	—	—
Low pH range	27.45	—	—	—	35.5	37.25	35.78	52.67	8.0	8.0

These tables show that, except for Fe^{2+} and Mn^{2+}, the equivalent masses are the same in both the ferrous and manganese reactions. Now, let

M_{CaOFe} = kilograms of lime used for the ferrous reaction
M_{NaOHFe} = kilograms of caustic used for the ferrous reaction
M_{Cl_2Fe} = kilograms of chlorine used for the ferrous reaction
$M_{NaClOFe}$ = kilograms of sodium hypochlorite used in the ferrous reaction
$M_{Ca(ClO)_2Fe}$ = kilograms of calcium hypochlorite used in the ferrous reaction
M_{KMnO_4Fe} = kilograms of potassium permanganate used in the ferrous reaction
M_{O_3Fe} = kilograms of ozone used in the ferrous reaction
M_{O_2Fe} = kilograms of dissolved oxygen used in the ferrous reaction
M_{CaOMn} = kilograms of lime used for the manganous reaction
M_{NaOHMn} = kilograms of caustic used for the manganous reaction
M_{Cl_2Mn} = kilograms of chlorine used for the manganous reaction
$M_{NaClOMn}$ = kilograms of sodium hypochlorite used in the manganous reaction
$M_{Ca(ClO)_2Mn}$ = kilograms of calcium hypochlorite used in the manganous reaction
M_{KMnO_4Mn} = kilograms of potassium permanganate used in the manganous reaction
M_{O_3Mn} = kilograms of ozone used in the manganous reaction
M_{O_2Mn} = kilograms of dissolved oxygen used in the manganous reaction
$\Delta[Fe]_{mg}$ = mg/L of ferrous to be removed
$\Delta[Mn]_{mg}$ = mg/L of manganous removed
V = cubic meters of water treated

Therefore,

$$M_{CaOFe} = \frac{28.05\Delta[Fe]_{mg}V}{1000(27.9)} = 0.0010\Delta[Fe]_{mg}V \qquad (13.38)$$

$$M_{NaOHFe} = \frac{40\Delta[Fe]_{mg}V}{1000(27.9)} = 0.0014\Delta[Fe]_{mg}V \qquad (13.39)$$

$$M_{Cl_2Fe} = \frac{35.5\Delta[Fe]_{mg}V}{1000(55.8)} = 0.00064\Delta[Fe]_{mg}V \qquad (13.40)$$

$$M_{NaClOFe} = \frac{37.25\Delta[Fe]_{mg}V}{1000(55.8)} = 0.00067\Delta[Fe]_{mg}V \qquad (13.41)$$

$$M_{Ca(ClO)_2Fe} = \frac{35.78\Delta[Fe]_{mg}V}{1000(55.8)} = 0.00064\Delta[Fe]_{mg}V \qquad (13.42)$$

$$M_{KMnO_4Fe} = \frac{52.67\Delta[Fe]_{mg}V}{1000(55.8)} = 0.00094\Delta[Fe]_{mg}V \qquad (13.43)$$

$$M_{O_3Fe} = \frac{8\Delta[Fe]_{mg}V}{1000(55.8)} = 0.00014\Delta[Fe]_{mg}V \qquad (13.44)$$

$$M_{O_2Fe} = \frac{8\Delta[Fe]_{mg}\cancel{V}}{1000(55.8)} = 0.00014\Delta[Fe]_{mg}\cancel{V} \qquad (13.45)$$

$$M_{CaOMn} = \frac{28.05\Delta[Mn]_{mg}\cancel{V}}{1000(27.45)} = 0.0010\Delta[Mn]_{mg}\cancel{V} \qquad (13.46)$$

$$M_{NaOHMn} = \frac{40.0\Delta[Mn]_{mg}\cancel{V}}{1000(27.45)} = 0.0015\Delta[Mn]_{mg}\cancel{V} \qquad (13.47)$$

$$M_{Cl_2Mn} = \frac{35.5\Delta[Mn]_{mg}\cancel{V}}{1000(27.45)} = 0.0013\Delta[Mn]_{mg}\cancel{V} \qquad (13.48)$$

$$M_{NaClOMn} = \frac{37.25\Delta[Mn]_{mg}\cancel{V}}{1000(27.45)} = 0.0014\Delta[Mn]_{mg}\cancel{V} \qquad (13.49)$$

$$M_{Ca(ClO)_2Mn} = \frac{35.78\Delta[Mn]_{mg}\cancel{V}}{1000(27.45)} = 0.0013\Delta[Mn]_{mg}\cancel{V} \qquad (13.50)$$

$$M_{KMnO_4Mn} = \frac{52.67\Delta[Mn]_{mg}\cancel{V}}{1000(27.45)} = 0.0019\Delta[Mn]_{mg}\cancel{V} \qquad (13.51)$$

$$M_{O_3Mn} = \frac{8.0\Delta[Mn]_{mg}\cancel{V}}{1000(27.45)} = 0.00029\Delta[Mn]_{mg}\cancel{V} \qquad (13.52)$$

$$M_{O_2Mn} = \frac{8.0\Delta[Mn]_{mg}\cancel{V}}{1000(27.45)} = 0.00029\Delta[Mn]_{mg}\cancel{V} \qquad (13.53)$$

Example 13.4   A raw water contains 2.5 mg/L Mn. Calculate the kilograms of lime per cubic meter needed to meet the limit concentration of 0.05 mg/L, if removal is to be done at the high pH range.

Solution:

$$M_{CaOMn} = 0.0010\Delta[Mn]_{mg}\cancel{V} = 0.0010(2.5 - 0.05)(1) = 2.45(10^{-3}) \text{ kg/m}^3 \text{ Ans}$$

Example 13.5   A raw water contains 40 mg/L iron. Calculate the kilograms of lime per cubic meter needed to meet the limit concentration of 0.3 mg/L, if removal is to be done at the high pH range.

Solution:

$$M_{CaOFe} = 0.0010\Delta[Fe]_{mg}\cancel{V} = 0.0010(40 - 0.3)(1) = 0.040 \text{ kg/m}^3 \quad \textbf{Ans}$$

13.8 ALKALINITY EXPRESSED IN OH⁻ AND ACIDITY EXPRESSED IN H⁺

Knowledge of the equivalent mass is needed in the succeeding discussions on the chemical requirements for pH adjustments, which entails adding acids or bases. As acids are added to water and wastewater to lower the pH, they need to react with existing alkalinities first before the pH changes. For the purpose of determining the equivalent masses of the acids as a result of the reaction due to this addition, the constituent species of the alkalinity need to be expressed in terms of a single unifying species. Once this unifying species has been established, the process of determining the equivalent mass becomes simpler. This particular species happens to be the hydroxide ion.

Similar statements also hold with respect to bases added to raise the pH. They need to react first with existing acidities before the pH can increase, and their equivalent masses also need to be determined by referring to a single unifying species. This single unifying species is the hydrogen ion. We will discuss the hydroxide ion first.

In the carbonate system, the total alkalinity species has been identified as composed of the constituent species carbonate ion, bicarbonate ion, and hydroxide ion. To express all these constituent species in terms of a single parameter, *total alkalinity*, they need to be expressed in terms of the number of equivalents. Thus, if the concentrations are in terms of equivalents per liter, they can all be added to sum up to the total alkalinity, of course, also in terms of equivalents per liter.

All the concentrations are expressed in terms of equivalents, so any one species may represent the total alkalinity. For example, the total alkalinity may be expressed in terms of equivalents of the carbonate ion, or in terms of equivalents of the bicarbonate ion, or in terms of the equivalents of the hydroxide ion. The number of equivalents of the hydroxide ion is equal to its number of moles, so it is convenient to expresses total alkalinity in terms of the equivalent hydroxide ion, and the reaction of total alkalinity may then be simply represented by the reaction of the hydroxide ion OH^-.

Without going into a detailed discussion, the total acidity may also be expressed in terms of the equivalent hydrogen ion. Thus, the reaction of total acidity may be represented by the H^+ ion.

13.9 CHEMICAL REQUIREMENTS FOR pH ADJUSTMENTS

The formulas for the kilograms of carbon dioxide, M_{CO_2pH}, needed to lower the pH and the kilograms of lime, M_{CaOpH}, needed to raise the pH were derived in Chapter 12. For convenience, they are reproduced below:

$$M_{CO_2pH} = 22\left\{[A_{cadd}]_{geq} = [A_{cur}]_{geq} + \frac{10^{-pH_{to}} - 10^{-pH_{cur}}}{\phi_a}\right\}V \qquad (13.54)$$

$$M_{CaOpH} = 28.05[A_{add}]_{geq}V = 28.05\left([A_{ccur}]_{geq} + \frac{10^{-pH_{cur}} - 10^{-pH_{to}}}{\phi_b}\right)V \qquad (13.55)$$

where,

$[A_{cadd}]$ = gram equivalents of acidity to be added
$[A_{add}]$ = gram equivalents of base to be added
$[A_{cur}]_{geq}$ = gram equivalents of current alkalinity
$[A_{ccur}]_{geq}$ = gram equivalents of current acicity
pH_{to} = the destination pH
pH_{cur} = current pH
V = cubic meters of water treated
ϕ_a = fractional dissociation of the hydrogen ion from the acid provided
ϕ_b = fractional dissociation of the hydroxide ion from the base provided

Let us derive the formulas for calculating the quantities of sulfuric acid, hydrochloric acid, and nitric acid and the formulas for calculating the quantities of caustic soda and soda ash that may be needed to lower and to raise the pH, respectively. To find the equivalent masses of the acids, they must be reacted with the hydroxyl ion. Reaction with this ion is necessary, since total alkalinity may be represented by the hydroxyl ion. Remember that the acids must first consume all the existing alkalinity represented in the overall by the OH⁻ before they can lower the pH. Thus, proceed as follows:

$$H_2SO_4 + 2OH^- \rightarrow 2H_2O + SO_4^{2-} \tag{13.56}$$

$$HCl + OH^- \rightarrow H_2O + Cl^- \tag{13.57}$$

$$HNO_3 + OH^- \rightarrow H_2O + NO_3^- \tag{13.58}$$

From these equations, the equivalent masses are sulfuric acid = $H_2SO_4/2$ = 49.05, hydrochloric acid = $HCl/1$ = 36.5, and nitric acid = $HNO_3/1$ = 63.01.

To find the equivalent masses of the caustic soda and soda ash, they must be reacted with an acid, which may be represented by the hydrogen ion. Again, remember that the bases specified previously must first consume all the existing acidity of the water represented in the overall by H^+ before they can raise the pH. Thus,

$$NaOH + H^+ \rightarrow H_2O + Na^+ \tag{13.59}$$

$$Na_2CO_3 + 2H^+ \rightarrow H_2CO_3 + 2Na^+ \tag{13.60}$$

From the equations, the equivalent masses are caustic soda = $NaOH/1$ = 40 and soda ash = $Na_2CO_3/2$ = 2 × 23 + 12 + 48/2 = 53.

Let $M_{H_2SO_4pH}$, M_{HClpH}, and M_{HNO_3pH} be the kilograms of sulfuric acid, hydrochloric acid, or nitric acid used to lower the pH from the current pH to pH_{to}. Gleaning from Equation (13.54) and the respective equivalent masses of H_2SO_4, HCl, and HNO_3 and the cubic meters, V, of water treated,

$$M_{H_2SO_4pH} = 49.05 \left\{ [A_{cadd}]_{geq} = [A_{cur}]_{geq} + \frac{10^{-pH_{to}} - 10^{-pH_{cur}}}{\phi_a} \right\} V \tag{13.61}$$

$$M_{HClpH} = 36.5 \left\{ [A_{cadd}]_{geq} = [A_{cur}]_{geq} + \frac{10^{-pH_{to}} - 10^{-pH_{cur}}}{\phi_a} \right\} V \qquad (13.62)$$

$$M_{HNO_3pH} = 63.01 \left\{ [A_{cadd}]_{geq} = [A_{cur}]_{geq} + \frac{10^{-pH_{to}} - 10^{-pH_{cur}}}{\phi_a} \right\} V \qquad (13.63)$$

And, also, gleaning from Equation (13.55) and the respective equivalent masses of NaOH and Na_2CO_3 and the cubic meters of water treated, V,

$$M_{NaOHpH} = 40.0 \left\{ [A_{ccur}]_{geq} + \frac{10^{-pH_{cur}} - 10^{-pH_{to}}}{\phi_b} \right\} V \qquad (13.64)$$

$$M_{Na_2CO_3pH} = 53.0 \left\{ [A_{ccur}]_{geq} + \frac{10^{-pH_{cur}} - 10^{-pH_{to}}}{\phi_b} \right\} V \qquad (13.65)$$

where M_{NaOHpH} and $M_{Na_2CO_3pH}$ are the kilograms of caustic soda or soda ash, respectively, used to raise the pH from the current pH to pH_{to}.

They are strong acids, therefore, the ϕ_a's for H_2SO_4, HCl, and HNO_3 are unity. The hydrogen ion resulting from the second ionization of sulfuric is very small so it can be neglected. Also, because NaOH is a strong base, its ϕ_b is equal to unity. The ϕ_b for Na_2CO_3 is not as straightforward, and we need to calculate it. Sodium carbonate ionizes completely, as follows:

$$Na_2CO_3 \rightarrow 2Na^+ + CO_3^{2-} \qquad (13.66)$$

The carbonate ion then proceeds to react with water to produce the hydroxide ion, as follows (Holtzclaw and Robinson, 1988):

$$CO_3^{2-} + HOH \rightleftharpoons HCO_3^- + OH^- \qquad K_b = 1.4(10^{-4}) \qquad (13.67)$$

Now, proceed to calculate the fractional dissociation of the hydroxide ion from sodium carbonate. Begin by assuming a concentration for the carbonate of 0.1 M. This will then produce, also, 0.1 M of the carbonate ion. By its subsequent reaction with water, however, its concentration at equilibrium will be smaller. Let x be the concentrations of HCO_3^- and OH^- at equilibrium. Then the concentration of CO_3^{2-} is $0.1 - x$ (at equilibrium). Thus,

$$K_b = 1.4(10^{-4}) = \frac{x^2}{0.1 - x}$$

$$x = 0.00367$$

TABLE 13.10
ϕ_a or ϕ_b for Acids and Bases Used in
Water and Wastewater Treatment

Acid or Base	ϕ_a or ϕ_b		
	0.1 M	0.5 M	1.0 M
H_2SO_4	1.0	1.0	1.0
HCl	1.0	1.0	1.0
HNO_3	1.0	1.0	1.0
H_2CO_3	0.00422	0.00094	0.000668
NaOH	1.0	1.0	1.0
Na_2CO_3	0.0367	0.0166	0.0118
$Ca(OH)_2$	0.125	0.025	0.0125

Therefore,

$$\phi_{b,\,Na_2CO_3} = \frac{0.00367}{0.1} = 0.0367$$

Note that, in the previous calculation, the activity coefficients have been ignored.

Lastly, find the ϕ_b for lime. The chemical reaction for lime after it has been slaked is

$$Ca(OH)_2 \rightleftharpoons Ca^{2+} + 2OH^- \qquad K_{sp,Ca(OH)_2} = 7.9(10^{-6}) \qquad (13.68)$$

Begin by assuming a concentration 0.1 M. Let x be the concentrations of Ca^{2+} at equilibrium; the corresponding concentration of OH^- will then be $2x$ at equilibrium. Thus,

$$K_{sp,Ca(OH)_2} = 7.9(10^{-6}) = x(2x)^2 = 4x^3$$

$$x = 0.0125$$

Therefore,

$$\phi_{b,\,Ca(OH)_2} = \frac{0.0125}{0.1} = 0.125$$

Table 13.10 shows some other values of ϕ_a and ϕ_b.

$$x = \frac{-1.4(10^{-4}) + \sqrt{[1.4(10^{-4})]^2 + 4(1.0)[1.4(10^{-4})]}}{2} = 0.0083$$

Example 13.6 A raw water containing 3 mg/L of manganese has a pH 4.0. To remove the manganese, the pH needs to be raised to 6.0. The current acidity is 30 mg/L as $CaCO_3$ $[Ca^{2+}] = 0.7$ mgmol/L. Assume the temperature is 25°C and ϕ_b of 0.0367. Calculate the amount of soda ash needed.

Solution:

$$M_{Na_2CO_3pH} = 53.0\left\{[A_{ccur}]_{geq} + \frac{10^{-pH_{cur}} - 10^{-pH_{to}}}{\phi_b}\right\}V$$

$$10^{-pH_{to}} = 10^{-6} \text{ gmols/L}; \qquad 10^{-pH_{cur}} = 10^{-4} \text{ gmols/L}; \quad V = 1.0 \text{ m}^3$$

Therefore,

$$M_{Na_2CO_3pH} = 53.0\left\{\frac{30}{1000(50)} + \frac{10^{-4} - 10^{-6}}{0.0367}\right\}(1) = 0.175 \text{ kg/m}^3 \quad \textbf{Ans}$$

13.10 SLUDGE PRODUCTION

For convenience, the chemical reactions responsible for the production of sludge solids are summarized next. These equations have been derived at various points.

$$Fe^{2+} + 2OH^- \rightarrow Fe(OH)_{2(s)}\downarrow \tag{13.69}$$

$$Fe^{2+} + \frac{1}{2}Cl_2 + 3H_2O \rightarrow Fe(OH)_{3(s)}\downarrow + Cl^- + 3H^+ \tag{13.70}$$

$$Fe^{2+} + \frac{1}{3}KMnO_4 + \frac{7}{3}H_2O \rightarrow Fe(OH)_{3(s)}\downarrow + \frac{1}{3}MnO_2\downarrow + \frac{1}{3}K^+ + \frac{5}{3}H^+ \tag{13.71}$$

$$Fe^{2+} + \frac{1}{6}O_3 + \frac{15}{6}H_2O \rightarrow Fe(OH)_{3(s)}\downarrow + 2H^+ \tag{13.72}$$

$$Fe^{2+} + \frac{5}{2}H_2O + \frac{1}{4}O_2 \rightarrow Fe(OH)_{3(s)}\downarrow + 2H^+ \tag{13.73}$$

$$Mn^{2+} + 2OH^- \rightarrow Mn(OH)_{2(s)}\downarrow \tag{13.74}$$

$$Mn^{2+} + Cl_2 + 2H_2O \rightarrow MnO_{2(s)}\downarrow + 2Cl^- + 4H^+ \tag{13.75}$$

$$Mn^{2+} + \frac{2}{3}KMnO_4 + \frac{2}{3}H_2O \rightarrow \frac{5}{3}MnO_{2(s)}\downarrow + \frac{2}{3}K^+ + \frac{4}{3}H^+ \tag{13.76}$$

TABLE 13.11
Summary of Equivalent Masses

	Fe^{2+}	Mn^{2+}	$Fe(OH)_{2(s)}$	$Fe(OH)_{3(s)}$	$Mn(OH)_{2(s)}$	$MnO_{2(s)}$
Equation (13.69)	27.9	—	44.9	—	—	—
Eqs. (13.70) through (13.73)	55.8	—	—	106.8	—	—
Eqs. (13.74) through (13.78)	—	27.45	—	—	44.45	43.45

$$Mn^{2+} + \frac{1}{3}O_3 + H_2O \rightarrow MnO_{2(s)}\downarrow + 2H^+ \tag{13.77}$$

The solids produced from the previous reactions are $Fe(OH)_{2(s)}$, $Fe(OH)_{3(s)}$, $Mn(OH)_{2(s)}$, and $MnO_{2(s)}$. From Equation (13.69), the equivalent mass of $Fe(OH)_{2(s)}$ is $Fe(OH)_2/2 = 44.9$ and that of Fe^{2+} is $Fe/2 = 27.9$. From Eqs. (13.70) through (13.73), the equivalent mass of $Fe(OH)_{3(s)}$ is $Fe(OH)_{3(s)}/1 = 106.8$; the corresponding equivalent mass of Fe^{2+} is $Fe/1 = 55.8$. Equations (13.74) through (13.77) produce the following equivalent masses: $Mn^{2+} = Mn/2 = 27.45$; $Mn(OH)_{2(s)} = Mn(OH)_{2(s)}/2 = 44.45$ and $MnO_{2(s)} = MnO_{2(s)}/2 = 43.45$. Table 13.11 summarizes the equivalent masses.

Let $M_{Fe(OH)_2FeRem}$ and $M_{Fe(OH)_3FeRem}$ be the kilogram mass of ferrous hydroxide and ferric hydroxide produced from $\Delta[Fe]_{mg}$ milligrams per liter of ferrous iron removed, respectively. $M_{Fe(OH)_2FeRem}$ is produced at the high pH range and $M_{Fe(OH)_3FeRem}$ is produced at the low pH range. Also, let $M_{Mn(OH)_2MnRem}$ and M_{MnO_2MnRem} be the kilogram mass of manganous hydroxide and manganic oxide produced from $\Delta[Mn]_{mg}$ milligrams per liter of manganese removed, respectively. Also, $M_{Mn(OH)_2MnRem}$ is produced at the high pH range and M_{MnO_2MnRem} is produced at the low pH range. Let the volume of water treated be V cubic meters. Using these information and the equivalent masses derived above produce the following:

$$M_{Fe(OH)_2FeRem} = \frac{44.9\Delta[Fe]_{mg}V}{1000(27.9)} = 0.0016\Delta[Fe]_{mg}V \tag{13.78}$$

$$M_{Fe(OH)_3FeRem} = \frac{106.8\Delta[Fe]_{mg}V}{1000(55.8)} = 0.0019\Delta[Fe]_{mg}V \tag{13.79}$$

$$M_{Mn(OH)_2MnRem} = \frac{44.45\Delta[Mn]_{mg}V}{1000(27.45)} = 0.0016\Delta[Mn]_{mg}V \tag{13.80}$$

$$M_{MnO_2MnRem} = \frac{43.45\Delta[Mn]_{mg}V}{1000(27.45)} = 0.0016\Delta[Mn]_{mg}V \tag{13.81}$$

Example 13.7 A raw water contains 2.5 mg/L Mn. Calculate the solids produced per cubic meter of water treated. The effluent is to contain 0.05 mg/L of manganese and the removal is to be done at the high pH range.

Solution:

$$M_{Mn(OH)_2MnRem} = 0.0016\Delta[Mn]_{mg}V$$

$$= 0.0016(2.5 - 0.05)(1) = 3.92(10^{-3}) \text{ kg/m}^3 \quad \textbf{Ans}$$

GLOSSARY

Complex ions—An ion formed in which a central atom or ion is bonded to a number of surrounding atoms, ions, or molecules.

Electronic configuration—The arrangement by which electrons occupy orbitals of an atom.

Equivalent mass—Mass divided by the number of reference species.

High pH range—The optimum pH range from 11.5 to 12.5.

Low pH range—The optimum pH range from 5.5 to 12.5.

Orbitals—A volume of space in an atom where electrons of specific energy occupy.

Oxidation—The removal of electrons from the valence shell of an atom.

Practical optimum pH range—A range of pH values at which unit process can be operated to produce satisfactory results.

Reduction—The addition of electrons to the valence shell of an atom.

Reference species—The number of moles of electrons involved or the number of positive (or negative) moles of charges or oxidation states in a reactant involved in a given chemical reaction.

Sludge—A mixture of solids and liquids in which the mixture behaves more like a solid than a liquid.

SYMBOLS

A_{add}	Alkalinity to be added to water
$[A_{add}]_{geq}$	Gram equivalents per liter of alkalinity to be added to water
A_{cadd}	Acidity to be added to water
$[A_{cadd}]_{geq}$	Gram equivalents per liter of acidity to be added to water
A_{cur}	Current alkalinity of water
$[A_{cur}]_{geq}$	Gram equivalents per liter of current alkalinity of water
$\Delta[Fe]mg$	mg/L of ferrous to be removed
$FeOH^+$	Monohydroxo Fe(II) complex ion
$FeOH^{2+}$	Monohydroxo Fe(III) complex ion
$Fe(OH)_2^+$	Dihydroxo Fe(III) complex ion
$Fe(OH)_3^-$	Trihydroxo Fe(II) complex ion
$Fe(OH)_4^-$	Tetrahydroxo Fe(III) complex ion

$Fe_2(OH)_2^{4+}$	Dihydroxo 2-Fe(III) complex ion
K_{FeOHc}	Equilibrium constant of $FeOH^+$ and $FeOH^{2+}$
$K_{Fe(OH)_2c}$	Equilibrium constant of $Fe(OH)_2^+$
$K_{Fe(OH)_3c}$	Equilibrium constant of $Fe(OH)_3^-$
$K_{Fe(OH)_4c}$	Equilibrium constant of $Fe(OH)_4^-$
$K_{Fe_2(OH)_2c}$	Equilibrium constant of $Fe_2(OH)_2^{4+}$
K_{MnOHc}	Equilibrium constant of $Mn(OH)^+$
$K_{Mn(OH)_2c}$	Equilibrium constant of $Mn(OH)_2^0$
$K_{Mn(OH)_3c}$	Equilibrium constant of $Mn(OH)_3^-$
$K_{sp,Fe(OH)_2}$	Solubility product constant of $Fe(OH)_{2(s)}$
$K_{sp,Fe(OH)_3}$	Solubility product constant of $Fe(OH)_{3(s)}$
$K_{sp,Mn(OH)_2}$	Solubility product constant of $Mn(OH)_{2(s)}$
$M_{Ca(ClO)_2Fe}$	Kilograms of calcium hypochlorite used in the ferrous reaction
$M_{Ca(ClO)_2Mn}$	Kilograms of calcium hypochlorite used in the manganous reaction
M_{CaOFe}	Kilograms of lime used for the ferrous reaction
M_{CaOMn}	Kilograms of lime used for the manganous reaction
M_{Cl_2Fe}	Kilograms of chlorine used for the ferrous reaction
M_{Cl_2Mn}	Kilograms of chlorine used for the manganous reaction
$M_{Fe(OH)_2FeRem}$	Kilograms of $Fe(OH)_2$ precipitated from removal of ferrous
$M_{Fe(OH)_3FeRem}$	Kilograms of $Fe(OH)_3$ precipitated from removal of ferrous
M_{KMnO_4Fe}	Kilograms of potassium permanganate used in the ferrous reaction
M_{KMnO_4Mn}	Kilograms of potassium permanganate used in the manganous reaction
$M_{Mn(OH)_2MnRem}$	Kilograms of $Mn(OH)_2$ precipitated from removal of manganese
M_{MnO_2MnRem}	Kilograms of MnO_2 precipitated from removal of manganese
$M_{NaClOFe}$	Kilograms of sodium hypochlorite used in the ferrous reaction
$M_{NaClOMn}$	Kilograms of sodium hypochlorite used in the manganous reaction
M_{NaOHFe}	Kilograms of caustic used for the ferrous reaction
M_{NaOHMn}	Kilograms of caustic used for the manganous reaction
$\Delta[Mn]_{mg}$	mg/L of manganous to be removed
M_{O_2Fe}	Kilograms of dissolved oxygen used in the ferrous reaction
M_{O_2Mn}	Kilograms of dissolved oxygen used in the manganous reaction
M_{O_3Fe}	Kilograms of ozone used in the ferrous reaction
M_{O_3Mn}	Kilograms of ozone used in the manganous reaction
$Mn(OH)^+$	Monohydroxo Mn(II) complex ion
$Mn(OH)_2^0$	Dihydroxo Mn(II) complex ion
$Mn(OH)_3^-$	Trihydroxo Mn(II) complex ion
M_{CaOpH}	Kilograms of lime needed to raise pH
M_{CO_2pH}	Kilograms of carbon dioxide needed to lower pH
M_{HClpH}	Kilograms of hydrochloric acid needed to lower pH
$M_{H_2SO_4pH}$	Kilograms of sulfuric acid needed to lower pH
$M_{Na_2CO_3pH}$	Kilograms of soda ash needed to raise pH
M_{NaOHpH}	Kilograms of caustic soda needed to raise pH
sp_{FeII}	Species containing the Fe(II) species

sp_{FeIII}	Species containing the Fe(III) species
sp_{Mn}	Species containing the Mn(II) species
$\forall$	Cubic meters of water treated
γ_H	Activity coefficient of the hydrogen ion
γ_{FeII}	Activity coefficient of the ferrous ion
γ_{FeIII}	Activity coefficient of the ferric ion
γ_{FeOHc}	Activity coefficient of $FeOH^+$ and $FeOH^{2+}$
$\gamma_{Fe(OH)_2c}$	Activity coefficient of $Fe(OH)_2^+$
$\gamma_{Fe(OH)_3c}$	Activity coefficient of $Fe(OH)_3^-$
$\gamma_{Fe(OH)_4c}$	Activity coefficient of $Fe(OH)_4^-$
$\gamma_{Fe_2(OH)_2c}$	Activity coefficient of $Fe_2(OH)_2^{4+}$
γ_{Mn}	Activity coefficient of Mn(II) ion
$\gamma_{Mn(OH)c}$	Activity coefficient of $Mn(OH)^+$
$\gamma_{Mn(OH)_3c}$	Activity coefficient of $Mn(OH)_3^-$

PROBLEMS

13.1 The concentration sp_{FeIII} is determined to be $1.59(10^{-6})$ mg/L. Assuming the dissolved solids are 140 mg/L, calculate the optimum pH.

13.2 The concentration sp_{FeII} is determined to be $6.981(10^{-3})$ mg/L. Assuming the dissolved solids are 140 mg/L, calculate the optimum pH.

13.3 The concentration sp_{FeII} is determined to be $6.981(10^{-3})$ mg/L. The dissolved solids are 140 mg/L and the temperature is 25°C. Calculate the optimum pH.

13.4 The concentration sp_{FeII} is determined to be $6.981(10^{-3})$ mg/L. The dissolved solids are 140 mg/L and the temperature is 25°C. Although the solubility product is known at this temperature, calculate it using values of the other parameters. The optimum pH is 11.95.

13.5 The concentration sp_{FeII} is determined to be $6.981(10^{-3})$ mg/L. The dissolved solids are 140 mg/L and the temperature is 25°C. Although the activity coefficient of the hydrogen ion can be obtained from the knowledge of the dissolved solids content, calculate it using values of the other parameters. The optimum pH is 11.95.

13.6 The concentration sp_{FeII} is determined to be $6.981(10^{-3})$ mg/L. The dissolved solids are 140 mg/L and the temperature is 25°C. Although the activity coefficient of $FeOH^+$ ion can be obtained from the knowledge of the dissolved solids content, calculate it using values of the other parameters. The optimum pH is 11.95.

13.7 The optimum pH for precipitating $Mn(OH)_2$ is 11.97 and the dissolved solids content of the water is 140 mg/L. The temperature is 25°C. Calculate the activity coefficient of the hydrogen ion. The data on dissolved solids may be used to calculate other parameters but not for the activity coefficient of the hydrogen ion.

13.8 The optimum pH for precipitating $Mn(OH)_2$ is 11.97 and the dissolved solids content of the water is 140 mg/L. The temperature is 25°C. Calculate the ion product of water. The given temperature may be used to calculate other parameters but not for the ion product.

13.9 The optimum pH for precipitating $Mn(OH)_2$ is 11.97 and the dissolved solids content of the water is 140 mg/L. The temperature is 25°C. Calculate the activity coefficient of the Mn(II) ion. The data on dissolved solids may be used to calculate other parameters but not for the activity coefficient of the manganese ion.

13.10 The optimum pH for precipitating $Mn(OH)_2$ is 11.97 and the dissolved solids content of the water is 140 mg/L. The temperature is 25°C. Calculate the equilibrium constant of $Mn(OH)_3^-$. The given temperature may used to calculate other parameters but not for the equilibrium constant of $Mn(OH)_3^-$.

13.11 The optimum pH for precipitating $Mn(OH)_2$ is 11.97 and the dissolved solids content of the water is 140 mg/L. The temperature is 25°C. Calculate the activity coefficient of the $Mn(OH)_3^-$ complex ion. The data on dissolved solids may be used to calculate other parameters but not for the activity coefficient of $Mn(OH)_3$.

13.12 The optimum pH for precipitating $Mn(OH)_2$ is 11.97 and the dissolved solids content of the water is 140 mg/L. The temperature is 25°C. Calculate the equilibrium constant of $Mn(OH)^+$. The given temperature may used to calculate other parameters but not for the equilibrium constant of $Mn(OH)^+$.

13.13 The optimum pH for precipitating $Mn(OH)_2$ is 11.97 and the dissolved solids content of the water is 140 mg/L. The temperature is 25°C. Calculate the activity coefficient of the $Mn(OH)^+$ complex ion. The data on dissolved solids may be used to calculate other parameters but not for the activity coefficient of $Mn(OH)^+$.

13.14 At a pH of 11.97 and a dissolved solids of 140 mg/L, the concentration of sp_{Mn} is 0.0179 mg/L. Calculate the solubility product constant of $Mn(OH)_{2(s)}$. The temperature is 25°C. This given temperature may be used to calculate other parameters but not for the solubility product constant.

13.15 At a pH of 11.97 and dissolved solids of 140 mg/L, the concentration of sp_{Mn} is 0.0179 mg/L. Calculate the activity coefficient of the hydrogen ion. The temperature is 25°C. The data on dissolved solids may be used to calculate other parameters but not to calculate the activity coefficient of the hydrogen ion.

13.16 At a pH of 11.97 and dissolved solids of 140 mg/L, the concentration of sp_{Mn} is 0.0179 mg/L. Calculate the equilibrium constant of $Mn(OH)^+$. The temperature is 25°C. This given temperature may be used to calculate other parameters but not the equilibrium constant of $Mn(OH)^+$.

13.17 At a pH of 11.97 and dissolved solids of 140 mg/L, the concentration of sp_{Mn} is 0.0179 mg/L. Calculate the activity coefficient of $Mn(OH)^+$. The temperature is 25°C. The data on dissolved solids may be used to calculate other parameters but not to calculate the activity coefficient of $Mn(OH)^+$.

13.18 At a pH of 11.97 and dissolved solids of 140 mg/L, the concentration of sp_{Mn} is 0.0179 mg/L. Calculate the equilibrium constant of $Mn(OH)_3^-$.

The temperature is 25°C. This given temperature may be used to calculate other parameters but not the equilibrium constant of $Mn(OH)_3^-$.

13.19 At a pH of 11.97 and dissolved solids of 140 mg/L, the concentration of sp_{Mn} is 0.0179 mg/L. Calculate the activity coefficient of $Mn(OH)_3^-$. The temperature is 25°C. The data on dissolved solids may be used to calculate other parameters but not to calculate the activity coefficient of $Mn(OH)_3^-$.

13.20 At a pH of 11.97 and dissolved solids of 140 mg/L, the concentration of sp_{Mn} is 0.0179 mg/L. Calculate the ion product of water. The temperature is 25°C. This given temperature may be used to calculate other parameters but not the ion product of water.

13.21 The kilograms of lime per cubic meter needed to meet the limit concentration of 0.05 mg/L is 0.00071. This removal is done at the high pH range. Calculate the concentration of manganese in the raw water.

13.22 The kilograms of lime per cubic meter needed to meet the limit concentration of 0.05 mg/L is 0.0014. This removal is done at the high pH range and the concentration of manganese in the raw water is 2.5 mg/L. Calculate the volume of water treated.

13.23 A raw water contains 2.5 mg/L Mn. Calculate the kilograms of caustic used per cubic meter needed to meet the limit concentration of 0.05 mg/L.

13.24 A raw water contains 2.5 mg/L Mn. Calculate the kilograms of chlorine per cubic meter needed to meet the limit concentration of 0.05 mg/L.

13.25 A raw water contains 2.5 mg/L Mn. Calculate the kilograms of sodium hypochlorite per cubic meter needed to meet the limit concentration of 0.05 mg/L.

13.26 A raw water contains 2.5 mg/L Mn. Calculate the kilograms of calcium hypochlorite per cubic meter needed to meet the limit concentration of 0.05 mg/L.

13.27 A raw water contains 2.5 mg/L Mn. Calculate the kilograms of ozone per cubic meter needed to meet the limit concentration of 0.05 mg/L.

13.28 A raw water contains 2.5 mg/L Mn. Calculate the kilograms of dissolved oxygen per cubic meter needed to meet the limit concentration of 0.05 mg/L.

13.29 A raw water contains 40 mg/L ferrous. Calculate the kilograms of caustic used per cubic meter needed to meet the limit concentration of 0.3 mg/L.

13.30 A raw water contains 40 mg/L ferrous. Calculate the kilograms of chlorine per cubic meter needed to meet the limit concentration of 0.3 mg/L.

13.31 A raw water contains 40 mg/L ferrous. Calculate the kilograms of sodium hypochlorite per cubic meter needed to meet the limit concentration of 0.3 mg/L.

13.32 A raw water contains 40 mg/L ferrous. Calculate the kilograms of calcium hypochlorite per cubic meter needed to meet the limit concentration of 0.3 mg/L.

13.33 A raw water contains 40 mg/L ferrous. Calculate the kilograms of ozone per cubic meter needed to meet the limit concentration of 0.3 mg/L.

13.34 A raw water contains 40 mg/L ferrous. Calculate the kilograms of dissolved oxygen per cubic meter needed to meet the limit concentration of 0.3 mg/L.

13.35 A raw water containing 3 mg/L of manganese has a pH 4.0. To remove the manganese, the pH needs to be raised to 6.0. The current acidity is 30 mg/L as $CaCO_3$. Calculate the amount of lime needed.

13.36 A raw water containing 3 mg/L of manganese has a pH 4.0. To remove the manganese, the pH needs to be raised to 6.0. The current acidity is 30 mg/L as $CaCO_3$. Calculate the amount of caustic needed.

13.37 A raw water contains 2.5 mg/L Mn. Calculate the solids produced per cubic meter of water treated. The effluent is to contain 0.05 mg/L of manganese and the removal is to be done at the low pH range using chlorine to oxidize the manganous ion.

13.38 A raw water contains 2.5 mg/L Mn. Calculate the solids produced per cubic meter of water treated. The effluent is to contain 0.05 mg/L of manganese and the removal is to be done at the low pH range using dissolved oxygen to oxidize the manganous ion.

13.39 A raw water contains 2.5 mg/L Mn. Calculate the solids produced per cubic meter of water treated. The effluent is to contain 0.05 mg/L of manganese and the removal is to be done at the low pH range using ozone to oxidize the manganous ion.

BIBLIOGRAPHY

Admakin, L. A., et al. (1998). Prospects for utilization of activated anthracites in technology for cleaning the waste waters. *Koks i Khimiya/Coke Chem.*, 8, 30–32.

American Water Works Association (1990). *Alternative Oxidants for the Removal of Soluble Iron and Manganese*. Am. Water Works Assoc., Denver.

American Water Works Association (1984). *Softening, Iron Removal, and Manganese Removal*. Am. Water Works Assoc., Denver.

AWI MARKETING. Anglian House, Ambury Road, Huntingdon, Cambs PE18 6NZ, England. As obtained from the Internet, July 1998.

Burns, F. L. (1997). Case study: Automatic reservoir aeration to control manganese in raw water Maryborough town water supply, Queensland, Australia. *Water Science Technol., Proc. 1997 1st IAWQ-IWSA Joint Specialist Conf. on Reservoir Manage. and Water Supply—An Integrated Syst.*, May 19–23, Prague, Czech Republic, 37, 2, 301–308. Elsevier Science Ltd., Exeter, England.

Clasen, J. (1997). Efficiency control of particle removal by rapid sand filters in treatment plants fed with reservoir water: A survey of different methods. *Water Science Technol., Proc. 1997 1st IAWQ-IWSA Joint Specialist Conf. on Reservoir Manage. and Water Supply—An Integrated Syst.*, May 19–23, Prague, Czech Republic, 37, 2, 19–26. Elsevier Science Ltd., Exeter, England.

Diz, H. R. and J. T. Novak (1997). Heavy metal removal in an innovative treatment system for acid mine drainage. *Hazardous Industrial Wastes, Proc. 1997 29th Mid-Atlantic Industrial and Hazardous Waste Conf.*, July 13–16, Blacksburg, VA, 183–192. Technomic Publ. Co. Inc., Lancaster, PA.

Gouzinis, A., et al. (1998). Removal of Mn and simultaneous removal of NH_3, Fe and Mn from potable water using a trickling filter. *Water Res.* 32, 8, 2442–2450.

Guo, X., et al. (1997). Iron removal from reduction leaching solution of Mn-nodule in dilute HCl. *Zhongguo Youse Jinshu Xuebao/Chinese J. Nonferrous Metals.* 7, 4, 53–56.

Hedberg, T. and T. A. Wahlberg (1997). Upgrading of waterworks with a new biooxidation process for removal of manganese and iron. *Water Science Technol., Proc. 1997 Int. Conf. on Upgrading of Water and Wastewater Syst.,* May 25–28, Kalmir, Sweden, 37, 9, 121–126. Elsevier Science Ltd., Exeter, England.

Holtzclaw Jr., H. F. and W. R. Robinson (1988). *General Chemistry.* D. C. Heath and Company, Lexington, MA, A7, 809.

Knocke, W. R. (1990). *Alternative Oxidants for the Removal of Soluble Iron and Manganese.* Am. Water Works Assoc., Denver.

McNeill, L. and M. Edwards (1997). Predicting as removal during metal hydroxide precipitation. *J. Am. Water Works Assoc.,* 89, 1, 75–86.

Perry, R. H. and C. H. Chilton (1973). *Chemical Engineers' Handbook.* p (3–138). McGraw-Hill, New York, 3–138.

Ranganathan, K. and C. Namasivayam (1998). Utilization of Waste Fe(III)/Cr(III) hydroxide for removal of Cr(VI) and Fe(II) by fixed bed system. *Hungarian J. Industrial Chem.* 26, 3, 169–172.

Rich, G. (1963). *Unit Process of Sanitary Engineering.* John Wiley & Sons, New York.

Robinson, R. B. (1990). *Sequestering Methods of Iron and Manganese Treatment.* Am. Water Works Assoc. Denver.

Snoeyink, V. L. and D. Jenkins (1980). *Water Chemistry.* John Wiley & Sons, New York, 210.

Sikora, F. J., L. L. Behrends, and G. A. Brodie (1995). Manganese and trace metal removal in successive anaerobic and aerobic wetland environments. *Proc. 57th Annu. Am. Power Conf., Part 2,* April 18–20, Chicago, IL, 57–2, 1683–1690. Illinois Institute of Technology, Chicago, IL.

Sincero, A. P. and G. A. Sincero (1996). *Environmental Engineering: A Design Approach.* Prentice Hall, Upper Saddle River, NJ.

14 Removal of Phosphorus by Chemical Precipitation

Phosphorus is a very important element that has attracted much attention because of its ability to cause eutrophication in bodies of water. For example, tributaries from as far away as the farmlands of New York feed the Chesapeake Bay in Maryland and Virginia. Because of the use of phosphorus in fertilizers for these farms, the bay receives an extraordinarily large amount of phosphorus input that has triggered excessive growths of algae in the water body. Presently, large portions of the bay are eutrophied. Without any doubt, all coves and little estuaries that are tributaries to this bay are also eutrophied. Thus, it is important that discharges of phosphorus be controlled in order to avert an environmental catastrophe. In fact, the eutrophication of the Chesapeake Bay and the clogging of the Potomac River by blue greens are two of the reasons for the passage of the Federal Water Pollution Control Act Amendments of 1972.

This chapter discusses the removal of phosphorus by chemical precipitation. It first discusses the natural occurrence of phosphorus, followed by a discussion on the modes of removal of the element. The chemical reactions of removal, unit operations of removal, chemical requirements, optimum pH range of operation, and sludge production are all discussed. The chemical precipitation method employed uses alum, lime, and the ferric salts, $FeCl_3$ and $Fe_2(SO_4)_3$.

14.1 NATURAL OCCURRENCE OF PHOSPHORUS

The element phosphorus is a nonmetal. It belongs to *Group VA* in the Periodic Table in the third period. Its electronic configuration is $[Ne]3s^23p^3$. [Ne] means that the neon configuration is filled. The valence configuration represented by the 3, the *M* shell, shows five electrons in the orbitals: 2 electrons in the *s* orbitals and 3 electrons in the *p* orbitals. This means that phosphorus can have a maximum oxidation state of +5. The commonly observed oxidation states are 3−, 3+, and 5+.

Phosphorus is too active a nonmetal to be found free in nature. Our interest in its occurrence is the form that makes it as fertilizer to plants. As a fertilizer, it must be in the form of orthophosphate. Phosphorus occurs in three phosphate forms: orthophosphate, condensed phosphates (or polyphosphates), and organic phosphates. Phosphoric acid, being triprotic, forms three series of salts: dihydrogen phosphates containing the $H_2PO_4^-$ ion, hydrogen phosphate containing the HPO_4^{2-} ion, and the phosphates containing the PO_4^{3-} ion. These three ions collectively are called orthophosphates. As orthophosphates, the phosphorus atom exists in its highest possible oxidation state of 5+. As mentioned, phosphorus can cause eutrophication in receiving streams. Thus, concentrations of orthophosphates should be controlled through removal before discharging the wastewater into receiving bodies of water. The orthophosphates of concern in wastewater are sodium phosphate (Na_3PO_4), sodium hydrogen phosphate (Na_2HPO_4), sodium dihydrogen phosphate (NaH_2PO_4), and ammonium hydrogen phosphate [$(NH_4)_2HPO_4$]. They cause the problems associated with algal blooms.

When phosphoric acid is heated, it decomposes, losing molecules of water forming the P–O–P bonds. The process of losing water is called *condensation*, thus the term condensed phosphates and, since they have more than one phosphate group in the molecule, they are also called *polyphosphates*. Among the acids formed from the condensation of phosphoric acid are *dipolyphosphoric acid* or *pyrophosphoric acid* ($H_4P_2O_7$, oxidation state = 5+), *tripolyphosphoric acid* ($H_5P_3O_{10}$, oxidation state = 5+), and *metaphosphoric acid* [$(HPO_3)_n$, oxidation state = 5+]. Condensed phosphates undergo hydrolysis in aqueous solutions and transform into the orthophosphates. Thus, they must also be controlled. Condensed phosphates of concern in wastewater are sodium hexametaphosphate ($NaPO_3)_6$, sodium dipolyphosphate ($Na_4P_2O_7$), and sodium tripolyphosphate ($Na_5P_3O_{10}$).

When organic compounds containing phosphorus are attacked by microorganisms, they also undergo hydrolysis into the orthophosphate forms. Thus, as with all the other phosphorus species, they have to be controlled before wastewaters are discharged.

Figure 14.1 shows the structural formulas of the various forms of phosphates. Note that the oxygen not bonded to hydrogen in orthophosphoric acid, trimetaphosphoric

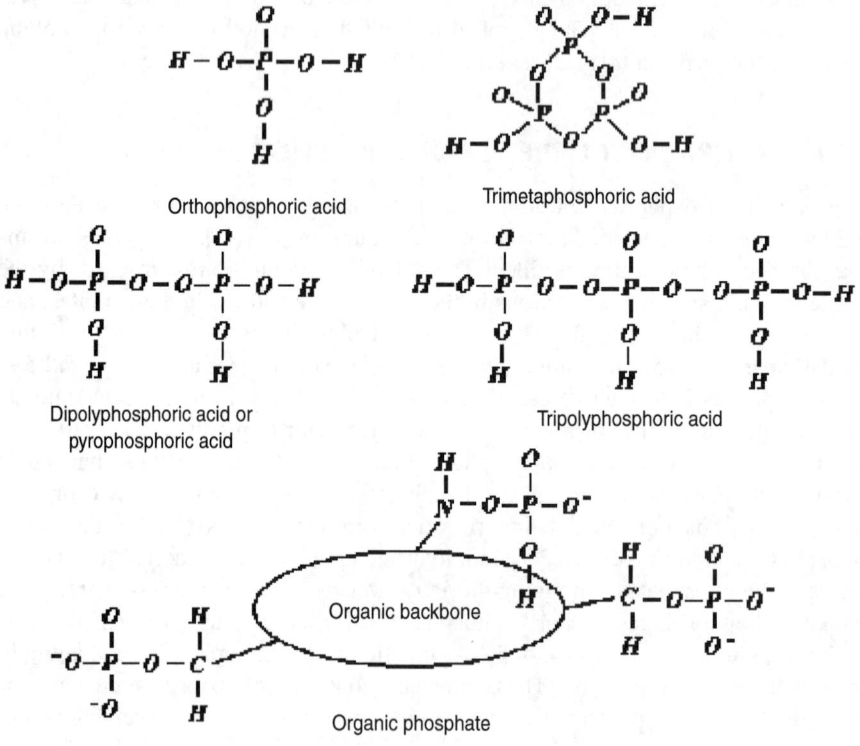

FIGURE 14.1 Structural formulas of various forms of phosphates.

acid, dipolyphosphoric acid, and tripolyphosphoric acid has a single bond with the central phosphorus atom. Oxygen has six electrons in its valence shell, therefore, this indicates that phosphorus has shared two of its own valence electrons to oxygen without oxygen sharing any of its electrons to phosphorus. The acceptance by oxygen of these phosphorus electrons, completes its valence orbitals to the required eight electrons for stability.

All of the hydrogen atoms are ionizable. This means that the largest negative charge for the complete ionization of orthophosphoric acid is 3–; that of trimetaphosphoric acid is also 3–. Dipolyphosphoric acid will have 4– as the largest negative charge and tripolyphosphoric acid will have 5– as the largest negative charge. The charges in organic phosphates depend upon the organic backbone the phosphates are attached to and how many of the phosphate radicals are being attached.

The phosphorus concentration in domestic wastewaters varies from 3 to 15 mg/L and that in lake surface waters, from 0.01 to 0.04 mg/L all measured as P. These values include all the forms of phosphorus.

14.2 MODES OF PHOSPHORUS REMOVAL

Again, as in previous chapters, the best place to investigate for determining the modes of removal is the table of solubility products constants as shown in Table 14.1. A precipitation product that has the lowest K_{sp} means that the substance is the most insoluble. As shown in the table, the phosphate ion can be precipitated using a calcium precipitant producing either $Ca_5(PO_4)_3(OH)_{(s)}$ or $Ca_3(PO_4)_2$. Of these two precipitates, $Ca_5(PO_4)_3(OH)_{(s)}$ has the smaller K_{sp} of $10^{-55.9}$; thus, it will be used as the criterion for the precipitation of phosphates. $Ca_5(PO_3)_3(OH)_{(s)}$ is also called *calcium hydroxy apatite*.

As shown in the table, the other mode of precipitation possible is through precipitating the phosphate ion as $AlPO_{4(s)}$ and $FePO_4$. The precipitant normally used in this instances are alum and the ferric salts (ferric chloride and ferric sulfate), respectively.

TABLE 14.1
Solubility Product Constants for Phosphate Precipitation

Precipitation Product	Solubility Product, K_{sp} at 25°C
$Ca_5(PO_3)_3(OH)_{(s)}$	$(10^{-55.9})$
$Ca_3(PO_4)_2$	(10^{-25})
$AlPO_{4(s)}$	(10^{-21})
$FePO_4$	$(10^{-21.9})$

14.3 CHEMICAL REACTION OF THE PHOSPHATE ION WITH ALUM

To precipitate the phosphate ion as aluminum phosphate, alum is normally used. The chemical reaction is shown next:

$$Al^{3+} + PO_4^{3-} \rightleftharpoons AlPO_{4(s)} \downarrow$$

$$AlPO_{4(s)} \downarrow \rightleftharpoons Al^3 + PO_4^{3-} \quad K_{sp,AlPO_4} = 10^{-21} \tag{14.1}$$

As shown in these reactions, the phosphorus must be in the phosphate form. The reaction occurs in water, so the phosphate ion originates a series of equilibrium orthophosphate reactions with the hydrogen ion. This series is shown as follows (Snoeyink and Jenkins):

$$PO_4^{3-} + H^+ \rightleftharpoons HPO_4^{2-}$$

$$\Rightarrow HPO_4^{2-} \rightleftharpoons PO_4^{3-} + H^+ \quad K_{HPO_4} = 10^{-12.3} \tag{14.2}$$

$$HPO_4^{2-} + H^+ \rightleftharpoons H_2PO_4^-$$

$$\Rightarrow H_2PO_4^- \rightleftharpoons HPO_4^{2-} + H^+ \quad K_{H_2PO_4} = 10^{-7.2} \tag{14.3}$$

$$H_2PO_4^- + H^+ \rightleftharpoons H_3PO_4$$

$$\Rightarrow H_3PO_4 \rightleftharpoons H_2PO_4^- + H^+ \quad K_{H_3PO_4} = 10^{-2.1} \tag{14.4}$$

Let sp_{PO_4Al} represent the species in solution containing the PO_4 species of the orthophosphates, using alum as the precipitant. Therefore,

$$[sp_{PO_4Al}] = [PO_4^{3-}] + [HPO_4^{2-}] + [H_2PO_4^-] + [H_3PO_4] \tag{14.5}$$

Express Equation (14.5) in terms of $[PO_4^{3-}]$ using Eqs. (14.2) through (14.4). This will enable $[sp_{PO_4Al}]$ to be expressed in terms of $[Al^{3+}]$ using Equation (14.1) and $K_{sp,AlPO_4}$. Proceed as follows:

$$[HPO_4^{2-}] = \frac{\{HPO_4^{2-}\}}{\gamma_{HPO_4}} = \frac{\{PO_4^{3-}\}\{H^+\}}{\gamma_{HPO_4}K_{HPO_4}} = \frac{\gamma_{PO_4}\gamma_H[PO_4^{3-}][H^+]}{\gamma_{HPO_4}K_{HPO_4}} \tag{14.6}$$

$$[H_2PO_4^-] = \frac{\{H_2PO_4^-\}}{\gamma_{H_2PO_4}} = \frac{\{HPO_4^{2-}\}\{H^+\}}{\gamma_{H_2PO_4}K_{H_2PO_4}} = \frac{\gamma_{PO_4}\gamma_H^2[PO_4^{3-}][H^+]^2}{\gamma_{H_2PO_4}K_{H_2PO_4}K_{HPO_4}} \tag{14.7}$$

$$[H_3PO_4] = \{H_3PO_4\} = \frac{\{H_2PO_4^-\}\{H^+\}}{K_{H_3PO_4}} = \frac{\gamma_{PO_4}\gamma_H^3[PO_4^{3-}][H^+]^3}{K_{H_3PO_4}K_{H_2PO_4}K_{HPO_4}} \qquad (14.8)$$

Equations (14.6) through (14.8) may now be substituted into Equation (14.5) to produce

$$[sp_{PO_4Al}] = [PO_4^{3-}] + \frac{\gamma_{PO_4}\gamma_H[PO_4^{3-}][H^+]}{\gamma_{HPO_4}K_{HPO_4}} + \frac{\gamma_{PO_4}\gamma_H^2[PO_4^{3-}][H^+]^2}{\gamma_{H_2PO_4}K_{H_2PO_4}K_{HPO_4}}$$

$$+ \frac{\gamma_{PO_4}\gamma_H^3[PO_4^{3-}][H^+]^3}{K_{H_3PO_4}K_{H_2PO_4}K_{HPO_4}} \qquad (14.9)$$

But, from Equation (14.1), $[PO_4^{3-}] = \{PO_4^{3-}\}/\gamma_{PO_4} = K_{sp,AlPO_4}/\gamma_{PO_4}\{Al^{3+}\} = K_{sp,AlPO_4}/\gamma_{PO_4}\gamma_{Al}[Al^{3+}]$. Substituting,

$$[sp_{PO_4Al}] = \frac{K_{sp,AlPO_4}}{\gamma_{PO_4}\gamma_{Al}[Al^{3+}]} + \frac{\gamma_H K_{sp,AlPO_4}[H^+]}{\gamma_{HPO_4}K_{HPO_4}\gamma_{Al}[Al^{3+}]} + \frac{\gamma_H^2 K_{sp,AlPO_4}[H^+]^2}{\gamma_{H_2PO_4}K_{H_2PO_4}K_{HPO_4}\gamma_{Al}[Al^{3+}]}$$

$$+ \frac{\gamma_H^3 K_{sp,AlPO_4}[H^+]^3}{K_{H_3PO_4}K_{H_2PO_4}K_{HPO_4}\gamma_{Al}[Al^{3+}]} \qquad (14.10)$$

γ_{PO_4}, γ_{Al}, γ_H, γ_{HPO_4}, and $\gamma_{H_2PO_4}$ are, respectively, the activity coefficients of the phosphate, aluminum, hydrogen, hydrogen phosphate, and dihydrogen phosphate ions. K_{HPO_4}, $K_{H_2PO_4}$, and $K_{H_3PO_4}$ are, respectively, the equilibrium constants of the hydrogen phosphate and dihydrogen phosphate ions and phosphoric acid.

[Al^{3+}] needs to be eliminated for the equation to be expressed solely in terms of [H^+]. When alum is added to water, it will unavoidably react with the existing natural alkalinity. For this reason, the aluminum ion will not only react with the phosphate ion to precipitate $AlPO_{4(s)}$, but it will also react with the OH^- to precipitate $Al(OH)_{3(s)}$. Also, Al^{3+} will form complexes $Al(OH)^{2+}$, $Al_7(OH)_{17}^{4+}$, $Al_{13}(OH)_{34}^{5+}$, $Al(OH)_4^-$, and $Al_2(OH)_2^{4+}$ in addition to the $Al(OH)_{3(s)}$. All these interactions complicate our objective of eliminating [Al^{3+}].

Consider, however, the equilibrium of $Al(OH)_{3(s)}$, which is as follows:

$$Al(OH)_{3(s)} \rightleftharpoons Al^{3+} + 3OH^- \qquad K_{sp,Al(OH)_3} = 10^{-33} \qquad (14.11)$$

$K_{sp,Al(OH)_3} = 10^{-33}$ may be compared with $K_{sp,AlPO_4} = 10^{-21}$. From $K_{sp,Al(OH)_3} = 10^{-33}$, the concentration of Al^{3+} needed to precipitate $Al(OH)_3$ may be calculated to be $2.0(10^{-9})$ gmol/L. Doing similar calculation from $K_{sp,AlPO_4} = 10^{-21}$, the concentration of Al^{3+} needed to precipitate $AlPO_4$ is $3.0(10^{-11})$ gmol/L. These two concentrations are so close to each other that it may be concluded that $Al(OH)_3$ and $AlPO_4$ are coprecipitating. This finding allows us to eliminate $[Al^{3+}]$. Thus,

$$[Al^{3+}] = \frac{\{Al^{3+}\}}{\gamma_{Al}} = \frac{K_{sp,Al(OH)_3}}{\gamma_{Al}\{OH^-\}^3} = \frac{K_{sp,Al(OH)_3}\gamma_H^3[H^+]^3}{\gamma_{Al}K_w^3}.$$

Substituting into Equation (14.10),

$$[sp_{PO_4Al}] = \frac{K_{sp,AlPO_4}K_w^3}{\gamma_{PO_4}K_{sp,Al(OH)_3}\gamma_H^3[H^+]^3} + \frac{K_{sp,AlPO_4}K_w^3}{\gamma_{HPO_4}K_{HPO_4}K_{sp,Al(OH)_3}\gamma_H^2[H^+]^2}$$

$$+ \frac{K_{sp,AlPO_4}K_w^3}{\gamma_{H_2PO_4}K_{H_2PO_4}K_{HPO_4}K_{sp,Al(OH)_3}\gamma_H[H^+]} + \frac{K_{sp,AlPO_4}K_w^3}{K_{H_3PO_4}K_{H_2PO_4}K_{HPO_4}K_{sp,Al(OH)_3}}$$

$$(14.12)$$

Equation (14.12) portrays the equilibrium relationship between $[sp_{PO_4}]$ and $[H^+]$.

Example 14.1   Calculate $[sp_{PO_4Al}]$ when the pH is 10. Assume the water contains 140 mg/L of dissolved solids.

Solution:

$$[sp_{PO_4Al}] = \frac{K_{sp,AlPO_4}K_w^3}{\gamma_{PO_4}K_{sp,Al(OH)_3}\gamma_H^3[H^+]^3} + \frac{K_{sp,AlPO_4}K_w^3}{\gamma_{HPO_4}K_{HPO_4}K_{sp,Al(OH)_3}\gamma_H^2[H^+]^2}$$

$$+ \frac{K_{sp,AlPO_4}K_w^3}{\gamma_{H_2PO_4}K_{H_2PO_4}K_{HPO_4}K_{sp,Al(OH)_3}\gamma_H[H^+]} + \frac{K_{sp,AlPO_4}K_w^3}{K_{H_3PO_4}K_{H_2PO_4}K_{HPO_4}K_{sp,Al(OH)_3}}$$

$$K_{sp,AlPO_4} = 10^{-21} \qquad \mu = 2.5(10^{-5})TDS \qquad \gamma = 10^{-\frac{0.5z_i^2(\sqrt{\mu})}{1+1.14(\sqrt{\mu})}}$$

$$\mu = 2.5(10^{-5})(140) = 3.5(10^{-3}) \qquad \gamma_{PO_4} = 10^{-\frac{0.5(3)^2[\sqrt{3.5(10^{-3})}]}{1+1.14[\sqrt{3.5(10^{-3})}]}} = 0.56$$

$$K_{sp,Al(OH)_3} = 10^{-33} \qquad \gamma_H = \gamma_{H_2PO_4} = 10^{-\frac{0.5(1)^2[\sqrt{3.5(10^{-3})}]}{1+1.14[\sqrt{3.5(10^{-3})}]}} = 0.94$$

$$\gamma_{HPO_4} = 10^{-\frac{0.5(2)^2[\sqrt{3.5(10^{-3})}]}{1+1.14[\sqrt{3.5(10^{-3})}]}} = 0.77 \quad K_{HPO_4} = 10^{-12.3} \quad K_{H_2PO_4} = 10^{-7.2} \quad K_{H_3PO_4} = 10^{-2.1}$$

Therefore,

$$
\begin{aligned}
[sp_{PO_4Al}] &= \frac{(10^{-21})(10^{-14})^3}{(0.56)(10^{-33})(0.94)^3[10^{-10}]^3} + \frac{(10^{-21})(10^{-14})^3}{(0.77)(10^{-12.3})(10^{-33})(0.94)^2[10^{-10}]^2} \\
&\quad + \frac{(10^{-21})(10^{-14})^3}{(0.94)(10^{-7.2})(10^{-12.3})(10^{-33})(0.94)[10^{-10}]} \\
&\quad + \frac{(10^{-21})(10^{-14})^3}{(10^{-2.1})(10^{-7.2})(10^{-12.3})(10^{-33})} \\
&= \frac{1.0(10^{-63})}{4.65(10^{-64})} + \frac{1.0(10^{-63})}{3.41(10^{-66})} + \frac{1.0(10^{-63})}{2.79(10^{-63})} + \frac{1.0(10^{-63})}{2.51(10^{-55})} \\
&= 295.76 \text{ gmols/L} = 9.17(10^6) \text{ mg/L as P}
\end{aligned}
$$

The answer of $9.17(10^{+6})$ mg/L emphasizes a very important fact: phosphorus cannot be removed at alkaline conditions. It will be shown in subsequent discussions that the solution conditions must be acidic for effective removal of phosphorus using alum.

14.3.1 DETERMINATION OF THE OPTIMUM pH RANGE

Equation (14.12) shows that $[sp_{PO_4Al}]$ is a function of the hydrogen ion concentration. This means that the concentration of the species containing the PO_4 species of the orthophosphates is a function of pH. If the equation is differentiated and the result equated to zero, however, an optimum value cannot be guaranteed to be found. A range of pH values can, however, be assigned and the corresponding values of sp_{PO_4Al} calculated. By inspection of the result, the optimum range can be determined. Tables 14.2 and 14.3 show the results of assigning this range of pH and values of sp_{PO_4Al} calculated using Equation (14.12). These tables show that optimum removal of phosphorus using alum results when the unit is operated at pH values less than 5.0.

Note: The dissolved solids content has only a negligible effect on the resulting concentrations.

14.4 CHEMICAL REACTION OF THE PHOSPHATE ION WITH LIME

Calcium hydroxy apatite contains the phosphate and hydroxyl groups. Using calcium hydroxide as the precipitant, the chemical reaction is shown below:

$$
5Ca^{2+} + 3PO_4^{3-} + OH^- \rightleftharpoons Ca_5(PO_4)_3OH_{(s)}\downarrow
$$

$$
Ca_5(PO_4)_3OH_{(s)}\downarrow \rightleftharpoons 5Ca^{2+} + 3PO_4^{3-} + OH^- \quad K_{sp,apatite} = 10^{-55.9} \quad (14.13)
$$

TABLE 14.2
Concentration of sp_{PO_4Al} as a Function of pH at 25°C (mg/L as P)

pH, Dissolved Solids = 140 mg/L	$[sp_{PO_4Al}]$ (mg/L)
0	$6.6(10^{-5})$
1	$7.1(10^{-5})$
2	$1.2(10^{-4})$
3	$6.5(10^{-4})$
4	$5.9(10^{-3})$
5	$5.9(10^{-2})$
6	$6.3(10^{-1})$
7	$1.1(10^{+1})$
8	$5.4(10^{+2})$
9	$4.8(10^{+4})$
10	$4.8(10^{+6})$
11	$5.8(10^{+8})$
12	$8.3(10^{+10})$
13	$4.0(10^{+13})$
14	$2.3(10^{+16})$
15	$3.5(10^{+19})$

TABLE 14.3
Concentration of sp_{PO_4Al} as a Function of pH at 25°C (mg/L as P)

pH, Dissolved Solids = 35,000 mg/L	$[sp_{PO_4Al}]$ (mg/L)
0	$6.6(10^{-5})$
1	$8.0(10^{-5})$
2	$2.1(10^{-4})$
3	$1.5(10^{-3})$
4	$1.5(10^{-2})$
5	$1.5(10^{-1})$
6	$2.0(10^{-0})$
7	$8.9(10^{+1})$
8	$7.6(10^{+3})$
9	$7.5(10^{+5})$
10	$8.2(10^{+7})$
11	$1.6(10^{+10})$
12	$9.2(10^{+12})$
13	$8.5(10^{+15})$
14	$8.5(10^{+18})$
15	$8.5(10^{+21})$

These equations also show that the phosphorus must be in the phosphate form. As in the case of alum, the phosphate ion produces the set of reactions given by Eqs. (14.2) through (14.4). Let sp_{PO_4Ca} be the species in solution containing the PO_4 species, using the calcium in lime as the precipitant. $[sp_{PO_4Ca}]$ will be the same as given by Equation (14.5) which, along with Eqs. (14.2) through (14.4), can be manipulated to produce Equation (14.9). From Equation (14.13),

$$[PO_4^{3-}] = \frac{\{PO_4^{3-}\}}{\gamma_{PO_4}} = \frac{\left(\dfrac{K_{sp,apatite}}{\{Ca^{2+}\}^5\{OH^-\}}\right)^{1/3}}{\gamma_{PO_4}} = \frac{K_{sp,apatite}^{1/3}\,\gamma_H^{1/3}[H^+]^{1/3}}{\gamma_{Ca}^{1/3}[Ca^{2+}]^{5/3}K_w^{1/3}\gamma_{PO_4}}$$

Substituting in Equation (14.9) and simplifying,

$$[sp_{PO_4Ca}] = \frac{K_{sp,apatite}^{1/3}\,\gamma_H^{1/3}[H^+]^{1/3}}{\gamma_{Ca}^{1/3}[Ca^{2+}]^{5/3}K_w^{1/3}\gamma_{PO_4}} + \frac{K_{sp,apatite}^{1/3}\,\gamma_H^{4/3}[H^+]^{4/3}}{\gamma_{HPO_4}K_{HPO_4}\gamma_{Ca}^{1/3}[Ca^{2+}]^{5/3}K_w^{1/3}}$$

$$+ \frac{K_{sp,apatite}^{1/3}\,\gamma_H^{7/3}[H^+]^{7/3}}{\gamma_{H_2PO_4}K_{H_2PO_4}K_{HPO_4}\gamma_{Ca}^{1/3}[Ca^{2+}]^{5/3}K_w^{1/3}}$$

$$+ \frac{K_{sp,apatite}^{1/3}\,\gamma_H^{10/3}[H^+]^{10/3}}{K_{H_3PO_4}K_{H_2PO_4}K_{HPO_4}\gamma_{Ca}^{1/3}[Ca^{2+}]^{5/3}K_w^{1/3}} \qquad (14.14)$$

γ_{Ca} and K_w are, respectively, the activity coefficient of the calcium ion and ion product of water.

The $[Ca^{2+}]$ in the denominator is analogous to $[Al^{3+}]$ in the case of alum; however, the present case is different in that $[Ca^{2+}]$ cannot be eliminated in a straightforward manner. $[Al^{3+}]$ was easily eliminated because $Al(OH)_{3(s)}$ was precipitating along with $AlPO_4$. $Ca(OH)_{2(s)}$, which could be precipitating, has a K_{sp} value of $7.9(10^{-6})$. This is much, much greater than $K_{sp,apatite} = 10^{-55.9}$ and calcium hydroxide will not be precipitating along with $Ca_5(PO_4)_3OH_{(s)}$. The other possible precipitate is $CaCO_3$ that is produced when lime reacts with the natural alkalinity of the water; however, calcium carbonate has a K_{sp} value of $4.8(10^{-9})$. Again, this value is much, much greater than $K_{sp,apatite}$ and calcium carbonate will not be precipitating along with $Ca_5(PO_4)_3OH_{(s)}$, either. We will, therefore, let the equation stand and express $[sp_{PO_4Ca}]$ as a function of $[Ca^{2+}]$, along with $[H^+]$ and the constants.

Example 14.2 Calculate $[sp_{PO_4Ca}]$ expressed as mg/L of P when the pH is 8. Assume the water contains 140 mg/L of dissolved solids and that $[Ca^{2+}] = 130$ mg/L as $CaCO_3$.

Solution:

$$[sp_{PO_4Ca}] = \frac{K_{sp,apatite}^{1/3}\,\gamma_H^{1/3}[H^+]^{1/3}}{\gamma_{Ca}^{1/3}[Ca^{2+}]^{5/3}K_w^{1/3}\gamma_{PO_4}} + \frac{K_{sp,apatite}^{1/3}\,\gamma_H^{4/3}[H^+]^{4/3}}{\gamma_{HPO_4}K_{HPO_4}\gamma_{Ca}^{1/3}[Ca^{2+}]^{5/3}K_w^{1/3}}$$

$$+ \frac{K_{sp,apatite}^{1/3}\,\gamma_H^{7/3}[H^+]^{7/3}}{\gamma_{H_2PO_4}K_{H_2PO_4}K_{HPO_4}\gamma_{Ca}^{1/3}[Ca^{2+}]^{5/3}K_w^{1/3}}$$

$$+ \frac{K_{sp,apatite}^{1/3}\,\gamma_H^{10/3}[H^+]^{10/3}}{K_{H_3PO_4}K_{H_2PO_4}K_{HPO_4}\gamma_{Ca}^{1/3}[Ca^{2+}]^{5/3}K_w^{1/3}}$$

$$K_{sp,apatite} = 10^{-55.9} \qquad \gamma_H = \gamma_{H_2PO_4} = 10^{-\frac{0.5(1)^2[\sqrt{3.5(10^{-3})}]}{1+1.14[\sqrt{3.5(10^{-3})}]}} = 0.94$$

$$\gamma_{Ca} = \gamma_{HPO_4} = 10^{-\frac{0.5(2)^2[\sqrt{3.5(10^{-3})}]}{1+1.14[\sqrt{3.5(10^{-3})}]}} = 0.77$$

$$[CA^{2+}] = \frac{130}{1000(40.1)} = 3.24(10^{-3})\,gmol/L$$

$$\gamma_{PO_4} = 10^{-\frac{0.5(3)^2[\sqrt{3.5(10^{-3})}]}{1+1.14[\sqrt{3.5(10^{-3})}]}} = 0.56$$

$$K_{HPO_4} = 10^{-12.3} \qquad K_{H_2PO_4} = 10^{-7.2} \qquad K_{H_3PO_4} = 10^{-2.1}$$

Therefore,

$$[sp_{PO_4Ca}] = \frac{(10^{-55.9})^{1/3}(0.94)^{1/3}[10^{-8}]^{1/3}}{(0.77)^{1/3}[3.24(10^{-3})]^{5/3}(10^{-14})^{1/3}(0.56)}$$

$$+ \frac{(10^{-55.9})^{1/3}(0.94)^{4/3}[10^{-8}]^{4/3}}{(0.77)(10^{-12.3})(0.77)^{1/3}[3.24(10^{-3})]^{5/3}(10^{-14})^{1/3}}$$

$$+ \frac{(10^{-55.9})^{1/3}(0.94)^{7/3}[10^{-8}]^{7/3}}{(0.94)(10^{-7.2})(10^{-12.3})(0.77)^{1/3}[3.24(10^{-3})]^{5/3}(10^{-14})^{1/3}}$$

$$+ \frac{(10^{-55.9})^{1/3}(0.94)^{10/3}[10^{-8}]^{10/3}}{(10^{-2.1})(10^{-7.2})(10^{-12.3})(0.77)^{1/3}[3.24(10^{-3})]^{5/3}(10^{-14})^{1/3}}$$

$$= \frac{4.91(10^{-22})}{7.85(10^{-10})} + \frac{4.62(10^{-30})}{5.41(10^{-22})} + \frac{4.34(10^{-38})}{4.16(10^{-29})} + \frac{4.08(10^{-46})}{3.52(10^{-31})}$$

$$= 9.58(10^{-9})\,gmol/L$$

$$= 6.25(10^{-13}) + 8.54(10^{-9}) + 1.04(10^{-9}) + 1.16(10^{-15})$$

$$= 2.97(10^{-4})\,mg/L \quad \textbf{Ans}$$

14.4.1 Determination of the Optimum pH and the Optimum pH Range

Analyzing Equation (14.14) shows that $[H^+]$'s are all written in the numerator. This means that if the equation is differentiated, the result will give positive terms on one side of the equation which will then be equated to zero to get the minimum. This kind of equation cannot guarantee a minimum. Thus, the optimum and the optimum pH range will be determined by preparing a table as was done in the case of alum. Tables 14.4 through 14.6 show the results. This table was prepared using Equation (14.14).

Tables 14.4 and 14.5 reveal very important information. The efficiency of removal of phosphorus increases as the concentration of calcium increases from 0 to 130 mg/L. Remember that we disallowed the use of the K_{sp} of $Ca(OH)_2$ because it was too large, and we concluded that the hydroxide would not be precipitating alongside with the apatite. It is not, however, preventable to add more lime in order to increase the concentration of the calcium ion to the point of saturation and, thus, be able to use $K_{sp,Ca(OH)_2}$ in calculations. In theory, before the hydroxide precipitation can happen, all the apatite particles would have already precipitated, resulting, indeed, in a very high efficiency of removal of phosphorus. Would this really happen? The answer would be yes, but this could be a good topic for applied research.

A concentration of zero for the calcium ion is, of course, nonexistent. Thus, Table 14.4 may be considered as purely imagined. Also, from the tables there is no single optimum pH; the range, however, may be determined by inspection as ranging

TABLE 14.4

Concentration of sp_{PO_4Ca} as a Function of pH at 25°C (mg/L as P)

pH, Dissolved Solids = 140 mg/L $[Ca^{2+}] = 0$ mg/L	$[sp_{PO_4Ca}]$ (mg/L)
0	$19.0(10^{+23})$
1	$3.4(10^{+21})$
2	$1.4(10^{+20})$
3	$6.6(10^{+18})$
4	$3.1(10^{+17})$
5	$1.4(10^{+16})$
6	$6.6(10^{+14})$
7	$3.1(10^{+13})$
8	$1.4(10^{+12})$
9	$6.6(10^{+10})$
10	$3.1(10^{+9})$
11	$1.4(10^{+8})$
12	$6.6(10^{+6})$
13	$3.1(10^{+5})$
14	$1.4(10^{+4})$
15	$6.6(10^{+2})$

TABLE 14.5
Concentration of sp_{PO_4Ca} as a Function of pH at 25°C (mg/L as P)

pH, Dissolved Solids = 140 mg/L $[Ca^{2+}]$ = 130 mg/L as $CaCO_3$	$[sp_{PO_4Ca}]$ (mg/L)
0	$7.7(10^{+16})$
1	$3.8(10^{+13})$
2	$3.1(10^{+10})$
3	$7.6(10^{+7})$
4	$3.2(10^{+5})$
5	$1.3(10^{+3})$
6	$7.4(10^{+0})$
7	$5.8(10^{-2})$
8	$2.97(10^{-4})$
9	$5.6(10^{-5})$
10	$2.6(10^{-6})$
11	$1.3(10^{-7})$
12	$9.7(10^{-9})$
13	$2.2(10^{-9})$
14	$9.0(10^{-10})$
15	$4.1(10^{-10})$

TABLE 14.6
Concentration of sp_{PO_4Ca} as a Function of pH at 25°C (mg/L as P)

pH, Dissolved Solids = 35,000 mg/L $[Ca^{2+}]$ = 130 mg/L as $CaCO_3$	$[sp_{PO_4Ca}]$ (mg/L)
0	$3.0(10^{+16})$
1	$1.7(10^{+13})$
2	$2.1(10^{+10})$
3	$7.1(10^{+7})$
4	$3.2(10^{+5})$
5	$1.6(10^{+3})$
6	$1.0(10^{+1})$
7	$1.9(10^{-1})$
8	$7.6(10^{-3})$
9	$3.5(10^{-4})$
10	$1.8(10^{-5})$
11	$1.6(10^{-6})$
12	$4.3(10^{-7})$
13	$1.9(10^{-7})$
14	$8.6(10^{-8})$
15	$4.0(10^{-8})$

from pH 7.0 and above. The effect of dissolved solids has reduced this range to 8 and above; however, a concentration of 35,000 mg/L of total solids is already very high and would not be encountered in the normal treatment of water and wastewater. This concentration is representative of the dissolved solids concentration of sea water.

14.5 CHEMICAL REACTION OF THE PHOSPHATE ION WITH THE FERRIC SALTS

The chemical reaction to precipitate the phosphate ion as ferric phosphate is shown next:

$$Fe^{3+} + PO_4^{3-} \rightleftharpoons FePO_4\downarrow \tag{14.15}$$

$$FePO_4\downarrow \rightleftharpoons Fe^{3+} + PO_4^{3-} \quad K_{sp,FePO_4} = 10^{-21.9}$$

As in the case of precipitation using alum and precipitation using lime, the phosphate ion produces the set of reactions given by Eqs. (14.2) through (14.4). Let $sp_{PO_4,FeIII}$ be the species in solution containing the PO_4 species of the orthophosphates using the ferric salts $FeCl_3$ and $Fe_2(SO_4)_3$ as the precipitants. Again, $[sp_{PO_4,FeIII}]$ will be the same as given by Equation (14.5) which, along with Eqs. (14.2) through (14.4), can be manipulated to produce Equation (14.9). From Equation (14.15),

$$[PO_4^{3-}] = \frac{\{PO_4^{3-}\}}{\gamma_{PO_4}} = \frac{K_{sp,FePO_4}}{\gamma_{PO_4}\{Fe^{3+}\}} = \frac{K_{sp,FePO_4}}{\gamma_{PO_4}\gamma_{FeIII}[Fe^{3+}]}.$$

Substituting in Equation (14.9) and simplifying,

$$[sp_{PO_4,FeIII}] = \frac{K_{sp,FePO_4}}{\gamma_{PO_4}\gamma_{FeIII}[Fe^{3+}]} + \frac{\gamma_H K_{sp,FePO_4}[H^+]}{\gamma_{HPO_4}K_{HPO_4}\gamma_{FeIII}[Fe^{3+}]}$$

$$+ \frac{\gamma_H^2 K_{sp,FePO_4}[H^+]^2}{\gamma_{H_2PO_4}K_{H_2PO_4}K_{HPO_4}\gamma_{FeIII}[Fe^{3+}]} + \frac{\gamma_H^3 K_{sp,FePO_4}[H^+]^3}{K_{H_3PO_4}K_{H_2PO_4}K_{HPO_4}\gamma_{FeIII}[Fe^{3+}]} \tag{14.16}$$

γ_{FeIII} is the activity coefficient of the ferric ion.

Investigate the possibility of eliminating $[Fe^{3+}]$ from the denominator of the above equation. This is, indeed, possible through the use of ferric hydroxide. The dissociation reaction is

$$Fe(OH)_{3(s)} \rightleftharpoons Fe^{3+} + 3(OH^-) \quad K_{sp,Fe(OH)_3} = 1.1(10^{-36}) \tag{14.17}$$

From $K_{sp,\text{Fe(OH)}_3} = 1.1(10^{-36})$, the concentration of Fe^{3+} needed to precipitate $Fe(OH)_3$ can be calculated to be equal to $4.5(10^{-10})$ gmol/L. And, from $K_{sp,\text{FePO}_4} = 10^{-21.9}$, the concentration of Fe^{3+} needed to precipitate $FePO_4$ can also be calculated to be equal to $1.1(10^{-11})$ gmol/L. Because these two concentrations are practically equal, $Fe(OH)_{3(s)}$ will definitely precipitate along with $FePO_4$. Therefore,

$$[Fe^{3+}] = \frac{\{Fe^{3+}\}}{\gamma_{\text{FeIII}}} = \frac{K_{sp,\text{Fe(OH)}_3}}{\gamma_{\text{FeIII}}\{OH^-\}^3} = \frac{K_{sp,\text{Fe(OH)}_3}\gamma_H^3[H^+]^3}{\gamma_{\text{FeIII}}K_w^3}.$$

Substituting in Equation (14.16),

$$[sp_{\text{PO}_4\text{FeIII}}] = \frac{K_{sp,\text{FePO}_4}K_w^3}{\gamma_{\text{PO}_4}K_{sp,\text{Fe(OH)}_3}\gamma_H^3[H^+]^3} + \frac{K_{sp,\text{FePO}_4}K_w^3}{\gamma_{\text{HPO}_4}K_{\text{HPO}_4}K_{sp,\text{Fe(OH)}_3}\gamma_H^2[H^+]^2}$$

$$+ \frac{K_{sp,\text{FePO}_4}K_w^3}{\gamma_{\text{H}_2\text{PO}_4}K_{\text{H}_2\text{PO}_4}K_{\text{HPO}_4}K_{sp,\text{Fe(OH)}_3}\gamma_H[H^+]} + \frac{K_{sp,\text{FePO}_4}K_w^3}{K_{\text{H}_3\text{PO}_4}K_{\text{H}_2\text{PO}_4}K_{\text{HPO}_4}K_{sp,\text{Fe(OH)}_3}}$$

$$(14.18)$$

Example 14.3   Calculate $[sp_{\text{PO}_4\text{FeIII}}]$ expressed as mg/L of P when the pH is 8. Assume the water contains 140 mg/L of dissolved solids.

Solution:

$$[sp_{\text{PO}_4\text{FeIII}}] = \frac{K_{sp,\text{FePO}_4}K_w^3}{\gamma_{\text{PO}_4}K_{sp,\text{Fe(OH)}_3}\gamma_H^3[H^+]^3} + \frac{K_{sp,\text{FePO}_4}K_w^3}{\gamma_{\text{HPO}_4}K_{\text{HPO}_4}K_{sp,\text{Fe(OH)}_3}\gamma_H^2[H^+]^2}$$

$$+ \frac{K_{sp,\text{FePO}_4}K_w^3}{\gamma_{\text{H}_2\text{PO}_4}K_{\text{H}_2\text{PO}_4}K_{\text{HPO}_4}K_{sp,\text{Fe(OH)}_3}\gamma_H[H^+]} + \frac{K_{sp,\text{FePO}_4}K_w^3}{K_{\text{H}_3\text{PO}_4}K_{\text{H}_2\text{PO}_4}K_{\text{HPO}_4}K_{sp,\text{Fe(OH)}_3}}$$

$$K_{sp,\text{FePO}_4} = 10^{-21.9} \quad \gamma_{\text{PO}_4} = 10^{-\frac{0.5(3)^2[\sqrt{3.5(10^{-3})}]}{1+1.14[\sqrt{3.5(10^{-3})}]}} = 0.56 \quad K_{sp,\text{Fe(OH)}_3} = 1.1(10^{-36})$$

$$\gamma_H = \gamma_{\text{H}_2\text{PO}_4} = 10^{-\frac{0.5(1)^2[\sqrt{3.5(10^{-3})}]}{1+1.14[\sqrt{3.5(10^{-3})}]}} = 0.94 \quad \gamma_{\text{HPO}_4} = 10^{-\frac{0.5(2)^2[\sqrt{3.5(10^{-3})}]}{1+1.14[\sqrt{3.5(10^{-3})}]}} = 0.77$$

$$K_{\text{HPO}_4} = 10^{-12.3} \quad K_{\text{H}_2\text{PO}_4} = 10^{-7.2} \quad K_{\text{HPO}_4} = 10^{-12.3} \quad K_{\text{H}_3\text{PO}_4} = 10^{-2.1}$$

$$[sp_{\text{PO}_4\text{FeIII}}] = \frac{(10^{-21.9})(10^{-14})^3}{(0.56)(1.1)(10^{-36})(0.94)^3[10^{-8}]^3}$$

$$+ \frac{(10^{-21.9})(10^{-14})^3}{(0.77)(10^{-12.3})(1.1)(10^{-36})(0.94)^2[10^{-8}]^2}$$

$$+\frac{(10^{-21.9})(10^{-14})^3}{(0.94)(10^{-7.2})(10^{-12.3})(1.1)(10^{-36})(0.94)[10^{-8}]}$$

$$+\frac{(10^{-21.9})(10^{-14})^3}{(10^{-2.1})(10^{-7.2})(10^{-12.3})(1.1)(10^{-36})}$$

$$=\frac{1.26(10^{-64})}{5.12(10^{-61})}+\frac{1.26(10^{-64})}{3.75(10^{-65})}+\frac{1.26(10^{-64})}{3.07(10^{-64})}+\frac{1.26(10^{-64})}{2.76(10^{-58})}$$

$$=2.46(10^{-4})+3.36+0.41+4.57(10^{-7})=3.77 \text{ gmols/L}$$

$$=1.17(10^5)\text{ mg/L} \quad \textbf{Ans}$$

14.5.1 Determination of the Optimum pH and the Optimum pH Range

As in the case of Eqs. (14.12) and (14.14), Equation (14.18) cannot guarantee an optimum if it is differentiated and equated to 0. Hence, the optimum and the optimum pH range will be determined by preparing a table as was done in the case of alum and lime. Tables 14.7 and 14.8 show the results. This table was prepared using Equation (14.18). As shown, the optimum range is equal to or less than 3.

TABLE 14.7
Concentration of $sp_{PO_4 FeIII}$ as a Function of pH at 25°C (mg/L as P)

pH, Dissolved Solids = 140 mg/L	$[sp_{PO_4 FeIII}]$ (mg/L)
0	$1.1(10^{-2})$
1	$1.2(10^{-2})$
2	$2.1(10^{-2})$
3	$1.1(10^{-1})$
4	$1.0(10^{-0})$
5	$1.0(10^{+1})$
6	$1.1(10^{+2})$
7	$1.8(10^{+3})$
8	$1.17(10^{+5})$
9	$8.4(10^{+6})$
10	$8.3(10^{+8})$
11	$8.9(10^{+10})$
12	$1.4(10^{+13})$
13	$6.9(10^{+15})$
14	$6.1(10^{+18})$
15	$6.0(10^{+21})$

TABLE 14.8

Concentration of sp_{PO_4FeIII} as a Function of pH at 25°C (mg/L as P)

pH, Dissolved Solids = 35,000 mg/L	$[sp_{PO_4FeIII}]$ (mg/L)
0	$1.2(10^{-2})$
1	$1.4(10^{-2})$
2	$3.7(10^{-2})$
3	$2.6(10^{-1})$
4	$2.6(10^{-0})$
5	$2.7(10^{+1})$
6	$3.8(10^{+2})$
7	$1.5(10^{+3})$
8	$1.3(10^{+6})$
9	$1.3(10^{+8})$
10	$1.4(10^{+10})$
11	$2.8(10^{+12})$
12	$1.6(10^{+15})$
13	$1.5(10^{+18})$
14	$1.5(10^{+21})$
15	$1.5(10^{+24})$

14.6 COMMENTS ON THE OPTIMUM pH RANGES

When the ferric salts are used to precipitate $FePO_4$, the PO_4^{3-} ion must compete with the OH^- ion. When the pH is high, there will be a large concentration of the OH^- and the phosphate ion can lose the competition and fail to precipitate; $Fe(OH)_3$ may precipitate instead of $FePO_4$. Thus, to precipitate the phosphate ion, the concentration of OH^- must be suppressed. This is done by adding more H^+, which reacts with the OH^-. If the OH^- is "busy" satisfying the hydrogen ion, the Fe^{3+} ion is now available for reaction with phosphate to precipitate $FePO_4$ instead of $Fe(OH)_3$. The addition of the hydrogen depresses the pH and the optimum pH is found to be equal to or less than 3.

For the case of the $AlPO_4$ precipitation, alum also produces the hydroxide $Al(OH)_3$ analogous to the ferric salts producing $Fe(OH)_3$. Thus, for the phosphate ion to outcompete the hydroxide ion, the pH must also be lowered in order to produce the desired precipitate of $AlPO_4$, instead of $Al(OH)_3$. Again, the pH is reduced by adding the hydrogen ion. The optimum pH range for $AlPO_4$ precipitation, we found to be equal to or less than 5.

In the case of the use of lime, the OH^- ion is a reactant for the production of the apatite, $Ca_5(PO_4)_3OH_{(s)}$. This is the reason why the pH must be raised, which is found to be equal to or greater than 7. In addition, no other precipitate competes with the precipitation of the apatite. Thus, overall, of the three precipitants, lime is the best.

14.7   EFFECT OF THE K_{sp}'s ON THE PRECIPITATION OF PHOSPHORUS

The worst of the precipitants are the ferric salts. Compare $K_{sp,\mathrm{Fe(OH)_3}} = 1.1(10^{-36})$ and $K_{sp,\mathrm{FePO_4}} = 10^{-21.9}$. These two K_{sp}'s produce comparable concentrations of the ferric ion to precipitate either $\mathrm{Fe(OH)_3}$ or $\mathrm{FePO_4}$. Thus, at high pH conditions, the phosphate ion would have a big competitor in the form of the hydroxide ion. An Fe^{3+} available in solution is grabbed by the OH^- ion to form the ferric hydroxide, leaving less amount of Fe^{3+} to precipitate ferric phosphate. The ferric salts are, therefore, a poor performer for removing phosphorus. Also, to be effective requires adjusting the pH to the pH of almost mineral acidity of less than 3.

Now, compare $K_{sp,\mathrm{Al(OH)_3}} = 10^{-33}$ and $K_{sp,\mathrm{AlPO_4}} = 10^{-21}$ for the case of using alum. This situation is similar to the case of the ferric salts. The concentrations of the aluminum ion to precipitate either the aluminum hydroxide or the aluminum phosphate are comparable. This means that the hydroxyl ion is also a competitor in the phosphate precipitation. The process operates poorly at high pH, although it has a better pH range of equal to or less than 5 compared to that of the ferric salts.

In the case of using lime to precipitate phosphorus, we know that calcium hydroxide has a much, much larger K_{sp} than that of the apatite, the required precipitate. The former has a K_{sp} of $K_{sp,\mathrm{Ca(OH)_2}} = 7.9(10^{-6})$. Compare this to that of apatite which is $K_{sp,apatite} = 10^{-55.9}$. Thus, adding lime to the water means that a large amount of the calcium ion is available to precipitate the apatite, $\mathrm{Ca_5(PO_4)_3OH_{(s)}}$. Also, no other precipitate competes with this precipitation. Therefore, using lime is the best method for removing phosphorus. Adding a quantity of lime to the point of saturation, $\mathrm{Ca(OH)_2}$ has the potential of removing phosphorus "100%." This process must, however, be investigated. In the overall, the process using lime entails adjusting the pH to greater than 7, which is much, much better than the range for the use of either alum or ferric salts.

14.8   UNIT OPERATIONS FOR PHOSPHORUS REMOVAL

The removal of phosphorus involves provision for the addition of chemicals both for adjusting the pH and for precipitating the phosphate, mixing them with the raw water, settling the resulting precipitates, and filtering those precipitates that escape removal by settling. The precipitates formed are still in suspension; thus, it is also necessary for them to flocculate by adding a flocculation tank.

14.9   CHEMICAL REQUIREMENTS

The chemicals required are alum, lime, and the ferric salts $\mathrm{FeCl_3}$ and $\mathrm{Fe_2(SO_4)_3}$. These are in addition to the acid or bases needed to adjust the pH. The formulas used to calculate the amounts of these chemicals for pH adjustment were already derived in Chapters 11, 12, and 13. The chemical requirements to be discussed here will only be for alum, lime, and the ferric salts.

The amount of alum needed to precipitate the phosphate is composed of the alum required to satisfy the natural alkalinity of the water and the amount needed to precipitate the phosphate. Satisfaction of the natural alkalinity will bring the equilibrium of aluminum hydroxide. Remember, however, that even if these quantities of alum were provided, the concentration of phosphorus that will be discharged from the effluent of the unit still has to conform to the equilibrium reaction that depends upon the pH level at which the process was conducted. The optimum pH, we have found, is equal to or less 5.

Consider the full alum formula in the chemical reaction,

$$Al_2(SO_4)_3 \cdot xH_2O + 2\,PO_4^{3-} \rightarrow 2\,AlPO_{4(s)}\downarrow + 3\,SO_4^{2-} + xH_2O \qquad (14.19)$$

Note: Only one arrow is used and not the forward and backward arrows. The reaction is not in equilibrium.

As mentioned before, the previous reaction for phosphate removal takes place in conjunction with the $Al(OH)_3$ equilibrium. Bicarbonate alkalinity is always present in natural waters, so this equilibrium is facilitated by alum reacting with calcium bicarbonate so that, in addition to the amount of alum required for the precipitation of phosphorus, more is needed to neutralize the bicarbonate. The reaction is represented by

$$Al_2(SO_4)_3 \cdot xH_2O + 3\,Ca(HCO_3)_2 \rightarrow 2\,Al(OH)_3\downarrow + 6\,CO_2 + 3\,CaSO_4 + xH_2O$$
$$(14.20)$$

From the previous reactions, the equivalent masses are alum = $Al_2(SO_4)_3 \cdot xH_2O/6 = 57.05 + 3x$ and calcium bicarbonate = $3\,Ca(HCO_3)_2/6 = 81.05$. The number of moles of phosphorus in PO_4^{3-} is one; thus, from the balanced chemical reaction, the equivalent mass of phosphorus is 2P/6 = 10.33.

Although the alkalinity found in natural waters is normally calcium bicarbonate, it is conceivable that natural alkalinities could also be associated with other cations. Thus, to be specific, as shown by Equation (14.20), the alkalinity discussed here is calcium bicarbonate, with the understanding that "other forms" of natural alkalinities may be equated to calcium bicarbonate alkalinity when expressed in equivalent concentrations. Let A_{Ca} be this calcium bicarbonate alkalinity.

The quantity of alum needed to react with the calcium bicarbonate alkalinity is exactly the quantity of alum needed to establish the aluminum hydroxide equilibrium. Let $[A_{Ca}]_{geq}$ be the gram equivalents per liter of this bicarbonate alkalinity in the raw water of volume V cubic meters and let M_{AlAlk} be the kilograms of alum at a fractional purity of P_{Al} required to react with it. Then,

$$M_{AlAlk} = \frac{[A_{Ca}]_{geq}(57.05 + 3x)V}{P_{Al}} \qquad (14.21)$$

Remember that alkalinity is also expressed analytically in terms of calcium carbonate in which the equivalent mass is considered 50. Let us digress from our main discussion and explore this matter further in order to determine how the various methods of expressing alkalinities are related. How will the analytical result be converted to an equivalent concentration to be used in Equation (14.21)?

Consider the following reactions:

$$CaCO_3 + 2H^+ \rightarrow H_2CO_3 + CA^{2+} \tag{14.22}$$

$$2Ca(HCO_3)_2 + 2H^+ \rightarrow H_2CO_3 + 2Ca^{2+} \tag{14.23}$$

From these reactions, the equivalent mass of calcium bicarbonate in relation to the equivalent mass of calcium carbonate of 50 is $2Ca(HCO_3)_2/2 = 162.1$ (much different in comparison with 81.05). Now, assume $[A_{Ca}]_{geqCaCO_3}$ gram equivalents per liter of calcium bicarbonate alkalinity being computed on the basis of Equation (14.23). This computation, in turn, is obtained from an analytical result based on Equation (14.22). The number of grams of the bicarbonate alkalinity is then equal to 162.1 $[A_{Ca}]_{geqCaCO_3}$. The quantity of mass remains the same in any system. Thus, referring to Equation (14.20), there will also be the same 162.1 $[A_{Ca}]_{geqCaCO_3}$ grams of the bicarbonate there. To find the number of equivalents, however, it is always equal to the mass divided by the equivalent mass. Therefore, based on Equation (14.20), the corresponding number of equivalents of the 162.1 $[A_{Ca}]_{geqCaCO_3}$ grams of calcium bicarbonate alkalinity is 162.1 $[A_{Ca}]_{geqCaCO_3}/81.05 = 2[A_{Ca}]_{geqCaCO_3}$. And,

$$[A_{Ca}]_{geq} = 2[A_{Ca}]_{geqCaCO_3} \tag{14.24}$$

Again, $[A_{Ca}]_{geq}$ is the calcium bicarbonate alkalinity referred to the reaction of alum with calcium bicarbonate and $[A_{Ca}]_{geqCaCO_3}$ is the same bicarbonate alkalinity referred to calcium carbonate with equivalent mass of 50. This equation converts analytical results expressed as calcium carbonate to a form that Equation (14.21) can use.

Now, let $[Phos]_{mg}$ be the milligram per liter of phosphorus as P in phosphate and let M_{AlPhos} be the kilograms of alum at fractional purity of P_{Al} required to precipitate this phosphate phosphorus. Letting f_P represent the fractional removal of phosphorus, the following equation is then produced:

$$M_{AlPhos} = \frac{f_P[Phos]_{mg}}{1000(10.33)P_{Al}}(57.05 + 3x) \, V = \frac{9.7(10^{-5})f_P[Phos]_{mg}(57.05 + 3x) \, V}{P_{Al}} \tag{14.25}$$

Letting M_{Al} be the total kilograms of alum needed for the removal of phosphorus,

$$M_{Al} = M_{AlAlk} + M_{AlPhos} = \frac{(57.05 + 3x)}{P_{Al}}([A_{Ca}]_{geq} + 9.7(10^{-5})f_P[Phos]_{mg}) \, V \tag{14.26}$$

Now, let us tackle the problem of calculating the chemical requirements for removing calcium apatite. Although calcium carbonate has a very high K_{sp} that it would not be precipitating with the apatite, the natural alkalinity would have to be neutralized nonetheless, producing the carbonate. The formation of the calcium carbonate solid will only occur, however, after the apatite formation has completed. Thus, the amount of lime needed to precipitate the phosphate is composed of the lime needed to precipitate the phosphate in apatite and the lime needed to neutralize the natural alkalinity of the water.

Note: Even if these quantities were provided, the concentration of phosphorus that would be discharged from the effluent of the unit still has to conform to the equilibrium reaction that depends upon the pH level at which the process was conducted.

The optimum pH, we have found, is equal to or greater than 7.
The applicable chemical reactions are:

$$Ca(OH)_2 + Ca(HCO_3)_2 \rightarrow H_2O + CaCO_{3(s)} \downarrow \qquad (14.27)$$

$$Ca(OH)_2 + \frac{3}{5} PO_4^{3-} \rightleftharpoons \frac{1}{5} Ca_5(PO_4)_3OH_{(s)} \downarrow + \frac{9}{5} OH^- \qquad (14.28)$$

To make the previous equations compatible, use 2 as the number of reference species, based on $Ca(OH)_2$. Therefore the equivalent masses are lime = CaO/2 = 37.05; calcium bicarbonate = $Ca(HCO_3)_2/2$ = 81.05; and phosphorus = 3/5P/2 = 9.3. In this particular case, because the equivalent mass of calcium bicarbonate is also equal to 81.05, $[A_{Ca}]_{geq} = 2[A_{Ca}]_{geqCaCO_3}$.

Let M_{CaOAlk} be the kilograms of lime at fractional purity of P_{CaO} required to react with the calcium carbonate alkalinity. Then,

$$M_{CaOAlk} = \frac{37.05[A_{Ca}]_{geq} V}{P_{CaO}} \qquad (14.29)$$

Now, letting $M_{CaOPhos}$ be the kilograms of lime required to react with the phosphorus,

$$M_{CaOPhos} = \frac{37.05 f_P[Phos]_{mg}}{1000(9.3)P_{CaO}} V = \frac{4.0(10^{-3}) f_P[Phos]_{mg} V}{P_{CaO}} \qquad (14.30)$$

Let M_{CaO} be the total kilograms of lime needed for the removal of phosphorus. Thus,

$$M_{CaO} = M_{CaOAlk} + M_{CaOPhos} = (37.05[A_{Ca}]_{geq} + 4.0(10^{-3}) f_P[Phos]_{mg}) \frac{V}{P_{CaO}} \qquad (14.31)$$

Last, let us tackle the calculation of the amount of ferric salts needed to precipitate the phosphate. This amount is composed of the amount needed to neutralize the natural alkalinity of the water and the amount needed to precipitate the phosphate. Again, remember, that even if these quantities were provided, the concentration of phosphorus that will be discharged from the effluent of the unit still has to conform to the equilibrium reaction that depends upon the pH level at which the process was conducted. The optimum pH, we have found, is equal to or less than 3.

The applicable respective chemical reactions for the satisfaction of the bicarbonate alkalinity are

$$2FeCl_3 + 3Ca(HCO_3)_2 \rightarrow 2Fe(OH)_3\downarrow + 6CO_2 + 3CaCl_2 \qquad (14.32)$$

$$Fe_2(SO_4)_3 + 3Ca(HCO_3)_2 \rightarrow 2Fe(OH)_3\downarrow + 6CO_2 + 3CaSO_4 \qquad (14.33)$$

Upon satisfaction of the previous reactions, the normal precipitation of the phosphate occurs. The respective reactions are

$$2FeCl_3 + 2PO_4^{3-} \rightarrow 2FePO_4\downarrow + 6Cl^- \qquad (14.34)$$

$$Fe_2(SO_4)_3 + 2PO_4^{3-} \rightarrow 2FePO_4\downarrow + 3SO_4^{2-} \qquad (14.35)$$

Note that the coefficients of Equation (14.34) have been adjusted to make the equation compatible with Equation (14.32) and all the rest of the above equations. From these equations, the equivalent masses are ferric chloride = $2FeCl_3/6$ = 54.1; calcium bicarbonate = $3Ca(HCO_3)_2/6$ = 81.05; ferric sulfate = $Fe_2(SO_4)_3/6$ = 66.65; and phosphorus = $2P/6$ = 15.5. Since the equivalent mass is still 81.05, $[A_{Ca}]_{geq}$ = 2 $[A_{Ca}]_{geqCaO_3}$.

Let $M_{FeIIIClAlk}$ and $M_{FeIIISO_4Alk}$ be the kilograms of ferric chloride and ferric sulfate, respectively, required to react with the calcium bicarbonate alkalinity. Also, let the respective purity be $P_{FeIIICl}$ and $P_{FeIIISO_4}$. Then,

$$M_{FeIIIClAlk} = \frac{54.1[A_{Ca}]_{geq}\,\Psi}{P_{FeIIICl}} \qquad (14.36)$$

$$M_{FeIIISO_4Alk} = \frac{66.65[A_{Ca}]_{geq}\,\Psi}{P_{FeIIISO_4}} \qquad (14.37)$$

Now, letting $M_{FeIIIClPhos}$ and $M_{FeIIISO_4Phos}$ be the kilograms of ferric chloride and ferric sulfate, respectively, required to react with the phosphorus,

$$M_{FeIIIClPhos} = \frac{54.1f_P[Phos]_{mg}}{1000(15.5)P_{FeIIICl}}\Psi = \frac{3.5(10^{-3})f_P[Phos]_{mg}\Psi}{P_{FeIIICl}} \qquad (14.38)$$

$$M_{FeIIISO_4Phos} = \frac{66.5f_P[Phos]_{mg}}{1000(15.5)P_{FeIIISO_4}}\Psi = \frac{4.3(10^{-3})f_P[Phos]_{mg}\Psi}{P_{FeIIISO_4}} \qquad (14.39)$$

Let $M_{FeIIICl}$ and $M_{FeIIISO_4}$ be the total kilograms of ferric chloride and ferric sulfate, respectively, needed for the removal of phosphorus. Thus,

$$M_{FeIIICl} = M_{FeIIIClAlk} + M_{FeIIIClPhos} = (54.1[A_{Ca}]_{geq} + 3.5(10^{-3})f_P[Phos]_{mg})\frac{\Psi}{P_{FeIIICl}}$$

$$(14.40)$$

$$M_{FeIIISO_4} = M_{FeIIISO_4Alk} + M_{FeIIISO_4Phos}$$

$$= (66.5[A_{Ca}]_{geq} + 4.3(10^{-3})f_P[Phos]_{mg})\frac{\Psi}{P_{FeIIISO_4}} \qquad (14.41)$$

14.10 SLUDGE PRODUCTION

Phosphorus is normally removed from wastewaters and, as such, the solids produced would include the original suspended solids or turbidity. Therefore, two sources contribute to the production of solids in phosphorus removal: the original suspended solids and the solids produced from chemical reaction. If V is the cubic meters of water treated and sp_{ss} are the solids removed at a fractional efficiency of f_{ss}, the kilograms of solids produced from the suspended solids is $(f_{ss}[sp_{ss}]_{mg}/1000)V$, where $[sp_{ss}]_{mg}$ is in mg/L.

For convenience, the chemical reactions responsible for production of chemical sludge are reproduced below. They were derived previously.

$$Al_2(SO_4)_3 \cdot xH_2O + 2PO_4^{3-} \rightarrow 2AlPO_{4(s)}\downarrow + 3SO_4^{2-} + xH_2O \qquad (14.42)$$

$$Al_2(SO_4)_3 \cdot xH_2O + 3Ca(HCO_3)_2 \rightarrow 2Al(OH)_3\downarrow + 6CO_2 + 3CaSO_4 + xH_2O \qquad (14.43)$$

$$Ca(OH)_2 + Ca(HCO_3)_2 \rightarrow H_2O + CaCO_{3(s)}\downarrow \qquad (14.44)$$

$$Ca(OH)_2 + \frac{3}{5}PO_4^{3-} \rightleftharpoons \frac{1}{5}Ca_5(PO_4)_3OH_{(s)}\downarrow + \frac{9}{5}OH^- \qquad (14.45)$$

$$2FeCl_3 + 3Ca(HCO_3)_2 \rightarrow 2Fe(OH)_3\downarrow + 6CO_2 + 3CaCl_2 \qquad (14.46)$$

$$Fe_2(SO_4)_3 + 3Ca(HCO_3)_2 \rightarrow 2Fe(OH)_3\downarrow + 6CO_2 + 3CaSO_4 \qquad (14.47)$$

$$2FeCl_3 + 2PO_4^{3-} \rightarrow 2FePO_4\downarrow + 6Cl^- \qquad (14.48)$$

$$Fe_2(SO_4)_3 + 2PO_4^{3-} \rightarrow 2FePO_4\downarrow + 3SO_4^{2-} \qquad (14.49)$$

Referring to the previous reactions, the solids produced from the removal of phosphorus are $AlPO_{4(s)}$, $Al(OH)_3$, $CaCO_3$, $Ca_5(PO_4)_3OH_{(s)}$, $Fe(OH)_3$, and $FePO_4$. The equivalent masses are aluminum phosphate $= 2AlPO_4/6 = 40.67$; aluminum hydroxide is $2Al(OH)_3/6 = 26.0$; calcium carbonate $= CaCO_3/2 = 50$; calcium hydroxyapatite $= 1/5Ca_5(PO_4)_3OH/2 = 50.25$; ferric hydroxide $= 2Fe(OH)_3/6 = 35.6$; and ferric phosphate $= 2FePO_4/6 = 50.27$. The equivalent mass of phosphorus varies. For Eqs. (14.42), (14.48), and (14.49), the equivalent mass of phosphorus $= 2P/6 = 10.33$; from Equation (14.45), equivalent mass $= 3/5P/2 = 9.3$. For calcium carbonate, the equivalent mass in all the reactions is $3Ca(HCO_3)_2/6$ or $Ca(HCO_3)_2/2 = 81.05$. Thus,

$$[A_{Ca}]_{geq} = 2[A_{Ca}]_{geqCaCO_3} \qquad (14.50)$$

From treating $V\,m^3$ of water with $[Phos]_{mg}$ mg/L of phosphate phosphorus being removed, let M_{AlPO_4} = the kilograms of aluminum phosphate solids produced;

$M_{Ca_5(PO_4)_3OH}$ = kilograms of apatite solids produced; and M_{FePO_4} = kilograms of ferric phosphate solids produced. Also, from reacting with $[A_{cur}]_{geq}$ gram moles per liter of natural alkalinity in V cubic meters of water, let $M_{Al(OH)_3}$ = kilograms of alum solids produced; M_{CaCO_3} = kilograms calcium carbonate solids produced; and $M_{Fe(OH)_3}$ = kilograms of ferric hydroxide solids produced. Finally, let M_{ss} = kilograms of solids produced from the original suspended solids. Considering the respective equivalent masses for the above equations, the following equations are produced:

$$M_{AlPO_4} = \frac{40.67 f_P[Phos]_{mg} V}{1000(10.33)} = 3.94\,(10^{-3}) f_P[Phos]_{mg} V \qquad (14.51)$$

$$M_{Ca_5(PO_4)_3OH} = \frac{50.25 f_P[Phos]_{mg} V}{1000(9.3)} = 5.40\,(10^{-3}) f_P[Phos]_{mg} V \qquad (14.52)$$

$$M_{FePO_4} = \frac{50.27 f_P[Phos]_{mg} V}{1000(10.33)} = 4.87(10^{-3}) f_P[Phos]_{mg} V \qquad (14.53)$$

$$M_{Al(OH)_3} = 26.0\,[A_{Ca}]_{geq} V \qquad (14.54)$$

$$M_{CaCO_3} = 50\,[A_{Ca}]_{geq} V \qquad (14.55)$$

$$M_{Fe(OH)_3} = 35.6[A_{Ca}]_{geq} V \qquad (14.56)$$

$$M_{ss} = \frac{f_{ss}[sp_{ss}]_{mg}}{1000} V \qquad (14.57)$$

Now, let $M_{AlPO_4 Solids}$ = total solids produced in removing phosphorus using alum; $M_{CaOPO_4 Solids}$ = total solids produced in removing phosphorus using lime; and $M_{FeIIIPO_4 Solids}$ = total solids produced in removing phosphorus using the ferric salts. $M_{AlPO_4 Solids} = M_{AlPO_4} + M_{Al(OH)_3} + M_{ss}$; $M_{CaOPO_4 Solids} = M_{Ca_5(PO_4)_3OH} + M_{CaCO_3} + M_{ss}$; and $M_{FeIIIPO_4 Solids} = M_{FePO_4} + M_{Fe(OH)_3} + M_{ss}$. Therefore,

$$M_{AlPO_4 Solids} = \left(3.94(10^{-3}) f_P[Phos]_{mg} + 26.0\,[A_{Ca}]_{geq} + \frac{f_{ss}[sp_{ss}]_{mg}}{1000}\right) V \qquad (14.58)$$

$$M_{CaOPO_4 Solids} = \left(5.40(10^{-3}) f_P[Phos]_{mg} + 50\,[A_{Ca}]_{geq} + \frac{f_{ss}[sp_{ss}]_{mg}}{1000}\right) V \qquad (14.59)$$

$$M_{FeIIIPO_4 Solids} = \left(4.87(10^{-3}) f_P[Phos]_{mg} + 35.6\,[A_{Ca}]_{geq} + \frac{f_{ss}[sp_{ss}]_{mg}}{1000}\right) V \qquad (14.60)$$

Example 14.4 A wastewater contains 10 mg/L of phosphorus expressed as P. The pH is 8.0. Assume the wastewater contains 100 mg/L of alkalinity expressed as $CaCO_3$. The phosphorus is to be removed using lime. If the wastewater flow is 0.75 m^3/sec, calculate the kilograms of solids produced per day. Assume the suspended solids content is 200 mg/L.

Solution:

$$M_{CaOPO_4 Solids} = \left(5.40(10^{-3}) f_P [Phos]_{mg} + 50 [A_{Ca}]_{geq} + \frac{f_{ss}[sp_{ss}]_{mg}}{1000} \right) V$$

Assume all of the phosphorus and suspended solids removed. Therefore, $f_P = 1; f_{ss} = 1$

$$[A_{Ca}]_{geq} = 2 [A_{Ca}]_{geqCaCO_3} = \frac{2(100)}{(1000)(50)} = 0.004 \text{ geq/L}$$

Therefore,

$$M_{CaOPO_4 Solids} = \left(5.40(10^{-3})(1)(10) + 50(0.004) + \frac{(1)(200)}{1000} \right)(0.75)(60)(60)(24)$$

$$= 29,419.2 \text{ kg/d} \quad \textbf{Ans}$$

GLOSSARY

Calcium hydroxy apatite—$Ca_5(PO_3)_3(OH)_{(s)}$.

Condensation—The formation of water molecules when substances react.

Condensed phosphates—Substances containing phosphates characterized by the presence of the P–O–P bonds formed when phosphoric acid combines in a process called condensation.

Metaphosphoric acid—A condensation product of phosphoric acid forming a chain of $(HPO_3)_n$.

Optimum pH—The pH at which maximum precipitation occurs or at which the concentration of the species to be removed is at the lowest.

Optimum pH range—A range of pH at which the quantity of precipitates produced or the concentration of the species to be removed remaining in solution is acceptable.

Orbitals—A volume of atomic space in which a maximum of two electrons occupy.

Orthophosphates—The group of species containing the of $H_2PO_4^-$, HPO_4^{2-}, and PO_4^{3-} groups.

Oxidation states—A condition of being an atom in relation to the extent of the atom losing or gaining electrons in its valence shell.

Polyphosphates—A compound containing two or more phosphate groups in the molecule.

Shell—A volume of atomic space holding electronic orbitals.

Solubility product constant—The product of the concentrations of ions, raised to appropriate powers, in equilibrium with their solid.

Valence configuration—The arrangement of electrons in an atomic shell that accounts for the chemical reactivity of the atom.

Valence shell—The shell of an atom responsible for chemical reactions.

SYMBOLS

A_{ca}	Calcium bicarbonate alkalinity
$[A_{Ca}]_{geq}$	Gram equivalents per liter of calcium bicarbonate alkalinity
$[A_{Ca}]_{geqCaCO_3}$	Gram equivalents per liter of alkalinity being computed on the basis of $CaCO_3$ having an equivalent mass of 50
$Al(OH)^{2+}$	Hydroxo Al(III) ion
$Al_7(OH)_{17}^{4+}$	17-hydroxo, 7-Al(III) complex ion
$Al_{13}(OH)_{34}^{5+}$	34-hydroxo, 13-Al(III) complex ion
$Al(OH)_4^-$	Tetrahydroxo Al(III) complex ion
$Al_2(OH)_2^{4+}$	Dihydroxo 2-Al(III) complex ion
f_{ss}	Fractional removal of original suspended of raw water
K_{HPO_4}	Equilibrium constant of the hydrogen phosphate ion
$K_{H_2PO_4}$	Equilibrium constant of the dihydrogen phosphate ion
$K_{H_3PO_4}$	Equilibrium constant of phosphoric acid
$K_{sp,Al(OH)_3}$	Solubility product constant of aluminum hydroxide
$K_{sp,AlPO_4}$	Solubility product constant of aluminum phosphate
$K_{sp,apatite}$	Solubility product constant of calcium hydroxy apatite phosphate
$K_{sp,Ca(OH)_2}$	Solubility product constant of calcium hydroxide
$K_{sp,Fe(OH)_3}$	Solubility product constant of ferric hydroxide
$K_{sp,FePO_4}$	Solubility product constant of ferric phosphate
K_w	Ion product of water
M_{Al}	Total kilograms of alum needed for the removal of phosphorus
M_{AlAlk}	Kilograms of alum that react with the calcium bicarbonate alkalinity
$M_{Al(OH)_3}$	Kilograms of alum solids produced in the removal of phosphorus
M_{AlPhos}	Kilograms of alum that react with phosphate phosphorus in raw water
M_{AlPO_4}	Kilograms of aluminum phosphate solids produced in the removal of phosphorus
$M_{AlPO_4Solids}$	Total solids produced in removing phosphorus using alum as the precipitant
$M_{CaOPO_4Solids}$	Total solids produced in removing phosphorus using lime as the precipitant (including suspended solids in raw water)
$M_{Ca_5(PO_4)_3OH}$	Kilograms of apatite solids produced in the removal of phosphorus
M_{CaCO_3}	Kilograms calcium carbonate solids produced in the removal of phosphorus
M_{CaO}	Kilograms of lime needed for the removal of phosphorus
M_{CaOAlk}	Kilograms of lime that react with the calcium bicarbonate alkalinity of the raw water
$M_{CaOPhos}$	Kilograms of lime that react with phosphate phosphorus in raw water
$M_{FeIIICl}$	Total kilograms of ferric chloride used for phosphate removal
$M_{FeIIIClAlk}$	Kilograms of ferric chloride that react with the calcium bicarbonate of the raw water

$M_{FeIIIClPhos}$	Kilograms of ferric chloride that reacts with the phosphate phosphorus of the raw water
$M_{Fe(OH)_3}$	Kilograms of ferric hydroxide solids produced in the removal of phosphorus
M_{FePO_4}	Kilograms of ferric phosphate solids produced in the removal of phosphorus
$M_{FeIIIPO_4Solids}$	Total solids produced in removing phosphorus using the ferric salts as precipitants (including suspended solids in raw water)
M_{ss}	Kilograms of solids produced from the original suspended solids in the removal of phosphorus
$M_{FeIIISO_4}$	Total kilograms of ferric sulfate used for phosphate removal
$M_{FeIIISO_4Alk}$	Kilograms of ferric sulfate that react with the calcium bicarbonate alkalinity of the raw water
$M_{FeIIISO_4Phos}$	Kilograms of ferric sulfate that react with the phosphate phosphorus of the raw water
p	The p orbital of the valence shell
P_{Al}	Fractional purity of alum
P_{CaO}	Fractional purity of lime
$P_{FeIIICl}$	Fractional purity of ferric chloride
$P_{FeIIISO_4}$	Fractional purity of ferric sulfate
$[Phos]_{mg}$	Milligram per liter of phosphate phosphorus in raw water
s	The s orbital of the valence shell
sp_{PO_4Al}	Species in solution containing the PO_4 species of the orthophosphates, using alum as the precipitant
sp_{PO_4Ca}	Species in solution containing the PO_4 species of the orthophosphates, using lime as the precipitant
sp_{PO_4FeIII}	Species in solution containing the PO_4 species of the orthophosphates, using the ferric salts $FeCl_3$ and $Fe_2(SO_4)_3$ as precipitants
$[sp_{ss}]_{mg}$	Milligrams per liter of suspended solids in raw water
Ψ	Cubic meters of water treated
x	Water of hydration of alum
γ_{Ca}	Activity coefficient of the calcium ion
γ_H	Activity coefficient of the hydrogen ion
γ_{HPO_4}	Activity coefficient of the hydrogen phosphate ion
$\gamma_{H_2PO_4}$	Activity coefficient of the dihydrogen phosphate ion
γ_{PO_4}	Activity coefficient of the phosphate ion

PROBLEMS

14.1   A plant is conducting phosphorus removal at a pH of 5.0. $[sp_{PO_4Al}]$ is calculated to be 0.112 mg/L as P and the unit is operated at approximately an average temperature of 25°C. The dissolved solids content of the raw water is 140 mg/L. Calculate $K_{sp,AlPO_4}$ assuming it cannot be determined from the given temperature.

14.2 A plant is conducting phosphorus removal at a pH of 5.0. $[sp_{PO_4Al}]$ is calculated to be 0.112 mg/L as P and the unit is operated at approximately an average temperature of 25°C. The dissolved solids content of the raw water is 140 mg/L. Calculate K_w assuming it cannot be determined from the given temperature.

14.3 A plant is conducting phosphorus removal at a pH of 5.0. $[sp_{PO_4Al}]$ is calculated to be 0.112 mg/L as P and the unit is operated at approximately an average temperature of 25°C. The dissolved solids content of the raw water is 140 mg/L. Calculate $K_{sp,Al(OH)_3}$ assuming it cannot be determined from the given temperature.

14.4 A plant is conducting phosphorus removal at a pH of 5.0. $[sp_{PO_4Al}]$ is calculated to be 0.112 mg/L as P and the unit is operated at approximately an average temperature of 25°C. The dissolved solids content of the raw water is 140 mg/L. Calculate γ_H assuming it cannot be determined from the given dissolved solids content of the raw water.

14.5 A plant is conducting phosphorus removal at an average temperature of approximately 25°C. $[sp_{PO_4Al}]$ is calculated to be 0.112 mg/L as P. The dissolved solids content of the raw water is 140 mg/L. Calculate the pH at which the unit is being operated.

14.6 A plant is conducting phosphorus removal at an average temperature of approximately 25°C and at a pH of 5.0. $[sp_{PO_4Al}]$ is calculated to be 0.112 mg/L as P. The dissolved solids content of the raw water is 140 mg/L. Calculate the γ_{PO_4} assuming it cannot be determined from the given dissolved solids content of the raw water.

14.7 A plant is conducting phosphorus removal at an average temperature of approximately 25°C and at a pH of 5.0. $[sp_{PO_4Al}]$ is calculated to be 0.112 mg/L as P. The dissolved solids content of the raw water is 140 mg/L. Calculate the γ_{HPO_4} assuming it cannot be determined from the given dissolved solids content of the raw water.

14.8 A plant is conducting phosphorus removal at an average temperature of approximately 25°C and at a pH of 5.0. $[sp_{PO_4Al}]$ is calculated to be 0.112 mg/L as P. The dissolved solids content of the raw water is 140 mg/L. Calculate the K_{HPO_4} assuming it cannot be determined from the given temperature.

14.9 A plant is conducting phosphorus removal at an average temperature of approximately 25°C and at a pH of 5.0. $[sp_{PO_4Al}]$ is calculated to be 0.112 mg/L as P. The dissolved solids content of the raw water is 140 mg/L. Calculate the $\gamma_{H_2PO_4}$ assuming it cannot be determined from the given dissolved solids content of the raw water.

14.10 A plant is conducting phosphorus removal at an average temperature of approximately 25°C and at a pH of 5.0. $[sp_{PO_4Al}]$ is calculated to be 0.112 mg/L as P. The dissolved solids content of the raw water is 140 mg/L. Calculate the $K_{H_2PO_4}$ assuming it cannot be determined from the given temperature.

14.11 A plant is conducting phosphorus removal at an average temperature of approximately 25°C and at a pH of 5.0. $[sp_{PO_4Al}]$ is calculated to be 0.112 mg/L as P. The dissolved solids content of the raw water is 140 mg/L. Calculate the $K_{H_3PO_4}$ assuming it cannot be determined from the given temperature.

14.12   Calculate $[sp_{PO_4Ca}]$ expressed as P when the pH is 7.0. Assume the water contains 140 mg/L of dissolved solids and that $[Ca^{2+}] = 130$ mg/L as $CaCO_3$. What is the precipitant used?

14.13   A plant is conducting phosphorus removal at a pH of 7.0. $[sp_{PO_4Ca}]$ is calculated to be 0.0126 mg/L as P and the unit is operated at approximately an average temperature of 25°C. The dissolved solids content of the raw water is 140 mg/L, and the concentration of the calcium ion is 130 mg/L. Calculate $K_{sp,apatite}$ assuming that it cannot be determined from the given temperature.

14.14   A plant is conducting phosphorus removal at a pH of 7.0. $[sp_{PO_4Ca}]$ is calculated to be 0.0126 mg/L as P and the unit is operated at approximately an average temperature of 25°C. The dissolved solids content of the raw water is 140 mg/L, and the concentration of the calcium ion is 130 mg/L. Calculate K_w assuming that it cannot be determined from the given temperature.

14.15   A plant is conducting phosphorus removal at a pH of 7.0. $[sp_{PO_4Ca}]$ is calculated to be 0.0126 mg/L as P and the unit is operated at approximately an average temperature of 25°C. The dissolved solids content of the raw water is 140 mg/L, and the concentration of the calcium ion is 130 mg/L. What is the precipitant used? Calculate γ_{Ca} assuming that it cannot be determined from the given dissolved solids content of the raw water.

14.16   A plant is conducting phosphorus removal at a pH of 7.0. $[sp_{PO_4Ca}]$ is calculated to be 0.0126 mg/L as P and the unit is operated at approximately an average temperature of 25°C. The dissolved solids content of the raw water is 140 mg/L. What is the precipitant used? Calculate [Ca].

14.17   A plant is conducting phosphorus removal at a pH of 7.0. $[sp_{PO_4Ca}]$ is calculated to be 0.0126 mg/L as P and the unit is operated at approximately an average temperature of 25°C. The dissolved solids content of the raw water is 140 mg/L, and the concentration of the calcium ion is 130 mg/L. Calculate γ_H assuming that it cannot be determined from the given dissolved solids content of the raw water.

14.18   A plant is conducting phosphorus removal. $[sp_{PO_4Ca}]$ is calculated to be 0.0126 mg/L as P and the unit is operated at approximately an average temperature of 25°C. The dissolved solids content of the raw water is 140 mg/L, and the concentration of the calcium ion is 130 mg/L. Calculate the pH.

14.19   A plant is conducting phosphorus removal at a pH of 7.0. $[sp_{PO_4Ca}]$ is calculated to be 0.0126 mg/L as P and the unit is operated at approximately an average temperature of 25°C. The dissolved solids content of the raw water is 140 mg/L, and the concentration of the calcium ion is 130 mg/L. Calculate γ_{PO_4} assuming that it cannot be determined from the given dissolved solids content of the raw water.

14.20   A plant is conducting phosphorus removal at a pH of 7.0. $[sp_{PO_4Ca}]$ is calculated to be 0.0126 mg/L as P and the unit is operated at approximately an average temperature of 25°C. The dissolved solids content of the raw water is 140 mg/L, and the concentration of the calcium ion is 130 mg/L.

Calculate γ_{HPO_4} assuming that it cannot be determined from the given dissolved solids content of the raw water.

14.21 A plant is conducting phosphorus removal at a pH of 7.0. $[sp_{PO_4Ca}]$ is calculated to be 0.0126 mg/L as P and the unit is operated at approximately an average temperature of 25°C. The dissolved solids content of the raw water is 140 mg/L, and the concentration of the calcium ion is 130 mg/L. Calculate K_{HPO_4} assuming that it cannot be determined from the given temperature.

14.22 A plant is conducting phosphorus removal at a pH of 7.0. $[sp_{PO_4Ca}]$ is calculated to be 0.0126 mg/L as P and the unit is operated at approximately an average temperature of 25°C. The dissolved solids content of the raw water is 140 mg/L, and the concentration of the calcium ion is 130 mg/L. Calculate $\gamma_{H_2PO_4}$ assuming that it cannot be determined from the given dissolved solids content of the raw water.

14.23 A plant is conducting phosphorus removal at a pH of 7.0. $[sp_{PO_4Ca}]$ is calculated to be 0.0126 mg/L as P and the unit is operated at approximately an average temperature of 25°C. The dissolved solids content of the raw water is 140 mg/L, and the concentration of the calcium ion is 130 mg/L. Calculate $K_{H_2PO_4}$ assuming that it cannot be determined from the given temperature.

14.24 A plant is conducting phosphorus removal at a pH of 7.0. $[sp_{PO_4Ca}]$ is calculated to be 0.0126 mg/L as P and the unit is operated at approximately an average temperature of 25°C. The dissolved solids content of the raw water is 140 mg/L, and the concentration of the calcium ion is 130 mg/L. Calculate $K_{H_3PO_4}$ assuming that it cannot be determined from the given temperature.

14.25 Calculate $[sp_{PO_4FeIII}]$ expressed as P when the pH is 3.0. Assume the water contains 140 mg/L of dissolved solids.

14.26 A plant is conducting phosphorus removal at a pH of 3.0. $[sp_{PO_4FeIII}]$ is calculated to be 0.141 mg/L as P and the unit is operated at approximately an average temperature of 25°C. The dissolved solids content of the raw water is 140 mg/L. What precipitant is used? Calculate $K_{sp,FePO_4}$ assuming that it cannot be determined from the given temperature.

14.27 A plant is conducting phosphorus removal at a pH of 3.0. $[sp_{PO_4FeIII}]$ is calculated to be 0.141 mg/L as P and the unit is operated at approximately an average temperature of 25°C. The dissolved solids content of the raw water is 140 mg/L. Calculate K_w assuming that it cannot be determined from the given temperature.

14.28 A plant is conducting phosphorus removal at a pH of 3.0. $[sp_{PO_4FeIII}]$ is calculated to be 0.141 mg/L as P and the unit is operated at approximately an average temperature of 25°C. The dissolved solids content of the raw water is 140 mg/L. Calculate $K_{sp,(FeOH)_3}$ assuming that it cannot be determined from the given temperature.

14.29 A plant is conducting phosphorus removal at a pH of 3.0. $[sp_{PO_4FeIII}]$ is calculated to be 0.141 mg/L as P and the unit is operated at approximately an average temperature of 25°C. The dissolved solids content of the raw water is 140 mg/L. Calculate γ_H assuming that it cannot be determined from the given dissolved solids content of the raw water.

14.30   A plant is conducting phosphorus removal. $[sp_{PO_4 FeIII}]$ is calculated to be 0.141 mg/L as P and the unit is operated at approximately an average temperature of 25°C. The dissolved solids content of the raw water is 140 mg/L. Calculate $[H^+]$.

14.31   A plant is conducting phosphorus removal at a pH of 3.0. $[sp_{PO_4 FeIII}]$ is calculated to be 0.141 mg/L as P and the unit is operated at approximately an average temperature of 25°C. The dissolved solids content of the raw water is 140 mg/L. Calculate γ_{PO_4} assuming that it cannot be determined from the given dissolved solids content of the raw water.

14.32   A plant is conducting phosphorus removal at a pH of 3.0. $[sp_{PO_4 FeIII}]$ is calculated to be 0.141 mg/L as P and the unit is operated at approximately an average temperature of 25°C. The dissolved solids content of the raw water is 140 mg/L. Calculate γ_{HPO_4} assuming that it cannot be determined from the given dissolved solids content of the raw water.

14.33   A plant is conducting phosphorus removal at a pH of 3.0. $[sp_{PO_4 FeIII}]$ is calculated to be 0.141 mg/L as P and the unit is operated at approximately an average temperature of 25°C. The dissolved solids content of the raw water is 140 mg/L. Calculate K_{HPO_4} assuming that it cannot be determined from the given temperature.

14.34   A plant is conducting phosphorus removal at a pH of 3.0. $[sp_{PO_4 FeIII}]$ is calculated to be 0.141 mg/L as P and the unit is operated at approximately an average temperature of 25°C. The dissolved solids content of the raw water is 140 mg/L. Calculate $\gamma_{H_2PO_4}$ assuming that it cannot be determined from the given dissolved solids content of the raw water.

14.35   A plant is conducting phosphorus removal at a pH of 3.0. $[sp_{PO_4 FeIII}]$ is calculated to be 0.141 mg/L as P and the unit is operated at approximately an average temperature of 25°C. The dissolved solids content of the raw water is 140 mg/L. Calculate $K_{H_2PO_4}$ assuming that it cannot be determined from the given temperature.

14.36   A plant is conducting phosphorus removal at a pH of 3.0. $[sp_{PO_4 FeIII}]$ is calculated to be 0.141 mg/L as P and the unit is operated at approximately an average temperature of 25°C. The dissolved solids content of the raw water is 140 mg/L. Calculate $K_{H_3PO_4}$ assuming that it cannot be determined from the given temperature.

14.37   A wastewater contains 10 mg/L of phosphorus expressed as P. The pH is 8.0. Assume the wastewater contains 140 mg/L of dissolved solids and 100 mg/L of alkalinity expressed as $CaCO_3$. The phosphorus is to be removed using lime. If the wastewater flow is 0.75 m^3/sec, calculate the kilograms of lime per day needed to react with the alkalinity.

14.38   A wastewater contains 10 mg/L of phosphorus expressed as P. The pH is 8.0. Assume the wastewater contains 140 mg/L of dissolved solids and 100 mg/L of alkalinity expressed as $CaCO_3$. The phosphorus is to be removed using lime. If the wastewater flow is 0.75 m^3/sec, calculate the kilograms of lime per day needed to react with the phosphorus.

14.39 A wastewater contains 10 mg/L of phosphorus expressed as P. The pH is 10.0. Assume the wastewater contains 100 mg/L of alkalinity expressed as $CaCO_3$. The phosphorus is to be removed using ferric ferric chloride. If the wastewater flow is 0.75 m^3/sec, calculate the kilograms of solids produced per day. Assume the suspended solids content is 200 mg/L.

BIBLIOGRAPHY

Baker, M. J., D. W. Blowes, and C. J. Ptacek (1998). Laboratory development of permeable reactive mixtures for the removal of phosphorus from onsite wastewater disposal systems. *Environ. Science Technol.* 32, 15, 2308–2316.

Banister, S. S., A. R. Pitman, and W. A. Pretorius (1998). Solubilisation of N and P during primary sludge acid fermentation and precipitation of the resultant P. *Water S.A.* 24, 4, 337–342.

Barralet, J., S. Best, and W. Bonfield (1998). Carbonate substitution in precipitated hydroxyapatite: An investigation into the effects of reaction temperature and bicarbonate ion concentration. *J. Biomedical Materials Res.* 41, 1, 79–86.

Battistoni, P., et al. (1998). Phosphate removal in real anaerobic supernatants: Modelling and performance of a fluidized bed reactor. *Water Science Technol. Wastewater: Nutrient Removal, Proc. 1998 19th Biennial Conf. Int. Assoc. on Water Quality, Part 1,* June 21–26, Vancouver, Canada, 38, 1, 275–283. Elsevier Science Ltd. Exeter, England.

Booker, N. A. and R. B. Brooks (1994). Scale-up of the rapid sewage treatment SIROFLOCTM process. *Process Safety Environ. Protection: Trans. Inst. Chem. Engineers, Part B,* 72, 2, 109–112. Inst. of Chem. Engineers, Rugby, England.

Fytianos, K., E. Voudrias, and N. Raikos (1998). Modelling of phosphorus removal from aqueous and wastewater samples using ferric iron. *Environ. Pollut.* 101, 1, 123–130.

Holtzclaw Jr., H. F. and W. R. Robinson (1988). *General Chemistry.* D. C. Heath and Company, Lexington, MA, A7.

Jiang, J. Q. and N. J. D. Graham (1998). Pre-polymerised inorganic coagulants and phosphorus removal by coagulation—A review. *Water S.A.* 24, 3, 237–244.

Jonsson, L. (1997). Experiences of nitrogen and phosphorus removal in deep-bed filters at Henriksdal Sewage Works in Stockholm. *Water Science Technol., Proc. 1997 Int. Conf. on Upgrading of Water and Wastewater Syst.,* May 25–28, Kalmar, Sweden, 37, 9, 193–200, Elsevier Science Ltd., Exeter, England.

Maurer, M., et al. (1999). Kinetics of biologically induced phosphorus precipitation in wastewater treatment. *Water Res.* 33, 2, 484–493.

Morse, G. K., et al. (1998). Review: Phosphorus removal and recovery technologies. *Science Total Environment.* 212, 1, 69–81.

Piotrowski, J. and R. Onderko (1998). Phosphorus control and reduction at a corrugating medium mill. *Proc. 1998 TAPPI Int. Environ. Conf. Exhibit, Part 1,* April 5–8, Vancouver, Canada, 1, 101–115. TAPPI Press, Norcross, GA.

Snoeyink, V. L. and D. Jenkins (1980). *Water Chemistry.* John Wiley & Sons, New York, 280, 301.

Strikland, J. (1998). Development and application of phosphorus removal from wastewater using biological and metal precipitation techniques. *J. Chartered Inst. Water Environment Manage.* 12, 1, 30–37.

Ugurlu, A. and B. Salman (1998). Phosphorus removal by fly ash. *Environment Int.* 24, 8, 911–918.

15 Removal of Nitrogen by Nitrification–Denitrification

We define a biological reaction as a reaction mediated by organisms. It encompasses both the organisms and the underlying chemical reactions. To fully apply the knowledge of biological reactions to the treatment of water and wastewater, the chemical nature of these reactions must be given center stage. In other words, to control the process of removing nitrogen by nitrification–denitrification, the intrinsic chemical reactions must be unraveled and fully understood. The organisms only serve as mediators (i.e., the producer of the enzymes needed for the reaction). Thus, on the most fundamental level, nitrogen removal is a chemical process (more accurately, a biochemical process), and the treatment for removal of nitrogen by nitrification–denitrification as used in this textbook is chemical in nature and the process is a chemical unit process. In fact, nitrification–denitrification removal of nitrogen can be effected by purely enzymatic means by providing the needed enzymes externally without ever using microorganisms.

Similar to phosphorus, nitrogen is a very important element that has attracted much attention because of its ability to cause eutrophication in bodies of water. As stated in the chapter on phosphorus removal, the Chesapeake Bay in Maryland and Virginia is fed by tributaries from farmlands as far away as New York. Because of the use of nitrogen in fertilizers for these farms, the bay receives an extraordinarily large amount of nitrogen input that has triggered excessive growths of algae in the water body. Presently, large portions of the bay are eutrophied.

This chapter discusses removal of nitrogen using the unit process of nitrification followed by denitrification. Half reactions are utilized in the discussion of the chemical reactions. Whether or not a particular reaction will occur can be determined by the free energy change of the reactants and products. Thus, half reactions are normally tabulated in terms of free energies. To understand the exact meaning of free energy as it relates to half reactions and thus to nitrogen removal, microbial thermodynamics is discussed. Carbon requirements, alkalinity dose requirements, and reaction kinetics as they apply to nitrogen removal are all discussed. A section on whether or not to remove nitrogen is also included.

15.1 NATURAL OCCURRENCE OF NITROGEN

The element nitrogen is a nonmetal. It belongs to Group VA in the Periodic Table in the second period. Its electronic configuration is $[He]2s^2 2p^3$. [He] means that the helium configuration is filled. The valence configuration represented by the 2, the L shell, shows five electrons in the orbitals: 2 electrons in the s orbitals and 3 electrons in the p orbitals. This means that, like phosphorus, nitrogen can have a maximum oxidation state of +5; its smallest oxidation state is 3–. Examples are nitrous oxide (N_2O, 1+); nitric oxide (NO, 2+); dinitrogen trioxide (N_2O_3, 3+); nitrogen dioxide (NO_2, 4+); dinitrogen tetroxide (N_2O_4, 4+); and dinitrogen pentoxide (N_2O_5, 5+). Our interest in nitrogen as it occurs in nature is in the form that makes it fertilizer to plants. These forms are the *nitrites, nitrates,* and *ammonia.* The nitrogen in ammonia exists in its smallest oxidation state of 3–; in nitrites, it exists as 3+, and in nitrates, it exists as 5+. These nitrogen species are utilized by algae as nutrients for growth. Also, because organic nitrogen hydrolyzes to ammonia, we will, in general, be concerned with this form of the nitrogen species.

15.2 TO REMOVE OR NOT TO REMOVE NITROGEN

The formula of algae is $(CH_2O)_{106}(NH_3)_{16}H_3PO_3$ (Sincero and Sincero, 1996). Gleaning from this formula, to curtail its production in any water body such as the Chesapeake Bay, it is necessary to control only any one of the elements of nitrogen, phosphorus, oxygen, hydrogen, or carbon. It must be stressed that only one needs to be controlled, because absence of any element needed for the construction of the algal body prevents the construction of the body. This is analogous to a car. To disable this car, you only need to remove one wheel and you can never drive the car.

Of course, oxygen, hydrogen, and carbon should never be controlled, because there are already plenty of them around. From the algae formula, the ratio of N to P is 16/1 = 16 mole for mole or 14(16)/31 = 7.2 mass for mass. Table 15.1 shows various values of nitrogen and phosphorus concentrations in the water column and the corresponding N/P ratios in some coastal areas of Maryland (Sincero, 1987). For those ratios greater than 7.2, phosphorus will run out first before nitrogen does. In these situations, phosphorus should be controlled first and nitrogen should be left alone in the discharge, until further investigation reveals that the ratio has reversed.

TABLE 15.1
Nitrogen and Phosphorus Ratios, Maryland Coastal Area

Organic N, mg/L	NH_3–N, mg/L	NO_2–N, mg/L	NO_3–N, mg/L	Total N, mg/L	Total P, mg/L	N/P Ratio
0.79	0.01	0.002	0.03	0.83	0.59	1.4
0.63	0.04	0.003	0.01	0.68	0.15	4.6
0.47	0.03	0.002	0.02	0.52	0.09	5.8
2.59	0.01	0.002	0.03	2.63	0.29	9.1
1.99	0.01	0.002	0.02	2.02	0.30	6.7
3.19	0.07	0.002	0.03	3.29	0.18	18.3
1.59	0.01	0.002	0.02	1.62	0.13	12.5
0.49	0.01	0.002	0.02	0.52	0.04	13.1
0.59	0.01	0.002	0.02	0.62	0.11	5.7
1.19	0.01	0.002	0.03	1.23	0.27	4.6

For those situations where the ratio is less than 7.2, nitrogen will run out first. In these cases, should nitrogen be controlled?

Certain forms of algae, the blue-greens, can synthesize nitrogen from the air into ammonia, which they need for growth (Sincero, 1984). These particular species are very resistant and can survive anywhere where there is a water body. Thus, if this is the case in a particular body of water, it may be a waste of money to remove nitrogen, because the alga could simply get the nitrogen it needs from the air. Phosphorus should be removed, instead. These situations can become very political, however, especially with some environmentalists. Some authorities even claim that it is still advisable to remove both nitrogen and phosphorus (D'Elia, 1977). It is in this situation that modeling of the effect of the discharge of the nutrients nitrogen and phosphorus on the eutrophication potential of the water body should be investigated accurately and in great detail.

15.3 MICROBIAL THERMODYNAMICS

The study of the relationships between heat and other forms of energy is called *thermodynamics*. All living things utilize heat, therefore, the science of thermodynamics may be used to evaluate life processes. An example of a life process is the growth of bacteria when wastewater is fed to them to treat the waste. Knowledge of microbial thermodynamics is therefore important to professionals involved in cleaning up wastewaters.

Variables involved in the study of the relationship of heat and energy are called *thermodynamic variables*. Examples of these variables are temperature, pressure, free energy, enthalpy, entropy, and volume. In our short discussion of thermodynamics, we will address enthalpy, entropy, and free energy. As mentioned, whether or not a particular reaction, such as a biological reaction, is possible can be determined by the free energy change between products and reactants. Free energy, in turn, is a function of the enthalpy and entropy of the reactants and products.

15.3.1 ENTHALPY AND ENTROPY

Let H represent the enthalpy, U the internal energy, P the pressure, and $\forall$ the volume of a particular system undergoing a process under study. The enthalpy H is defined as

$$H = U + P\forall \qquad (15.1)$$

Internal energy refers to all the energies that are present in the system such as kinetic energies of the molecules, ionization energies of the electrons, bond energies, lattice energies, etc. The system possesses all these energies by virtue of its being and are all integral (that is, internal) with the system.

Let us derive the relationship between enthalpy and the heat exchange during a biological reaction, where *biological reaction* is a chemical reaction mediated by organisms. Biological reactions are carried out at constant pressure; hence, the heat exchange is a heat exchange at constant pressure. Designate this exchange as Q_p. The first law of thermodynamics states that any heat added to a system minus any work W that the system is doing at the same time manifests itself in the form of an increase of the internal energy. In differential form,

$$dU = dQ_p - dW = dQ_p - Pd\Psi \tag{15.2}$$

The only work done in biological reactions is the work of pushing the surroundings (the atmosphere) in which the reaction is occurring. This is a pressure-volume work; hence, the $Pd\Psi$ term.

Because the biological reaction is at constant pressure, differentiate the enthalpy equation at constant pressure. This produces

$$dH = dU + Pd\Psi \tag{15.3}$$

This may be combined with Equation (15.2) to eliminate dU producing

$$dH = dQ_p \tag{15.4}$$

This equation concludes that change in enthalpy is a heat exchange at constant pressure between the system under study and its surroundings.

Before we discuss entropy, define reversible process and reversible cycle. A *reversible process* is a process in which the original state or condition of a system can be recovered back if the process is done in the opposite direction from that in which it is currently being done. To perform a reversible process, the steps must be conducted very, very slowly, in an infinitesimal manner, and without friction. From the definition of a reversible process, the definition of a reversible cycle follows. A *reversible cycle* is a cycle in which the reversible process is applied in every step around the cycle.

Heat added to a system causes its constituent particles to absorb the energy resulting in the system being more chaotic than it was before. If the heat is added reversibly, the ratio of the infinitesimal heat added to the temperature T during the infinitesimal time that the heat is added defines the *change in entropy*. If this addition is done around a reversible cycle, the state or condition of the system at the end of the cycle will revert back to its original state or condition at the beginning of the cycle. This must be so, since the whole process is being done reversibly in every step along the way around the cycle. Hence, the change in entropy around a reversible cycle is zero.

Let S be the entropy and Q_{rev} be the reversible heat added. In a given differential step, the heat added is dQ_{rev}. The differential change in entropy in every differential step is therefore $dS = dQ_{rev}/I$. Around the cycle, the change in entropy is the integral, thus

$$\oint dS = S_e - S_b = \Delta S = \oint \frac{dQ_{rev}}{T} = 0 \tag{15.5}$$

The symbol $\oint$ means that the integrand is to be integrated around the cycle and subscripts e and b refer to the end and the beginning of the cycle, respectively. If the process is not around a cycle, the previous subscripts simply mean the end and beginning of the process. In this case, the integral will not be zero and the equation is written as the integral

$$\int_b^e dS = S_e - S_b = \Delta S = \int_b^e \frac{dQ_{rev}}{T} \tag{15.6}$$

$$= \frac{Q_{rev}}{T} \quad T = constatnt \tag{15.7}$$

Interpretations of enthalpy and entropy. The heat absorbed by the system causes more agitation of its constituent elements. This increased agitation and chaos is the entropy increase and is calculated by Eqs. (15.6) and (15.7). The entropy increase is an increase in disorder of the constituents of the system. The energy state of the system is increased, but because the energy supporting this state is nothing more than supporting chaos, this energy is a wasted energy. The equations therefore calculate the loss in energy of the system as a result of increased chaos or disorder.

Consider a fuel such as coal, and burn it in a furnace. The burning of the coal occurs under constant atmospheric pressure. As the coal burns, heat is released; this heat is energy Q_p, which may be used to produce electricity by using a boiler and a turbine generator. From Equation (15.4), Q_p is equal to the enthalpy change ΔH. We therefore conclude that before the coal was burned, it possessed an enthalpy H which, by virtue of Equation (15.4), is its energy content. By the entropy change during the process of burning, however, all this energy is not utilized as useful energy but is subtracted by the change in entropy. The electrical energy that is ultimately delivered to the consumer is less by an amount equal to the overall entropy change in the transformation of coal to electricity.

In biological reactions, the fuel is the food. In biological nitrogen removal, nitrogen in its appropriate form is fed to microorganisms to be utilized as food. This food possesses enthalpy as does coal; and, similar to coal, its energy content cannot all be utilized as useful energy by the microorganism as a result of the inevitable entropy inefficiency that occurs in the process of consuming food.

15.3.2 Free Energy

Will a certain food provide energy when utilized by microorganisms? If the answer is yes, then the food will be eaten; and if it is in a wastewater, the wastewater will be cleaned up. The answer to this question can now be quantified by the combination of the concept of enthalpy and entropy. This combination is summed up in a term called *free energy*. *Free energy* G is defined as

$$G = H - TS \tag{15.8}$$

Because H is an energy content and S is a wasted energy, G represents the useful energy (or, alternatively, the maximum energy) the fuel can provide after the wasteful energy (represented by S) has been subtracted from the energy content. Thus, the term free energy.

Biological processes are carried out at a given constant temperature as well as constant pressure. Thus, differentiating the free-energy equation at constant temperature,

$$dG = dH - TdS = dH - Q_{rev} \tag{15.9}$$

Note: In order for G to be a maximum (i.e., to be a free energy), Q must be the Q_{rev} as depicted in the equation.

15.4 OXIDATION–REDUCTION REACTIONS OF NITROGEN FOODS

Life processes involve electron transport. Specifically, the mitochondrion and the chloroplast are the sites of this electron movement in the eucaryotes. In the procaryotes, this function is embedded in the sites of the cytoplasmic membrane. As far as electron movement is concerned, life processes have similarity to a battery cell. In this cell, electrons move because of electrical pressure, the voltage difference. By the same token, electrons move in an organism because of the same electrical pressure, the voltage difference. In a battery cell, one electrode is oxidized while the other is reduced; that is, oxidation–reduction occurs in a battery cell. Exactly the same process occurs in an organism.

In an oxidation–reduction reaction, a mole of electrons involved is called the *faraday*, which is equal to 96,494 *coulombs*. A mole of electrons is equal to one equivalent of any substance. Therefore, a faraday is equal to one equivalent.

Let n be the number of faradays of charge or mole of a substance participating in a reaction, and let the general reaction be represented by the half-cell reaction of Zn as follows:

$$Zn \rightarrow Zn^{2+} + 2e^- \tag{15.10}$$

The couple Zn/Zn^{2+} has an electric pressure between them. Now, take another couple such as Mg/Mg^{2+} whose half-cell reaction would be similar to that of Equation (15.10). The couples Zn/Zn^{2+} and Mg/Mg^{2+} do not possess the same voltage potential. If the two couples are connected together, they form a cell. Their voltage potentials are not the same, so a voltage difference would be developed between their electrodes and electric current would flow.

Let the voltage between the electrodes of the above cell be measured by a potentiometer. Designate this voltage difference by ΔE. In potentiometric measurements, no electrons are allowed to flow, but only the voltage tendency of the electrons to flow is measured. No electrons are allowed to flow, therefore, no energy is dissipated due to friction of electrons "rubbing along the wire." Thus, any energy associated with this no-electron-to-flow process represents the maximum energy available. Because voltage is energy in joules per coulomb of charge, the energy associated can be calculated from the voltage difference. This associated energy corresponds to a no-friction loss process; it is therefore a maximum energy—the change in free energy—after the entropic loss has been deducted.

Let n, the number of faradays involved in a reaction, be multiplied by F, the number of coulombs per faraday. The result, nF, is the number of coulombs involved in the reaction. If nF is multiplied by ΔE, the total associated energy obtained from the previous potentiometric experiment results. Because the voltage measurement was done with no energy loss, by definition, this associated energy represents the free energy change of the cell (i.e., the maximum energy change in the cell). In symbols,

$$\Delta G = \pm nF \, \Delta E \qquad (15.11)$$

The $\pm$ sign has been used. A convention used in chemistry is that if the sign is negative, the process is spontaneous and if the sign is positive, the process is not spontaneous.

As mentioned, the battery cell process is analogous to the living cell process of the mitochondrion, the chloroplast, and the electron-transport system in the cytoplasmic membrane of the procaryotes. Thus, Equation (15.11) can represent the basic thermodynamics of a microbial system.

In the living cell, organic materials are utilized for both energy (oxidation) and synthesis (reduction). Microorganisms that utilize organic materials for energy are called heterotrophs. Those that utilize inorganics for energy are called *autotrophs*. Autotrophs utilize CO_2 and HCO_3^- for the carbon needed for cell synthesis; the heterotrophs utilize organic materials for their carbon source. Autotrophs that use inorganic chemicals for energy are called *chemotrophs*; those that use sunlight are *phototrophs*. The bacteria that consumes the nitrogen species in the biological removal of nitrogen are chemotrophic autotrophs. Algae are phototrophic autotrophs.

Somehow, in life processes, the production of energy from release of electrons does not occur automatically but through a series of steps that produce a high energy-containing compound. This high energy-containing compound is ATP (adenosine triphosphate). Although ATP is not the only high energy-containing compound, it is by far the major one that fuels synthesis in the cell. ATP is the energy currency that the cell relies upon for energy supply.

The energy function of ATP is explained as follows: ATP contains two high-energy bonds. To form these bonds, energy must be obtained from an energy source through electron transfer. The energy released is captured and stored in these bonds. On demand, hydrolysis of the bond releases the stored energy which the cell can then use for synthesis and cell maintenance.

ATP is produced from ADP (adenosine diphosphate) by coupling the release of electrons to the reaction of organic phosphates and ADP producing ATP. ATP has two modes of production: *substrate-level phosphorylation* and *oxidative phosphorylation*. In the former, the electrons released by the energy source are absorbed by an *intermediate product* within the system. The electron absorption is accompanied by an energy release and ATP is formed. The electron-transport system is simple.

Fermentation is an example of a substrate-level phosphorylation process that uses intermediate absorbers such as formaldehyde. Substrate-level phosphorylation is inefficient and produces only a few molecules of ATP.

In the oxidative phosphorylation mode, the electron moves from one electron carrier to another in a series of complex reduction and oxidation steps. The difference between substrate-level and oxidative phosphorylation is that in the former, the transport is *simple*, while in the latter, it is *complex*. For a hydrogen-containing energy source, the series starts with the initial removal of the hydrogen atom from the molecule of the source. The hydrogen carries with it the electron it shared with the original source molecule, moving this electron through a series of intermediate carriers such as NAD (nicotinamide adenine dinucleotide). The intermediate that receives the electron-carrying hydrogen becomes reduced. The reduction of NAD, for example, produces $NADH_2$. The series continues on with further reduction and oxidation steps. The whole line of reduction and oxidation constitutes the electron-transport system. At strategic points of the transport system, ATP is produced from ADP and inorganic phosphates.

The other version of oxidative phosphorylation used by autotrophs involves the release of electrons from an inorganic energy source. An example of this is the release of electrons from NH_4^-, oxidizing NH_4^+ to NO_2^-, and the release of electrons from NO_2^-, oxidizing NO_2^- to NO_3^-.

The transported electrons emerge from the system to reduce a final external electron acceptor. The type of the final acceptor depends upon the environment on which the electron transport is transpiring and may be one of the following: for aerobic environments, the acceptor is O_2; for anaerobic environments, the possible acceptors are NO_3^-, SO_4^{-2}, and CO_2. When the acceptor is NO_3^-, the system is said to be *anoxic*.

The values of free energy changes are normally reported at standard conditions. In biochemistry, in addition to the requirement of unit activity for the concentrations of reactants and products, pressure of one atmosphere, and temperature of 25°C, the hydrogen ion concentration is arbitrarily set at pH 7.0. Following this convention, Equation (15.11) may be written as

$$\Delta G_o' = \pm nF\Delta E_o' \tag{15.12}$$

The primes emphasize the fact that the standard condition now requires the $\{H^+\}$ to be 10^{-7} moles per liter. The subscript o signifies conditions at standard state.

In environmental engineering, it is customary to call the substance oxidized as the electron donor and the substance reduced as the electron acceptor. The electron donor is normally considered as food. In the context of nitrogen removal, the foods are the nitrites, nitrates, and ammonia. Equation (15.10) is an example of an electron donor reaction. Zn is the donor of the electrons portrayed on the right-hand side of the half-cell reaction. On the other hand, the reverse of the equation is an example of an electron acceptor reaction. Zn^{+2} would be the electron acceptor. McCarty (1975) derived values for free energy changes of half-reactions for various electron donors and acceptors utilized in a bacterial systems. The ones specific for the nitrogen species removal are shown in Table 15.2.

TABLE 15.2
Half-Cell Reactions for Bacterial Systems in Nitrogen Removal

$$\Delta G_o'$$
kcal/electron–mol

Reactions for cell synthesis:

Ammonia as nitrogen source:

$$\frac{1}{5}CO_2 + \frac{1}{20}HCO_3^- + \frac{1}{20}NH_4^+ + H^+ + e^- \rightarrow \frac{1}{20}C_5H_7NO_2 + \frac{9}{20}H_2O \qquad —$$

Nitrate as nitrogen source:

$$\frac{1}{28}NO_3^- + \frac{5}{28}CO_2 + \frac{29}{28}H^+ + e^- \rightarrow \frac{1}{28}C_5H_7NO_2 + \frac{11}{28}H_2O \qquad —$$

Reactions for electron acceptors:

Oxygen as acceptior:

$$\frac{1}{4}O_2 + H^+ + e^- \rightarrow \frac{1}{2}H_2O \qquad -18.68$$

Nitrate as acceptor:

$$\frac{1}{5}NO_3^- + \frac{6}{5}H^+ + e^- \rightarrow \frac{1}{10}N_2 + \frac{3}{5}H_2O \qquad -17.13$$

Nitrite as acceptor:

$$\frac{1}{3}NO_2^- + \frac{4}{3}H^+ + e^- \rightarrow \frac{1}{6}N_2 + \frac{2}{3}H_2O \qquad —$$

Reactions for electron donors:

Domestic wastewater as donor (heterotrophic reaction):

$$\frac{1}{50}C_{10}H_{19}NO_3 + \frac{9}{25}H_2O \rightarrow \frac{9}{50}CO_2 + \frac{1}{50}NH_4^+ + \frac{1}{50}HCO_3^- + H^+ + e^- \qquad -7.6$$

Nitrite as donor:

$$\frac{1}{2}NO_2^- + \frac{1}{2}H_2O \rightarrow \frac{1}{2}NO_3^- + H^+ + e^- \qquad +9.43$$

Ammonia as donor:

$$\frac{1}{6}NH_4^+ + \frac{1}{3}H_2O \rightarrow \frac{1}{6}NO_2^- + \frac{4}{3}H^+ + e^- \qquad +7.85$$

15.4.1 CRITERION FOR SPONTANEOUS PROCESS

It is a law of nature that things always go in the direction of creating greater chaos—this is the second law of thermodynamics. Any system, except those at temperature equals absolute zero, is always disordered.* The energy required to maintain this disorder, we have found, is called entropy. As mentioned, any system possesses free energy at any instant, this energy being the net energy remaining after the entropy required to maintain the current disorder has been subtracted from the enthalpy (energy content).

When the system goes from state 1 (current state) to state 2, its free energy at the latter state may or may not be the same as the former. If the free energy at state 2

---

* At absolute zero, all particles practically cease to move and are therefore structured and orderly.

is the same as that in state 1, the system must be in *equilibrium*. If the free energy at state 2 is greater than that at state 1, then some outside free energy must have been added to the system. In Table 15.2, this is the case of nitrite as a donor and ammonia as a donor. External free energies of 9.43 kcal/electron-mol and 7.85 kcal/electron-mol, respectively, has been added to the system; these values are indicated by the plus signs. External sources of energy are being required, so these half-cell reactions cannot occur spontaneously; they are said to be *endothermic* (i.e., requiring external energies to effect the reaction).

On the one hand, when the free energy at state 2 is less than that at state 1, some energy must have been released by the system to the surroundings, thus manifesting in the decrease of free energy. A decrease in free energy is indicated in the table with a negative sign. This energy has been released "voluntarily" by the system without some form of "coercion" from the surroundings. The release is spontaneous, and therefore the reaction is spontaneous.

Note: Thus, this is the criterion for a spontaneous process: When the free energy change is negative, the process is spontaneous.

Judging from Table 15.2, when the electrons that travel through the electron transport system are finally accepted by oxygen, a large amount of energy equal to 18.68 kcal/electron-mol is liberated. This liberated energy is then captured in the bonds of ATP. The same statement holds for the others whose free energy changes have negative signs. Thus, any material in wastewater, edible by organisms, will release energy, resulting in their destruction. The more energy that can be released, the easier it is to be treated using microorganisms.

15.5 MODES OF NITROGEN REMOVAL

The physical removal of nitrogen using the unit operation of stripping was discussed in a previous chapter. The present chapter concerns only the removal of nitrogen by biochemical means, as mediated by microorganisms. The technique of the unit process is to release the nitrogen in the form of the gas N_2 to the atmosphere. This will first entail nitrifying the nitrogens using the species of bacteria *Nitrosomonas* and *Nitrobacter. Nitrosomonas* oxidizes the ammonium ion into nitrites, deriving from this oxidation the energy it needs. *Nitrobacter* then oxidizes the nitrites into nitrates, also deriving from this oxidation the energy that it needs. These oxidations into nitrites and nitrates is called *nitrification*. Nitrification is an aerobic process.

After the nitrogen has been nitrified, the second unit process of *denitrification* is then applied. The denitrifying bacteria, which are actually heterotrophs, convert the nitrates into nitrogen gas, thus ridding the wastewater of nitrogen. Denitrification is an anaerobic process.

15.6 CHEMICAL REACTIONS IN NITROGEN REMOVAL

In biochemical nitrogen removal, BNR, two steps are required: oxidation of nitrogen to nitrate and subsequent reduction of the nitrate to gaseous nitrogen, N_2. The oxidation steps are mediated by *Nitrosomonas* and by *Nitrobacter*, as mentioned previously. The reduction step is mediated by the normal heterotrophic bacteria. We will now discuss the chemical reactions involved in these oxidations and reduction.

15.6.1 NITRIFICATION: *NITROSOMONAS* STAGE

From Table 15.2, the generalized donor reaction, acceptor reaction, and synthesis reaction mediated by *Nitrosomonas* are, respectively, shown by the following half-cell reactions:

$$\frac{1}{6} NH_4^+ + \frac{1}{3} H_2O \rightarrow \frac{1}{6} NO_2^- + \frac{4}{3} H^+ + e^- \quad \text{donor reaction} \quad (15.13)$$

$$\frac{1}{4} O_2 + H^+ + e^- \rightarrow \frac{1}{2} H_2O \quad \text{acceptor reaction} \quad (15.14)$$

$$\frac{1}{5} CO_2 + \frac{1}{20} HCO_3^- + \frac{1}{20} NH_4^+ + H^+ + e^-$$

$$\rightarrow \frac{1}{20} C_5H_7NO_2 + \frac{9}{20} H_2O \quad \text{synthesis reaction} \quad (15.15)$$

The table shows three possibilities for a donor reaction: domestic wastewater, nitrite, and ammonia.

Note: Ammonia and ammonium are interchangeable, since one converts to the other.

We are nitrifying ammonia, so the ammonium is the donor. The table also shows other possibilities for the electron acceptor. Because nitrification is aerobic, however, oxygen is the acceptor. Lastly, the synthesis reaction has two possibilities: one using ammonia and the other using nitrate. Of these two possible species, organisms prefer the ammonium ion to the nitrate ion. Only when the ammonium ions are consumed will the nitrates be used in synthesis.

If the previous equations are simply added without modifications, electrons will remain free to roam in solution; this is not possible. In actual reactions, the previous equations must be modified to make the electrons given up by Equation (15.13) be balanced by the electrons accepted by Equations (15.14) and (15.15). Starting with one gram-equivalent of NH_4–N, based on Equation (15.15), assume m equivalents are incorporated into the cell of *Nitrosomonas*, $C_5H_7NO_2$. Thus, the NH_4–N equivalent remaining for the donor reaction of Equation (15.13) is $1 - m$, equals $(1 - m)$ $[((1/20)N)/1] (1/N) = \frac{1}{20}(1 - m)$ moles. [The 1/20 came from Equation (15.15) used to compute the equivalent mass of NH_4–N.] Thus, $\frac{1}{20}(1 - m)$ moles of NH_4–N is available for the donor reaction to donate electrons. By Equation (15.13), the modified donor reaction is then

$$\frac{1/6}{1/6}\left(\frac{1}{20}\right)(1 - m) NH_4^+ + \frac{1/3}{1/6}\left(\frac{1}{20}\right)(1 - m)H_2O$$

$$\rightarrow \frac{1/6}{1/6}\left(\frac{1}{20}\right)(1 - m)NO_2^- + \frac{4/3}{1/6}\left(\frac{1}{20}\right)(1 - m)H^+ + \frac{1}{1/6}\left(\frac{1}{20}\right)(1 - m)e^- \quad (15.16)$$

From Equation (15.15), the m equivalents of NH_4–N is $m[((1/20)N)/1] (1/N) = \frac{m}{20}$ moles. Thus, the synthesis reaction becomes

$$\frac{1/5}{1/20}\left(\frac{m}{20}\right)CO_2 + \frac{1/20}{1/20}\left(\frac{m}{20}\right)HCO_3^- + \frac{1/20}{1/20}\left(\frac{m}{20}\right)NH_4^+ + \frac{1}{1/20}\left(\frac{m}{20}\right)H^+ + \frac{1}{1/20}\left(\frac{m}{20}\right)e^-$$

$$\rightarrow \frac{1/20}{1/20}\left(\frac{m}{20}\right)C_5H_7O_2N + \frac{9/20}{1/20}\left(\frac{m}{20}\right)H_2O \tag{15.17}$$

From Eqs. (15.16) and (15.17), the electron-moles left for the acceptor reaction is

$$\frac{1}{1/6}\left(\frac{1}{20}\right)(1-m) - \frac{1}{1/20}\left(\frac{m}{20}\right) = \frac{3-13m}{10}$$

Hence, the acceptor reaction, Equation (15.14), modifies to

$$\frac{1}{4}\left(\frac{3-13m}{10}\right)O_2 + \frac{3-13m}{10}H^+ + \frac{3-13m}{10}e^- \rightarrow \frac{1}{2}\left(\frac{3-13m}{10}\right)H_2O \tag{15.18}$$

Adding Eqs. (15.16), (15.17), and (15.18) produces the overall reaction for the *Nitrosomonas* reaction shown next. Note that, after addition, the electrons e^- are gone. The overall reaction should indicate no electrons in the equation, since electrons cannot just roam around in the solution. They must be taken up by some atom. The overall reaction is

$$\frac{1}{20}NH_4^+ + \frac{(3-13m)}{40}O_2 + \frac{m}{20}HCO_3^- + \frac{m}{5}CO_2$$

$$\rightarrow \frac{m}{20}C_5H_7NO_2 + \frac{1}{20}(1-m)NO_2^- + \frac{1-2m}{20}H_2O + \frac{1-m}{10}H^+ \tag{15.19}$$

15.6.2 Nitrification: *Nitrobacter* Stage

Nitrobacter utilizes the nitrites produced by *Nitrosomonas* for energy. The unmodified donor, acceptor, and synthesis reactions are written below, as taken from Table 15.2:

$$\frac{1}{2}NO_2^- + \frac{1}{2}H_2O \rightarrow \frac{1}{2}NO_3^- + H^+ + e^- \qquad \text{donor reaction} \tag{15.20}$$

$$\frac{1}{4}O_2 + H^+ + e^- \rightarrow \frac{1}{2}H_2O \qquad \text{acceptor reaction} \tag{15.21}$$

$$\frac{1}{5}CO_2 + \frac{1}{20}HCO_3^- + \frac{1}{20}NH_4^+ + H^+ + e^-$$

$$\rightarrow \frac{1}{20}C_5H_7NO_2 + \frac{9}{20}H_2O \qquad \text{synthesis reaction} \tag{15.22}$$

In the BNR process, the *Nitrobacter* reaction follows the *Nitrosomonas* reaction. From Equation (15.19), $1/20(1 - m)$ moles of NO_2–N have been produced from the original one equivalent of ammonia nitrogen based on Equation (15.15). These nitrites serve as the elector donor for *Nitrobacter*. Thus, the donor reaction of Equation (15.20) becomes

$$\frac{1/2}{1/2}\left[\frac{1}{20}(1-m)\right]NO_2^- + \frac{1/2}{1/2}\left[\frac{1}{20}(1-m)\right]H_2O$$

$$\rightarrow \frac{1/2}{1/2}\left[\frac{1}{20}(1-m)\right]NO_3^- + \frac{1}{1/2}\left[\frac{1}{20}(1-m)\right]H^+ + \frac{1}{1/2}\left[\frac{1}{20}(1-m)\right]e^- \quad (15.23)$$

Let n, based on Equation (15.22), be the equivalents of *Nitrobacter* cells produced. This quantity, n equivalents, is equal to

$$n\left(\frac{\frac{C_5H_7O_2N}{20}}{1}\right)\bigg/ C_5H_7O_2N = n/20 \text{ moles.}$$

Modifying the synthesis reaction,

$$\frac{1/5}{1/20}\left(\frac{n}{20}\right)CO_2 + \frac{1/20}{1/20}\left(\frac{n}{20}\right)HCO_3^- + \frac{1/20}{1/20}\left(\frac{n}{20}\right)NH_4^+ + \frac{1}{1/20}\left(\frac{n}{20}\right)H^+ + \frac{1}{1/20}\left(\frac{n}{20}\right)e^-$$

$$\rightarrow \frac{1/20}{1/20}\left(\frac{n}{20}\right)C_5H_7O_2N + \frac{9/20}{1/20}\left(\frac{n}{20}\right)H_2O \quad (15.24)$$

Subtracting the electrons used in Equation (15.24), $1/(1/20)(n/20)$, from the electrons donated in Equation (15.23), $1/(1/2)[1/20(1 - m)]$, produces the amount of electrons available for the acceptor (energy) reaction, $(1 - m - 10n)/10$. Modifying the acceptor reaction,

$$\frac{1}{4}\left[\frac{1-m-10n}{10}\right]O_2 + \left[\frac{1-m-10n}{10}\right]H^+ + \left[\frac{1-m-10n}{10}\right]e^- \rightarrow \frac{1}{2}\left[\frac{1-m-10n}{10}\right]H_2O$$

$$(15.25)$$

Adding Equations (15.23), (15.24), and (15.25) produces the overall reaction for *Nitrobacter*,

$$\frac{(1-m)}{20}NO_2^- + \frac{n}{20}NH_4^+ + \frac{n}{5}CO_2 + \frac{n}{20}HCO_3^- + \frac{n}{20}H_2O + \left[\frac{1-m-10n}{40}\right]O_2$$

$$\rightarrow \frac{n}{20}C_5H_7O_2N + \frac{1-m}{20}NO_3^- \quad (15.26)$$

15.6.3 OVERALL NITRIFICATION

The *Nitrosomonas* and the *Nitrobacter* reactions may now be added to produce the overall nitrification reaction as shown next.

$$\frac{1+n}{20}NH_4^+ + \frac{m+n}{20}HCO_3^- + \frac{m+n}{5}CO_2 + \frac{2-7m-5n}{20}O_2$$

$$\rightarrow \frac{m+n}{20}C_5H_7NO_2 + \frac{1-m}{10}H^+ + \frac{1-m}{20}NO_3^- + \frac{1-2m-n}{20}H_2O \quad (15.27)$$

The literature reports cell yields (productions) for *Nitrosomonas* of 0.04 to 0.29 milligrams of the bacteria per milligram of NH_4–N destroyed and cell yields for *Nitrobacter* of 0.02 to 0.084 milligrams of the bacteria per milligrams of NO_2–N destroyed (Mandt and Bell, 1982). *Yield* or *specific yield* simply means the amount of organisms produced per unit amount of substrate consumed. Also, for nitrification to be the dominant reaction, the BOD_5/TKN ratios should be less than 3. At these ratios, the nitrifier population is about 10% and higher. For BOD_5/TKN ratios of greater than 5.0, the process may be considered combined carbonaceous-nitrification reaction. At these ratios, the nitrifier population is less than 4%. In addition, to ensure complete nitrification, the dissolved oxygen concentration should be, at least, about 2.0 mg/L.

15.6.4 DENITRIFICATION: HETEROTROPHIC SIDE REACTION STAGE

Aside from the normal anoxic reaction, two side reactions must be considered in denitrification: the continued oxidation using the leftover dissolved oxygen from the nitrification reaction stage, and nitrite reduction. Immediately after nitrification is stopped, a large concentration of dissolved oxygen still exists in the reactor—this would be around 2.0 mg/L. In nitrification, the nitrifiers are mixed with the heterotrophic bacteria. The heterotrophic bacteria are fast growers compared to *Nitrosomonas* and *Nitrobacter*, so they overwhelm the reaction and the overall chemical process is the normal heterotrophic carbonaceous reaction in the last stage of oxidation. By control of the process, the growth rates of the nitrifiers and the heterotrophs are balanced during nitrification and the two types of bacteria grow together. As soon as the oxygen supply is cut off, however, the nitrifiers cannot compete against the heterotrophs for the ever decreasing concentration of dissolved oxygen. Thus, the activities of the nitrifiers "fade away," and the heterotrophs predominate in the last stage of oxidation after aeration is cut off.

The other side reaction is the reduction of nitrite to the nitrogen gas. Although the process is aimed at oxidizing nitrogen to the nitrate stage, some nitrite can still be found. In the absence of oxygen, after the heterotrophs have consumed all the remaining oxygen, no other reaction can occur except for the reduction of nitrite. This is discussed further, after the discussion on the regular denitrification reaction.

Now, derive the chemical reaction for the heterotrophic stage. Let r be the equivalents of O_2 (based on the oxygen acceptor reaction) used at this stage. The removal of nitrogen is done in conjunction with the treatment of sewage. Thus, sewage ($C_{10}H_{19}NO_3$) must be the electron donor. Using sewage, the ammonium ion is produced, see Table 15.2. As mentioned before, given NO_3^- and NH_4^+ in solution, organisms prefer to use NH_4^+ first, before NO_3^- for synthesis. Thus, of the two competing synthesis reactions, the one using the ammonium is then the one favored by the bacteria. Letting q be the equivalents of cells, based on the synthesis reaction, produced during the last stage of the aerobic reaction, the synthesis reaction may be modified as follows:

$$\frac{q}{5}CO_2 + \frac{q}{20}HCO_3^- + \frac{q}{20}NH_4^+ + qH^+ + qe^- \rightarrow \frac{q}{20}C_5H_7NO_2 + \frac{9q}{20}H_2O \qquad (15.28)$$

From the r equivalents of O_2 (based on the oxygen acceptor reaction), the acceptor reaction is

$$\frac{r}{4}O_2 + rH^+ + re^- \rightarrow \frac{r}{2}H_2O \qquad (15.29)$$

From these equations, the total electron moles needed from the donor is $r + q$. Thus, the donor reaction is modified as follows:

$$\frac{r+q}{50}C_{10}H_{19}NO_3 + \frac{9(r+q)}{25}H_2O \rightarrow \frac{9(r+q)}{50}CO_2 + \frac{r+q}{50}NH_4^+$$

$$+ \frac{r+q}{50}HCO_3^- + (r+q)H^+ + (r+q)e^- \qquad (15.30)$$

Adding Eqs. (15.28), (15.29), and (15.30), the overall aerobic reaction is produced:

$$\frac{r+q}{50}C_{10}H_{19}NO_3 + \frac{r}{4}O_2 \rightarrow \frac{q}{20}C_5H_7NO_2 + \frac{9r-q}{50}CO_2$$

$$+ \frac{2r-3q}{100}NH_4^+ + \frac{2r-3q}{100}HCO_3^- + \frac{14r+9q}{100}H_2O \qquad (15.31)$$

From Equation (15.31), let Y_c be the cell yield in moles per unit mole of sewage. Then, $Y_c = ((q/20)/(r + q/50)) = 5q/2(r + q)$. And,

$$q = \frac{2rY_c}{5 - 2Y_c} \qquad (15.32)$$

The reaction involving sewage is called a *carbonaceous reaction*, because it is the reaction where organisms utilize the carbon of sewage for synthesis of the cells. Thus, Y_c is also called a *carbonaceous cell yield*.

15.6.5 Denitrification: Normal Anoxic Stage

Let s, based on the synthesis reaction, be the equivalents of cells produced for the regular anoxic denitrification reaction. The organisms in denitrification are heterotrophic, which must use sewage because, as mentioned, removal of nitrogen is done in conjunction with the treatment of sewage. Again, the ammonium ion is produced in the process, making it the source of nitrogen in the synthesis reaction. Thus, modifying the ammonium synthesis reaction, using the s equivalents of cells, produces the following reaction:

$$\frac{s}{5}CO_2 + \frac{s}{20}HCO_3^- + \frac{s}{20}NH_4^+ + sH^+ + se^- \rightarrow \frac{s}{20}C_5H_7NO_2 + \frac{9s}{20}H_2O \qquad (15.33)$$

Denitrification is an anaerobic process; thus, it will not be using oxygen as the electron acceptor. From Table 15.2, two possibilities exist for the electron acceptor: nitrite or nitrate. In nitrate, the oxidation state of nitrogen is 5+; in nitrite, the oxidation state of nitrogen is 3+. The nitrate ion is at the higher oxidation state, so it is easier for it to be reduced than the nitrite ion. Thus, nitrate is the electron acceptor.

Let p, based on the NO_3^- acceptor reaction, be the equivalents of NO_3–N utilized. The revised acceptor reaction is

$$\frac{p}{5}NO_3^- + \frac{6p}{5}H^+ + pe^- \rightarrow \frac{p}{10}N_2 + \frac{3p}{5}H_2O \qquad (15.34)$$

From Eqs. (15.33) and (15.34), the total e^- needed from the donor reaction is $s + p$ electron moles. Thus, the donor reaction using sewage is

$$\frac{s+p}{50}C_{10}H_{19}NO_3 + \frac{9(s+p)}{25}H_2O \rightarrow \frac{9(s+p)}{50}CO_2 + \frac{s+p}{50}NH_4^+$$

$$+ \frac{s+p}{50}HCO_3^- + (s+p)H^+ + (s+p)e^- \qquad (15.35)$$

Adding Eqs. (15.33), (15.34), and (15.35), the overall reaction for denitrification is produced,

$$\frac{p}{5}NO_3^- + \frac{p}{5}H^+ + \frac{s+p}{50}C_{10}H_{19}NO_3$$

$$\rightarrow \frac{s}{20}C_5H_7NO_2 + \frac{p}{10}N_2 + \frac{2p-3s}{100}NH_4^+$$

$$+ \frac{2p-3s}{100}HCO_3^- + \frac{9p-s}{50}CO_2 + \frac{24p+9s}{100}H_2O \qquad (15.36)$$

From Equation (15.36), let Y_{dn} be the cell yield in moles per unit moles of NO_3–N. Then,

$$s = 4pY_{dn} \qquad (15.37)$$

15.6.6 DENITRIFICATION: NO_2-REDUCTION SIDE REACTION STAGE

Now, derive the overall reaction for the reduction of nitrite. NO_2^- can go only one way: conversion to the nitrogen gas. Although right after cutting aeration, some dissolved oxygen still remains in water, the concentration is not sufficient to oxidize the nitrite to nitrate. As soon as the heterotrophic side reaction is complete, anaerobic conditions set in and the environment becomes a reducing atmosphere. The NO_2^- with nitrogen having an oxidation state of 3+ must then be reduced. The nitrogen atom has three possible reduction products: reduction to nitrous oxide, N_2O, reduction to N_2, and reduction to the ammonium ion.

The nitrogen in N_2O has an oxidation state of 1+; that in N_2 has an oxidation state of 0; and that in the ammonium ion has an oxidation state of 3–. Because the environment is now severely reducing, the reduction process takes an extra step. After reducing the nitrogen atom from 3+ in nitrite to 1+ in nitrous oxide, the process continues one more step to the nitrogen molecule. In theory, it is still possible to proceed with the reduction down to the formation of the ammonium ion. This, however, will require the formation of bonds between the nitrogen atom and three atoms of hydrogen (to form NH_3), plus the hydrogen bond between the ammonia molecule and the hydrogen proton (to form the ammonium ion, NH_4^+). This needs extra energy. Thus, the process stops at the liberation of the nitrogen gas. This whole reduction process makes the nitrite ion the electron acceptor for it to be reduced. Of course, the N_2O could also be formed; but again, the atmosphere is severely reduced and the nitrogen gas must be the one produced.

Let t, based on the NO_2^- acceptor reaction, be the equivalents of NO_2–N utilized. From Table 15.2, the revised acceptor reaction is

$$\frac{t}{3}NO_2^- + \frac{4t}{3}H^+ + te^- \rightarrow \frac{t}{6}N_2 + \frac{2t}{3}H_2O \qquad (15.38)$$

As in the normal anoxic process, sewage is used as the electron donor, because it is the wastewater being treated. Again, as an electron donor, the ammonium ion is produced serving as the nitrogen source for synthesis. Thus, let u, based on the synthesis reaction, be the equivalents of cells formed. The synthesis reaction then becomes

$$\frac{u}{5}CO_2 + \frac{u}{20}HCO_3^- + \frac{u}{20}NH_4^+ + uH^+ + ue^- \rightarrow \frac{u}{20}C_5H_7NO_2 + \frac{9u}{20}H_2O \qquad (15.39)$$

From Eqs. (15.38) and (15.39), the total moles of e^- needed from the electron donor is $t + u$. The modified donor reaction is

$$\frac{t+u}{50}C_{10}H_{19}NO_3 + \frac{9(t+u)}{25}H_2O \rightarrow \frac{9(t+u)}{50}CO_2 + \frac{t+u}{50}NH_4^+$$

$$+ \frac{t+u}{50}HCO_3^- + (t+u)H^+ + (t+u)e^- \qquad (15.40)$$

Adding Eqs. (15.38), (15.39), and (15.40) produces the overall reaction for the nitrite reduction,

$$\frac{t}{3}NO_2^- + \frac{t}{3}H^+ + \frac{t+u}{50}C_{10}H_{19}NO_3$$

$$\rightarrow \frac{u}{20}C_5H_7NO_2 + \frac{2t-3u}{100}NH_4^+$$

$$+ \frac{2t-3u}{100}HCO_3^- + \frac{9t-u}{50}CO_2 + \frac{t}{6}N_2 + \frac{92t+27u}{300}H_2O \qquad (15.41)$$

Let Y_{dc} be the cell yield in moles per unit moles of sewage. Then,

$$Y_{dc} = \frac{\dfrac{u}{20}}{\dfrac{t+u}{50}} = \frac{5u}{2(t+u)}$$

and,

$$u = \frac{2tY_{dc}}{5-2Y_{dc}} \qquad (15.42)$$

Table 15.3 shows some values of Y_c, Y_{dn}, and Y_{dc}. These values, however, may only be used for very rough estimates. For an actual firm design, a laboratory or pilot plant study for a given waste should be conducted.

TABLE 15.3
Values of Y_c, Y_{dn}, and Y_{dc}

Carbon Source	Y_c	Y_{dn}	Y_{dc}
Domestic waste, mg VSS/mg BOD$_5$	0.40	—	—
Domestic waste, mg VSS/mg COD	0.56	—	—
Soft drink waste, mg VSS/mg COD	0.35	—	—
Skim milk, mg VSS/mg BOD$_5$	0.38	—	—
Pulp and paper, mg VSS/mg BOD$_5$	0.36	—	—
Shrimp processing, mg VSS/mg BOD$_5$	0.35	—	—
Methanol, mg VSS/mg NO$_3$–N	—	0.7–1.5	—
Domestic sludge, mg VSS/mg BOD$_5$	—	—	0.040–0.100
Fatty acid, mg VSS/mg BOD$_5$	—	—	0.040–0.070
Carbohydrate, mg VSS/mg BOD$_5$	—	—	0.020–0.040
Protein, mg VSS/mg BOD$_5$	—	—	0.050–0.090

Note: VSS = Volatile suspended solids

15.7 TOTAL EFFLUENT NITROGEN

Although the idea behind denitrification is to remove nitrogen, due to the use of sewage as a carbon source for synthesis in the heterotrophic side reaction, normal anoxic denitrification, and nitrite reduction, some ammonia is produced. These productions are indicated in Eqs. (15.31), (15.36), and (15.41). For convenience, these reactions are reproduced next:

$$\frac{r+q}{50}C_{10}H_{19}NO_3 + \frac{r}{4}O_2 \rightarrow \frac{q}{20}C_5H_7NO_2 + \frac{9r-q}{50}CO_2$$

$$+ \frac{2r-3q}{100}NH_4^+ + \frac{2r-3q}{100}HCO_3^- + \frac{14r+9q}{100}H_2O \tag{15.43}$$

$$\frac{p}{5}NO_3^- + \frac{p}{5}H^+ + \frac{s+p}{50}C_{10}H_{19}NO_3 \rightarrow \frac{s}{20}C_5H_7NO_2 + \frac{p}{10}N_2 + \frac{2p-3s}{100}NH_4^+$$

$$+ \frac{2p-3s}{100}HCO_3^- + \frac{9p-s}{50}CO_2 + \frac{24p+9s}{100}H_2O \tag{15.44}$$

$$\frac{t}{3}NO_2^- + \frac{t}{3}H^+ + \frac{t+u}{50}C_{10}H_{19}NO_3 \rightarrow \frac{u}{20}C_5H_7NO_2 + \frac{2t-3u}{100}NH_4^+$$

$$+ \frac{2t-3u}{100}HCO_3^- + \frac{9t-u}{50}CO_2 + \frac{t}{6}N_2 + \frac{92t+27u}{300}H_2O \tag{15.45}$$

From Equation (15.43), the number of moles of ammonia nitrogen produced per mole of oxygen is

$$\frac{\frac{2r-3q}{100}}{\frac{r}{4}} = \frac{2r-3q}{25r}$$

The units of r and q are all in equivalents. In the derivation, although the unit of equivalent was used, there is no restriction for using equivalents per liter instead. Let r' expressed in mmol/L units be the concentration corresponding to r, which is now in meq/L. r' is the concentration of dissolved oxygen left after the aeration is cut off. For substitution into $2r - 3q/25r$, these must be converted into units of milligram equivalents per liter as follows:

$$r = \frac{r'(20)}{\frac{r}{4}O_2} = 4r'$$

$(r/4)O_2/r$ is the equivalent mass of oxygen. This r may be substituted into $q = 2rY_c/(5 - 2Y_c)$ such that $q = 8r'Y_c/(5 - 2Y_c)$. Substituting these newfound values of r and q into $(2r - 3q)/25r$, we have

$$\frac{2r - 3q}{25r} = \frac{8r' - 3\frac{8r'Y_c}{5 - 2Y_c}}{100r'} = \frac{2(1 - Y_c)}{5(5 - 2Y_c)} \quad (15.46)$$

$2(1 - Y_c)/5(5 - 2Y_c) = (2r - 3q)/25r$ is the number of mmol/L of ammonia nitrogen produced per mmol/L of dissolved oxygen. We have said the r' is the concentration of dissolved oxygen left after the aeration is cut off. This may be assumed as all consumed, since the process is now going toward being completely anaerobic. Therefore, the mmol/L of ammonia nitrogen produced in the heterotrophic side reaction is

$$\frac{\text{milligram moles per liter of ammonia nitrogen produced}}{r' \text{ milligram moles per liter of oxygen used}} = \frac{2(1 - Y_c)}{5(5 - 2Y_c)}r' \quad (15.47)$$

A similar procedure is applied to find the ammonia nitrogen produced from the normal anoxic denitrification. From Equation (15.44), the number of moles of ammonia nitrogen produced per mole of nitrate nitrogen destroyed is

$$\frac{\frac{2p - 3s}{100}}{\frac{p}{5}} = \frac{2p - 3s}{20p}$$

The units of p and s are all in equivalents. Again, in the derivation, although the unit of equivalent was used, there is no restriction for using equivalents per liter instead. Let p' expressed in mmol/L units be the concentration corresponding to p, which is now in meq/L. p' is the concentration of nitrate nitrogen destroyed in the normal anoxic denitrification stage. For substitution into $2p - 3s/20p$, the conversion into units of milligram equivalents per liter is as follows:

$$p = \frac{p'(N)}{\frac{p}{5}N} = 5p'$$

This p may be substituted into $s = 4pY_{dn}$ such that $s = 20p'Y_{dn}$.
Substituting these newfound values of p and s into $2p - 3s/20p$, we have

$$\frac{2p - 3s}{20p} = \frac{1 - 6Y_{dn}}{10} \quad (15.48)$$

$(1 - 6Y_{dn})/10 = (2p - 3s)/20p$ is the number of mmol/L of ammonia nitrogen produced per mmol/L of nitrate nitrogen destroyed. The p' mmol/L of nitrate nitrogen constitutes a fraction of the total concentration of nitrate nitrogen produced during the nitrification process. Let this concentration be p'' mmol/L and let $f_{p''}$ be the fraction of p'' destroyed. Then, $p' = f_{p''} p''$. Therefore, the mmol/L of ammonia nitrogen produced during the normal anoxic denitrification is

$$\frac{\text{milligram moles per liter of ammonia nitrogen produced}}{p' \text{milligram moles per liter of nitrate nitrogen destroyed}}$$

$$= \frac{1 - 6Y_{dn}}{10} p' = \frac{1 - 6Y_{dn}}{10} f_{p''} p'' \tag{15.49}$$

The equation to calculate the amount of ammonia nitrogen produced from the nitrite-reduction side reaction may be derived in a similar manner. We will no longer go through the steps but simply write the answer at once:

$$\frac{\text{milligram moles per liter of ammonia nitrogen produced}}{t' \text{milligram moles per liter of nitrite nitrogen destroyed}} = \frac{3(1 - Y_{dc})}{10(5 - 2Y_{dc})} t' \tag{15.50}$$

t' is the mmol/L of nitrite nitrogen that corresponds to the t meq/L of nitrite nitrogen destroyed or used during the nitrite-reduction stage of the denitrification process. It is assumed that all of the nitrites that appeared after the aeration is shut off are being converted to the nitrogen gas.

Adding Eqs. (15.47), (15.49), and (15.50) gives the concentration of NH_4–N in milligram moles per liter, $[NH_4-N]_{mmol}$, that will appear in the effluent of the nitrification–denitrification process:

$$[NH_4-N]_{mmol} = \frac{2(1 - Y_c)}{5(5 - 2Y_c)} r' + \frac{1 - 6Y_{dn}}{10} f_{p''} p'' + \frac{3(1 - Y_{dc})}{10(5 - 2Y_{dc})} t' \tag{15.51}$$

The total nitrogen concentration that will appear at the effluent of the nitrification–denitrification process as a result of the consumption of the residual oxygen from nitrification, destruction of nitrate, and destruction of nitrite in the denitrification step is the sum of the ammonia nitrogen in the above equation plus the nitrate nitrogen not destroyed in the denitrification step. Let $[TN]_{mmol}$ be the milligram moles per liter of total nitrogen in the effluent. Thus,

$$[TN]_{mmol} = [NH_4-N]_{mmol} + (1 - f_{p''}) p''$$

$$= \frac{2(1 - Y_c)}{5(5 - 2Y_c)} r' + \frac{1 - 6Y_{dn}}{10} f_{p''} p'' + \frac{3(1 - Y_{dc})}{10(5 - 2Y_{dc})} t' + (1 - f_{p''}) p'' \tag{15.52}$$

Again, t' = mmol/L of nitrite nitrogen appearing after the nitrification step, assumed totally destroyed; p'' = mmol/L of nitrate nitrogen appearing after the nitrification step, with $f_{p''}$ fraction destroyed; and r' = mmol/L of dissolved oxygen remaining right after aeration is cut off, assumed totally consumed.

Note that the production of effluent nitrogen depends upon the values of the cell yields. For Y_c equal to or greater than 1, no effluent ammonia nitrogen is produced from the heterotrophic side of the denitrification; for $6Y_{dn}$ equal to or greater than 1, no effluent ammonia nitrogen is produced from the normal anoxic denitrification;

and for Y_{dc} equal to or greater than 1, no effluent ammonia nitrogen is produced from the nitrite-reduction side of the denitrification. These facts should be considered in the calculation by setting the values equal to zero when the respective conditions are met.

15.7.1 Units of Cell Yields

Three types of cell yields are used in the previous derivations: Y_c, Y_{dn}, and Y_{dc}. The units of Y_c and Y_{dc} are in terms of moles of organisms per unit mole of sewage. The units of Y_{dn} are in terms of moles of organism per unit mole of nitrate nitrogen. The units normally used in practice for Y_c and Y_{dc}, however, are either in terms of mass of organisms or cells (approximated by the volatile suspended solids value, VSS) per unit mass of BOD_5 or mass of organisms or cells per unit mass of COD. For the case of Y_{dn}, the units used in practice are in terms of mass of the organisms or cells per unit of mass of the nitrate nitrogen. Unlike the conversion of Y_c and Y_{dc} from mass basis to mole basis which is harder, the conversion of Y_{dn} from mass basis to mole basis is very straightforward; thus, we will address the conversion of the former.

Let the cells yielding Y_c and Y_{dc} be designated collectively as Y_{BOD_5} when expressed in terms of mass cells per mass BOD_5 and let them be designated collectively as Y_{COD} when expressed in terms of mass cells per mass COD. To make the conversion from Y_{BOD_5} or Y_{COD} to Y_c or Y_{dc}, the electrons released by sewage as the electron donor must be assumed to be all taken up by the oxygen electron acceptor. This is because we want the conversion of one to the other—partial taking up of the electrons does not make the conversion. The two half reactions are reproduced below for convenience.

$$\frac{1}{50}C_{10}H_{19}NO_3 + \frac{9}{25}H_2O \rightarrow \frac{9}{50}CO_2 + \frac{1}{50}NH_4^+ + \frac{1}{50}HCO_3^- + H^+ + e^- \quad (15.53)$$

$$\frac{1}{4}O_2 + H^+ + e^- \rightarrow \frac{1}{2}H_2O \quad (15.54)$$

Judging from these two equations, the mole ratio of oxygen to sewage is $(1/4)/(1/50) = (25/2)$. Oxygen is the same as the ultimate carbonaceous biochemical oxygen demand, CBOD. Let f_{BOD_5} be the mole ratio of BOD_5 to CBOD, which would be the same as the mole ratio of BOD_5 to oxygen. The mole ratio of BOD_5 to sewage is then $f_{BOD_5}(25/2)$.

$$Y_{BOD_5}\frac{\text{mass of cells}}{\text{mass of BOD}_5} \text{ implies } Y_{BOD_5}\frac{\dfrac{\text{mass of cells}}{C_5H_7NO_2}}{\dfrac{\text{mass of BOD}_5}{f_{BOD_5}^{(32)}}} = \frac{1}{113}Y_{BOD_5}\frac{\text{mol cells}}{\text{mol BOD}_5}$$

$$= \frac{1}{113}Y_{BOD_5}\frac{\text{mol cells}}{f_{BOD_5}\left(\dfrac{25}{2}\right)\text{mol sewage}}$$

Therefore,

$$Y_c \text{ or } Y_{dc} = \frac{1}{113} Y_{BOD_5} \frac{\text{mol cells}}{f_{BOD_5} \left(\frac{25}{2}\right) \text{mol sewage}}$$

$$= 7.08(10^{-4}) \left(\frac{Y_{BOD_5}}{f_{BOD_5}}\right) \frac{\text{mol cells}}{\text{mol sewage}} \qquad (15.55)$$

Although not correct exactly, for practical purposes, COD may considered equal to CBOD or oxygen:

$$Y_{COD} \frac{\text{mass of cells}}{\text{mass of COD}} \text{ implies } Y_{COD} \frac{\overset{\text{mass of cells}}{\overset{C_5H_7NO_2}{\text{mass of COD}}}}{\underset{(32)}{\text{mass of COD}}} = \frac{1}{113} Y_{COD} \frac{\text{mol cells}}{\text{mol COD}}$$

$$= \frac{1}{113} Y COD \frac{\text{mol cells}}{\left(\frac{25}{2}\right) \text{mol sewage}}$$

Therefore,

$$Y_c \text{ or } Y_{dc} = 7.08(10^{-4}) Y_{COD} \frac{\text{mol cells}}{\text{mol sewage}} \qquad (15.56)$$

Example 15.1 A domestic wastewater with a flow of 20,000 m^3/d is to be denitrified. The effluent from the nitrification tank contains 30 mg/L of NO_3–N, 2.0 mg/L of dissolved oxygen, and 0.5 mg/L of NO_2–N. If the total nitrate nitrogen is to be destroyed by 95%, calculate the total ammonia nitrogen produced from the reactions and total nitrogen in the effluent.

Solution:

$$[NH_4\text{–N}]_{mmol} = \frac{2(1-Y_c)}{5(5-2Y_c)} r' + \frac{1-6Y_{dn}}{10} f_{p''} p'' + \frac{3(1-Y_{dc})}{10(5-2Y_{dc})} t'$$

$$[TN]_{mmol} = \frac{2(1-Y_c)}{5(5-2Y_c)} r' + \frac{1-6Y_{dn}}{10} f_{p''} p'' + \frac{3(1-Y_{dc})}{10(5-2Y_{dc})} t' + (1-f_{p''}) p''$$

Referring to Table 15.3, assume:

$$Y_c = 0.35 \text{ mg VSS/mg BOD}_5 = 7.08(10^{-4}) \left(\frac{Y_{BOD_5}}{f_{BOD_5}}\right) \frac{\text{mmol cells}}{\text{mmol sewage}}$$

and assume $f_{BOD_5} = 0.67$.

Therefore,

$$Y_c = 7.08(10^{-4})\left(\frac{0.35}{0.67}\right)\frac{\text{mmol cells}}{\text{mmol sewage}} = 3.7(10^{-4})\frac{\text{mmol cells}}{\text{mmol sewage}}$$

$$r' = \frac{2}{32} = 0.0625 \text{ mmol/L}$$

$Y_{dn} = 0.9$ mg VSS/mg NO_3–N   (from Table 15.3)

$$Y_{dn} = \frac{0.9/113}{1/14} = 0.112 \qquad f_{p''} = 0.95 \qquad p'' = 30/14 = 2.14 \text{ mmol/L}$$

$Y_{dc} = 0.05$ mg VSS/mg BOD_5 (from Table 15.3)

$$Y_{dc} = 7.08(10^{-4})\left(\frac{0.05}{0.67}\right) = 5.28(10^{-5}) \qquad t' = \frac{0.5}{4} = 0.0357 \text{ mmol/L}$$

Therefore,

$$[NH_4\text{–N}]_{\text{mmol}} = \frac{2[1 - 3.7(10^{-4})]}{5[5 - 2(3.7)(10^{-4})]}(0.0625) + \frac{1 - 6(0.112)}{10}(0.95)(2.14)$$

$$+ \frac{3[1 - 5.28(10^{-5})]}{10[5 - 2(5.28)(10^{-5})]}(0.037)$$

$$= 5.0(10^{-3}) + 0.067 + 2.22(10^{-3}) = 0.074 \text{ mmol/L} \Rightarrow 1.036 \text{ mg/L} \quad \textbf{Ans}$$

$$[TN]_{\text{mmol}} = 0.074 + (1 - 0.95)(2.14) = 0.181 \text{ mmol/L} \Rightarrow 2.53 \text{ mg/L} \quad \textbf{Ans}$$

Example 15.2 A domestic wastewater with a flow of 20,000 m^3/d is to be denitrified. The effluent from the nitrification tank contains 30 mg/L of NO_3–N, 2.0 mg/L of dissolved oxygen, and 0.5 mg/L of NO_2–N. If the total nitrate nitrogen in the effluent is limited by the state agency to 3.0 mg/L, calculate the percent destruction of the nitrate nitrogen and the ammonia nitrogen concentration in the effluent.

Solution:

$$[TN]_{\text{mmol}} = \frac{2(1 - Y_c)}{5(5 - 2Y_c)}r' + \frac{1 - 6Y_{dn}}{10}f_{p''}p'' + \frac{3(1 - Y_{dc})}{10(5 - 2Y_{dc})}t' + (1 - f_{p''})p''$$

$$= 5.0(10^{-3}) + \frac{1 - 6(0.112)}{10}f_{p''}(2.14) + 2.22(10^{-3}) + (1 - f_{p''})(2.14)$$

$$[TN]_{\text{mmol}} = 3/14 = 0.214 \text{ mmol/L}$$

$$0.214 = 5.0(10^{-3}) + \frac{1 - 6(0.112)}{10}f_{p''}(2.14) + 2.22(10^{-3}) + (1 - f_{p''})(2.14)$$

$$= 5.0(10^{-3}) + 0.070 f_{p''} + 2.22(10^{-3}) + 2.14 - 2.14 f_{p''}$$

$$2.07 f_{p''} = 1.93 f_{p''} = 0.93 \quad \textbf{Ans}$$

$$[\mathrm{NH_4-N}]_{\mathrm{mmol}} = \frac{2(1-Y_c)}{5(5-2Y_c)}r' + \frac{1-6Y_{dn}}{10}f_{p''}p'' + \frac{3(1-Y_{dc})}{10(5-2Y_{dc})}t'$$

$$= 5.0(10^{-3}) + \frac{1-6(0.112)}{10}(0.93)(2.14) + 2.22(10^{-3})$$

$$= 0.072 \ \mathrm{mmol/L} \Rightarrow 1.008 \ \mathrm{mg/L} \quad \textbf{Ans}$$

15.8 CARBON REQUIREMENTS FOR DENITRIFICATION

As shown in the derivations, carbon is necessary during denitrification. Because the organisms responsible for this process are heterotrophs, sewage was provided as the source. The pertinent reactions for the heterotrophic side reaction, normal anoxic denitrification reaction, and the nitrite-reduction side reaction are given, respectively, in Eqs. (15.43), (15.44), and (15.45).

From Equation (15.43), the number of moles of sewage needed per mole of oxygen is

$$\frac{\frac{r+q}{50}}{\frac{r}{4}} = \frac{2(r+q)}{25r}$$

Substituting $r = 4r'$ and $q = 8r'Y_c/(5 - 2Y_c)$, we have

$$\frac{2(r+q)}{25r} = \frac{2\left(4r' + \frac{8r'Y_c}{5-2Y_c}\right)}{25(4r')} = \frac{2}{5(5-2Y_c)} \qquad (15.57)$$

$2/5(5 - 2Y_c) = 2(r + q)/25r$ is the number of mmol/L of sewage needed per mmol/L of dissolved oxygen. Therefore, the total mmol/L of sewage needed during the heterotrophic side reaction is

$$\frac{\text{milligram moles per liter of sewage needed}}{r'\,\text{milligram moles per liter of oxygen used}} = \frac{2}{5(5-2Y_c)}r' \qquad (15.58)$$

From Equation (15.44), the number of moles of sewage needed per mole of nitrate nitrogen is

$$\frac{\frac{s+p}{50}}{\frac{p}{5}} = \frac{s+p}{10p}$$

Substituting $p = 5p'$ and $s = 20p'\,Y_{dn}$, we have

$$\frac{s+p}{10p} = \frac{20p'Y_{dn}+5p'}{10(5p')} = \frac{20Y_{dn}+5}{50} = \frac{4Y_{dn}+1}{10} \qquad (15.59)$$

From this equation, the total mmol/L of sewage needed during the normal anoxic denitrification is

$$\frac{\text{milligram moles per liter of sewage needed}}{p' \text{milligram moles per liter of nitrate nitrogen destroyed}} = \frac{4Y_{dn}+1}{10}p'$$

$$= \frac{4Y_{dn}+1}{10}f_{p''}p'' \qquad (15.60)$$

The relationship of t' (in mmol/L) and t was obtained as follows:

$$t = \frac{t'(N)}{\frac{t}{3}N} = 3t'$$

This t may be substituted into

$$u = \frac{2tY_{dc}}{5-2Y_{dc}}$$

such that

$$u = \frac{2tY_{dc}}{5-2Y_{dc}} = \frac{6t'Y_{dc}}{5-2Y_{dc}}$$

From Equation (15.45), the number of moles of sewage needed per mole of nitrite nitrogen is

$$\frac{\frac{t+u}{50}}{\frac{t}{3}} = \frac{3(t+u)}{50t}$$

Substituting these newfound values of t and u into $3(t+u)/50\,t$, we have

$$\frac{3(t+u)}{50t} = \frac{3(3t'+\frac{6t'Y_{dc}}{5-2Y_{dc}})}{50(3t')} = \frac{3}{10(5-2Y_{dc})} \qquad (15.61)$$

From this equation, the total mmol/L of sewage needed during the nitrite-reduction stage of the denitrification is

$$\frac{\text{milligram moles per liter of sewage needed}}{t' \text{milligram moles per liter of nitrite nitrogen destroyed}} = \frac{3}{10(5-2Y_{dc})}t' \qquad (15.62)$$

Adding Eqs. (15.58), (15.60), and (15.62) gives the overall milligram moles per liter of sewage, $[\text{SEW}_{denit}]_{\text{mmol}}$, needed in the denitrification process:

$$[\text{SEW}_{denit}]_{\text{mmol}} = \frac{2r'}{5(5 - 2Y_c)} + \frac{(4Y_{dn} + 1)f_{p''}p''}{10} + \frac{3t'}{10(5 - 2Y_{dc})} \quad (15.63)$$

$[\text{SEW}_{denit}]_{\text{mmol}}$ may be converted into units in terms of BOD_5 in mg/L. From Eqs. (15.53) and (15.54), the ratio of milligrams BOD_5 to millimole of sewage is f_{BOD_5} $((1/4)\text{O}_2)/(1/50) = 400 f_{\text{BOD}_5}$. Therefore, the corresponding concentration of $[\text{SEW}_{denit}]_{\text{mmol}}$ in terms of BOD_5 in mg/L, $[\text{BOD}_{5\,denit}]_{\text{mg}}$, is

$$[\text{BOD}_{5\,denit}]_{\text{mg}} = 400 f_{\text{BOD}_5}[\text{SEW}_{denit}]_{\text{mmol}}$$

$$= 400 f_{\text{BOD}_5}\left[\frac{2r'}{5(5 - 2Y_c)} + \frac{(4Y_{dn} + 1)f_{p''}p''}{10} + \frac{3t'}{10(5 - 2Y_{dc})}\right] \quad (15.64)$$

Example 15.3 After cutting off the aeration to a nitrification plant, the dissolved oxygen concentration is 2.0 mg/L. How much sewage is needed for the carbonaceous reaction in this last stage of aerobic reaction before denitrification sets in? How much NH_4–N is produced from this carbonaceous reaction?

Solution:

$$\frac{\text{milligram moles per liter of sewage needed}}{r'\text{milligram moles per liter of oxygen used}} = \frac{2}{5(5 - 2Y_c)}r'$$

$$r' = \frac{2}{32} = 0.0625 \text{ mmol/L} \qquad Y_c = 3.7(10^{-4})\frac{\text{mmol cells}}{\text{mmol sewage}}$$

Therefore,

$$\frac{\text{milligram moles per liter of sewage needed}}{r'\text{milligram moles per liter of oxygen used}} = \frac{2}{5[5 - 2(3.7)(10^{-4})]}(0.0625)$$

$$= 0.0050 \quad \textbf{Ans}$$

$$\frac{\text{milligram moles per liter of ammonia nitrogen produced}}{r'\text{milligram moles per liter of oxygen used}} = \frac{2(1 - Y_c)}{5(5 - 2Y_c)}r'$$

$$= \frac{2[1 - 3.7(10^{-4})]}{5[5 - 2(3.7)(10^{-4})]}(0.0625) = 0.005 \quad \textbf{Ans}$$

Example 15.4 A domestic wastewater with a flow of 20,000 m^3/d is to be denitrified. The effluent from the nitrification tank contains 30 mg/L of NO_3–N, 2.0 mg/L of dissolved oxygen, and 0.5 mg/L of NO_2–N. Assume that an effluent total nitrogen not to exceed 3.0 mg/L is required by the permitting agency. **(a)** Calculate the quantity of domestic sewage to be provided to satisfy the carbon requirement in mmoles per liter. **(b)** What is the corresponding BOD_5 to be provided?

Solution:

(a) $[SEW_{denit}]_{mmol} = \dfrac{2r'}{5(5-2Y_c)} + \dfrac{(4Y_{dn}+1)f_{p''}p''}{10} + \dfrac{3t'}{10(5-2Y_{dr})}$

$r' = \dfrac{2}{32} = 0.0625\ \text{mmol/L}$ $Y_c = 3.7(10^{-4})\dfrac{\text{mmol cells}}{\text{mmol sewage}}$

$f_{p''} = 0.93$ from Example 5.2

$Y_{dn} = 0.9$ mg VSS/mg NO$_3$–N (from Table 15.3)

$f_{p''} = 0.93$ from Example 5.2

$Y_{dn} = \dfrac{0.9/113}{1/14} = 0.112$ $p'' = 30/40 = 2.14$ mmol/L

$Y_{dc} = 0.05$ mg VSS/mg BOD$_5$ (from Table 15.3)

$Y_{dc} = 7.08(10^{-4})\left(\dfrac{0.05}{0.67}\right) = 5.28(10^{-5})$ $t' = \dfrac{0.5}{14} = 0.0357$ mmol/L

Therefore,

$$[SEW_{denit}]_{mmol} = \dfrac{2(0.0625)}{5[5-2(3.7)(10^{-4})]} + \dfrac{[4(0.112)+1](0.93)(2.14)}{10}$$

$$+ \dfrac{3(0.0357)}{10[5-2(5.28(10^{-5}))]} = 0.30\ \text{mmol/L}\quad\textbf{Ans}$$

(b) $[BOD_{5\,denit}]_{mg} = 400f_{BOD_5}\left[\dfrac{2r'}{5(5-2Y_c)} + \dfrac{(4Y_{dn}+1)f_{p''}p''}{10} + \dfrac{3t'}{10(5-2Y_{dr})}\right]$

Assume $f_{BOD_5} = 0.67$

Therefore,

$$[BOD_{5\,denit}]_{mg} = 400(0.67)(0.30) = 80.4\ \text{mg/L}\quad\textbf{Ans}$$

15.9 ALKALINITY PRODUCTION AND ASSOCIATED CARBON REQUIREMENT

If the nitrification reactions are inspected (see the overall reaction, in particular, reproduced next), it will be found that bicarbonate alkalinity is needed:

$$\dfrac{1+n}{20}\,NH_4^+ + \dfrac{m+n}{20}\,HCO_3^- + \dfrac{m+n}{5}\,CO_2 + \dfrac{2-7m-5n}{20}\,O_2$$

$$\rightarrow \dfrac{m+n}{20}\,C_5H_7NO_2 + \dfrac{1-m}{10}\,H^+ + \dfrac{1-m}{20}\,NO_3 + \dfrac{1-2m-n}{20}\,H_2O \quad (15.65)$$

On the other hand, inspection of Table 15.2 for the half reaction of sewage as an electron donor reproduced as Equation (15.66) shows that bicarbonate alkalinity is produced from sewage.

$$\frac{1}{50}C_{10}H_{19}NO_3 + \frac{9}{25}H_2O \rightarrow \frac{9}{50}CO_2 + \frac{1}{50}NH_4^+ + \frac{1}{50}HCO_3^- + H^+ + e^- \quad (15.66)$$

A good process design should consider accounting for alkalinity requirements and production.

Equation (15.65) portrays the requirement of alkalinity in nitrification. From this equation, the moles of alkalinity required per mole of nitrate nitrogen produced is

$$\frac{\frac{m+n}{20}}{\frac{1-m}{20}} = \frac{m+n}{1-m}$$

The equivalents m and n may be replaced by the corresponding yields of *Nitrosomonas* and *Nitrobacter*, respectively. Let these yields be Y_m and Y_n in mol/mol units, respectively. From Equation (15.19), the moles of *Nitrosomonas* per unit mole of ammonia nitrogen is $(m/20)/(1/20) = m = Y_m$, and from Equation (15.26), the moles of *Nitrobacter* per unit mole of nitrite nitrogen is

$$\frac{\frac{n}{20}}{\frac{(1-m)}{20}} = \frac{n}{1-m} = \frac{n}{1-Y_m} = Y_n$$

from which $n = Y_n(1 - Y_m)$. Substituting these new values into $\frac{m+n}{1-m}$, we have

$$\frac{m+n}{1-m} = \frac{Y_m + Y_n(1-Y_m)}{1-Y_m}$$

Therefore, the mmol/L of alkalinity required, $[A_{nit}]_{mmol}$, for the p'' mmol/L of nitrate nitrogen produced in the nitrification stage is

$$[A_{nit}]_{mmol} = \frac{Y_m + Y_n(1-Y_m)}{1-Y_m}p'' \quad (15.67)$$

Equation (15.31) may be used to derive the production of alkalinity from sewage. To signify that the equation pertains to the regular heterotrophic secondary treatment of sewage as opposed to the heterotrophic side reaction, change r and q to r_s and q_s, respectively, where s stands for secondary in the regular heterotrophic secondary treatment reaction. The overall manipulations are indicated next. Note that all the terms of the equation are divided by the coefficient of HCO_3^- and then multiplied by 1/20.

$$\frac{\frac{r_s+q_s}{50}}{\frac{2r_s-3q_s}{100}}\left(\frac{1}{20}\right)C_{10}H_{19}NO_3 + \frac{\frac{r_s}{4}}{\frac{2r_s-3q_s}{100}}\left(\frac{1}{20}\right)O_2 \rightarrow \frac{\frac{q_s}{20}}{\frac{2r_s-3q_s}{100}}\left(\frac{1}{20}\right)C_5H_7NO_2$$

$$+ \frac{\frac{9r_s-q_s}{50}}{\frac{2r_s-3q_s}{100}}\left(\frac{1}{20}\right)CO_2 + \left(\frac{1}{20}\right)NH_4^+ + \frac{1}{20}HCO_3^- + \frac{\frac{14r_s+9q_s}{100}}{\frac{2r_s-3q_s}{100}}\left(\frac{1}{20}\right)H_2O \quad (15.68)$$

The previous result shows that 1/20 mole of HCO_3^- came from

$$\frac{\frac{r_s+q_s}{50}}{\frac{2r_s-3q_s}{100}}\left(\frac{1}{20}\right) = \frac{(r_s+q_s)}{2r_s-3q_s}\left(\frac{1}{10}\right) \quad \text{moles of sewage}$$

This is equivalent to

$$\frac{\frac{(r_s+q_s)}{2r_s-3q_s}\left(\frac{1}{10}\right)}{\frac{1}{20}} = \frac{2(r_s+q_s)}{2r_s-3q_s}$$

moles of sewage required per mole of the bicarbonate produced.

As in the case of r being replaced by r', replace r_s by r'_s, which is the mmol/L of dissolved oxygen at which the secondary treatment of sewage is being operated. From Equation (15.29),

$$r_s = \frac{r'_s (2O)}{\frac{r_s}{4}O_2/r_s} = 4r'_s$$

This expression for r_s may be substituted into

$$q_s = \frac{2r_s Y_c}{5-2Y_c} = \frac{8r'_s Y_c}{5-2Y_c}$$

Substitute these newfound values of r_s and q_s into

$$\frac{2(r_s+q_s)}{2r_s-3q_s}$$

to produce

$$\frac{2(4r'_s + \frac{8r'_s Y_c}{5-2Y_c})}{8r'_s - 3\frac{8r'_s Y_c}{5-2Y_c}} = \frac{5}{5+Y_c}$$

The mmol/L of sewage required, $[SEW_{nit}]_{mmol}$, for the $[A_{nit}]_{mmol}$ milligram moles per liter of alkalinity required for nitrificationis then equal

$$[SEW_{nit}]_{mmol} = \frac{5}{5+Y_c}[A_{nit}]_{mmol}$$

$$= \frac{5}{5+Y_c}\frac{Y_m + Y_n(1-Y_m)}{1-Y_m}p'' \tag{15.69}$$

$[SEW_{nit}]_{mmol}$ may be written in terms of BOD_5 as follows:

$$[BOD_{5nit}]_{mg} = 400 f_{BOD_5} \left[\frac{5}{5+Y_c} \frac{Y_m + Y_n(1-Y_m)}{1-Y_m} p'' \right] \quad (15.70)$$

Example 15.5 A domestic wastewater with a flow of 20,000 m³/d is to be nitrified. After settling in the primary clarifier, its BOD_5 was reduced to 150 mg/L. The concentration of nitrate nitrogen after nitrification is 30 mg/L. **(a)** Calculate the amount of bicarbonate alkalinity required for nitrification. **(b)** Determine if the primary treated sewage can produce enough alkalinity to satisfy nitrification requirement. Assume that 97% of the nitrate nitrogen is destroyed in denitrification.

Solution:

(a) $[A_{nit}]_{mmol} = \dfrac{Y_m + Y_n(1-Y_m)}{1-Y_m} p''$

The literature reports cell yields Y_m for *Nitrosomonas* of 0.04–0.29 mg of the bacteria per mg of NH_4–N destroyed and cell yields for *Nitrobacter* Y_n of 0.02–0.084 mg of the bacteria per milligrams of NO_2–N destroyed.

$$\text{Let } Y_m = 1.5 \text{ mg/mg} = \frac{1.5/113}{1/14} = 0.186; \ Y_n = 0.05 \text{ mg/mg}$$

$$= \frac{0.05/113}{1/14} = 0.0062$$

$$p'' = 30/14 = 2.14 \text{ mmol/L}$$

Therefore,

$$[A_{nit}]_{mmol} = \frac{0.186 + 0.0062(1-0.186)}{1-0.186}(2.14) = 0.50 \text{ mmol/L} \quad \textbf{Ans}$$

(b) $[SEW_{nit}]_{mmol} = \dfrac{2(5-Y_c)}{10-7Y_c} \dfrac{Y_m + Y_n(1-Y_m)}{1-Y_m} p''$

$$\text{let } Y_c = 3.7(10^{-4}) \frac{\text{mmol cells}}{\text{mmol sewage}} \qquad \text{from a previous example.}$$

Therefore,

$$[SEW_{nit}]_{mmol} = \frac{10 - 7(3.7)(10^{-4})}{2[5 - 3.7(10^{-4})]}(0.50) = 0.50 \quad \textbf{Ans}$$

15.10 REACTION KINETICS

The discussion so far has simply dealt with material balances. For example, they predict the concentration of nitrogen in the effluent, alkalinity requirement and production, and carbon requirement. None, however, would be able to compute the size of the reactor and how fast or how long the sewage should be treated; and, although the concentrations of the nitrogen in the effluent have been predicted, the actual resulting value would depend on how long the reaction was allowed to take place. The subject that deals with studying how fast reactions proceed to completion is called *reaction kinetics*. As a result of knowing how long a reaction proceeds, the size of the reactor can then be determined. In general, there are three types of kinetics involved in the removal of nitrogen: nitrification kinetics, denitrification kinetics, and carbon kinetics.

15.10.1 KINETICS OF GROWTH AND FOOD UTILIZATION

The kinetics of the microbial process may be conveniently divided into kinetics of growth and kinetics of food (or substrate) utilization. The kinetics of growth was derived previously in the Background Prerequisites chapter, in connection with the derivation of the Reynolds transport theorem. The result was

$$\frac{d[X]}{dt} = \mu_m \frac{[S]}{K_s + [S]}[X] - k_d[X] \qquad (15.71)$$

where $[X]$ is the concentration of mixed population of microorganisms utilizing the organic waste; t is the time of reaction; μ_m is the maximum *specific growth rate* of the mixed population in units of per unit time; $[S]$ is the concentration of substrate; K_s is called the *half-velocity constant*; and k_d is the rate of decay.

Substrate kinetics may be established by noting that as organisms grow, substrates are consumed. Therefore, the rate of decrease of the concentration of the substrate is proportional to the rate of increase of the concentration of the organisms. The first term on the right-hand side of Equation (15.71) is the rate of increase of microorganism that corresponds to the rate of decrease of the substrate. Thus, the rate of decrease of the substrate is

$$-dS/dt = U\mu_m \frac{[S]}{K_s + [S]}[X] = \frac{1}{Y}\mu_m \frac{[S]}{K_s + [S]}[X] \qquad (15.72)$$

U is a proportionality constant called the specific substrate utilization rate; it is the reciprocal of Y, the specific yield of organisms. The *specific substrate utilization rate* simply means the amount of substrate consumed per unit amount of organisms produced.

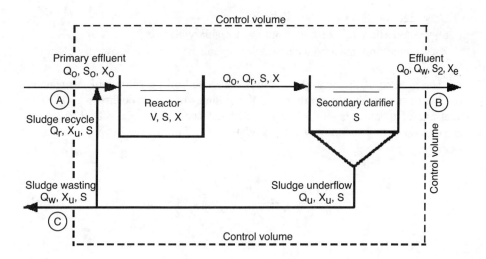

FIGURE 15.1 Schematic of the activated sludge process.

15.10.2 MATERIAL BALANCE AROUND THE ACTIVATED SLUDGE PROCESS

Biological nitrogen removal is basically a part of the general activated sludge process. Figure 15.1 shows the schematic of the process. Primary effluent is introduced into the reactor where air is supplied for consumption by the aerobic microorganisms. In the reactor, the solubles and colloids are transformed into microbial masses. The effluent of the reactor then goes to the secondary clarifier where the microbial masses are settled and separated from the clarified water. The clarified water is then discharged as effluent to some receiving stream. The settled microbial masses form the sludge. Some of the settled sludges are waste and some are recycled to the reactor. When once again exposed to the air in the reactor, the organisms become reinvigorated or activated, thus the term *activated sludge*. The wasting of the sludge is necessary so as not to cause buildup of mass in the reactor.

Develop the mathematics of the activated sludge process by performing a material balance around the control volume as indicated by the dashed lines. The symbols depicted in the figure mean as follows:

Q_o = influent flow to the process from the effluent of the primary clarifier
S_o = substrate remaining after settling in the primary clarifier and introduced into the reactor
X_o = influent mass concentration of microorganisms
Q_r = recycle flow from the secondary clarifier
S = substrate remaining after consumption by organisms in the reactor
X = mass concentration of organisms developed in the reactor
X_e = mass concentration of organisms exiting from the secondary clarifier

Q_w = rate of flow of sludge wasting
Q_u = underflow rate of flow from the secondary clarifier
X_u = underflow mass concentration of organisms from the secondary clarifier.

The volume of the control volume is comprised of the volume of the reactor, volume of the secondary clarifier, and the volumes of the associated pipings. The total rate of increase is $d\int_V [X] d\bar{V}/dt$; the local rate of increase is $\partial \int_V [X] d\bar{V}/\partial t$; and the convective rate of increase is $\oint_A [X]\vec{v}\cdot\hat{n}\, dA$. Thus, by the Reynolds transport theorem,

$$\frac{d\int_V [X] d\bar{V}}{dt} = \frac{\partial\int_V [X]\, d\bar{V}}{\partial t} + \oint_A [X]\vec{v} \cdot \hat{n} dA \qquad (15.73)$$

where
$[X]$ = the concentration of X
$\bar{V}$ = control volume
A = the bounding surface area of the control volume
$\hat{n}$ = outward-pointing normal vector
$\vec{v}$ = velocity vector.

Because aeration is done only in the reactor, this is the place where growth of organisms occurs. Thus, although it is said that $\bar{V}$ is the volume of the control volume, in reality, this volume is simply the volume of the reactor.

$d[X]/dt$ was previously given in Equation (15.71); combining this with Equation (15.73) after integrating over the volume of the reactor, produces

$$\frac{d\int_V [X] d\bar{V}}{dt} = \frac{d[X]}{dt}\bar{V} = \mu_m \frac{[S]}{K_s + [S]}[X]\bar{V} - k_d[X]\bar{V} \qquad (15.74)$$

The convective rate of increase of $[X]$ is

$$\oint_A [X]\vec{v} \cdot \hat{n} dA = (-Q_o[X_o] + (Q_o - Q_w)[X_e] + Q_w[X_u]) \qquad (15.75)$$

Note that $Q_o[X_o]$ is preceded by a negative sign. This is because at point A, the direction of the velocity vector is opposite to the direction of the unit normal vector. Now, assuming the operation of the reactor is at steady state, the local derivative is equal to zero. Substituting everything into Equation (15.73),

$$\mu_m \frac{[S]}{K_s + [S]}[X]\bar{V} - k_d[X]\bar{V} = 0 + (-Q_o[X_o] + (Q_o - Q_w)[X_e] + Q_w[X_u]) \qquad (15.76)$$

After some algebraic manipulation,

$$\mu_m \frac{[S]}{K_s + [S]} = \frac{Q_w[X_u] + (Q_o - Q_w)[X_e] - Q_o[X_o]}{[X]\mathbb{V}} + k_d \qquad (15.77)$$

The same mathematical operations may be applied to the substrate. These will not be repeated but the result will be written at once, which is

$$-\mathbb{V}\frac{d[S]}{dt} = \frac{\mu_m}{Y}\frac{[S]}{K_s + [S]}[X]\mathbb{V} = -\{0 + (-Q_o[S_o] + (Q_o - Q_w)[S] + Q_w[S])\} \qquad (15.78)$$

Note that the total derivative and Eulerian derivative are prefixed by a negative sign. This is because they represent a rate of decrease of the mass of the substrate as opposed to the rate of increase of the mass of the microorganisms. Equation (15.78) may be manipulated to produce

$$\mu_m \frac{[S]}{K_s + [S]} = \frac{Q_o}{\mathbb{V}}\frac{Y}{[X]}([S_o] - [S]) \qquad (15.79)$$

Equating the right-hand sides of Eqs. (15.77) and (15.79) and rearranging,

$$\frac{Q_w[X_u] + (Q_o - Q_w)[X_e] - Q_o[X_o]}{[X]\mathbb{V}} = \frac{Q_o}{\mathbb{V}}\frac{Y}{[X]}([S_o] - [S]) - k_d \qquad (15.80)$$

$[X]\mathbb{V}$ is the mass of biomass (organisms) contained in the volume of the reactor at any time. The is also called the *mixed liquor volatile suspended solids*. The volatile suspended is used to estimate the biomass in the reactor. $Q_w[X_u] + (Q_o - Q_w)[X_e] - Q_o[X_o]$ is the net rate of biomass wasting. Since the mass is wasted to prevent buildup of solids, it represents the mass that has accumulated over some interval of time in the reactor. Hence, if the quantity of biomass in the reactor is divided by the rate of biomass wasting, the ratio obtained is the time it takes to accumulate the biomass solids inside the reactor. This ratio $\theta_c = [X]\mathbb{V}/Q_w[X_u] + (Q_o - Q_w)[X_e] - Q_o[X_o]$ is called by various names: *mean cell residence time* (*MCRT*), *sludge retention time* (*SRT*), and *sludge age*. This ratio is different from the ratio obtained by dividing the quantity of water in the reactor by the net rate of discharge from the reactor. This latter ratio, $\mathbb{V}/Q_o = \theta$, is called the *nominal hydraulic retention time* (*NHRT*). The word nominal is used here because θ is not the actual detention time of the tank. The actual detention time is $\mathbb{V}/(Q_o + Q_r)$, where Q_r is the recycled or recirculated flow. Using θ_c and θ in Equation (15.81) and solving for $[X]$ and $[S]$,

$$[X] = \frac{\theta_c Y([S_o] - [S])}{\theta(1 + k_d \theta_c)} \tag{15.81}$$

$$[S] = [S_o] - \frac{[X]\theta(1 + k_d \theta_c)}{\theta_c Y} \tag{15.82}$$

Y, k_d, μ_m, etc. are called kinetic constants.

Note: All example values of kinetic constants reported in succeeding discussions are at 20°C.

15.10.3 NITRIFICATION KINETICS

The kinetics of nitrification are a function of several factors the most important of which include pH, temperature, and the concentrations of ammonia and dissolved oxygen. Experience has shown that the optimum pH of nitrification lies between 7.2 and 8.8. Outside this range, the rate becomes limited. As shown in the nitrification reaction, acidity is produced. If this acidity is not buffered by addition of sufficient alkalinity, the pH could control the process and the kinetics become pH limited. We have not addressed the mathematics of this issue.

Temperature affects the half-velocity constant K_s of the Monod equation. It also affects maximum growth rates. Growth rates for *Nitrosomonas* and *Nitrobacter* are affected differently by change in temperature. At elevated temperatures *Nitrosomonas* growth rates are accelerated while those of *Nitrobacter* are depressed. *Nitrosomonas* is, however, slow growing compared to *Nitrobacter* that the kinetics is controlled by *Nitrosomonas*.

The value of the concentration of dissolved oxygen in nitrification does not promote unhampered growth of organisms. Aeration equipment are not that efficient; they cannot provide an unlimited amount of dissolved oxygen. In fact, the reason why sludge is wasted is to avoid exhaustion of the dissolved oxygen as a result of too large a concentration of organisms consuming oxygen in the reactor. Thus, the nitrification process is always limited by dissolved oxygen.

Expressing the effects of these predominantly limiting factors on the specific growth rate μ_n of the nitrifiers, the following Monod equation is obtained:

$$\mu_n = \mu_{nm}\left(\frac{[S_{NH_4}]}{K_{sNH_4} + [S_{NH_4}]}\right)\left(\frac{[S_{O_2}]}{K_{sO_2} + [S_{O_2}]}\right)(C_{pH}) \tag{15.83}$$

where μ_{nm} is the maximum μ_n; $[S_{NH_4}]$ is the $[S]$ concentration of NH_4–N.

Note: Because TKN hydrolyzes to produce NH_4^+, $[S_{NH_4}]$ may also represent [TKN]); K_{sNH_4} is K_s for nitrification; $[S_{O_2}]$ is the concentration of dissolved oxygen; K_{sO_2} is the half-velocity constant for oxygen; C_{pH} and is a fractional correction factor due to the effect of pH.

K_{sNH_4}, μ_{nm}, and C_{pH} have been determined experimentally and are given, respectively, by (Mandt and Bell, 1982):

$$K_{sNH_4} = 10^{0.051T-1.158} \qquad (15.84)$$

$$\mu_{nm} = 0.47e^{0.098(T-15)} \qquad (15.85)$$

$$C_{pH} = 1 - 0.833(7.2 - pH) \qquad (15.86)$$

The values of K_{sO_2} ranges from 0.25–2.46 mg/L. T is in °C, K_{sNH_4} is in mg/L, and μ_{nm} is per day.

Applied to nitrification, Equation (15.77) in conjunction with Equation (15.83) may be written as

$$\mu_{nm}\left(\frac{[S_{NH_4}]}{K_{sNH_4} + [S_{NH_4}]}\right)\left(\frac{[S_{O_2}]}{K_{sO_2} + [S_{O_2}]}\right)(C_{pH})$$
$$= \frac{Q_w[X_{nu}] + (Q_o - Q_w)[X_{ne}] - Q_o[X_{no}]}{[X_n]\mathcal{V}} + k_{dn} \qquad (15.87)$$

where subscripts n are appended to signify nitrification.

However,

$$\mu_{nm}\left(\frac{[S_{NH_4}]}{K_{sNH_4} + [S_{NH_4}]}\right)\left(\frac{[S_{O_2}]}{K_{sO_2} + [S_{O_2}]}\right)(C_{pH}) = \mu_n$$

and

$$\frac{Q_w[X_{nu}] + (Q_o - Q_w)[X_{ne}] - Q_o[X_{no}]}{[X_n]\mathcal{V}} = \frac{1}{\theta_{cn}}$$

where μ_n is the specific growth rate and θ_{cn} is the mean cell retention time (MCRT) for the nitrifiers. Thus, the equation becomes

$$\mu_n = \frac{1}{\theta_{cn}} + k_{dn} \qquad (15.88)$$

Typical range of values for k_{dn} is 0.03–0.06 per day; typical value of $\mu_n = 0.05/d$.

Our interest is in being able to determine the length of time that the nitrifiers are to be grown so that the size of the reactor can be calculated. This means determining the equation for θ_{cn}. From Equation (15.88), $\theta_{cn} = 1/(\mu_n - k_{dn})$ which, from Equation (15.83),

$$\theta_{cn} = \frac{1}{\mu_{nm}\left(\frac{[S_{NH_4}]}{K_{sNH_4} + [S_{NH_4}]}\right)\left(\frac{[S_{O_2}]}{K_{sO_2} + [S_{O_2}]}\right)(C_{pH}) - k_{dn}} \qquad (15.89)$$

Using Eqs. (15.84) through (15.86), Equation (15.89) becomes

$$\theta_{cn} = \cfrac{1}{0.47 e^{0.098(T-15)}\left(\cfrac{[S_{NH_4}]}{10^{0.051T-1.158} + [S_{NH_4}]}\right)\left(\cfrac{[S_{O_2}]}{K_{sO_2} + [S_{O_2}]}\right)[1 - 0.833(7.2 - pH)] - k_{dn}}$$

(15.90)

In Equation (15.88), the theoretical limit of μ_n may be approached when θ_{cn} approaches infinity. At this limiting condition, μ_n is equal to k_{dn}. Combining this condition with Equation (15.83) and Eqs. (15.84) through (15.86) and solving for $[S_{NH_4}]$, the limit concentration $[S_{NH_4 lim}]$ is

$$[S_{NH_4 lim}] = \cfrac{k_{dn} 10^{0.051T-1.158}}{0.47 e^{0.098(T-15)}\left(\cfrac{[S_{O_2}]}{K_{sO_2} + [S_{O_2}]}\right)[1 - 0.833(7.2 - pH)] - k_{dn}}$$

(15.91)

15.10.4 DENITRIFICATION KINETICS

In denitrification, the parameters that can limit the kinetics are the concentrations of the carbon source $[S_c]$ and the nitrate nitrogen $[S_{NO_3}]$. The Monod equation may be written as

$$\mu_{dn} = \mu_{dnm}\left(\frac{[S_{NO_3}]}{K_{sNO_3} + [S_{NO_3}]}\right)\left(\frac{[S_c]}{K_{sc} + [S_c]}\right)$$

(15.92)

where μ_{dn} is the specific growth rate, μ_{dnm} is the maximum μ_{dn}, S_{NO_3} is the concentration of nitrate nitrogen, K_{sNO_3} is the half-velocity constant for denitrification of nitrates, $[S_c]$ is the concentration of the carbon source, and K_{sc} is the half-velocity constant for the carbon source. Values for μ_{dnm} range from 0.3 to 0.9 per day, those for K_{sNO_3} range from 0.06 to 0.14 mg/L, and those for K_{sc} range from 25–100 mg/L as BOD_5 (Mandt and Bell, 1982).

By analogy with nitrification kinetics,

$$\mu_{dn} = \frac{1}{\theta_{cdn}} + k_{ddn}$$

(15.93)

$$\theta_{cdn} = \cfrac{1}{\mu_{dnm}\left(\cfrac{[S_{NO_3}]}{K_{sNO_3} + [S_{NO_3}]}\right)\left(\cfrac{[S_c]}{K_{sc} + [S_c]}\right) - k_{ddn}}$$

(15.94)

$$[S_{NO_3 lim}] = \cfrac{k_{ddn} K_{sNO_3}}{\mu_{dnm}\left(\cfrac{[S_{NO_3}]}{K_{sNO_3} + [S_{NO_3}]}\right)\left(\cfrac{[S_c]}{K_{sc} + [S_c]}\right) - k_{ddn}}$$

(15.95)

where θ_{cdn} = MCRT for the denitrifiers, $k_{ddn} = k_d$ for denitrifiers, and $[S_{NO_3 lim}]$ = limiting value of $[S_{NO_3}]$. Values for k_{ddn} range from 0.04–0.08 per day, typical = 0.04/d.

15.10.5 CARBON KINETICS

As mentioned, three types of kinetics are involved in the removal of nitrogen: nitrification kinetics, denitrification kinetics, and carbon kinetics. Carbon kinetics refer to the kinetics of the heterotrophic aerobic reactions. By analogy with the nitrification or denitrification kinetics,

$$\mu_c = \frac{1}{\theta_{cc}} + k_{dc} \tag{15.96}$$

$$\theta_{cc} = \frac{1}{\mu_{cm}\left(\dfrac{[S_c]}{K_{sc} + [S_c]}\right) - k_{dc}} \tag{15.97}$$

$$[S_{clim}] = \frac{k_{dc}K_{sc}}{\mu_{cm}\left(\dfrac{[S_c]}{K_{sc} + [S_c]}\right) - k_{dc}} \tag{15.98}$$

where μ_c = specific growth rate of heterotrophs, $k_{dc} = k_d$ of heterotrophs, $\theta_{cc} = \theta_c$ of heterotrophs, μ_{cm} = maximum μ_c, $K_{sc} = K_s$ of heterotrophs, $[S_c]$ = concentration of carbonaceous limiting nutrient, and $[S_{clim}]$ = limiting value of $[S_c]$. Values for k_{dc} range from 0.04–0.075 per day, typical = 0.06/d. Values for μ_{cm} range from 0.8–8 per day, typical = 2/d.

Example 15.6 An activated sludge process is used to nitrify an influent with the following characteristics to 2.0 mg/L ammonia: O_0 = 20,000 m^3/d, BOD$_5$ = 200 mg/L, TKN = 60 mg/L, and average operating temperature = 22°C. The effluent is to have a BOD$_5$ of 5.0 mg/L. The DO is to be maintained at 2.0 mg/L. Assume K_{sO_2} = 1.3 mg/L. Calculate the limiting concentration of ammonia and the MCRT needed to nitrify to this limiting concentration.

Solution:

$$[S_{NH_4 lim}] = \frac{k_{dn}10^{0.051T-1.158}}{0.47e^{0.098(T-15)}\left(\dfrac{[S_{O_2}]}{K_{sO_2} + [S_{O_2}]}\right)[1 - 0.833(7.2 - pH)] - k_{dn}}$$

Let k_{dn} = 0.05 per day

Therefore,

$$[S_{NH_4 lim}] = \frac{(0.05)10^{0.051(22)-1.158}}{0.47e^{0.098(22-15)}\left(\frac{2}{1.3+2}\right)[1-0.833(7.2-7.2)]-0.05}$$

$$= \frac{0.046}{0.52} = 0.088 \text{ mmol/L} \quad \textbf{Ans}$$

At the limiting concentration, MCRT $= \theta_{cn} = \infty$ **Ans**

15.10.6 REACTOR SIZING

Two possible calculations can be used to determine the volume of the denitrification reactor: calculation in which denitrification is allowed to occur in the same tank where nitrification is also occurring and calculation in which denitrification is done in a separate tank or in a separate portion of the same tank as in the oxidation ditch reactor. Where denitrification and nitrification occurs in the same tank, the process is a batch process, an example of which is the sequencing batch reactor (SBR).

In the SBR process, the timing of the nitrification and denitrification processes are properly sequenced at appropriate time intervals in a fill-and-draw mode consisting of four phases: fill, react, settle, and draw. In the fill phase, influent is introduced to fill the reactor vessel. In the react phase, the nitrification reaction is first allowed to take place by aerating the tank. Aeration is then stopped to allow denitrification to set in. In the settling phase, the microorganisms are allowed to settle, followed by the draw phase, where the supernatant formed during the settling phase is decanted. After decanting, the cycle starts all over again. Each of these phases has a predetermined period of time.

In the operation of the nitrification unit, the nitrifiers and the heterotrophs are coexisting. This means that there are two mean cell retention times that must be satisfied. The nitrifiers are slow growers compared to the heterotrophs, therefore, their mean cell retention time, θ_{cn}, is greater. Therefore, in the design of the reactor, θ_{cc} must be made equal to θ_{cn}. In this case, the retention time for the heterotrophs is satisfied as well as the retention time for the nitrifiers. If the reverse is made equal, that is, if θ_{cn} is made equal to θ_{cc}, the time will be very short and the nitrifiers will not have sufficient time to grow and get established and will simply be washed out into the effluent. In addition, a safety factor, SF, should be applied to obtain the design $\theta_c (\theta_{cc}^d = \theta_{cn}^d)$. A value for the safety factor of 2.0 may be used. In equation form, this is written as

$$SF = \frac{\theta_{cn}^d}{\theta_{cn}} = \frac{\theta_{cc}^d}{\theta_{cn}} \tag{15.99}$$

Once the MCRT has been established, the volume of the reactor may be determined. Substituting $\theta = V/Q_o$ into Equation (15.81), the reactor volume required is

$$V = \frac{\theta_{cn}^d Y_c \{Q_o([S_{co}]-[S_c])\}}{(1+k_{dc}\theta_{cn}^d)[X]} \tag{15.100}$$

$[X]$ is the combined concentration of the heterotrophs and nitrifiers.

As noted in Equation (15.100), the design is based on the heterotrophs with the mean cell retention of the nitrifiers being the one used. Thus, the process is considered as simply the normal activated sludge process, with nitrification considered incidental to the scheme. Also, note that in the equation, concentrations are found both in the numerator and the denominator. This means that any unit of measurement for concentrations can be used.

The proportion of the respective population of heterotrophs and nitrifiers varies with the ratio of the influent BOD_5 to influent TKN as shown in Table 15.4. Based on this table, the concentrations of the respective species may be calculated. Also, the design total volatile solids (MLVSS, for mixed liquor volatile suspended solids) vary depending upon the type of activated sludge process being used. Table 15.5 shows some representative values. Note that the air supplied is measured at normal air temperatures.

TABLE 15.4
Relationships between the Fraction of Nitrifiers and the BOD_5/TKN Ratio

BOD_5/TKN	Nitrifier Fraction
0.5	0.40
1	0.20
2	0.15
3	0.08
4	0.06
5	0.05
6	0.04
7	0.04
8	0.03
9	0.03

A. P. Sincero and G. A. Sincero (1996). *Environmental Engineering: A Design Approach.* p. 443. Prentice Hall, Upper Saddle River, NJ.

TABLE 15.5
Some Design Parameters for Variations of the Activated Sludge Process

Type of Process	θ_c (days)	θ (hours)	MLVSS (mg/L)	Air Supplied (m³/kg BOD_5)
Tapered aeration	4–15	5	1500–3000	50
Conventional	4–15	6	1500–3000	50
Step aeration	4–15	4	2000–3500	50
Pure oxygen	8–20	2	6000–8000	—
Extended aeration	20–30	20	3000–6000	100

Same reactor. After the nitrifiers and the heterotrophs, the denitrifiers will now also be considered. Considering the denitrification is to occur in the same vessel as for the nitrification–heterotrophic reaction step, a second volume will be compared with the first volume computed previously; the larger of the two controls the design. The applicable equation for the second case (denitrification) is

$$V = \frac{\theta_{cdn}^d Y_{dn} \{Q_o([S_{NO_3o}] - [S_{NO_3}])\}}{(1 + k_{ddn}\theta_{cdn}^d)[X_{dn}]} \qquad (15.101)$$

$[X_{dn}]$ is the concentration of the denitrifiers. Since the denitrifiers are basically heterotrophs, $[X_{dn}]$ is equal to $[X_c]$. As in previous equations, concentrations are found both in the numerator and the denominator. Again, this means that any unit of measurement for concentrations can be used.

Separate reactors. Where denitrification and nitrification are done in separate tanks or in a separate portion of the same tank, as in the oxidation ditch process, each will require its own reactor volume. The volume of the reactor for denitrification will be given by Equation (15.101) and that for nitrification will be given by Equation (15.100).

Incorporating the SF, the modified forms of Eqs. (15.100) and (15.101) are, respectively, as follows:

$$V_{carbnit} = \frac{\theta_{cn}(SF)Y_c\{Q_o([S_{co}] - [S_c])\}}{(1 + k_{dc}\theta_{cn}SF)[X_c]} \qquad (15.102)$$

$$V_{denit} = \frac{\theta_{cdn}(SF)Y_{dn}\{Q_o([S_{NO_3o}] - [S_{NO_3}])\}}{(1 + k_{ddn}\theta_{cdn}SF)[X_{dn}]} \qquad (15.103)$$

$V_{carbnit}$ is the volume of the activated sludge reactor, where the carbonaceous and the nitrification reactions occur. V_{denit} is the volume for denitrification.

Example 15.7 A domestic wastewater with a flow of 4000 m^3/d is nitrified. The influent to the reactor tank contains 60 mg/L of TKN, 25 mg/L of ammonia nitrogen, and 200 mg/L of BOD_5. The effluent from the tank contains 2.0 mg/L of dissolved oxygen, 0.5 mg/L of NO_2–N, 2.0 mg/L of ammonia nitrogen, and 10 mg/L of BOD_5. **(a)** Calculate the volume of the reactor. **(b)** How long does it take to attain the limiting nitrate ion concentration?

Solution:

(a) $$V_{carbnit} = \frac{\theta_{cn}(SF)Y_c\{Q_o([S_{co}] - [S_c])\}}{(1 + k_{dc}\theta_{cn}SF)[X_c]}$$

$$\theta_{cc} = \frac{1}{\mu_{cm}\left(\dfrac{[S_c]}{K_{sc} + [S_c]}\right) - k_{dc}}$$

$$\theta_{cn} = \frac{1}{0.47e^{0.098(T-15)}\left(\dfrac{[S_{NH_4}]}{10^{0.051T-1.158} + [S_{NH_4}]}\right)\left(\dfrac{[S_{O_2}]}{K_{sO_2} + [S_{O_2}]}\right)[1 - 0.833(7.2 - pH)] - k_{dn}}$$

Let $\mu_{cm} = 2$ per day $K_{sc} = 50$ mg/L $k_{dc} = 0.05$ per day

Therefore,

$$\theta_{cc} = \frac{1}{2\left(\frac{10}{50+10}\right) - 0.05} = 3.53d$$

Let $T = 10°C$ $pH = 7.2$ $k_{dn} = 0.06$ per day $K_{sO_2} = 1.3$ mg/L
Therefore,

$$\theta_{cn} = \frac{1}{0.47 e^{0.098(10-15)}\left(\frac{2}{10^{0.051(10)-1.158}+2}\right)\left(\frac{2}{1.3+2}\right)[1-0.833(7.2-7.2)]-0.06}$$

$$= \frac{1}{0.29(0.54)-0.06} = 10.35 \text{ days} > 3.53 \text{ days}$$

Let SF = 2 Y_c = 0.35 mg VSS/mg BOD$_5$. Assume MLVSS = 3000 mg/L composed of hetrotrophs and nitrifiers.

$$\frac{BOD_5}{TKN} = \frac{200}{60} = 3.33$$

From Table 15.4, nitrifier population:

$$\frac{x-0.08}{0.06-0.08} = \frac{3.33-3}{4-3} \qquad x = 0.073$$

therefore, heterotrophs = $(1 - 0.073)(3000) = 2781$
Therefore,

$$V_{carbnit} = \frac{10.35(2)(0.35)\{(4000(200-10)\}}{[1+0.05(10.35)(2)](2781)} = \frac{5,506,200}{5659.335} = 973 \text{ m}^3 \quad \textbf{Ans}$$

(b) It will take an infinite time to attain the limiting nitrite concentration. **Ans**

15.10.7 DETERMINATION OF KINETIC CONSTANTS

Various equations derived earlier require the values of the kinetic constants Y, k_d, K_s, and μ_m. These constants are characteristics of a given waste and must be determined uniquely for that waste. This section will develop a method for obtaining these values from experimental data. This method is based on the fact that if an equation can be transformed into a straight-line equation, only two data points are needed to determine the values of the parameters.

Equation (15.81) may be rearranged to produce

$$\frac{[S_o]-[S]}{[X]\theta} = \frac{1}{\theta_c Y} + \frac{k_d}{Y} \tag{15.104}$$

This equation is an equation of a straight line with $1/Y$ as the slope and k_d/Y as the y-intercept. Thus, two pairs of values of $1/\theta_c$ and $[S_o] - [S]/[X]\theta$ are all that are needed to obtain the parameters Y and k_d. Let m_Y be the slope of the line. Thus,

$$Y = \frac{1}{m_Y} \qquad (15.105)$$

Although only two data points are needed, experiments for the determination of the parameters are normally conducted to produce more data points. Let there be a total of n data points and divide them into two portions. Call the first portion as n_1; the second portion n_2 would be $n - n_1$. For the functional relationship of a straight line in the xy plane, the slope m (equal to m_Y for the present case) is

$$m = \frac{y_2 - y_1}{x_2 - x_1} \qquad (15.106)$$

where the 1 and 2 are indices for the two points of the straight line. An actual experimentation for the determination of the kinetic parameters uses no recycle. Thus, θ_c is equal to θ. And, for the straight-line equation of Equation (15.104) letting θ_c be equal to θ, y is equal to $[S_o] - [S]/[X]\theta$ and x is equal to $1/\theta$. Using the n experimental points, we produce

$$x_2 = \frac{\sum_{n_1+1}^{n} \frac{1}{\theta}}{n - n_1} \qquad (15.107)$$

$$x_1 = \frac{\sum_{1}^{n_1} \frac{1}{\theta}}{n_1} \qquad (15.108)$$

$$y_2 = \frac{\sum_{n_1+1}^{n} \frac{[S_o] - [S]}{[X]\theta}}{n - n_1} \qquad (15.109)$$

$$y_1 = \frac{\sum_{1}^{n_1} \frac{[S_o] - [S]}{[X]\theta}}{n_1} \qquad (15.110)$$

Therefore,

$$m = \frac{y_2 - y_1}{x_2 - x_1} = m_Y = \frac{\dfrac{\sum_{n_1+1}^{n} \frac{[S_o] - [S]}{[X]\theta}}{n - n_1} - \dfrac{\sum_{1}^{n_1} \frac{[S_o] - [S]}{[X]\theta}}{n_1}}{\dfrac{\sum_{n_1+1}^{n} \frac{1}{\theta}}{n - n_1} - \dfrac{\sum_{1}^{n_1} \frac{1}{\theta}}{n_1}} \qquad (15.111)$$

Equation (15.105) becomes

$$Y = \frac{1}{m_Y} = \frac{\dfrac{\sum_{n_1+1}^{n}\dfrac{1}{\theta}}{n-n_1} - \dfrac{\sum_{1}^{n_1}\dfrac{1}{\theta}}{n_1}}{\dfrac{\sum_{n_1+1}^{n}\dfrac{[S_o]-[S]}{[X]\theta}}{n-n_1} - \dfrac{\sum_{1}^{n_1}\dfrac{[S_o]-[S]}{[X]\theta}}{n_1}} \tag{15.112}$$

Equation (15.104) may be solved for k_d. Doing this and using the new expression for Y and the experimental data points, produces

$$k_d = \left\{ \frac{\dfrac{\sum_{n_1+1}^{n}\dfrac{1}{\theta}}{n-n_1} - \dfrac{\sum_{1}^{n_1}\dfrac{1}{\theta}}{n_1}}{\dfrac{\sum_{n_1+1}^{n}\dfrac{[S_o]-[S]}{[X]\theta}}{n-n_1} - \dfrac{\sum_{1}^{n_1}\dfrac{[S_o]-[S]}{[X]\theta}}{n_1}} \right\} \left\{ \dfrac{\sum_{1}^{n_1}\dfrac{[S_o]-[S]}{[X]\theta}}{n_1} \right\} - \dfrac{\sum_{1}^{n_1}\dfrac{1}{\theta}}{n_1} \tag{15.113}$$

Now, derive the equation for determining the kinetic constants K_s and μ_m. Start by transforming Equation (15.79) to

$$\frac{\theta[X]}{[S_o]-[S]} = \frac{K_s Y}{\mu_m[S]} + \frac{Y}{\mu_m} \tag{15.114}$$

which is also an equation of a straight line. $K_s Y/\mu_m$ is the slope and Y/μ_m is the y-intercept. The process of deriving the equations for K_s and μ_m is similar to that in the cases of Y and k_d. Thus, the derivation will no longer be shown. The final results are

$$\mu_m = \frac{Y\left(\dfrac{\sum_{n_1+1}^{n}\dfrac{1}{[S]}}{n-n_1} - \dfrac{\sum_{1}^{n_1}\dfrac{1}{[S]}}{n_1} \right)}{\left(\dfrac{\sum_{1}^{n_1}\dfrac{\theta[X]}{[S_o]-[S]}}{n_1} \right)\left(\dfrac{\sum_{n_1+1}^{n}\dfrac{1}{[S]}}{n-n_1} - \dfrac{\sum_{1}^{n_1}\dfrac{1}{[S]}}{n_1} \right) - \left(\dfrac{\sum_{n_1+1}^{n}\dfrac{\theta[X]}{[S_o]-[S]}}{n-n_1} - \dfrac{\sum_{1}^{n_1}\dfrac{\theta[X]}{[S_o]-[S]}}{n_1} \right)\left(\dfrac{\sum_{1}^{n_1}\dfrac{1}{[S]}}{n_1} \right)} \tag{15.115}$$

$$K_s = \frac{\dfrac{\sum_{n_1+1}^{n}\dfrac{\theta[X]}{[S_o]-[S]}}{n-n_1} - \dfrac{\sum_{1}^{n_1}\dfrac{\theta[X]}{[S_o]-[S]}}{n_1}}{\dfrac{\sum_{n_1+1}^{n}\dfrac{1}{[S]}}{n-n_1} - \dfrac{\sum_{1}^{n_1}\dfrac{1}{[S]}}{n_1}} \left(\frac{\mu_m}{Y} \right) \tag{15.116}$$

Note that in the previous methods, the values were obtained through the calculation of slopes. Depending on the relationships of the independent and dependent

TABLE 15.6
Variables Represented by
Variables in Kinetic Equations

Variable in Equation	Variables Represented		
Y	Y_c	Y_n	Y_{dn}
k_d	k_{dc}	k_{dn}	k_{ddn}
μ_m	μ_{cm}	μ_{nm}	μ_{dnm}
K_s	K_{sc}	K_{sNH_4}	K_{sNO_3}
S	S_c	S_{NH_4}	S_{NO_3}
S_o	S_{co}	$S_{NH_{4_o}}$	$S_{NO_{3_o}}$

variables, these slopes could be positive or negative. Thus, values of the computed parameters could be negative. Whether the slopes are positive or negative, however, is immaterial. The sign of the values of the parameters obtained should be taken as positive. Table 15.6 presents the kinetic variables represented by the variables in the kinetic equations.

Example 15.8 A bench-scale kinetic study was conducted on a particular carbonaceous waste producing the result below. Calculate Y_c, k_{dc}, μ_{cm}, and K_{sc}.

Unit No.	S_{co} (mg/L, BOD$_5$)	S (mg/L, BOD$_5$)	θ (days)	X_c (mg/L)
1	300	8	3.3	130
2	300	14	2.1	126
3	300	17	1.5	132
4	300	30	1.1	129
5	300	40	1.0	122

Solution:

$$Y = \frac{1}{m_Y} = \frac{\dfrac{\sum_{n_1+1}^{n} \dfrac{1}{\theta}}{n-n_1} - \dfrac{\sum_1^{n_1} \dfrac{1}{\theta}}{n_1}}{\dfrac{\sum_{n_1+1}^{n} \dfrac{[S_o]-[S]}{[X]\theta}}{n-n_1} - \dfrac{\sum_1^{n_1} \dfrac{[S_o]-[S]}{[X]\theta}}{n_1}}$$

$$k_d = \left\{ \frac{\dfrac{\sum_{n_1+1}^{n} \dfrac{1}{\theta}}{n-n_1} - \dfrac{\sum_1^{n_1} \dfrac{1}{\theta}}{n_1}}{\dfrac{\sum_{n_1+1}^{n} \dfrac{[S_o]-[S]}{[X]\theta}}{n-n_1} - \dfrac{\sum_1^{n_1} \dfrac{[S_o]-[S]}{[X]\theta}}{n_1}} \right\} \left\{ \dfrac{\sum_1^{n_1} \dfrac{[S_o]-[S]}{[X]\theta}}{n_1} \right\} - \dfrac{\sum_1^{n_1} \dfrac{1}{\theta}}{n_1}$$

$$\mu_m = \frac{Y\left(\dfrac{\sum_{n_1+1}^{n}\frac{1}{[S]}}{n-n_1} - \dfrac{\sum_{1}^{n_1}\frac{1}{[S]}}{n_1}\right)}{\left(\dfrac{\sum_{1}^{n_1}\frac{\theta[X]}{[S_o]-[S]}}{n_1}\right)\left(\dfrac{\sum_{n_1+1}^{n}\frac{1}{[S]}}{n-n_1} - \dfrac{\sum_{1}^{n_1}\frac{1}{[S]}}{n_1}\right) - \left(\dfrac{\sum_{n_1+1}^{n}\frac{\theta[X]}{[S_o]-[S]}}{n-n_1} - \dfrac{\sum_{1}^{n_1}\frac{\theta[X]}{[S_o]-[S]}}{n_1}\right)\left(\dfrac{\sum_{1}^{n_1}\frac{1}{[S]}}{n_1}\right)}$$

$$K_s = \frac{\dfrac{\sum_{n_1+1}^{n}\frac{\theta[X]}{[S_o]-[S]}}{n-n_1} - \dfrac{\sum_{1}^{n_1}\frac{\theta[X]}{[S_o]-[S]}}{n_1}}{\dfrac{\sum_{n_1+1}^{n}\frac{1}{[S]}}{n-n_1} - \dfrac{\sum_{1}^{n_1}\frac{1}{[S]}}{n_1}}\left(\frac{\mu_m}{Y}\right)$$

S_{co} (mg/L, BOD$_5$)	S (mg/L, BOD$_5$)	θ (days)	X_c (mg/L)	$1/\theta$	$[S_o]-[S]/[X]\theta$	$1/[S]$	$\theta[X]/[S_o]-[S]$
300	8	3.3	30	0.303	0.681	0.0077	1.47
300	14	2.1	126	0.476	1.081	0.0079	0.93
300	17	1.5	132	0.667	1.429	0.0076	0.67
300	30	1.1	129	0.909	1.903	0.0078	0.53
300	40	1.0	122	1.000	2.131	0.0082	0.47

Let $n_1 = 3$

$$\sum_{1}^{n_1}\frac{1}{\theta} = 0.303 + 0.476 + 0.667 = 1.446 \qquad \sum_{n_1+1}^{n}\frac{1}{\theta} = 0.909 + 1.000 = 1.909$$

$$\sum_{1}^{n_1}\frac{[S_o]-[S]}{[X]\theta} = 0.681 + 1.081 + 1.429 = 3.191$$

$$\sum_{n_1+1}^{n}\frac{[S_o]-[S]}{[X]\theta} = 1.903 + 2.131 = 4.034$$

$$\sum_{1}^{n_1}\frac{1}{[S]} = 0.0077 + 0.0079 + 0.0076 = 0.0232$$

$$\sum_{n_1+1}^{n}\frac{1}{[S]} = 0.0078 + 0.0082 = 0.016$$

$$\sum_{1}^{n_1}\frac{\theta[X]}{[S_o]-[S]} = 1.47 + 0.93 + 0.67 = 3.07$$

$$\sum_{n_1+1}^{n}\frac{\theta[X]}{[S_o]-[S]} = 0.53 + 0.47 = 1.00$$

Therefore,

$$Y_c = \frac{\dfrac{1.909}{2} - \dfrac{1.446}{3}}{\dfrac{4.034}{2} - \dfrac{3.191}{3}} = 0.496 \text{ mg/mg BOD}_5 \quad \textbf{Ans}$$

$$k_{dc} = \left\{ \left[\frac{\dfrac{1.909}{2} - \dfrac{1.446}{3}}{\dfrac{4.034}{2} - \dfrac{3.191}{3}} \right] \left\{ \frac{3.191}{3} \right\} \right\} - \frac{1.446}{3} = 0.046/\text{d} \quad \textbf{Ans}$$

$$\mu_{mc} = \frac{0.496 \left(\dfrac{0.016}{2} - \dfrac{0.0232}{3} \right)}{\left(\dfrac{3.07}{3} \right) \left(\dfrac{0.016}{2} - \dfrac{0.0232}{3} \right) - \left(\dfrac{1.00}{2} - \dfrac{3.07}{3} \right) \left(\dfrac{0.0232}{3} \right)}$$

$$= \frac{1.32(10^{-4})}{2.73(10^{-4}) + 4.05(10^{-3})} = 0.031/\text{d} \quad \textbf{Ans}$$

$$K_{sc} = \frac{\dfrac{1}{2} - \dfrac{3.07}{3}}{\dfrac{0.016}{2} - \dfrac{0.0232}{3}} \left(\frac{0.031}{0.496} \right) = -122.7 \text{ say } 122.7 \text{ mg/L} \quad \textbf{Ans}$$

GLOSSARY

Activated sludge—Microorganisms sludge subjected to reaeration.

Anoxic—An anaerobic environment where the final electron acceptor is the nitrate ion.

Anoxic denitrification, normal—The actual conversion of nitrates to nitrogen gas.

Autotrophs—Microorganisms that utilize inorganics for energy.

Biological reaction—Chemical reaction mediated by an organism.

Carbonaceous reaction—The reaction of heterotrophs utilizing the carbon content of the substrate for both energy and synthesis.

Carbonaceous cell yield—The cell yield of heterotrophs, Y_c.

Cell yield—Mass of microbial cells produced per unit mass of substrate.

Chemotrophs—Autotrophs that use inorganic chemicals for energy.

Denitrification—The process that converts nitrates to nitrogen gas

Endogenous growth rate—The condition of growth in which food is in short supply and in which organisms may cannibalize each other.

Enthalpy—Internal energy plus the product of pressure and volume. The change in enthalpy is equal to the heat exchange between the system and its surroundings at constant pressure.

Entropy—Measure of disorder of a system.

Entropy change—The ratio of the infinitesimal heat added to the temperature T during the infinitesimal time that the heat is added, if the heat is added reversibly.

Eulerian derivative—The rate of change of a property when the system is open plus the rate of change of the property as the masses cross the boundary of the system.

Faraday—One mole of electrons.

Free energy—Represents the useful energy (or, alternatively, the maximum energy) a system fuel can provide after the wasted energy due to entropy has been subtracted from the energy content.

Half-velocity constant—Concentration of the substrate that makes the specific growth rate equal to one-half the maximum growth rate.

Heterotrophs—Microorganisms that utilize organic materials for energy.

Heterotrophic side reaction stage—The initial stage of denitrification where heterotrophs consume the last amounts of dissolved oxygen before anaerobic conditions finally set in.

Internal energy—Refers to all the energies that are present in the system such as kinetic energies of the molecules, ionization energies of the electrons, bond energies, lattice energies, etc.

Lagrangian derivative—The rate of change of a property when the system is closed.

Logarithmic growth rate—Unlimited rate of growth rate.

Mean cell residence time (MCRT), sludge retention time (SRT), or sludge age—The biomass in the reactor divided by the net rate of biomass wasting.

Mixed liquor volatile suspended solids—The volatile solids suspended in an activated sludge reactor.

Nitrification—The oxidation of ammonia into nitrites and nitrates.

Nitrobacter—Species of bacteria that converts nitrites into nitrates

Nitrosomonas—Species of bacteria that converts ammonia into nitrates.

Nominal hydraulic retention time (NHRT)—The ratio obtained by dividing the quantity of water in the reactor by the net rate of discharge of the water from the reactor.

NO_2-Reduction side reaction stage—The stage of denitrification where nitrates are converted to nitrogen gas.

Oxidative phosphorylation—A mode of electron transport where the electron released by the energy source moves through a series intermediates trapping energy in more ATP molecules than is possible in substrate-level phosphorylation.

Phototrophs—Autotrophs that use sunlight for energy.

Reaction kinetics—The subject that deals with studying how fast reactions proceed to completion.

Reversible cycle—Cycle in which a reversible process is applied in every step around the cycle

Reversible process—Process in which the original state or condition of a system can be recovered if the process is done in the opposite direction from that in which it is being currently done.

Reynolds transport theorem—A theorem that states that the rate of change of a property of a system when the boundary of the system is closed is equal to the rate of change of the property when the boundary is open plus the rate of change of the property as the masses cross the boundary of the system.

Specific growth rate—Proportional constant in the rate of change of the concentration of organisms with respect to time modeled as a first order process.

Specific substrate utilization rate—Amount of substrate consumed per unit amount of organisms produced.

Spontaneous process—A process in which the change in free energy is less than zero.

Substrate-level phosphorylation—An electron transport system where the electrons released by the energy source is absorbed by a single intermediate product within the system trapping energy in only a few molecules of ATP.

Thermodynamics—The study of the relationships between heat and other forms of energy.

Thermodynamic variables—Variables involved in the study of the relationship of heat and energy.

SYMBOLS

$[A_{nit}]_{mmol}$	Milligram moles per liter of alkalinity required in nitrification
BOD_u	Ultimate oxygen demand, mg/L
BOD_5	mg/L of oxygen demand at end of five days
$[BOD_{5nit}]_{mg}$	mg/L of BOD_5 corresponding to $[SEW_{nit}]_{mmol}$
$[BOD_{5denit}]_{mg}$	mg/L of BOD_5 corresponding to $[SEW_{denit}]_{mmol}$
COD	Chemical oxygen demand, considered approximately equal to BOD_u
C_{pH}	Fractional correction factor due to the effect of pH
e^-	Mole of electron
E	Electromotive force
E'_o	Electromotive force at $\{H^+\} = 10^{-7}$ moles per liter
f_{BOD_5}	Ratio of BOD_5 to $BOD_u Y_{BOD_5}$ mole cells/mole sewage
$f_{p''}$	Fraction of p'' destroyed
F	Number of coulombs per *farady*
G	Free energy
G'_o	Standard free energy at $\{H^+\} = 10^{-7}$ moles per liter
H	Enthalpy
k_d	Rate of decay
k_{dc}	k_d of heterotrophs, per day
k_{dn}	k_d of nitrifiers, per day
k_{ddn}	k_d of denitrifiers, per day
K_s	Half-velocity constant
K_{sc}	K_s for carbon source, mg/L
K_{sNH_4}	K_s for nitrification, mg/L
K_{sNO_3}	K_s for denitrification, mg/L
K_{sO_2}	K_s for dissolved oxygen, mg/L

m	Equivalents or equivalents per liter of *Nitrosomonas* produced from one equivalent or one equivalent per liter of NH_4-N based on Equation (18.15)
n	The number of *faradays* or moles of electrons involved in a reaction; equivalents or equivalents per liter of *Nitrobacter* cells produced based on Equation (18.22) per equivalent or equivalent per liter of NH_4-N based on Equation (18.15)
p	Equivalents or equivalents per liter of NO_3-N, based on the NO_3^- acceptor reaction, destroyed in the denitrification step
p'	Millimoles per liter of nitrate nitrogen destroyed in the normal anoxic denitrification stage
p''	Total milligram moles per liter of nitrate nitrogen produced in nitrification
P	Pressure
q	Equivalents or equivalents per liter of cells, based on the synthesis reaction, produced during the last stage of the aerobic reaction
Q_o	Influent flow to the process from the effluent of the primary clarifier
Q_p	Heat exchange between system and surroundings at constant presure
Q_r	Recycle flow from the secondary clarifier
Q_{rev}	Heat exchange between system and surroundings in a reversible process
Q_u	Underflow rate of flow from the secondary clarifier
Q_w	Rate of flow of sludge wasting
r	Equivalents or equivalents per liter of O_2, based on the oxygen acceptor reaction, used in the heterotrophic stage of the denitrification reaction
r'	Millimoles per liter of dissolved oxygen left after the aeration is cut off
s	Equivalents or equivalents per liter of cells, based on the synthesis reaction, produced during the regular anoxic denitrification reaction
S	Entropy
$[S]$	mg/L of substrate remaining after consumption by organisms in the reactor
S_b	Entropy at beginning of process
$[S_{co}]$	mg/L of the carbon source influent to reactor
$[S_c]$	mg/L of the carbon source in reactor
$[S_{clim}]$	Limiting value of $[S_c]$
S_e	Entropy at end of process
$[SEW_{denit}]_{mmol}$	Milligram moles per liter of sewage needed in the denitrification process
$[SEW_{nit}]_{mmol}$	Milligram moles per liter of sewage required in nitrification
SF	Safety factor
$[S_{NH_4}]$	mg/L of ammonia nitrogen in reactor
$[S_{NH_4 lim}]$	Limit concentration of $[S_{NH_4}]$
$[S_{NH_{4o}}]$	mg/L of ammonia nitrogen influent to the reactor
$[S_{NO_3}]$	mg/L of nitrate nitrogen in reactor

$[S_{NO_3o}]$ mg/L of nitrate nitrogen influent to the reactor

$[S_o]$ Substrate remaining after settling in the primary clarifier, mg/L

$[S_{O_2}]$ mg/L of dissolved oxygen

t Equivalents or equivalents per liter of NO_2-N, based on the NO_2^- acceptor reaction, utilized in the nitrite reduction stage of the denitrification; time of reaction

t' Milligram moles per liter of nitrite nitrogen utilized in the nitrite reduction stage of the denitrification process equals the milligram moles per liter of nitrite nitrogen appearing after the nitrification step

T Temperature

u Equivalents or equivalents per liter of cells, based on the synthesis reaction, produced in the nitrite reduction stage

U Internal energy; specific substrate utilization rate, per day

V Volume of reactor considered equal to the volume of control volume

$V_{carbnit}$ Volume of activated sludge reactor

V_{denit} Volume of reactor based on denitrification reaction

W Work done by system on surroundings

$[X]$ Milligram per liter of mixed population of microorganisms utilizing organic waste.

$[X_e]$ Milligrams per liter of organisms exiting from the secondary clarifier

$[X_o]$ Influent mass concentration of microorganisms to the reactor, mg/L

$[X_{ne}]$ mg/L of nitrifiers at effluent of secondary clarifier

$[X_{nu}]$ mg/L of nitrifiers at underflow of secondary clarifier

$[X_n]$ mg/L nitrifiers in reactor

$[X_u]$ Underflow mass concentration of organisms from the secondary clarifier, mg/L

Y Specific yield of organisms

Y_c Heterotrophic cell yield in moles per unit mole of sewage

Y_{BOD_5} Cell yield of heterotrophs, moles of cells per mole of BOD_5

Y_{COD} Cell yield of heterotrophs, moles of cells per mole of COD

Y_{dn} Cell yield of denitrifiers in moles per unit moles of NO_3-N

Y_{dc} Cell yield of carbonaceous anaerobic cells in moles per unit moles of sewage with nitrite as electron acceptor

μ Specific growth rate of mixed population of organisms, per day

μ_c Specific growth rate of heterotrophs, per day

μ_{cm} Maximum μ_c

μ_{dn} Specific growth rate of denitrifiers

μ_{dnm} Maximum μ_{dn}

μ_m Maximum μ

μ_n Specific growth of nitrifiers, per day

μ_{nm} Maximum μ_n

θ_{cc} θ_c of heterotrophs, per day

θ_{cn} Mean cell retention time (MCRT) for the nitrifiers

θ_{cdn} Mean cell retention time (MCRT) for the denitrifiers

PROBLEMS

15.1   A domestic wastewater with a flow of 20,000 m^3/d is to be denitrified. The effluent from the nitrification tank contains 30 mg/L of NO_3-N, 2.0 mg/L of dissolved oxygen, and 0.5 mg/L of NO_2-N. The total nitrate nitrogen is destroyed by 95% and the concentration of ammonia nitrogen in the effluent is 1.039 mg/L. Assuming all kinetic parameters are known except Y_c, calculate Y_c.

15.2   A domestic wastewater with a flow of 20,000 m^3/d is to be denitrified. The effluent from the nitrification tank contains 30 mg/L of NO_3-N, 2.0 mg/L of dissolved oxygen, and 0.5 mg/L of NO_2-N. The total nitrate nitrogen is destroyed by 95% and the concentration of total nitrogen in the effluent is 2.53 mg/L. Assuming all kinetic parameters are known except Y_c, calculate Y_c.

15.3   A domestic wastewater with a flow of 20,000 m^3/d is to be denitrified. The effluent from the nitrification tank contains 30 mg/L of NO_3-N, and 0.5 mg/L of NO_2-N. The total nitrate nitrogen is destroyed by 95% and the concentration of total nitrogen in the effluent is 2.53 mg/L. Assume all kinetic parameters. Calculate the concentration of oxygen right after the aeration is cut off.

15.4   A domestic wastewater with a flow of 20,000 m^3/d is to be denitrified. The effluent from the nitrification tank contains 30 mg/L of NO_3-N and 0.5 mg/L of NO_2-N. The total nitrate nitrogen is destroyed by 95% and the concentration of ammonia nitrogen in the effluent is 1.039 mg/L. Assume all kinetic parameters are known. Calculate the concentration of oxygen right after the aeration is cut off.

15.5   A domestic wastewater with a flow of 20,000 m^3/d is to be denitrified. The effluent from the nitrification tank contains 30 mg/L of NO_3-N, 2.0 mg/L of dissolved oxygen, and 0.5 mg/L of NO_2-N. The concentration of total nitrogen in the effluent is 2.53 mg/L. Assuming all kinetic parameters are known, calculate the fraction of nitrate nitrogen destroyed in the denitrification process.

15.6   A domestic wastewater with a flow of 20,000 m^3/d is to be denitrified. The effluent from the nitrification tank contains 30 mg/L of NO_3-N, 2.0 mg/L of dissolved oxygen, and 0.5 mg/L of NO_2-N. The total nitrate nitrogen is destroyed by 95% and the concentration of total nitrogen in the effluent is 2.53 mg/L. Assuming all kinetic parameters are known except Y_{dn}, calculate Y_{dn}.

15.7   A domestic wastewater with a flow of 20,000 m^3/d is to be denitrified. The effluent from the nitrification tank contains 30 mg/L of NO_3-N, 2.0 mg/L of dissolved oxygen, and 0.5 mg/L of NO_2-N. The total nitrate nitrogen is destroyed by 95% and the concentration of ammonia nitrogen in the effluent is 1.039 mg/L. Assuming all kinetic parameters are known except Y_{dn}, calculate Y_{dn}.

15.8   A domestic wastewater with a flow of 20,000 m^3/d is to be denitrified. The effluent from the nitrification tank contains 30 mg/L of NO_3-N and 0.5 mg/L of NO_2-N. The total nitrate nitrogen is destroyed by 95% and the concentration of total nitrogen in the effluent is 2.53 mg/L. Assume all kinetic parameters. The concentration of oxygen right after the aeration is cut off is 2.0 mg/L. Calculate the concentration of ammonia nitrogen in the effluent.

15.9 A domestic wastewater with a flow of 20,000 m^3/d is to be denitrified. The effluent from the nitrification tank contains 30 mg/L of NO_3-N and 0.5 mg/L of NO_2-N. The total nitrate nitrogen is destroyed by 95% and the concentration of ammonia nitrogen in the effluent is 1.039 mg/L. Assume all kinetic parameters are known. The concentration of oxygen right after the aeration is cut off is 2.0 mg/L. Calculate the concentration of total nitrogen in the effluent.

15.10 A domestic wastewater with a flow of 20,000 m^3/d is to be denitrified. The effluent from the nitrification tank contains 30 mg/L of NO_3-N and 2.0 mg/L of dissolved oxygen. The concentration of total nitrogen in the effluent is 2.53 mg/L. Assuming all kinetic parameters are known and that the nitrate nitrogen is destroyed by 95%, calculate the concentration of nitrite nitrogen that appeared right after aeration is stopped.

15.11 After cutting off the aeration to a nitrification plant, the dissolved oxygen concentration is 2.0 mg/L. Calculate the millimoles per liter of sewage needed for the carbonaceous reaction in this last stage of aerobic reaction before denitrification sets in.

15.12 After cutting off the aeration to a nitrification plant, the dissolved oxygen concentration is 2.0 mg/L. The BOD_5 sewage needed for the carbonaceous reaction in this last stage of aerobic reaction before denitrification sets in is 1.34 mg/L. Assuming all kinetic parameters are known except Y_c, calculate Y_c.

15.13 The BOD_5 sewage needed for the carbonaceous reaction in the last stage of aerobic reaction before denitrification sets in is 1.34 mg/L. Assuming all kinetic parameters are known, calculate the dissolved oxygen concentration right after the aeration is cut off.

15.14 The millimoles per liter of sewage needed for the carbonaceous reaction in the last stage of aerobic reaction before denitrification sets in is 0.005. Assuming all kinetic parameters are known, calculate the dissolved oxygen concentration right after the aeration is cut off.

15.15 After cutting off the aeration to a nitrification plant, the dissolved oxygen concentration is 2.0 mg/L. The millimoles per liter of sewage needed for the carbonaceous reaction in the last stage of aerobic reaction before denitrification sets in is 0.005. Assuming all kinetic parameters are known except Y_c, calculate Y_c.

15.16 A domestic wastewater with a flow of 20,000 m^3/d is to be denitrified. The effluent from the nitrification tank contains 30 mg/L of NO_3-N, 2.0 mg/L of dissolved oxygen, and 0.5 mg/L of NO_2-N. Assume that an effluent total nitrogen not to exceed 3.0 mg/L is required by the permitting agency. The BOD_5 required to satisfy the carbon requirement is 79.15 mg/L. Assuming all kinetic parameters are known except Y_c, calculate Y_c.

15.17 A domestic wastewater with a flow of 20,000 m^3/d is to be denitrified. The effluent from the nitrification tank contains 30 mg/L of NO$_3$-N and 0.5 mg/L of NO$_2$-N. Assume that an effluent total nitrogen not to exceed 3.0 mg/L is required by the permitting agency. The BOD$_5$ required to satisfy the carbon requirement is 79.15 mg/L. Assuming all kinetic parameters are known, calculate the dissolved oxygen concentration remaining right after the aeration is cut off.

15.18 A domestic wastewater with a flow of 20,000 m^3/d is to be denitrified. The effluent from the nitrification tank contains 30 mg/L of NO$_3$-N and 2.0 mg/L of dissolved oxygen. Assume that an effluent total nitrogen not to exceed 3.0 mg/L is required by the permitting agency. The BOD$_5$ required to satisfy the carbon requirement is 79.15 mg/L. Assuming all kinetic parameters are known, calculate the nitrite nitrogen concentration right after the aeration is cut off.

15.19 A domestic wastewater with a flow of 20,000 m^3/d is to be denitrified. The effluent from the nitrification tank contains 30 mg/L of NO$_3$-N, 2.0 mg/L of dissolved oxygen, and 0.5 mg/L of NO$_2$-N. Assume that an effluent total nitrogen not to exceed 3.0 mg/L is required by the permitting agency. The BOD$_5$ required to satisfy the carbon requirement is 79.15 mg/L. Assuming all kinetic parameters are known except Y_{dn}, calculate Y_{dn}.

15.20 A domestic wastewater with a flow of 20,000 m^3/d is to be denitrified. After settling in the primary clarifier, its BOD$_5$ was reduced to 150 mg/L. From previous experience, the concentration of nitrate nitrogen after nitrification is 30 mg/L. The BOD$_5$ required for nitrification was found to be 4.21 mmol/L. Assuming all kinetic parameters are known except Y_c, calculate Y_c.

15.21 A domestic wastewater with a flow of 20,000 m^3/d is to be denitrified. After settling in the primary clarifier, its BOD$_5$ was reduced to 150 mg/L. From previous experience, the concentration of nitrate nitrogen after nitrification is 30 mg/L. The BOD$_5$ required for nitrification was found to be 4.21 mmol/L. Assuming all kinetic parameters are known except Y_m, calculate Y_m.

15.22 A domestic wastewater with a flow of 20,000 m^3/d is to be denitrified. After settling in the primary clarifier, its BOD$_5$ was reduced to 150 mg/L. From previous experience, the concentration of nitrate nitrogen after nitrification is 30 mg/L. The BOD$_5$ required for nitrification was found to be 4.21 mmol/L. Assuming all kinetic parameters are known except Y_n, calculate Y_n.

15.23 A domestic wastewater with a flow of 20,000 m^3/d is to be denitrified. After settling in the primary clarifier, its BOD$_5$ was reduced to 150 mg/L. The BOD$_5$ required for nitrification was found to be 4.21 mmol/L. Assuming all kinetic parameters are known, calculate the concentration of nitrate nitrogen subjected to denitrification.

15.24 A domestic wastewater with a flow of 20,000 m^3/d is to be denitrified. After settling in the primary clarifier, its BOD$_5$ was reduced to 150 mg/L. From previous experience, the concentration of nitrate nitrogen after nitrification is 30 mg/L. The BOD$_5$ required for nitrification was found to be

134.72 mg/L. Assuming all kinetic parameters are known except Y_c, calculate Y_c. Make appropriate assumptions for the ratio of BOD_5 to BOD_u.

15.25 A domestic wastewater with a flow of 20,000 m^3/d is to be denitrified. After settling in the primary clarifier, its BOD_5 was reduced to 150 mg/L. From previous experience, the concentration of nitrate nitrogen after nitrification is 30 mg/L. The BOD_5 required for nitrification was found to be 134.72 mg/L. Assuming all kinetic parameters are known except Y_m, calculate Y_m. Make appropriate assumptions for the ratio of BOD_5 to BOD_u.

15.26 A domestic wastewater with a flow of 20,000 m^3/d is to be denitrified. After settling in the primary clarifier, its BOD_5 was reduced to 150 mg/L. From previous experience, the concentration of nitrate nitrogen after nitrification is 30 mg/L. The BOD_5 required for nitrification was found to be 134.72 mg/L. Assuming all kinetic parameters are known except Y_n, calculate Y_n. Make appropriate assumptions for the ratio of BOD_5 to BOD_u.

15.27 A domestic wastewater with a flow of 20,000 m^3/d is to be denitrified. After settling in the primary clarifier, its BOD_5 was reduced to 150 mg/L. The BOD_5 required for nitrification was found to be 134.72 mg/L. Assuming all kinetic parameters are known, calculate the concentration of nitrate nitrogen subjected to denitrification. Make appropriate assumptions for the ratio of BOD_5 to BOD_u.

15.28 A domestic wastewater with a flow of 20,000 m^3/d is to be denitrified. After settling in the primary clarifier, its BOD_5 was reduced to 150 mg/L. From previous experience, the concentration of nitrate nitrogen after nitrification is 30 mg/L. The BOD_5 required for nitrification was found to be 134.72 mg/L. Assuming all kinetic parameters are known, calculate the ratio of BOD_5 to BOD_u.

15.29 A domestic wastewater with a flow of 20,000 m^3/d is to be denitrified. After settling in the primary clarifier, its BOD_5 was reduced to 150 mg/L. From previous experience, the concentration of nitrate nitrogen after nitrification is 30 mg/L. The amount of bicarbonate alkalinity required for nitrification is 25.0 mg/L as calcium carbonate. Assuming all kinetic parameters are known except Y_m, calculate Y_m.

15.30 A domestic wastewater with a flow of 20,000 m^3/d is to be denitrified. After settling in the primary clarifier, its BOD_5 was reduced to 150 mg/L. From previous experience, the concentration of nitrate nitrogen after nitrification is 30 mg/L. The amount of bicarbonate alkalinity required for nitrification is 25.0 mg/L as calcium carbonate. Assuming all kinetic parameters are known except Y_n, calculate Y_n.

15.31 A domestic wastewater with a flow of 20,000 m^3/d is to be denitrified. After settling in the primary clarifier, its BOD_5 was reduced to 150 mg/L. The amount of bicarbonate alkalinity required for nitrification is 25.0 mg/L as calcium carbonate. Assuming all kinetic parameters are known, calculate the nitrate nitrogen concentration subjected to denitrification.

15.32 An activated sludge process is used to nitrify an influent with the following characteristics to 2.0 mg/L ammonia: $Q_0 = 1$ mgd, $BOD_5 = 200$ mg/L, TKN = 60 mg/L, and minimum operating temperature = 10°C. The effluent is to have a BOD_5 of 5.0 mg/L. The DO is to be maintained at 2.0 mg/L. Assume $K_{sO_2} = 1.3$ mg/L. The MCRT needed to attain the limiting concentration of ammonia nitrogen concentration is $1.34(10^8)$ days. What do you conclude about this MCRT? Calculate the limiting ammonia nitrogen concentration.

15.33 An activated sludge process is used to nitrify an influent with the following characteristics to 2.0 mg/L ammonia: $Q_0 = 1$ mgd, $BOD_5 = 200$ mg/L, and TKN = 60 mg/L. The effluent is to have a BOD_5 of 5.0 mg/L. The DO is to be maintained at 2.0 mg/L. Assume $K_{sO_2} = 1.3$ mg/L. The MCRT needed to attain the limiting concentration of ammonia nitrogen is $1.34(10^8)$ days and the limiting ammonia nitrogen concentration is 1.26 mg/L. At what temperature was the unit operated?

15.34 An activated sludge process is used to nitrify an influent with the following characteristics to 2.0 mg/L ammonia: $Q_0 = 1$ mgd, $BOD_5 = 200$ mg/L, TKN = 60 mg/L, and minimum operating temperature = 10°C. The effluent is to have a BOD_5 of 5.0 mg/L. Assume $K_{sO_2} = 1.3$ mg/L. The MCRT needed to attain the limiting concentration of ammonia nitrogen is $1.34(10^8)$ days and the limiting ammonia nitrogen concentration is 1.26 mg/L. Calculate the dissolved oxygen concentration at which the unit was operated.

15.35 An activated sludge process is used to nitrify an influent with the following characteristics to 2.0 mg/L ammonia: $Q_0 = 1$ mgd, $BOD_5 = 200$ mg/L, TKN = 60 mg/L, and minimum operating temperature = 10°C. The effluent is to have a BOD_5 of 5.0 mg/L. The DO is to be maintained at 2.0 mg/L. Assume $K_{sO_2} = 1.3$ mg/L. Calculate the MCRT needed to attain the effluent ammonia nitrogen concentration from the nitrification tank of 2.0 mg/L.

15.36 An activated sludge process is used to nitrify an influent with the following characteristics to around 2.0 mg/L ammonia: $Q_0 = 1$ mgd, $BOD_5 = 200$ mg/L, TKN = 60 mg/L, and minimum operating temperature = 10°C. The effluent is to have a BOD_5 of 5.0 mg/L. The DO is to be maintained at 2.0 mg/L. Assume $K_{sO_2} = 1.3$ mg/L. If the MCRT is 20 days, calculate the ammonia nitrogen concentration from the nitrification tank.

15.37 A domestic wastewater with a flow of 20,000 m^3/d is nitrified. The effluent from the nitrification tank contains 30 mg/L of NO_3-N, 2.0 mg/L of dissolved oxygen, and 0.5 mg/L of NO_2-N. Assume an effluent total nitrogen not to exceed 3.0 mg/L and an effluent BOD_5 not to exceed 10 mg/L. Calculate the volume of the reactor based on the heterotrophic reaction.

15.38 A domestic wastewater with a flow of 20,000 m^3/d is nitrified. The effluent from the nitrification tank contains 30 mg/L of NO_3-N, 2.0 mg/L of dissolved oxygen, and 0.5 mg/L of NO_2-N. Assume an effluent total nitrogen not to exceed 3.0 mg/L and an effluent BOD_5 not to exceed 10 mg/L. Calculate the volume of the reactor based on the nitrifiers.

15.39 A domestic wastewater with a flow of 20,000 m^3/d is nitrified. The effluent from the nitrification tank contains 30 mg/L of NO_3-N, 2.0 mg/L of dissolved oxygen, and 0.5 mg/L of NO_2-N. Assume an effluent total nitrogen not to exceed 3.0 mg/L and an effluent BOD_5 not to exceed 10 mg/L. Calculate the volume of the reactor based on the denitrifiers.

15.40   A bench-scale kinetic study was conducted on a particular carbonaceous waste producing the result below. Calculate Y_c, k_{dc}, μ_{cm}, and K_{sc}.

Unit No.	S_{co}, mg/L	S, mg/L	θ, days	X_c, mg/L
1	300	10	3.5	150
2	300	15	2.4	146
3	300	19	1.8	138
4	300	32	1.2	130
5	300	44	1.1	125

BIBLIOGRAPHY

Barros, P. R. and B. Carlsson (1998). Iterative Design of a Nitrate Controller Using an External Carbon Source in an Activated Sludge Process. *Water Science Technol., Proc. 7th Int. Workshop on Instrumentation, Control and Automation of Water and Wastewater Treatment and Transport Syst.*, July 6–9, 1997, Brighton, England, 37, 12, 95–102. Elsevier Science Ltd., Exeter, England.

Cuevas-Rodriguez, G., O. Gonzalez-Barcelo, and S. Gonzalez-Martinez (1998). Wastewater fermentation and nutrient removal in sequencing batch reactors. *Water Science Technol., Wastewater: Nutrient Removal, Proc. 1998 19th Biennial Conf. Int. Assoc. on Water Quality.* Part 1, June 21–26, Vancouver, Canada, 38, 1, 255–264. Elsevier Science Ltd., Exeter, England.

D'Elia, C. (1977). University of Maryland Chesapeake Bay Laboratory, personal communication.

Furukawa, S., et al. (1998). New operational support system for high nitrogen removal in oxidation ditch process. *Water Science Technol., Proc. 7th Int. Workshop on Instrumentation, Control and Automation of Water and Wastewater Treatment and Transport Syst.*, July 6–9, 1997, Brighton, England, 37, 12, 63–68. Elsevier Science Ltd., Exeter, England.

Gupta, A. B. and S. K. Gupta (1999). Simultaneous carbon and nitrogen removal in a mixed culture aerobic RBC biofilm. *Water Res.*, 33, 2, 555–561.

Harremoes, P., et al. (1998). Six years of pilot plant studies for design of treatment plants for nutrient removal. *Water Science Technol., Wastewater: Nutrient Removal, Proc. 1998 19th Biennial Conf. Int. Assoc. on Water Quality. Part 1,* June 21–26, Vancouver, Canada, 38, 1, 219–226. Elsevier Science Ltd., Exeter, England.

Koch, G., et al. (1999). Potential of denitrification and solids removal in the rectangular clarifier. *Water Res.*, 33, 2, 309–318.

Mandt, M. G. and B. A. Bell (1982). *Oxidation Ditches in Wastewater Treatment.* Ann Arbor Science Publishers. Ann Arbor, MI, 48, 51, 64.

McCarty, P. L. (1975). Stoichiometry of biological reactions, in *Progress in Water Technology.* Pergamon Press, London.

Monod, J. (1949). The growth of bacterial cultures. *Ann Rev. Microbiol.* Vol. 3.

Orhon, D., et al. (1998). Unified basis for the design of nitrogen removal activated sludge process—the Braunschweig exercise. *Water Science Technol., Wastewater: Nutrient Removal, Proc. 1998 19th Biennial Conf. Int. Assoc. on Water Quality. Part 1,* June 21–26, Vancouver, Canada, 38, 1, 227–236. Elsevier Science Ltd., Exeter, England.

Piirtola, L., B. Hultman, and M. Lowen (1998). Effects of detergent zeolite in a nitrogen removal activated sludge process. *Water Science Technol., Solid Wastes: Sludge and Landfill Manage., Proc. 1998 19th Biennial Conf. Int. Assoc. on Water Quality, Part 2,* June 21–26, Vancouver, Canada, 38, 2, 41–48. Elsevier Science Ltd., Exeter, England.

Sen, P. and S. K. Dentel (1998). Simultaneous nitrification–denitrification in a fluidized bed reactor. *Water Science Technol., Wastewater: Nutrient Removal, Proc. 1998 19th Biennial Conf. Int. Assoc. on Water Quality, Part 1,* June 21–26, Vancouver, Canada, 38, 1, 247–254. Elsevier Science Ltd., Exeter, England.

Siegrist, H., et al. (1998). Nitrogen loss in a nitrifying rotating contactor treating ammonium-rich wastewater without organic carbon. *Water Science Technol., Wastewater: Biological Processes, Proc. 1998 19th Biennial Conf. Int. Assoc. on Water Quality, Part 7,* June 21–26, Vancouver, BC, Canada, 38, 8–9, 241–248. Elsevier Science Ltd., Exeter, England.

Sincero, A. P. and G. A. Sincero (1996). *Environmental Engineering: A Design Approach.* Prentice Hall, Upper Saddle River, NJ, 44, 443.

Sincero, A. P. (1987). Chlorophyl-a relationships in the coastal areas of Maryland. A dissertation submitted to the faculty of the School of Engineering, George Washington University, Washington, as a partial fulfillment for the degree of Doctor of Science.

Sincero, A. P. (1984). Eutrophication and the fallacy of nitrogen removal. in *Pollution Engineering.* Pudvan Publishing Co. Northbrook, IL.

Thomsen, H. A., et al. (1997). Load dependent control of BNR-WWTP by dynamic changes of aeration volumes. *Water Science Technol., Proc. 7th Int. Workshop on Instrumentation, Control and Automation of Water and Wastewater Treatment and Transport Syst.,* July 6–9, Brighton, England, 37, 12, 157–164. Elsevier Science Ltd., Exeter, England.

Tonkovic, Z. (1998). Aerobic stabilisation criteria for BNR biosolids. *Water Science Technol., Solid Wastes: Sludge and Landfill Manage., Proc. 1998 19th Biennial Conf. Int. Assoc. on Water Quality, Part 2,* June 21–26, Vancouver, Canada, 38, 2, 133–141. Elsevier Science Ltd., Exeter, England.

Tonkovic, Z. (1998). Energetics of enhanced biological phosphorus and nitrogen removal processes. *Water Science Technol., Wastewater: Nutrient Removal, Proc. 1998 19th Biennial Conf. Int. Assoc. on Water Quality, Part 1,* June 21–26, Vancouver, Canada, 38, 1, 177–184. Elsevier Science Ltd., Exeter, England.

Woodbury, B. L., et al. (1998). Evaluation of reversible fixed-film static-bed bio-denitrification reactors. *Water Science Technol., Wastewater: Nutrient Removal Proc. 1998 19th Biennial Conf. Int. Assoc. on Water Quality, Part 1,* June 21–26, Vancouver, Canada, 38, 1, 311–318. Elsevier Science Ltd., Exeter, England.

16 Ion Exchange

This chapter discusses the unit process of ion exchange, including topics such as ion exchange reactions, unit operations of ion exchange, sodium and hydrogen cycles, regeneration, and design of ion exchangers. Some of these topics have already been discussed in the various sections of the unit operations in Part II and will only be incorporated into this chapter by reference.

16.1 ION EXCHANGE REACTIONS

Ion exchange is the displacement of one ion by another. The displaced ion is originally a part of an insoluble material, and the displacing ion is originally in solution. At the completion of the process, the two ions are in reversed places: the displaced ion moves into solution and the displacing ion becomes a part of the insoluble material.

Two types of ion exchange materials are used: the cation exchange material and the anion exchange material. The *cation exchange material* exchanges cations, while the *anion exchange material* exchanges anions. The insoluble part of the exchange material is called the *host*. If R^{-n} represents the host part and C^{+m} the exchangeable cation, the cation exchange material may be represented by $(R^{-n})_r(C^{+m})_{rn/m}$, where r is the number of active sites in the insoluble material, rn/m is the number of charged exchangeable particles attached to the host material, $-n$ is the charge of the host, and $+m$ is the charge of the exchangeable cation. On the other hand, if R^{+o} represents the host part of the anion exchange material and A^{-p} its exchangeable anion, the exchange material may be represented by $(R^{+o})_r(A^{-p})_{ro/p}$, where the subscripts and superscripts are similarly defined as those for the cation exchange material. Letting C_s^{+q} be the displacing cation from solution, the cation exchange reaction is

$$(R^{-n})_r(C^{+m})_{rn/m} + \frac{rn}{q}C_s^{+q} \rightleftharpoons (R^{-n})_r(C_s^{+q})_{rn/q} + \frac{rn}{m}C^{+m} \qquad (16.1)$$

Also, letting A_s^{-t} be the displacing anion from solution, the anion exchange reaction may be represented by

$$(R^{+o})_r(A^{-p})_{ro/p} + \frac{ro}{t}A_s^{-t} \rightleftharpoons (R^{+o})_r(A_s^{-t})_{ro/t} + \frac{ro}{p}A^{-p} \qquad (16.2)$$

As shown by the previous equations, ion exchange reactions are governed by equilibrium. For this reason, effluents from ion exchange processes never yield pure water.

TABLE 16.1
Displacement Series
for Ion Exchange

La^{3+}	SO_4^{2-}
Y^{3+}	CrO_4^{2-}
Ba^{2+}	NO_3^-
Pb^{2+}	AsO_4^{-3}
Sr^{2+}	PO_4^{-3}
Ca^{2+}	MoO_4^{-2}
Ni^{2+}	I^-
Cd^{2+}	Br^-
Cu^{2+}	Cl^-
Zn^{2+}	F^-
Mg^{2+}	OH^-
Ag^+	—
Cs^+	—
Rb^+	—
K^+	—
NH_4^+	—
Na^+	—
Li^+	—
H^+	—

Table 16.1 shows the *displacement series* for ion exchange materials. When an ion species high in the table is in solution, it can displace ion species in the insoluble material below it in the table and, thus, be removed from solution. As noted in this table, to remove any cation in solution, the displaceable cation must be the proton H^+; and to remove any anion, the displaceable anion must be the hydroxyl ion OH^-.

Originally, natural and synthetic alumino silicates, called *zeolites*, were the only ones used as exchange materials. Presently, they have been largely replaced by synthetic resins. *Synthetic resins* are insoluble polymers to which are added, by certain chemical reactions, acidic and basic groups called *functional groups*. These groups are capable of performing reversible exchange reactions with ions in solution. The total number of these groups determines the *exchange capacity* of the exchange material, while the type of functional group determines ion selectivity. When the exchange capacity of the exchange material is exhausted, the exchanger may be regenerated by the reverse reactions above. The principles of regeneration are discussed in the section on "Sodium, Hydrogen Cycle, and Regeneration."

16.2 UNIT OPERATIONS OF ION EXCHANGE

Figure 16.1 shows the schematics of the unit operations of ion exchange. Figure 16.1a shows a cation exchanger and Figure 16.1b shows an anion exchanger. In both units, the influent is introduced at the top of the vessel. The bed of ion exchanger materials would be inside the vessels, where, as the water to be treated passes through, exchange

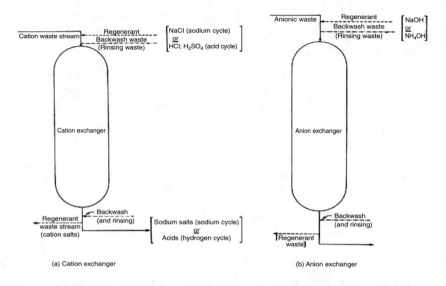

FIGURE 16.1 Unit operations of ion exchange.

of ions takes place. This exchange of ions is the chemical reaction of the unit process of ion exchange; the mere physical passing through of the water with the attendant head loss and pumping consideration is the unit operation of ion exchange. The unit operations of head losses are similar to those of granular filtration discussed in Part II. The unit operation of pumping was also discussed in Part II. After ion exchange, product waters are withdrawn at the respective bottoms of the vessels.

16.3 SODIUM, HYDROGEN CYCLE, AND REGENERATION

As shown in Table 16.1, sodium, lithium, and hydrogen are the logical choices for the exchangeable ions. In practice, however, sodium and hydrogen are the ions of choice. The cation exchange resin using sodium may be represented by $(R^{-n})_r(Na^+)_{rn}$. Its exchange reaction with Ca^{+2} and similar cations is shown below:

$$(R^{-n})_r(Na^+)_{rn} + \frac{rn}{2}Ca^{+2} \rightleftharpoons (R^{-n})_r(Ca^{+2})_{rn/2} + rnNa^+ \qquad (16.3)$$

As shown, Ca^{+2} has become embedded in the resin, thus removed from solution, and Na^+ has become solubilized. Similar reactions may be formulated for the rest of the ions in Table 16.1.

As soon as the resin is exhausted, it may be regenerated. As shown in Equation (16.3), by the *Law of Mass Action*, the reaction may be driven to the left by increasing the concentration of the sodium ion on the right. In practice, this is what is actually done. The resin is regenerated by using a concentration of NaCl of about 5 to 10%, thus, driving the reaction to the left. Operations where regeneration is done using NaCl is said to run on the *sodium cycle*. Regeneration may also be made using acids,

TABLE 16.2
General Properties of Ion Exchangers

Exchanger, cycle	Exchange Capacity, $\dfrac{geq}{m^3}$	Regenerant	Regenerant Requirement, $\dfrac{geq}{m^3}$
Cation exchangers:			
Natural zeolite, Na	175–350	NaCl	3–6
Synthetic zeolite, Na	350–700	NaCl	2–3
Resin, Na	350–1760	NaCl	1.8–3.6
Resin, H	350–1760	H_2SO_4	2–4
Anion exchanger:			
Resin, OH	700–1050	NaOH	5–8

such as H_2SO_4. When regeneration is through the use of acids, the cycle is called the *hydrogen cycle* (from the proton or hydrogen ion content of acids).

Table 16.2 shows approximate exchange capacities and regeneration requirements for ion exchangers. As shown, the values have great ranges. Thus, in practice, one must have to perform an actual experiment or obtain data from the manufacturer for a particular ion exchanger to determine the exchange capacity and regeneration requirement. The capacity of an ion exchanger in terms of volume of influent treated varies with the nature and concentration of ions in solution. This is much the same as the characteristics of activated carbon. Hence, the experimental procedure is practically the same as that of activated carbon.

Tables 16.3 and 16.4 show some additional properties of exchangers. The acidic exchangers are cationic exchangers. They are called acidic because their exchange sites are negatively charged to which the H^+ ion can attach, hence, acidic. The strongly acidic cation exchangers readily remove cations from solutions, while the weakly acidic exchangers will remove ions such as calcium and magnesium but have limited ability to remove sodium and potassium, which are way down the table in the displacement series. The basic exchangers, on the other hand, are anionic. They have positively charged exchange sites to which the hydroxyl ion can attach and other basic species such the quaternary and amine groups. The strongly basic exchanger can readily remove all anions. The weakly basic ones remove mainly the anions of strong acids such as SO_4^{2-}, Cl^-, and NO_3^-.

16.4 PRODUCTION OF "PURE WATER"

Theoretically, it would seem possible to produce pure water by combining the cation exchanger operating on the hydrogen cycle and the anion exchanger operating on the OH cycle. This is shown in the following discussions. Let Equation (16.1) be written specifically for the hydrogen cycle. The resulting equation is

$$(R^{-n})_r(H^+)_{rn} + \frac{rn}{q}C_s^{+q} \rightleftharpoons (R^{-n})_r(C_s^{+q})_{rn/q} + rnH^+ \tag{16.4}$$

TABLE 16.3
Some Additional Properties of Cation Exchangers

Material	Exchange Capacity, $\dfrac{\text{dry meq}}{\text{gm}}$, Average	Packed Density, $\dfrac{\text{kg}}{\text{m}^3}$, Average	Particle Shape
Strongly Acidic:			
Sulfonated polystyrene:			
Homogeneous resin:			
1% cross-linked	5.4	750	Spherical
2% cross-linked	5.5	720	Spherical
4% cross-linked	5.2	800	Spherical
5–6% cross-linked	5.0	810	Spherical
8–10% cross-linked	4.9	855	Spherical
12% cross-linked	5.1	840	Spherical
14% cross-linked	4.6	940	Spherical
16% cross-linked	4.9	860	Spherical
20% cross-linked	3.9	840	Spherical
Macroporous	4.8	790	Spherical
Sulfonated phenolic resins	2.4	800	Granular
Resins from phenol methylene sulfuric acid	2.9	730	Granular
Sulfonated coal	1.6	430	Granular
Weakly Acidic:			
Acrylic or meta acrylic:			
Homogeneous resin:			
5% cross-linked	10.0	720	Spherical
10% cross-linked	6.5	750	Spherical
Macroporous	8.0	745	Spherical
Phenolic and related condensation products	2.5	720	Granular
Polystyrene phosphonic acid	6.6	735	Granular
Polystyrene iminodiacetate	2.9	735	Spherical
Inorganic materials:			
Greensand	0.14	1280	Granular
Aluminum silicate	1.4	800	Granular
Celluloses:			
Phosphonic, low capacity	1.0	—	Fiber
Phosphonic, high capacity	7.0	—	Granular
Methyl carboxylic	0.7	—	Fiber

From this equation, the number of reference species is $rn/q(q)$, based on the cation in solution; and the equivalent mass of species C_s^{+q} is

$$\frac{\dfrac{rn}{q}C_s^{+q}}{\dfrac{rn}{q}(q)} = \frac{C_s^{+q}}{q}$$

TABLE 16.4
Some Additional Properties of Anion Exchangers

Material	Exchange Capacity, $\dfrac{\text{dry meq}}{\text{gm}}$, Average	Packed Density, $\dfrac{\text{kg}}{\text{m}^3}$, Average	Particle Shape
Strongly Basic:			
Polystyrene matrix:			
Trimethyl benzene ammonium:			
1% cross-linked	3.2	700	Spherical
2% cross-linked	3.5	700	Spherical
4% cross-linked	4.0	670	Spherical
8% cross-linked	3.5	720	Spherical
Dimethyl hydroxyethyl benzyl ammonium			
1–4% cross-linked	3.2	705	Spherical
6% cross-linked	3.1	705	Spherical
8% cross-linked	3.4	705	Spherical
10–12% cross-linked	3.0	705	Spherical
Condensation products with pyridium quaternary amine	4.0	800	Spherical
Weakly Basic:			
Aminopolystyrene	5.6	690	Spherical
Mixed aliphatic amine and quaternary ammonium	3.7	900	Granular
Epoxy polyamine	8.5	740	Spherical

Letting the molar concentration of C_s^{+q} be $[C_s^{+q}]$ gmol/L, the corresponding concentration in geq/L is

$$\frac{[C_s^{+q}]C_s^{+q}}{\dfrac{C_s^{+q}}{q}} = q[C_s^{+q}]$$

Note: From $q[C_s^{+q}]$, the units of q is equivalents per mole.

Therefore, the total concentration in gram equivalents per liter of removable cations in solution, $[CatT]_{eq}$, is the sum of all the cations. Let there be a total of i cations. Thus,

$$[CatT]_{eq} = \sum_{i=1}^{i=m} q_i[C_{s_i}^{+q_i}] \tag{16.5}$$

As $[CatT]_{eq}$ of cations is removed from solution, a corresponding number of equivalent concentrations of anions pair with the H^+ ions displaced from the cation bed.

Let $[AnionT]_{eq}$ and $[HT]_{eq}$ be the total anions and the hydrogen ions displaced, respectively. Since the number of equivalents of one substance in a reaction is equal to the number of equivalents of all the other substances participating in the reaction,

$$[AnionT]_{eq} = [HT]_{eq} = [CatT]_{eq} \qquad (16.6)$$

Let the $[AnionT]_{eq}$ from the effluent of the cation exchanger be introduced into an anion exchanger. For the anion exchanger operating under the OH cycle, the total equivalents of OH^- released from the anion bed is equal to that of the anions, $[AnionT]_{eq}$, removed from solution. Let $[OHT]_{eq}$ be this total OH^-. Since $[AnionT]_{eq}$ is equal to $[HT]_{eq}$, $[OHT]_{eq}$ must be equal to $[HT]_{eq}$. This means that all the acids produced in the cation exchanger are neutralized in the anion exchanger, and all ions in the water have been removed by using the combination of cation exchanger followed by anion exchanger.

On the surface, the combination of cation exchanger and anion exchanger would mean that pure water is produced. As shown in Equations (16.1) and (16.2), however, the unit process of ion exchange is governed by equilibrium constants. The values of these constants depend upon how tightly the removed ions from solution are bound to the bed exchanger sites. In general, however, by the nature of equilibrium constants, the concentrations of the affected solutes in solution are extremely small. Practically, then, we may say that "pure water" has been produced.

By analogy with Equation (16.5),

$$[AnionT]_{eq} = \sum_{i=1}^{i=m} t_i[A_{s_i}^{-t_i}] \qquad (16.7)$$

As with q_i, the units of t_i are equivalents per mole.

Example 16.1   A wastewater contains the following ions: $CrO_4^{2-} = 120$ mg/L, $Cu^{2+} = 30$ mg/L, $Zn^{2+} = 15$ mg/L, and $Ni^{2+} = 20$ mg/L. Calculate the total equivalents of cations and anions, assuming the volume of the wastewater is 450 m³.

Solution:

Ions (mg/L)	Equiv. Mass	Cations (meq/L)	Anions (meq/L)
$CrO_4^{2-} = 120$	58[a]	—	2.069[b]
$Cu^{2+} = 30$	31.75	0.945	—
$Zn^{2+} = 15$	32.7	0.469	—
$Ni^{2+} = 20$	29.35	0.681	—
		$\Sigma = 2.395$	$\Sigma = 2.069$

Note: The ions are not balanced, meaning error in analysis.

[a] Equiv. mass $CrO_4^{2-} = [52 + 4(16)]/2 = 58$. [b] $120/58 = 2.069$

Total equivalents of cations = 2.395(450) = 1077.75   **Ans**
Total equivalents of cations = 2.069(450) = 931.05   **Ans**

16.5 ACTIVE OR EXCHANGE ZONE

Figure 16.2 is the same figure illustrated in a previous chapter under carbon adsorption. The length of the active zone was derived in that chapter and is reproduced next.

$$\delta = \frac{2\left\{(\mathcal{V}_x - \mathcal{V}_b)[C_o] - \Sigma(\mathcal{V}_{n+1} - \mathcal{V}_n)\left(\frac{[C_{n+1}] + [C_n]}{2}\right)\right\}}{A_s \rho_p \left(\frac{X}{M}\right)_{ult}} \tag{16.8}$$

where
δ = length of active zone
$\mathcal{V}_x$ = total volume of water or wastewater treated at complete exhaustion of bed
$\mathcal{V}_b$ = volume treated at breakthrough
$[C_o]$ = influent concentration to δ
$\mathcal{V}_{n+1}$ = total volume treated at time t_{n+1}
$\mathcal{V}_n$ = total volume treated at time t_n
$[C_{n+1}]$ = concentration of solute at effluent of δ at time t_{n+1}
$[C_n]$ = concentration of solute at effluent of δ at time t_n
A_s = surficial area of exchanger bed

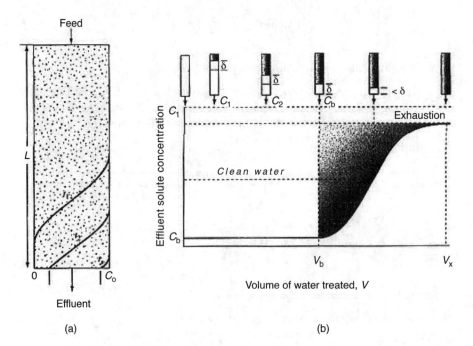

(a)   (b)

FIGURE 16.2 Active zones at various times during adsorption and the breakthrough curve.

ρ_p = pack density of ion exchange material

$(X/M)_{ult}$ = ultimate exchange capacity of the bed or simply, the exchange capacity of the bed.

It should be emphasized that to use the equation $[C_o]$, $[C_{n+1}]$, and $[C_n]$ should be the total concentration of ions. For example, if the influent is composed of the ions $Ca^{2+} = 50$ mg/L, $Mg^{2+} = 60$ mg/L, and $Zn^{2+} = 2$ mg/L, then $[C_o]_{meq}$ in meq/L is $50/(Ca/2) + 60/(Mg/2) + 2/(Zn/2)$.

Example 16.2   A breakthrough experiment is conducted for a wastewater producing the results below. Determine the length δ of the active zone. The diameter of the column used is 2.5 cm, and the packed density of the bed is 750 kg/m³. $[C_o]$ is equal to 2.2 meq/L. $(X/M)_{ult} = 6.5$ meq/g.

C, meq/L	V, L
0.06	1.0
0.08	1.20
0.09	1.30
0.10	1.40
0.20	1.48
0.46	1.58
1.30	1.70
1.80	1.85
2.10	2.00

Solution:

$$\delta = \frac{2\left\{(V_x - V_b)[C_o] - \Sigma(V_{n+1} - V_n)\left(\frac{[C_{n+1}] + [C_n]}{2}\right)\right\}}{A_s \rho_p \left(\frac{X}{M}\right)_{ult}}$$

C, meq/L	V, L	$(V_{n+1} - V_n)$	$\left(\frac{[C_{n+1}] + [C_n]}{2}\right)$	$(V_{n+1} - V_n)\left(\frac{[C_{n+1}] + [C_n]}{2}\right)$
0.06	1.0	0.20	0.07	0.014
0.08	1.20	0.10	0.085	0.0085
0.09	1.30	0.10	0.095	0.0095
0.10	1.40	0.08	0.15	0.012
0.20	1.48	0.10	0.33	0.033
0.46	1.58	0.12	0.88	0.1056
1.30	1.70	0.15	1.55	0.2325
1.80	1.85	0.15	1.95	0.2925
2.10	2.00			$\Sigma = 0.7076$

$$A_s = \frac{\pi(0.025)^2}{4} = 0.00049 \text{ m}^2$$

Therefore,

$$\delta = \frac{2\{(2-1)[2.2]-0.7076\}}{(0.00049)(750)(1000)(6.5)} = 0.0012 \text{ m} = 1.2 \text{ mm}$$

16.6 DESIGN OF ION EXCHANGERS

Generally, designs of ion exchangers should include the following: quantity of exchange materials and regenerants; dimension of the bed (volume of bed); interval of bed regeneration, backwash, and rinse water requirements. The amount of exchange materials determines the dimension of the bed. The interval of regeneration may be arbitrarily set from which the quantity of exchange bed material may be calculated. Regeneration, backwash, and rinse waters may pose pollution problems.

16.6.1 QUANTITY OF EXCHANGE MATERIALS

Before discussing quantities of exchange materials, a method of expressing exchange capacity in terms of calcium carbonate is addressed. This method of expressing capacity is very troublesome, and it should not have been adopted; nonetheless, it is used and we must know it. As shown in Tables 16.2, 16.3, and 16.4, equivalents, among other units, are used to express exchange capacities. This is appropriate because reactants react in equivalent amounts; but to express this in terms of calcium carbonate is a bit unusual. As addressed in previous chapters, however, arbitrarily adopt $CaCO_3/2 = 50$ as the equivalent mass of calcium carbonate. From this, the exchange capacity, expressed in equivalents, may be obtained by dividing the exchange capacity expressed in calcium carbonate by 50.

Let us first derive the formula for the exchange materials for the cation bed. The amount of exchange bed materials required can be determined by calculating first the amount of displacing ions in solution to be removed. Let the exchange capacity of the bed be $(X/M)_{ult}$ meq/g of bed. The equivalents of ion displaced from the bed is equal to the equivalents of displacing ion in solution; therefore, the mass of bed material *CatTBedMass* in kilograms is

$$CatTBedMass = \frac{([CatT]_{eq})(Q)(t_{int})}{\left(\dfrac{X}{M}\right)_{ult}}\left(\frac{1000}{24}\right)$$

$$= \frac{(\sum_{i=1}^{i=m} q_i[C_{s_i}^{+q_i}])(Q)(t_{int})}{\left(\dfrac{X}{M}\right)_{ult}}\left(\frac{1000}{24}\right) \qquad (16.9)$$

Q is the m³/d of flow and t_{int} is the interval of regeneration in hours. In concept, the interval of regeneration may be arbitrarily set. A value of 8 hours is not unreasonable. The factor (1000/24) is used so that the unit of *CatTBedMass* will be in kilograms.

By analogy, the mass of bed material for the anion exchanger, *AnionTBedMass*, in kilograms is

$$AnionTBedMass = \frac{([Anion]_{eq})(Q)(t_{int})}{\left(\frac{X}{M}\right)_{ult}}\left(\frac{1000}{24}\right)$$

$$= \frac{(\sum_{i=1}^{i=m}t_i[A_{s_i}^{-t_i}])(Q)(t_{int})}{\left(\frac{X}{M}\right)_{ult}}\left(\frac{1000}{24}\right) \qquad (16.10)$$

Using the packed density of bed material ρ_p kg/m^3, the volume of bed can be calculated. Let *CatTBedVol* and *AnionTBedVol* be the respective volumes in m^3. Thus,

$$CatTBedVol = \frac{\dfrac{(\sum_{i=1}^{i=m}q_i[C_{s_i}^{+q_i}])(Q)(t_{int})}{\left(\frac{X}{M}\right)_{ult}}}{\rho_p}(1+swell)\left(\frac{1000}{24}\right)$$

$$= \frac{(\sum_{i=1}^{i=m}q_i[C_{s_i}^{+q_i}])(Q)(t_{int})(1+swell)}{\rho_p\left(\frac{X}{M}\right)_{ult}}\left(\frac{1000}{24}\right) \qquad (16.11)$$

$$AnionTBedVol = \frac{(\sum_{i=1}^{i=m}t_i[A_{s_i}^{-t_i}])(Q)(t_{int})(1+swell)}{\rho_p\left(\frac{X}{M}\right)_{ult}}\left(\frac{1000}{24}\right) \qquad (16.12)$$

A very important property of exchanger beds is percentage of *swell*. This actually determines the final size of the tank into which the material is to be put when the unit is put into operation. Thus, we use the parameter swell, which is the fraction of swelling of the bed. Values of this parameter vary over a wide range from zero for zeolites (greensand) to 1.0 for polystyrene iminodiacetate. Thus, an experiment should be performed on any given exchanger to determine percent swell or the manufacturer should be consulted.

Example 16.3 Using a bed exchanger, 75 m^3 of water per day is to be treated for hardness removal between regenerations having intervals of 8 h. The raw water contains 400 mg/L of hardness as $CaCO_3$. The exchanger is a resin of exchange capacity of 1412.8 geq/m^3. Assume that the packed density of the resin is 720 kg/m^3. Calculate the mass of exchanger material to be used and the resulting volume when the exchanger is put into operation.

Solution: Assume cation exchanger:

$$CatTBedMass = \frac{(\sum_{i=1}^{i=m}q_i[C_{s_i}^{+q_i}])(Q)(t_{int})}{\left(\frac{X}{M}\right)_{ult}}\left(\frac{1000}{24}\right)$$

Also, assume that all of the cations are removed.

$$\sum_{i=1}^{i=m} q_i [C_{s_i}^{+q_i}] = \frac{400}{50} = 8 \text{ meq/L} = 0.004 \text{ geq/L}$$

$$\left(\frac{X}{M}\right)_{ult} = 1412.8 \ \frac{\text{geq}}{\text{m}^3} = 1412.8 \ \frac{(1000) \text{ meq}}{720(1000) \text{ g}} = 1.96 \ \frac{\text{meq}}{\text{g}}$$

Therefore,

$$CatTBedMass = \frac{0.004(75)(8)}{1.96} \left(\frac{1000}{24}\right) = 51.02 \text{ kg} \quad \textbf{Ans}$$

$$CatTBedVol = \frac{(\sum_{i=1}^{i=m} q_i [C_{s_i}^{+q_i}])(Q)(t_{int})(1 + swell)}{\rho_p \left(\frac{X}{M}\right)_{ult}} \left(\frac{1000}{24}\right)$$

Assume swell = 0.8

$$CatTBedVol = \frac{51.02}{720}(1 + 0.8) = 0.13 \text{ m}^3 \quad \textbf{Ans}$$

16.6.2 QUANTITY OF REGENERANT

Let *CatRegenerant* in kilogram equivalents be the quantity of regenerant required and R be the regenerant requirement in equivalents per equivalent of solute removed. The concentration of removable cations in gram equivalents per liter is $[CatT]_{eq} = \sum_{i=1}^{i=m} q_i [C_{s_i}^{+q_i}]$; therefore,

$$CatRegenerant = \sum_{i=1}^{i=m} q_i [C_{s_i}^{+q_i}](R)(Q)(t_{int})\left(\frac{1}{24}\right) \qquad (16.13)$$

This equation represents the regenerant required between intervals of regeneration. The kilograms of regenerant depend upon the regenerant used. For example, if the regenerant is common table salt, then *CatRegenerant* should be multiplied by NaCl to obtain the kilograms of the salt.

By analogy with cation exchangers, the kilogram equivalents of regenerant, *AnionRegenerant*, used to regenerate anion exchangers is

$$AnionRegenerant = \sum_{i=1}^{i=m} t_i [A_{s_i}^{+t_i}](R)(Q)(t_{int})\left(\frac{1}{24}\right) \qquad (16.14)$$

Example 16.4   Using a bed exchanger, 75 m³ of water per day is to be treated for hardness removal between regenerations having intervals of 8 hours. The raw water contains 80 mg/L of Ca^{2+} and 15 mg/L of Mg^{2+}. The exchanger is a resin of exchange capacity of 1412.8 geq/m³. Assume that the packed density of the resin is 720 kg/m³. Calculate the kilograms of sodium chloride regenerant required assuming $R = 2$ and that all of the cations were removed.

Solution:

$$CatRegenerant = \sum_{i=1}^{i=m} q_i[C_{s_i}^{+q_i}](R)(Q)(t_{int})\left(\frac{1}{24}\right)$$

$$\sum_{i=1}^{i=m} q_i[C_{s_i}^{+q_i}] = \frac{80}{(40.1/2)(1000)} + \frac{15}{(24.3/2)(1000)} = 5.22(10^{-3})\frac{geq}{L}$$

$$= 5.22(10^{-3})\frac{keq}{m^3}$$

Therefore,

$$CatRegenerant = 5.22(10^{-3})(2)(75)(8)\left(\frac{1}{24}\right)$$

$$= 0.26 \text{ kg per interval of regeneration }\textbf{Ans}$$

16.6.3 WASTEWATER PRODUCTION

In the operation of ion exchangers wastewaters are produced. These come from the solvent water used to dissolve the regenerant and the backwash and rinse requirements. To illustrate the method for the production of wastewater from the solvent water, let us use an example calculation. In the sodium cycle, the concentration of NaCl is about 5 to 10% for an average of 7.5%. This means that, for example, if the quantity of regenerant required is 0.26 kg, the volume of wastewater produced from regeneration can be calculated as follows: the total mass of regenerant solution is 0.26/0.075 = 3.47 kilograms; the corresponding volume is 3.47/1000 = 0.0035 m³. For an interval of regeneration of 8 h and assuming a rate of flow for the water treated of 75 m³/d, the volume of water treated is 75/24(8) = 25 m³. Thus, the wastewater produced is 0.0035/25 (100) = 0.014% by volume.

The other wastewater produced as a result of regeneration is the backwash and rinse waters. As soon as the bed is exhausted, it must be backwashed to remove debris accumulated during the service cycle. In addition, after regeneration, the bed should be rinsed to remove any residual regenerant. Backwash and rinse water requirements should be determined by experiment on an actual exchanger bed to be used in design.

The quantity of backwash and rinse requirement is best expressed as a function of bed volume. Let *CatBackwashRinseVol* be the m^3 of backwash and rinse waters required for the cation exchanger and *BackwashRinseV* be the corresponding m^3 of backwash and rinse waters per cubic meter of bed. For the cation exchanger, the volume of bed was previously derived as

$$CatBedVol = \frac{(\sum_{i=1}^{i=m} q_i[C_{s_i}^{+q_i}])(Q)(t_{int})}{\rho_p\left(\frac{X}{M}\right)_{ult}}\left(\frac{1000}{24}\right) m^3$$

with the swelling not being considered. Thus,

$$CatBackwashRinseVol = \frac{(\sum_{i=1}^{i=m} q_i[C_{s_i}^{+q_i}])(Q)(t_{int})(BackwashRinseV)}{\rho_p\left(\frac{X}{M}\right)_{ult}}\left(\frac{1000}{24}\right)$$

$$(16.15)$$

By analogy, the corresponding equation for the anion exchanger is

$$AnionBackwashRinseVol = \frac{(\sum_{i=1}^{i=m} t_i[A_{s_i}^{+t_i}])(Q)(t_{int})(BackwashRinseV)}{\rho_p\left(\frac{X}{M}\right)_{ult}}\left(\frac{1000}{24}\right)$$

$$(16.16)$$

AnionBackwashRinseVol is the m^3 of backwash and rinse waters required for the anion exchanger. An example backwash and rinse waters requirement is 18 m^3/m^3 of bed volume.

Example 16.5   Using a bed exchanger, 75 m^3 of water per day is to be treated for hardness removal between regenerations having intervals of 8 hours. The raw water contains 80 mg/L of Ca^{2+} and 15 mg/L of Mg^{2+}. The exchanger is a resin of exchange capacity of 1412.8 geq/m^3. Assume that the packed density of the resin is 720 kg/m^3. Calculate the total volume of rinse and backwash requirement if the backwash and rinse per unit volume of bed is 18 m^3/m^3.

Solution:

$$CatBackwashRinseVol = \frac{(\sum_{i=1}^{i=m} q_i[C_{s_i}^{+q_i}])(Q)(t_{int})(BackwashRinseV)}{\rho_p\left(\frac{X}{M}\right)_{ult}}\left(\frac{1000}{24}\right)$$

$$\left(\sum_{i=1}^{i=m} q_i[C_{s_i}^{+q_i}]\right) = 5.22(10^{-3})\frac{keq}{m^3}$$

$$\left(\frac{X}{M}\right)_{ult} = 1412.8\frac{geq}{m^3} = 1412.8\frac{(1000)\ meq}{720(1000)\ g} = 1.96\frac{meq}{g}$$

Therefore,

$$CatBackwashRinseVol = \frac{5.22(10^{-3})(75)(8)(18)}{720(1.96)}\left(\frac{1000}{24}\right) = 1.66\,\text{m}^3 \quad \textbf{Ans}$$

16.7 HEAD LOSSES IN ION EXCHANGERS

During ion exchange, the water or wastewater is allowed to flow through the bed. In order for this to happen, head loss allowances must be provided between influent and effluent. Ion exchanger media are actually granular, so head loss calculations are the same as those in granular filters. In addition, the mechanics of backwashing are also exactly the same as in granular filters. Rinsing may be done in downflow mode or in upflow mode. If it is done in a downflow mode, then it is the same as regular filtration. If rinsing is done in an upflow mode, then the mechanics are the same as backwashing. Rinsing in an upflow mode is more effective as more intimate contact between bed grains and regenerant are facilitated by the grains being suspended and agitated by the upward flow of the regenerant solution. Granular head loss calculation formulas were already derived in a previous chapter on filtration and will not be repeated here. Example calculations were also presented in that chapter, which should be consulted for a review of the material.

GLOSSARY

Active zone—A segment of exchanger bed engaged in exchanging ions.
Anion exchange bed material—Exchanges anions in solution for the anions in the bed.
Cation exchange bed material—Exchanges cations in solution for the cations in the bed.
Displacement series—A table of ions indicating that ions at the top of the table can displace ions below them in the table.
Exchange capacity—The ultimate quantity of ions that an ion exchanger can remove from solution.
Hydrogen cycle—A mode of operation of anion exchangers in which regeneration is done using acid.
Ion exchange—The displacement of one ion by another.
Sodium cycle—A mode of operation of cation exchangers in which regeneration is done using NaCl.
Synthetic resins—Insoluble polymers to which are added, by certain chemical reactions, acidic and basic groups called functional groups.
Zeolites—Natural and synthetic alumino silicates.

SYMBOLS

A^{-p}	Represents the exchangeable anion in the anion exchanger
$[Anion]_{eq}$	Gram equivalents per liter of total displacing anion in solution

CatBackwashRinseVol	Cubic meters of backwash and rinse waters required for the anion exchanger
AnionTBedMass	Kilograms of anion bed material
AnionTBedVol	Cubic meters of bed volume
AnionRegenerant	Kilogram equivalents of regenerant required for anion exchangers
A_s	Surficial area of exchanger bed
A_s^{-t}	Displacing anion in solution
$[A_{s_i}^{-t_i}]$	Gram moles per liter of ith displacing anion in solution
C^{+m}	Represents the exchangeable cation in the cation exchanger
$[C_n]$	Concentration of solute at effluent of δ at time t_n
$[C_{n+1}]$	Concentration of solute at effluent of δ at time t_{n+1}
$[C_o]$	Influent concentration to δ
$[C_{s_i}^{+q_i}]$	Gram moles per liter of ith displacing cation in solution
$[CatT]_{eq}$	Gram equivalents per liter of total removable cations in solution
BackwashRinseV	Cubic meters of backwash and rinse waters required for the cation or anion exchanger per cubic meter of bed
CatBackwashRinseVol	Cubic meters of backwash and rinse waters required for the cation exchanger
CatTBedMass	Kilograms of cation exchange material
CatTBedVol	Cubic meters of bed volume
CatRegenerant	Kilogram equivalents of regenerant required for cation exchangers
C_s^{+q}	Displacing cation in solution
$[HT]_{eq}$	Equivalent concentration of displaced hydrogen ion from anion exchanger
$+m$	Charge of the exchangeable cation in the cation exchanger, equivalents per mole
$-n$	Charge of the cation exchanger host, equivalents per mole
$+o$	Charge of the anion exchanger host, equivalents per mole
$-p$	Charge of the exchangeable anion in the anion exchanger, equivalents per mole
$+q$	Charge of displacing cation in solution, equivalents per mole
q_i	Charge of the ith displacing cation in solution, equivalents per mole
Q	Cubic meters of flow per day
r	Number of active sites in the insoluble cation or anion host material
R	Regenerant requirement in equivalents per equivalent of solute removed
swell	Fraction swelling of bed material
t	Charge of displacing anion in solution, equivalents per mole
t_i	Charge of ith displacing anion in solution, equivalents per mole
t_{int}	Time interval of regeneration, hours

R^{-n}	Represents the host part of a cation exchanger
$(R^{+o})_r(A^{-p})_{ro/p}$	Representation of anion exchanger material
$(R^{-n})_r(C^{+m})_{rn/m}$	Representation of cation exchanger material
R^{+o}	Represents the host part of an anion exchanger
V_b	Volume treated at breakthrough
V_n	Total volume treated at time t_n
V_{n+1}	Total volume treated at time t_{n+1}
V_x	Total volume of water or wastewater treated at complete exhaustion of bed
$(X/M)_{ult}$	Ultimate exchange capacity of the bed or simply, the exchange capacity of the bed. For ion exchangers, $(X/M)_{ult}$ has the units of milligram equivalents of solute exchanged per gram of exchanger material
δ	Length of active zone
ρ_p	Pack density of ion exchange material, kilograms per cubic meter

PROBLEMS

16.1 A wastewater contains the following ions: CrO_4^{2-} = 120 mg/L, Cu^{2+} = 30 mg/L, Zn^{2+}= 15 mg/L, and Ni^{2+} = 20 mg/L. Calculate the total gmols of cations and anions, assuming the volume of the wastewater is 450 m³.

16.2 A breakthrough experiment is conducted for a wastewater producing the results in the following table. Determine the packed density of the bed if δ = 0.0012 m. The diameter of the column used is 2.5 cm. $[C_o]$ is equal to 2.20 meq/L. $(X/M)_{ult}$ = 6.5 meq/g.

C (meq/L)	V_x (L)
0.06	1.0
0.08	1.20
0.09	1.30
0.10	1.40
0.20	1.48
0.46	1.58
1.30	1.70
1.80	1.85
2.10	2.00

16.3 Using a bed exchanger, a volume of water is treated for hardness removal between regenerations having intervals of 8 h. The raw water contains 400 mg/L of hardness as $CaCO_3$. The exchanger is a resin of exchange capacity of 1412.8 geq/m³. Assume that the packed density of the resin is 720 kg/m³. The mass of exchanger material is 51.02 kg and its volume is 0.13 m³ at a swell of 0.8. Calculate the volume of water treated.

16.4   Using a bed exchanger, a volume of water is treated for hardness removal. The raw water contains 400 mg/L of hardness as $CaCO_3$. The exchanger is a resin of exchange capacity of 1412.8 geq/m^3. Assume that the packed density of the resin is 720 kg/m^3. The mass of exchanger material is 51.02 kg and its volume is 0.072 m^3. Calculate the volume of water treated if regeneration is to take place every 8 h.

16.5   Using a bed exchanger, a volume of water is treated for hardness removal. The raw water contains 400 mg/L of hardness as $CaCO_3$. Assume that the packed density of the resin is 720 kg/m^3. The mass of exchanger material is 51.02 kg and its volume is 0.13 m^3 at a swell of 0.8. Calculate the exchange capacity if the volume of water treated is 75 m^3/d and if regeneration is to take place every 8 h.

16.6   Using a bed exchanger, a volume of water is treated for hardness removal. The raw water contains 400 mg/L of hardness as $CaCO_3$. Assume that the packed density of the resin is 720 kg/m^3. The mass of exchanger material is 51.02 kg and its volume is 0.13 m^3 at a swell of 0.8. The exchange capacity is 1412.8 geq/m^3 and the volume of water treated is 75 m^3/d. Determine the interval of regeneration.

16.7   Using a bed exchanger, 75 m^3 of water per day is to be treated for hardness removal between regenerations having intervals of 8 h. The raw water contains 80 mg/L of Ca^{2+} and 15 mg/L of Mg^{2+}. The exchanger is a resin of exchange capacity of 1412.8 geq/m^3. Assume that the packed density of the resin is 720 kg/m^3. The kilograms of sodium chloride regenerant required is 0.26. Calculate R assuming that all of the cations were removed.

16.8   Using a bed exchanger, a volume of water is to be treated for hardness removal between regenerations having intervals of 8 h. The raw water contains 80 mg/L of Ca^{2+} and 15 mg/L of Mg^{2+}. The exchanger is a resin of exchange capacity of 1412.8 geq/m^3. Assume that the packed density of the resin is 720 kg/m^3. The kilograms of sodium chloride regenerant required is 0.26. Calculate the volume of water treated assuming $R = 2$ and that all of the cations were removed.

16.9   Using a bed exchanger, 75 m^3 of water per day is treated for hardness removal. The raw water contains 80 mg/L of Ca^{2+} and 15 mg/L of Mg^{2+}. The exchanger is a resin of exchange capacity of 1412.8 geq/m^3. Assume that the packed density of the resin is 720 kg/m^3. The kilograms of sodium chloride regenerant required is 0.26. Calculate the interval of regeneration assuming $R = 2$ and that all of the cations were removed.

16.10 Using a bed exchanger, a volume of water is to be treated for hardness removal between regenerations having intervals of 8 hours. The raw water contains 80 mg/L of Ca^{2+} and 15 mg/L of Mg^{2+}. The exchanger is a resin of exchange capacity of 1412.8 geq/m^3. Assume that the packed density of the resin is 720 kg/m^3. The total volume of the rinse and backwash requirement is 1.66 m^3. If the backwash and rinse per unit volume of the bed is 18 m^3/m^3, calculate the volume of water treated.

16.11   Using a bed exchanger, 75 m^3 of water per day is to be treated for hardness removal. The raw water contains 80 mg/L of Ca^{2+} and 15 mg/L of Mg^{2+}. The exchanger is a resin of exchange capacity of 1412.8 geq/m^3. Assume that the packed density of the resin is 720 kg/m^3. The total volume of the rinse and backwash requirement is 1.66 m^3. If the backwash and rinse per unit volume of the bed is 18 m^3/m^3, calculate the interval of regeneration.

16.12   Using a bed exchanger, 75 m^3 of water per day is to be treated for hardness removal. The raw water contains 80 mg/L of Ca^{2+} and 15 mg/L of Mg^{2+}. The exchanger is a resin of exchange capacity of 1412.8 geq/m^3. Assume that the packed density of the resin is 720 kg/m^3. The total volume of the rinse and backwash requirement is 1.66 cubic meters and the interval of regeneration is 8 h. Calculate the backwash and rinse per unit volume of the bed.

16.13   Using a bed exchanger, 75 m^3 of water per day is to be treated for hardness removal. The raw water contains 80 mg/L of Ca^{2+} and 15 mg/L of Mg^{2+}. The exchanger is a resin of exchange capacity of 1412.8 geq/m^3. The total volume of the rinse and backwash requirement is 1.66 m^3, the interval of regeneration is 8 h, and the backwash and rinse per unit volume of the bed is 18 m^3/m^3. Calculate the packed density of the bed.

16.14   Using a bed exchanger, 75 m^3 of water per day is to be treated for hardness removal. The raw water contains 80 mg/L of Ca^{2+} and 15 mg/L of Mg^{2+}. The total volume of the rinse and backwash requirement 1.66 m^3, the interval of regeneration is 8 h, and the backwash and rinse per unit volume of the bed is 18 m^3/m^3. The packed density of the bed is 720 kg/m^3. Calculate the exchange capacity of the bed.

BIBLIOGRAPHY

Ahmed, S., S. Chughtai, and M. A. Keane (1998). Removal of cadmium and lead from aqueous solution by ion exchange with Na-Y zeolite. *Separation Purification Technol.*, 13, 1, 57–64.

Bishkin, B. (1998). PCB wastewater treatment. *Printed Circuit Fabrication.* 21, 11.

Brown, C. J., et al. (1998). Chloride removal from Kraft liquors using ion exchange technology. *Technical Assoc. Pulp and Paper Industry, Proc. 1998 Pulping Conf., Part 3* Oct. 25–29, Montreal, Canada, 3, 12p, TAPPI Press, Norcross, GA.

Gromov, S. L. (1998). Technological advantages of the countercurrent regeneration process for ion-exchange resins: UPCORE backwash rinsing. *Teploenergetika.* 45, 3, 230–233.

Heller, T., et al. (1997). Properties and performance of Type III strong base anion exchange resin: Purolite A555. *Proc. 1997 Int. Conf. on Power Generation, POWER-GEN,* Dec. 9–11, Dallas, TX, 158. PennWell Publ. and Exhib., Houston. TX.

Ibrahim, M., et al. (1998). Photoactive ion exchange resins. *Proc. 1998 TAPPI Int. Environ. Conf. Exhibit, Part 1,* April 5–8, Vancouver, Canada, 1, 215–216. TAPPI Press, Norcross, GA.

Kats, B. M., et al. (1998). Exchange of copper, lead, cadmium and manganese ions on carboxylic ion-exchange fiber. *Ukrainskii Khimicheskii Zhurnal.* 64, 1–2, 30–34.

Keane, M. A. (1998). Removal of copper and nickel from aqueous solution using Y zeolite ion exchangers. *Colloids Surfaces A: Physicochem. Eng. Aspects.* 138, 1, 11–20.

Mamchenko, A. V. and M. S. Novozhenyuk (1997). Sorption of humus substances by ion exchange resin in water softening. *Khimiya i Tekhnologiya Vody.* 19, 3, 242–253.

Perry, R. H. and C. H. Chilton (Eds.) (1969). *Chemical Engineers' Handbook.* McGraw-Hill, New York, 16-2–16-12.

Petruzzelli, D., et al. (1998). Aluminum recovery from water clarifier sludges by ion exchange. *Reactive Functional Polymers.* 38, 2–3, 227–236.

Ramalho, R. S. (1977). *Introduction to Watewater Treatment Prcess.* Academic Press, New York, 359–367.

Rich, L. G. (1963). *Unit Processes of Sanitary Engineering.* John Wiley & Sons, New York, 117–125.

Sanks, L. R. (1972). Ion exchange, in *Water Quality Engineering, New Concepts and Developments,* E. L. Thackston and W. W. Eckenfelder (Eds.). The Jenkins Book Publishing Company, New York, 147–172.

Sawicki, J. A., P. J. Sefranek, and S. Fisher (1998). Depth distribution and chemical form of iron in low cross-linked crud-removing resin beds. *Nuclear Instruments Methods in Physics Res., Section B: Beam Interactions with Materials and Atoms,* 142, 1–2, 122–132.

Sincero, A. P. and G. A. Sincero (1996). *Environmental Engineering: A Design Approach.* Prentice Hall, Upper Saddle River, NJ, 410–413.

Sylvester, P. and A. Clearfield (1998). Removal of strontium and cesium from simulated Hanford groundwater using inorganic ion exchange materials. *Solvent Extraction Ion Exchange.* 16, 6, 1527–1539.

Tahir, H., et al. (1998). Estimation and removal of chromium ions from tannery wastes using Zeolite-3A. *Adsorption Science Technol.* 16, 3, 153–161.

Wang, H. and G. Zhang (1998). Development of potable pure water solution. *Desalination, Proc. 1998 Conf. on Membranes in Drinking and Industrial Water Production, Part 3,* Sep. 21–24, Amsterdam, Netherlands, 119, 1–3, 353–354. Elsevier Science B.V., Amsterdam, Netherlands.

Wisniewski, J. and G. Wisniewska (1997). Application of electrodialysis and cation exchange technique to water and acid recovery. *Environ. Protection Eng.,* 23, 3–4, 35–45.

17 Disinfection

Disinfection is a unit process involving reactions that render pathogenic organisms harmless. A companion unit process is sterilization. *Sterilization* refers to the killing of all organisms. Sterilization is not often practiced in the treatment of water and wastewater. Thus, this chapter will only discuss the unit process of disinfection. This discussion will include methods of disinfection, factors affecting disinfection, and the various disinfectants that have been used. Because chlorine is the most widely used disinfectant, its chemistry will be discussed at length. The design of chlorination unit operations equipment will also be discussed. The following disinfectants will also be specifically addressed: ozone and ultraviolet light.

17.1 METHODS OF DISINFECTION AND DISINFECTANT AGENTS USED

Generally, two methods of disinfection are used: chemical and physical. The chemical methods, of course, use chemical agents, and the physical methods use physical agents. Historically, the most widely used chemical agent is chlorine. Other chemical agents that have been used include ozone, ClO_2, the halogens bromine and iodine and bromine chloride, the metals copper and silver, $KMnO_4$, phenol and phenolic compounds, alcohols, soaps and detergents, quaternary ammonium salts, hydrogen peroxide, and various alkalis and acids.

As a strong oxidant, ClO_2 is similar to ozone. (Ozone will be discussed specifically later in this chapter.) It does not form trihalomethanes that are disinfection by-products and suspected to be carcinogens. Also, ClO_2 is particularly effective in destroying phenolic compounds that often cause severe taste and odor problems when reacted with chlorine. Similar to the use of chlorine, it produces measurable residual disinfectants. ClO_2 is a gas and its contact with light causes it to photooxidize, however. Thus, it must be generated on-site. Although its principal application has been in wastewater disinfection, chlorine dioxide has been used in potable water treatment for oxidizing manganese and iron and for the removal of taste and odor. Its probable conversion to chlorate, a substance toxic to humans, makes its use for potable water treatment questionable.

The physical agents of disinfection that have been used include ultraviolet light (UV), electron beam, gamma-ray irradiation, sonification, and heat (Bryan, 1990; Kawakami et al., 1978; Hashimoto et al., 1980). Gamma rays are emitted from radioisotopes, such as cobalt-60, which, because of their penetrating power, have been used to disinfect water and wastewater. The electron beam uses an electron generator. A beam of these electrons is then directed into a flowing water or wastewater to be disinfected. For the method to be effective, the liquid must flow in thin layers.

Several theories have been proposed as to its mechanics of disinfection, including the production of intermediates and free radicals as the beam hits the water. These intermediates and free radicals are very reactive and are thought to possess the disinfecting power. In sonification, high-frequency ultrasonic sound waves are produced by a vibrating-disk generator. These waves rattle microorganisms and break them into small pieces. Ultraviolet light will be addressed specifically later in this chapter.

In general, the effect of disinfectants is thought to occur as a result of damage to the cell wall, alteration of cell permeability, alteration of the protoplasm, and inhibition of enzymatic activities. Damage to the cell wall results in cell lysis and death. Some agents such as phenolic compounds and detergents alter the permeability of the cytoplasmic membrane. This causes the membrane to lose selectivity to substances and allow important nutrients such as phosphorus and nitrogen to escape the cell. Heat will coagulate protoplasm and acids and alkali will denature it causing alteration of the structure and producing a lethal effect on the microorganism. Finally, oxidants, such as chlorine, can cause the rearrangement of the structure of enzymes. This rearrangement will inhibit enzymatic activities.

17.2 FACTORS AFFECTING DISINFECTION

The effectivity of disinfectants are affected by the following factors: time of contact between disinfectant and the microorganism and the intensity of the disinfectant, age of the microorganism, nature of the suspending liquid, and temperature. Each of these factors are discussed next.

17.2.1 TIME OF CONTACT AND INTENSITY OF DISINFECTANT

In the context of how we use the term, intensity refers to the intensive property of the disinfectant. Intensive properties, in turn, are those properties that are independent of the total mass or volume of the disinfectant. For example, concentrations are expressed as mass *per unit volume*; the phrase "per unit volume" makes concentration independent of the total volume. Hence, concentration is an intensive property and it expresses the intensity of the disinfectant. Another intensive property is radiation from an ultraviolet light. This radiation is measured as power impinging upon a *square unit of area*. The "per unit area" here is analogous to the "per unit volume." Thus, radiation is independent of total area and is, therefore, an intensive property that expresses the intensity of the radiation, which, in this case, is the intensity of radiation of the ultraviolet light.

It is a universal fact that the time needed to kill a given percentage of microorganisms decreases as the intensity of the disinfectant increases, and the time needed to kill the same percentage of microorganisms increases as the intensity of the disinfectant decreases, therefore, the time to kill and the intensity are in inverse ratio to each other. Let the time be t and the intensity be I. Thus, mathematically,

$$t \propto \frac{1}{I^m}$$

                          (17.1)

Note: I has been raised to the power m, which is a constant. This is to make the relationship more general.

Letting k be the proportionality constant, the equation becomes

$$t = \frac{k}{I^m} \tag{17.2}$$

In this equation, if I^m is multiplied by t, and if I is expressed as the concentration of the disinfectant C in mg/L, the equation is the famous Ct concept with m equal to 1 and t in minutes. Ct values at given temperatures and pH are tabulated in Ct tables used by regulating authorities and by the U.S. Environmental Protection Agency. The time to kill t is synonymous with the time of inactivation of the microorganisms.

The constants may be obtained from experimental data by converting the above equation first into an equation of a straight line. Taking the logarithms of both sides,

$$ln t = ln k - m ln I \tag{17.3}$$

This equation is the equation of the straight line with y-intercept ln k and slope m. The constants may then be solved using experimental data.

Assume n experimental data points, and divide them into two groups. Let there be l data points in the first group; the second group would have $m - l$ data points. From analytic geometry,

$$m = -\frac{\dfrac{\sum_{l+1}^{n} ln t_i}{n-l} - \dfrac{\sum_{1}^{l} ln t_i}{l}}{\dfrac{\sum_{l+1}^{n} ln I_i}{n-l} - \dfrac{\sum_{1}^{l} ln I_i}{l}} \tag{17.4}$$

Substituting Equation (17.4) into Equation (17.2) and solving for k produces

$$k = \exp\left[\frac{\sum_{1}^{l} ln t_i}{l} + m\left\{\frac{\sum_{1}^{l} ln I_i}{l}\right\}\right] \tag{17.5}$$

Having obtained m and k, the time t can be solved using Equation (17.2) from a knowledge of the value of I. This time is called the *contact time* for disinfection, and the intensity I is called the *lethal dose*. From Equation (17.2) any reasonable amount of dose is lethal when administered in a sufficient amount of contact time as calculated from the equation. We call Equation (17.2) the *Universal Law of Disinfection.*

Example 17.1 It is desired to design a bromide chloride contact tank to be used to disinfect a secondary-treated sewage discharge. To determine the contact time, an experiment was conducted producing the following results:

BrCl Dosage (mg/L)	Contact Time (min/residual fecal coliforms) (No./100 mL)		
	15	30	60
3.6	10,000	4,000	600
15.0	800	410	200
47.0	450	200	90

Determine the contact time to be used in design, if it is desired to have a log 2 removal efficiency for fecal coliforms. Calculate the Ct value. The original concentration of fecal coliforms is 40,000 per/100 mL.

Solution: The percentage corresponding to a log removal can be obtained by first assuming any original value of the concentration of the microorganisms x_1, computing the next value x_2 based on the given log removal, and computing the corresponding percentage. Thus, let $x_1 = 8888888$. Then,

$$\log 8888888 - \log x_2 = 2 \text{ and } x_2 = 88888.88$$

$$\% = \frac{8888888 - 88888.88}{8888888} = 99$$

Note: Any number could have been assumed for x_1 and the answer would still be 99. Thus, log 2 removal is equal to 99% removal or 99% inactivation.

BrCl Dosage (mg/L)	15 min	% Inactivation	30 min	% Inactivation	60 min	% Inactivation
3.6	10,000	0.75[a]	4000	0.90	600	0.985
15.0	800	0.98	410	0.98975	200	0.995
47.0	450	0.98875	200	0.995	90	0.99775

[a] $0.75 = (40{,}000 - 10{,}000)/40{,}000$

From the previous table, we produce the following table for the time to effect a 99% inactivation:

BrCl Dosage (mg/L)	Time to 99% Inactivation (min)
3.6	61.76[a]
15.0	31.43[b]
47.0	18

30	0.90	$\dfrac{x-60}{60-30} = \dfrac{0.99-0.985}{0.985-0.90}$
60	0.985	[a] $x = 61.76$
x	0.99	
30	0.98975	$\dfrac{x-30}{30-60} = \dfrac{0.99-0.98975}{0.98975-0.995}$
x	0.99	[b] $x = 31.43$
60	0.995	

Use a contact time of 30 minutes and find the corresponding lethal dosage.

$$t = \frac{k}{I^m}; \qquad m = -\frac{\dfrac{\sum_{l+1}^n \ln t_i}{n-l} - \dfrac{\sum_1^l \ln t_i}{l}}{\dfrac{\sum_{l+1}^n \ln I_i}{n-l} - \dfrac{\sum_1^l \ln I_i}{l}}; \qquad k = \exp\left[\frac{\sum_1^l \ln t_1}{l} + m\left\{\frac{\sum_1^l \ln I_i}{l}\right\}\right]$$

BrCl Dosage (mg/L)	Time to 99% Inactivation (min)	In I_i	In t_i
3.6	61.76	1.28	4.1233
15.0	31.43	2.708	3.4477
47.0	18	3.850	2.8904

let $l = 1$; $n = 3$

$$m = -\frac{\dfrac{(3.4477 + 2.8904)}{3-1} - \dfrac{(4.1233)}{1}}{\dfrac{(2.708 + 3.850)}{3-1} - \dfrac{1.28}{1}} = -\frac{\dfrac{6.3381}{2} - 4.1233}{\dfrac{6.558}{2} - 1.28} = 0.477$$

$$k = \exp\left[\frac{(4.1233)}{1} + 0.477\{1.28\}\right] = \exp[4.733] = 113.60$$

Therefore,

$$t = \frac{k}{I^m}; \qquad 30 = \frac{113.60}{I^{0.477}}; \qquad I = 16.3 \text{ mg/L } \textbf{Ans}$$

Therefore,

$$Ct = 16.3(30) = 489 \text{ mg/L} \cdot \text{min} \quad \textbf{Ans}$$

17.2.2 AGE OF THE MICROORGANISM

The effectiveness of a disinfectant also depends upon the age of the microorganism. For example, young bacteria can easily be killed, while old bacteria are resistant. As the bacterium ages, a polysaccharide sheath is developed around the cell wall; this contributes to the resistance against disinfectants. For example, when using 2.0 mg/L of applied chlorine dosage, for bacterial cultures of about 10 days old, it takes 30 min of contact time to produce the same reduction as for young cultures of about one day old dosed with one minute of contact time. In the extreme case are the bacterial spores; they are, indeed, very resistant and many of the chemical disinfectants normally used have little or no effect on them.

17.2.3 Nature of the Suspending Fluid

In addition to the time of contact and age of the microorganism, the nature of the suspending fluid also affects the effectiveness of a given disinfectant. For example, extraneous materials such ferrous, manganous, hydrogen sulfide, and nitrates react with applied chlorine before the chlorine can do its job of disinfecting. Also, the turbidities of the water reduces disinfectant effectiveness by shielding the microorganism. Hence, for most effective kills, the fluid should be free of turbidities.

17.2.4 Effect of Temperature

We have learned from previous chapters that equilibrium and reaction constants are affected by temperature. The length of time that a disinfection process proceeds is a function of the constants of the underlying reaction between the microorganism and the disinfectant; thus, it must also be a function of temperature. The variation of the contact time to effect a given percentage kill with respect to temperature can therefore be modeled by means of the *Van't Hoff equation*. This equation was derived for the equilibrium constants in Chapter 11, which is reproduced next:

$$K_{T2} = K_{T1} exp\left[\frac{\Delta H_{298}^o}{RT_1 T_2}(T_2 - T_1)\right] \tag{17.6}$$

K_{T1} and K_{T2} are the equilibrium constants at temperatures T_1 and T_2, respectively. ΔH_{298}^o is the standard enthalpy change of the reaction and R is the universal gas constant. If K_{T1} is replaced by contact time t_{T1} at temperature $T1$ and K_{T2} is replaced by contact time t_{T2} at temperature $T2$, the resulting equation would show that as the temperature increases, the contact time to kill the same percentage of microorganisms also increases. Of course, this is not true. Thus, the replacement should be the other way around. Doing this is the same as interchanging the places in the difference term between T_1 and T_2 inside the *exp* function. Thus, doing the interchanging,

$$t_{T2} = t_{T1} exp\left[\frac{\Delta H_{298}^o}{RT_1 T_2}(T_1 - T_2)\right] \tag{17.7}$$

Table 17.1 shows the standard enthalpy change as a function of pH for both aqueous chlorine and chloramines, and Table 17.2 shows the various possible values of the universal gas constant.

Example 17.2   The contact time for a certain chlorination process at a pH of 7.0 and a temperature of 25°C is 30 min. What would be the contact time to effect the same percentage kill if the process is conducted at a temperature 18°C?

Solution:

$$t_{T2} = t_{T1} exp\left[\frac{\Delta H_{298}^o}{RT_1 T_2}(T_1 - T_2)\right]$$

TABLE 17.1
Standard Enthalpy Changes at 25°C

Compound	pH	ΔH^o_{298} (J/mol)
Aqueous chlorine	7.0	34,332
	8.5	26,796
	9.8	50,242
	10.7	62,802
Chloramines	7.0	50,242
	8.5	58,615
	9.5	83,736

TABLE 17.2
Values and Units of R

R Value	R Units	K Concentration Units Used	$\Delta H°$ Units	T Units
0.08205	$\dfrac{L \text{ atm}}{\text{gmmole } K°}$	$\dfrac{\text{gmmoles}}{L}$	—	°K
8.315	$\dfrac{J}{\text{gmmole } K°}$	$\dfrac{\text{gmmoles}}{L}$	$\dfrac{J}{\text{gmmole}}$	°K
1.987	$\dfrac{\text{cal}}{\text{gmmole } K°}$	$\dfrac{\text{gmmoles}}{L}$	$\dfrac{\text{cal}}{\text{gmmole}}$	°K
82.05	$\dfrac{\text{atm.cm}^3}{\text{gmmole} \cdot K°}$	$\dfrac{\text{gmmoles}}{L}$	—	°K

From Table 17.1,

$$\Delta H^o_{298} = 34{,}332 \text{ j/mol} = 34{,}332 \text{ N} \cdot \text{m/mol}$$

From Table 17.2,

$$R = 8.315 \frac{J}{\text{gmmole} \cdot K°} = 8.315 \frac{N \cdot m}{\text{gmmole} \cdot K°}$$

Therefore,

$$t_{T2} = 30\,exp\left[\frac{34{,}332}{8.315(298)(291)}(298 - 291)\right] = 30\,exp[0.333] = 41.87 \text{ min} \quad \textbf{Ans}$$

17.3 OTHER DISINFECTION FORMULAS

The literature reveals other disinfection formulas. These include *Chick's law* for contact time, modifications of Chick's law, and relationship between concentration of disinfectant and concentration of microorganisms reduced in a given percentage kill. Chick's law and its modification called the *Chick–Watson model*, however, are not useful formulas, because they do not incorporate either the concentration of the disinfectant that is needed to kill the microorganisms or the incorporation of the concentration is incorrect. The relationship of the concentration of disinfectant and the concentration of the microorganisms is also not a useful formula, since it does not incorporate the contact time required to kill the microorganisms. It must be noted that for a formula to be useful, it must incorporate both the concentration (intensity) of the disinfectant and the contact time corresponding to this concentration effecting a given percentage kill. For these reasons, these other disinfection formulas are not discussed in this book.

The Chick–Watson model needs to be addressed further. Watson explicitly expressed the constant k in Chick's law in terms of the concentration of disinfectant C as αC^n, where α is an activation constant and n is another constant termed the constant of dilution. Chick's Law, thus, became $dN/dt = -\alpha C^n dt$, where N is the concentration of microorganisms and t is time. Note that C is a function of time. When this equation was integrated, however, it was assumed constant, thus producing the famous Chick–Watson model,

$$ln\frac{N}{N_o} = -\alpha\, C^n t$$

where N_o is the initial concentration of microorganisms. Because the concentration C was assumed constant with time during integration, this equation is incorrect and, therefore, not used in this book.

17.4 CHLORINE DISINFECTANTS

The first use of chlorine as a disinfectant in America was in New Jersey in the year 1908 (Leal, 1909). At that time George A. Johnson and John L. Leal chlorinated the water supply of Jersey City, NJ.

The principal compounds of chlorine that are used in water and wastewater treatment are the molecular chlorine (Cl_2), calcium hypochlorite [$Ca(OCL)_2$], and sodium hypochlorite [NaOCl]. Sodium hypochlorite is ordinary bleach. Chlorine is a pale-green gas, which turns into a yellow-green liquid when pressurized. Both the aqueous and liquid chlorine react with water to form hydrated chlorine. Below 9.4°C, liquid chlorine forms the compound $Cl_2 \cdot 8H_2O$.

Chlorine gas is supplied from liquid chlorine that is shipped in pressurized steel cylinders ranging in size from 45 kg and 68 kg to one tonne containers. It is also shipped in multiunit tank cars that can contain fifteen 1-tonne containers and tank cars having capacities of 15, 27, and 50 tonnes.

In handling chlorine gas, the following points are important to consider:

- Chlorine gas is very poisonous and corrosive. Therefore, adequate ventilation should be provided. In the construction of the ventilation system,

the capturing hood vents should be placed at floor level, because the gas is heavier than air.
- The storage area for chlorine should be walled off from the rest of the plant. There should be appropriate signs posted in front of the door and back of the building. Gas masks should be provided at all doors and exits should be provided with clearly visible signs.
- Chlorine solutions are very corrosive and should therefore be transported in plastic pipes.
- The use of calcium hypochlorite or sodium hypochlorite as opposed to chlorine gas should be carefully considered when using chlorination in plants located near residential areas. Accidental release of the gas could endanger the community. Normally, small plants that usually lack well-trained personnel, should not use gaseous chlorine for disinfection.

Calcium hypochlorite is available in powder or granular forms and compressed tablets or pellets. Depending upon the source of the chemical, a wide variety of container sizes and shapes are available. Because it can oxidize other materials, calcium hypochlorite should be stored in a cool dry place and in corrosion-resistant containers. High-test calcium hypochlorite, HTH, contains about 70% chlorine. (Available chlorine will be defined later.)

Sodium hypochlorite is available in solution form in strengths of 1.5 to 15% with 3% the usual maximum strength. The solution decomposes readily at high concentrations. Because it is also affected by heat and light, it must be stored in a cool dry place and in corrosion-resistant containers. The solution should be transported in plastic pipes. Sodium hypochlorite can contain 5 to 15% available chlorine.

17.4.1 CHLORINE CHEMISTRY

The chemistry of chlorine discussed in this section includes hydrolysis and optimum pH range of chlorination, expression of chlorine disinfectant concentration, reaction mediated by sunlight, reactions with inorganics, reactions with ammonia, reactions with organic nitrogen, breakpoint reaction, reactions with phenols, formation of trihalomethanes, acid generation, and available chlorine.

Chlorine has the electronic configuration of $[Ne]3s^23p^5$ and is located in Group VIIA of the Periodic Table in the third period. [Ne] means that this element has the electronic configuration of the noble gas neon. The letters p and s refer to the p and s orbitals; the superscripts indicate the number of electrons that the orbitals contain. Thus, the p orbital contains 5 electrons and the s orbital contains two electrons, making a total of seven electrons in its valence shell. This means that the chlorine atom needs to acquire only one more electron to attain the neon configuration for stability. This makes chlorine a very good oxidizer. In fact, it is a characteristic of Group VIIA to attain a charge of −1 when the members of this group oxidizes other substances. The members of this group starting from the strongest oxidizer to the least are fluorine, chlorine, bromine, iodine, and astatine. This group forms the family of elements called the *halogen family.*

All the chlorine disinfectants reduce to the chloride ion (Cl^-) when they oxidize other substances, which must, of course, be reducing substances. The chlorine starts with an oxidation state of zero and ends up with a −1; it only needs one reduction step. One the other hand, the hypochlorites start with oxidation states of +1 and end up with also a −1; thus, they need two reduction steps. Because the chlorine atom only needs one reduction step, while the hypochlorites need two, the chlorine atom is a stronger oxidizer than the hypochlorites. As a stronger oxidizer, it is also a stronger disinfectant.

Hydrolysis and optimum pH range of chlorination. As previously mentioned, chlorine is supplied in the form of liquefied chlorine. The liquid must then be evaporated into a gas. As the gas, $Cl_{2(g)}$, is applied into the water or wastewater, it dissolves into aqueous chlorine, $Cl_{2(aq)}$, as follows:

$$Cl_{2(g)} \rightleftharpoons Cl_{2(aq)} \qquad K_{Claq} = 6.2(10^{-2}) \tag{17.8}$$

$Cl_{2(aq)}$ then hydrolyzes, one of the chlorine atoms being oxidized to +1 and the other reduced to −1. This reaction is called *disproportionation*. The reaction is as follows:

$$Cl_{2(aq)} + H_2O \rightleftharpoons HOCl + H^+ + Cl^{-1} \qquad K_H = 4.0(10^{-4}) \tag{17.9}$$

From Equation (17.9), the hypochlorous acid, HOCl, is formed, which is one of the chlorine disinfectants. If its formula is analyzed, it will be found that the chlorine has an oxidation state of +1, as we mentioned before. Note also that hydrochloric acid is formed. This is a characteristic in the use of the chlorine gas as a disinfectant. The water becomes acidic. Also, as we have mentioned, the chlorine molecule is a much stronger oxidizer than the hypochlorite ion and, hence, a stronger disinfectant. From Equation (17.9), if the water is intentionally made acidic, the reaction will be driven to the left, producing more of the chlorine molecule. This condition will then produce more disinfecting power. As will be shown later, however, this condition, where the chlorine molecule will exist, is at a very low pH hovering around zero. This makes the chlorine molecule useless as a disinfectant.

HOCl further reacts to produce the following dissociation reaction:

$$HOCl \rightleftharpoons H^+ + OCl^- \qquad K_a = 10^{-7.5} \tag{17.10}$$

Using Equation (17.9), let us calculate the distribution of $Cl_{2(aq)}$ and HOCl. Expressing in the form of equilibrium equation,

$$K_H = 4.0(10^{-4}) = \frac{[HOCl][H^+][Cl^{-1}]}{[Cl_{2(aq)}]} \tag{17.11}$$

Taking logarithms, rearranging, and simplifying,

$$\frac{[Cl_{2(aq)}]}{[HOCl]} = 10^{\{pK_H - pH + \log[Cl^{-1}]\}} = 10^{\{3.40 - pH + \log[Cl^{-1}]\}} \tag{17.12}$$

pK_H is the negative logarithm to the base 10 of K_H.

Table 17.3 shows the ratios of $[Cl_{2(aq)}]/[HOCl]$ and $[HOCl]/[Cl_{2(aq)}]$ as functions of pH and the chloride concentrations, using Equation (17.12). The concentration of 1.0 gmmole/L of chloride is 35,500 mg/L. This will never be encountered in the normal treatment of water and wastewater. Disregarding this entry in the table, the

TABLE 17.3

Ratios of $\dfrac{[Cl_{2(aq)}]}{HOCl}$ and $\dfrac{[HOCl]}{[Cl_{2(aq)}]}$ as Functions pH and Chloride Concentration

[Cl⁻] (gmmoles/L)	pH	$\dfrac{[Cl_{2(aq)}]}{HOCl}$	$\dfrac{[HOCl]}{[Cl_{2(aq)}]}$	pH	$\dfrac{[Cl_{2(aq)}]}{HOCl}$	$\dfrac{[HOCl]}{[Cl_{2(aq)}]}$
1.0	0	$2.51(10^3)$	$3.98(10^{-4})$	1	251	$3.98(10^{-3})$
	2	25.12	$3.98(10^{-2})$	3	2.51	0.398
	4	0.251	3.98	5	0.0251	39.81
	6	$2.51(10^{-3})$	398	7	$2.51(10^{-4})$	$3.98(10^3)$
10^{-1}	0	251	$3.98(10^{-3})$	1	25.12	0.0398
	2	2.51	0.398	3	0.251	3.98
	4	0.025	40	5	$2.51(10^{-3})$	398
	6	$2.51(10^{-4})$	$3.98(10^3)$	7	$2.51(10^{-5})$	$3.98(10^4)$
10^{-3}	0	2.51	0.398	1	0.251	3.98
	2	0.025	40	3	$2.51(10^{-3})$	398.11
	4	$2.51(10^{-4})$	$3.98(10^3)$	5	$2.51(10^{-5})$	$3.98(10^4)$
	6	$2.51(10^{-6})$	$3.98(10^5)$	7	$2.51(10^{-7})$	$3.98(10^6)$
10^{-5}	0	0.0251	39.81	1	$2.51(10^{-3})$	398
	2	$2.51(10^{-4})$	$3.98(10^3)$	3	$2.51(10^{-5})$	$3.98(10^4)$

concentration of $Cl_{2(aq)}$ is already practically nonexistent at around pH 4.0 and above. In fact, it is even practically nonexistent at pH's less than 4, except when the pH is close to zero and chloride concentration of 0.1 gmmol/L; but, 0.1 gmmol/L is equal to 3,500 mg/L, which is already a very high chloride concentration and will not be encountered in the treatment of water and wastewater. Practically, then, for conditions encountered in practice, at pH's greater than 4.0, [HOCl] predominates over $Cl_{2(aq)}$.

Now, using Equation (17.10), let us calculate the distribution of HOCl and OCl⁻. Note that from the previous result, HOCl predominates over $Cl_{2(aq)}$ above pH 4.0, $Cl_{2(aq)}$ being practically zero. Thus, above this pH, the distribution of the chlorine disinfectant species will simply be for HOCl and OCl⁻. Expressing Equation (17.10) in the form of the equilibrium equation,

$$K_a = 10^{-7.5} = \frac{[H^+][OCl^-]}{[HOCl]} \qquad (17.13)$$

Taking logarithms, rearranging, and simplifying,

$$\frac{[OCl^-]}{[HOCl]} = 10^{pH-pK_a} = 10^{pH-7.5} \qquad (17.14)$$

pK_a is the negative logarithm to the base 10 of K_a.

Table 17.4 shows the ratios of [OCl⁻]/[HOCl] and [HOCl]/[OCl⁻] as functions of pH using Equation (17.14). This table shows that HOCl predominates over OCl⁻ at pH's less than 7.5. Also considering Table 17.3, we make the conclusion that for all practical purposes, HOCl predominates over all chlorine disinfectant species

TABLE 17.4

Ratios of $\frac{[OCl^-]}{[HOCl]}$ and $\frac{[HOCl]}{[OCl^q]}$ as Functions of pH

pH	$\frac{[OCl^-]}{[HOCl]}$	$\frac{[HOCl]}{[OCl^-]}$	pH	$\frac{[OCl^-]}{[HOCl]}$	$\frac{[HOCl]}{[OCl^-]}$
0	$3.16(10^{-8})$	$3.16(10^7)$	1	$3.16(10^{-6})$	$3.16(10^6)$
2	$3.16(10^{-6})$	$3.16(10^5)$	3	$3.16(10^{-5})$	$3.16(10^4)$
4	$3.16(10^{-4})$	$3.16(10^3)$	5	$3.16(10^{-3})$	316.2
6	$3.16(10^{-2})$	31.62	7	0.316	3.16
7.5	1.0	1.0	8	3.16	0.316
9	31.62	$3.16(10^{-2})$	10	316.2	$3.16(10^{-3})$
11	$3.16(10^3)$	$3.16(10^{-4})$	12	$3.16(10^4)$	$3.16(10^{-5})$
13	$3.15(10^5)$	$3.16(10^{-6})$	14	$3.16(10^6)$	$3.16(10^{-7})$

in all pH ranges up to less than 7.5. At exactly pH 7.5, the concentrations of HOCl and OCl⁻ are equal and above this pH, OCl⁻ predominates over all chlorine disinfectant species. This reality is more than just a theoretical interest, because HOCl is 80 to 100% more effective than OCl⁻ as a disinfectant (Snoeyink and Jenkins, 1980). We now conclude that the optimum pH range for chlorination is up to 7.0. Beyond this range, OCl⁻ predominates and the disinfection becomes less effective.

The three species $Cl_{2(aq)}$, HOCl, and OCl⁻ are called *free chlorine*. Although $Cl_{2(aq)}$ is a stronger oxidizer than the other two, it is not really of much use, unless the chlorination is done under a very acidic condition.

Example 17.3   At pH 6.0, calculate the mole fraction of HOCl.

Solution: At pH 6.0 the mole fraction of $Cl_{2(aq)}$ is practically zero. From Table 17.4, the mole ratio of HOCl to OCl⁻ at this pH is 31.62.

Therefore,

$$\text{mole fraction of HOCl} = \frac{[HOCl]}{[HOCl] + [OCl^-] + [Cl_{2(aq)}]}$$

$$= \frac{31.62}{31.62 + 1 + 0} = 0.97 \quad \textbf{Ans}$$

Expression of chlorine disinfectant concentration.   Now that we have detailed the various reactions of the chlorine disinfectants, it is time to unify the concentrations of the chlorine species. By convention, the concentrations of the three species are expressed in terms of the molecular chlorine, Cl_2. The pertinent reactions are written as follows:

$$Cl_{2(aq)} + H_2O \rightleftharpoons HOCl + H^+ + Cl^{-1} \tag{17.15}$$

$$NaOCl \rightleftharpoons Na^+ + OCl^- \tag{17.16}$$

$$Ca(OCl)_2 \rightleftharpoons Ca^{2+} + 2OCl^- \tag{17.17}$$

From these reactions,

$$1.0 \text{ mg/L HOCl} = \frac{Cl_2}{HOCl} = \frac{2(35.5)}{1.008 + 16 + 35.5} = 1.35 \text{ mg/L Cl}_2 \quad (17.18)$$

$$1.0 \text{ mg/L NaOCl} = \frac{Cl_2}{NaOCl} = \frac{2(35.5)}{23 + 16 + 35.5} = 0.95 \text{ mg/L Cl}_2 \quad (17.19)$$

$$1.0 \text{ mg/L Ca(OCl)}_2 = \frac{2Cl_2}{Ca(OCl)_2} = \frac{4(35.5)}{40.1 + 2(16 + 35.5)} = 0.99 \text{ mg/L Cl}_2 \quad (17.20)$$

Whenever a concentration of a chlorine disinfectant is mentioned, the above equations are implicitly referred to, and this concentration is the equivalent Cl_2 concentration. Of course, the equivalent Cl_2 concentration of the chlorine gas disinfectant is Cl_2.

Reaction mediated by sunlight. Aqueous chlorine is not stable in the presence of sunlight. Sunlight contains ultraviolet light. This radiation provides the energy that drives the chemical reaction for breaking up the molecule of hypochlorous acid. The water molecule breaks up, first releasing electrons that are then needed to reduce the chlorine atom in HOCl to chloride. The overall reaction is as follows:

$$2HOCl \rightarrow 2H^+ + 2Cl^- + O_2 \quad (17.21)$$

The O_2 comes from the break up of the water molecule, oxidizing its oxygen atom to the molecular oxygen.

The previous reaction in the presence of sunlight is very significant. If the disinfectant is to be stored in plastic containers, then this container must be made opaque; otherwise, the chlorine gas will be converted to hydrochloric acid, and the hypochlorites will be converted into the corresponding salts of calcium and sodium.

Example 17.4 A solution of sodium hypochlorite containing 50 kg of NaOCl is stored in a transparent plastic container. It had been stored outdoors for quite some time and then used to disinfect a swimming pool. How effective is the disinfection? What material is used instead for the disinfection? And how many kilograms is it?

Solution: Because the solution is stored outdoors and in a transparent container, the following reaction occurs:

$$2NaOCl \rightarrow 2Na^+ + 2Cl^- + O_2$$

From this reaction, no disinfectant exists in the container and the disinfection is not effective. **Ans**

The material used instead for disinfection is the salt NaCl. **Ans**

The amount of sodium chloride used to disinfect is

$$\frac{NaCl}{NaOCl}(50) = \frac{23 + 35.5}{23 + 16 + 35.5}(50) = 39.26 \text{ kg} \quad \textbf{Ans}$$

Reactions with inorganics. Reducing substances that could be present in the raw water and raw wastewater and treated water and treated wastewater are ferrous, manganous, nitrites, and hydrogen sulfide. Thus, these are the major substances that can interfere with the effectiveness of chlorine as a disinfectant. The interfering reactions are written as follows:

with ferrous:

$$2Fe^{2+} \rightarrow 2Fe^{3+} + 2e^- \tag{17.22}$$

$$\frac{HOCl + H^+ + 2e^- \rightarrow Cl^- + H_2O}{} \tag{17.23}$$

$$2Fe^{2+} + HOCl + H^+ \rightarrow 2Fe^{3+} + Cl^- + H_2O \tag{17.24}$$

with manganous:

$$Mn^{2+} \rightarrow Mn^{4+} + 2e^- \tag{17.25}$$

$$\frac{HOCl + H^+ + 2e^- \rightarrow Cl^- + H_2O}{} \tag{17.26}$$

$$Mn^{2+} + HOCl + H^+ \rightarrow Mn^{4+} + Cl^- + H_2O \tag{17.27}$$

with nitrites:

$$NO_2^- + H_2O \rightarrow NO_3^- + 2H^+ + 2e^- \tag{17.28}$$

$$\frac{HOCl + H^+ + 2e^- \rightarrow Cl^- + H_2O}{} \tag{17.29}$$

$$NO_2^- + HOCl \rightarrow NO_3^- + Cl^- + H^+ \tag{17.30}$$

with hydrogen sulfide:

$$H_2S + 4H_2O \rightarrow SO_4^{2-} + 10H^+ + 8e^- \tag{17.31}$$

$$\frac{4HOCl + 4H^+ + 8e^- \rightarrow 4Cl^- + 4H_2O}{} \tag{17.32}$$

$$H_2S + 4HOCl \rightarrow SO_4^{2-} + 4Cl^- + 6H^+ \tag{17.33}$$

For activated sludge plants that produce only partial nitrification, it is a common complaint of operators that a residual chlorine cannot be obtained at the effluent. The reason for this is the reaction of nitrites with the chlorine disinfectant producing nitrates as shown in Reaction (17.30).

Example 17.5 An activated sludge plant of a small development is out of order, and a decision has been made following approval from a state agency to discharge raw sewage to a river. The effluent was found to contain 8 mg/L of ferrous, 3 mg/L manganous, 20 mg/L nitrite as nitrogen, and 4 mg/L hydrogen sulfide. Calculate the mg/L of HTH needed to be dosed before actual disinfection is realized. What is the chlorine concentration?

Solution: Examining Reactions (17.22) to (17.33) reveals that the number of reference species is equal to two moles of electrons except Reaction (17.33), which has eight moles of electrons. By considering all the other reactions, the number of milliequivalents of HOCl needed

$$= \frac{8}{2Fe/2} + \frac{3}{Mn/2} + \frac{20}{NO_2/2} + \frac{4}{H_2S/8} = \frac{8}{2(55.8)/2} + \frac{3}{54.9/2} + \frac{20}{(14+32)/2}$$

$$+ \frac{4}{[2(1.008)+32.1]/8}$$

$$= 2.06 \text{ meq/L} \Rightarrow 2.06\left(\frac{HOCl}{2}\right) = 2.06\left(\frac{1.008+16+35.5}{2}\right) = 54.08 \text{ mg/L}$$

$$Ca(OCl)_2 \rightarrow Ca^{2+} + 2OCl^-$$

$$2OCl^- + 2H_2O \rightleftharpoons 2HOCl + 2OH^-$$

Therefore,

$$mg/L \text{ of HTH} = 54.08\left(\frac{Ca(OCl)_2}{2HOCl}\right) = 54.08\left[\frac{40.1+2(16+35.5)}{2(1.008+16+35.5)}\right]$$

$$= 73.69 \text{ mg}/L \quad \textbf{Ans}$$

From Equation (17.20), the chlorine concentration $= 73.69(0.99) = 72.95$ mg/L

Ans

Reactions with ammonia and optimum pH range for chloramine formation. Effluents from sewage treatment plants can contain significant amounts of ammonia that when disinfected, instead of finding free chlorine, substitution products of ammonia called *chloramines* are found. In addition, in water treatment plants, ammonia are often purposely added to chlorine. This, again, also forms the chloramines. Chloramines are disinfectants like chlorine, but they are slow reacting, and it is this slow-reacting property that is the reason why ammonia is used. The purpose is to provide residual disinfectant in the distribution system. In other words, the formation of chloramines assures that when the water arrives at the tap of the consumer, a certain amount of disinfectant still exists.

The formation of chloramines is a stepwise reaction sequence. When ammonia and chlorine are injected into the water that is to be disinfected, the following reactions occur, one after the other in a stepwise manner.

$$NH_3 + HOCl \rightarrow NH_2Cl(\text{monochloramine}) + H_2O \tag{17.34}$$

$$NH_2Cl + HOCl \rightarrow NHCl_2(\text{dichloramine}) + H_2O \tag{17.35}$$

$$NHCl_2 + HOCl \rightarrow NCl_3(\text{trichloramine}) + H_2O \tag{17.36}$$

First, it is to be noted that the reaction is expressed in terms of HOCl. By the equivalence of reactions, however, the above reactions can be manipulated if the equivalent amount of the other two species is desired to be known. In monochloramines and dichloramines, therefore, the chlorine is combined in ammonia; they are called *combined chlorines*. As will be shown in subsequent discussions, the concentration of trichloramine is practically zero during disinfection; thus, it is not included in the definition of combined chlorine.

Reaction (17.34) indicates that at the time when one mole of HOCl is added to one mole of NH_3, the conversion into monochloramine is essentially complete. In view of the relationship of HOCl and OCl^- as a function of pH, however, this statement is not exactly correct. From previous discussions, at pH 7.5, hypochlorous acid and the hypochlorite ion exist in equal mole concentrations, but beyond pH 7.5, the hypochlorite ion predominates. OCl^- does not directly react with NH_3 to form the monochloramine, but must first hydrolyze to produce the HOCl before Reaction (17.34) proceeds. Thus, when the pH is above 7.5, addition of one mole of HOCl to one mole of ammonia does not guarantee complete conversion into NH_2Cl. At these pH values, the one mole of HOCl added becomes lesser, because of the predominance of the hypochlorite ion. HOCl, however, exists at practically 100 concentrations at pH's below 7.0; hence, at this range, a mole for mole addition would essentially guarantee the aforementioned conversion into monochloramine.

Now, Reactions (17.35) and (17.36) indicate that by adding two moles of HOCl and three moles of HOCl, the conversion into dichloramine and the trichloramine are, respectively, essentially complete. For the same reason as in the case of the conversion into monochloramine, these two and three moles are not really these values, because the resulting concentrations depend upon the pH of the solution. Above pH 7.5, the conversions are not complete.

Let us have more discussion regarding the formation of dichloramine. The oxidation state of nitrogen in NH_2Cl from where the dichloramine comes from is -1. The oxidation state of the nitrogen in $NHCl_2$, itself, is $+1$. This means that in order to form the dichloramine, two electrons must be abstracted from the nitrogen atom. Now, the other substances that have been observed to occur, as the amount of hypochlorous acid added is increased, are the nitrogen gas and nitrates. Take the case of the nitrogen gas. The oxidation state of the N atom in the N_2 molecule is zero. This means that in order to form the nitrogen gas from NH_2Cl only one electron needs to be abstracted from the nitrogen atom; this is an easier process than abstracting two electrons. It must then be concluded that before the dichloramine is formed, the gas must have already been forming, and that for the dichloramine to be formed, more HOCl is needed than is needed for the formation of the gas.

The reaction for the formation of the nitrogen gas may be written as follows:

$$2NH_2Cl \rightarrow N_{2(g)} + 2Cl^- + 4H^+ + 2e^- \tag{17.37}$$

$$HOCl + H^+ + 2e^- \rightarrow Cl^- + H_2O \tag{17.38}$$

$$\overline{2NHCl_2 + HOCl \rightarrow N_{2(g)} + 3Cl^- + 3H^+ + H_2O} \tag{17.39}$$

Let us discuss the formation of the monochloramine versus the formation of the nitrogen gas. The oxidation state of the nitrogen atom in ammonia is −3. And, again, its oxidation state in NH_2Cl is −1. Thus, forming the monochloramine from ammonia needs the abstraction of two electrons from the nitrogen atom. Now, again, the oxidation state of nitrogen in the nitrogen gas is zero, which means that to form the nitrogen gas from ammonia needs the abstraction of three electrons; this is harder than abstracting two electrons. Thus, in the reaction of HOCl and NH_3, the monochloramine is formed rather than the nitrogen gas, and the gas is formed only when the conversion into monochloramine is complete by more additions of HOCl.

Consider the formation of the nitrate ion. The oxidation state of nitrogen in the nitrate ion is +5. Thus, this ion would not be formed from ammonia, because this would need the abstraction of eight electrons. If it is formed from the monochloramine, it would need the abstraction of six electrons, and if formed from the dichloramine, it would need the abstraction of four electrons. Thus, in the chloramine reactions with HOCl, the nitrate is formed from the dichloramine. We will, however, compare which formation forms first from the dichloramine: trichloramine or the nitrate ion. The oxidation state of the nitrogen atom in trichloramine is +3. Thus, to form the trichloramine, two electrons need to be abstracted from the nitrogen atom. This may be compared to the abstraction of four electrons from the nitrogen atom to form the nitrate ion. Therefore, the trichloramine forms first before the nitrate ion does.

The reaction for the formation of the nitrate ion may be written as follows:

$$NHCl_2 + 3H_2O \rightarrow NO_3^- + 2Cl^- + 7H^+ + 4e^- \tag{17.40}$$

$$2HOCl + 2H^+ + 4e^- \rightarrow 2Cl^- + 2H_2O \tag{17.41}$$

$$\overline{NHCl_2 + 2HOCl + H_2O \rightarrow NO_3^- + 2Cl^- + 5H^+} \tag{17.42}$$

Now, let us discuss the final fate of trichloramine during disinfection. In accordance with the chloramine reactions [Reactions (17.34) to (17.36)], by the time three moles of HOCl have been added, a mole of trichloramine would have been formed. This, however, is not the case. As mentioned, while the monochloramine decomposes in a stepwise fashion to convert into the dichloramine, its destruction into the nitrogen gas intervenes. Thus, the eventual formation of the dichloramine would be less; in fact, much, much less, since, as we have found, formation of the gas is favored over the formation of the dichloramine. In addition, monochloramine and dichloramine, themselves, react with each other along with HOCl to form another gas N_2O [$NH_2Cl + NHCl_2 + HOCl \rightarrow N_2O + 4H^+ + 4Cl^-$]. Also, there may be more other side reactions that could occur before the eventual formation of the dichloramine from monochloramine. Overall, as soon as the step for the conversion of the dichloramine to the trichloramine is reached, the concentration of dichloramine is already very low and the amount of trichloramine produced is also very low. Thus, if, indeed, trichloramine has a disinfecting power, this disinfectant property is useless, since the concentration is already very low in the first place. This is the reason why combined chlorine is only composed of the monochloramine and the dichloramine. Also, it follows that since dichloramine is, itself, simply decomposed, it is not the important combined chlorine disinfectant; the monochloramine is. If the objective is the formation of the disinfecting chloramines, it is only necessary to add chlorine to a level of a little more than one mole of chlorine to one mole of ammonia in order to simply form monochloramine. Beyond this is a waste of chlorine.

Now, let us determine the optimum pH range for the formation of the mono-chloramine. The key to the determination of this range is the predominance of HOCl. Hypochlorous acid predominates over the pH range of less the 7.0; therefore, the optimum pH range for the formation of monochloramine is also less than 7.0.

Example 17.6 In order to provide residual disinfectant in the distribution system, chloramination is applied to the treated water. If two moles of HOCl have been added per mole of ammonia, calculate the moles of nitrogen gas produced.

Solution: So many intervening reactions may be occurring during chlorami-nation that it is not possible to determine exactly the amount of resulting species. Experimentally, a sample may be put in a closed jar and chloramination performed. The liberated nitrogen gas may then be measured; but in this problem the moles of nitrogen produced simply cannot be calculated. **Ans**

Reactions with organic nitrogen. Chlorine reacts with organic amines to form organic chloramines. Examples of the organic amines are those with the groups $-NH_2$, $-NH-$, and $-N =$. Parallel to its reaction with ammonia, HOCl also reacts with organic amines to form organic monochloramines and organic dichloramines by the chloride atom simply attaching to the nitrogen atom in the organic molecule. For example, methyl amine reacts with HOCl as follows:

$$CH_3Cl + HOCl \rightarrow CH_3NHCl \text{ (an organic monochloramine,}$$
$$\text{monochloromethyl amine)} + H_2O \tag{17.43}$$

As in the conversion of monochloramine to dichloramine, monochloromethyl amine converts to dichloromethyl amine in the second step reaction as follows:

$$CH_3NHCl + HOCl \rightarrow CH_3NCl_2 \text{ (an organic dichloramine,}$$
$$\text{dichloromethyl amine)} + H_2O \tag{17.44}$$

Other nitrogen-containing organic compounds are the amides which contain the group $-OCNH_2$ and $-CNH-$. The ammonia and organic amine molecules have basic properties. They react readily with HOCl, which is acidic. The organic amides, on the hand, are less basic than the amines are; thus, they do not react as readily to form organic chloramides with hypochlorous acid. They consume chlorine, however, so organic amides as well as organic amines are important in chloramination. Although the organic chloramides and organic chloramines have some disinfecting power, they are not as potent as the ammonia chloramines; thus, their formation is not beneficial. Organic chloramides and organic chloramines are also combined chlorines.

Example 17.7 Show the half reaction that will exhibit the property of organic chloramines as disinfectants.

Solution: A characteristic property of chlorine disinfectant is the conversion of the chlorine atom in the disinfectant into the chloride ion. Thus, in portraying the chemical reaction, the formation of the chloride should always be shown. Let CH_3NHCl represent the organic chloramines. Therefore, its half reaction as a disinfectant is as follows:

$$CH_3NHCl + 4H^+ + 4e^- \text{ (from organism disinfected)} \rightarrow Cl^- + NH_4^+ + CH_4$$

As this half reaction shows, the disinfectant grabs four mole electrons from the organism disinfected per mole of the disinfectant, disabling the organism. **Ans**

Breakpoint reactions. Figure 17.1 shows the status of chlorine residual as a function of chlorine dosage. From zero chlorine applied at the beginning to point A, the applied chlorine is immediately consumed. This consumption is caused by reducing species such as Fe^{2+}, Mn^{2+}, H_2S, and NO_2^-. The reactions of these substances on HOCl have been discussed previously. As shown, no chlorine residual is produced before point A.

In waters and wastewaters, organic amines and their decomposition products such as ammonia may be present. In addition, ammonia may be purposely added for chloramine formation to produce chlorine residuals in distribution systems. Also, other organic substances such as organic amides may be present as well. Thus, from point A to B, chloro-organic compounds and organic chloramines are formed. Ammonia will be converted to monochloramine at this range of chlorine dosage.

Beyond point B, the chloro-organic compounds and organic chloramines break down. Also, at this range of chlorine dosage, the monochloramine starts to convert to the dichloramine, but, at the same time, it also decomposes into the nitrogen gas and, possibly, other gases as well. These decomposition reactions were addressed previously.

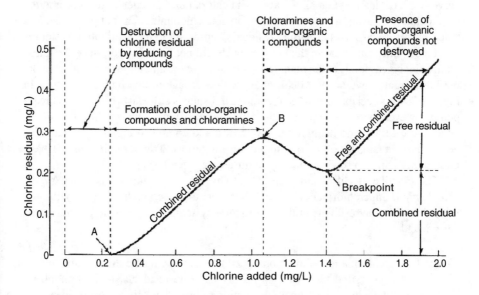

FIGURE 17.1 Chlorine residual versus applied chlorine.

As the curve continues to go "downhill" from point B, the dichloramine converts to the trichloramine, the conversion being complete at the lowest point indicated by "breakpoint." As shown, this lowest point is called the *breakpoint*. In addition, nitrates will also be formed from the dichloramine before reaching the breakpoint. In fact, other substances would have been formed as decomposition products from monochloramine and dichloramine, as well as other substances would have been formed as decomposition products from the chloro-organic compounds and organic chloramines.

As shown by the downward swing of the curve, the reactions that occur between point B and the breakpoint are all breakdown reactions. Substances that have been formed before reaching point B are destroyed in this range of dosage of chlorine. In other words, the chloro-organics that have been formed, the organic chloramines that have been formed, the ammonia chloramines that have formed, and all other substances that have been formed from reactions with compounds such as phenols and fulvic acids are all broken down within this range. These breakdown reactions have been collectively called *breakpoint reactions*.

The breakpoint reactions only break down the decomposable fractions of the respective substances. All the nondecomposables will remain after the breakpoint. This will include, among other nondecomposables, the residual organic chloramines, residual chloro-organic compounds, and residual ammonia chloramines. As we have learned, the trichloramine fraction that comes from ammonia chloramines has to be very small at this point to be of value as a disinfectant. All the substances that could interfere with disinfection and all decomposables would have already been destroyed, therefore, any amount of chlorine applied beyond the breakpoint will appear as free chlorine residual.

Important knowledge is gained from this "chlorine residual versus applied chlorine" curve. We have learned that all the ammonia chloramines practically disappear at the breakpoint. We have also learned that the organic chloramines are not good disinfectants. Therefore, as far as providing residual disinfectant in the distribution system is concerned, chlorination up to the breakpoint should not be practiced. Since the maximum point corresponds to the maximum formation of the ammonia monochloramine, the ideal practice would be to chlorinate with a dosage at this point. Note that, in Figure 17.1, appreciable amounts of combined residuals still exists beyond the breakpoint; however, these combined residuals mainly consist of combined chloro-organics, which have little or no disinfecting properties, and combined organic chloramines, which have, again, little or no disinfecting properties. Trichloramine, as we have mentioned, will be present at a very minuscule concentration.

The practice of chlorinating up to and beyond the breakpoint is called *super-chlorination*. Superchlorination ensures complete disinfection; however, it will only leave free chlorine residuals in the distribution system, which can simply disappear very quickly.

Note: If superchlorination is to be practiced to ensure complete disinfection and it is also desired to have long-lasting chlorine residuals, then ammonia should be added after superchlorination to bring back the chlorine dosage to the point of maximum monochloramine formation.

Example 17.8   Referring to Figure 17.1, if a dosage of 1.8 mg/L is administered, determine the amount free chlorine residual that results, the amount of combined residual that results, and the amount of combined ammonia chloramine residual that results. Also, determine the amount of organic chloramine residual that results.

Solution: From the figure, the concentration of residual chlorine at a dosage of 1.8 mg/L = 0.38 mg/L. The concentration of the residual at the breakpoint = 0.20 mg/L. Therefore,

free chlorine residual = 0.38 − 0.20 = 0.18 mg/L   **Ans**
amount of combined residual = 0.20 mg/L   **Ans**
amount of combined ammonia chloramine ≃ 0   **Ans**
amount of organic chloramine cannot be determined   **Ans**

Reactions with phenols. Chlorine reacts readily with phenol and organic compounds containing the phenol group by substituting the hydrogen atom in the phenol ring with the chlorine atom. These chloride substitution products are extremely odorous. Because phenols and phenolic groups of compounds can be present in raw water supplies as a result of discharges from industries and from natural decay of organic materials, the formation of these odorous substances is a major concern of water treatment plant operators.

Figure 17.2 shows the threshold odor as a function of pH and the concentration of chlorine dosage. Figure 17.2a uses a concentration of 0.2 mg/L and, at a pH of 9.0, the maximum threshold odor concentration is around 28 μg/L. When the pH is reduced to 8.0 this threshold worsens to around 20 μg/L, and when the pH is further reduced to 7.0, the threshold concentration becomes worst at around 13 μg/L. Thus, chlorination at acidic conditions would produce very bad odors compared to chlorination at high pH values. This is very unfortunate, because HOCl predominates at the lower pH range, which is the effective range of disinfection.

In Figure 17.2b the concentration of chlorine has been increased to 1.0 mg/L. For the same adjustments of pH, the maximum threshold concentrations are about the same as in Figure 17.2a; however, in the cases of pH's 7.0 and 8.0, the threshold odors practically vanish at approximately 3 to 5 h after contact as opposed to greater than 60 h when the dosage was only 0.2 mg/L. Thus, increasing the dosage produces the worst nightmare for odor production.

Figure 17.3 shows the reaction scheme for the breakdown of phenol to odorless low molecular weight decomposition products using HOCl. The threshold odor concentrations of the various chloride substituted phenolic compounds are also indicated in brackets. Note that the worst offenders are 2-monochlorophenol and 2,4-dichlorophenol, which have an odor threshold of 2.0 μg/L. In order to effect these breakdown reactions, superchlorination would be necessary, which would also mean that the odor had increased before it disappeared.

Example 17.9   In the reaction scheme of Figure 17.3, what atom has been displaced in ortho chlorophenol by the chlorine atom to form 2,6-dichlorophenol?

Solution: The hydrogen atom in the phenol ring has been displaced by the chlorine.   **Ans**

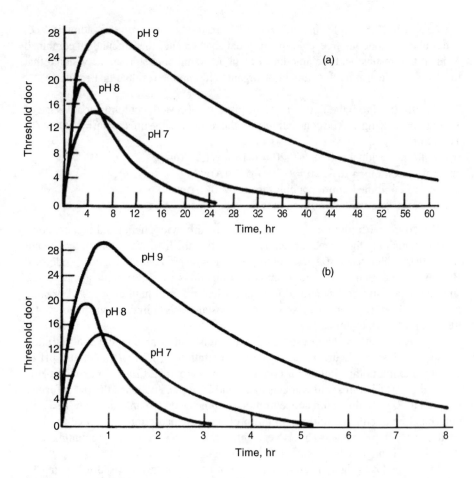

FIGURE 17.2 Threshold odor from chlorination of phenol: (a) chlorine 0.2 mg/L, initial phenol 5.0 mg/L; (b) chlorine 1.0 mg/L, initial phenol 5.0 mg/L; all at 25°C and threshold odors are concentrations in μg/L. (From Lee, G. F. (1967). *Principles and Applications of Water Chemistry*. S. D. Faust and J. V. Hunters (Eds.). John Wiley & Sons, New York. With permission.)

Formation of trihalomethanes. Reactions of chlorine with organic compounds such as fulvic and humic acids and humin produce undesirable by-products. These by-products are known as *disinfection by-products, DBPs*. Examples of DBPs are chloroform and bromochloromethane; these DBPs are suspected carcinogens. Snoeyink and Jenkins (1980) wrote a series of reactions that demonstrate the basic steps by which chloroform may be formed from an acetyl-group containing organic compounds. These reactions are shown in Figure 17.4.

Note that the initial reaction involves the splitting of the hydrogen atom from the methyl group using the hydroxyl ion. The hydroxyl ion is again used in (3), (5), and (7). Because the hydroxyl is involved, this would mean that chloroform formation is enchanced at high pH. To prevent formation of the chloroform, all that is necessary

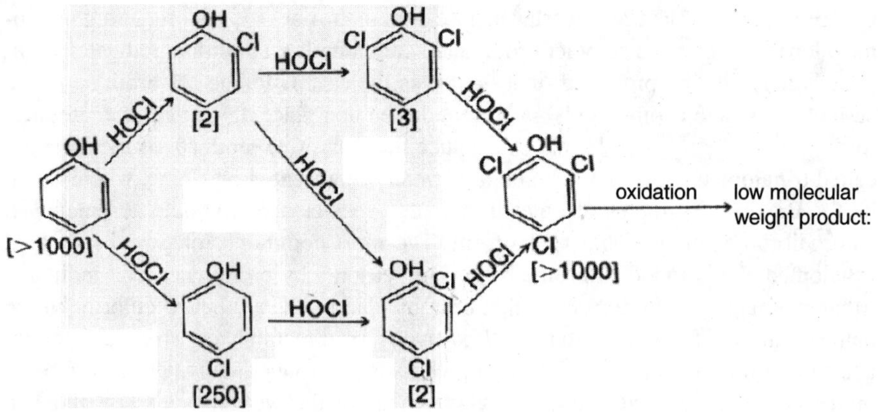

FIGURE 17.3 Chlorination for the breakdown of phenol; numbers in brackets are odor threshold concentrations in $\mu g/L$.

$$(1) \quad R-\overset{\overset{O}{\|}}{C}-CH_3 \overset{OH^-}{\rightleftharpoons} R-\overset{\overset{O^-}{|}}{C}=CH_2 + H^+$$

$$(2) \quad R-\overset{\overset{O^-}{|}}{C}=CH_2 + HOCl \rightarrow R-\overset{\overset{O}{\|}}{C}-CH_2Cl + OH^-$$

$$(3) \quad R-\overset{\overset{O}{\|}}{C}-CH_2Cl \overset{OH^-}{\rightleftharpoons} R-\overset{\overset{O^-}{|}}{C}=CHCl + H^+$$

$$(4) \quad R-\overset{\overset{O^-}{|}}{C}=CHCl + HOCl \rightarrow R-\overset{\overset{O}{\|}}{C}-CHCl_2 + OH^-$$

$$(5) \quad R-\overset{\overset{O}{\|}}{C}-CHCl_2 \overset{OH^-}{\rightleftharpoons} R-\overset{\overset{O^-}{|}}{C}=CCl_2 + H^+$$

$$(6) \quad R-\overset{\overset{O^-}{|}}{C}=CCl_2 + HOCl \rightarrow R-\overset{\overset{O}{\|}}{C}-CCl_3 + OH^-$$

$$(7) \quad R-\overset{\overset{O}{\|}}{C}-CCl_3 + H_2O \overset{OH^-}{\rightarrow} R-\overset{\overset{O}{\|}}{C}-OH + CHCl_3$$

FIGURE 17.4 Proposed scheme for chloroform formation.

then is simply to chlorinate at low pH, and this in fact, would be fortunate, since HOCl predominates at low pH values instead of at high pH values. Based on this fact, if superchlorination is to be conducted, it should be done at low pH values. Further research, however, should be performed to establish the accuracy of this assumption.

Chlorinated waters and wastewaters can contain not only chloroform and bromochloromethane but also other brominated compounds. In addition, iodinated compounds may also be produced; that is, this is the case if iodine (or bromine in the case of brominated compounds) is present, in the first place. In general, the products formed from the halogen family to produce the derivative products of methane are called trihalomethanes. The formula is normally represented by CHX_3, where X can be Cl, Br, or I. Examples of other brominated species are bromodichloromethane, chlorodibromomethane, and bromoform. The most commonly observed iodinated trihalomethane is iododichloromethane. The reason why brominated and iodinated trihalomethanes can be formed is that bromine and iodine are below chlorine in the halogen family of the periodic table. It is an observed fact in chemistry that stronger acids drive the weaker acids. The acid precursor of stronger acids (Cl in HOCl) are higher in the series than those of the precursor of the weaker acids (Br and I in HOBr and HOI, respectively). For this reason, HOCl drives the weaker acids HOBr and HOI. These two acids then react in the same way as HOCl when it produces the brominated and iodinated trihalomethanes.

Example 17.10   In Figure 17.4, when hydrogen is abstracted from the methyl group, what happens to the double between carbon and oxygen?

Solution: As shown, the double bond is ruptured making the oxygen end negative and single bonded. The double bond switches to become a carbon-to-carbon double bond. As indicated in subsequent reactions, this flip-flopping of the double bond continues until the formation of chloroform.   **Ans**

Acid generation. Whether or not acid will be produced depends upon the form of chlorine disinfectant used. Using chlorine gas will definitely produce hydrochloric acid. Sodium hypochlorite and calcium hypochlorite will not produce any acid; on the contrary, it can result in the production of alkalinity. Superchlorination using HOCl will definitely produce acids.

As shown in Equation (17.9), a mole of hydrochloric acid is produced per mole of chlorine gas that reacts. Chlorination uses up the disinfectant, so this reaction would be driven to the right and any mole of chlorine gas added will be consumed. Thus, if a mmol/L of the gas is dosed, this will produce a mmol/L of HCl. This is equivalent to one mgeq of the acid, which must also be equivalent to a mgeq of alkalinity. The analytical equivalent mass of alkalinity in terms of $CaCO_3$ is 50 mg $CaCO_3$ per mgeq. Thus, the mmol/L of hydrochloric acid produced will need 50 mg/L of alkalinity expressed as $CaCO_3$ for its neutralization. Or, simply, one mmol of hydrochloric acid requires 50 mg of alkalinity expressed as $CaCO_3$ for its neutralization.

In superchlorination, breakpoint reactions like Eqs. (17.37) to (17.42) will transpire. A host of other reactions may also occur such that all these, as shown by the preceding reactions, produce acids. The number of reactions are many, therefore, it is not possible to predict the amount of acids produced by using simple stoichiometry. The only way to determine this amount is to run a jar test as is done in the determination of the optimum alum dose. Metcalf & Eddy, Inc. wrote that in practice it is found that 15.0 mg/L of alkalinity is needed per mg/L of ammonia nitrogen (Metcalf & Eddy, Inc., 1972). But, again, the best method would be to run the jar test.

Example 17.11 A flow of 25,000 m^3/d of treated water is to be disinfected using chlorine in pressurized steel cylinders. The raw water comes from a reservoir where the water from the watershed has a very low alkalinity. With this low raw-water alkalinity, coupled with the use of alum in the coagulation process, the alkalinity of the treated water when it finally arrives at the chlorination tank is practically zero. Calculate the amount of alkalinity required to neutralize the acid produced during the addition of the chlorine gas.

Solution: Because the dose of the chlorine is not given, assume it to be 1.0 mg/L, which is equal to:

$$\frac{20}{2(35.5)} = 0.0142 \text{ millimol/L of Cl}_2$$

Therefore,

alkalinity needed $= 0.0142(50) = 0.71$ mg/L as $CaCO_3$ **Ans**

Available chlorine. The strength of a chlorine disinfectant is measured in terms of available chlorine. *Available chlorine* is defined as the ratio of the mass of chlorine to the mass of the disinfectant that has the same unit of oxidizing power as chlorine. The unit of disinfecting power of chlorine may be found as follows in terms of one mole of electrons:

$$Cl_2 + 2e^- \rightarrow 2Cl^- \tag{17.45}$$

From this equation, the unit of oxidizing power of Cl_2 is $Cl_2/2 = 35.5$. Consider another chlorine disinfectant such as NaOCl. To find its available chlorine, its unit of disinfecting power must also, first, be determined.

$$NaOCl + 2e^- + 2H^+ \rightarrow Cl^- + Na^+ + H_2O \tag{17.46}$$

From this equation, the unit of disinfecting power of NaOCl is NaOCl/2 = 37.24. Therefore, the available chlorine of NaOCl is the ratio of the mass of chlorine to the mass of NaOCl that has the same unit of oxidizing power as chlorine, or available chlorine of NaOCl = 35.5/37.24 = 0.95 or 95%. In other words, NaOCl is 95% effective compared with chlorine.

Example 17.12 What is the available chlorine in dichloramine?

Solution: When dichloramine oxidizes a substance, its chlorine atom is reduced to chloride; And, as gleaned from its formula, the nitrogen must be converted to the NH_4^+ ion. Thus, the oxidation-reduction reaction using only half the reaction is

$$NHCl_2 + 4e^- + 3H^+ \rightarrow 2Cl^- + NH_4^+$$

Therefore,

$$\text{the unit of oxidizing power of dichloramine} = NHCl_2/4 = 21.48$$

and, available chlorine = 35.5/21.48 = 1.65 or 165%   **Ans**

17.4.2 DESIGN OF CHLORINATION UNIT OPERATIONS FACILITIES

Important parameters to be considered in the design of chlorination unit operations facilities should include chlorine feeders, dosage control, chlorine injection and initial mixing, contact time and chlorine dosage, and maintenance of self-cleaning velocities through the chlorine contact tank. Each of these will be discussed in succession.

Chlorine feeders. Chlorine feeders may be categorized as feeders for chlorine gas and feeders for hypochlorite solutions. Schematics of chlorine gas feeder systems are shown in Figures 17.5, 17.6, and 17.7, as well as a drawing of a hypochlorite feeder is shown in Figure 17.8. Hypochlorite feeders are also called *hypochlorinators*.

Figure 17.5 is a chlorine gas feeder system that utilizes an evaporator. Evaporation of liquid chlorine speeds up the dosing delivery process. When dosage rates exceed 680 kg/d, evaporators must be used. Lower dosage rates may also require the use of evaporators when space is limited.

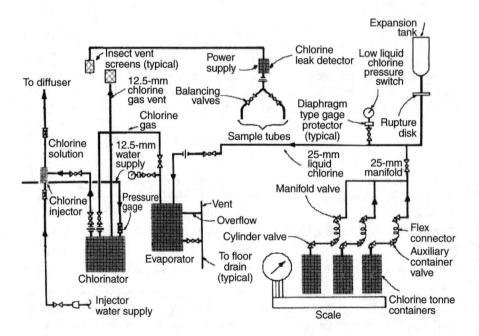

FIGURE 17.5 Schematic of chlorine feeder system using evaporators and tonne or larger containers.

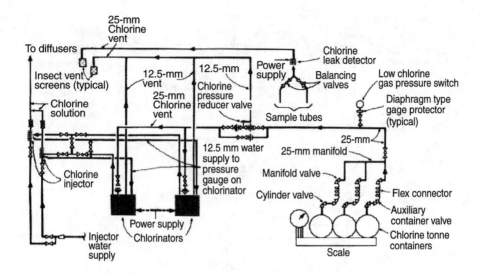

FIGURE 17.6 Schematic of chlorine feeder system using tonne containers.

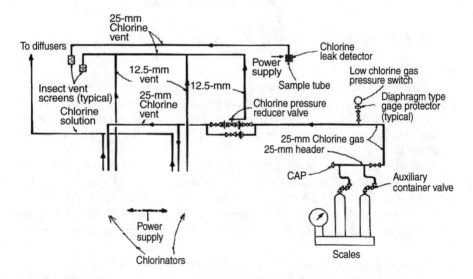

FIGURE 17.7 Schematic of chlorine feeder system using chlorine cylinders.

Following the schematic, tonne containers containing liquid chlorine are put on top of a scale in order to measure usage. Through a suitable piping, liquid chlorine under pressure flows into the evaporator. The chlorine is then evaporated into its gaseous state and conveyed into the chlorinator.

The driving mechanism that abstracts the chlorine gas from the chlorinator is the chlorine injector. This injector is powered by water pressure. As the water passes through

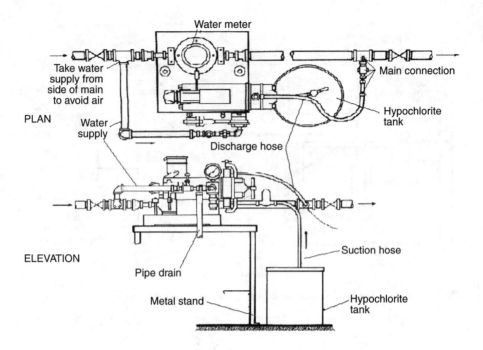

FIGURE 17.8 Hypochlorinator paced by mainline meter.

it, a suction vacuum is created through the piping leading to the chlorinator thereby abstracting the chlorine gas. The chlorinator serves as the metering device. The gas mixes with the water in the injector producing the HOCl chlorine solution. This solution is then conveyed into diffusers at the point of application to the flowing water to be disinfected.

Note the chlorine gas detector. The liquid chlorine is under pressure and could break pipe connections, so the location at the container section is appropriate. Also note the expansion tank and the rupture disk. Again, because the liquid chlorine is under a very high pressure, any leakage that may occur in the piping may be stopped by rupturing the disk and allowing the chlorine to expand into the expansion tank.

Figure 17.6 shows the chlorine gas system that also uses tonne containers but without the evaporator, while Figure 17.7 shows the system that uses the smaller chlorine cylinders. In these systems, the liquid chlorine is evaporated into gas through the pressure reducing value, instead of evaporators. The operating principle is the same: the gaseous chlorine is abstracted from the chlorinator, which serves as the metering device, through vacuum created by the chlorine injector. The resulting chlorine solution produced at the injector is then conveyed to diffusers at the point of application. Also note the chlorine detector. Liquid chlorine is always under pressure, therefore, chlorine detectors must be provided; however, the system is a little smaller than the one using evaporators, so the rupture disk and the expansion tank are not provided.

Hypochlorinators are most suited for smaller installations. For installations serving about ten people, it is possible to use drip-fed hypochlorinators. For up to 100 people, orifice-controlled feeders using a constant head tank mounted overhead and fed by gravity can be used successfully. The most satisfactory means of feeding hypochlorite solutions is through the use of low capacity proportioning pumps.

Figure 17.8 shows the hypochlorinator using low capacity proportioning pumps. As indicated in the elevation drawing, hypochlorite solution is sucked from the hypochlorite tank by the proportioning pump. The amount of dosage is proportioned according to the reading of the meter shown in the plan drawing. The hypochlorite solution is then injected into the main at the discharge side of the chlorinator through the discharge hose.

Example 17.13   Specify the chlorine dosing system to be used for a wastewater treatment plant processing an average of 25,000 m³/d of wastewater. The peaking factor is 3.0 and the regulatory agency requires a dosage of 20.0 mg/L.

Solution:

$$\text{Chlorine dosage} = \frac{25,000(3.0)(20)}{1000} = 1,500 \text{ kg/d} > 680 \text{ kg/d}$$

Therefore, use a system that uses evaporators.   **Ans**

Dosage control. There are five ways of providing dosage control. The first one is *manual control*. This is the simplest and involves the operator adjusting the flow rate of chlorine to match requirements. The chlorine residual is checked at intervals of time such as 15 min and dosage adjusted accordingly. The residual desired may be in the vicinity of 0.5 mg/L. This method of control is obviously used in small facilities. The second method is *program control*. The program control is a selected set pattern of dosage that must have already been determined to effect the desired disinfection. Program control is the cheapest way to attain automatic control. The third method is *flow-portioned control*. Flow-proportioned control proportions dosage according to the flow rate of the water to be disinfected. This is the case of the hypochlorite dosing system of Figure 17.8. The rate of suction of the solution is proportioned to the reading of the meter. Flow-proportioned control is also called *flow-paced control*. The fourth method is *residual-proportioned control*. Residual-proportioned control proportions dosage according to the amount of chlorine residual desired. This system requires an automatic residual chlorine analyzer at the effluent and a signal transmitter. The signal is sent to a controller that then changes valve settings for proper dosage. The fifth and final method is the combination of the flow-proportioned and residual-proportioned control. In this setup, the two signals coming from the flow meter and the residual chlorine analyzer are transmitted to a controller that calculates the resulting valve setting according to these signals' input.

Figure 17.9 shows the fifth method of control. As indicated, the signal from the flow meter, labeled "pacing control signal," and the signal from the chlorine residual analyzer, labeled "residual trim control signal," are sent to the chlorine feeder. Inside the chlorine

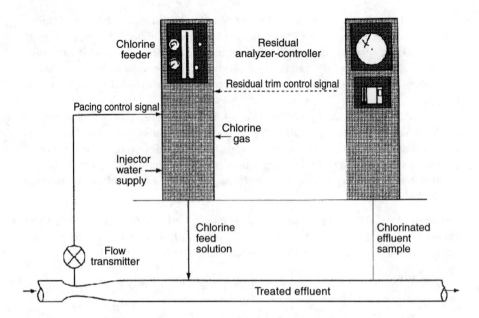

FIGURE 17.9 Automatic residual chlorine control using combined flow-proportioned and residual-proportioned control.

feeder would be the controller that combines the two input signals to generate a command to adjust the valves to the correct setting. The correct valve setting then dispenses chlorine solution as shown by the arrow labeled "chlorine feed solution."

Example 17.14   The residual chlorine analyzer transmits a signal correspond-ing to 0.7 mg/L of chlorine and the flow transmitter transmits a signal corresponding to 0.01 m³/s. If the desired chlorine residual is 0.5 mg/L, at what rate is the disin-fectant dosed into the treated effluent?

Solution: To solve this problem, the chlorine demand must be given. Assume this to be 0.4 mg/L. Thus, the total dose is 0.7 + 0.4 = 1.1 mg/L, or

$$\text{total chlorine dose} = \frac{0.01(1.1)}{1000}(60)(60)(24) = 0.95 \text{ kg/d} \quad \textbf{Ans}$$

Chlorine injection and initial mixing.   To ensure effective disinfection, chlo-rine should be completely mixed at the point of application. As learned in the discussion of chlorine chemistry, several intervening substances are present in water and wastewater that exert chlorine demand before the disinfectant can do its job of disinfecting. Thus, a rapid mix should be instituted at the point of application so that the disinfectant can immediately act on the microorganisms rather than wasting time reacting with extraneous intervening substances. The unit operation of mixing has already been discussed under the chapter on mixing and flocculation.

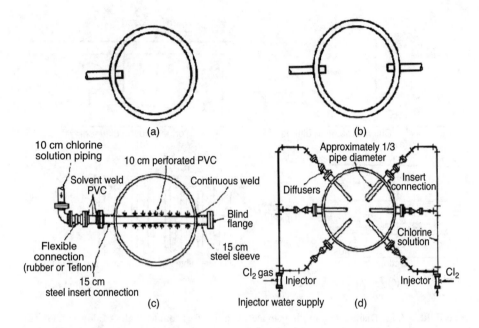

FIGURE 17.10 Disinfectant mixing devices. (From Metcalf & Eddy, Inc. (1972). *Wastewater Engineering: Treatment, Disposal, Reuse.* McGraw-Hill, New York, 385. With permission.)

Other mixing devices that have not been discussed in that chapter are shown in Figures 17.10 and 17.11.

The arrangement in Figure 17.10a is used to inject disinfectant into waters and wastewaters flowing in small pipes. Figure 17.10b is also an arrangement used for small pipes but with dual injection points. As the flow moves downstream, the disinfectant is mixed with the liquid due to turbulence. For injection into larger pipes, the arrangement in Figures 17.10c and 17.10d may be used. In Figure 17.10c, a PVC pipe is perforated along its length. The perforations are for the disinfectant to diffuse through. The process of diffusing through the perforations causes turbulence and mixing. Figure 17.10d shows multiple injection ports converging into a point. The turbulence caused by this arrangement and the turbulence of the flowing water mixes the disinfectant.

Figure 17.11a shows a mixing induced by the turbulence produced by a submerged baffle. The arrangement in Figure 17.11b produces turbulence over and under the baffles. The arrangement shown in 17.11c is similar to the one shown in Figure 17.10c, except that in the former, the diffuser is installed in a closed conduit; the latter is in an open channel. The arrangement in Figure 17.11d is also in an open channel and is equipped with a hanging-nozzle type diffuser.

Contact time and chlorine dosage. The two most important parameters used in the design of chlorine contact tanks is the contact time and dosage of chlorine. These parameters have already been discussed; the equation is given by the universal law of disinfection, Equation (17.2). Figure 17.12 shows a contact tank used to disinfect treated sewage.

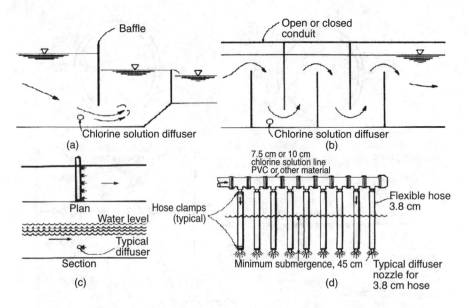

FIGURE 17.11 Disinfectant mixing devices, continued. (From Metcalf & Eddy, Inc. (1972). *Wastewater Engineering: Treatment Disposal, Reuse.* McGraw-Hill, New York, 386. With permission.)

FIGURE 17.12 A chlorine contact tank in serpentine configuration. (From Metcalf & Eddy, Inc. (1972). *Wastewater Engineering: Treatment, Disposal, Reuse.* McGraw-Hill, New York, 387. With permission.)

As shown in Figure 17.12, the flow of water is in a serpentine configuration; the direction of the arrows indicates the direction of flow. This configuration allows the satisfaction of the contact time. Since the channel flow cross section is compressed into a smaller cross-section area, this scheme approximates a plug flow. Note that the figure shown is only one of the contact tank compartments in this plant. One of the other compartments is partly shown at the top of the figure.

Example 17.15 A chlorine disinfection study was conducted to determine the constants of Equation (17.2). For a log 2 removal efficiency, the value of m is found to be 0.35 and the value of k is found to be 100. Calculate the contact time if the regulatory agency requires a chlorine dose of 20 mg/L.

Solution:

$$t = \frac{100}{20^{0.35}} = 35 \text{ min} \quad \textbf{Ans}$$

Maintenance of self-cleaning velocity. In the case of sewage treatment plants, some solids would have escaped settling. In the case of water treatment plants, however, the effluent should be very clear with no danger of solids depositing on the chlorine contact tank. As in any design of open channels, the velocity through the cross section should be self-cleaning. We have seen this requirement in design of sewers. The design of chlorine contact tanks is no exception. Self-cleaning velocities of 2.0 to 4.5 m/min have been mentioned in the literature.

Example 17.16 A total flow of 1000 m^3/d is to be disinfected. What should be the cross-sectional area of the serpentine channel in order to maintain self-cleaning velocity? What would be the total combined length of the channel if the contact time has been calculated to be 35 min?

Solution: Cross-sectional area = 1000/24(60)(2.0) = 0.34 m^2, assuming a self-cleaning velocity of 2.0 m/min. **Ans**
Length of channel = 35(2) = 70 m. The flow in the tank will not be exactly plug flow, so increase channel length by 10%. Thus, length of channel = 70(1.10) = 77 m **Ans**

17.5 DECHLORINATION

Effluents from sewage treatment plants are not allowed to contain residual chlorine in excess of tolerable values as determined by water quality standards. For example, in discharges to trout streams, the residual chlorine should not exceed 0.02 mg/L. Thus, chlorinated effluents should be dechlorinated. Sulfur dioxide, sodium sulfite, sodium metabisulfite, and activated carbon have been used for dechlorination. Because sulfur dioxide, sodium sulfite, and sodium metabisulfite contain sulfur, we will call them sulfur dechlorinating agents. Dechlorination is an oxidation-reduction reaction. The chemical reactions involved in dechlorination are discussed next.

17.5.1 Chemical Reactions Using Sulfur Dechlorinating Agents

Irrespective of the type of residual chlorine present, all the species can be represented by a single species by means of the equivalence of chemical reactions. It is convenient to use HOCl to represent all these species. Thus, the chemical reactions using sulfur dioxide as the dechlorinating agent with HOCl representing the residual chlorine species are as follows:

$$SO_2 + H_2O \rightarrow HSO_3^- + H^+ \tag{17.47}$$

$$HSO_3^- + H_2O \rightarrow SO_4^{2-} + 3H^+ + 2e^- \tag{17.48}$$

$$\frac{HOCl + H^+ + 2e^- \rightarrow Cl^- + H_2O}{SO_2 + HOCl + H_2O \rightarrow SO_4^{2-} + Cl^- + 3H^+} \tag{17.49}$$

$$ \tag{17.50}$$

Sodium sulfite is Na_2SO_3. Thus, the chemical reactions using sodium sulfite as the dechlorination agent are as follows:

$$Na_2SO_3 + H_2O \rightarrow SO_4^{2-} + 2Na^+ + 2H^+ + 2e^- \tag{17.51}$$

$$\frac{HOCl + H^+ + 2e^- \rightarrow Cl^- + H_2O}{Na_2SO_3 + HOCl \rightarrow SO_4^{2-} + 2Na^+ + Cl^- + H^+} \tag{17.52}$$

$$ \tag{17.53}$$

Sodium metabisulfite is $Na_2S_2O_5$. Thus, the chemical reactions using sodium metabisulfite as the dechlorination agent are as follows:

$$Na_2S_2O_5 + 3H_2O \rightarrow 2SO_4^{2-} + 2Na^+ + 6H^+ + 4e^- \tag{17.54}$$

$$\frac{2HOCl + 2H^+ + 4e^- \rightarrow 2Cl^- + 2H_2O}{Na_2S_2O_5 + 2HOCl + H_2O \rightarrow 2SO_4^{2-} + 2Na^+ + 2Cl^- + 4H^+} \tag{17.55}$$

$$ \tag{17.56}$$

17.5.2 Chemical Reactions Using Activated Carbon

Carbon is a reducing agent; thus, it follows that when in contact with chlorine it will reduce it to the chloride state. Carbon, itself, will be oxidized to carbon dioxide. The chemical reactions follow:

$$C + 2H_2O \rightarrow CO_2 + 4H^+ + 4e^- \tag{17.57}$$

$$\frac{2HOCl + 2H^+ + 4e^- \rightarrow 2Cl^- + 2H_2O}{C + 2HOCl \rightarrow CO_2 + 2Cl^- + 2H^+} \tag{17.58}$$

$$ \tag{17.59}$$

The unit operation of carbon adsorption is discussed in Part II of this book. The use of activated carbon in dechlorination will use the same unit operation, except that in the present case, the carbon will be consumed by chemical reaction. In the unit operation as discussed in Part II, the operation is purely physical, and carbon can be regenerated, which is also the same case in dechlorination except that there is a large loss of carbon between regenerations.

Example 17.17   A total flow of 25,000 m^3/d is to be dechlorinated after disinfection using chlorine. Sulfur dioxide is to be used for dechlorination. If the total residual chlorine (TRC) is 0.5 mg/L, how many kilograms per day of sulfur dioxide are needed?

Solution:

$$1.0 \text{ mg/L HOCl} = \frac{Cl_2}{HOCl} = \frac{2(35.5)}{1.008 + 16 + 35.5} = 1.35 \text{ mg/L } Cl_2$$

Therefore,

$$0.5 \text{ mg/L TRC} = \frac{0.5}{1.35} = 0.37 \text{ mg/L HOCl}$$

And, from Equation (17.50),

$$\text{kg/d of sulfur dioxide} = \frac{\frac{SO_2}{HOCl}(0.37)}{1000}(25,000)$$

$$= \frac{\frac{32+32}{1.008+16+35.5}(0.37)}{1000}(25,000) = 11.27 \text{ kg/d Ans}$$

Example 17.18   A total flow of 25,000 m^3/d is to be dechlorinated after disinfection using chlorine. An activated carbon bed is to be used for dechlorination. If the total residual chlorine (TRC) is 0.5 mg/L, how many kilograms of activated carbon are lost from the bed per day?

Solution:

$$1.0 \text{ mg/L HOCl} = \frac{Cl_2}{HOCl} = \frac{2(35.5)}{1.008 + 16 + 35.5} = 1.35 \text{ mg/L } Cl_2$$

Therefore,

$$0.5 \text{ mg/L TRC} = \frac{0.5}{1.35} = 0.37 \text{ mg/L HOCl}$$

And, from Equation (17.59),

$$\text{kg/d of activated carbon} = \frac{\dfrac{C}{2HOCl}(0.37)}{1000}(25{,}000)$$

$$= \frac{\dfrac{12}{2(1.008+16+35.5)}(0.37)}{1000}(25{,}000) = 1.06 \text{ kg/d} \quad \textbf{Ans}$$

17.5.3 EFFECT OF DECHLORINATED EFFLUENTS ON DISSOLVED OXYGEN OF RECEIVING STREAMS

Although precise control may be achieved in dosing the sulfur dechlorinating agents, some residual will exist. This residual should be minimized not only because of waste of chemicals but also because of its effect on the dissolved oxygen of receiving streams. The reactions of these residuals on oxygen are as follows:

for sulfur dioxide:

$$2HSO_3^- + 2H_2O \rightarrow 2SO_4^{2-} + 6H^+ + 4e^- \qquad (17.60)$$

$$\underline{O_2 + 4H^+ + 4e^- \rightarrow 2H_2O} \qquad (17.61)$$

$$2HSO_3^- + O_2 \rightarrow 2SO_4^{2-} + 2H^+ \qquad (17.62)$$

for sodium sulfite:

$$2Na_2SO_3 + 2H_2O \rightarrow 2SO_4^{2-} + 4Na^+ + 4H^+ + 4e^- \qquad (17.63)$$

$$\underline{O_2 + 4H^+ + 4e^- \rightarrow 2H_2O} \qquad (17.64)$$

$$2Na_2SO_3 + O_2 \rightarrow 2SO_4^{2-} + 4Na^+ \qquad (17.65)$$

for sodium metabisulfite:

$$Na_2S_2O_5 + 3H_2O \rightarrow 2SO_4^{2-} + 2Na^+ + 6H^+ + 4e^- \qquad (17.66)$$

$$\underline{O_2 + 4H^+ + 4e^- \rightarrow 2H_2O} \qquad (17.67)$$

$$Na_2S_2O_5 + O_2 + H_2O \rightarrow 2SO_4^{2-} + 2Na^+ + 2H^+ \qquad (17.68)$$

The effects of the above reactions on the receiving streams are a reduction of dissolved oxygen and an increase in BOD and COD. In addition, the right-hand sides of the reactions, except those for sodium sulfite, indicate production of the hydrogen ion. Thus, there is the possibility of a decrease in pH of the receiving streams.

Example 17.19   Calculate the residual $Na_2S_2O_5$ in a dechlorination process that will result in a decrease of pH to 5.0 in a receiving stream assuming the original pH is 7.5.

Solution:

Increase in hydrogen concentration $= (10^{-5} - 10^{-7.5}) = 9.97(10^{-6})\,\text{gmols/L}$

Therefore, from Equation (17.68),

residual $Na_2S_2O_5 = \dfrac{9.97(10^{-6})}{2} = 4.99(10^{-6})\,\text{gmols/L} = 4.99(10^{-6})(Na_2S_2O_5)$

$= 4.99(10^{-6})[2(23) + 2(32.1) + 5(16)] = 9.49(10^{-4})\,\text{gm/L} = 0.95\,\text{mg/L}$ **Ans**

17.5.4 UNIT OPERATIONS IN DECHLORINATION

The unit operations equipment for dechlorination are similar to that for chlorination. In fact, dechlorination equipment is interchangeable with chlorination equipment. The contact tank, however, is not used in dechlorination using the sulfur dechlorinating agents. The reactions are so very fast that no contact tanks are necessary. As in chlorination, the key process control parameters are precise dosage and monitoring of residual chlorine and adequate mixing at the point of application in the process stream. The unit operation of dechlorination using activated carbon is similar to the one used in the unit operation of carbon adsorption, with the difference, as mentioned before, of the loss of large amounts of activated carbon in the case of dechlorination.

17.6 DISINFECTION USING OZONE

Ozone is a very strong oxidizer and has been found to be superior to chlorine in inactivating resistant strains of bacteria and viruses. It is very unstable, however, having a half-life of only 20 to 30 min in distilled water at 20°C. It is therefore generated on site before use.

Ozone may be produced by first refrigerating air to below the dew point to remove atmospheric humidity. The dehumidified air is then passed through desiccants such as silica gel and activated silica to dry to −40 to −60°C. The dried and dehumidified air is then introduced between two electrically and oppositely charged plates or through tubes where an inner core and the inner side of the tube serve as the oppositely charged plates. Passage through these plates converts the oxygen in the air into ozone according to the following reaction:

$$3O_2 \rightarrow 2O_3 \tag{17.69}$$

Typical ozone dosage is 1.0 to 5.3 kg/1000 m^3 of treated water at a power consumption of 10 to 20 kW/kg of ozone. It has been observed that complete destruction of poliovirus in distilled water is accomplished with a residual of 0.3 mg/L of ozone in 3 min. As in the case of chlorination where a chlorine demand is first exerted before the actual disinfection process can take place, an ozone demand in ozonation is also first exerted before the actual disinfection process can take place. The ozone demand can be variable; therefore, the contact time for ozonation is best determined by a pilot plant study. Depending upon the nature of the ozone demand, a contact time of 20 min is not unreasonable. A residual ozone concentration of 0.4 mg/L has been found to be effective (Rice et al., 1979).

The immediate ozone demand parallels that of the immediate chlorine demand. Recall that this demand is due to ferrous, manganous, nitrites, and hydrogen sulfide. The immediate demand reactions with ozone are as follows:

with ferrous:

$$2Fe^{2+} \rightarrow 2Fe^{3+} + 2e^-$$ (17.70)

$$O_3 + e^- \rightarrow O^- + O_2$$ (17.71)

$$O^- + 2H^+ + e^- \rightarrow H_2O$$ (17.72)

$$\overline{2Fe^{2+} + O_3 + 2H^+ \rightarrow 2Fe^{3+} + O_2 + H_2O}$$ (17.73)

with manganous:

$$Mn^{2+} \rightarrow Mn^{4+} + 2e^-$$ (17.74)

$$O_3 + e^- \rightarrow O^- + O_2$$ (17.75)

$$O^- + 2H^+ + e^- \rightarrow H_2O$$ (17.76)

$$\overline{Mn^{2+} + O_3 + 2H^+ \rightarrow Mn^{4+} + O_2 + H_2O}$$ (17.77)

with nitrites:

$$NO_2^- + H_2O \rightarrow NO_3^- + 2H^+ + 2e^-$$ (17.78)

$$O_3 + e^- \rightarrow O^- + O_2$$ (17.79)

$$O^- + 2H^+ + e^- \rightarrow H_2O$$ (17.80)

$$\overline{NO_2^- + O_3 \rightarrow NO_3^-}$$ (17.81)

with hydrogen sulfide:

$$H_2S + 4H_2O \rightarrow SO_4^{2-} + 10H^+ + 8e^-$$ (17.82)

$$4O_3 + 4e^- \rightarrow 4O^- + 4O_2$$ (17.83)

$$4O^- + 8H^+ + 4e^- \rightarrow 4H_2O$$ (17.84)

$$\overline{H_2S + 4O_3 \rightarrow SO_4^{2-} + 4O_2 + 2H^+}$$ (17.85)

As indicated in all the previous reactions, an intermediate O^- is first produced. This is called nascent oxygen and is the one responsible for the potent property of ozone as a strong oxidant and, hence, as a strong disinfectant.

Note: All in all, a mole of ozone grabs 2 moles of electrons, making it a strong oxidizer. Also, all the previous reactions show that molecular O_2 has been produced from the decomposition of ozone.

This is one of the advantages in the use of ozone; the effluent is saturated with dissolved oxygen. Of course, the previous reactions are simply for the immediate ozone demand and have nothing to do with disinfection. As mentioned, these reactions must be satisfied first before the actual act of disinfecting commences.

17.6.1 UNIT OPERATIONS IN OZONATION

In its simplest form, the unit operations of ozonation involve the production of ozone and the mechanics of dissolving and mixing the ozone in the water or wastewater. The unit operations of dissolving and mixing have already been discussed in Part II of this book. The design of the contact tank is similar to that of chlorination, the contact time and dosage being best determined by a pilot plant study. As mentioned before, a contact time of 20 min is not unreasonable, and a residual of 0.4 mg/L of ozone has been found to be effective. The equation for ozone disinfection follows the universal law of disinfection given by Equation (17.2).

Example 17.20   A total flow of 1000 m^3/d is to be disinfected using ozone. What should be the cross-sectional area of the serpentine channel in order to maintain self-cleaning velocity? What would be the total combined length of the channel if the contact time has been determined by a pilot study to be 20 min?

Solution: Cross-sectional area = 1000/24(60)(2.0) = 0.34 m^2, assuming a self-cleaning velocity of 2.0 m/min.   **Ans**
Length of channel = 20(2) = 40 m. The flow in the tank will not be exactly plug flow, so increase channel length by 10%. Thus, length of channel = 40(1.10) = 44 m.   **Ans**

17.7 DISINFECTION USING ULTRAVIOLET LIGHT

Water, air, and foodstuff can be disinfected using ultraviolet light, UV. This radiation destroys bacteria, bacterial spores, molds, mold spores, viruses, and other microorganisms. Radiation at a wavelength of around 254 nm penetrates the cell wall and is absorbed by the cell materials including DNA and RNA stopping cell replication or causing death. The use of UV radiation for disinfection dates back to the 1900s when it was used to a limited extent in disinfecting water supplies.

The low-pressure mercury arc lamp is the principal means of producing ultraviolet light. This is used because about 85% of the light output is monochromatic at a wavelength of 253.7 nm which is within the optimum range for germicidal effect. It should be noted, however, that the optimum range for germicidal effect is also within the UV-B ultraviolet radiation. UV-B is less than or equal to 320 nm and is the dangerous range for causing skin cancer. Thus, it is important that workers be not unduly exposed to this radiation.

FIGURE 17.13 Installation of UV lamp (left) and a UV lamp assembly (right). (Courtesy of Trojan Technologies, Inc. With permission.)

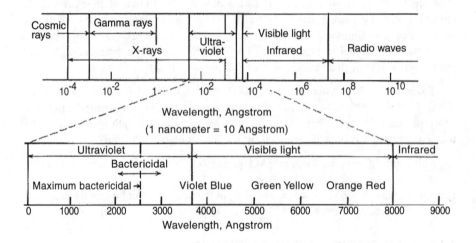

FIGURE 17.14 Electromagnetic spectrum.

Lamp bulbs are typically 0.75 m to 1.5 m in length and 15–20 mm in diameter. Mercury vapor is sealed inside this bulb. To produce the radiation, electric current, which generates an electric arc, is passed through the bulb. In actual use in the treatment of water and wastewater, the lamps are either suspended over the liquid to be disinfected or immersed in it. When immersed, the lamps are encased in quartz tube to prevent the cooling effect of the liquid on the lamps. The right-hand side portion of Figure 17.13 shows an ultraviolet lamp assembly; the left-hand side shows an actual installation located below floor level.

Figure 17.14 shows the electromagnetic spectrum. From the figure, the ultraviolet light occupies a band along the spectrum extending from 30 Å to 3650 Å (3 nm– 365 nm). The bactericidal wave range extends from 2000 Å to 2950 Å (200 nm–295 nm);

the optimum wave range extends from 2500 Å to 2700 Å (250 nm–270 nm). As mentioned before, the UV-B range is ≤ 320 nm. The other "half" of the UV spectrum is called the UV-A range and extends from 320–365 nm. UV-A is not carcinogenic.

Note: A very small "window" of spectrum is available for visible light.

17.7.1 Unit Operations in UV Disinfection

The unit operations in UV disinfection, among other things, involve the generation of the ultraviolet radiation and its insertion into or above the flowing water or wastewater to be disinfected. The contact time is very short, being in the range of seconds to a few minutes. Thus, the contact tank for UV disinfection is short relative to the contact tanks for chlorination and ozonation which require the serpentine configuration to satisfy the contact time required. The contact time and UV intensity is best determined by a pilot plant study. Intensity is normally expressed as milliwatts or microwatts per square centimeter of projected area. To be effective, the "sheet" of flowing liquid should be thin so that the radiation can penetrate. The equation for UV disinfection follows the universal law of disinfection given by Equation (17.2).

Example 17.21   A total flow of 1000 m^3/d is to be disinfected using UV. What should be the cross-sectional area of the channel in order to maintain self-cleaning velocity? What would be the total combined length of the channel if the contact time has been determined by a pilot study to be 1.0 min?

Solution: Cross-sectional area = $1000/24(60)(2.0) = 0.34$ m², assuming a self-cleaning velocity of 2.0 m/min.   **Ans**
Length of channel = $1.0(2) = 2.0$ m. The flow in the tank will not be exactly plug flow, so increase channel length by 10%. Thus, length of channel = $2.0(1.10) = 2.2$ m.   **Ans**

Example 17.22   A study was conducted to determine the optimum UV intensity and contact time to be used to design a UV contact tank. For a 99% kill, the results are as follows:

Contact Time (sec)	Intensity (mW/cm²)
20	0.4
66	0.045

Determine the dosage intensity and the contact time.

Solution:

$$t = \frac{k}{I^m}$$

$$m = -\frac{\dfrac{\sum_{l+1}^{m} \ln t_i}{m-l} - \dfrac{\sum_{1}^{l} \ln t_i}{l}}{\dfrac{\sum_{l+1}^{m} \ln I_i}{m-l} - \dfrac{\sum_{1}^{l} \ln I_i}{l}} = -\frac{4.1896 - 2.996}{-3.101 - (-0.9163)} = 0.546$$

$$k = \exp\left[\frac{\sum_1^l \ln t_i}{l} + m\left\{\frac{\sum_1^l \ln I_i}{l}\right\}\right] = \exp[2.996 + 0.546\{-0.9163\}]$$

$$= \exp[2.496] = 12.13$$

Therefore,

$$t = \frac{k}{I^m} = \frac{12.13}{I^{0.546}}$$

Let $I = 0.25$ mW/cm^2; thus,

$$\text{contact time} = t = \frac{12.13}{I^{0.546}} = \frac{12.13}{0.25^{0.546}} = 25.86 \text{ sec} \quad \textbf{Ans}$$

GLOSSARY

Available chlorine—The ratio of the mass of chlorine to the mass of the disinfectant that has the same unit of oxidizing power as chlorine.

Breakpoint—The endpoint of the breakpoint reactions.

Breakpoint reactions—Decomposition reactions before the breakpoint.

Chloramines—Reaction products of chlorine with ammonia and organic amines.

Chlorination—Application of chlorination for disinfection and for other purposes.

Combined chlorine—Chloramines composed of monochloramines and dichloramines.

Dechlorination—A unit process of removing residual chlorine disinfectants.

Disinfection—A unit process involving reactions that render pathogenic organisms harmless.

Disinfection by-products (DBPs)—By-products formed during chlorination.

Disproportionation—A chemical reaction in which an element is oxidized as well as reduced.

Flow-proportioned or flow-paced control—A method of controlling dosage that proportions the dosage according to the flow rate of the water to be disinfected.

Free chlorine—The three species of $Cl_{2(aq)}$, HOCl, and OCl$^-$.

Halogen family—Group of elements in the periodic table consisting of fluorine, chlorine, bromine, iodine, and astatine.

Hypochlorinators—Hypochlorite feeders.

Log removal efficiency—Efficiency of removal corresponding to the reduction of one logarithm of the concentration of organisms.

Intensity—An intensive property of a disinfectant.

Intensive property—Property independent of the total mass or volume of the disinfectant.

Ozonation—Application of ozone for disinfection and other purposes.

Program control—Selected set of pattern of dosage that must have already been determined to effect the desired disinfection.

Residual-proportioned control—A method of controlling dosage that proportions the dosage according to the amount of chlorine residual desired.

Sterilization—A unit process involving reactions that kill all organisms.

Superchlorination—The practice of chlorinating up to and beyond the breakpoint.

Threshold concentration—The concentration at which the odor is barely detectable.

Trihalomethane—Halogen derivative of methane.

SYMBOLS

ΔH^o_{298}	Standard enthalpy change
k	Constant of proportionality in the universal disinfection equation
K_a	Ionization constant for HOCl
K_{Claq}	Equilibrium constant for dissolution of gaseous chlorine into water
K_H	Hydrolysis equilibrium constant for $Cl_{2(g)}$
K_{T1}	Equilibrium constants at temperatures T_1
K_{T2}	Equilibrium constant at temperatures T_2 enthalpy change of the reaction and R is the universal gas constant.
I	Intensity of disinfectant
m	Power of Intensity in the universal disinfection equation
pK_a	Negative logarithm to the base 10 of K_a
pK_H	Negative logarithm to the base 10 of K_H
R	Universal gas constant
t	Contact time of disinfection
T_1	Initial temperature in the Van't Hoff equation
T_2	Final temperature in the Van't Hoff equation

PROBLEMS

17.1 It is desired to design a bromide chloride contact tank to be used to disinfect a secondary-treated sewage discharge. To determine the contact time, an experiment was conducted producing the following results:

	Contact Time (min)/Residual Fecal Coliforms (No./100 mL)		
BrCl Dosage (mg/L)	15	30	60
3.6	10,000	4,000	600
15.0	800	410	200
47.0	450	200	90

Determine the contact time to be used in design, if it is desired to have a log 3 removal efficiency for fecal coliforms. The original concentration of fecal coliforms is 40,000 per/100 mL.

17.2   It is desired to design a bromide chloride contact tank to be used to disinfect a secondary-treated sewage discharge. To determine the contact time, an experiment was conducted producing the results below:

BrCl dosage (mg/L)	Contact Time (min)/Residual Fecal Coliforms (No./100 mL)		
	15	30	60
3.6	10,000	4,000	600
47.0	450	200	90

Determine the contact time to be used in design, if it is desired to have a log 3 removal efficiency for fecal coliforms. The original concentration of fecal coliforms is 40,000 per/100 mL.

17.3   It is desired to design a bromide chloride contact tank to be used to disinfect a secondary-treated sewage discharge. To determine the contact time, an experiment was conducted producing the following results:

BrCl Dosage (mg/L)	Contact Time (min)/Residual Fecal Coliforms (No./100 mL)		
	15	30	60
3.6	10,000	4,000	600
15.0	800	410	200

Determine the contact time to be used in design, if it is desired to have a log 3 removal efficiency for fecal coliforms. The original concentration of fecal coliforms is 40,000 per/100 mL.

17.4   The contact time for a certain chlorination process at a pH of 7.0 and a temperature of 25°C is 30 min. What would be the contact time to effect the same percentage kill if the process is conducted at a temperature 8°C?

17.5   The contact time for a certain chlorination process at a pH of 7.0 and a temperature of 25°C is 30 min. What would be the contact time to effect the same percentage kill if the process is conducted at a temperature 30°C.

17.6   The contact time for a certain chlorination process at a pH of 7.0 and a temperature of 25°C is 30 min. What would be the contact time to effect the same percentage kill if the process is conducted at a temperature 20°C.

17.7   At pH 8.0, calculate the mole fraction of HOCl.

17.8   At pH 2.0, calculate the mole fraction of HOCl.

17.9   At pH 6.0, calculate the mole fraction of OCl^-.

17.10 A container containing 50 kg of $Ca(OCl)_2$ is stored in a transparent plastic container. It had been stored outdoors for quite some time and then used to disinfect a swimming pool. How effective is the disinfection? What material is used instead for the disinfection? And how many kilograms is it?

17.11 A container containing 50 kg of $Ca(OCl)_2$ is stored in an opaque plastic container. It had been stored outdoors for quite some time and then used to disinfect a swimming pool. How effective is the disinfection? What material is used instead for the disinfection? And how many kilograms is it?

17.12 An activated sludge plant of a small development is out of order, and a decision has been made following approval from a state agency to discharge raw sewage to a river. The effluent was found to contain 8 mg/L of ferrous, 3 mg/L manganous, 20 mg/L nitrite as nitrogen, and 4 mg/L hydrogen sulfide. Calculate the mg/L of bleach needed to be dosed before actual disinfection is realized. What is the chlorine concentration?

17.13 An activated sludge plant of a small development is out of order, and a decision has been made following approval from a state agency to discharge raw sewage to a river. The effluent was found to contain 8 mg/L of ferrous and 4 mg/L hydrogen sulfide. Calculate the mg/L of HTH needed to be dosed before actual disinfection is realized. What is the chlorine concentration?

17.14 An activated sludge plant of a small development is out of order, and a decision has been made following approval from a state agency to discharge raw sewage to a river. The effluent was found to contain 8 mg/L of ferrous and 4 mg/L hydrogen sulfide. Calculate the mg/L of bleach needed to be dosed before actual disinfection is realized. What is the chlorine concentration?

17.15 In order to provide residual disinfectant in the distribution system, chloramination is applied to the treated water. If three moles of HOCl have been added per mole of ammonia, calculate the moles of nitrogen gas produced.

17.16 In order to provide residual disinfectant in the distribution system, chloramination is applied to the treated water. If three moles of HOCl have been added per mole of ammonia, calculate the moles of trichloramine produced.

17.17 In order to provide residual disinfectant in the distribution system, chloramination is applied to the treated water. If three moles of HOCl have been added per mole of ammonia, calculate the moles of laughing gas produced.

17.18 Show the half reaction that will exhibit the property of organic chloramides as disinfectants.

17.19 Referring to Figure 17.1, if a dosage of 2.0 mg/L is administered, determine the amount of free chlorine residual that results, the amount of combined residual that results, and the amount of combined ammonia chloramine residual that results. Also, determine the amount of organic chloramine residual that results.

17.20 In the reaction scheme of Figure 17.3, how many hydrogen atoms have been displaced in ortho-chlorophenol by the chlorine atom to form 2,6,-dichlorophenol?

17.21 In the reaction scheme of Figure 17.3, what is the name of the compound before the phenol ring is finally splintered.

17.22 In Figure 17.4, at what reaction number is the chloroform finally formed?

17.23 In Figure 17.4, at what reaction number is an alcohol being formed?

17.24 In Figure 17.4, considering all the reactions, what would the resulting pH be?

17.25 A flow of 25,000 m³/day of treated water is to be disinfected using chlorine in pressurized steel cylinders. The raw water comes from a reservoir where the water from the watershed has a very low alkalinity. With this low raw-water alkalinity, coupled with the use of alum in the coagulation process, the alkalinity of the treated water when it finally arrives at the chlorination tank is reduced to 10 mg/L. Calculate the amount of alkalinity required to neutralize the acid produced during the addition of the chlorine gas.

17.26 What is the available chlorine in monochloramine?

17.27 Calculate the available chlorine in $Ca(OCl)_2$.

17.28 Specify the chlorine dosing system to be used for a wastewater treatment plant processing an average of 5000 m³/d of wastewater. The peaking factor is 3.0 and the regulatory agency requires a dosage of 20.0 mg/L.

17.29 Specify the chlorine dosing system to be used for a wastewater treatment plant processing an average of 800 m³/d of wastewater. The peaking factor is 3.0 and the regulatory agency requires a dosage of 20.0 mg/L.

17.30 The residual chlorine analyzer transmits a signal corresponding to 0.5 mg/L of chlorine and the flow transmitter transmits a signal corresponding to 0.1 m³/s. If the desired chlorine residual is 0.5 mg/L, at what rate is the disinfectant dosed into the treated effluent?

17.31 A chlorine disinfection study was conducted to determine the constants of Equation (17.2). For a log 2 removal efficiency, the value of m is found to be 0.35 and the value of k is found to be 115. Calculate the contact time if the regulatory agency requires a chlorine dose of 20 mg/L.

17.32 A total flow of 25,000 m³/d is to be dechlorinated after disinfection using chlorine. Sodium metabisulfite is to be used for dechlorination. If the total residual chlorine (TRC) is 0.5 mg/L, how may kilograms per day of the metabisulfite is needed?

17.33 A total flow of 25,000 m³/d is to be dechlorinated after disinfection using chlorine. Sodium sulfite is to be used for dechlorination. If the total residual chlorine (TRC) is 0.5 mg/L, how may kilograms per day of the sulfite is needed?

17.34 A total flow of 25,000 m³/d is to be dechlorinated after disinfection using chlorine. An activated carbon bed is to be used for dechlorination. If the total residual chlorine (TRC) is 0.7 mg/L, how may kilograms of activated carbon is lost from the bed per day?

17.35 Calculate the residual $Na_2S_2O_5$ in a dechlorination process that will result in a decrease of pH to 6.0 in a receiving stream assuming the original pH is 7.5.

17.36 Calculate the residual sulfur dioxide in a dechlorination process that will result in a decrease of pH to 5.0 in a receiving stream assuming the original pH is 7.5.

17.37 Calculate the residual sodium sulfite in a dechlorination process that will result in a decrease of pH to 5.0 in a receiving stream assuming the original pH is 7.5.

17.38 A total flow of 2,000 m^3/d is to be disinfected using ozone. What should be the cross-sectional area of the serpentine channel in order to maintain self-cleaning velocity? What would be the total combined length of the channel if the contact time has been determined by a pilot study to be 24 min?

17.39 A total flow of 1000 m^3/d is to be disinfected using UV. What should be the cross-sectional area of the channel in order to maintain self-cleaning velocity? What would be the total combined length of the channel if the contact time has been determined by a pilot study to be 1.09 min?

BIBLIOGRAPHY

Bodzek, M. and K. Konieczny (1998). Comparison of various membrane types and module configurations in the treatment of natural water by means of low-pressure membrane methods. *Separation Purification Technol.,* 14, 1–3, 69–78.

Bryan, E. H. (1990). The National Science Foundation's support of research on uses of ionizing radiation in treatment of water and wastes. *Environ. Eng., Proc. 1990 Specialty Conf., ASCE.* Arlington, VA, 47–54.

Buisson, H., et al. (1997). Use of immersed membranes for upgrading wastewater treatment plants. *Water Science Technol., Proc. 1997 Int. Conf. on Upgrading of Water and Wastewater Syst.,* May 25–28, 37, 9, 89–95. Elsevier Science Ltd., Exeter, England.

Burch, J. D. and K. E. Thomas (1998). Water disinfection for developing countries and potential for solar thermal pasteurization. *Solar Energy.* 64, 1–3, 87–97.

Burttschell, R. H., et al. (1959). Chlorine derivatives of phenol causing taste and odor. *J. Am. Water Works Assoc.* 205–214.

Camel, V. and A. Bermond (1998). Use of ozone and associated oxidation processes in drinking water treatment. *Water Res.* 32, 11, 3208–3222.

Chapman, J. S. (1998). Characterizing bacterial resistance to preservatives and disinfectants. *Int. Biodeterioration Biodegradation.* 41, 3–4, 241–245.

Cicek, N., et al. (1998). Using a membrane bioreactor to reclaim wastewater. *J. Am. Water Works Assoc.* 90, 11, 105–113.

Clark, R. M. and M. Sivaganesan (1998). Predicting chlorine residuals and formation of TTHMs in drinking water. *J. Environ. Eng.* 124, 12, 1203–1210.

Denyer, S. P. and G. S. A. B. Stewart (1998). Mechanisms of action of disinfectants. *Int. Biodeterioration Biodegradation.* 41, 3–4, 261–268.

Downey, D., D. K. Giles, and M. J. Delwiche (1998). Finite element analysis of particle and liquid flow through an ultraviolet reactor. *Computers Electronics in Agriculture.* 21, 2, 81–105, 0168–1699.

Edzwald, J. K. and M. B. Kelley (1997). Control of cryptosporidium: From reservoirs to clarifiers to filters. *Water Science Technol., Proc. 1997 1st IAWQ-IWSA Joint Specialist Conf. on Reservoir Manage. Water Supply—An Integrated Syst.,* May 19–23, 37, 2, 1–8, Elsevier Science Ltd., Exeter, England.

Engelhardt, N., W. Firk, and W. Warnken (1998). Integration of membrane filtration into the activated sludge process in municipal wastewater treatment. *Water Science Technol., Wastewater: Industrial Wastewater Treatment, Proc. 1998 19th Biennial Conf. Int. Assoc. on Water Quality, Part 4,* June 21–26, 38, 4–5, 429–436, Elsevier Science Ltd., Exeter, England.

Fair, G. M., et al. (1948). The behavior of chlorine as a water disinfectant. *J. Am. Water Works Assoc.* 1051.

Gadgil, A. (1998). Drinking water in developing countries. *Annu. Review of Energy and the Environment.* 23, 253–286.

Hashimoto, T. Miyata, and W. Kawakami (1980). Radiation-induced decomposition of phenol in flow system. *Radiat. Phys. Chem.* 16, 59–65.

Hayashi, T., O. K. Kikuchi, and T. Dohino (1998). Electron beam disinfestation of cut flowers and their radiation tolerance. *Radiat. Physics Chem.* 51, 2, 175–179.

Hunter, G. F. and K. L. Rakness (1997). Start-up and optimization of the ozone disinfection process at the Sebago Lake water treatment facility. *Ozone: Science Eng.* 19, 3, 255–272.

Johnson, P., et al. (1998). Determining the optimal theoretical residence time distribution for chlorine contact tanks. *Aqua (Oxford).* 47, 5, 209–214.

Kawakami, W., et al. (1978). Electron-beam oxidation treatment of a commercial dye by use of a dual-tube bubbling column reactor. *Environ. Science Technol.* 12, 189–194.

Lawand, T. A., J. Ayoub, and H. Gichenje (1997). Solar disinfection of water using transparent plastic bags. *RERIC Int. Energy J.* 19, 1, 37–44.

Leal, J. L. (1909). The sterilization plant of the Jersey City Water Supply Company at Boonton, NJ. *Proc. Am. Water Works Assoc.*

Lee, G. F. (1967). *Principles and Applications of Water Chemistry.* S. D. Faust and J. V. Hunters (Eds.) John Wiley & Sons, New York.

Lee, Seockheon, M. Nakamura, and S. Ohgaki (1998). Inactivation of phage $Q\beta$ by 254 nm UV light and titanium dioxide photocatalyst. *J. Environ. Science Health, Part A: Toxic/Hazardous Substances Environ. Eng.* 33, 8, 1643–1655.

Madaeni, S. S. (1999). Application of membrane technology for water disinfection. *Water Res.* 33, 2, 301–308.

Magara, Y., S. Konikane, and M. Itoh (1998). Advanced membrane technology for application to water treatment. *Water Science Technol., Proc. 1997 Workshop on IAWQ-IWSA Joint Group on Particle Separation,* July 1–2, 1997, 37, 10, Elsevier Science Ltd., Exeter, England.

Marks, D. (1998). Automated chlor/dechlor control protects fishery, reduces chemical and labor costs. *Water/Eng. Manage.* 145, 11.

Metcalf & Eddy, Inc. (1972). *Wastewater Engineering: Treatment, Disposal, Reuse.* McGraw-Hill, New York, 297, 351, 385, 386, 387.

Momba, M. N. B., et al. (1998). Evaluation of the impact of disinfection processes on the formation of biofilms in potable surface water distribution systems. *Water Science Technol. Wastewater: Biological Processes, Proc. 1998 19th Biennial Conf. Int. Assoc. on Water Quality. Part 7,* June 21–26, 38, 8–9, 283–289. Elsevier Science Ltd., Exeter, England.

Plummer, J. D. and J. K. Edzwald (1997). Effect of ozone on disinfection by-product formation of algae. *Water Science Technol., Proc. 1997 1st IAWQ-IWSA Joint Specialist Conf. on Reservoir Manage. and Water Supply—An Integrated Syst.,* May 19–23, 37, 2, 49–55, Elsevier Science Ltd., Exeter, England.

Prokopov, V. A., G. V. Tolstopyatova, and E. D. Maktaz (1997). Hygienic aspects of dioxide chlorine use in drinking water supply. *Khimiya i Tekhnologiya Vody.* 19, 3, 275–288.

Potapchenko, N. G., et al. (1997). Disinfection action of ozone in water on *Escherichia coli* 1257. *Khimiya i Tekhnologiya Vody.* 19, 3, 315–320.

Rice, R. G., et al. (1979). Ozone utilization Europe. *AIChE 8th Annual Meeting,* Houston, TX.

Rodriguez, M. J. and J.-B. Serodes (1999). Assessing empirical linear and non-linear modelling of residual chlorine in urban drinking water systems. *Environ. Modelling Software.* 14, 1, 93–102.

Schmidt, W., B. Hambsch, and H. Petzoldt (1997). Classification of algogenic organic matter concerning its contribution to the bacterial regrowth potential and by-products formation. *Water Science Technol., Proc. 1997 1st IAWQ-IWSA Joint Specialist Conf. on Reservoir Manage. and Water Supply—An Integrated Syst.,* May 19–23, 37, 2, 91–96. Elsevier Science Ltd., Exeter, England.

Schulz, S. and H. H. Hahn (1997). Generation of halogenated organic compounds in municipal waste water. *Water Science Technol., Proc. 1997 2nd IAWQ Int. Conf. on the Sewer as a Physical, Chemical and Biological Reactor,* May 25–28, 37, 1, 303–309. Elsevier Science Ltd., Exeter, England.

Siddiqui, M., et al. (1998). Modeling dissolved ozone and bromate ion formation in ozone contactors. *Water, Air Soil Pollut.* 108, 1–2, 1–32.

Sincero, A. P. and G. A. Sincero (1996). *Environmental Engineering: A Design Approach.* Prentice Hall, Upper Saddle River, NJ.

Snoeyink, V. L. and D. Jenkins (1980). *Water Chemistry.* John Wiley & Sons, New York, 402.

Tosa, K. and T. Hirata (1999). Photoreactivation of enterohemorrhagic *Escherichia coli* following UV disinfection. *Water Res.* 33, 2, 361–366.

Vera, L., et al. (1998). Can microfiltration of treated wastewater produce suitable water for irrigation? *Water Science Technol., Wastewater: Industrial Wastewater Treatment, Proc. 1998 19th Biennial Conf. Int. Assoc. on Water Quality, Part 4,* June 21–26, 38, 4–5, 395–403. Elsevier Science Ltd., Exeter, England.

Appendices and Index

Appendix 1
Density and Viscosity of Water

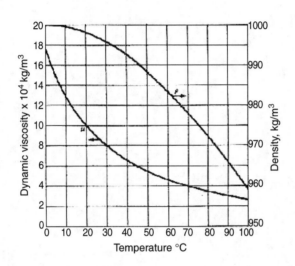

Appendix 2
Atomic Masses of the Elements Based on C-12

Name	Symbol	Atomic Number	Atomic Mass
Actinium	Ac	89	227.0278
Aluminum	Al	13	26.98154
Americium	Am	95	(243)
Antimony	Sb	51	121.75
Argon	Ar	18	39.948
Arsenic	As	33	74.9216
Astatine	At	85	(210)
Barium	Ba	56	137.33
Berkelium	Bk	97	(247)
Beryllium	Be	4	9.01218
Bismuth	Bi	83	208.9804
Boron	B	5	10.81
Bromine	Br	35	79.904
Cadmium	Cd	48	112.41
Calcium	Ca	20	40.08
Californium	Cf	98	(251)
Carbon	C	6	12.011
Cerium	Ce	58	140.12
Cesium	Cs	55	132.9054
Chlorine	Cl	17	35.453
Chromium	Cr	24	51.996
Cobalt	Co	27	58.9332
Copper	Cu	29	63.546
Curium	Cm	96	(247)
Dysprosium	Dy	66	162.50
Einsteinium	Es	99	(252)
Erbium	Er	68	167.26
Europium	Eu	63	151.96
Fermium	Fm	100	(257)
Fluorine	F	9	18.99840
Francium	Fr	87	(223)
Gadolinium	Gd	64	157.25

(continued)

Name	Symbol	Atomic Number	Atomic Mass
Gallium	Ga	31	69.72
Germanium	Ge	32	72.59
Gold	Au	79	196.9665
Hafnium	Hf	72	178.49
Helium	He	2	4.00260
Holmium	Ho	67	164.9304
Hydrogen	H	1	1.0079
Indium	In	49	114.82
Iodine	I	53	126.9045
Iridium	Ir	77	192.22
Iron	Fe	26	55.847
Krypton	Kr	36	83.80
Lanthanum	La	57	138.9055
Lawrencium	Lr	103	(260)
Lead	Pb	82	207.2
Lithium	Li	3	6.941
Lutetium	Lu	71	174.967
Magnesium	Mg	12	24.305
Manganese	Mn	25	54.9380
Mendelevium	Md	101	(258)
Mercury	Hg	80	200.59
Molybdenum	Mo	42	95.94
Neodymium	Nd	60	144.24
Neon	Ne	10	20.179
Neptunium	Np	93	237.0482
Nickel	Ni	28	58.69
Niobium	Nb	41	92.9064
Nitrogen	N	7	14.0067
Nobelium	No	102	(259)
Osmium	Os	76	190.2
Oxygen	O	8	15.9994
Palladium	Pd	46	106.42
Phosphorus	P	15	30.97376
Platinum	Pt	78	195.08
Plutonium	Pu	94	(244)
Polonium	Po	84	(209)
Potassium	K	19	39.0983
Praseodymium	Pr	59	140.9077
Promethium	Pm	61	(145)
Protactinium	Pa	91	231.0359
Radium	Ra	88	226.0254
Radon	Ra	86	(222)
Rhenium	Re	75	186.207
Rhodium	Rh	45	102.9055
Rubidium	Rb	37	85.4678
Ruthenium	Ru	44	101.07

(continued)

Name	Symbol	Atomic Number	Atomic Mass
Samarium	Sm	62	150.36
Scandium	Sc	21	44.9559
Selenium	Se	34	78.96
Silicon	Si	14	28.0855
Silver	Ag	47	107.868
Sodium	Na	11	22.98977
Strontium	Sr	38	87.62
Sulfur	S	16	32.06
Tantalum	Ta	73	180.9479
Technetium	Tc	43	(98)
Tellurium	Te	52	127.60
Terbium	Tb	65	158.9254
Thallium	Tl	81	204.383
Thorium	Th	90	232.0381
Thulium	Tm	69	168.9342
Tin	Sn	50	118.71
Titanium	Ti	22	47.88
Tungsten	W	74	183.85
Unnilennium	Une	109	(266)
Unnilhexium	Unh	106	(263)
Unniloctium	Uno	108	(265)
Unnilpentium	Unp	105	(262)
Unnilquadium	Unq	104	(261)
Unnilseptium	Uns	107	(262)
Uranium	U	92	238.029
Vanadium	V	23	50.9415
Xenon	Xe	54	131.29
Ytterbium	Yb	70	173.04
Yttrium	Y	39	88.9059
Zinc	Zn	30	65.39
Zirconium	Zr	40	91.224

Note: Values in parentheses (isotope atomic masses of longest half-life) are used for radioactive elements where atomic weights cannot be quoted precisely without knowledge of the origin of the elements.

Appendix 3
Saturation Values of Dissolved Oxygen Exposed to Saturated Atmosphere at One Atmosphere Pressure at Given Temperatures

Temperature (°C)	Chloride Concentration (mg/L)					Saturated H_2O Vapor Pressure (kPa)
	0	5,000	10,000	15,000	20,000	
0	14.62	13.79	12.97	12.14	11.32	0.6108
1	14.23	13.41	12.61	11.82	11.03	0.6566
2	13.84	13.05	12.28	11.52	10.76	0.7055
3	13.48	12.72	11.98	11.24	10.50	0.7575
4	13.13	12.41	11.69	10.97	10.25	0.8129
5	12.80	12.09	11.39	10.70	10.01	0.8719
6	12.48	11.79	11.12	10.45	9.78	0.9347
7	12.17	11.51	10.85	10.21	9.57	1.0013
8	11.87	11.24	10.61	9.98	9.36	1.0722
9	11.59	10.97	10.36	9.76	9.17	1.1474
10	11.33	10.73	10.13	9.55	8.98	1.2272
11	11.08	10.49	9.92	9.35	8.80	1.3119
12	10.83	10.28	9.72	9.17	8.62	1.4017
13	10.60	10.05	9.52	8.98	8.46	1.4969
14	10.37	9.85	9.32	8.80	8.30	1.5977
15	10.15	9.65	9.14	8.63	8.14	1.7054
16	9.95	9.46	8.96	8.47	7.99	1.8173
17	9.74	9.26	8.78	8.30	7.84	1.9367
18	9.54	9.07	8.62	8.15	7.70	2.0630
19	9.35	8.89	8.45	8.00	7.56	2.1964
20	9.17	8.73	8.30	7.86	7.42	2.3373
21	8.99	8.57	8.14	7.71	7.28	2.4861
22	8.83	8.42	7.99	7.57	7.14	2.6430

(continued)

Temperature (°C)	Chloride Concentration (mg/L)					Saturated H$_2$O Vapor Pressure (kPa)
	0	5,000	10,000	15,000	20,000	
23	8.68	8.27	7.85	7.43	7.00	2.8086
24	8.53	8.12	7.71	7.30	6.87	2.9831
25	8.38	7.96	7.56	7.15	6.74	3.1671
26	8.22	7.81	7.42	7.02	6.61	3.3608
27	8.07	7.67	7.28	6.88	6.49	3.5649
28	7.92	7.53	7.14	6.75	6.37	3.7796
29	7.77	7.39	7.00	6.62	6.25	4.0055
30	7.63	7.25	6.86	6.49	6.13	4.2430

Appendix 4
SDWA Acronyms

BAT	Best available technology
BTGA	Best technology generally available
CWS	Community water systems
DWEL	Drinking water equivalence level
EMSL	EPA Environmental Monitoring and Support Laboratory (Cincinnati)
GAC	Granular activated carbon
IOC	Inorganic chemical
IPDWR	Interim primary drinking water regulation
LOAEL	Lowest observed adverse effect level
LOQ	Limit of quantitation
MCL	Maximum contaminant level
MCLG	Maximum contaminant level goal
MDL	Method detection limit
NOAEL	No observed adverse effect level
NPDWR	National primary drinking water regulation
NTNCWS	Non-transient noncommunity water system
PAC	Powdered activated carbon
PHS	Public Health Service
POET	Point-of-entry technology
POUT	Point-of-use technology
PQL	Practical quantitation level
PTA	Packed tower aeration
RFD	Reference dose
RIA	Regulatory impact analysis
RMCL	Recommended maximum contaminant level
RPDWR	Revised primary drinking water regulation
RSC	Relative source contribution
SDWA	Safe drinking water act
SMCL	Secondary maximum contaminant level
SNARL	Suggested no adverse response level
SOC	Synthetic organic chemical
TNCWS	Transient noncommunity water system
UIC	Underground injection control
URTH	Unreasonable risk to health
VOC	Volatile organic chemical

Appendix 5
Sample Drinking Water VOCs

Contaminant	MCL (mg/L)	BAT
Benzene	0.005–zero	PTA or GAC
Carbon tetrachloride	0.005–zero	PTA or GAC
p-Dichlorobenzene	0.075	PTA or GAC
1,2-Dichloroethane	0.005–zero	PTA or GAC
1,1-Dichloroethylene	0.007	PTA or GAC
1,1,1-Trichloromethane	0.20	PTA or GAC
Trichloroethylene	0.005–zero	PTA or GAC
Vinyl chloride	0.002–zero	PTA
Bromobenzene	—	—
Bromodichloromethane	—	—
Bromoform	—	—
Bromomethane	—	—
Chlorobenzene	—	—
Chlorodibromomethane	—	—
Chloroethane	—	—
Chloroform	—	—
Chloromethane	—	—
o-Chlorotoluene	—	—
p-Chlorotoluene	—	—
Dibromomethane	—	—
m-Dichlorobenzene	—	—
o-Dichlorobenzene	—	—
cis-1,2-Dichloroethylene	—	—
trans-1,2-Dichloroethylene	—	—
Dichloromethane	—	—
1,1-Dichloroethane	—	—
1,2-Dichloropropane	—	—
1,3-Dichloropropane	—	—
2,2,-Dichloropropane	—	—
1,1-Dichloropropene	—	—
1,3-Dichloropropene	—	—
Ethylbenzene	—	—
Styrene	—	—

(continued)

Contaminant	MCL (mg/L)	BAT
1,1,1,2-Tetrachloroethane	—	—
1,1,2,2-Tetrachloroethane	—	—
Tetrachloroethylene	—	—
Toluene	—	—
1,1,2-Trichloroethane	—	—
1,2,3-Trichloropropane	—	—
m-Xylene	—	—
o-Xylene	—	—
p-Xylene	—	—

Appendix 6
Sample Drinking Water SOCs and IOCs

Contaminants	MCL (mg/L)
SOC:	
Acrylamide	treatment technique
Alachlor	0.002
Aldicarb	0.01
Aldicarb sulfone	0.04
Aldicarb sulfoxide	0.01
Atrazine	0.003
Carbofuran	0.04
Chlordane	0.002
2,4-D	0.07
Dibromochloropropane	0.0002
o-Dichlorobenzene	0.6
cis-1,2-Dichloroethylene	0.07
trans-1,2-Dichloroethylene	0.1
1,2-Dichloropropane	0.005
Epichlorohydrin	treatment technique
Ethylbenzene	0.7
Ethylene dibromide (EDB)	0.00005
Heptachlor	0.0004
Heptachlor epoxide	0.0002
Lindane	0.0002
Methoxychlor	0.4
Monochlorobenzene	0.1
PCBs (as decachlorobiphenyls)	0.0005
Pentachlorophenol	0.2
Styrene	0.005
Tetrachloroethylene	0.005
Toluene	2.0
Toxaphene	0.005
2,4,5-TP (Silvex)	0.05
Xylene	10

(continued)

Contaminants	MCL (mg/L)
IOC:	
Arsenic	0.05
Asbestos	7 million fibers/L (longer than 10 μm)
Barium	5.0
Cadmium	0.005
Chromium	0.1
Mercury	0.002
Nitrate	10 as N
Nitrite	1 as N
Selenium	0.05
Silver	0.05

Appendix 7
Secondary MCLs for a Number of Substances

Contaminant	SMCL (mg/L)
Chloride	250
Color	15 color units
Copper	1
Corrosivity	Non-corrosive
Foaming agents	0.5
Hydrogen sulfide	0.05
Iron	0.3
Manganese	0.05
Odor	3 TON
pH	6.5–8.5
Sulfate	250
Total dissolved solids	500
Zinc	5

Appendix 8
Some Primary Drinking-Water Criteria

Contaminants	Concentration (mg/L, unless otherwise noted)
Arsenic	0.05
Barium	1.00
Cadmium	0.010
Chromium	0.05
Lead	0.05
Mercury	0.002
Nitrate as N	10.00
Selenium	0.01
Silver	0.05
Chlorinated hydrocarbon:	
Endrin (1,2,3,4,10,10-hexachloro-6,7-epoxy-1,4,4a,5,6,7,8,8a-octo-hydro-1,4-endo,endo-5,8-dimetha-nonaphthalene)	0.0002
Lindane (1,2,3,4,5,6-hexachlorocyclohexane, gamma isomer)	0.004
Methoxychlor (1,1,1-trichloro-2,2-bis{p-methoxyphenyl}ethane)	0.1
Toxaphene ($C_{10}H_{10}Cl_8$-technical chlorinated camphene, 67–69% chlorine)	0.005
Chlorophenoxys:	
2,4-D (2,4-dichlorophenoxyacetic acid)	0.1
2,4,5-TP (2,4,5-trichlorophenoxy propionic acid)	0.01
Turbidity	Based on monthly average: 1 TU or up to 5 TU if the water supplier can demonstrate that the higher turbidity does not interfere with disinfection.

(continued)

Contaminants	Concentration (mg/L, unless otherwise noted)
Microbiological contaminants	Based on average of two consecutive days: 5 TU Membrane filter technique: not to exceed 1/100 mL on monthly basis; not to exceed 4/100 mL of coliforms in more than one sample in fewer than 20 samples per month on an individual sample basis; not to exceed 4/100 mL of coliforms in more than 5% of sample in more than 20 samples per month on an individual basis. Fermentation tube method: **1. 10-mL standard portions:** coliforms shall not be present in more than 10% of the portions on monthly basis; coliforms shall not be present in three or more portions in more than one sample in fewer than 20 samples per month on an individual basis; coliforms shall not be present in three or more portions in more than 5% of samples in more than 20 samples per month on an individual basis. **2. 100-mL standard portions:** coliforms shall not be present in more than 60% of the portions on monthly basis; coliforms shall not be present in five portions in more than one sample in fewer than 20 samples per month on an individual basis; coliforms shall not be present in five portions in more than 20% of samples in more than 20 samples per month on an individual basis.

Appendix 9
Some Secondary Drinking-Water Criteria

Contaminants	Concentration (mg/L, unless otherwise noted)
Chloride	250
Color	15 CU (color units)
Copper	1
Corrosivity	Noncorrosive
Foaming agents	0.5
Hydrogen sulfide	0.05
Iron	0.3
Manganese	0.05
Odor	<3 TON
pH	6.5–8.5
Sulfate	250
Total dissolved solids	500
Zinc	5

Appendix 10
Physical Constants

Physical constant	Value
Atomic mass unit (amu)	$\frac{1}{12}$ the mass of $C^{12} = 1.6605402(10^{-27})$ kg
Acceleration due to gravity (g)	9.80665 m/s^2 and 32.174 ft/s^2 at sea level
Avogadro's number	$6.0221367(10^{23})$ per gmol
Boltzmann constant	$1.380658(10^{-23})$ J/K°
Charge-to-mass ratio for electrons	$1.75881962(10^{11})$ coulomb per kg
Electrical permittivity constant	$8.85(10^{-12})$ coulomb/V-m
Electron charge	$1.60217733(10^{-19})$ coulomb
Electron rest mass	$9.109390(10^{-31})$ kg
Faraday constant	$9.6485309(10^4)$ coulomb per equivalent
gmole	22.4 liters at STP of 0°C and 1 atm
lbmole	359 ft^3 ideal gas at STP of 0°C and 1 atm
Neutron rest mass	$1.6749286(10^{-27})$ kg
Planck constant	$6.6260755(10^{-34})$ J-s
Proton rest mass	$1.6726231(10^{-27})$ kg
Speed of light in vacuum	$2.99792458(10^8)$ m/s
Standard temperature and pressure (STP)	1 atm and 0°C
Standard room temperature	25°C
Universal gas constant	$8.205784(10^{-2})$ L-atm-°K/gmol, 8.314510 J/gmol-°K, 1.987 cal/gmol-°K, 82.05 atm-cm^3/gmol-°K, $4.968(10^4)$ lb$_m$-ft^2/(lbmole)-°R, 49,720 ft-lb$_f$/slug-°R

Appendix 11
Conversion Factors

From	To
Å, angstrom	10^{-8} cm
acre	43,560 ft^2, 0.00156 mi^2, 0.405 ha
amp	coulomb/s
atm	101.325 kN/m^2, 14.696 lb_f/in^2, 101.325 kPa, 1.013 bar, 29.92 in Hg and 760 mm Hg at 0°C, 760 torr, 33.936 ft of H_2O (60°F), 2116.2 lb_f/ft^2
barrel	42 gal
bu/ha	0.4047 bu/acre
btu	252.2 cal, 778.2 ft-lb_f
cal	4.1868 J
cal/gmol	1.8 Btu/lbmol
centipoise	0.000672 lb_m/s-ft, 10^{-3} Pa-s
cm	0.3937 in
cm^3	1 mL
coulomb	$6.2(10^{18})$ electrons, 1 amp-s
dyne	10^{-5} N
erg	1 dyne-cm, 10^{-7} J
esu	$1.59(10^{-19})$ amp-s, $1.59(10^{-19})$ coulomb
ft	0.305 m
ft^3	7.481 gal, 28.32 L, 0.0283 m^3
°F	[1.8(°C) + 32]
g	0.0353 oz, 0.0022 lb_m
gal	3.785 L
gpm	0.227 m^3/h
grain	$6.480(10^{-2})$ g
grain/ft^3	2.29 g/m^3
g/m^3	8.3454 lb_m/Mgal
ha	2.4711 acre, 10^4 m^2
hp	746 W, 33000 ft-lb_f/min, 2545 Btu/h, 0.0738 boiler hp
hp-hr	$1.98(10^6)$ ft-lbf
hertz	1 cycle/s
inch	2.54 cm
J	$2.7778(10^{-7})$ kW-h, 0.7376 ft-lb_f, 1.0 W-s, 1.0 N-m, 0.2388 cal, 10^7 ergs
kg	2.2046 lb_m

(*continued*)

From	To
kg/ha	0.8922 lb_m/acre
kg/ha-d	0.8922 lb_m/acre-d
kg/m^2-d	0.2048 lb_m/ft^2-d
kg/m^3-d	62.428 lb_m/10^3 ft^3-d
kg/m^3	8345.4 lb_m/Mgal
kg/kW-h	1.6440 lb_m/hp-h
kJ	0.9478 Btu
kJ/kg	0.4303 Btu/lb_m
km	0.6214 mi
kPa	0.0099 atm
kW	0.9478 Btu/s, 1.3410 hp
kW-hr	3412 Btu, 1.341 hp-hr
kW/m^3	5.0763 hp/10^2-gal
kW/10^3-m^3	0.0380 hp/10^3-ft^3
°K	°C + 273.16
lb_f	[lb_m/(g_c lb_m/slug)]g (g_c = 32.174; g is the gravitational acceleration)
lb_m	453.6 g, 7000 grains, 16 oz
long ton	2000 lb_m
L	0.2642 gal, 0.0353 ft^3, 33.8150 oz, 1.057 quarts, 10^{-3} m^3
L/m^2-min	35.3420 gal/ft^2-d
L/m^2	2.4542(10^{-2}) gal/ft^2
m	3.2808 ft, 39.3701 in, 1.0936 yd
m^2	2.471(10^{-4}) acre
m^3	35.314 ft^3, 1000 L
micron	10^{-6} m, 10^{-3} mm
mile	5280 ft, 1609 km
m/s^2	3.2808 ft/s^2, 39.3701 in/s^2
m^3	264.1720 gal, 8.1071(10^{-4}) acre-ft
m^3/10^3-m^3	133.6806 ft^3/Mgal
m^3/m-min	10.7639 ft^3/ft-min
m^3/m-d	80.5196 gal/ft^2-d
m^3/m^2-h	589.0173 gal/ft^2-d
m^3/m^2-d	24.5424 gal/ft^2-d
m^3/m^3	0.1337 ft^3/gal
Mg	1.1023 short ton, 0.9842 long ton
MJ	0.3725 hp-h
newton, N	0.2248 lb_f, 1 kg-m/s^2
N/m^2	1Pa
poise	1 g/cm-s, 0.1 Pa-s
Pa	1 N/m^2, 1.4504(10^{-4}) lb_f/in^2
°R	°F + 459.49
short ton	2240 lb_m
stoke	cm^2/s
ton of refrigeration	200 Btu/min
W	0.7376 ft-lb_f/s, 1 J/s, 10^7 ergs/s
W/m^2-C°	0.1763 Btu/ft^2-F°-h

Index

A

ABS, *see* Alkyl benzene sulfonate (ABS)
Absolute pressure, 232, 304
Absorption, *see also* Aeration; Mass transfer;
     Stripping
    ammonia stripping, 450–453
    basics, 419
    operating line, 446–447
    tower height and sizing, 446–450
ACF, *see* Activated carbon fiber (ACF)
Acidity
    basics, 54–56, 143–144
    coagulation, 578–579
    expressed as $CaCO_3$, 578–579
    pH levels, 577–578, 613
Acids, chlorine disinfection, 762–763
Acids and bases, 30, 51–56
Activated carbon
    activation techniques, 392–393
    dechlorination, 772–774
Activated carbon fiber (ACF), 392
Activated sludge, 690
Activate silica, 552
Activation techniques, carbon adsorption,
    392–393
Active concentration, 41, 514
Active site, molecules, 551
Active zone
    carbon adsorption, 399–404
    ion exchange, 726–728
Actual oxygen rate/requirement (AOR), 430–431,
    439–442
Adenosine diphosphate (ADP), 665–666
Adenosine triphosphate (ATP), 665–666
ADP, *see* Adenosine diphosphate (ADP)
Adsorption capacity, carbon, 393–395
Aeration, *see also* Absorption; Mass transfer;
     Stripping
    actual oxygen requirement, 439–442
    basics, 419–420, 430
    basin sizing, 443–444
    bubble aerators, 444–446
    contact time, 443–444
    equipment specification, 430–432
    parameters, 432–439
    sizing, 433

Aerobacter species, 151
Age
    bacteria and other microorganisms, 743
    sludge, 693
Air saturation tank pressure, 281
Air temperature, influence, 128
Alcohols, 144
Algae, 627, 660–661
Algal blooms, 143
Alkalinity
    basics, 54–56, 143–144
    bicarbonate, 686–687
    calcium carbonate, 530–531
    coagulation, 575, 578–579
    expressed as $CaCO_3$, 578–579
    hydroxyl ions, 614
    natural water, 534
    nitrogen removal, 686–689
    pH adjustment, 572, 577–578, 613
    water softening, 505–506
    water stabilization, 522–523
Alkyl benzene sulfonate (ABS), 145
Allegheny County, Maryland, 434
Alum, *see also* Aluminum
    coagulation, 553–559, 568–570
    phosphorus removal, 37, 630–633, 643–644
Aluminum, 43–46, 554, *see also* Alum
Amebiasis, 168, 170
Amine group, 546
Ammonia
    chlorine disinfection, 753–756
    free, 141
    nitrification, 669, 677–680
    stripping, 450–453
Amphoteric substances, 54, 144
Analogy (Jose, Pedro, and Maria), 579
Animals, 166, *see also* specific animal
Anion exchange material, 719
Anion membranes, 373
Anne Arundel County, 265
Anoxic stage, 674
AOR, *see* Actual oxygen rate/requirement (AOR)
Apolarity, membranes, 383
Approach channel, 245
Approach velocity, 183
Apron aerators, 433, *see also* Aeration
Area meters, 185